Ordered Intermetallics –
Physical Metallurgy and Mechanical Behaviour

NATO ASI Series

Advanced Science Institutes Series

A Series presenting the results of activities sponsored by the NATO Science Committee, which aims at the dissemination of advanced scientific and technological knowledge, with a view to strengthening links between scientific communities.

The Series is published by an international board of publishers in conjunction with the NATO Scientific Affairs Division

A **Life Sciences**	Plenum Publishing Corporation
B **Physics**	London and New York
C **Mathematical**	Kluwer Academic Publishers
and Physical Sciences	Dordrecht, Boston and London
D **Behavioural and Social Sciences**	
E **Applied Sciences**	
F **Computer and Systems Sciences**	Springer-Verlag
G **Ecological Sciences**	Berlin, Heidelberg, New York, London,
H **Cell Biology**	Paris and Tokyo
I **Global Environmental Change**	

NATO-PCO-DATA BASE

The electronic index to the NATO ASI Series provides full bibliographical references (with keywords and/or abstracts) to more than 30000 contributions from international scientists published in all sections of the NATO ASI Series.
Access to the NATO-PCO-DATA BASE is possible in two ways:

– via online FILE 128 (NATO-PCO-DATA BASE) hosted by ESRIN,
Via Galileo Galilei, I-00044 Frascati, Italy.

– via CD-ROM "NATO-PCO-DATA BASE" with user-friendly retrieval software in English, French and German (© WTV GmbH and DATAWARE Technologies Inc. 1989).

The CD-ROM can be ordered through any member of the Board of Publishers or through NATO-PCO, Overijse, Belgium.

Series E: Applied Sciences - Vol. 213

Ordered Intermetallics –
Physical Metallurgy
and Mechanical Behaviour

edited by

C. T. Liu
Metals and Ceramics Division,
Oak Ridge National Laboratory,
Oak Ridge, Tennessee, U.S.A.

R. W. Cahn
Department of Materials Science and Metallurgy,
University of Cambridge,
Cambridge, U.K.

and

G. Sauthoff
Max-Planck-Institut für Eisenforschung GmbH,
Düsseldorf, Germany

Springer-Science+Business Media, B.V.

Proceedings of the NATO Advanced Research Workshop on
Ordered Intermetallics – Physical Metallurgy and Mechanical Behaviour
Irsee, Germany
23–28 June 1991

Library of Congress Cataloging-in-Publication Data

Ordered intermetallics--physical metallurgy and mechanical behaviour
  proceedings of the NATO advanced research workshop, Irsee, Germany,
  23-28 June 1991 / edited by C.T. Liu and R.W. Cahn and G. Sauthoff.
       p.   cm. -- (NATO ASI series. Series E, Applied sciences ; vol.
  213)
     ISBN 978-94-010-5119-4   ISBN 978-94-011-2534-5 (eBook)
     DOI 10.1007/978-94-011-2534-5
     1. Intermetallic compounds--Mechanical properties--Congresses.
  2. Physical metallurgy--Congresses.   I. Liu, C. T. (chain Tsuan),
  1937-   . II. Cahn, R. W. (Robert W.), 1924-   . III. Sauthoff, G.
  IV. Series NATO ASI series. Series E, Applied sciences , no. 213.
  TA483.073 1992
  669'.94--dc20   92-10360

ISBN 978-94-010-5119-4

CONTENTS

SECTION 1. Electronic Structure and Phase Stability

SECTION 2. Deformation and Dislocation Structures

SECTION 3. Ductility and Fracture

SECTION 4. Kinetic Processes and Creep Behavior

SECTION 5. Research Programs and Highlights

PREFACE

Ordered intermetallics constitute an unique class of metallic materials which have the potential to be developed as new-generation materials for structural use at high temperatures in hostile environments. At present, there is a worldwide interest in intermetallics, and extensive efforts have been devoted to intermetallic research and development in the U.S., Japan, European countries, and other nations. As a result, significant advances have been made in all areas of intermetallic research. The objective of this NATO Advanced Workshop on ordered intermetallics is (1) to review the recent progress and (2) to assess the future direction of intermetallic research in the areas of electronic structure and phase stability, deformation and fracture, and high-temperature properties. Emphasis has been placed on bringing experts from different fields into close contact with members of the mainline intermetallic community, with the expectation of mutual benefit and increased collaboration in the future.

The NATO Advanced Workshop on Ordered Intermetallics — Physical Metallurgy and Mechanical Behavior was held at Irsee, Germany, on June 23-28, 1991. The five-day workshop consisted of six sessions: (1) Electronic Structure and Phase Stability, (2) Deformation and Dislocation Structures, (3) Ductility and Fracture, (4) Kinetic Processes and Creep Behavior, (5) Research Programs and Highlights, and (6) Assessment of Current Research and Recommendation for Future Work. The first four sessions reviewed the recent advances in the three focused areas. The fifth session provided highlights of the intermetallic research under major programs and in different institutes and countries. A total of 35 featured papers and 9 highlight talks were presented in the workshop, with 59 participants from 11 countries. The last session provided a forum to discussion research areas for future studies.

This volume is a record of the proceedings of this NATO Advanced Workshop. A total of 43 invited papers were included, and all of the papers were reviewed by peers. We are grateful to Prof. D. deFontaine who voluntarily wrote the article "Why First-Principles Calculations for Alloys," which is a brief overview of first-principles calculations with emphasis on prediction of phase equilibrium. Merits and drawbacks of various calculation methods are briefly discussed in this paper. Unfortunately, because of the malfunction of the record system used in the workshop, the discussions in the last session could not be recorded and presented in this proceedings. Instead, a very brief summary of the workshop, including a general consensus of the discussions, is given at the end of this proceedings.

This NATO Workshop was co-sponsored by the U.S. Department of Energy, the Commission of European Communities, and the U.S. Naval Research European Office. We are grateful for their financial support. We are pleased to acknowledge the graduate students Mr. B. Zeumer, D. Letzig, and R. Yang for their excellent on-site arrangement of the workshop, and Ms. Connie Dowker and Shirin Badlani for their diligent secretarial services. Thanks are also due to Oak Ridge National Laboratory and Max-Planck Institut für Eisenforschung for their general support of this workshop.

C. T. Liu
R. W. Cahn
G. Sauthoff

PHASE STABILITY, AND COHESIVE, ELECTRONIC AND MECHANICAL PROPERTIES OF INTERMETALLIC COMPOUNDS

A. J. FREEMAN, J.-H. XU, T. HONG, W. LIN
Department of Physics and Astronomy
Northwestern University
Evanston, IL 60208-3112

ABSTRACT. Current sophisticated electronic structure simulations are at the forefront of understanding and predicting a variety of materials properties of intermetallic compounds. Several examples are given here that illustrate how first principles total energy local density methods have addressed the problems of (i) phase stability and the effect of ternary additions, (ii) anti-phase boundaries (APB's) in B2 NiAl, FeAl and RuAl aluminides and other faults in determining their structural and bonding character. A key objective has been to attempt to understand, at the electronic level, fundamental quantities that may be related to the crucial ductility issue in high temperature intermetallics. Differences between observed ductility properties of related systems may relate to their differing electronic and bonding properties, particularly the nature of p-d hybridization and the directional charge distributions of the states near the Fermi energy.

1. Introduction

As will be apparent from the proceedings of this workshop, the last five years have witnessed a rapid growth in the study and understanding of intermetallic compounds. Much of this progress may be seen to be the direct result of the introduction of computational theory, notably electronic structure theory, to tackle the urgent problems of the experimentalists and the incessant call of industry for new and better materials for aerospace applications. This reflects the present state of the field and the reaching of a threshold: advanced theoretical–computational techniques combined with the power of supercomputers provide an understanding of matter at the atomic-scale with an unprecedented level of detail and accuracy. This capability has given birth to a new branch of scientific endeavor: computational materials science. In contrast to an analytic–theoretical approach, which isolates and idealizes real systems to unravel fundamental relations and laws, the computational approach is synthetic: its goal is to simulate more and more details of the system studied including as much of the environment as possible.

Designing new materials with specific mechanical, thermal chemical and electronic properties hinges on one basic assumption: the properties of the macroscopic ensemble are related to and can be derived from the properties of individual molecules and atomic building blocks (such as crystallographic unit cells in solids). But which atomic-scale physical quantities

1

C. T. Liu et al. (eds.), Ordered Intermetallics – Physical Metallurgy and Mechanical Behaviour, 1–14.
© 1992 Kluwer Academic Publishers.

Table 1. Observed variation of the crystal structure in transition metal-trialuminides across the early-transition metal series. HT and RT denote the high temperature and room temperature phases, respectively.

IIIB	IVB	VB
$ScAl_3$	$TiAl_3$	VAl_3
($L1_2$)	(DO_{22})	(DO_{22})
YAl_3	$ZrAl_3$	$NbAl_3$
($L1_2$ above 950 °C)	(DO_{23})	(DO_{22})
(BaPb above 640 °C)	($L1_2$ metastable)	
(DO_{19} RT)		
$LaAl_3$	$HfAl_3$	$TaAl_3$
(DO_{19})	(DO_{22} HT)	(DO_{22})
	(DO_{23} RT)	

and observables need to be looked at? This is the important question today and the key to successful materials by design in the future. Thus, much of what will be discussed at this workshop focusses on bridging the gap between the predictive capabilities of these first principles theoretical computations and the macroscopic properties that determine the success or failure of a given alloy system.

In the following, as illustration, we limit ourselves to discussing only a few selected examples taken from several different materials and problems we have been studying. Our aim is to provide a general overview of progress made to date as well as to bring out the complexity of the different issues involved and, at the same time to illustrate the kind of answers that our theoretical–computational approach can offer to the understanding of these issues by yielding well-defined results and making precise predictions. Various examples have demonstrated that it is possible not only to make quantitative predictions for real systems, but more importantly, to gain insights into the underlying physics of these materials and the phenomena addressed.

2. Electronic Structure and Phase Stability in Transition-metal Intermetallics

2.1. TRANSITION-METAL TRIALUMINIDES

The studies of early transition metal (TM) trialuminides, (TM = Sc, Ti, V, Y, Zr, Nb, etc.) have both technological and scientific significance: they are attractive as potential structural materials for use in high temperature environments[1], or as thermally stable (i.e., low coarsening rate) precipitates for developing so-called super-"alumalloys"[2]. Recently, thin film TM aluminides have also become of great interest in the microelectronics industry[3]. Furthermore, it is scientifically interesting to note the observed variation of the crystal structure of TM trialuminides across each of the early transition-metal series[4] (cf., Table 1). Note that the stable cubic $L1_2$ structure for $ScAl_3$ at room temperature

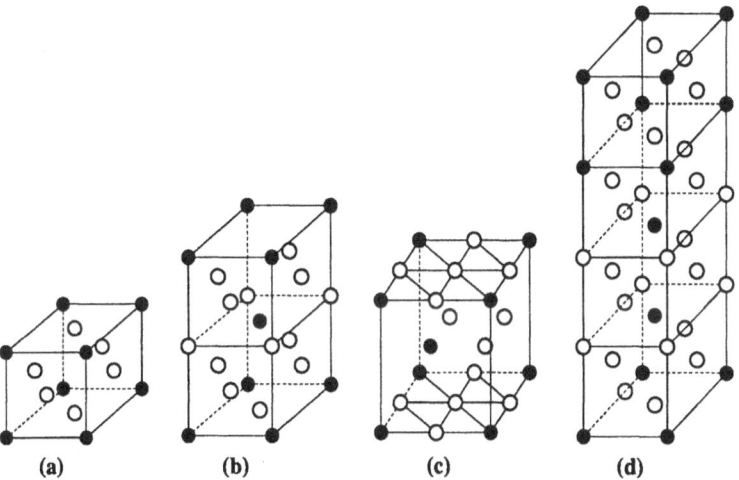

Fig.1 The unit cell of (a) $L1_2$, (b) $D0_{22}$, (c) $D0_{19}$ and (d) $D0_{23}$ structures.

appears uniquely in the left top corner of Table 1. The stability of the cubic $L1_2$ phase decreases from the left (top) to the right (bottom) of the transition metals; on the other hand, the stability of the tetragonal structure increases from left to right. In the middle, $ZrAl_3$ possesses both the cubic $L1_2$ and the tetragonal $D0_{23}$ structures as metastable and stable forms. In the whole last column, all trialuminides from VAl_3 to $NbAl_3$ to $TaAl_3$ only crystallize in the tetragonal $D0_{22}$ form.

We have studied the electronic structure and the structural stability of $TMAl_3$ (TM = Sc, Ti, V, Y, Zr and Nb) in the $L1_2$ (Fig.1(a)), the $D0_{22}$ (Fig.1(b)) and the naturally stable form (i.e., $D0_{19}$ and $D0_{23}$ (Figs. 1(c) and (d)) phases for YAl_3 and $ZrAl_3$, respectively) using the total energy local-density approach[5].

Fig.2 exhibits the total energy as a function of the Wigner-Seitz (WS) sphere radius ($r^0{}_{ws}$) for $ZrAl_3$ in the $L1_2$, $D0_{22}$ and $D0_{23}$ structures. Note that the locations of the equilibrium $r^0{}_{ws}$ are very close to each other. The calculated equilibrium total energy, E_{tot}, and the density of states at the Fermi level, $N(E_F)$, for these TM trialuminides in $L1_2$, $D0_{22}$ and in their stable forms ($D0_{19}$ for YAl_3 and $D0_{23}$ for $ZrAl_3$) are listed in Table 2. For all these trialuminides, the calculated energetically favored structure are completely in agreement with the experimentally observed stable structures. It is interesting to note that the differences in E_{tot} (14-19 mRy/f.u.) for the group-IV TM trialuminides ($TiAl_3$ and $ZrAl_3$) between the stable and hypothetical phases are much smaller than those ($\sim$ 72 mRy/f.u.) for both group-III TM ($ScAl_3$ and YAl_3) and group-V TM (VAl_3 and $NbAl_3$) trialuminides. This indicates a crossover situation between the stable and the hypothetical phases. Therefore, ternary additions in the group-IV TM trialuminides may alleviate the phase transition among the different phases. This proves indeed to be the case: the additions of Cu,Ni,Fe,Mn,Cr,Zn and Ag to $TiAl_3$ yield $L1_2$ structural compounds (such as, Al_5CuTi_2)[6], and partial substitution of V for Zr in $ZrAl_3$ increases the thermal stability of the metastable $L1_2$ phase in

4

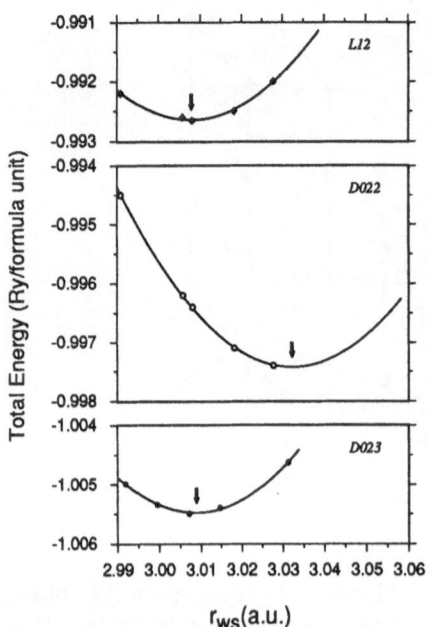

Fig.2 Total energy (subtracted by -8641 Ry/f.u.) vs. WS radius for $ZrAl_3$ in $L1_2$, DO_{22}, and DO_{23} structures.

Table 2. The total energy, E_{tot} (in mRy/f.u.), and the total density of states at E_F, $N(E_F)$ (in states/eV f.u.), of TMAl$_3$ (TM = Sc, Ti, V, Y, Zr, Nb) in the $L1_2$, DO_{22} and DO_{19} (for YAl$_3$) or DO_{23} (for ZrAl$_3$) structures. Here the total energy of the stable structure is taken as the energy zero.

	YAl$_3$		ZrAl$_3$		NbAl$_3$	
	E_{tot}	$N(E_F)$	E_{tot}	$N(E_F)$	E_{tot}	$N(E_F)$
$L1_2$	37.0	1.11	18.3	1.68	72.3	2.45
DO_{22}	71.7	2.26	13.5	1.80	0.0	0.15
DO_{19}	0.0	1.47	-	-	-	-
DO_{23}	-	-	0.0	1.09	-	-
	ScAl$_3$		TiAl$_3$		VAl$_3$	
	E_{tot}	$N(E_F)$	E_{tot}	$N(E_F)$	E_{tot}	$N(E_F)$
$L1_2$	0.0	1.15	14.0	1.88	58.0	3.07
DO_{22}	32.3	2.00	0.0	1.79	0.0	0.04

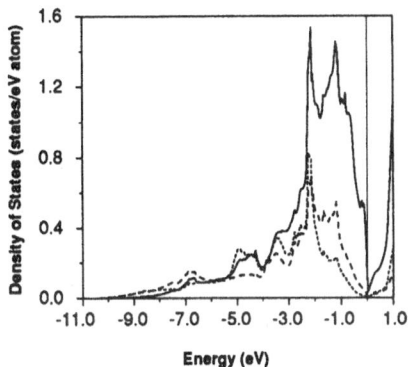

Fig.3 Partial DOS for NbAl$_3$ in the D0$_{22}$ structure: solid, dashed and broken lines denote Nb 4d, Al(1) 3p and Al(2) 3p states, respectively. Al(1) and Al(2) denote the Al atom sitting in the same and below the (001) plane of the Nb layer.

dilute alloys[7]. By contrast, it has not been possible to convert D0$_{22}$ into L1$_2$ by means of ternary additions[8] to NbAl$_3$, which can be understood from the large E$_{tot}$ difference ($\sim$ 72 mRy/f.u.) between the stable (D0$_{22}$) and hypothetical (L1$_2$) phase. In particular, note that the calculated results provide a very clear trend of increasing (decreasing) structural stability of the D0$_{22}$ (L1$_2$) phase on going from YAl$_3$ to ZrAl$_3$ to NbAl$_3$. Likewise, we have seen the same variation of the structural stability from ScAl$_3$ to TiAl$_3$ to VAl$_3$. In addition, note that the stable phase is often accompanied with a lower N(E$_F$) (cf., Table 2).

In order to gain insight at the microscopic level into the phase stability of the TM trialuminides and its variation trend with the TM constituent, we investigated the electronic structures of the three trialuminides YAl$_3$, ZrAl$_3$, and NbAl$_3$ in the D0$_{22}$ structure. For example, an analysis of the Nb-4d and Al-3p partial density of states (DOS) for D0$_{22}$ NbAl$_3$, shown in Fig.3, demonstrates that the hybridization between the Nb-d and Al-p is so substantial that the Nb-d and Al-p states almost overlap each other completely in the whole energy region from the bottom of the band to high above E$_F$. Further, a broad and prominent peak formed at about -1.1 eV for YAl$_3$, -1.5 eV for ZrAl$_3$, and -2.1 eV for NbAl$_3$ arises mainly from the TM-d – Al-p bonding states and a deep valley (or pseudogap[9]) between the bonding and the nonbonding states is located at about 1.6 eV for YAl$_3$, 1.1 eV for ZrAl$_3$ and near E$_F$ for NbAl$_3$. We find that the overall features of the total DOS for these three trialuminides resemble each other. Similarly, we have nearly the same DOS curve for ScAl$_3$, TiAl$_3$ and VAl$_3$. In other words, the rigid band approximation holds well for these trialuminides when we move from group-III to -IV to -V TM. As a result, the valence electrons gradually fill the bonding states, the Fermi level for YAl$_3$ (ScAl$_3$) shifts up for ZrAl$_3$ (TiAl$_3$), and finally moves close to the pseudogap in the DOS for NbAl$_3$ (VAl$_3$) (maximizing the bonding states[10]). Thus, one is led to a phase transition from the L1$_2$ phase for YAl$_3$ (ScAl$_3$) to D0$_{22}$ for NbAl$_3$ (VAl$_3$). In other words, the packing ratio of the valence electrons in reciprocal (equivalently k) space determines the crystal structure in real space.

2.2. ORDERED PSEUDOBINARY (Ni,Co,Fe)$_3$V ALLOY

The ordered pseudobinary alloy (Ni,Co,Fe)$_3$V may be an another example of the valence

6

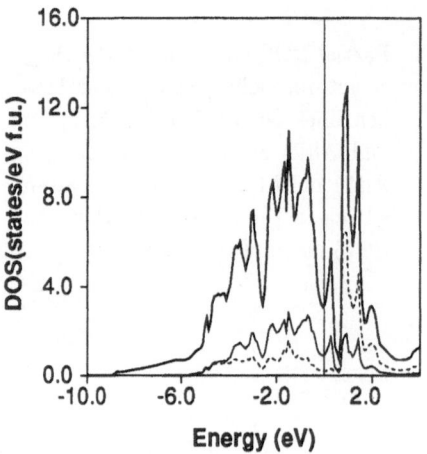

Fig.4 DOS of L1$_2$ Co$_3$V. Thick solid line denotes the total DOS. Thin solid and dotted lines denote the Co d and V d states, respectively.

electron packing in reciprocal space leading to a structural transition in real space. Sinha[11], and Liu and Inouye[12] found that the ordered crystal structure in pseudobinary Ni$_3$V-Co$_3$V-Fe$_3$V alloy can be correlated with electron concentration e/a (i.e., the number of valence electrons per atom). We inspected the DOS for the "constituent" compounds Ni$_3$V, Co$_3$V and Fe$_3$V of the pseudobinary alloy (Ni,Co,Fe)$_3$V. Fig. 4. shows the Co-d and V-d partial DOS for cubic L1$_2$ Co$_3$V. It is seen clearly that the d-d hybridization is so strong that the Co-d and V-d states overlap over the whole energy region from -8.0 eV up to high above E$_F$. A broad and prominent peak at about -1.5 eV arises from the Co-d and V-d bonding states, and a deep valley separates the bonding and antibonding states located at about 0.5 eV. Fig. 5 exhibits the total DOS for all three L1$_2$ intermetallics Fe$_3$V, Co$_3$V and Ni$_3$V. Note that the overall features of the DOS for these three compounds resemble each other; if the bottom of the valence band for Fe$_3$V (Ni$_3$V) is shifted to coincide with that of Co$_3$V, the three DOS curves nearly fall onto the same contour; in other words, the rigid band approximation holds well for these three compounds in the same (L1$_2$) structure. As stated before, the filling bonding or antibonding states will increase or decrease the cohesion (or stability). Indeed, the structural stability of the L1$_2$ structure increases from Fe$_3$V to Co$_3$V, because Co$_3$V has more valence electrons than Fe$_3$V (therefore, filling a larger occupied portion of the bonding states), and with an increase of the number of valence electrons continuously from Co$_3$V to Ni$_3$V the valence electrons now occupy the antibonding region in Ni$_3$V, leading to a phase transition to the D0$_{22}$ structure[13].

We can understand that the magic number e/a (equivalently, phase stability boundary) is associated with the fill-up of the bonding states. As stated above, the electronic structures for the Ni$_3$V, Co$_3$V and Fe$_3$V have a common feature, i.e., a deep valley separates the d-d bonding and antibonding regions. The number of electrons accommodated in the bonding region is surprisingly nearly invariant from 33.83, 33.92 to 33.75 electrons (equivalently, e/a from 8.46, 8.48 to 8.44) for Ni$_3$V, Co$_3$V and Fe$_3$V, respectively, i.e., the bonding region for the L1$_2$ phase can accomodate at most 33.9 electrons (or e/a ∼8.5). Similarly, the e/a boundary (8.50, 8.49 and 8.53 for Ni$_3$V, Co$_3$V and Fe$_3$V, respectively) is found

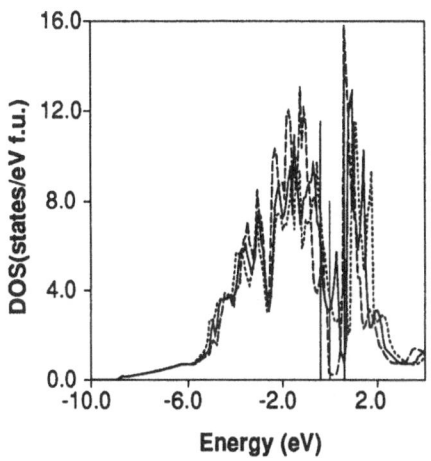

Fig. 5 Comparison of the total DOS of L1$_2$ Ni$_3$V, Co$_3$V and Fe$_3$V; the bottom of the DOS of Ni$_3$V and Fe$_3$V has been shifted to coincide that of Co$_3$V. Dashed, solid and dotted lines denote Ni$_3$V, Co$_3$V and Fe$_3$V, respectively.

for the hexagonal D0$_{19}$ phase[14]. This means that for any TM$_3$V (TM= Ni,Co and Fe) alloy possessing the electron concentration that exceeds the boundary of e/a = 8.5 - 8.53, the extra valence electrons stacked into the antibonding region destabilize both the L1$_2$ and D0$_{19}$ structures, and results in a phase transition to the D0$_{22}$ structure. This coincides with Ref.[12] 's observation that when e/a exceeds 8.54, there is a resultant change in the basic layer structure from T (triangular) type to R (rectangular) type[15] (i.e, the D0$_{19}$ structure is converted to the tetragonal D0$_{22}$ structure).

2.3. EFFECT OF TERNARY ADDITIONS: Al$_3$Ti + Cu

The titanium trialuminide, Al$_3$Ti, that crystallizes in the D0$_{22}$ structure, is considered as a promising candidate for aerospace applications. It can be converted to the cubic L1$_2$ structure by ternary additions (such as Cu, Ni, Fe and Zn) which could offer more slip systems and, consequently, might make the material more ductile.

Since Cu is the prototype element with wide composition range[16] in Al$_3$Ti that was reported to form the compound Al$_5$CuTi$_2$[6], we chose Al$_3$Ti+Cu for our first principles calculations[17]. Further, we only considered a composition of 12.5 % Cu to substitute either Al ot Ti in Al$_3$Ti, because it makes the computing tractable. By doing the substitution for both D0$_{22}$ and L1$_2$ Al$_3$Ti, we obtained four different unit cells (configurations): L1$_2$- and D0$_{22}$-like Al$_5$CuTi$_2$, and L1$_2$- and D0$_{22}$-like Al$_6$CuTi.

Our results indicate (as shown in Fig. 6 for the configurations with c/a = 2) that (i) Cu as a ternary addition strongly favors the Al site (by about 0.2 eV/atom) over the Ti site; (ii) the site preference and relative stability of the L1$_2$- and D0$_{22}$-like structures all agree with experiment; (iii) the site preference dominates over the structural factor.

We inspected the density of states for Al$_5$CuTi$_2$. As we expected from an earlier paper[18], the rigid band model works very well for Al$_5$CuTi$_2$. The profile of the total DOS for

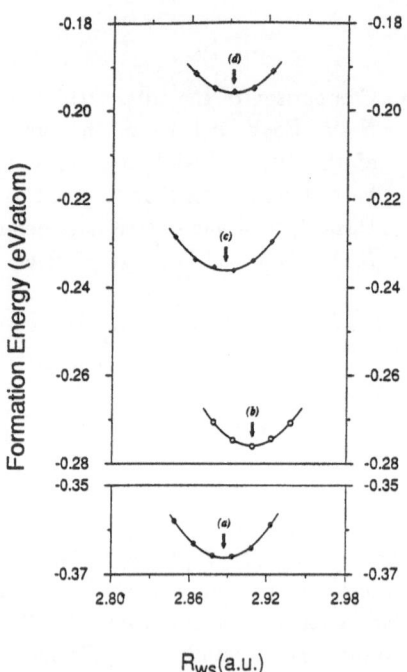

Fig. 6 Formation energy vs. WS radius with $c/a=2.0$ for: (a) $L1_2$-like Al_5CuTi_2, (b) $D0_{22}$-like Al_5CuTi_2, (c) $D0_{22}$-like Al_6CuTi, and (d) $L1_2$-like Al_6CuTi.

Al_5CuTi_2 is very much like that in pure Al_3Ti except for a prominent peak associated with Cu d electrons which are basically not hybridized with other components. Since Cu has fewer electrons participating in bonding than either Ti or Al, the Fermi energy is shifted towards higher binding energy in both $L1_2$- and $D0_{22}$-like structures compared to their counterparts in pure Al_3Ti. The number of Cu valence electrons brought E_F into a deep valley separating bonding from non-bonding and antibonding regions for the $L1_2$ structure. On the other hand, in the $D0_{22}$-like structure the E_F is located near a major peak due to the hybridization of Al-p and Ti-d electrons results in a kind of instability.

When comparison of the charge density between the $L1_2$ and the $D0_{22}$ structures for pure Al_3Ti was made, it was found that the charge density in the region between the first nearest neighbor Ti-Al and the second nearest neighbor Ti-Ti or Al-Al on the (001) plane with equal numbers of Ti and Al atoms is generally larger in the $D0_{22}$ structure than in the $L1_2$ structure. Meanwhile, the charge density in the $D0_{22}$ structure along the directions connecting the atoms on different (001) planes (denoted as non-planar bonds) is comparable with that in the $L1_2$ structure, resulting in a stronger hybridization for $D0_{22}$ Al_3Ti when one counts all the bonds together. It is interesting to note that for $L1_2$-like Al_5CuTi_2, the charge density (with $c/a=2.0$) on the (001) plane between the neighbors is larger than that in pure Al_3Ti (with $c/a=2.0$), thus surprisingly showing a similar effect to that found as one introduces tetragonal elongation into pure Al_3Ti. On the other hand, the non-planar bonds in the $L1_2$-like Al_5CuTi_2 are still comparable to those in the $L1_2$ Al_3Ti, or at most, are slightly weakened from the latter.

Table 3. Comparison of calculated and observed cohesive properties and calculated APB energies of NiAl, FeAl and RuAl.

		Latt. const. (Å)	B (Mbar)	$E_{Formation}$ (kcal/mol)	E_{APB} (erg/cm^2)
NiAl	calc.	2.89	2.1	33.8	880
	expt.	2.886[4]	1.89[23]	28.3[29]	-
RuAl	calc.	3.03	2.1	35.8	580
	expt.	3.03[4]	-	-	-
FeAl	calc.	2.87	2.0	14.0	490
	expt.	2.909[4]	1.52[30]	12.2[29]	110[30]^a

a: Measured in FeAl with 26-36 at.% Al.

3. Electronic Structure and Mechanical Properties

3.1. ANTI-PHASE BOUNDARY ENERGIES AND BONDING CHARACTERISTICS IN NiAl, FeAl and RuAl

The B2 aluminides, FeAl, NiAl and recently RuAl, have been studied as promising high temperature structural materials due to their intriguing properties[19]-[24]. From a number of experimental studies, the major deformation mode in stoichiometric B2 NiAl was found to be < 001 > slip[19] with only three independent slip systems. Since this could not meet the von Mises[25] criterion for ductile deformation in polycrystalline materials, NiAl is very brittle. On the other hand, stoichiometric FeAl deforms mainly by < 111 > slip[21,26], which could support ductile deformation. Further studies showed that the brittleness in FeAl could be caused to interstitial impurities[27]. In RuAl, some degree of ductility under compression was also reported[24]. The slip vectors of RuAl were found to be in the < 100 >, < 110 > and < 111 > directions. In addition, it is also speculated that the absence of the < 111 > slip vectors in NiAl could be attributed to a slight excess of charge in < 111 > direction[23], and a covalent-like bonding between Ni and Al was proposed[28]. The activation of < 111 > slip in NiAl is expected to change its brittle nature. A comparative study of these compounds helps the understanding of their similarities and differences.

In Table 3, cohesive properties and APB energies of NiAl, FeAl and RuAl are listed. The equilibrium lattice constants calculated for these compounds were found to be in good agreement with the experimental values (within 1%). The results show both NiAl and RuAl to have a large bulk modulus. For NiAl it is about 10 % larger than experiment. NiAl has been experimentally found to have larger formation energy (28.3 kcal/mol) than FeAl (12.2 kcal/mol). Our calculated results for these compounds are within 15 % of experiment. The formation energy for RuAl was calculated to be fairly large (35.8 kcal/mol; also much larger than that for FeAl). No experimental result is available for comparison.

The APB energy is thought to be an important factor in controlling slip behavior[31]. Application of the first principles calculations to these problems becomes a natural step. In the

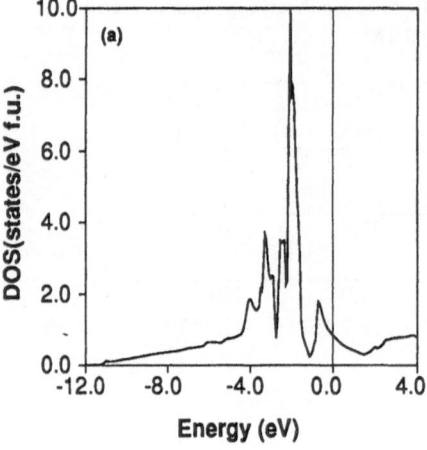

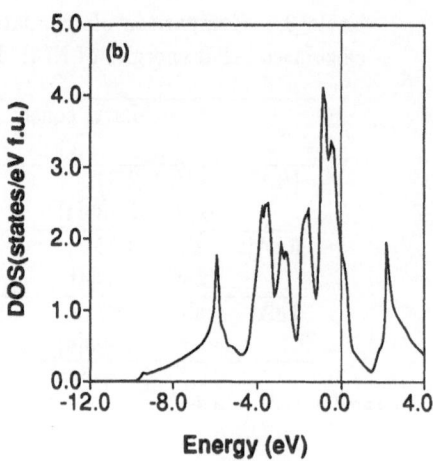

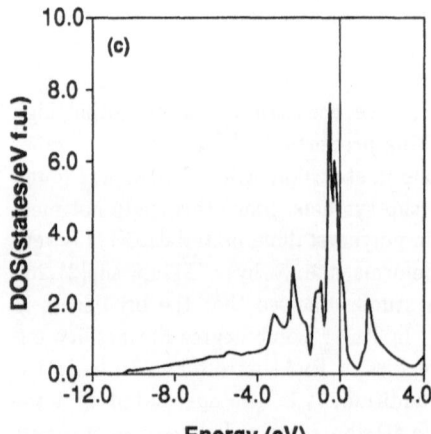

Fig. 7 Total DOS for: (a) NiAl, (b) RuAl and (c) FeAl.

$\frac{1}{2} < 111 > \{110\}$ direction, extremely high calculated APB energies for stoichiometric NiAl (1000 erg/cm^2) were obtained. The calculated APB energies for FeAl is about 490 erg/cm^2 (56 % that of NiAl, cf., Table 3), and the APB energy of RuAl is only 580 erg/cm^2, or 66 % that of NiAl. Thus, it is understandable that FeAl and RuAl can have the < 111 > slip vector while NiAl does not.

The DOS characteristics of RuAl and FeAl are similar to those of NiAl (cf. Fig. 7). It is found that the TM-d and Al-p hybridization is predominant in these compounds. In FeAl and RuAl, the Fermi energy lies on the main peak in the bonding region while in NiAl, E_F is located on the secondary peak. The introduction of an APB will greatly enhance the secondary peak. Thus, the density of states below E_F is more severely affected by an APB in NiAl than in FeAl and RuAl. The APB energy of NiAl is expected to be higher than those of FeAl and RuAl

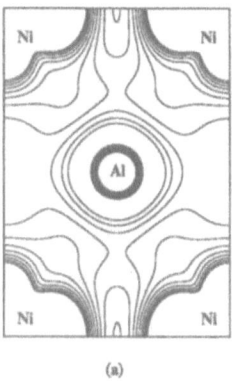 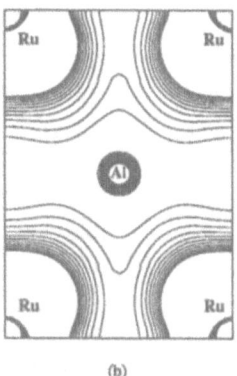 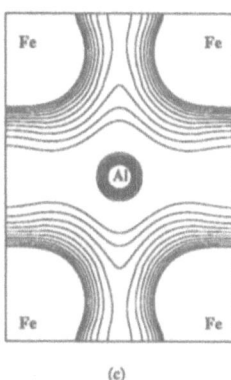

(a)   (b)   (c)

Fig. 8 Charge density near E_F (in unit of $1/(\text{a.u.})_3$) in the $\{110\}$ plane for (a) NiAl, (b) RuAl and (c) FeAl.

The charge density on the $\{110\}$ plane from the states near-the-Fermi-energy for these compounds are presented in Fig. 8. The results for the NiAl $\{110\}$ plane (Fig. 8(a)) show a directional charge distribution between Ni and Al in the $< 111 >$ direction. In RuAl (Fig. 8(b)) and FeAl (Fig. 8(c)), the charge connection is not directional in $< 111 >$. Between the nearest neighbor Ni atoms in NiAl, the charge connection is seen to be extremely weak compared to d-d bonding between Ru-Ru and Fe-Fe. A relatively strong bond between Al-Al in NiAl is also observed. On the other hand, the Al-Al bond is weak in RuAl and FeAl. The NiAl results are consistent with the report that some covalent bonding may exist and be superimposed upon the metallic bond[32].

3.2. INTERFACIAL ENERGY : TWIN AND STACKING FAULT ENERGIES OF Al AND Pd

Twin (TW) and stacking fault (SF) energies (in the broad sense, interfacial energies) are of significance for determining many physical properties. Thus, for example, it is well known that the mechanical properties are closely related to the existence of faults in alloys[33]. Despite the fact that a large amount of experimental data about fault energies has been accumulated for pure metals[34], there are still large uncertainties in the measured values of such interfacial energies. For example, the value of the TW energy for pure Cu ranges from 20 to 160 erg/cm^2. Direct measurements of these interfacial energies require very delicate techniques. Moreover, the measured value of such an interfacial energy depends sensitively upon many factors, such as the existence of internal stresses, impurities in the matrix, temperature, etc.[33]

We studied the (111) twin boundary and stacking fault energies of Al and Pd [35] using the all-electron total energy self-consistent linear muffin-tin orbital (LMTO) method within the framework of density functional theory. Fault energies are determined by subtracting the total energies with fault from without one, both obtained for the same size supercells.

Table 5.   Twin and stacking fault energies for Al and Pd (in erg/cm^2)

		Al			Pd	
		TW	ISF	ESF	TW	ISF
	Ref.[35]	130±15	280±40	~ 260	97±5	-
calc.	Ref.[36]	118	-	-	-	-
	Ref.[37]	61	160	133	-	-
	Ref.[33]	75	166	-	-	180
expt.	Ref.[38]	120	200	-	-	180
	Ref.[39]	75-100	135	180	-	-
	Ref.[34]	100-125	160	200	-	-

The calculated TW and SF energies for Al and Pd are listed in Table 4. The calculated intrinsic fault energy (280± 40 erg/cm^2) for Al has approximately the same value as the extrinsic one (~ 260 erg/cm^2). This is expected from their geometrical arrangement, i.e., an intrinsic (extrinsic) SF a layer is removed (inserted) into the otherwise normal sequence along the [111] direction. We still do not understand the experimental finding, i.e., the extrinsic fault energy for Al is approximately 25% larger than that of the intrinsic one (cf., Table 5). Note that the calculated TW energy (130±15 erg/cm^2) is in fairly good agreement with the value 118 erg/cm^2 obtained from the LKKR method[36]; however, it is in general 30% larger than that observed value (extrapolated to 0 K) for Al. A plausible explanation for this discrepancy between the calculated and experimental values might be attributed to (i) neglect of the relaxation around the fault region, (ii) effect of the (finite) size of the supercell (which includes, incorrectly, the interactions between faults), and/or (iii) use of the local density approximation (we noted a similar overestimated value for the calculated antiphase boundary energy for NiAl[40]). Nevertheless, our calculated results appear to provide support for the experimental finding that 2 times the TW energy $\simeq$ the ESF energy for both Al and Pd.

Acknowledgements

This work was supported by the Air Force Office of Scientific Research (grant Nos. 88-0346 and F49620-88-C-0052 under Dr. A. Rosenstein's program). We thank Drs. R. Darolia, D. Dimiduk and R. Field for close interactions.

References

[1] Lipsitt, H. A. (1986), in S. M. Allen, R. M. Pelloux and R. Widmer (eds.), Properties of Advanced High Temperature Alloys, American Society of Metals, Metal Park, OH, pp. 157; Kumar, K. S. (1990), Inter. Mater. Rev 35, 293.

[2] Fine, M. E. (1975), Metall. Trans. A6, 625; Kubel, E. J. (1986), Adv. Mater. Pro. Metal Prog. 130, 43.

[3] Colgan, E. G. (1990), Mater. Sci. Reports 5, 1.

[4] Pearson, W. B. (1967), A Handbook of Lattice Spacing and Structures of Metals and

Alloys Vol.2, Pergamon, Oxford; Pearson, W. B. (1967), in W. B. Pearson (ed.), Structure Reports, International Union of Crystallography, Utrecht 32A, pp. 14.

[5] Xu, J.-h. and Freeman, A. J. (1989), Phys. Rev. B40, 11927; Xu, J.-h. and Freeman, A. J. (1991), J. Mater. Res. 6, 1188.

[6] Maeland, A. J. and Narasimhan, D. (1989), in High-Temperature Ordered Intermetallic Alloys III, Materias Research Society, Pittsburgh, PA, pp. 723.

[7] Zedalis, M. S. (1986), Ph.D. thesis, Northwestern University; Chen, Y. N., Fine, M. E., Weertman, J. R. and Lewis, R. E. (1987), Scripta Metall. 21, 1003.

[8] Subramanian, P. R. and Simmons, J. P. (1991), Scripta Metall. 25, 231.

[9] Pasturel, A., Colinet, C. and Hicter, P. (1985), Physica B+C 132, 177.

[10] Hoffmann, R. (1988), Rev. Mod. Phys. 60, 601.

[11] Sinha, A. K. (1969), Trans. Metall. Soc. AIME 245, 911.

[12] Liu, C. T. and Inouye, H. (1979), Metall. Trans. A10, 1515.

[13] Lin, W., Xu, J.-h. and Freeman, A. J. (1991), in High-Temperature Ordered Intermetallic Alloys IV, Materias Research Society, Pittsburgh, PA, pp. 131.

[14] Lin, W., Xu, J.-h. and Freeman, A. J. (1991), to appear in Phys. Rev. B.

[15] Liu, C. T. (1989), in G. M. Stocks and A. Gonis (eds.), Alloy Phase Stability, Kluwer Academic, Norwell, Massachusetts, pp. 7.

[16] Mazdiyasni, S., Miracle, D. B., Dimiduk, D. M., Mendiratta, M. G. and Subramanian, P. R. (1989), Scripta Metall. 23, 327.

[17] Hong, T. and Freeman, A. J. (1991), J. Mater. Res. 6, 330.

[18] Hong, T., Watson-Yang, T. J., Freeman, A. J. and Xu, J.-h. (1990), Phys. Rev. B41, 12462.

[19] Ball, A. and Smallman, R. E. (1966), Acta Metall. 14, 1349; (1968) ibid. 16, 233.

[20] Stephens, J. R. (1987), in High-Temperature Ordered Intermetallic Alloys, Materias Research Society, Pittsburgh, PA, pp. 381; and references therein.

[21] Vedula, K. and Stephens, J. R. (1987), in High-Temperature Ordered Intermetallic Alloys II, Materias Research Society, Pittsburgh, PA, pp. 381; and references therein.

[22] Baker, I. and Munroe, P. R. (1989), in High Temperature Aluminides and Intermetallics, Metallurgical Society of AIME, Warrendale, PA, pp. 425.

[23] Fleischer, R. L., Dimiduk, D. M. and Lipsitt, H. A. (1989), Ann. Rev. Mater. Sci. 19, 231; and references therein.

[24] Fleischer, R. L., Field, R. D. and Briant, C. L. (1991), Metall. Trans. A22, 403.

[25] von Mises, R. (1928), Z. Angew. Math. Mech. 8, 161.

[26] Yamagata, T. and Yoshida, H. (1973), Mater. Sci. and Eng. 12, 95.

[27] Dimiduk, D. M. and Miracle, D. B. (1989), in High-Temperature Ordered Intermetallic Alloys III, Materias Research Society, Pittsburgh, PA, pp. 349.

[28] Cooper, M. J. (1963), Phil. Mag. 89, 811.

[29] Hultgren, R., Desai, P. D., Hawkings, D. T., Gleiser, M. and Kelley, K. K. (1973), Selected Values of the Thermodynamic Properties of Binary Alloys, American Society for Metals, Metals Park, OH.

[30] Yoo, M. H., Takasuga, T., Hanada, S. and Izumi, O. (1990), Mater. Trans. JIM 31, 435.

[31] Marcinkowski, M. J. (1974), in Treatise on Materials Science and Technology, Academic Press, New York, pp. 333.

[32] Noebe, R. D., Bowman, R. R., Kim, J. T., Larsen, M. and Gibala, R. (1989), in High Temperature Aluminides and Intermetallics, Metallurgical Society of AIME, Warrendale, PA, pp. 271.

[33] Hirth, J. P. and Lothe, J. (1982), Theory of Dislocations, 2nd ed., John Wiley & Sons, New York.

[34] Murr, L. E. (1975), Interfacial Phenomena in Metals and Alloys, Addison Wesley, New York.

[35] Xu, J.-h., Lin, W. and Freeman, A. J. (1991), Phys. Rev. B43, 2018.

[36] Maclaren, J. M., Crampin, S., Vvedensky, D. D. and Eberhart, M. E. (1989), Phys. Rev. Lett. 63, 2586.

[37] Simon, J. P. (1979), J. Phys. F9, 1425.

[38] Reed-Hill, R. E. (1973), in Physical Metallurgy Principles, 2nd ed., D. Van Norstrand, New York, pp. 892.

[39] Smallman, R. E. and Dobson, P. S. (1970), Metal. Trans. 1, 2383.

[40] Freeman, A. J., Hong, T. and Xu, J.-h. (1989), in V. Vitek and D. J. Srolovitz (eds.), Atomic Simulation of Materials, Plenum, New York, pp. 41.

FIRST PRINCIPLES THEORY OF ALLOY PHASE STABILITY: ORDERING AND PRE-MARTENSITIC PHENOMENA IN β-PHASE NiAl

G. M. Stocks * , W. A. Shelton * , D. M. Nicholson * , F. J. Pinski ** , B. Ginatempo † , A. Barbieri ‡ , B. L. Györffy ‡ , D. D. Johnson § , J. B. Staunton ¶ , P. E. A. Turchi ‖ and M. Sluiter ‖

*Metals and Ceramics Division, Oak Ridge National Laboratory, Oak Ridge, Tennessee, 37831-6114, USA.
**Department of Physics, University of Cincinnati, Cincinnati, Ohio 45221, USA.
†Istituto di Fisica Teorica, Universta di Messina, Messina, Italy.
‡H.H. Wills Physics Laboratory, University of Bristol, Bristol BS8 1TL, U.K.
§Sandia National Laboratories, Livermore, California 94551-0969, USA.
¶Department of Physics, University of Warwick, Coventry CV4 7AL, UK.
‖Lawrence Livermore National Laboratory, Livermore, California 94550, USA.

ABSTRACT

We review, briefly, the first principles KKR–CPA theory of the electronic structure and energetics of alloys in which compositional disorder plays a role. We also review the first principles theories of ordering and alloy phase stability that are built on the KKR–CPA description of the disordered state. We point to a number of underlying electronic driving mechanisms of ordering and clustering that have been uncovered. Specifically, we emphasize the important role that Fermi surface nesting plays in driving specific instabilities . Using a newly developed method for treating the effects of disorder in alloys that have complex lattices we present results for the ordering energies of beta phase $Ni_cAl_{(1-c)}$ alloys. Finally, we show alloy "Fermi surfaces" in β–phase $Ni_{0.625}Al_{0.375}$ that support the notion that the pre-martensitic phenomena observed in this alloy are Fermi surface driven.

INTRODUCTION

Although the central theme of this Advanced Research Workshop is ordered intermetallic alloys, it is quite clear that a detailed understanding of their properties requires consideration of the effects of disorder. The need to study these effects arises in a variety of ways. Firstly, by studying the energetics of nearby disordered phases one may uncover the electronic mechanisms that give rise to the formation of the specific ordered phases observed. Such an understanding can provide insights into possible ways of obtaining desirable phases or suppressing undesirable ones. Secondly, except at precise stoichiometry, some form of disorder is present even in alloys that are nominally described as ordered. Thirdly, substitutional and interstitial alloying additions again result in disorder. Indeed, most of the ordered intermetallic alloys that are currently under development and that are the primary subjects of this ARW fall into these latter two categories. The whole question of phase stability is bound up in the first. The point of the discussion then becomes to what extent disorder plays an important role in determining the properties of these materials.

C. T. Liu et al. (eds.), Ordered Intermetallics – Physical Metallurgy and Mechanical Behaviour, 15–36.

During the last few years significant progress has been made towards developing first principles theories of the properties of alloys. These advances are built upon the foundation provided by the local density approximation to density functional theory (LDA-DFT) [1]. For systems that have underlying periodicity, the equations of density functional theory can now be solved to a high degree of precision using standard band theory methods. The effect to which this advance can be put, within the context of ordered intermetallics, is amply demonstrated by other contributors to this workshop [2], [3]. Here, we will consider aspects of the first principles theory af alloys that specifically involve some aspect of substitutional disorder.

Systems and problem areas that we have in mind are illustrated in figs. 1 and 2. In fig.

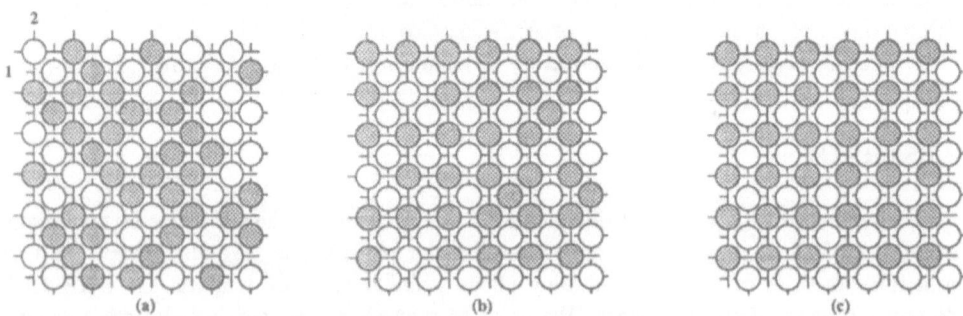

Figure 1: Two dimensional depiction of an equiatomic AB alloy. We have identified two sublattices (1,2) and two atomic species ($A \equiv$ unshaded; $B \equiv$ shaded), (a) disordered $A_{0.5}B_{0.5}$ solid solution for which the long range order parameter $\eta = 0$, (b) a partially ordered alloy $0 \leq \eta \leq 1$ (c) ordered AB intermetallic compound $\eta = 1$.

1 we indicate three distinct situations. From the computational standpoint, the completely ordered state (c) is the most straight forward. Because it possesses translational symmetry, standard band theory methods can applied to solving the LDA–DFT equations. In (b) some of the A-atoms on sublattice 1 have been interchanged with B-atoms on sub-lattice 2 i.e. $0 \leq \eta \leq 1$ where η is the long range order parameter $\eta = 2c_1 - 1$ and c_1 is the concentration of the A-species on sub-lattice 1. In (a) $\eta = 0$ i.e. the A- and B-species are randomly distributed across the two sub-lattices, at this point it is natural to drop the distinction between the two sublattices and consider the alloy has a single atom per unit cell random solid solution. In fig.2 we consider two more circumstances where disorder is endemic to the problem. In (a) we consider an ordered intermetallic that is hyper-stoichiometric in the A-species. Clearly, many situations are possible, for example the excess A-atoms can substitute onto the B sub-lattice creating antisite defects as is pictured in fig.2. Other possibilities include, the creation of constitutional vacancies and the creation combinations of antisite defects and constitutional vacancies. Finally, in fig.2 (b) we picture the addition of a third element. We have assumed that this element substitutes randomly for both the A- and B-species. Once again this is only one of many possibilities, others include preferential substitution and the possibility that introduction of a third element will induce randomization of the original compound.

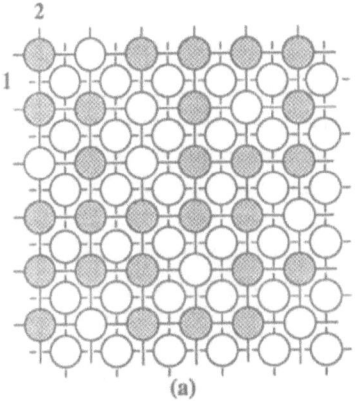
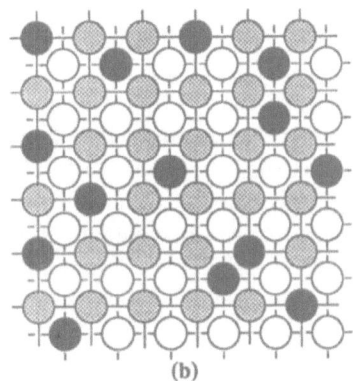

Figure 2: Two dimensional depiction of *(a)* off-stoichiometric AB ordered intermetallic, *(b)* ternary addition to binary AB ordered intermetallic

The Korringa–Kohn–Rostoker coherent potential approximation (KKR-CPA) [4] is a first principles theory of the electronic structure and energetics of substitutionally disordered alloys. The KKR-CPA has mostly been applied to binary solid solutions [5],[6] [7]. However, the method is not limited to such cases [8] [9] and a new implementation of a fully self-consistent total energy complex lattice version has been developed [10] that can treat systems having multiple sublattices and that are only partially disordered as in the examples given above.

In the next two sections we outline the first principles LDA-KKR-CPA method for calculating the electronic structure and energetics of disordered alloys and briefly discuss the first principles theories of ordering and phase stability that have been built on the foundation provided by the KKR–CPA and we point to some examples of their successful application. In the remaining sections, we turn our attention β-phase NiAl alloys. We present calculations of the ordering energy of B2 structure NiAl based both on generalized perturbation method (GPM) [11], [12] pair-wise interchange potentials and on direct calculation of the energy difference between a hypothetical disordered bcc phase and the B2 structure ordered intermetallic compound. We argue that the ordering is dominated by the first neighbor interaction and that, to the extent that the lack of ductility in this system is related to this observation, it will be difficult to overcome by macro-alloying. In the next section, we show, in some detail, results of electronic structure calculations for the martensitic composition $Ni_{0.625}Al_{0.375}$ which are consistent with the notion that the pre-martensitic phonon softening observed in this system [13], [14] is Fermi surface driven as has recently been suggested by Zhao and Harmon [15]. The final section contains some concluding remarks.

LDA–KKR–CPA

The density functional theory (DFT) is, in principle, an exact method for calculating the energetics of an electron system in the field of the atomic nuclei [1]. The local density approximation (LDA) converts density functional theory into a practical computational method. For ordered solids, such as pure metals and intermetallic compounds, the LDA has proven sufficiently reliable to allow one to contemplate problems of metallurgical interest, e.g. calculation of the small energy changes associated with allotropic transformations and intermetallic compound formation. However, the task of obtaining a complete understanding of phase stability in alloys requires theories of the energetics of both ordered phases and disordered, solid solution, phases. It was for treating the latter that the KKR-CPA was developed.

The straightforward application of LDA-DFT to treating substitutionally disordered alloys is not a practical proposition since, in principle, it involves calculation of the energy of each configuration of the alloy and then averaging these over the ensemble of configurations. In order to see this more clearly let us outline what such an application would entail. We begin by specifying the configuration of the alloy by a set of occupation variables $\{\xi_i\}$ where ξ_i takes on the value 1 if there is an A atom at the lattice site labelled i and 0 if the site is occupied by a B atom. LDA-DFT involves solving the Hartree self-consistent field like, Kohn–Sham equations

$$\left[-\nabla^2 + \sum_i v_i(\vec{r})\right] \psi_n(\vec{r}) = \epsilon_n \psi_n(\vec{r}) \tag{1}$$

where the crystal potential $v_i(\vec{r})$ takes the form

$$v_i(\vec{r}) = v\left(\vec{r} - \vec{R}_i; [n(\vec{r}; \{\xi_i\})]\right) \tag{2}$$

and the functional $v(\vec{r} - \vec{R}_i; [n])$ is given by the LDA, $\vec{R}$ is the position vector of the i-th nucleus, and $n(\vec{r}; \{\xi_i\})$ is the charge density for a given configuration $\{\xi_i\}$. The charge density $n(\vec{r}; \{\xi_i\})$ is given in terms of the eigen-solutions $\psi_n(\vec{r})$ of eq. 1 as

$$n(\vec{r}; \{\xi_i\}) = \sum_n |\psi_n(\vec{r})|^2 f(\epsilon_n) \tag{3}$$

For a specific configuration, the self consistent solution of eqs. 1, 2, and 3 involves making an initial guess of $n(\vec{r}; \{\xi_i\})$ (say from the superposition of neutral atom charge densities), calculation of $v(\vec{r} - \vec{R}_i; [n])$ using the LDA, solving eq. 1 and recalculation of $n(\vec{r}; \{\xi_i\})$ from eq. 3. If the new $n(\vec{r}; \{\xi_i\})$ is not equal to the original $n(\vec{r}; \{\xi_i\})$ to within some prescribed accuracy then a new guess of $n(\vec{r}; \{\xi_i\})$ is made and the process is repeated until convergence is obtained. The total energy for that configuration $E_0^{\text{LDA}}(\{\xi_i\})$ can then be calculated from the the self-consistent $n(\vec{r}; \{\xi_i\})$ and the band energies ϵ_n which are the eigen-values of eq. 1.

The final step in this *gedanken* calculation is to calculate the configurationally averaged total energy by averaging over the ensemble of configurations

$$\overline{E} = \sum_{\{\xi_i\}} P(\{\xi_i\}) E_0^{\text{LDA}}(\{\xi_i\}) \tag{4}$$

where $P(\{\xi_i\})$ is the probability distribution.

Clearly, this is an impossible proposition for two major reasons. Firstly, it is not possible to solve eq. 1 for an arbitrary configuration that has no translational symmetry. Secondly, the set of configurations $\{\xi_i\}$ involved is prohibitively large. For a homogeneously random alloy $P(\{\xi_i\})$ is given by

$$P(\{\xi_i\}) = \prod_i p(\xi_i) \tag{5}$$

where for a $A_cB_{(1-c)}$ binary solid solution $p(\xi_i)$ may be parameterised in terms of the concentration c of the A species as

$$p(\xi_i) = c\xi_i + (1-c)(1-\xi_i). \tag{6}$$

In a straight ahead application of LDA–DFT the above procedure is clear, calculate the energy for each configuration then average over configurations.

In order to obtain a practical, albeit approximate, method for carrying out total energy calculations we eschew the above, direct, approach and advocate the use of the so called Korringa-Kohn-Rostoker Coherent Potential Approximation (KKR-CPA) method [6]. In the LDA-KKR-CPA two primary assumptions are made. Firstly, that the system can be represented by the homogeneous random probability distribution eqs. 5, 6. Secondly, that the processes of averaging and self-consistency can be inverted. This latter step involves replacing the crystal potential in eq. 2 by $\langle v_i(\vec{r} - \vec{R}_i; [n(\vec{r}; \{\xi_i\})])\rangle_{i,\alpha}$, its average over all the occupation variables save the one referring to the site i,

$$
\begin{aligned}
\langle v_i(\vec{r} - \vec{R}_i; [n(\vec{r}; \{\xi_i\})])\rangle_{i,\alpha} &= \bar{v}(\vec{r} - \vec{R}_i; \xi_i) \\
&= \xi_i \bar{v}^A(\vec{r} - \vec{R}_i; \bar{n}^A, \bar{n}_0, \bar{n}) + \\
&\quad (1 - \xi_i)\bar{v}^B(\vec{r} - \vec{R}_i; \bar{n}^B, \bar{n}_0, \bar{n})
\end{aligned}
\tag{7}
$$

where the potentials functions $\bar{v}^\alpha(\vec{r} - \vec{R}_i; \bar{n}^\alpha, \bar{n}_0, \bar{n})$ $\alpha = A, B$ are the LDA potential functionals evaluated at the partially averaged densities. The configurationally averaged single site potentials depend upon the partially averaged single site charge density

$$\bar{n}^\alpha = \langle n(\vec{r}; \{\xi_i\})\rangle_{i,\alpha} \tag{8}$$

inside the muffin-tin sphere surrounding the site i, $\bar{n}_0$ it's compositional and volume average in the region between the the muffin-tin sphere and the Wigner-Seitz cell, and $\bar{n} = \sum_\alpha c^\alpha \bar{n}^\alpha$, the compositionally averaged density, elsewhere. Thus, in the scheme that we are advocating here for treating the effects of disorder within the LDA-DFT, the potential function and charge densities involved in the self-consistent solution of eqs. 1, 2, and 3 are the partially averaged quantities of eqs. 7 and 8. Though much simplified, the scheme is still not practical since eq. 1 still has to solved for the random potential $\langle v_i(\vec{r} - \vec{R}_i; [n(\vec{r}; \{\xi_i\})])\rangle_{i,\alpha}$.

This latter problem is solved employing the KKR-CPA method [6]. Rather than dealing with eq. 1 we deal with the corresponding equation for the single particle Green function $G(\vec{r}, \vec{r}'; \epsilon)$

$$\left[-\epsilon - \nabla^2 + \sum_i \bar{v}(\vec{r} - \vec{R}_i; \xi_i) \right] G(\vec{r}, \vec{r}'; \epsilon) = \delta(\vec{r} - \vec{r}') \tag{9}$$

The KKR–CPA is a method that allows the direct calculation of the partially averaged Greens function $\langle G(\vec{r}, \vec{r}'; \epsilon)\rangle_{i,\alpha}$ [6] [16]. The partially averaged charge densities $\bar{n}^\alpha$ can then be obtained in terms of the partially averaged Greens function $\langle G(\vec{r}, \vec{r}'; \epsilon) \rangle_{i,\alpha}$ from the usual relation [16]

$$\bar{n}^\alpha(\vec{r}) = -\frac{2}{\pi} \int f(\epsilon) \Im \langle G(\vec{r}, \vec{r}; \epsilon) \rangle_{i,\alpha} d\epsilon. \tag{10}$$

where $\Im$ indicates the imaginargy part. Whilst the KKR–CPA method only gives an approximation to the exact configurationally averaged Green function, the approximation has proven reliable in a wide variety of circumstances. These range from calculations of the electronic structure itself to detailed comparison with experimental probes of the electronic structure such as photoelectron spectroscopies, positron annihilation measurements of Fermi surfaces, and transport properties to cite but a few [6],[17],[5], [18], and [19].

Once the $\bar{n}^\alpha(\vec{r})$ have been obtained within the KKR-CPA all that remains is to calculate the configurationally averaged total energy $\overline{E}$. For a random $A_c b_{(1-c)}$ alloy this takes the form [7]

$$\overline{E} = cE_J[\bar{n}^A, \bar{n}_0, \bar{n}] + (1-c)E_J[\bar{n}^B, \bar{n}_0, \bar{n}] \tag{11}$$

here E_J is the functional derived by Janak [20] for ordered systems with crystal potential in the muffin-tin form. A remarkable feature of eq. 11 is that, thanks to the use of CPA, $\overline{E}$ retains the variational properties characteristic of $E[n]$ for pure systems. Namely

$$\frac{\partial \overline{E}}{\partial \bar{n}^A} = 0 \quad \text{and} \quad \frac{\partial \overline{E}}{\partial \bar{n}^B} = 0. \tag{12}$$

This latter is clearly one of the reasons for the success of the KKR–CPA theory for the total energy [7].

Before closing this section we wish to define two other quantities that will be of use later. The first is the averaged densities of states given

$$\bar{n}^\alpha(\epsilon) = -\frac{1}{\pi} \int_{\Omega_i} \Im \langle G(\vec{r}, \vec{r}; \epsilon) \rangle_{i,\alpha} d\vec{r} \tag{13}$$

where the integral is over the i-th Wigner–Seitz cell in which there is an α-type atom. Major structure in this quantity can be measured in a number of the experimental probes mentioned above. Its first moment,

$$\overline{E}_{BS}^\alpha = \int f(\epsilon)\epsilon\bar{n}^\alpha d\epsilon \tag{14}$$

gives the band structure contribution to the energy $E_J[\bar{n}^\alpha, \bar{n}_0, \bar{n}]$ in eq. eq:total-energy and provides the basis for present application of the Gyorffy-Stocks [21] and generalized perturbation method (GPM) [12] approaches to ordering and phase stability discussed in the next section.

The second quantity of interest is the Bloch spectral function $A^B(\vec{k}, \epsilon)$

$$A^B(\vec{k}, \epsilon) = \sum_{i,j} e^{i\vec{k}\cdot(\vec{R}_i - \vec{R}_j)} \int_{\Omega_i} \Im \langle G(\vec{r} + \vec{R}_i; \vec{r} + \vec{R}_j; \epsilon) \rangle d\vec{r}. \tag{15}$$

The Bloch spectral function contains a complete description of the electronic structure of a random alloy. It is the generalization to a disordered system of the band structure of pure metals and ordered compounds. Indeed, for an ordered system it reduces to

$$A^B(\vec{k}, \epsilon) = \sum_\nu \delta(\epsilon - \epsilon_{\vec{k},\nu}) \tag{16}$$

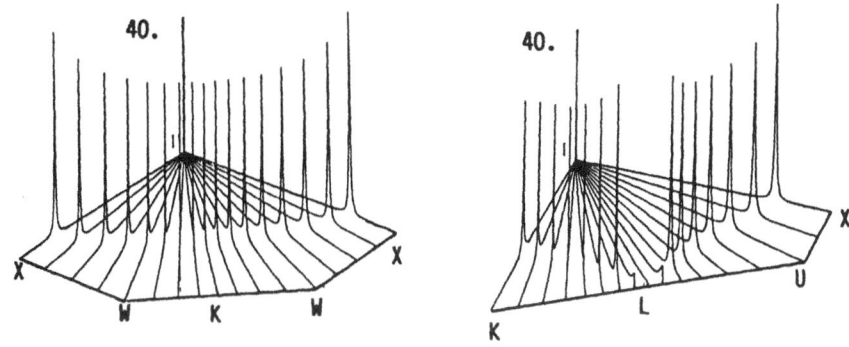

Figure 3: The Bloch spectral density function $A^B(\vec{k}, \epsilon)$ at the Fermi energy $\epsilon = \epsilon_F$ along various directions emanating from the center of the fcc Brillouin zone for a disordered $Cu_{0.75}Pd_{0.25}$ alloy.

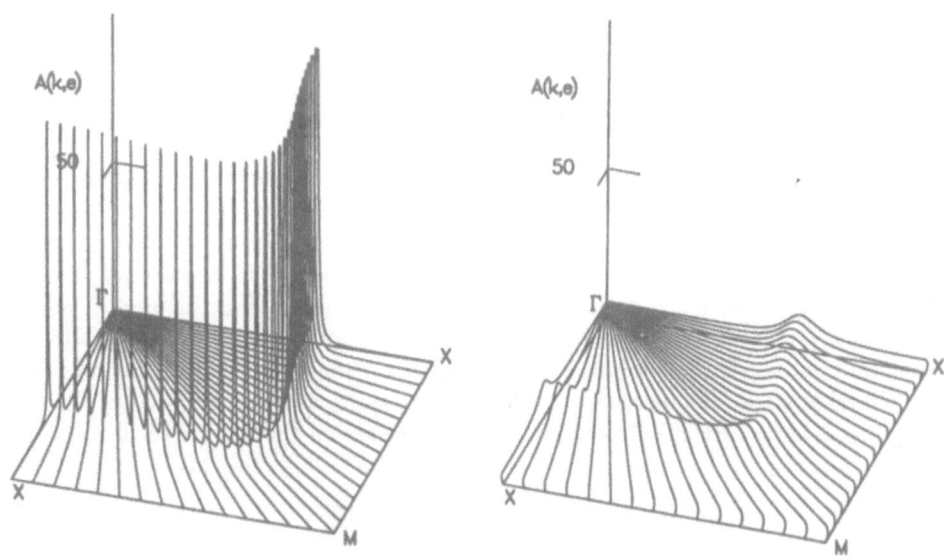

Figure 4: The Bloch spectral density function $A^B(\vec{k}, \epsilon)$ at the Fermi energy $\epsilon = \epsilon_F$ along various directions emanating from the center of the bcc Brillouin zone for ordered NiAl (left) and hypothetical disordered bcc $Ni_{0.5}Al_{0.5}$ alloy (right). For the ordered alloy the Fermi surface is calculated at a slightly complex energy ($\Im(\epsilon_F) = 0.005$ Ry.) in order to introduce a sufficient broadening to allow plotting

where $\vec{k}$ is the Bloch wave-vector and ν is the band index. For disordered alloys the δ-function peaks of eq. 16 broaden into peaks with finite width and finite heights whose position in $\vec{k}$, ϵ space trace out the band structure. The finite widths are interpretable as finite inverse life-times if viewed in energy space or inverse mean free paths if viewed in $\vec{k}$ space. It should be stressed that while $\vec{k}$ is not a good quantum number for a given configuration $\{\xi_i\}$ it does correspond to the translational symmetry of the average lattice. Specifically, $A^B(\vec{k},\epsilon)$ is periodic and has the Brillouin zone of the unoccupied lattice. As examples of systems in which the effects of disorder on the electronic structure are very different, in figs. 3 and 4 we show the Fermi energy Bloch spectral functions for disordered fcc $Cu_{0.75}Pd_{0.25}$ and hypothetical disordered bcc $Ni_{0.5}Al_{0.5}$. In fig. 4 we also plot $A^B(\vec{k},\epsilon)$ for ordered B2-structure NiAl at a slightly complex energy. The imaginary part in the energy gives rise to a small amount of broadening in $A^B(\vec{k},\epsilon)$. For $Cu_{0.75}Pd_{0.25}$ the effects of disorder on $A^B(\vec{k},\epsilon)$ are small, the peaks are well defined, and the corresponding mean free path is long. Under such circumstances, it is possible to view the surface in $\vec{k}$-space defined by the locus of the peak positions of $A^B(\vec{k},\epsilon)$ as being the "Fermi surface" of the disordered alloy. For $Ni_{0.5}Al_{0.5}$ the effects of disorder on $A^B(\vec{k},\epsilon)$ are large, the weight in $A^B(\vec{k},\epsilon)$ is spread throughout the Brillouin zone and it is no longer possible to speak of this alloy having a Fermi surface in any meaningful sense.

FIRST PRINCIPLES THEORIES OF ORDERING

Currently, two rather different approaches for obtaining a first principles theory of compositional ordering in metallic alloys are being actively pursued. The first of these is the first principles mean field theory method of Gyorffy and Stocks (GS) [21]. In this approach the concentration-concentration direct correlation function $S^{(2)}(\vec{q})$ is evaluated directly in terms of the electronic structure. At the present time such a direct approach cannot be implemented without serious approximation. In the Gyorffy and Stocks approach the most significant of these is the mean field approximation for the configurational entropy. One of the central results of the theory is that the Warren-Cowley short range order (SRO) parameter, that is measured in diffuse scattering experiments on the disordered phase, is given by

$$\alpha(\vec{q}) = \frac{1}{1 - \beta\bar{c}(1 - \bar{c})S^{(2)}(\vec{q})} \tag{17}$$

It should be noted that the above formula is exact and can be used as the definition of the direct correlation function and, therefore, provides a useful basis for comparisons between different theories and experiments. For instance, for a pairwise Ising model $S^{(2)}(\vec{q}) = v(\vec{q})$ where $v(\vec{q})$ is the lattice Fourier transform of the pair potential and eq. 17 becomes the usual Krivoglaz–Clapp–Moss expression [22], [23], [24].

The power of the above approach is that $S^{(2)}(\vec{q})$ may be calculated from the results of a LDA–KKR–CPA calculation of the electronic structure in the homogeneously disordered phase [21] and as a result involves no arbitrary or adjustable parameters. Thus, the predictions of the theory regarding the nature of the ordering process are couched in terms of the underlying electronic structure. Furthermore, the method is implemented in reciprocal space and is able to treat interactions of arbitrary range and complexity. This is particularly important for systems in which the interactions that govern the ordering process are long ranged in real space as in the $Cu_cPd_{(1-c)}$ alloys discussed below [25] where the ordering is driven by Fermi surface nesting.

The Gyorffy-Stocks method has been applied to a variety of alloy systems in order to identify the electronic mechanisms driving the ordering processes. All of the KKR–CPA calculations of $S^{(2)}(\vec{q})$ and $\alpha(\vec{q})$ that have been published to date rest on the the the further assumption that the full expression for the electronic energy used in the formal derivation of the KKR-CPA theory of $S^{(2)}(\vec{q})$ [21] can be approximated by the band structure energy $\overline{E}_{BS}^{\alpha}$ of eq. 14 alone, i.e that double counting corrections to the total energy can be ignored. This approximation is expected to be valid for systems where charge transfer is small as is the case for the systems discussed below.

In $Pd_cRh_{(1-c)}$ alloys band filling or electron to atom ratio (e/a) is the dominant factor [25], [26] in deciding that the alloy clusters in the disordered phase and phase separates at low temperature. The calculations on $Pd_cRh_{(1-c)}$ put on a first principles basis the picture obtained using a simple tight binding description of the electronic structure that transition metal/transition metal alloys for which the d-band is either almost empty $(e/a \approx 0)$ or almost full $(e/a \approx 10$; as is the case for $Pd_cRh_{(1-c)})$ should phase separate whilst for half filling $(e/a \approx 5)$ alloys should order. In $Ni_cPt_{(1-c)}$ alloys two effects related to atomic size [27],[28] have been identified has being responsible for driving ordering, over coming the tendency towards phase separation that one would expect based on band filling arguements (for NiPt $e/a = 10$).

In Cu-rich $Cu_cPd_{(1-c)}$ alloys ordering is driven by Fermi surface nesting [21], [25], [29]. In the disordered phase this gives rise to peaks in the SRO diffuse scattering that are incommensurate with the underlying lattice. The electron diffraction patterns from various

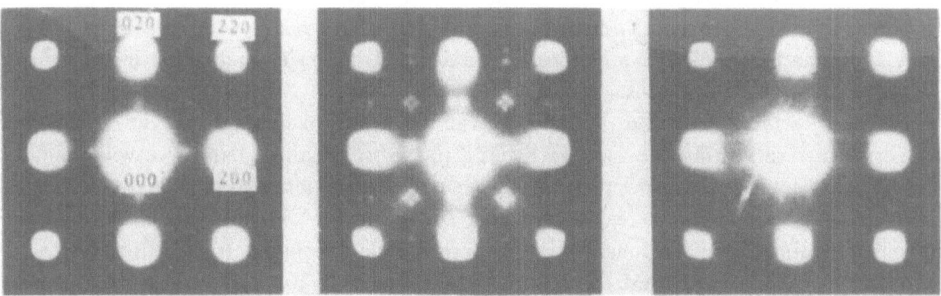

Figure 5: Electron diffraction patterns from various Cu_cPd_{1-c} alloys taken from the work of Oshima and Watanabe.

$Cu_cPd_{(1-c)}$ alloys taken from the work of Oshima and Watanabe (fig. 5) clearly show the four fold SRO diffuse scattering peaks split about the (110)-superlattice position. Furthermore, they show the rapid concentration dependence of this splitting. The large white areas in fig. 5 are the Bragg peaks associated with the underlying fcc symmetry of the disordered phase. Fig. 6 shows the calculated SRO $\alpha(\vec{q})$ corresponding to the measurements of fig. 5. The concentration dependent splitting of the SRO diffuse peaks seen in the experiments is very well described by the theoretical results. As was pointed out by Gyorffy and Stocks [21] the positions of the peaks in the SRO diffuse scattering patterns are determined by

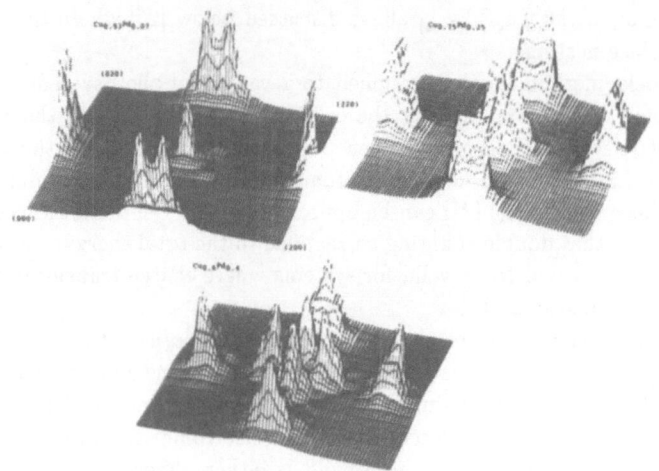

Figure 6: Calculated Warren-Cowley SRO parameter $\alpha(\vec{q})$ in the plane containing the reciprocal lattice points 000, 020, 022, and 002 for various Cu_cPd_{1-c} alloys.

spanning vectors connecting well defined, flat, parallel sheets of the alloys Fermi surface. The particular part of the Fermi surface that gives rise to this Fermi surface nesting mechanism is normal to the Γ-K direction and can be clearly seen in fig. 3. The specific nesting vector connects the flat part of the Fermi surface in one Brillouin zone with its equivalent in an adjacent one.

At low temperature Cu-rich Cu_cPd_{1-c} alloys order into a series of long period ordered structures (LPOS) [30] many of which can be predicted on the basis of the calculated $S^{(2)}(\vec{q})$ [31] by assuming that $S^{(2)}(\vec{q})$ can be interpreted as a pair potential and using this in a mean field theory of the LPOS phases.

In general terms, the $Cu_cPd_{(1-c)}$ alloy systems is a nice example of the powerful effect that Fermi surface nesting can have in driving phase transitions. Later, we will point to another example when discussing pre-martensitic phenomena observed in β-phase $Ni_cAl_{(1-c)}$

An alternate approach to treating ordering and phase stability in alloys is to map the electronic bonding forces responsible for compositional ordering onto an effective lattice Hamiltonian. If we assume that the interactions are entirely pairwise such a Hamiltonian takes the form

$$H = \frac{1}{2} \sum_{ij} [v_{ij}^{AA} \xi_i \xi_j + v_{ij}^{AB} \xi_i (1 - \xi_j) + v_{ij}^{BA} (1 - \xi_i) \xi_j + v_{ij}^{BB} (1 - \xi_i)(1 - \xi_j)] \qquad (18)$$

where $v_{ij}^{\alpha\beta}$ is the interaction energy between an α (=A or B) and a β (=A or B) type atom at the sites i and j respectively. In this language an ordering tendency is due to AB bonds being more attractive than AA or BB bonds, i.e. $v_{ij} \equiv v_{ij}^{AA} + v_{ij}^{BB} - 2v_{ij}^{AB} > 0$. By contrast $v_{ij} < 0$ implies clustering. The principle drawback of this class of methods is that there is no *apriori* guarantee that a suitable lattice Hamiltonian exists that reproduces the complex, non-local, and often long-ranged energetics of the real electron-ion system.

In first principles applications of this approach the multi-site interchange potentials are obtained either by fitting to the LDA total energies calculated for some finite set of ordered structures, as in the Connolly-Williams scheme [32], or, as in the generalized perturbation method (GPM) [33], [12] and the embedded cluster method (ECM) [11], are obtained on the bases of the KKR–CPA. In the GPM and ECM approaches, it is convenient to expand about the energy of the homogeneously disordered state, thus the energy may be partitioned as

$$E(\{c_i\}) = \overline{E} + \frac{1}{2} {\sum_{ij}}' V_{ij}^{(2)} \delta c_i \delta c_j + \frac{1}{3} {\sum_{ijk}}' V_{ijk}^{(3)} \delta c_i \delta c_j \delta c_k + \ldots \tag{19}$$

where $\delta c_i = \xi_i - c$ and $\overline{E}$ is the energy of the random alloy eq. 11. The the many-site interchange potentials $V_{ij}^{(2)}$, $V_{ijk}^{(3)}$ etc. are then given in terms of quantities that are available at the end of a KKR–CPA calculation of the random solid solution [12], [11]. Once the appropriate interaction potentials have been calculated they may be used in either cluster variation method (CVM) or Monte-Carlo simulations. This approach has been used to treat ordering and phase stability in a number of alloy systems. Again, in the calculations to date the energy expression used in deriving the relevant GPM–KKR–CPA expressions has been approximated by the band structure energy. Therefore, the same caveats apply to the GPM results as apply to the $S^{(2)}(\vec{q})$ results described above. A recent application to the phase diagram of $Cu_c Zn_{1-c}$ alloys [34] has yielded a good account of both order-disorder phenomena and structural transformations over most of the phase diagram. In a forthcoming paper we will present the results of a similar study of $Ni_c Al_{1-c}$ alloys. In the next section of this paper we use the calculated multi-site interchange potentials to discuss ordering energies in $Ni_c Al_{1-c}$ alloys and to comment on the nature of the bonding.

ORDER–DISORDER ENERGIES AND BONDING IN β–PHASE NiAl

Although the energy difference between the ideal disordered phase of an alloy and a related ordered phase is not generally experimentally measurable it does provide an important measure of the strength of the ordering interactions present in the system. The order–disorder energy ΔE^{o-d} for a binary $A_c B_{(1-c)}$ may be define as

$$\Delta E^{o-d}(X, x) = E^o(X) - E^d(x) \tag{20}$$

where x refers to the structure of the disordered lattice and X refers to some ordered structure based on this lattice and $E^d(x)$ and $E^o(X)$ are the corresponding ground state $(T = 0°K)$ energies. Clearly, when ΔE^{o-d} is large the alloy is strongly ordering and the order–disorder temperature is high.

From the forgoing sections it is clear that there are a number of ways to proceed in calculating $\Delta E^{o-d}(X, x)$. It can obtained from the GPM interchange potentials calculated in the disordered state. It can also be obtained by direct subtraction of LDA ground state energies of the ordered and disordered states. In the following we use both these approaches in order to clarify the nature of the interactions that give rise to the strongly ordered B2 phase found in the $Ni_c Al_{(1-c)}$ alloy system.

In fig. 7 we show calculated GPM values of the first three near neighbor pairwise interchange potentials in $Ni_{0.5} Al_{0.5}$. More distant pairwise interactions are found to be

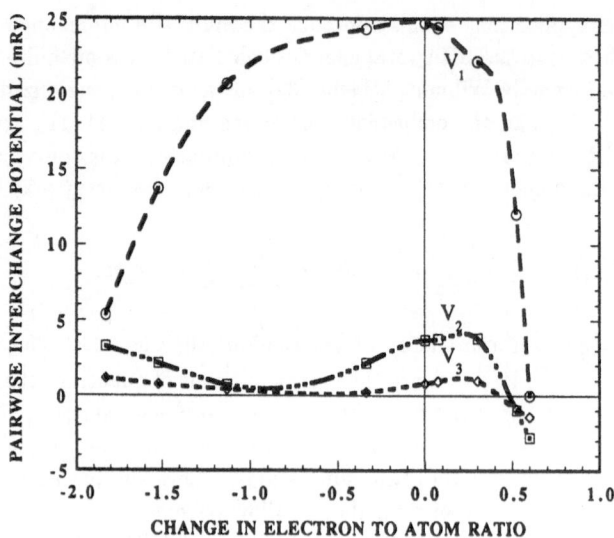

Figure 7: Pairwise interchange potentials for $Ni_{0.5}Al_{0.5}$. The first, second and third neighbor interactions are label accordingly

small, as are higher order $(3, 4, \ldots$-site) interactions. In fig. 7 the V_{ij} are plotted as a function of the departure of e/a from that of $Ni_{0.5}Al_{0.5}$. Thus, the values of V_{ij} at the zero of the abscissa are those appropriate to stoichiometric $Ni_{0.5}Al_{0.5}$. Clearly this system is strongly ordering. Furthermore, the ordering is dominated by first neighbor bonding between Ni and Al atoms ($V_1 \approx 25$ mRy/atom). For pairwise interactions the ordering energy for B2 is given by [35]

$$\Delta E^{o-d}(B2, bcc) \approx -V_1 + \frac{3}{4}V_2 + \frac{3}{2}V_3 \qquad (21)$$

Using the values of fig.7 yields $E_{ord}^{B2} = -20.3$ mRy/atom. Used in an appropriate statistical treatment these interactions give rise to an order disorder temperature in excess of 5,000K [36].

In fig. 8 we show the energies of disordered bcc $Ni_{0.5}Al_{0.5}$ and of ordered B2 structure NiAl obtained by direct calculation of the ground state energies of the ordered and disordered states using the LDA–KKR–CPA method. In order to minimize the relative error we treat the disordered phase of $Ni_{0.5}Al_{0.5}$ as comprising two interpenetrating simple cubic lattices each with the composition $Ni_{0.5}Al_{0.5}$. The ideal stoichiometric B2 ordered phase is treated similarly except that the sub-lattice compositions are $Ni_{1.0}Al_{0.0}$ and $Ni_{0.0}Al_{1.0}$. In fig. 8 the minima of the energy versus lattice spacing curves give the ground state energies, $E^o(B2)$ and $E^d(bcc)$ respectively the difference then gives the order–disorder energy $\Delta E^{o-d}(B2, bcc)$. The minima in the two curves occur at different lattice spacings. The predicted equilibrium lattice spacings are 5.45 a.u. and 5.40 a.u. for the disordered and ordered phases respectively corresponding to a volume expansion of approximately 3% on disordering. The experimentally determined lattice spacing for stoichiometric NiAl is 5.455 a.u. [37]. Interestingly, the order–disorder energy obtained from this direct calculation, $\Delta E^{o-d}(B2, bcc)$=-46.7mRy/atom, is larger by a factor of two than that obtained from the

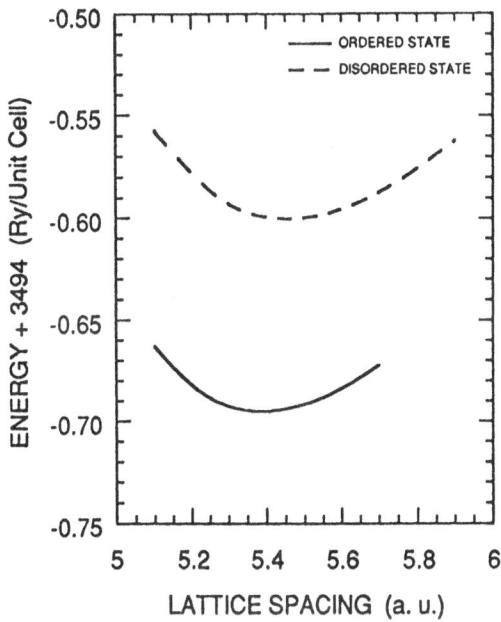

Figure 8: Ground state energies of ordered and disordered NiAl

GPM interactions. Since the GPM interactions were calculated on the basis of the band structure contribution $\overline{E}_{BS}^{\alpha}$ to the total energy alone, neglecting double counting corrections, it is not necessarily surprising that the two calculations differ. That the two values differ so much is presumably due to the relatively large charge transfer found in this alloy system. Indeed the amount of charge transfer changes substantially upon ordering. The charges associated with Wigner-Sietz cells surrounding Ni(Al)-sites are 10.2(2.8) electrons and 10.4(2.6) electrons for the disordered and ordered phases respectively. As a result of the large charge transfer the Madelung energy obtained from the ordered array of Ni and Al sites in the B2 compound makes a significant contribution to the ordering energy.

In fig. 9 we show the calculated electronic densities of states (DOS) of both ordered and disordered NiAl. Because of the large differences in the scattering cross-sections of Ni and Al sites throughout the whole range of band energies the DOS curve for the disordered alloy is relatively structureless. Clearly, upon ordering the weight in the DOS close to ϵ_F is reduced and the weight at low energies is increased resulting in the stabilization of the B2 structure.

That the interactions giving rise to the strong ordering tendency are short ranged can also be deduced from the total energy calculations. For, hyper-stoichiometric $Ni_{0.625}Al_{0.375}$, if we assume that the excess Ni goes substitutionally onto the Al-sublattice forming a random array of antisite defects, we find $\Delta E^{o-d}(B2, bcc)$=-24.5mRy/atom. In B2 NiAl all the nearest neighbor bonds are "correct" (i.e. are Ni-Al bonds); in disordered $Ni_{0.5}Al_{0.5}$ on average half of the nearest neighbor bonds are correct and half are "wrong"; for the model used above for ordered $Ni_{0.625}Al_{0.375}$ three quarters of the bonds are correct. If we assume the ordering energy to be dominated by the first neighbor interaction we expect the ordering energy of $Ni_{0.625}Al_{0.375}$ to be half that of NiAl. This is close to what was found in the full

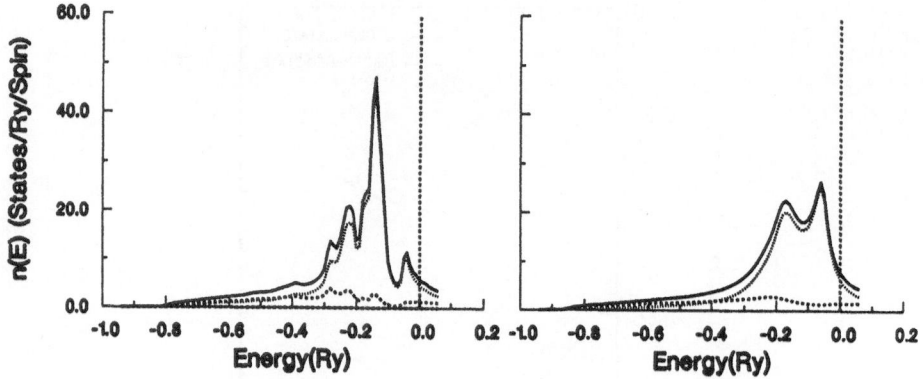

Figure 9: Densities of states for ordered B2 structure NiAl (left) and disordered bcc $Ni_{0.625}Al_{0.375}$ (right). The total density of states is given by the solid line, the contribution from Ni-sites by the dotted line and from Al-sites by the dashed line.

blown calculations.

Having obtained some insights into the nature of the interactions that drive the ordering process in NiAl, we close this section by making some comments on the implications that these insights have for attempts to modify the mechanical properties of this system by macro-alloying. An interesting feature of the interactions displayed in fig. 7 is their relative insensitivity to e/a. The observation that slip in NiAl alloys occurs in the 100-plane rather than in the 111-plane as in FeAl is generally cited as the central reason for the brittle behavior on NiAl. That the 111-slip is not favored in NiAl is, presumably, related to the very strong first neighbor bonding since 111-slip results in the creation of "wrong" bonds. Clearly, one way to make 111-slip more favorable is to reduce the first neighbor bond strength. If we assume that the addition of a third element results only in a change in e/a (the rigid band model) it is clear that it will require large additions to accomplish a substantial reduction. In this regard addition of elements that increase e/a have will have a more beneficial effect than those that reduce it. However, even here this does not appear to be a very promising route. It should also be remembered that the high melting point of NiAl is also related to the strong bonding, thus any global reduction in bond strength that would favor 111-slip would also likely reduce the melting point.

Of course, it is dangerous to base predictions of the effects on the bonding of the addition of third elements on the rigid band model since it is notoriously unreliable. A much better way to proceed is to treat the effects of alloying additions on the basis of the complex lattice LDA–KKR–CPA method. Such calculations would allow direct calculation of the energetics and predictions of site selectivities etc..

FERMI SURFACE NESTING IN MARTENSITIC β-PHASE $Ni_c Al_{(1-c)}$ ALLOYS

In the concentration range $0.60 < c < 0.68$ β–phase $Ni_c Al_{(1-c)}$ transform martensitically from the bcc-based B2 structure into fcc-derived structures. Most notably, $Ni_{0.625} Al_{0.375}$ alloys transform into the 7R structure, (Zhadnov notation [38],[39]), lower Ni-content alloys transform into the 3R or Ll_0 structure. The martensitic transition temperature T_M is strongly concentration dependent varying from near $0°K$ at low Ni-content to well above room temperature for high Ni-content [40], [41].

Above T_M a number of so called pre-martensitic phenomena are observed. Chief amongst these are "tweed" observed in transmission electron microscopy, the associated "streaking" observed in selected area electron diffraction [41], and the phonon anomalies observed in neutron scattering experiments [13], [14].

The inelastic neutron scattering experiments of Shapiro et al [13],[14] revealed a concentration and temperature dependent dip in the $\langle \varsigma\varsigma 0 \rangle$ transverse acoustic branch of the phonon spectrum. At $c = 0.50$ the dip is very shallow and is at a distance $q_0 \approx 0.12(2\pi/a)$ along the $\langle \varsigma\varsigma 0 \rangle$ direction. The dip moves to larger values of q_0 with increasing concentration. At $c = 0.625$ the dip is at $q_0 \approx 0.16(2\pi/a)$. For $Ni_{0.625} Al_{0.375}$ the dip becomes more pronounced as the temperature is lowered towards T_M. Since the phase transformation is first order, the phonon frequency does not go to zero at $T_M \approx 80K$ as it would in an ideal soft phonon transformation. However, it is clear that this anomaly is closely connected with the phase transformation. Indeed, for $Ni_{0.625} Al_{0.375}$ the reciprocal of q_0 is close to the wavelength of the martensitic 7R structure into which it transforms. Thus, understanding the mechanism that gives rise to the pre-martensitic phonon softening can be expected to provide important clues into the mechanism that drives the martensitic transformation.

Recently, Zhao and Harmon [15] have performed a detailed study of the phonon spectra of Ni-rich B2-phase NiAl alloys and have identified a Fermi surface nesting mechanism as giving rise to the phonon anomaly. The calculations of Zhao and Harmon are based on first principles calculations of the electronic structure of B2 NiAl. The band structure obtained in these calculations is then fitted to a non-orthogonal Slater-Koster tight binding Hamiltonian. This model Hamiltonian is then used to evaluate the electron phonon matrix elements using the method of Varma and Weber [42], [43]. In this method the total dynamical matrix, D, is written as sum of three terms, $D = D_0 + D_1 + D_2$, the first two of which are grouped together and are fitted to the measured phonon spectrum using a Born-von Karman force constant model. The final term, D_2 term involves the electron phonon matrix elements and is calculated directly from the fitted electronic structure. Alone, the Born-von Karman fit yields the overall structure in the phonon spectrum, however, it does not reproduce the phonon softening. This requires inclusion of the third term involving the electron phonon matrix elements. Since D_2 is calculated on the basis of the electronic structure it is possible to identify the specific features in the electronic structure that give rise to the softening.

Within the Varma-Weber theory D_2 is given by

$$D_2(\kappa\alpha, \kappa'\beta|\vec{q}) = - \sum_{\vec{k},\mu,\nu} \frac{f_{\vec{k},\mu}(1 - f_{\vec{k}+\vec{q},\mu})}{\epsilon_{\vec{k}+\vec{q},\mu} - \epsilon_{\vec{k},\mu}} g^{\kappa\alpha}_{\vec{k},\mu;\vec{k}+\vec{q},\mu} g^{\kappa'\beta}_{\vec{k},\mu;\vec{k}+\vec{q},\mu} \tag{22}$$

where $\epsilon_{\vec{k},\mu}$ is the energy of the μ-th band at point $\vec{k}$ in the Brillouin zone, $f_{\vec{k},\mu}$ is the Fermi function, and the $g^{\kappa\alpha}_{\vec{k},\mu;\vec{k}+\vec{q},\mu}$ are electron-phonon matrix elements. Structure in D_2 arising from Fermi surface nesting occurs when the phonon wave-vector $\vec{q}$ is such that it connects parallel, flat sheets of Fermi surface i.e. the denominator of eq. 22 for $\epsilon = \epsilon_F$, the Fermi energy, is near zero over a substantial fraction of $\vec{k}$-space. Zhao and Harmon found nesting in the $\vec{k}_x, \vec{k}_y$ plane for values of $\vec{k}_z$ in the range $0.12(2\pi/a) \leq \vec{k}_z \leq 0.24(2\pi/a)$. This nesting, together with a strong electron-phonon matrix element for the transverse mode, gives rise to the anomaly.

For pure metals and for ordered systems the methodology followed by Zhao and Harmon is quite sound since it is based on standard band theory technology. Unfortunately, for disordered alloys and off-stoichiometric compounds it is much less well founded. In order to be able to treat the off-stoichiometric $Ni_{0.625}Al_{0.375}$ compound Zhao and Harmon made use of the rigid band model. They based their calculations on the electronic structure of stoichiometric NiAl assuming that the only effect on the electronic structure of adding excess Ni is to change the position of the Fermi energy. They choose valences of 1 and 3 for Ni and Al respectively, (arguing that in Ni the other 9 that we normally take as valence electrons are in a fixed d^9 core like configuration), modify ϵ_F accordingly, and use this modified ordered compound band structure to perform the calculation of the phonon spectra. The problem with this approach is that the effect of going off-stoichiometry is not only to modify the position of ϵ_F but also to smear out the electronic structure in a complex energy and $\vec{k}$ dependent manner. Thus the whole question of the existence of a band structure, in the sense that is required by eq. 22, is thrown into question.

As we have pointed out earlier a much better way to calculate the electronic structure of off-stoichiometric compounds is, at the outset, to face up to the fact that they are disordered and to use the LDA-KKR-CPA method. In this theory the band structure encapsulated in the $\epsilon_{\vec{k},\mu}$ relation is replaced by the Bloch spectral function $A^B(\vec{k}, \epsilon)$. At the present time, there is no first principles theory of the dynamical matrix for disordered alloys similar to that outlined above. However, since the mechanism identified by Zhao and Harmon rests critically upon the picture of the electronic structure and Fermi surface obtained using the rigid band model we can enquire into its validity. Specifically, if the effect of adding disorder into the problem is to smear out those parts of the Fermi surface responsible for the nesting this would negate the mechanism identified by Zhao and Harmon.

We have calculated the electronic structure of $Ni_{0.625}Al_{0.375}$ using the complex lattice LDA–KKR–CPA method described earlier. We assume that the structure is B2 and that the excess Ni goes onto the Al-sublattice. Thus the system is describe by two interpenetrating simple cubic lattices, one is occupied entirely by Ni, i.e. has the composition $Ni_{1.0}Al_{0.0}$, the second has Ni and Al randomly distributed with composition $Ni_{0.25}Al_{0.75}$. Once the self-consistent crystal potentials have been obtained the $A^B(\vec{k}, \epsilon)$ was calculated using a generalization of eq. 15. $A^B(\vec{k}, \epsilon)$ along a number of directions in the $\vec{k}_x, \vec{k}_y$ plane for an number of values of $\vec{k}_z$ is displayed in fig. 10. For $\vec{k}_z = 0$, $A^B(\vec{k}, \epsilon)$ is quite sharp in all directions, although, it is sharpest along the Γ-X direction and less sharp along Γ-M. As $\vec{k}_z$ is increased $A^B(\vec{k}, \epsilon)$ remains sharp along $\langle 100 \rangle$-directions for $\vec{k}_z < 0.3(2\pi/a)$ whilst along $\langle 110 \rangle$-directions it is rapidly smeared out by the disorder on the Al-sublattice. If we interpret the the locus of the peaks in $A^B(\vec{k}, \epsilon)$ as the Fermi surface and the width as the inverse mean free path of the electrons, clearly the Fermi surface is well define around the X-points in the Brillouin zone and much less so elsewhere.

In fig. 11 we show a summary plot of the locus of the peaks in $A^B(\vec{k}, \epsilon)$ of fig.10. What

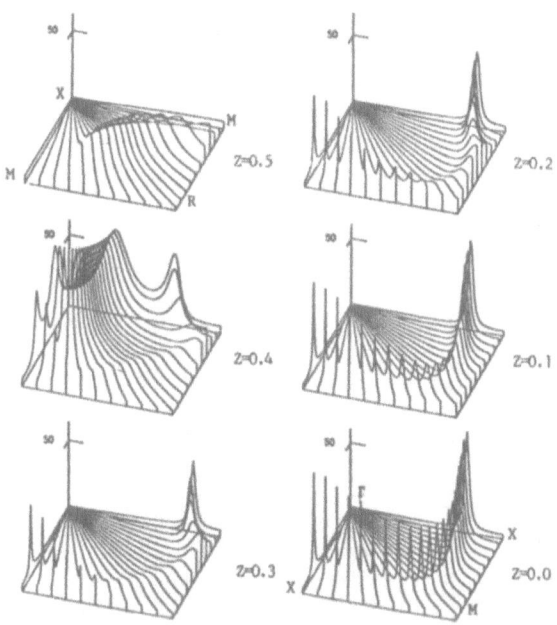

Figure 10: $A^B(\vec{k}, \epsilon)$ along various directions in $\vec{k}_z = constant$ planes in one octant of the simple cubic Brillouin zone for off-stoichiometric $Ni_{0.625}Al_{0.375}$.

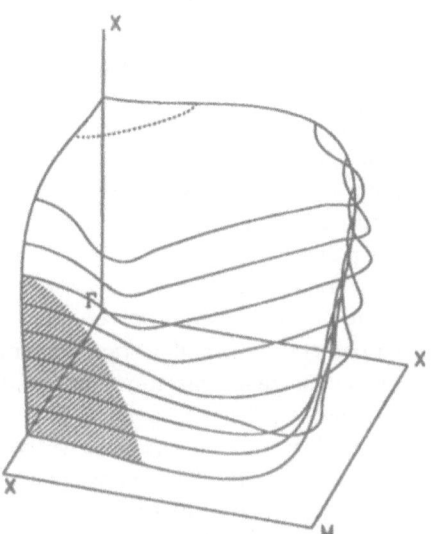

Figure 11: The locus of the peaks in $A^B(\vec{k}, \epsilon)$ of fig. 10 in the various $\vec{k}_z = constant$ planes. Starting from the basal plane the curves are for increments in $\vec{k}_z$ of 0.05 i.e. twice the density of fig. 10. The two uppermost curves are both for $\vec{k}_z = 0.4$ since, has may be seen from fig. 10, $A^B(\vec{k}, \epsilon)$ has two peaks. For values of $\vec{k}_z > 0.4$ the peaks are so ill defined that we did not plot their locus. The shaded area around the X-point indicates the region over which the Fermi surface is well defined.

is clear from this plot is that the Fermi surface is quite well defined over a substantial region of the Brillouin zone around the ⟨100⟩-direction. Since, it is the Fermi surface in the neighborhood of this direction that, according to Zhao and Harmon, is responsible for the nesting this is an important observation. Furthermore, in this region the Fermi surface is quite flat normal to the $\vec{k}_z = constant$ plane for values of $\vec{k}_z$ around $0.15(2\pi/a)$.

In fig. 12 we plot the Fermi surface in the $\vec{k}_z = 0.15$ plane in the extended zone scheme.

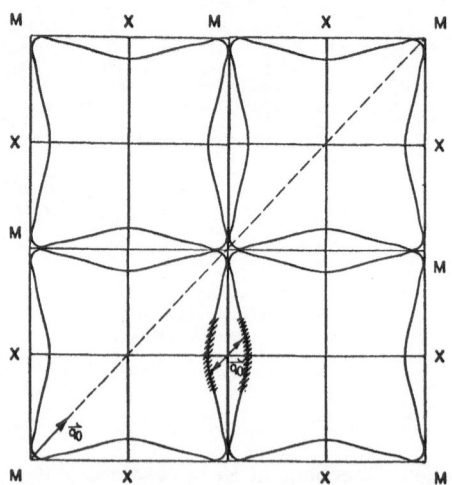

Figure 12: Fermi surface of off-stoichiometric B2-structure $Ni_{0.625}Al_{0.375}$ in the $\vec{k}_z = 0.15$ plane over four adjacent Brillouin zones. The shaded region indicates where the Fermi surface is well defined. We have marked by ↔ a possible nesting vector.

We have also indicated the part of the Fermi surface that is well defined and a spanning vector that can give rise to nesting. The spanning vector is $\sim 0.18\langle 110\rangle(2\pi/a)$, this is identical to that of Zhao and Harmon. Both of which are somewhat larger than the actual lock in vector for the 7R structure of $1/7\langle 110\rangle(2\pi/a)$.

Even though we do not have a first principles theory of the phonon spectra in disordered off-stoichiometric compounds our results lend strong support to the conclusions of Zhao and Harmon that the pre-martensitic phonon softening observed in B2 phase $Ni_{0.625}Al_{0.375}$ is Fermi surface driven. Such an observation has important consequences for future theories in that is implies that the interactions giving rise to the pre-martensitic behavior are long ranged. Thus we can conjecture that it will not be possible to describe this behavior on the basis of simple pair and short ranged interatomic potentials without the explicit inclusion of these subtle band structure effects. Since the effects of disorder on the electronic structure are extremely energy and $\vec{k}$ dependent, it is also clear that its effects will have to be considered at the outset. The extent to which such conclusions apply to the martensitic phase transformation itself remains to be seen. That disorder is important in understanding the nature of the atomic displacements that precede the martensitic transformation has already been provided by molecular dynamics studies based on embedded atom potentials [44]. In this work it was demonstrated that the characteristic "tweed" diffraction pattern is obtained for random positioning of the Ni and Al atoms on the disordered sublattice but tweed is not seen if the NiAl atoms are placed in some ordered array. Thus, it appears that a theory that includes both compositional and displacement fluctuations will be required in order to fully understand these complex phenomena.

CONCLUSIONS

We have reviewed the first principles theory of the electronic structure and energetics of disordered metallic alloys based on the LDA–KKR–CPA method. We have also outlined two approaches to describing ordering and phase stability and have given examples of their successful application to understanding the underlying electronic mechanisms that drive ordering. We have used a newly developed complex-lattice LDA–KKR–CPA code to make direct calculations of the ordering energy in NiAl. This system is very strongly ordering and is dominated by the first neighbor interaction. Based on a rigid band argument it is unlikely that this bond strength can be reduced by small additions of a third element. Thus, attempts to improve the ductility of NiAl by changing the dislocation slip plane from 100 to 111 as a result of global reductions in the near neighbor bond strength are unlikely to succeed.

Calculation of the Fermi surfaces of β–phase $Ni_{0.625}Al_{0.375}$ show that the disorder caused by the random distribution of Ni-antisite defects results in a complex $\vec{k}$-dependent smearing of the Fermi surface. Around the X-point point of the simple cubic Brillouin zone the Fermi surface remains relatively sharp, but elsewhere, and in particular around the M-point, the integrity of the Fermi surface is lost. Fortunately, the Fermi surface remains sharp over those parts that were identified by Zhao and Harmon [15] as giving rise to the pre-martensitic phonon softening observed in this material. Furthermore, Fermi surface nesting occurs with a nesting vector almost identical to that found by Zhao and Harmon on the basis of their rigid band model calculations. Thus, supporting the view that this pre-martensitic phenomenon is Fermi surface driven. Clearly, given the detailed information about the topology and smearing contained in the calculated Bloch spectral density, 2-dimensional ACAR positron annihilation measurements of the Fermi surface would be extremely valuable.

Using the complex lattice LDA–KKR–KPA methods outlined in this paper it is possible to address many questions regarding the properties of intermetallic alloys in which disorder plays a role. Along these lines studies of site selectivities of finite concentrations of third alloying elements are currently being undertaken.

ACKNOWLEDGMENTS

GMS: Work supported by Division of Materials Science Office of Basic Energy Sciences, U.S. Department of Energy under subcontractEAC05-84OR21400 with Martin-Marietta Energy Systems,Inc., and by a grant of computer time at NERSC from DOE-BES-DMS. DMN: Work supported by U.S. Department of Energy Assistant Secretary of Conservation and Renewable Energy, Office of Industrial Technologies, Advanced Industrial Concepts Materials Program, under subcontract DEAC05-84OR21400 with Martin-Marietta Energy Systems, Inc.. FJP: Work partially supported by Cray Research, Inc., and the Ohio Supercomputer Center. BG: Work supported by Consiglio Nazionale Delle Ricerche (Italy). DDJ: Work supported by Department of Energy, Basic Energy Sciences, Division of Material Sciences.

MS and PEAT: Work performed under the auspices of the U.S. Department of Energy by the Lawrence Livermore National Laboratory under contract number W-7405-ENG-48 WAS: Present address; Computational Physics Inc., P.O. Box 788, Annandale, Virginia, USA.

REFERENCES

1. S. Lundqvist and N. H. March, editors, *Theory of the Inhomogeneous Electron Gas*, Plenum, New York, 1983.

2. A. J. Freeman, (1992), This volume.

3. M. H. Yoo and C. L. Fu, (1992), This volume.

4. G. M. Stocks, W. M. Temmerman, and B. L. Gyorffy, Phys. Rev. Letters **41**, 339 (1978).

5. H. Winter and G. M. Stocks, Phys. Rev. **B27**, 882 (1983).

6. G. M. Stocks and H. Winter, The electronic structure of complex systems, page 463, New York, 1984, Plenum.

7. D. D. Johnson, D. M. Nicholson, F. J. Pinski, G. B. L., and G. M. Stocks, Phys. Rev. **41**, 9701 (1990).

8. W. M. Temmerman and A. J. Pindor, Joural of Physics F: Metal Physics **13**, 1869 (1983).

9. A. Pindor, B. L. Gyorffy, and W. M. Temmerman, Joural of Physics F: Metal Physics **13**, 1627 (1983).

10. W. A. Shelton, G. M. Stocks, D. M. Nicholson, F. J. Pinski, and B. Ginatempo, (1990), to be published.

11. A. Gonis et al., Phys. Rev. B **36**, 4630 (1987).

12. P. E. A. Turchi, G. M. Stocks, W. H. Butler, D. M. Nicholson, and A. Gonis, Phys. Rev. B **37**, 5982 (1988).

13. S. M. Shapiro, J. Z. Larese, Y. Noda, S. C. Moss, and L. E. Tanner, Phys. Rev. Letters **57**, 3199 (1986).

14. S. M. Shapiro, B. X. Yang, G. Shirane, Y. Noda, and L. E. Tanner, Phys. Rev. Letters **62**, 1298 (1989).

15. G. L. Zhao and B. N. Harmon, (1991), To be published. Receipt of a prepint is gratefully acknowledged.

16. J. S. Faulkner and G. M. Stocks, Phys. Rev. B **21**, 3222 (1980).

17. P. J. Durham, The electronic structure of complex systems, page 709, New York, 1984, Plenum.

18. H. Winter, P. J. Durham, W. M. Temmerman, and G. M. Stocks, Phys. Rev. B **33**, 2370 (1986).

19. J. C. Swihart, W. H. Butler, G. M. Stocks, D. M. Nicholson, and R. C. Ward, Phys. Rev. Letters **57**, 1181 (1986).

20. J. F. Janak, Phys. Rev. B **9**, 3985 (1974).

21. B. L. Gyorffy and G. M. Stocks, Phys. Rev. Letters **50**, 374 (1983).

22. M. A. Krivoglaz, *Theory of X-ray and Thermal Neutron Scattering by Real Crystals*, Plenum, New York, 1969.

23. R. C. Clapp and S. C. Moss, Phys. Rev. **142**, 418 (1966).

24. R. C. Clapp and S. C. Moss, Phys. Rev. **171**, 754 (1968).

25. B. L. Gyorffy, D. D. Johnson, F. J. Pinski, D. M. Nicholson, and G. M. Stocks, in *Alloy Phase Stability*, edited by G. Stocks and A. Gonis, volume 163, page 421, Boston, 1989, NATO-ASI, Kluwer.

26. D. M. Nicholson et al., (1991), To be published.

27. F. Pinski et al., Phys. Rev. Letters **66**, 766 (1991).

28. B. L. Gyorffy et al., (1991), To be published.

29. B. L. G. A. Barbieri et al., in *Alloy Phase Stability and Design*, edited by G. Stocks, D. P. Pope, and A. F. Giamei, volume 186, page 3, Pittsburgh, 1991, Materials Research Society, Materials Research Society.

30. H. Sato and R. S. Toth, *Alloying Behaviour and Effects in Concentrated Solid Solutions*, Gordon and Breach, New York, 1965.

31. G. Ceder et al., in *Alloy Phase Stability and Design*, edited by G. Stocks, D. P. Pope, and A. F. Giamei, volume 186, page 65, Pittsburgh, 1991, Materials Research Society, Materials Research Society.

32. J. W. D. Connolly and A. R. Williams, Phys. Rev. **B27**, 5169 (1983).

33. F. Ducastelle and F. Gautier, J. Phys. F **6**, 2039 (1976).

34. P. E. A. Turchi et al., Phys. Rev. Letters **67**, 1779 (1991).

35. P. E. A. Turchi, (1984), PhD Thesis: Univerite Pierre et Marie Curie, Paris VI.

36. P. E. A. Turchi, M. Sluiter, F. J. Pinski, and D. D. Johnson, in *Alloy Phase Stability and Design*, edited by G. Stocks, D. P. Pope, and A. F. Giamei, volume 186, page 59, Pittsburgh, 1991, Materials Research Society, Materials Research Society.

37. P. Villars and L. D. Calvert, *Pearson's Handbook of Crystallographic Data for Inter-metallic Phases*, volume 3, American Society for Metals, Metals Park, Ohio, 1985.

38. M. Ahlers, Prog. Mater. Sci. **30**, 135 (1988).

39. Z. Nishiyama, *Martensitic Transformations*, Academic Press, New York, 1978.

40. S. Ochai and M. Ueno, Japan Inst. Met **52**, 157 (1988).

41. L. E. Tanner, D. Schryvers, and S. M. Shapiro, Mat. Sci. and Eng. **A127**, 205 (1990).

42. C. M. Varma and W. Weber, Phys. Rev. Letters **39**, 1094 (1977).

43. C. M. Varma and W. Weber, Phys. Rev. B **19**, 6142 (1979).

44. C. Becquart, P. C. Clapp, and J. A. Rifkin, (1991), in press.

WHY FIRST-PRINCIPLES CALCULATIONS FOR ALLOY PHASE EQUILIBRIA?

D. de FONTAINE

Department of Materials Science and Mineral Engineering, University of California, Berkeley, CA 94720, USA, and
Materials Sciences Division, Lawrence Berkeley Laboratory, Berkeley, CA 94720, USA

ABSTRACT. A brief non-technical overview is presented for first-principles calculations of alloy phase equilibria. Merits and drawbacks of various methods are briefly discussed.

1. Introduction

What are *first-principles calculations* and what can be done with them? In view of the importance of this topic and of the misunderstandings which it has generated, it may be useful to try to answer these questions at least qualitatively. Such is the purpose of this brief paper. The literature on the subject is now quite vast, so that no attempt will be made here to review the field. Bibliography is cited only sparingly. For more details of electronic structure calculations, the reader is referred to the review by Heine with the marvellous title "LCAO, from under a cloud to out in the sun." [1]

Ultimately all materials problems are quantum mechanical in nature. Does that mean, however, that we should try to improve the ductility of, say, high-temperature alloys by solving the Schrödinger (or Dirac) equation? Surely not. Should we construct a *supercell* with a grain boundary in it, and let a dislocation run through it while we compute all energies by electronic band structure calculations? At present, such an undertaking is not practical, but perhaps later generations of computers will make such huge projects feasible.

In the meantime, we should continue to use classical, macroscopic, continuum models which have been around a long time and which are currently taught in Materials Science and Physics Departments: elasticity, plasticity, dislocation theory, fracture mechanics, bulk and surface thermodynamics, and so on. In the past, these classical theories have not been used to full advantage because, very often, materials parameters required "to make the equations go" were unavailable. What first-principles calculations can now provide, however, are values for these elusive parameters. In other words, quantum mechanical calculations can generate the required *data bases*, so that, for the first time in the relatively young history of Materials Science, theory can become truly predictive, with first-principles calculations serving to establish a continuous computational path from purely atomic phenomena all the way to practical problems concerning, say, mechanical properties.

By *first-principles* or *ab initio* calculations we mean, in the present context, more than just calculating cohesive energies of crystals knowing the atomic numbers and locations of atoms in the unit cell; we also mean calculating entropies, hence free energies. Hence, good statistical mechanical models, which are also tractable, must be available, along with interaction energies which determine the system's thermodynamics. It follows that such *ab initio* calculations require

C. T. Liu et al. (eds.), Ordered Intermetallics – Physical Metallurgy and Mechanical Behaviour, 37–45.

the solution of both quantum and statistical mechanical problems. Moreover, computations must be carried out with a high degree of accuracy since important parameters generally result from small differences of large numbers.

Some of the electronic energy calculation techniques currently in use for perfect crystals at absolute zero of temperature will be briefly described in Section 2, then the extension to disordered systems will be discussed in Section 3. Applications to phase diagrams will be mentioned in Section 4. For reviews of the subject matter treated here the interested reader may consult Refs. [2]-[4].

2. Perfect Crystals at Absolute Zero

Let the term "perfect crystal" denote an elemental crystal or a crystalline stoichiometric compound free from imperfections. The problem is to calculate the *cohesive energy* of the compound in a given crystal structure, i.e., the difference in energy between the crystalline aggregate and the isolated atoms. It is necessary to solve the Schrödinger (or Dirac) equation for the electronic states of each free atom and for the periodic crystalline structure. As by-products, one obtains electronic densities of state and, when the wave functions are calculated, the electron density in the unit cell.

The problem is impossible to solve exactly, since all electrons interact dynamically with themselves and with atomic nuclei, some 10^{22} in number. Hence, drastic simplifications are required. Since the nuclei are so much more massive than the electrons, one first assumes that the nuclei are stationary and occupy fixed lattice positions. By this decoupling procedure, nuclei dynamics can be handled subsequently at non-zero temperatures by lattice dynamics methods.

The remaining problem is still very difficult: that of electrons interacting with the field of fixed nuclei and with all other electrons. Here, one generally makes the important one-electron approximation, i.e., one solves self-consistently for a single electron moving in the effective potential field of the nuclei and the charge density of all other electrons. Various methods have been proposed for carrying out these self-consistent calculations, the most frequently used being the local density functional method (LDF) [5] which has made first-principles calculations feasible on fast computers.

The time-independent Schrödinger equation applied to solids reduces to an extremely large eigenvalue problem which can be solved in principle by two types of mathematical techniques: variational methods or Green's function methods. In either case, wave functions may be expanded in orthonormal sets of functions, the choice of which depends on computational convenience but also on the nature of the simplified potential adopted.

Atomic potentials have a spherically symmetric 1/r dependence near the nucleus but lose that symmetry nearer to the Wigner-Seitz cell boundaries. One simplification consists of replacing the true potential in each WS cell by a central spherically symmetric one terminating, near cell edges, in a perfectly flat potential. Such is the so-called muffin-tin (MT) potential. This potential is adopted by various computational methods which then differ from each other by the choice of basis functions: APW (augmented plane waves), ASW (augmented spherical waves), and KKR (Korringa-Kohn-Rostoker, a "multiple scattering" method). A linearized version of the latter, obtained by expansions about fixed energy values, goes by the name of LMTO (linear muffin-tin orbital). A further simplification is possible, that which consists of eliminating the "interstitial" region between MT spheres altogether by allowing them to overlap. Such is the atomic sphere approximation. The resulting method, the LMTO-ASA, is extremely efficient but works best for close-packed structures. Other methods can be "linearized" as well, resulting in the LAPW method, for example.

The trouble with the spherically symmetric muffin-tin potential is that departure from simple local symmetry cannot be handled properly. Hence, in order to handle, say, c/a relaxation in tetragonal unit cells, or distortions required to compute elastic constants, full (F) potentials must be retained in the formalism. Corresponding electronic structure methods go by the acronym FLAPW, FLMTO.

For molecules, one often uses as basis functions the atomic orbitals themselves. Wave functions are then built up by linear combinations of atomic orbitals (LCAO). That method can of course be used for (infinite) periodic structures as well, but the matrix elements of the Hamiltonian are very difficult to compute *ab initio* in this basis. It is then customary to obtain these elements from fits to the band structure or as derived from the LMTO [6]. The ingredients of matrix elements are then treated as parameters of the model and electronic self-consistency is handled in an approximate manner. One then has the very convenient, physically meaningful, but sometimes computationally unreliable tight-binding approximation (TB), which nevertheless works rather well for transition metals. Finally, let us mention pseudopotential techniques, which are particularly well suited to covalently bonded structures [7].

Computer codes required to perform the calculations are of such complexity that non-initiates are well advised not to attempt to write their own versions. Fortunately, ready-made, quasi "black box" LMTO-ASA, APW, and ASW codes are available and, to a lesser degree, KKR and FLMTO codes. Tight-binding codes are simple enough to generate, but Cambridge University puts out a nice package which includes the clever recursion algorithm for calculating electronic densities of state for disordered systems [8]. Still, as for all sophisticated techniques, one must know the principles of the method quite well in order to use the tool to full advantage. To borrow an analogy familiar to all materials scientists, it is not necessary to build an electron microscope oneself, but one must know the principles of operation very thoroughly in order to operate the instrument effectively. Also, the more powerful the instrument, the longer must be the learning stage for successful utilization.

3. Disordered Systems

3.1 DEFECTS

The simplest type of configurational disorder is the isolated point defect, a vacant lattice site, for example. The introduction of a defect breaks translational symmetry so that the great simplification offered by application of the Bloch Theorem, valid for periodic structure, can no longer be used. To restore translational symmetry, what is usually done is to set up a periodic lattice of supercells, each one containing identical copies of the defect configuration considered. Care must be taken, however, to insure that the lattice parameter of the supercell array be significantly larger than the linear dimensions of the defect(s) to be calculated. If the defects are (relatively) too close together in supercell space, then the calculation yields not just the self-energy of the defects that one wishes to calculate, but also the interaction energies which are an artifact of the repeating nature of the defects in their supercells. It is clear, therefore, that the N^3 computational rule referred to above, where now N is the number of atoms in the supercell, severely limits the application of the method to very simple defect structures.

Next in order of complexity are extended defects such as dislocations, surfaces, interfaces, antiphase boundaries, etc. Again, the supercell technique may be used, but what is gained in accuracy by performing detailed first-principles calculations is often lost by the necessity of solving somewhat oversimplified problems. Rather than carrying out brute force calculations, it is usually preferable to use semi-empirical potentials, obtained from "embedded atom" techniques [9], and to model actual atomic configurations, relaxation included, by molecular dynamics. Gross features of extended defect structures and estimates of formation energies can be obtained by these methods, which, however, appear not to be sufficiently accurate at present to predict the thermodynamics of ordered or partially ordered (or disordered) structures. This problem is taken up in the next section. A very promising method of calculating cluster energies for use in extended defect problems has been introduced recently by Pettifor [10]. Green's functions techniques are also being developed for isolated defects in essentially infinite solids.

3.2 DISORDERED ALLOYS

As a first step, consider concentrated alloys in a state of complete disorder. Here, the supercell idea is not applicable: the unit cell is the whole crystal itself. The term "complete disorder" means a distribution of, say, A and B atoms, of specified average concentration c, where atomic site occupations are completely uncorrelated. With that understanding, a method for calculating average electronic structures and cohesive energies naturally suggests itself: calculate the required properties of a particular configuration in a given crystalline region, repeat the calculation for all configurations which conserve the chosen average concentration, and take averages of the results. Even in a small region and for a small number of sample configurations, such a computation requires prohibitively long computational times, unless a simple electronic structure approach is used, such as the tight binding method. This procedure, labeled DCA (direct configurational averaging), has been used with success for certain transition metal alloys [11].

If more elaborate and accurate electronically self-consistent methods are required, configurational averaging must be replaced by a technique which, by creating as it were identical "average atoms," restores translational symmetry to disordered systems. That method, known as the "coherent potential approximation" (CPA) [12], defines an average atomic potential (in a binary alloy, say) by requiring that the replacement of it, on the average, by an A or a B potential causes no additional electronic scattering. The average potential is then determined self-consistently. The CPA is a mean field technique, unlike the DCA, since the properties of an (artificial) average structure are computed rather than the average of properties. Nevertheless, the CPA has given very reliable results, say, for the density of states of completely disordered systems [13]. Let us note, finally, that the CPA relies on the calculation of an average Green's function, hence must be used in conjunction with the tight binding method, the KKR, or, very recently, with the LMTO-ASA.

3.3 PARTIAL ORDER

Perfectly ordered or completely disordered structures at absolute zero cannot describe real alloys, and cannot deal with the complexities of phase stability as a function of temperature, for example. Then the properties of phases, stable or metastable, must be calculated as a function of temperature and concentration c (or "chemical field" μ, a difference of chemical potentials). Better yet, what is sought is a way of determining properties as a function of alloy configuration. In the context of this section, it is understood that "disorder" designates atomic (A, B, ...) disorder on the sites of a *given lattice*. Complete order then produces *superstructures* of the given lattice (usually fcc or bcc).

A general and rigorous method exists for expanding in a set of basis functions any alloy property which depends on configuration:

$$f(\sigma) \;=\; \sum_{\alpha} f_{\alpha}\, \varphi_{\alpha}(\sigma) \quad. \tag{1}$$

In this equation, first derived by Sanchez, Ducastelle and Gratias [14], σ denotes an arbitrary crystal configuration (location of A and B atoms on lattice sites, in the binary A-B case), $\varphi_{\alpha}(\sigma)$ are *cluster functions* which may be chosen to form an orthonormal set, f_{α} are the coefficients of the expansion, and the index α denotes clusters of lattice sites: point, pairs, triplets, quadruplets, etc. Energies (formation, ordering, ...), elastic moduli and other properties may be thus expanded. What is required in macroscopic systems is the expectation values of the properties $f(\sigma)$, averaged over configurations:

$$<f> \;=\; \sum_{\alpha} f_{\alpha}\, \xi_{\alpha} \tag{2}$$

where, at least for binary systems, the ξ_{α} are correlation functions: point (average concentration), pair, triplet, ..., in general, multisite correlation functions. The expansion has been shown to be valid for both concentration-dependent and -independent coefficients f_{α} [15].

Eq. (2) shows very clearly how the statistical and quantum mechanical calculations are decoupled, in some sense: the f_{α} coefficients are obtained from electronic band structure calculations at absolute zero, the ξ_{α} correlations are obtained by minimizing a free energy functional at the temperatures and average concentrations of interest. The cluster variation method (CVM) [16] free energy functional is most commonly used as it is naturally expressed as a function of multisite correlation functions, the ξ_{α}, and, when used with large enough clusters, has proved to be highly accurate and reliable.

The problem, of course, is to calculate the coefficients f_{α}. As was seen in Section 2, quantum mechanical techniques exist for calculating energies of perfectly ordered or completely disordered crystals for which the considerable simplification of translational symmetry is available, either because it is naturally present in stoichiometric compounds, or because it is imposed by the CPA on disordered crystals. There are thus basically two methods for obtaining the f_{α} expansion coefficients: that based on perfect order calculations, and those based on complete disorder.

The former, which may be called the "structure inversion method," was first used by Connolly and Williams [17] and often goes by the names of these two authors: one decides which clusters α are going to contribute importantly, say, to the energies of partially ordered structures. Then a number of ordered superstructures are selected, at least equal in number to the non-equivalent clusters considered, and total energy *ab initio* calculations are performed on each superstructure. Since the ξ_{α} correlations can be calculated exactly for perfect superstructures, hence are considered "known," Eqs. (2), written for all structures calculated, form a linear system in the f_{α} unknown, which can be solved by matrix inversion or least squares optimization. This method has many advantages since it is generally simpler and more accurate to compute properties of perfect crystals than of disordered ones, and since more flexibility is available in the choice of band structure computational techniques. All of the electronic structure methods mentioned in Section 2, and more, may be used. The disadvantage of the inversion method is that the resulting values of the f_{α}

coefficients depend to some extent on the choice of superstructures. This undesirable feature is attenuated by considering a large enough set of structures to calculate [18].

The other methods, based on disorder, consist in calculating the f_α directly as responses of the disordered system to the perturbation caused by the corresponding cluster function φ_α. If the perturbing function is a harmonic concentration wave in the CPA medium, we have the $S^{(2)}$ method of Györffy and Stocks [19] or, equivalently, the original k-space generalized perturbation method (GPM) of Ducastelle and Gautier [20]. Today, the real-space GPM method is generally used [21], or the embedded cluster method (ECM) [22], both consisting of cluster perturbations of the single-site CPA medium. As mentioned above, the multiple scattering formalism must be used (KKR) or, in simplified form, the tight binding approximation. In this way, each cluster coefficient f_α is computed independently of the others and the calculation is pursued until the f_α for large clusters become very small. The series (1) or (2) are known to converge, but no general theoretical convergence criterion is available at present. The KKR-CPA is unfortunately a highly computationally-intensive method and at present rather unwieldy.

An alternative method has already been mentioned: the method of direct configurational averaging (DCA). Small clusters are embedded in a medium which is averaged by repeated choices of random configurations, that of the embedded cluster remaining fixed. This technique, which does not rely on the mean field approximation, is very efficient when applied in the tight binding formalism used in conjunction with the so-called "orbital peeling algorithm" [23]. The disadvantage of this technique is that only simplified band structure methods, such as the TB, have been used up to now, hence are not applicable to all alloy systems.

4. Phase Diagram Calculations

There are several reasons for wishing to calculate *ab initio* temperature-composition phase diagrams. First of all, if the phase diagram is not known empirically, it is of course useful to determine theoretically what the diagram may look like, even approximately. Such was the case for the oxygen-ordering $YBa_2Cu_3O_x$ phase diagram: the diagram was calculated from first principles [24] and, subsequently, experimental phase transition points were found to fall almost perfectly on the calculated phase boundaries [25]. Secondly, even when, as is usually the case, binary phase diagrams have been empirically determined, it is useful to perform the calculations for the following reasons: if calculated and experimentally determined phase diagrams agree reasonably well, it means that the calculations, which are very complex ones, are accurate and that the methods are sound. It is found indeed that the nature of the phases present at equilibrium and the location of phase boundaries depend critically on the accuracy of the calculations. Hence, the phase diagram is an extremely sensitive test of the computations. Also, calculations give information not only about equilibrium but also about metastable phases, which can be of considerable interest to alloy designers. Finally, and perhaps most importantly, if one can calculate a phase diagram, one has at one's disposal all quantitative information concerning formation energies, entropies, free energies, states of order, lattice parameters, elastic moduli, etc., of stable and metastable structures in all concentration and temperature ranges.

The calculations proceed roughly as follows: first, the appropriate lattices are selected, i.e., those which are expected to be parent to the terminal solutions and superstructures found in the system. Then the f_α *for each lattice* are determined by one of the methods alluded to above. With these sets of values, one can then perform, for each lattice, a ground state analysis along the lines described elsewhere in these workshop proceedings [26]. In some cases, it is possible to predict truly *ab initio* which will be the stable superstructures, without actually guessing at a number of competing structures and calculating, *a posteriori*, which are the ones with lowest energy at absolute zero.

Statistical mechanical methods (CVM, Monte Carlo simulation) are then used to determine free energies as a function of temperature and chemical field for the various lattices and their relevant superstructures. Common tangents are constructed in the usual way and phase boundaries plotted. Up to now, only crystalline phases can be handled in this first-principles way. The liquid can be included by means of a fitted free energy curve.

The degree of success and degree of difficulty encountered in this type of undertaking depend critically on the system envisaged. For example, in the Al-Li case [27], relevant intermetallic compounds were almost all superstructures of fcc and bcc lattices, and were stabilized by effective cluster interactions (ECI, f_α coefficients pertaining to configurational energy) limited to first neighbor coordination shell for fcc and first and second for bcc. Hence, small clusters could be used in the CVM free energy functional, and only a few structures needed to be calculated for the Connolly and Williams inversion. The resulting phase diagram and calculated parameters, such as lattice parameter, formation energies, and bulk moduli agreed closely with experimental values [27].

For Al-Ti, the situation is entirely different. Superstructures of fcc, bcc and also hcp are expected, and some of these structures can only be stabilized by fairly distant effective interactions [28]. Moreover, tetragonal distortions of non-cubic equilibrium phases are important and have to be taken into account in the perfect-structure calculations, which means that the structures must be "relaxed" not only with respect to atomic volume (or lattice parameter a), but also with respect to c/a and even b/a ratios and unit cell angles. That, in turn, means that only full potential codes may be used, turning the complete set of calculations into a very elaborate undertaking indeed. In addition, local elastic relaxations are expected to be important in the disordered states as well, and vibrational entropy will have to be included to stabilize bcc over hcp on the Ti-rich side of the phase diagram. Neither of these *displacive* forms of disorder has received adequate theoretical formulations. Needless to say, no one has yet "tamed" the Al-Ti system computationally, although it is technologically a highly relevant one.

Often, several different techniques must be brought to bear on the problem. For example, in a very recent work [29], FLAPW was used to calculate very precisely the cohesive energies of pure fcc and bcc Cu and Zn, the KKR-CPA-GPM combination was used to obtain the energies of complete fcc and bcc disorder in Cu-Zn and the effective cluster interactions as a function of concentration for both lattices, and an empirical Debye correction was used to stabilize bcc Cu at high temperature (above the melting point). In that way, the solid-state (no liquid) Cu-rich portion of the Cu-Zn phase diagram was calculated, including fcc and bcc superstructures, some inaccessible in practice due to sluggish kinetics at low temperatures. Predictions about diffuse scattering in α-brass were made by the equivalent of the $S^{(2)}$ method.

5. Conclusion

The full dynamical problem of interacting electrons and nuclei is rendered tractable by successive decoupling procedures: electronic motion is decoupled from nuclear (or ionic) motion by the Born-Oppenheimer approximation, and electronic structure calculations at absolute zero are decoupled from equilibrium configurational calculations by performing cluster expansions, Eqs. (1) or (2).

The decoupling of vibrational and configurational entropy contributions then allows phase equilibrium determination on a given lattice to be handled as an Ising model problem [30].

Although much remains to be done, the various methods which have been developed to perform the required calculations have produced very encouraging results. It is therefore hoped that, in the near future, a true First-Principles Thermodynamics of Alloys will become available.

Acknowledgements

This work was supported by the Director, Office of Energy Research, Office of Basic Energy Sciences, Materials Sciences Division of the U.S. Department of Energy under Contract No. DE-AC03-76SF00098. The author is indebted to Mr. Mark Asta, Dr. Gerbrand Ceder, Dr. Prabhakar P. Singh and Mr. Christopher Wolverton for many useful discussions.

References

1. Heine, V. (1980), 'LCAO: From under a Cloud to Out in the Sun', in H. Ehrenreich, F. Seitz and D. Turnbull (eds), Solid State Physics, Vol. 35, Academic Press, pp. 1-127.
2. de Fontaine, D. (1989) 'The Cluster Variation Method and the Calculation of Alloy Phase Diagrams', in G. M. Stocks and A. Gonis (eds.), Alloy Phase Stability, Kluwer Academic Publishers, Dordrecht, pp. 177-203.
3. Gonis, A., Sluiter, M., Turchi, P. E. A., Stocks, G. M. and Nicholson, D. M. (1991), J. Less-Common Metals 168, 127.
4. Lu, Z. W., Wei, S.-H., Zunger, A., Frotapessou, S. and Ferreira, L. G. (1991) Phys. Rev. B 44, 512; Wei, S.-H., Ferreira, L. G. and Zunger, A. (1990) Phys. Rev. B 41, 8240.
5. Hohenberg, P. and Kohn, W. (1964) Phys. Rev. B 136, 864; Kohn, W. and Sham, L. J. (1965) Phys. Rev. A 140, 1133.
6. Andersen, O. K., Jepsen, O. and Glötzel, D. (1985) 'Canonical Description of the Band Structure of Metals', Highlights of Condensed Matter Theory, pp. 59-176.
7. Ihm, J., Zunger, A. and Cohen, M. L. (1979), J. Phys. C 12, 4409.
8. Nex, C. N. M. (1984) Comput. Phys. Commun. 34, 101.
9. Daw, M. S. and Baskes, M. I. (1984) Phys. Rev. B 29, 6443.
10. Pettifor, D. G. and Aoki, M. (1991), Phil. Trans. R. Soc. (London) A, and this workshop.
11. Dreyssé, H., Berera, A., Wille, L. T. and D. de Fontaine (1989) Phys. Rev. B 39, 2442.
12. Faulkner, J. S. (1982), in J W. Christian, P. Haasen and T. B. Massalski (eds.), Prog. Mat. Sc. Vol. 27, pp. 1-187.
13. Winter, H., Durham, P. J., Temmerman, W. M. and Stocks, G. M. (1986) Phys. Rev. B 33, 2370.
14. Sanchez, J. M., Ducastelle, F., and Gratias, D. (1984) Physica (Utrecht) 128A, 334.
15. Asta, M., Wolverton, C., de Fontaine, D., and Dreyssé, H. (1991) Phys. Rev. B 44, 4907-4913; Wolverton, C., Asta, M., Dreyssé, H., and de Fontaine, D. (1991) Phys. Rev. B 44, 4914-4924.
16. Kikuchi, R. (1951) Phys. Rev. 81, 988-1003.
17. Connolly, J. W. D. and Williams, A. R. (1983) Phys. Rev. B 27, 5169.
18. Ferreira, L. G., Wei, S.-H. and Zunger, A. (1989), Phys. Rev. B 40, 3197.
19. Györffy, B. L. and Stocks, G. M. (1983), Phys. Rev. Lett. 50, 374.

20. Ducastelle, F. and Gautier, F. (1976) J. Phys. F.6, 2039.

21. Tréglia, G., Ducastelle, F. and Gautier, F. (1978) J. Phys. F 8, 1437; Ducastelle, F. and Tréglia, G. (1980) J. Phys. F 10, 2137; Bieber, A., Ducastelle, F., Gautier, F., Tréglia, G. and Turchi, P. (1983) Solid State Commun. 45, 585.

22. Gonis, A., Zhang, X.-G., Freeman, A. J., Turchi, P., Stocks, G. M. and Nicholson, D. M. (1987) Phys. Rev. B 36, 4630.

23. Haydock, R. (1980) Solid State Phys. 35, 215.

24. Ceder, G., Asta, M. Carter, W. C., Kraitchman, M., de Fontaine, D., Mann, M. E. and Sluiter, M. (1990) Phys. Rev. B 41, 8698.

25. Andersen, N. H., Lebech, B. and Poulsen, H. F. (1990) J. Less-Common Metals 164, 124.

26. de Fontaine, D., Wolverton, C., Ceder, G., and Dreyssé, H. (1991) this workshop.

27. Sluiter, M., de Fontaine, D., Guo, X. Q., Podloucky, R., and Freeman, A. J. (1990) Phys. Rev. B 42, 10460-10476.

28. Asta, M., unpublished work at University of California, Berkeley.

29. Turchi, P. E. A., Sluiter, M., Pinski, F. J., Johnson, D. D., Nicholson, D. M., Stocks, G. M. and Staunton, J. B. (1991) Phys. Rev. Lett. 67, 1779.

30. Ceder, Gerbrand (1991) 'Alloy Theory and Its Applications to Long Period Superstructure Ordering in Metallic Alloys and High Temperature Superconductors', Ph.D. dissertation, University of California, Berkeley.

STRUCTURE MAPS FOR ORDERED INTERMETALLICS

D.G. PETTIFOR
Department of Mathematics
Imperial College of Science, Technology and Medicine
London SW7 2BZ.

ABSTRACT. The search for new pseudobinary intermetallics with a required structure type and mechanical properties can be guided at the outset by the use of *phenomenological* two-dimensional structure maps which order the known structural data base on binary compounds. This paper reviews the progress made and problems encountered in using the maps in developing new cubic transition metal aluminides. In particular, it emphasizes the need for a *microscopic* quantum mechanical understanding of why, for example, Al_3Nb cannot be stabilized in the cubic form whereas Al_3Ti can, and why the latter remains brittle even though single crystals of Ni_3Al with the same cubic $L1_2$ crystal structure are ductile.

1. INTRODUCTION

The importance of crystal structure in determining the mechanical properties of intermetallics was beautifully illustrated by the work of C.T. Liu and colleagues [1] which was reported at the 1984 Bethesda meeting on *High Temperature Alloys: Theory and Design*. They showed that Co_3V, which is hexagonal with an e/a ratio of 8.0, could be transformed to the cubic $L1_2$ phase by alloying with iron so that e/a < 7.9. Whereas the hexagonal phase is brittle with a tensile elongation of only 1%, the cubic phase is ductile with an elongation of 40%. It was natural, therefore, that alloy developers should ask theorists which alloying elements must be added to the brittle non-cubic transition-metal aluminides such as Ti_3Al, $TiAl_3$, and $NbAl_3$ in order to transform them to the cubic $L1_2$ phase [2, 3]. The hope was that the recently proposed phenomenological structure maps [4] would help guide the alloy developers in their search for such cubic pseudo-binary intermetallics. We will see that this hope has only been partially

47

C.T. Liu et al. (eds.), Ordered Intermetallics – Physical Metallurgy and Mechanical Behaviour, 47–59.
© 1992 *Kluwer Academic Publishers.*

48

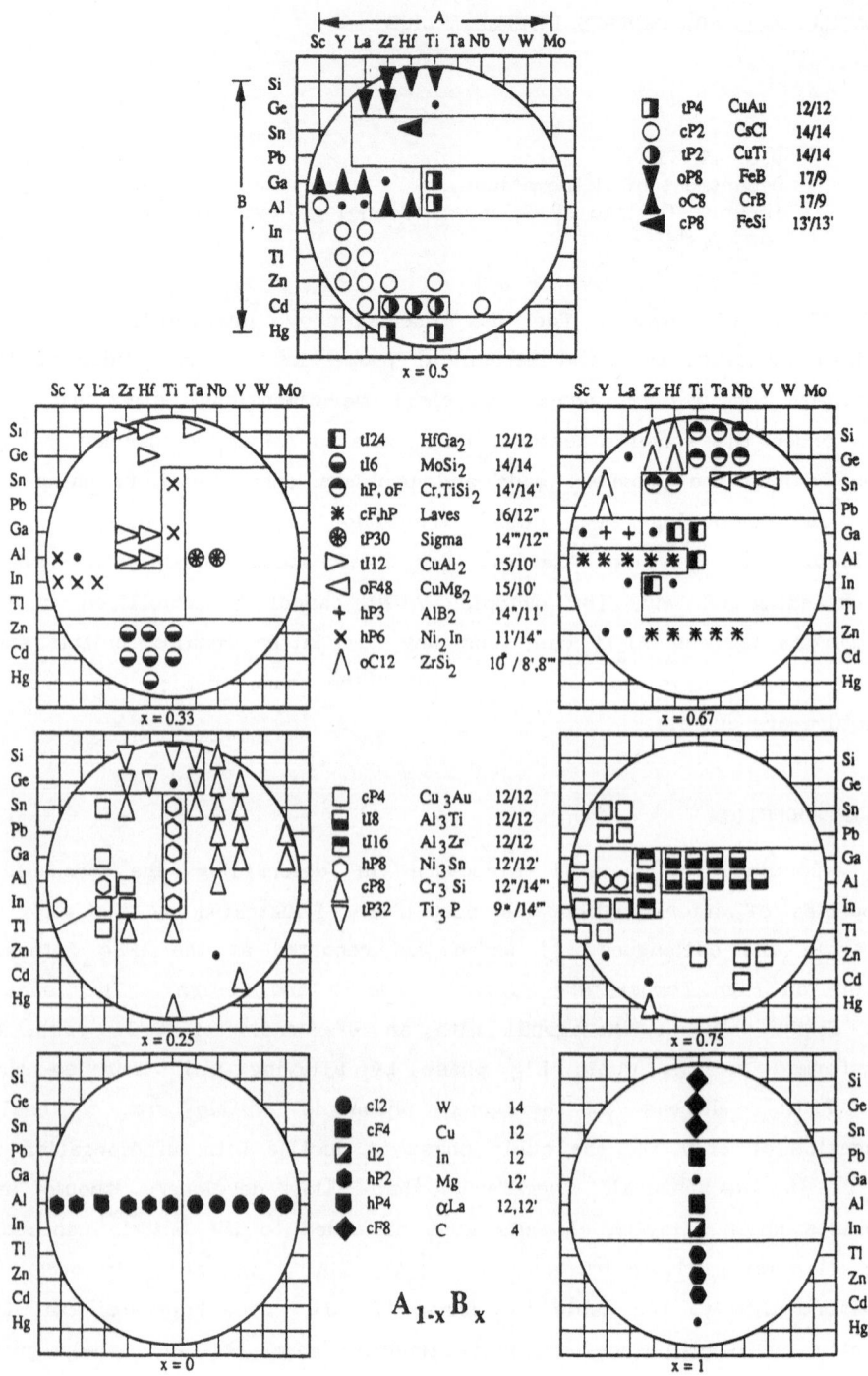

Fig. 1 Titanium-Aluminium neighbourhood maps [6].

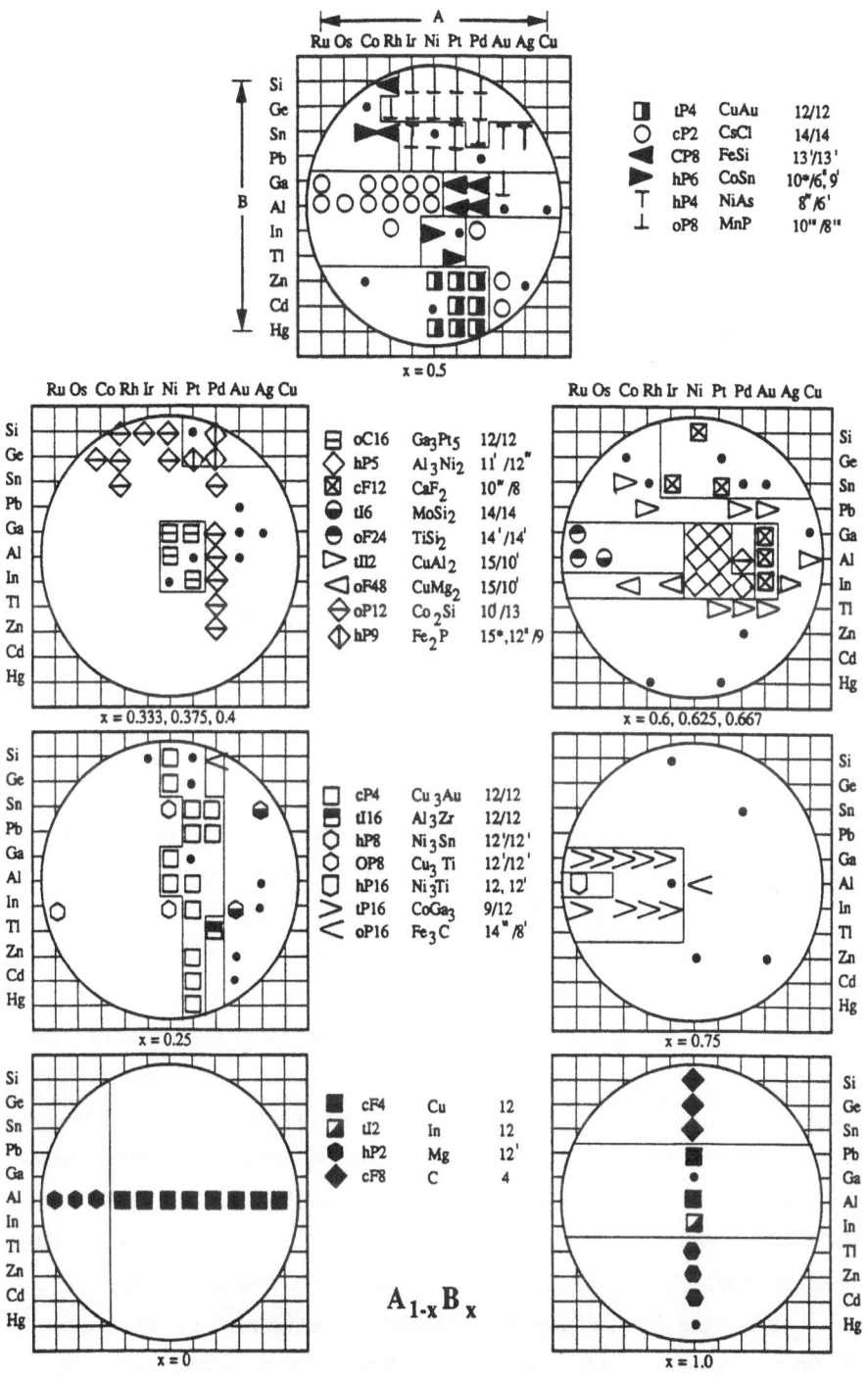

Fig. 2 Nickel-Aluminium neighbourhood maps [6].

realized and that a microscopic quantum mechanical understanding of the bonding at the atomistic level is required.

2. STRUCTURE MAPS

Figs. 1 and 2 show the relevant part of the structure maps in the neighbourhood of the titanium and nickel aluminides respectively [5, 6]. The ordering of the elements along the axes was obtained by running a one-dimensional string through the two-dimensional periodic table in a certain way, thereby placing all the elements in sequential order, termed the Mendeleev number $\mathfrak{M}$ (see fig 1 of [4]). All the known data on the ground state structure of *binary* compounds $A_{1-x}B_x$ could then be presented within a single three-dimensional plot $(\mathfrak{M}_A, \mathfrak{M}_B, x)$. The neighbourhood maps in fig. 1 are two-dimensional cross-sections of the three-dimensional cylinder whose axis is centred on $(\mathfrak{M}_A = 51, \mathfrak{M}_B = 80)$ corresponding to the co-ordinates of titanium-aluminium. Similarly, fig. 2 is centred on $(\mathfrak{M}_A = 67, \mathfrak{M}_B = 80)$ corresponding to nickel-aluminium. The 2p bonded elements Be, B, C, N and O are excluded from such neighbourhood plots because their small size and high electronegativity give them unique properties. The key in figs. 1 and 2 gives not only the Pearson symbol, but also the recently proposed generalized Jensen notation for the local co-ordination polyhedra about the A and B sites (see table 5 of [7] and figs. 3 and 4 of [8]).

We see that whereas Ni_3Al and $NiAl$ sit in cubic domains, the intermetallic Ti_3Al and the tri-aluminides $HfAl_3$, $TiAl_3$, $TaAl_3$, $NbAl_3$ and VAl_3 sit in hexagonal DO_{19} (hP8) and tetragonal DO_{22} (tI8) domains respectively. However, the latter are adjacent to cubic $L1_2$ (cP4) domains, which suggested that it might be possible to stabilize the cubic structure by alloying so that the resultant average Mendeleev co-ordinates $(\bar{\mathfrak{M}}_A, \bar{\mathfrak{M}}_B)$ fall in the cubic domain. This hope has only been partially realized. Schneibel and Porter [9] have indeed succeeded in stabilizing cubic $ZrAl_3$ by alloying to take the average Mendeleev number $\bar{\mathfrak{M}}_B$ down into the cubic domain. However, attempts to stabilize cubic Ti_3Al from hexagonal DO_{19} or cubic $NbAl_3$ from tetragonal DO_{22} have failed [2], [3]. Moreover, the cubic tri-aluminide pseudo-binaries based on $ZrAl_3$ and $TiAl_3$ still remain brittle, cleaving transgranularly, even though they have the same crystal structure as ductile single crystals of Cu_3Au or Ni_3Al. Theorists are, therefore, faced with two immediate problems. Firstly, why can some non-cubic transition metal aluminides be stabilized with the cubic structure by alloy addition (e.g. $TiAl_3$) but others cannot (e.g. $NbAl_3$). Secondly, why if the cubic close packed phase of the tri-aluminides can be stabilized, does it still remain brittle. We shall see that the answer to both questions will require a proper quantum mechanical understanding of the bonding at the atomistic level.

3. FIRST PRINCIPLES PREDICTIONS

First principles quantum mechanical calculations, which are based on the Local Density Functional (LDF) approximation, can now predict routinely the relative energies of different (simple) structure types. For example, fig. 3 shows the heat of formation of the titanium aluminides Ti_3Al, $TiAl$, and $TiAl_3$ which were calculated by van Schilfgaarde et al [10]. We see that the LDF approximation predicts correctly the hexagonal DO_{19} structure of Ti_3Al and the tetragonal DO_{22} structure of $TiAl_3$. But even more importantly it predicts the proximity in energy of competing metastable phases. It is at once apparent that the cubic $L1_2$ form of $TiAl_3$ is only about $0.02eV/atom$ or $2kJ/mole$ less stable than the tetragonal phase. In fact, as had been noted earlier by Nicholson et al [11], $TiAl_3$ would order as $L1_2$ rather than DO_{22} if the phase is constrained to order with respect to an underlying face centred cubic lattice with no tetragonal distortion. This has been observed experimentally to be the case for $HfAl_3$ and $ZrAl_3$ precipitates constrained within a cubic aluminium matrix [12]. The LDF calculations, therefore, explain why $TiAl_3$ can be stabilized in the cubic form whereas to date Ti_3Al has not been, since in the latter case the metastable $L1_2$ phase is energetically further away from the ground state.

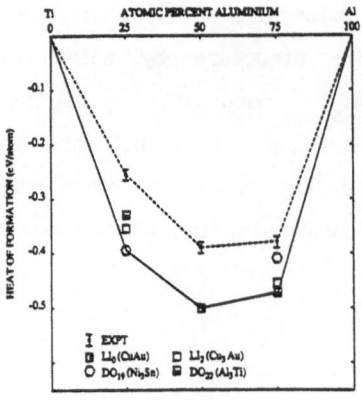

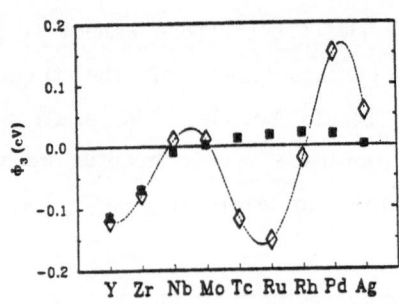

Fig. 3 Heat of formation of titanium aluminides (after [10]).

Fig. 4 The nearest-neighbour three-body cluster interaction $\Phi_3^{(1)}$ for the 4d transition metal aluminides (after [13]). Squares denote the predictions of Miedema's model.

The heat of formation of the titanium aluminides is skewed towards the aluminium rich end in fig. 3. This deviation from regular solution behaviour is controlled by the effective *first*-nearest-neighbour three-body cluster interaction $\Phi_3^{(1)}$. For a disordered alloy $A_{1-x}B_x$ the heat of formation may be written

$$\Delta H_{dis}(x) = -4x(1-x)[\Phi_2^{(1)} + (1-2x)\Phi_3^{(1)} + \dots], \qquad (1)$$

so that if $\Phi_3^{(1)}$ is positive (negative) the curve is skewed towards the A(B) rich end. Carlsson [13] has evaluated the effective *first*-nearest-neighbour cluster interactions up to $\Phi_4^{(1)}$ for the *cubic* transition metal aluminides $T_{1-x}Al_x$ by computing the energy of fcc (T), $L1_2$ (T_3Al), $L1_0$ (TAl), $L1_2$(TAl_3), and fcc (Al) within the LDF approximation. He finds that $\Phi_3^{(1)}$ *oscillates* across the transition metal aluminide series as shown for the 4d series in fig. 4. The 3d series shows a similar oscillatory behaviour (compare figs. 4(a) and 4(b) of [13]). As expected from fig. 3, $\Phi_3^{(1)}$ is negative for the titanium (zirconium) aluminides, corresponding to $TiAl_3$ ($ZrAl_3$) being more stable than Ti_3Al (Zr_3Al) as is known experimentally [14].

The three-body cluster interaction $\Phi_3^{(1)}$ has minima in the vicinity of Y(Sc) and Ru(Fe), maxima in the vicinity of Mo(Cr) and Pd(Ni). This is consistent with the observed *asymmetry* between the T_3Al and TAl_3 neighbourhood structure maps in figs 1 and 2. $ScAl_3$, YAl_3 and $RuAl_3$ exist as close-packed phases whereas Sc_3Al, Y_3Al and Ru_3Al do not exist. On the other hand, cubic $L1_2$ Ni_3Al exists whereas there is no corresponding close-packed phase $NiAl_3$. Of course, predictions about the occurrence of a given phase can only be made after comparison with all other competing phases. Thus although the zirconium aluminides have a negative $\Phi_3^{(1)}$, the cubic 3:1 stoichiometric phase Zr_3Al is found in addition to tetragonal $ZrAl_3$, which is similar to the titanium aluminides except for the reversal of the cubic-hexagonal stability of the 3:1 stoichiometry (see fig. 3).

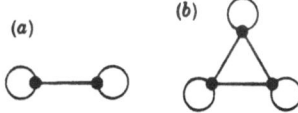

(a)   *(b)*

Fig. 5 The two-body (a) and three-body (b) diagrams which contribute to Φ_2 and Φ_3 respectively [15]. The bubbles represent the electron hopping on the same site in order to investigate whether it is occupied by an A or B atom.

The origin of the oscillations in $\Phi_3^{(1)}$ is entirely quantum mechanical. We see in fig. 4 that the Miedema "macroscopic atom" model does not predict such rapid variations across the series [13]. Instead these oscillations may be traced directly to wave mechanical interference effects around three-membered rings of atoms as illustrated in fig. 5 [15]. Similar oscillations as a function of the average number of valence electrons per atom ratio e/a have been predicted for the relative DO_{22} versus $L1_2$ phase stability of transition metal intermetallics [16]. The rapid oscillatory variation accounts for the fact that although $TiAl_3$ and $NbAl_3$ both lie in the same DO_{22} domain adjacent to a cubic $L1_2$ domain in fig. 1, the energy required to transform $NbAl_3$ to the cubic phase is nearly an order of magnitude larger than that for $TiAl_3$ (see fig. 2 of Carlsson and Meschter [17]). It is, therefore, extremely unlikely that cubic $NbAl_3$ will ever be stabilized by alloying additions, supporting recent experimental evidence that the 1964 report of the $L1_2$ phase $Nb_2(Al_5Ni)$ is mistaken [3]. Instead it is not surprising from the neighbourhood maps in fig. 1 that the alloy $Nb_2(Al_5Ni)$ comprises the Laves phase $Nb(Al_{0.8}Ni_{0.2})_2$ and $B2(NiAl)$ in addition to the parent phase $DO_{22}(NbAl_3)$ (Darolia, private communication 1987).

A clue as to why the cubic scandium and titanium tri-aluminides are brittle even though they have the same crystal structure as the ductile single crystals of Ni_3Al may be provided by their elastic constants. Fu [18, 19] has recently calculated these within first principles Local Density Functional (LDF) Theory. The predicted values for the cubic intermetallics are given in table 1 together with the experimental values for nickel, aluminium, and silicon for comparison. The ratio of the shear modulus to the bulk modulus, namely μ/K, is given in the last column as this has proved an effective criterion for deciding whether a sharp crack cleaves or blunts [20]. Cottrell [21] has shown that ductile fcc and bcc metals generally have $\mu/K \lesssim 0.4$, whereas brittle cubic metals have $\mu/K \gtrsim 0.5$. We see from table 1 that the intermetallic Ni_3Al satisfies the ductile criterion $\mu/K \lesssim 0.4$, whereas the tri-aluminides $TiAl_3$ and $ScAl_3$ do not [22].

4. SIMPLIFICATIONS

The ratio μ/K provides only an indication of the possible crack tip response in intermetallics. More detailed understanding of the nature and mobility of dislocation cores is required for predicting mechanical behaviour [23]. Atomistic simulations of dislocations, however, require further theoretical simplifications as solving the LDF Schrödinger equation directly is computationally very time consuming and requires high precision. The total energy of the γ' phase Ni_3Al, say, is of the order of 10^5 eV per formula unit whereas the energy difference between the stable *cubic* close packed structure $L1_2$ and the metastable *hexagonal* close packed structure DO_{19} is of the order of 10^{-1} eV per formula unit. If we are to believe the computational prediction that cubic is more stable that hexagonal we had better have converged to a precision of 1

in 10^7.   Moreover, theoretical science is not just about predicting
numbers but also about providing concepts and insight.   This implies
simplification.

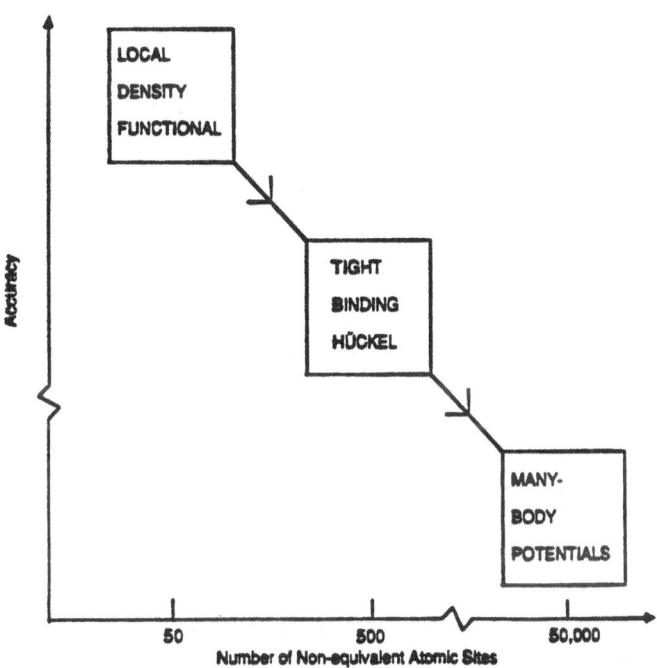

Fig. 6   Schematic representation of the decrease in accuracy with
         increasing number of non-equivalent atomic sites which can be
         treated computationally by different levels of approximation.
         The downward arrows linking the boxes indicate the progression
         from the first principles LDF theory to Tight Binding Hückel
         theory to many-body potentials through the application of well-
         understood approximations [24].

    Fig. 6 shows schematically three different levels of approximation
[24].   At the first level, favoured by physicists, the cohesive and
structural properties are obtained by solving the LDF Schrödinger
equation.   Because of the high precision required current computers can
treat a maximum of only about one hundred non-equivalent atomic sites.
At the second level, favoured by chemists, the bonding is described by
Tight-Binding Hückel theory [25] in terms of the σ, π and δ bond

integrals which result from the overlap of the angularly dependent
valence orbitals. Because of the simplifications introduced by chemical
intuition, current computers can handle up to about a thousand
non-equivalent atomic sites, but with a lower accuracy than LDF theory.
At the third level, favoured by metallurgists, the atoms are regarded as
fuzzy balls which intereact via pair potentials such as that of
Lennard-Jones. The simplicity of pair-potentials and recently developed
many-body embedded atom potentials [26] allows current computers to
simulate the behaviour of hundreds of thousands of atoms.

However, if we are to simulate trends in materials' properties
reliably, then we must take cognisance of Albert Einstein's caveat:
"make it as simple as possible, but not simpler." A many-body potential
which regards atoms as fuzzy spherical balls without any angular
character would not be able to explain, for example, why a single
crystal of close-packed cubic $TiAl_3$ is brittle whereas that of Ni_3Al is
ductile. This can be seen in the behaviour of the Cauchy pressure
$C_{12}-C_{44}$ in table 1: whereas nickel, aluminium, and Ni_3Al have *positive*
values of the Cauchy pressure, the tri-aluminides have *negative* values
which are comparable to that of silicon. This reflects the nature of
the bonding at the atomistic level. If the bonding is describeable by
simple pairwise potentials such as Lennard-Jones, then the Cauchy
pressure would be zero. If the bonding is more metallic in that
spherical atoms are embedded in the electron gas of the surrounding
neighbours, then the Cauchy pressure would be positive [27]. A negative
Cauchy pressure requires angular character to the bonding.

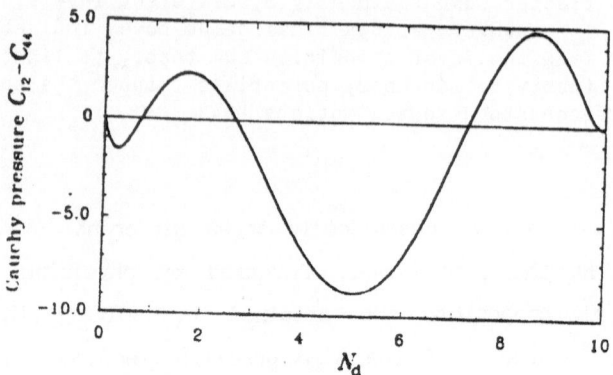

Fig. 7. The d bond Cauchy pressure $C_{12} - C_{44}$ as a function
of the number of valence d electrons per atom N_d
for the bcc lattice [15].

Very recently a new many body potential has been derived within Tight Binding Hückel theory which explicitly includes the angular character of the bonding orbitals [15, 28, 29, 30]. It can lead to both positive or negative values of the Cauchy pressure as is illustrated in fig. 7 for bcc transition metals [15]. It remains to apply these new bond order potentials to the atomistic simulation of dislocation cores and crack tips in order to explore why isostructural phases can display such a wide variety of mechanical response.

5. CONCLUSION

The *phenomenological* structure maps are very useful as an initial guide in the search for new pseudobinary alloys with a required structure type. However, a fundamental *microscopic* theory of the bonding at the atomistic level is required for the detailed prediction of crystal structure and defect behaviour.

ACKNOWLEDGEMENTS

I should like to thank the US Department of Energy, Energy Conversion and Utilization Technologies (ECUT) Materials Program, for financial support under subcontract no. 19X-55992V through Martin Marietta Energy Systems Inc.

REFERENCES

[1]   Liu, C.T. (1984) in J.O. Stiegler (ed.), High-Temperature Alloys: Theory and Design, AIME, New York, pp. 289-308.
[2]   Liu, C.T., Horton, J.A. and Pettifor, D.G. (1989) Mat. Res. Soc. Symp. Proc. 133, 37-43.
[3]   Subramanian, P.R., Simmons, J.P., Mendiratta, M.G., and Dimiduk, D.M. (1989) Mat. Res. Soc. Symp. Proc., 133 51-56.
[4]   Pettifor, D.G. (1988) Mat. Science and Technol., 4 675-691.
[5]   Pettifor, D.G. (1991) in High Temperature Intermetallics, Institute of Metals, London, pp. 93-99.
[6]   Pettifor, D.G. (1991) in O. Izumi (ed.), Intermetallic Compounds: Structure and Mechanical Properties, Japan Institute of Metals, Sendai, pp. 149-156.
[7]   Jensen, W.B. (1989) in F.R. de Boer and D.G. Pettifor, (eds.) The Structures of Binary Compounds, Cohesion and Structure Vol. 2, North-Holland, Amsterdam, pp. 105-146.
[8]   Villars, P., Mathis, K., and Hulliger, F. (1989) in F.R. de Boer and D.G. Pettifor (eds.) The Structures of Binary Compounds, Cohesion and Structure Vol. 2, North-Holland, Amsterdam, pp. 1-104.
[9]   Schneibel, J.H., and Porter, W.D. (1989) Mater. Res. Soc. Symp. Proc. 133, 335-341.
[10]   van Schilfgaarde, M., Paxton, A.T., Pasturel, A., and Methfessel, M. (1990) Mat. Res. Soc. Symp. Proc. on Alloy Phase Stability and

58

Design.

[11] Nicholson, D.M., Stocks, G.M., Temmerman, W.M., Sterne P., and Pettifor, D.G. (1989) Mat. Res. Soc. Symp. Proc. 133, 17-22.

[12] see, for example, Levoy, N.F., and Vandersande, J.B. (1989) Met. Trans. 20A, 999-1019.

[13] Carlsson, A.E. (1989) Phys Rev. B40, 912-923.

[14] de Boer, F.R., Boom, R., Mattens, W.C.M., Miedema, A.R. and Niessen, A.K. (1988) Cohesion in Metals: Transition Metal Alloys, Cohesion and Structure Vol. 1, North-Holland, Amsterdam, p. 130.

[15] Pettifor, D.G., and Aoki, M. (1991) Phil. Trans. R. Soc. Lond. A334 439-449.

[16] Bieber, A., and Gautier, F. (1989) Solid State Commun., 38, 1219-22.

[17] Carlsson, A.E., and Meschter, P.J. (1989) J. Mater. Res. 4, 1060-63.

[18] Fu, C.L., and Yoo, M.H. (1989) Mat. Res. Soc. Symp. Proc. 133, 81-86.

[19] Fu, C.L. (1990) J. Mater. Res. 5, 971.

[20] Kelly, A., Tyson, W.R., and Cottrell, A.H. (1967) Phil. Mag. 15, 567-586.

[21] Cottrell, A.H. (1989) Mat. Science and Technol. 5, 1165.

[22] Cottrell, A.H. (1991) in Proceedings of 2nd European Conference on Advanced Materials and Processes, 22-24 July 1991, Cambridge, U.K.

[23] Vitek, V. (1985) in M.H. Loretto (ed.), Dislocations and Properties of Real Materials, Institute of Metals, London, pp. 30-50.

[24] Pettifor, D.G. (1991) in G.A.D. Briggs (ed.), New Materials for the Next Century: a Scientific, Technological and Industrial Revolution, Blackwells, Oxford.

[25] Hoffmann, R. (1988) Solids and Surfaces: a Chemist's View of Bonding in Extended Structures, V.C.H., New York.

[26] Daw, M.S., and Baskes, M.I. (1984) Phys. Rev. B29 6443.

[27] Johnson, R.A. (1988) Phys. Rev., B37, 3924-3931.

[28] Pettifor, D.G. (1989) Phys. Rev. Lett., 63 2480-2483.

[29] Pettifor, D.G. (1990) Springer Proc. Phys. 48 64-84.

[30] Pettifor, D.G. and Aoki, M. (1991) in J.L. Morán-López and J.M. Sanchez (eds.), Adriatico Conference on Structural and Phase Stability of Alloys, Plenum, New York.

Table 1: Elastic Constants

(Experimental values in square brackets)

10^{11}N/m^2	C_{11}	C_{12}	C_{44}	Aniso-tropy C_{44}/C'	Cauchy Press. $C_{12}-C_{44}$	μ/K
Ni	[2.61]	[1.51]	[1.32]	[2.41]	[0.19]	[0.36]
Al	[1.14]	[0.62]	[0.32]	[1.23]	[0.30]	[0.35]
Si	[1.66]	[0.64]	[0.80]	[1.57]	[-0.16]	[0.59]
Ni$_3$Al	2.35	1.45	1.32	2.93	0.13	0.33
	[2.30]	[1.49]	[1.32]	[3.25]	[0.17]	[0.30]
TiAl$_3$	1.77	0.77	0.85	1.70	-0.08	0.53
ScAl$_3$	1.89	0.43	0.66	0.90	-0.23	0.77

CLUSTER EXPANSION OF fcc Pd-V INTERMETALLICS

D. de FONTAINE[1,2], C. WOLVERTON[1,3], G. CEDER[1,2*] and H. DREYSSE[4]
[1]Lawrence Berkeley Laboratory, Berkeley, CA 94720, USA
[2]Department of Materials Science and Mineral Engineerir ؛, University of California, Berkeley,
   CA 94720, USA
[3]Department of Physics, University of California, Berkeley, CA 94720, USA
[4]Laboratoire de Physique du Solide, Université de Nancy, Vandoeuvre-les-Nancy, France

ABSTRACT. A cluster expansion is used to compute fcc ground states from first principles for the Pd-V
system. Intermetallic structures are not assumed but derived rigorously by minimizing the configurational
energy subject to linear constraints. A large number of concentration-independent interactions are calculated
by the method of direct configurational averaging. Agreement with the fcc-based portion of the
experimentally-determined Pd-V phase diagram is quite satisfactory.

1. Introduction

In recent years, first-principles total electronic energy calculations have been remarkably successful
in predicting heats of formation, lattice parameters, and elastic moduli of simple intermetallic
compounds at their stoichiometric compositions and at zero Kelvin. Performing similar calculations
for off-stoichiometric compositions, in disordered or partiall؛ ordered states, has obviously not
progressed as rapidly, as it is required to solve combined quantum and statistical problems at a high
level of accuracy.

One aim of such calculations is to derive, virtually from first principles, reasonable temperature-
composition phase diagrams for binary metallic systems, say. Thermodynamic quantities, such as
free energies, entropies, enthalpies, states of order may then be deduced as by-products of the
calculations.

Over the last ten years or so, it has become apparent that the preferred way of investigating
alloys computationally (ordered or disordered) is through the medium of expansions in cluster
functions. The theoretical framework is rigorous, flexible and completely general, and is essential
for formulating both the energy (E) and the configurational entropy (S), hence the free energy
(F = E –TS).

The cluster expansion method can also be used to tackle the difficult problem of ground state
determination. Predicting, without guesswork, which superstructures of a given lattice have
minimum energy, is essential and is a topic which has perhaps not received as much attention lately
as it deserves. In this article, we shall address precisely this problem, and, after having outlined the
basic theory, we shall apply the cluster methods to the ground state determination of fcc
superstructures in the Pd-V system.

---

*Present address: Department of Materials Science and Eng..neering, Massachusetts Institute of
Technology, Cambridge, MA 02139, USA.

C. T. Liu et al. (eds.), Ordered Intermetallics – Physical Metallurgy and Mechanical Behaviour, 61–71.
© 1992 *Kluwer Academic Publishers.*

2. Cluster Expansions

It was back in 1951 that Kikuchi [1] introduced the idea of *clusters* in the statistical thermodynamics of the Ising model as a way of improving systematically on the currently known approximations of the configurational entropy. Since the free energy was obtained, in Kikuchi's method, by minimizing a functional with respect to cluster variables, he called his hierarchy of approximations the *cluster variation method* (CVM). Its application to the calculation of phase diagrams was suggested by Van Baal [2] in 1973, and the use of the method has expanded considerably ever since.

The cluster method is now viewed as far more general than the early practitioners of the CVM probably envisaged. Today, clusters (on a lattice) are considered as the essential building blocks for any description of alloy properties which depend on configuration. This approach was first described in 1982, in a remarkable paper by Sanchez, Ducastelle and Gratias [3]. Although these authors treated the general multicomponent case, we shall here summarize results for binary systems only. An alternative and very elegant method of treating multicomponent systems was suggested by Finel [4] and was very recently described in a very clear and comprehensive review article by Inden and Pitsch [5].

In a binary alloy (AB) let the pseudo-spin variable $\sigma_p = +1$ (-1) stand for an A (B) atom at lattice site p. Consider now a set of lattice points $\{p,p',p''\dots\}_\alpha$ which we shall denote as "the cluster α." It was shown [6,7] that *cluster functions* $\varphi_\alpha(\sigma)$ can be constructed so as to form an orthonormal set in the space of 2^N configurations, N being the total number of lattice points. A convenient choice [3] is the direct product of σ variables on the cluster points:

$$\varphi_\alpha(\sigma) = \sigma_p \, \sigma_{p'} \, \sigma_{p''} \dots \tag{1}$$

The set $\{\varphi_\alpha\}$ is orthonormal in the sense that the scalar product $<\varphi_\alpha(\sigma), \varphi_\beta(\sigma)>$, defined as the normalized sum of the product $\varphi_\alpha\varphi_\beta$ over all configurations, is unity if the two clusters α and β coincide, zero otherwise. It follows that any function of configuration, $f(\sigma)$, say, can be expanded in the set of cluster functions

$$f(\sigma) = \sum_\alpha f_\alpha \, \varphi_\alpha(\sigma) \tag{2}$$

with generalized Fourier coefficients given by

$$f_\alpha = <\varphi_\alpha(\sigma), f(\sigma)> \tag{3}$$

Of particular interest is the expectation value of the function $f(\sigma)$, obtained by taking an ensemble average of Eq. (2) at given T and chemical field μ (difference of chemical potentials $\mu_B - \mu_A$):

$$<f> = \sum_\alpha f_\alpha \xi_\alpha \tag{4}$$

where the ξ_α, denoted *multisite correlation functions*, are ensemble averages of the corresponding cluster functions. Eq. (4) is important in that, in principle, it shows how to express any macroscopic alloy property as an expansion in cluster correlation functions with coefficients calculated by Eq. (3). This cluster expansion technique, applied to the case of the energy E, has led to various methods for calculating macroscopic alloy parameters directly from quantum mechanical computations. One such method will now be described.

3. Effective Cluster Interactions

Let us apply the formalism of Eq. (4) to the expectation value of the internal energy E. It is necessary to calculate the expansion coefficients E_α, generally called *effective cluster interactions* (ECI). These parameters are obtained from Eq. (3), with cluster functions given by Eq. (1). It would appear that the formalism requires the computation of energies as a function of alloy configuration, $E(\sigma)$, for all possible configurations. Actually, the sum of (4) converges fairly rapidly [7], so that only a few ECI's need to be calculated. As an example, let α represent a pair of lattice sites, p and q, say. In that case, when the summation implied by the scalar product (3) is written out explicitly, the following single expression for effective pair ($\alpha \equiv pq$) interactions is obtained [6-9]:

$$E_{pq} = \frac{1}{4}(W_{AA} + W_{BB} - W_{AB} - W_{BA}) \ .$$

(5)

In Eq. (5), W_{IJ} represents the average energy of all configurations of the system with atom of Type I at p and J at q. The formalism can be extended to any reasonable size cluster: triplets (p,q,r), quadruplets, etc. The physical meaning of the ECI's is thus quite clear: E_α represents a linear combination of average energies of systems containing cluster α in all of its possible configurations.

This definition of ECI's is perfectly rigorous, and, in this approach, leads to the following important properties:

(a) Since the embedding medium is averaged over all possible configurations, and since all configurations of the embedded cluster are considered, the ECI's are *concentration independent*.

(b) Since the ECI's, as per Eq. (5), consist of sums of very similar energies (W), with equal number of positive and negative contributions, the E_α are much smaller in magnitude than the "potentials" W.

(c) As already anticipated in the pioneering work of Gautier and Ducastelle [10], the magnitude of the E_α decreases rapidly with distance, in the case of pairs, and with the number of points in cluster [7]. This convergence property is responsible for making the whole notion of cluster expansion into a practical reality. It is important to note that no such convergence holds for the "potentials" W themselves. However, in taking sums and differences, as in Eq. (5), the long-range portions of the energies tend to cancel out.

This is not the only way to proceed: it has been shown recently [6,7] that orthonormal sets of functions on configuration space could be constructed by means of summations over configurations which conserve average concentration. The ECI's produced in this manner are thus necessarily concentration-dependent, but it was proved that, somewhat surprisingly, both concentration-dependent and -independent expansion schemes are in fact completely equivalent [6,7]. Of course, the coefficients of the expansions will have different numerical values, and the convergence rate of the two series may well differ.

The concentration-independent scheme presents definite advantages, not the least of which is the simplification of the search for ground states. This subject will be treated in Section 5, but first let us describe a method for actually computing concentration-independent effective cluster interactions.

4. Direct Configurational Averaging

The most obvious way of computing ECI's (E_α) is by taking sums and differences of W's, as in Eq. (5). The energies W themselves can be calculated by selecting an arbitrary configuration σ in a finite potion of the crystal (containing N atoms), computing the energy by suitable electronic structure techniques, then repeating the procedure over and over, with different configurations selected at random, keeping that on the chosen α-cluster fixed. It has been shown that convergence to a true "random medium" is rather rapid [11]; usually about 20 to 50 configurations suffice [7].

We have used the tight-binding method with parameters obtained from LMTO calculations for the pure elements of the binary alloy considered. On-site energies were determined by disallowing charge transfer, a reasonable assumption for transition metals, and the Shiba prescription was used for evaluating off-diagonal two-center hopping integrals [7,11]. Because the recursion method is used to calculate the density of states, it then turns out that the W potentials need not be calculated individually: the technique of "orbital peeling" [12] produces E_α directly, so that taking a (small) difference of large numbers is not required. Generally, 10 recursion levels are used with a quadratic terminator.

The average medium, in the present case, was practically an fcc solution of 50% Pd and 50% V [13]. The tight-binding parameters were derived from pure fcc Pd and fcc V at a molar volume midway between that of equilibrium fcc Pd and fcc V at zero Kelvin. Effective cluster interactions were calculated for all pair and triplet and eight quadruplet clusters which belong to the set of subclusters of the 14-point fcc cube. Fig. 1 is a plot of the logarithm of the magnitude of the ratio of ECI E_α to the first neighbor pair interaction for pairs (circles), triangles (crosses) and quadruplets (squares), as a function of the order of the largest pair in the cluster. Several distinct triplets and some quadruplets have the same largest pair, hence their representative points appear on the same vertical in Fig. 1. It is apparent that ECI's converge fairly rapidly with pair separation and with number of points in the cluster, although not monotonically. This set of 26 effective interactions represents, to date, the largest set of ECI's yet calculated for any alloy system. It is becoming increasingly clear, however, that large numbers of ECI's are required to describe adequately the properties of most alloy systems.

5. Ground State Analysis

Predicting, for a given binary system, which intermetallic structures will have lowest energy, for all concentrations, at zero Kelvin, is an impossible task. Fortunately, most intermetallics of interest are superstructures of either fcc, bcc or hcp. Then, the problem of determining the lowest-energy *superstructures* of a given lattice is a simpler one which, in favorable cases, can be solved exactly. Each lattice must of course be handled separately: the ECI's calculated on different lattices will have different values. As for other intermetallic compounds, those which are *not* superstructures, they must be treated differently. For these "interloper" phases, their total energies must be calculated directly by appropriate electronic structure codes and compared to other, possibly competitive structures. Here, we shall investigate only the minimum-energy fcc-based superstructures of the Pd-V system, for the calculated set of ECI's, over the whole Pd-V concentration interval. This set of structures constitute the set of *fcc ground states of order* for the system in question.

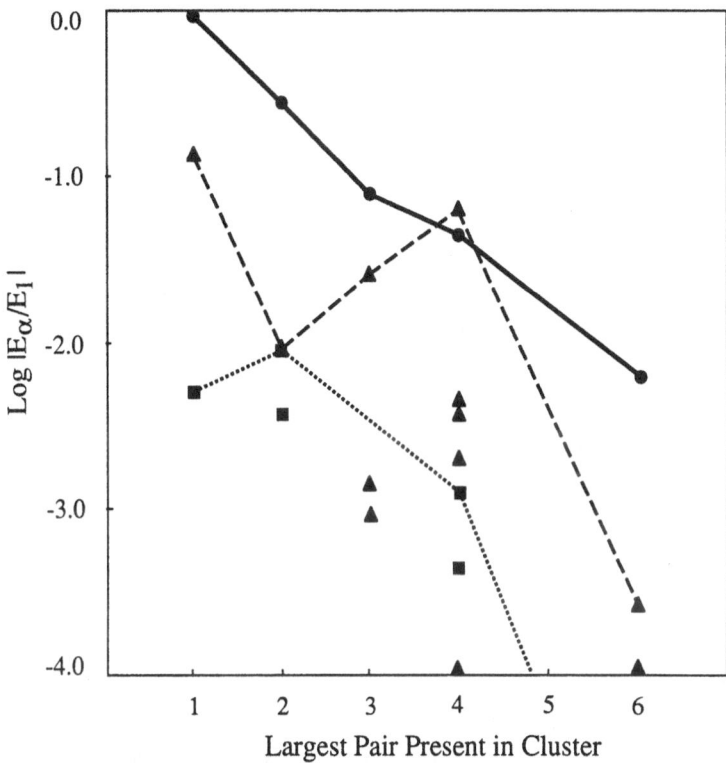

Figure 1. Logarithm of ECI's (normalized to nn pair interaction E_1), as a function of largest pair distance in cluster, for three types of clusters: pairs (circle symbols), triplets (triangles), quadruplets (squares). Lines connect highest-energy ECI's for the three classes of clusters. Geometrically distinct triplets and quadruplets which have same largest pair are located on the same vertical.

Eq. (2), written for the energy $E(\sigma)$, is the one to minimize, but it must first be rewritten in a more convenient form. Many of the clusters (α) appearing in the summation are equivalent through the space group symmetry operations of the underlying lattice. The set of such clusters equivalent to a given one by symmetry is known as the orbit of the given cluster. Each distinct orbit (or its generating cluster) will be denoted by the index j. The total number of clusters in orbit j is then the total number of lattice translational symmetry operations times the number of equivalent clusters per lattice point, or multiplicity m_j. Let us also denote the "empty cluster" by the index j = 0. Then, the energy of a given stoichiometric superstructure, per lattice point, is, by Eq. (4), given by the linear form

$$e = e_0 + \sum_{j=1}^{J} m_j E_j \xi_j \tag{6}$$

where the brackets have been removed from $< e >$ since, at absolute zero of temperature, the expectation value is just the energy of the perfect structure. The variables ξ_j here are not strictly ensemble averages, but "orbit averages" of cluster functions; such averaging process must be taken into account since the symmetry of the ordered superstructure is generally lower than that of the parent lattice. The summation in Eq. (6) extends from the "point" cluster to some maximal cluster(s), denoted by the index J.

Simply minimizing the linear function (6) with given ECI's E_j will not do, since the parameters ξ_j must describe a real structure, or mixture of structures, on the lattice. Hence, a number of constraints (i.e., linear inequalities) on the domain of ξ_j must be imposed. The required constraints are usually derived from considerations of clusters (see Refs. [14] and [15] and references cited therein), but the most straightforward method is probably that suggested by Sanchez and one of the present authors [16] and described fully, for the case of pair interactions, by Finel [15] and in a recent review [5]. The handling of combinatorics of large clusters was treated even more recently in the Ph.D. dissertation of one of the present authors [17] and a more detailed application to the Pd-V system will be published elsewhere [13].

Briefly, the idea is the following: denote the probability of finding a given cluster, say a nearest-neighbor triangle of lattice points (equilateral triangle in fcc) populated by atoms in a certain configuration ($\sigma \equiv$ AAA, AAB, ...) by the symbol $x_j(\sigma)$. This probability, or "dressed" cluster concentration, being a function of configuration, can be expanded in a set of cluster functions, as in Eq. (2) [17]. For simplicity, let distinct configurations on a given cluster be labeled by the index k. For the maximal cluster J, the concentrations of various configurations k are then given by [18]:

$$x_J(\sigma_k) = \rho_J^o \left(1 + \sum_{j=1}^{J} C_{kj} \xi_j \right) \tag{7}$$

where ρ_J^o is a normalization factor given by the reciprocal of the number of configurations on the cluster, i.e., 2^{-J}. The summation is over all subclusters j of the maximal cluster J and the coefficients C_{kj}, calculated by means of Eq. (3), are elements of a rectangular matrix, the so-called *configuration matrix* (or C-matrix). Often, more than one "maximal cluster" is used, J, J', J", ..., neither one being a subcluster of any other.

Since the x_J are probabilities, their values must be constrained to lie between 0 and 1. Then, only the lower constraint needs to be considered, since the upper one is guaranteed by the fact that cluster averages lie between -1 and $+1$. Hence, from Eq. (7), we must have, for all maximal clusters, and for all cluster configurations k,

$$\sum_j C_{kj} \xi_j \geq -1 \ . \tag{8}$$

These linear inequalities define a convex region in multidimensional ξ-space, the so-called configurational polyhedron, which contains all realizable configurations on the lattice. The determination of ground states then consists of minimizing the energy (objective) function (6), under the constraint of inequalities (8). This is a standard problem in linear programming and can be solved, when the ECI's are given, by the simplex algorithm. It follows that the vertices of the configuration polyhedron are the solutions sought, i.e, the ordered ground state superstructures, different vertices corresponding to different stoichiometries.

The C-matrix, which has more rows (configurations k) than columns (subcluster types j) contains all the geometric properties of the problem, and is used to transfer that information (lattice type, largest cluster(s), subclusters, symmetry equivalence) to both ground state and CVM codes. Unfortunately, the number of (sub)clusters and the number of configurations tend to increase exponentially with the number of points in the largest cluster retained in the energy (or entropy) approximation chosen. For example, in the 13-, 14-point fcc approximation (central lattice point and its twelve nearest neighbors, fcc cube itself), there are 742 distinct clusters, 554 configurations on the 14-point cluster and 288 on the 13-point cluster. Hence the C-matrix has 842×742 elements! Clearly, the enumeration of all variables and constraints must be obtained by a suitable computer algorithm based on group theoretic considerations. One such algorithm has recently been developed by one of the present authors [17].

Despite the computer automation provided by the C-matrix code and the simplex algorithm, the 13-, 14-point fcc approximation lies pretty much at the limit of what can be done practically at present. We have used this approach to determine the fcc ground states of fcc Pd-V with, as input, the 26 effective cluster interactions calculated by the DCA method, as described above. The resulting ground state map is shown in Fig. 2, in which the *formation energies* (in eV/atom) are plotted as a function of concentration. Formation energies are defined as the actual ordered ground state energy compared to the linear combination of pure fcc Pd and fcc V at the same concentration. The ordered states of minimum energy are indicated by filled squares joined together by a dotted line representing the "convex hull" for this problem. Open squares represent ordered superstructures which narrowly miss being ground states. Their energies were calculated separately, through the use of Eq. (6), using known values of the ξ_j structural variables. The full curve in Fig. 2 is the calculated energy of the completely disordered Pd-V fcc solid solution, theoretically resulting from the infinite-temperature fcc solution quenched infinitely rapidly to zero Kelvin.

All other structures indicated have been derived rigorously and are guaranteed to be the true and only fcc ground states for the given set of interactions. The nomenclature used to describe the superstructures is a hybrid one, consisting of standard Strukturbericht (such as DO_{22}) and prototype designations (such as $MoPt_2$). The structures are determined as follows: the simplex algorithm automatically zeroes in on a vertex of the configuration polyhedron and returns all vertex coordinates (ξ_j) appropriate to that structure. The investigator's task then consists in constructing an actual superstructure on the fcc lattice which has these ξ_j as structural variables. If a structure can be constructed from the vertex coordinates then it is guaranteed to be an absolute minimum energy structure, hence a true ground state.

In the past, ground state searches have often produced "non-constructible structures," which necessarily indicated that the set of constraints was too "loose," in some sense, i.e., incomplete. In the present search, fortunately (i.e., with the 13-, 14-point approximation and the 26 interactions chosen), all vertices turned out to correspond to "constructible" structures, including the very large-unit-cell $HfGa_2$ structure, which could certainly not have been guessed at. For points on the convex hull (dotted line), a mixture of two phases, with structures given by the square symbols on either side, is the stable state.

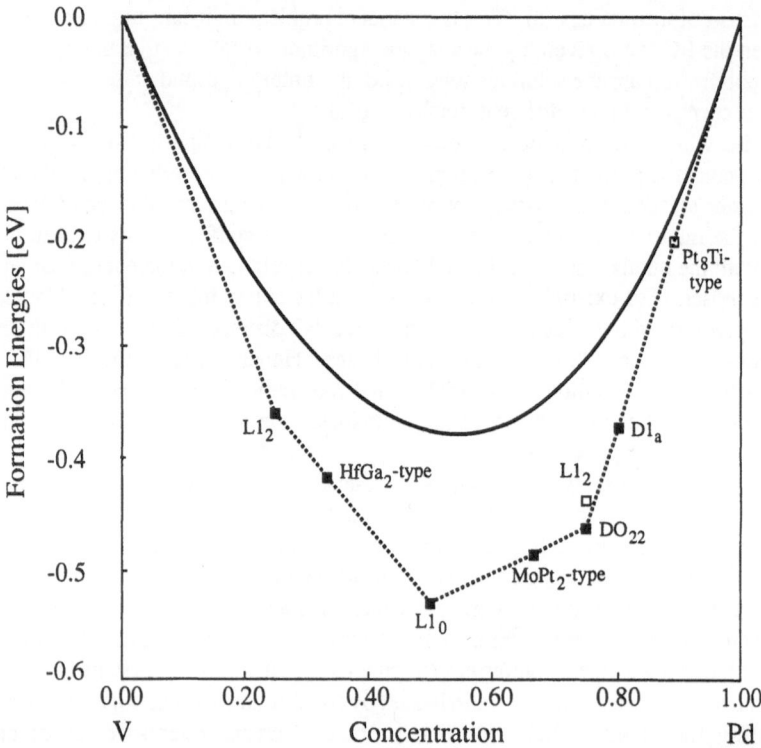

Figure 2. Formation energies for Pd-V fcc ground states as a function of Pd concentration.

6. Discussion

It must be emphasized that what has just been presented is a *true first-principles determination of lowest-energy structures; not a single adjustable or experimentally-derived parameter has been introduced.* Of course, this does not mean that the calculations were exact: local density and atomic sphere approximations were used to calculate tight-binding parameters of pure fcc Pd and V, the tight-binding Hamiltonian was used for the calculation of the ECI's, the Shiba prescription was used to determine off-diagonal hopping parameters, a limited number of recursion levels was used in the density of states calculation, a limited number of configurations was considered in the DCA method, elastic interactions and vibrational entropy were ignored, a limited number (though large, by current standards) of ECI's was calculated. However, the ground state structures were *derived* from these first-principles calculations without any preconceived notion of what the unit cells might be, only that they must be fcc superstructures. The same sort of calculation will also be performed with bcc superstructures at a later date. Then the two sets of ground states will be compared with one another.

Let us now investigate how well the present calculation compares to a previous one and to experimental evidence. Other ECI calculations on the fcc-based Pd-V system are those of Turchi et al. [19], who used the KKR-CPA-GPM scheme (Korringa-Kohn-Rostoker, Coherent Potential Approximation, Generalized Perturbation Method). The first four pair interactions calculated at concentration 0.5 by the latter method [19] and by the present one (DCA) agree closely. Both studies show the importance of describing this system by interactions extending beyond the second-neighbor pair: truncating the inter-action set after the second-neighbor pair would stabilize the $L1_2$ structure for Pd_3V, instead of the correct DO_{22}.

Ordered superstructures can be classified according to the dominant *special-point ordering wave* [14,20]. In fcc, there are three ordering wave families $<100>$, $<1\frac{1}{2}0>$ and $<\frac{1}{2}\frac{1}{2}\frac{1}{2}>$. The $L1_2$ and $L1_0$ structures belong to the $<100>$ family, the $MoPt_2$, DO_{22} and $D1_a$ belong to the $<1\frac{1}{2}0>$ family [20]. A glance at the ground state diagram of Fig. 2 clearly shows that the V-side is dominated by the $<100>$ wave, the Pd-side by the $<1\frac{1}{2}0>$. Competition is close, however: the $L1_2$ and DO_{22} V_3Pd structures have almost the same formation energies. Also, there is experimental evidence [21] that the short-range order above the Pd_3V transition temperature is of $<100>$ type, whereas the long-range order (DO_{22}) is of $<1\frac{1}{2}0>$ type. For a given binary system to belong to more than one special-point family, either concentration-dependent interactions must be used, or, as in the present case, multisite interactions must be taken into account. The latter approach allows for a rigorous ground state determination to be made through the linear programming algorithm, as explained above.

Obviously, the ground state diagram of Fig. 2 cannot do full justice to the actual Pd-V phase diagram [22]. In reality, vanadium is bcc and that side of the phase diagram must be dominated by that lattice and its superstructures or other non-superstructure phases. The Pd side is dominated by the fcc lattice, as expected, and it is found that the DO_{22} and $MoPt_2$-type structures are predicted correctly, respectively for Pd_3V and Pd_2V. The $L1_0$ structure is not observed in the experimentally-determined phase diagram, but a B19 structure has been reported, and described as a "distorted $L1_0$ phase" [23]. In Fig. 2, the $D1_a$ (Pd_4V) and Pd_8V structure points are located practically on the tie-line between DO_{22} and pure Pd so that it is difficult to predict whether those superstructures or phase mixtures will be the true observed ground states. Somewhat surprisingly, the $D1_a$ is not observed experimentally at temperatures for which atomic mobility is high enough to produce equilibrium states, but Pd_8V has been observed by Cheng and Ardell [24] in high-energy proton-irradiated samples. The $D1_a$ structure is the equilibrium one for Ni_4Mo, however. In Fig. 2, the vertical distance between a structure's formation energy and the full curve of the disordered-state energy represents the *ordering energy* (at zero Kelvin) and therefore gives a rough idea of the corresponding order-disorder transition temperature. It is therefore anticipated, from consideration of Fig. 2, that the ordering temperatures of the Pd_4V and, especially, Pd_8V, will be quite low, perhaps therefore unobservable, except under accelerated kinetic conditions such as those produced by irradiation. These two structures depend, for their stabilization, on interactions beyond the second neighbor pair.

7. Conclusion

We have presented a rigorous, first-principles ordered ground state determination of fcc-based Pd-V intermetallic structures. The structures were not selected *a priori*, then compared energetically with one another, as is usually done, but actually derived by an algorithm which guarantees minimum energy in the given context. The cluster expansion method was used here both for calculating the configurational energy and for obtaining the inequalities required by the linear programming technique.

The cluster method is an offshoot of the original cluster variation method but is now considered to be the fundamental technique for describing alloy thermodynamics in general: in principle, any function of configuration can be expressed as an expansion in orthonormal cluster functions. This basic property establishes the link between quantum mechanics and statistical mechanics, hence with classical thermodynamics itself. Here, only ground state applications were discussed in detail, but phase diagrams can be calculated by the CVM, and all derived quantities such as long- and short-range order parameters, diffuse intensity [24], elastic moduli as a function of atomic configuration, etc.

Many practical problems remain to the solved, however. The Pd-V case shows clearly that reliable results can only be obtained if a fairly large number of effective cluster interactions is calculated. The number of cluster functions then tends to increase exponentially, rapidly making the problem completely intractable. Computer algorithms for deriving the configuration matrix are now available [17], without which the present computations could not have been performed. Still, the complexity of the problem can be overwhelming.

It is interesting to note that, originally, the cluster approach was proposed as a method to improve the reliability of the configurational *entropy* [1]. Today, the emphasis has shifted to the configurational *energy*: it is primarily the latter contribution to the free energy that dictates which cluster approximation is required, hence whether or not the problem is tractable by present theoretical means. Undoubtedly, better algorithms will be developed in future, thereby ushering in a true first-principles thermodynamics of materials.

Acknowledgements

This work was supported by the Director, Office of Energy Research, Office of Basic Energy Sciences, Materials Sciences Division of the U.S. Department of Energy under Contract No. DE-AC03-76SF00098, and by grants from NATO (No. 0512/88) and from the National Science Foundation (No. INT-8815493). The authors wish to thank Mr. Mark Asta and Dr. Prabhakar P. Singh for many helpful discussions, and Dr. Mark Van Schilfgaarde for generously making available to them an LMTO-ASA code.

References

1. Kikuchi, R. (1951) Phys. Rev. 81, 988-1003.
2. Van Baal, D. M. (1973) Physica (Utrecht) 64, 571.
3. Sanchez, J. M., Ducastelle, F. and Gratias, D. (1984) Physica (Utrecht) 128A, 334.
4. Finel, A., private communication to D. de Fontaine.
5. Inden, G. and Pitsch, W. (1991) 'Atomic Ordering', in Peter Haasen (ed.) Materials Science and Technology, Vol. 5, VCH Press, Weinheim, pp. 497-552.
6. Asta, M., Wolverton, C., de Fontaine, D. and Dreyssé, H. (1991) Phys. Rev. B (in press).
7. Wolverton, C., Asta, M., Dreyssé, H. and de Fontaine, D. (1991) Phys. Rev. B (in press).
8. de Fontaine, D. (1985) 'On the Feasibility of *ab Initio* Calculations of Ordering Alloy Phase Diagrams', in C. C. Koch, C. T. Liu and N. S. Stoloff (eds.), High Temperature Ordered Intermetallic Alloys, MRS Symposia Proceedings, Vol. 39, pp. 43-64.
9. de Fontaine, D. (1989) 'The Cluster Variation Method and the Calculation of Alloy Phase Diagrams', in G. M. Stocks and A. Gonis (eds.), Alloy Phase Stability, Kluwer Academic Publishers, pp. 177-203.
10. Ducastelle, F. and Gautier, F. (1976) J. Phys. F 6, 2039.

11. Dreyssé, H., Berera, A., Wille, L. T., and de Fontaine, D. (1989) Phys. Rev. B 39, 2442-2452.

12. Burke, N. R. (1976) Surf. Sci. 58, 349.

13. Wolverton, C., Ceder, G., de Fontaine, D., and Dreyssé, H., to be published.

14. de Fontaine, D. (1979) 'Configurational Thermodynamics of Solid Solutions', in H. Ehrenreich, F. Seitz and D. Turnbull (eds.), Solid State Physics, Vol. 34, Academic Press, pp. 73-294.

15. Finel, A. (1987) 'Contribution à l'Etude des Effets d'Ordre dans le Cadre du Modèle d'Ising: Etats de Base et Diagrammes de Phase', doctoral dissertation, Université Pierre et Marie Curie, Paris (unpublished).

16. Sanchez, J. M. and de Fontaine, D. (1981) 'Theoretical Prediction of Ordered Super-structures in Metallic Alloys', in Michael O'Keeffe and Alexandra Navrotsky (eds.), Structure and Bonding in Crystals, Vol. II, Academic Press, pp. 117-132.

17. Ceder, Gerbrand (1991) 'Alloy Theory and Its Applications to Long Period Superstructure Ordering in Metallic Alloys and High Temperature Superconductors', Ph.D. dissertation, University of California, Berkeley (unpublished).

18. Sanchez, J. M. and de Fontaine, D. (1978) Phys. Rev. B 17, 2926-2936.

19. Turchi, P., Stocks, G., Butler, W., Nicholson, D. and Gonis, A. (1988) Phys. Rev. B 37, 5982.

20. de Fontaine, D. (1975) Acta Metall. 23, 553-571.

21. Solal, F., Caudron, R., Ducastelle, F., Finel, A. and Loiseau, A. (1987) Phys. Rev. Lett. 58, 2245-2248.

22. Massalski, T. B., Okamoto, H., Subramanian, P. R. and Kacprzak, L. (eds.) (1990) Binary Alloy Phase Diagrams, Second Edition, ASM International.

23. Maldonado, C. and Schubert, K. (1964) Z. Metallkde. 55, 619.

24. Cheng, J. and Ardell, A. J. (1988) J. Less-Common Met. 141, 45.

25. Sanchez, J. M. (1982) Physica (Utrecht) 111A, 200.

Solubility and Phase Stability in Ordered Intermetallics

J.M. Sanchez, J.D. Becker, and J.K. Tien
Center for Materials Science and Engineering
The University of Texas
Austin, TX 78712
USA

ABSTRACT. Structural and thermomechanical properties which include lattice constants, bulk moduli, Debye temperatures, and Grüneisen constants for the elements and ordered compounds can be obtained accurately using total energy electronic structure calculations within the local density approximation. Examples of these calculations are given using the linearized muffin tin orbital method (LMTO) for the Nb-Ru and Ru-Zr binary systems. Volume dependent pair- and many-body chemical interactions are extracted from the total energy results. These are used with the cluster variation method (CVM) to calculate the solid state equilibrium phase diagram by incorporating local volume relaxation and vibrational free energy using the Debye-Grüneisen approximation.

1.   Introduction

The use of *ab initio* total energy calculations within the local density functional theory has been shown (Mohri et al (1988), Carlsson and Sanchez (1988), and Wei et al (1987) for example) to provide a viable approach to the description of partially ordered alloys with long and short range order. The approach, generically known as the Connolly-Williams (Connolly and Williams (1983)) method, is based on a rigorous cluster expansion of the energy in terms of multisite correlation functions (Sanchez *et al* (1984)). Knowledge of the total energy of a selected set of compounds allows the determination of effective short range interactions which map the alloy into a generalized Ising model. Thermodynamic properties of the system, including the phase diagram, are usually obtained by incorporating the configurational entropies using the Cluster Variation Method (CVM) (Kikuchi(1952)). More recently, vibrational entropy in the

C. T. Liu et al. (eds.), Ordered Intermetallics – Physical Metallurgy and Mechanical Behaviour, 73–88.
© 1992 *Kluwer Academic Publishers.*

Debye-Grüneisen approximation has been included in the total free energy of solid state phases (Sanchez *et al* (1991)).

In some systems local volume relaxation has a significant effect upon phase stability. A method proposed recently (Becker et al (1990)), which allows for partial local volume relaxation, is seen to improve estimates of ordering temperatures and disordered alloy solubility.

Here we review and give examples of a statistical thermodynamic theory of alloys that includes electronic structure clalculations of ordered compounds using the local density approximation, long and short-range order using the CVM, vibrational free energies in the Debye-Grüneisen approximation, and local volume relaxation. The extent to which this theory, requiring only atomic numbers as input, may be used to calculate complex alloy phase diagrams is explored with the ruthenium-niobium and ruthenium-zirconium systems. These systems are selected because RuZr is a cubic structure (B2) with a high melting temperature, and NbRu$_3$ forms a metastable L1$_2$ compound with a high disordering temperature. Possibly, the addition of zirconium to the the niobium-ruthenium system will stablize the L1$_2$ compound-- yielding a high temperature intermetallic with the ductility inherent in the L1$_2$ structure. This study investigates the accuracy of the phase diagram calculation for the binary systems as a preliminary step to the ternary phase diagram calculations.

2.   Total energy calculations

The linear muffin tin orbital (LMTO) method within the atomic sphere approximation (ASA), together with the local density approximation, allows for efficient calculation of self-consistent electronic total energies using a minimal basis set (Andersen(1975)). The method used (Skriver(1986)) includes the Hedin and Lundquist exchange-correlation potential and corrections to first-order in energy. The mesh density of the reciprocal lattice is chosen such that extrapolation of calculated total energies at each volume to an infinite density changes the total energy by no more than 0.001 Ry.

The total energy is calculated for each of five crystal structures for the fcc and hcp lattices and each of six structures for the bcc lattice at fifteen different average atomic volumes near the equilibrium volume in the Nb-Ru and Zr-Ru binary systems.

3.   Morse potential

The volume dependence of the electronic binding energy for each compound is parameterized using a four parameter Morse potential of the form:

$$E(r) = A - 2Ce^{-\lambda(r-r_0)} + Ce^{-2\lambda(r-r)}, \tag{1}$$

In (1), C is the cohesive energy, and r_0 is the equilibrium Wigner-Seitz radius.

Furthermore, the bulk modulus is given in terms of the fitting parameters by:

$$B = \frac{-C\lambda^2}{6\pi r_0} \tag{2}$$

The four parameters are fit to the LMTO generated data using a least-squares fit. Furthermore, using the approach proposed by Moruzzi et al (1988) the Debye temperature, Θ_D, and Grüneisen constant, γ, at the equilibrium radius are given by:

$$(\Theta_D)_0 = A\left(\frac{r_0 B}{M}\right)^{\frac{1}{2}} \tag{3}$$

$$\gamma = \frac{\lambda r_0}{2} \tag{4}$$

where M is the atomic mass, and A is a proportionality constant. The fitting parameters for the elements and compounds studied are shown in Table 1.

Table 1. Morse parameter fits

Compound	Structure	r0[a.u.]	a[Ry/atom]	C[Ry/atom]	λ[a.u.$^{-1}$]
Zr	fcc	3.412	0.112	0.717	0.837
	hcp	3.409	0.121	0.732	0.836
	bcc	3.399	0.143	0.749	0.809
Ru	fcc	2.884	-0.112	0.62	1.403
	hcp	2.872	-0.108	0.637	1.395
	bcc	2.907	-0.106	0.596	1.386
Nb	fcc	3.193	-0.005	0.746	1.023
	hcp	3.182	-0.011	0.742	1.024
	bcc	3.166	-0.01	0.767	1.026
Zr3Ru	L12	3.271	-0.021	0.624	1.013
	DO3	3.245	-0.032	0.633	1.026
	DO19	3.27	0.018	0.667	0.967
ZrRu	B2	3.114	-0.078	0.651	1.147
	B32	3.111	-0.075	0.626	1.13
	L10	3.119	-0.074	0.626	1.13
	Pmma	3.13	-0.071	0.627	1.122
ZrRu3	L12	3.001	-0.104	0.619	1.256
	DO3	3.011	-0.101	0.616	1.248
	DO19	3.008	-0.099	0.625	1.241
Nb3Ru	L12	3.109	-0.0745	0.691	1.156
	DO3	3.081	-0.079	0.712	1.152
	DO19	3.1	-0.065	0.699	1.125
NbRu	B2	3.025	-0.099	0.68	1.23
	B32	3.018	-0.01	0.668	1.227
	L10	3.016	-0.097	0.677	1.219
	Pmma	3.021	-0.092	0.68	1.21
NbRu3	L12	2.951	-0.11	0.654	1.305
	DO3	2.961	-0.136	0.613	1.344
	DO19	2.943	-0.118	0.654	1.316

Figures 1 (a and b) are the ground-state diagrams calculated using the Morse potential fits to the LMTO results. For Zr-Ru the B2 phase is the only stable intermediate compound as is expected from the experimental phase diagram. The ground-state NbRu3 phase is predicted to be a DO19 structure. Villars and Calvert(1985) cite a high-pressure phase with the L12 structure; however, no such transition is expected from the total energy calculations.

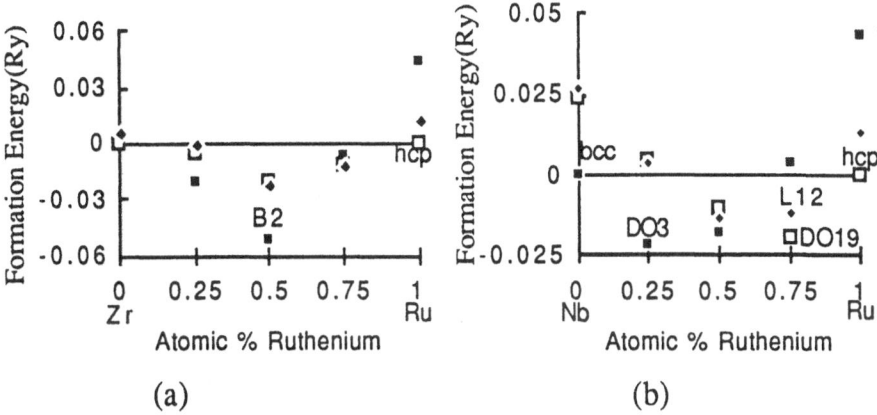

Figure 1: Ground-state energies of the elements and binary compounds underlain by the bcc, hcp, and fcc lattices for the (a) Zirconium-Ruthenium and (b) Niobium-Ruthenium systems.

4. Vibrational free energy

The vibrational free energy of each ordered compound at given atomic volume and temperature is given by the Debye-Grüneisen approximation as:

$$F(r,T) = \frac{9}{8}k_B\Theta_D + E(r) - k_BT\left[D\left(\frac{\Theta_D}{T}\right) - 3\ln\left(1-e^{-\Theta_D/T}\right)\right] \tag{5}$$

where D is the Debye function, and the Debye temperature is given by:

$$\Theta_D = \Theta_D^0\left(\frac{r_0}{r}\right)^{3\gamma}$$

Implicit in Moruzzi's calculation of the Debye temperature from the Morse potential parameters is the assumption of an effective speed of sound, and of a constant Poisson's ratio approximately equal to $1/3$. These assumptions are reasonable for most of the cubic elements, although they are, in general, not applicable for hexagonal elements. The assumptions considered by Moruzzi *et al* (1988) leads to the the value , A=41.63 $(K^2$-g/mole-a.u.-kbar$)^{1/2}$ of the proportionality constant in (3) (see Fig. 2). This expression is rescaled for hexagonal elements by fitting straight lines to measured values of the Debye temperature, bulk moduli, and atomic masses for all cubic and hexagonal transition metals. An empirical constant A=49.96 $(K^2$-g/mole-a.u.-kbar$)^{1/2}$ (see (3)) yields a good fit with experimental data.

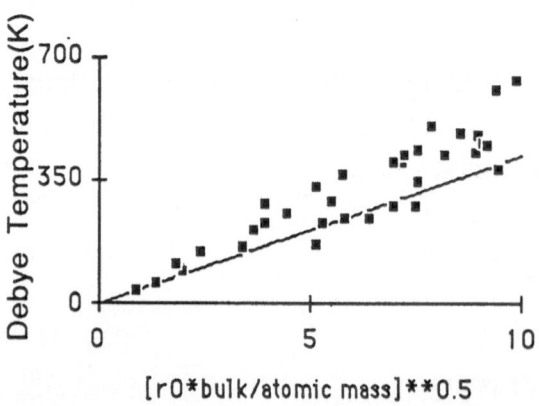

[r0*bulk/atomic mass]**0.5

Figure 2: Experimental (Kittel (1986)) measurements of Debye temperature versus square root of bulk modulus times equilibrium Wigner-Seitz radius divided by the atomic mass for each of the cubic transition metals. The superimposed line has a slope of 41.63.

5.   Cluster Variation Method

The cluster expansion of Sanchez *et al*(1984) provides the basis for the description of the effective chemical interactions of disorderd alloys due to electronic and vibrational free energies and of the configurational entropy. Here, the general method is specialized to binary crystals and summarized.

Let each configuration of the binary crystal with N sites be denoted by the N-dimensional vector $\sigma = \{\sigma_1, \sigma_2, ... \sigma_N\}$, where σ_i is the spin operator having values +1 or -1 for A- or B-atoms, respectively. The two polynomials in σ_i, $\phi_0(\sigma_i)=1$ of order 0 and $\phi_1(\sigma_i)=\sigma_i$ of order 1, form a complete orthonormal set. The set of orthonormal characteristic functions in the N-dimensional space spanned by the vector σ is the direct product of $\{\phi_0(\sigma_i),\phi_1(\sigma_i)\}$ with $i=1,2,..N$, or:

$$\Phi_\alpha(\sigma) = \prod_{i\in\alpha} \sigma_i = \sigma_{i_1}\sigma_{i_2}\cdots\sigma_{i_n} \qquad (6)$$

where α is a cluster of n atoms. Thus, there is a one to one correspondence between the set of orthogonal functions, $\Phi_\alpha(\sigma)$ and the set of all clusters α in the crystal.

Since, $\Phi_\alpha(\sigma)$ form a complete orthonormal set, i.e.

$$\frac{1}{2^N}\sum_\sigma \Phi_\alpha(\sigma)\Phi_\beta(\sigma) = \delta_{\alpha,\beta} \qquad (7)$$

any function of the configuration, $F(\sigma)$, may be written as:

$$F(\sigma) = \sum_\alpha F_\alpha\Phi_\alpha(\sigma) \qquad (8)$$

where the sum is over all clusters including the empty cluster. The projections of $F(\sigma)$ on the orthogonal cluster basis, F_α, are:

$$F_\alpha = \langle F(\sigma)\cdot\Phi_\alpha(\sigma)\rangle = \frac{1}{2^N}\sum_\sigma F(\sigma)\Phi_\alpha(\sigma) \qquad (9)$$

The symmetry of the space group of the crystal requires the cluster projections F_α to be invariant for all clusters α which are related by a symmetry operation of that group. Thus,

$$F(\sigma) = \sum_n F_n\Theta_n(\sigma), \qquad (10a)$$

$$\Theta_n(\sigma) = \sum_{\alpha\in n} \Phi_\alpha(\sigma), \qquad (10b)$$

and, due to the orthogonality of $\Phi_\alpha(\sigma)$,:

$$\frac{1}{2^N}\sum_\sigma \Theta_n(\sigma)\Theta_m(\sigma) = z_n N \delta_{n,m} \tag{11}$$

where $z_n N$ is the total number of n-type clusters in the crystal. Using the notation $\xi_n = <\Phi_\alpha(\sigma)>$ for the expectation value of the characteristic functions, the expectation value of a function of the configuration is:

$$\overline{F} = \langle F(\sigma) \rangle = N \sum_{n=0}^{N} z_n F_n \xi_n \tag{12}$$

5.1 Chemical interactions

Connolly and Williams (1983) used this cluster expansion, truncated to small cluster sizes, to obtain effective chemical interactions of disordered alloys from the formation energies of ordered compounds. In any such truncated expansion the maximum interaction range is chosen *a priori*.

In this study, interactions are limited to nearest neighbors for fcc- and hcp-lattices and to next-nearest neighbors for bcc-lattices. For a given choice of maximum cluster size, the set of relevant crystal structures is determined uniquely by the vertices of the associated configuration polyhedron (Sanchez and Sigli(1985) and de Fontaine (1981)). Thus, for the tetrahedron approximation used here, five high symmetry structures are required for the fcc-lattice and six for the bcc-lattice. So (12) becomes:

$$F_k = \sum_{n=0}^{M} z_n V_n \xi_{k,n} \tag{13}$$

for each ordered structure k, where M=4 for fcc-lattice and M=5 for bcc-lattice, and n labels the subcluster types, i.e. the empty (n=0), point (n=1), pair(n=2), next-nearest pair (n=3; for bcc only), triangle (n=3; n=4 for bcc), and tetrahedron (n=4; =5 for bcc). Equation (13) may be inverted and generalized to allow for volume and temperature dependence of the effective chemical interactions, V_n, to:

$$V_n(r,T)= \frac{N_n}{2^4 z_n} \sum_{k=0}^{M} \omega_k \xi_{k,n} F_k(r,T) \tag{14}$$

where N_n is the number of n-type clusters in a tetrahedron and ω_k is the number of equivalent configurations for the tetrahedron cluster associated with ordered structure, k. Here, $V_k(r,T)$ denotes the vibratioanal free energy given by (5), and r is the set of Wigner-Seitz radii at which each ordered compound energy is evaluated. The coefficients required for the inversion of (13) are given in Table 2. Figure 3 shows the chemical interactions as a function of Wigner-Seitz radius, r ($r_i=r$), for each of the three lattice types in the Nb-Ru and Zr-Ru systems.

Table 2.   Correlation functions for fcc and bcc lattices

System	ω_k	$\xi_{k,0}$	$\xi_{k,1}$	$\xi_{k,2}$	$\xi_{k,3}$	$\xi_{k,4}$	$\xi_{k,5}$
A(fcc)	1	1	1	1	1	1	-
A_3B(L1$_2$)	4	1	1/2	0	-1/2	-1	-
AB(L1$_0$)	6	1	0	-1/3	0	1	-
AB_3(L1$_2$)	4	1	-1/2	0	1/2	-1	-
B(fcc)	1	1	-1	1	-1	1	-
A(bcc)	1	1	1	1	1	1	1
A_3B(DO$_3$)	4	1	1/2	0	0	-1/2	-1
AB(B2)	2	1	0	-1	1	0	1
AB(B32)	4	1	0	0	-1	0	1
AB_3(DO$_3$)	1	1	-1/2	0	0	1/2	-1
B(bcc)	1	1	-1	1	1	-1	1

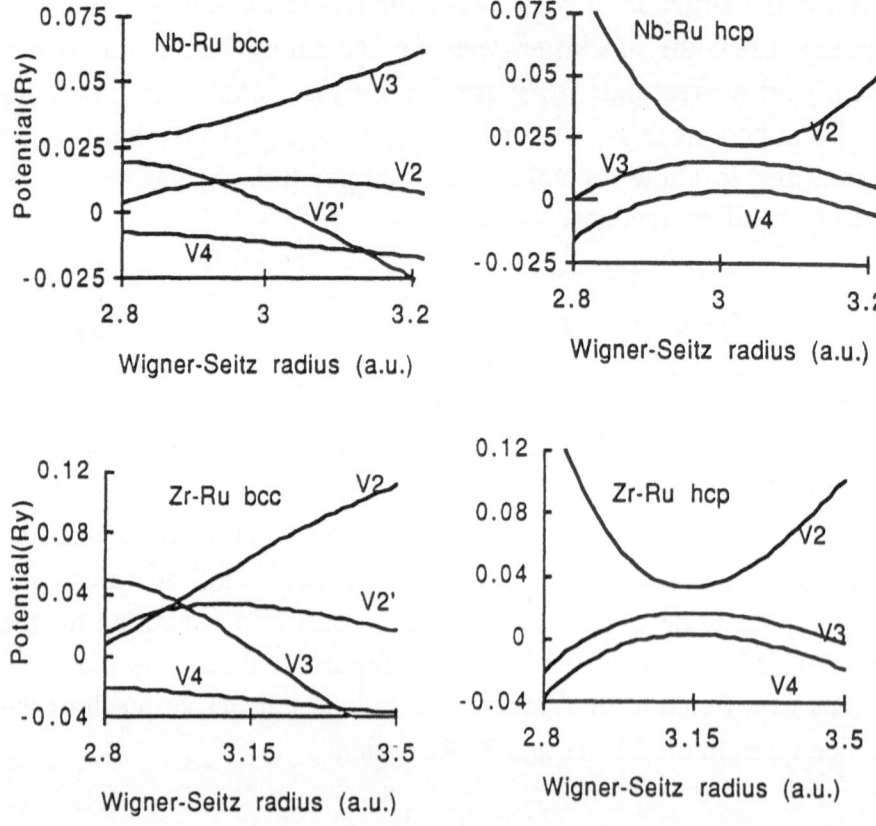

Figure 3: Ground-state pair and many body chemical interactions for (a) the Niobium-Ruthenium system and (b) the Zirconium-Ruthenium system, for bcc and hcp lattices.

5.2 Local volume relaxation

The set of atomic radii, r, at which the compound energies are evaluated may be "relaxed" according to different constraints. Total relaxation is achieved by minimization of the free energy with respect to each r_k, which is equvalent to minimization of the free energies of each of the compounds independently. Here we use a "partial" relaxation in which the total free energy is minimized with respect to the Wigner-Seitz radius of each atom type, A and B, constraining each ordered compound to an average Wigner-Seitz radius prescribed by Vegard's law:

$$r_k=[(1-c_k)r_A{}^3+c_k r_B{}^3]^{1/3} \qquad (15)$$

where c_k is the concentration of B in ordered compound k.

5.3   Configurational entropy

For a given probability distribution $X(\sigma)$, the configurational entropy is:

$$S = -k_B \sum X(\sigma)\ln X(\sigma) \qquad (16)$$

where the sum is over all 2^N configurations of the crystal. A sequence of cluster entropies defined by:

$$S_\alpha = -k_B \sum X_\alpha(\sigma_\alpha)\ln X_\alpha(\sigma_\alpha) \qquad (17)$$

clearly converges to the exact configurational entropy as the cluster size α approaches N. Sanchez et al (1984) use a Mobius transformation to write the cluster entropies, S_α, in terms of a set of irreducibles cluster contributions, $\hat{S}_\alpha$:

$$S_\alpha=\sum_{\beta \subseteq \alpha} \hat{S}_\beta \qquad (18)$$

where the sum extends over all subclusters of α, including α but excluding the empty cluster. The key approximation of the CVM is to neglect the contributions $\hat{S}_\alpha$ for clusters larger than a maximum cluster. Due to the symmetry of the space group of the crystal, (18) can be written as:

$$S = N \sum_{n=1}^{m} z_n \hat{S}_n \qquad (19)$$

where m labels the maximum cluster. This equation, in turn, may be rewritten in terms of cluster entropies:

$$S = N \sum_{n=1}^{m} z_n a_n S_n = -N k_B \sum_{n=1}^{m} z_n a_n \sum_{\sigma_n} X_n(\sigma_n) \ln X_n(\sigma_n) \qquad (20)$$

where the coefficients a_n are determined using:

$$\sum_{\beta \supseteq \alpha} a_\beta = 1 \qquad (21)$$

valid for each subcluster α of the maximum cluster, and with the sum extending over all subclusters β of the maximum clusters that contain α.

6.   Phase diagram calculation

The total free energy functional of the disordered alloy is:

$$F_{tot} = \sum_{n=0}^{M} z_n \xi_n V_n(r,T) + T k_B \sum_{n=1}^{m} z_n a_n \sum_{\sigma_n} X_n(\sigma_n) \ln X_n(\sigma_n) \qquad (22)$$

where the cluster probability distributions are given in terms of an independent set of multisite correlation functions by:

$$X_m(\sigma_m) = \frac{1}{2^n} \left[1 + \sum_{n=1}^{m} \Theta_n(\sigma_n) \xi_n \right] \qquad (23)$$

where Θ_n are the characterisitic functions defined by the sums of products of the configurational variables σ_i for lattice sites i belonging to cluster n. At a given temperature and concentration (determined by the point correlation function, ξ_1) the functional is minimized with respect to the Wigner-Seitz radii, **r**, and the remaining correlation functions to determine the equilibrium free energy.

Figures 4(a and b) are the calculated and experimental phase diagrams for the Nb-Ru and the Zr-Ru systems, respectively. The predicted ordering temperature of the NbRu$_3$ phase, 1999K, is approximately 9% above the experimentally determined temperature, 1813K. The major discrepancy with the experimental data is found in the high temperature solubility in Ru-rich alloys.

The predicted eutectoid temperature at the Zr-rich end of the Zr-Ru phase diagram is 1020K, 10% above the measured value 928K, and the calculated hcp-bcc transition temperature, 1259K, is 11% above the measured temperature, 1136K. This calculated transition temperature is very sensitive to the Debye temperature and Grüneisen constant. Thus, the *ad hoc* approach used to estimate the Debye temperature for the hcp compounds requires some scrutiny.

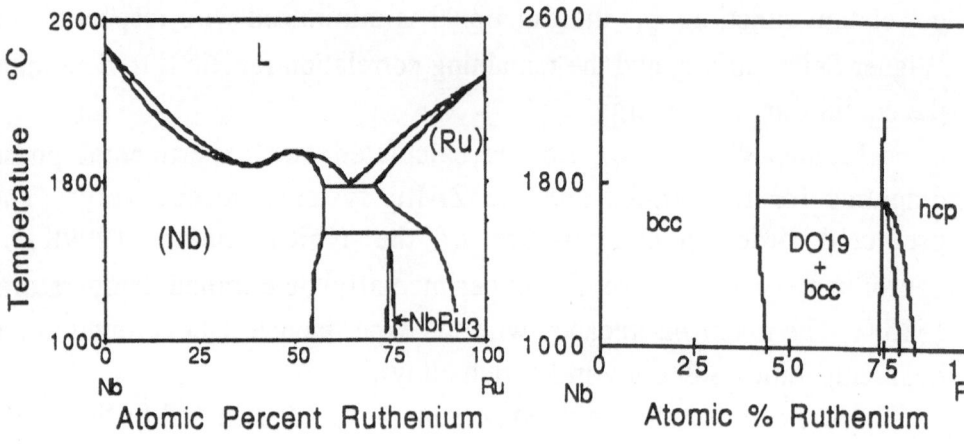

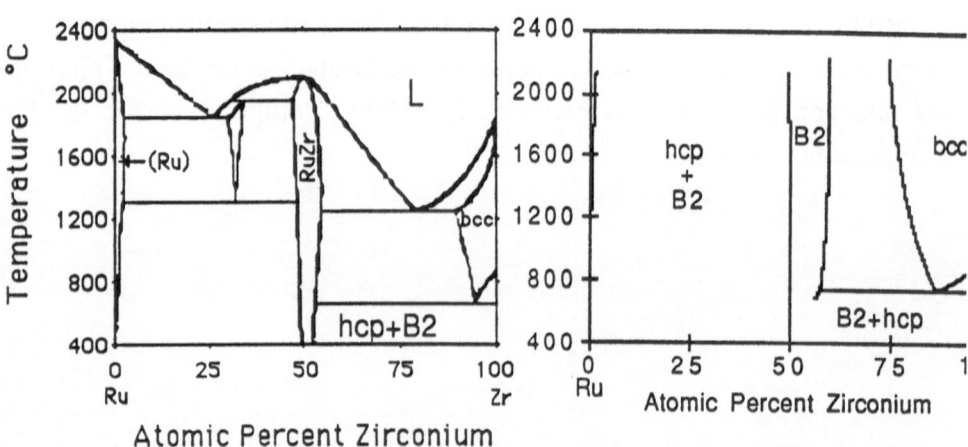

Figure 4: Experimental [Massalski] and calculated phase diagrams for (a) Niobium-Ruthenium and (b) Zirconium-Ruthenium;

7. Conclusions

The LMTO method used in this study predicts well the relative stabilities of cubic structures and, qualitatively, hexagonal structures. The discrepancies of the predicted solubilites of the (Ru) phase in the Nb-Ru system may indicate the need for longer range chemical interactions.

Variation of the Debye temperature and Grüneisen constant can reproduce the experimental Zr-Ru phase diagram accurately, but a truly first-principles calculation will require the implementationof more reliable models for the computation of vibrational entropies as well as refinements of the relaxation schemes. Experimental characterization of binary and ternary phases in the Zr-Nb-Ru system is currently in progress.

Acknowledgements:

This work was supported by the Office of Naval Research, contract number NOOO14-90-J-1185

References

Anderson,O.K., Phys Rev B **12**, 3060 (1975)

Becker, J.D., Sanchez, J.M., and Tien, J.K., Symposium Q, *Proceedings of Materials Research Society Fall Meeting* 1990

Carlsson, A.E. and Sanchez, J.M., *Solid State Comm* **65**, 527 (1988)

Connolly, J.W.D. and Williams,A.R., Phys Rev B **27**, 5169 (1983)

Kikuchi, R., Phys Rev **81**, 988 (1951)

Kittel, C., *Introduction to Solid State Physics*, Wiley (1986)

Massalski,T.B., Murray, J.L., Bennet, L.H., and Baker, H., ed., *Binary Alloy Phase Diagrams*, ASM (1986)

Mohri, T., Terakura, K., Ogudi, T.O., and Wotanebe, K., *Acta Met* **36**, 2239 (1988)

Sanchez, J.M. and de Fontaine, D., *Structure and Bonding in Crystals*, ed., O'Keefe, M. and Navrotsky, A., Academic Press, New York, Vol. II, P.117 (1981)

Sanchez, J.M., Ducastelle, F., and Gratias, D., Physica **128A**, 334 (1984)

Sanchez, J.M., Stark, J.P., and Moruzzi, V.L., submitted to Physical Review (1991)

Skriver, H.L., *The LMTO Method*, Springer-Verlag (1984)

Sluiter, M. and de Fontaine, D., Phys Rev B **42**, 10460 (1990)

Moruzzi, V.L., Janak, J.F., and Schwarz, Phys Rev B **37**, 790 (1988)

Villar's, P. and Calvert, L.D., ed., *Pearson,s Handbook of Crystallographic Data for Intermetallic Phases*, ASM (1988)

Wei, S.-H., Mbaye, A.A., Ferreira, L.G., and Zunger, A., Phys Rev B **36** 4163 (1987)

Computer Calculation of Intermetallic Phase Diagrams

G. INDEN
Max-Planck-Institut für Eisenforschung GmbH
P.O.Box 140 444
D-4000 Düsseldorf 1
Germany

1. Summary

The computer calculation of phase diagrams has been developed in the last two decades starting with the pioneering work of Kaufman et al.. His approach was applied and extended by a number of research groups known under the logo of CALPHAD. Most of their work can be found in the CALPHAD journal and at the occasion of their annual CALPHAD meetings. The purpose of these activities was and still is to provide a tool for estimating the phase equilibria in multicomponent systems. Usually the amount of experimental data is meagre and comes from different areas, e.g. phase diagram data, calorimetric and general thermophysical measurements, structural data. The computer programs are constructed in such a way that they can bundle all these different kinds of informations in an optimizer module to arrive at numerical values for adjustable parameters which define the thermodynamic functions. These functions have been developed in the past and this activity goes on. We may distinguish three types of approaches: (I) the pure formal mathematical approach consisting in a polynomial representation of the thermodynamic functions (CALPHAD approach), (II) the modeling of thermodynamic functions, (III) statistical thermodynamics treatments.

The presented paper tried to give an overview of these techniques with emphasis on the special aspects of materials development based on intermetallic phases. Special aspects were: How good is the description of intermetallic phases? What about the homogeneity range of intermetallic phases? What about the predictive power of the techniques, e.g. what can be predicted in a ternary system from existing knowledge in the binary subsystems? These aspects were discussed and illustrated with particular examples out of the huge amount of existing data in the literature as well as unpublished new results.

The interested reader is refered to the references listed below. This list cannot be complete. A selection has been made such that the main strategies can be found and that typical aspects of phase diagrams of intermetallic systems are treated.

2. References

2.1. CALPHAD TECHNIQUES

2.1.1 *General Aspects*

L. Kaufman and H. Bernstein, Computer Calculation of Phase Diagrams, Academic Press, New York 1970

L. Kaufman, Computer Based Thermochemical Modeling of Multicomponent Phase Diagrams, in Alloy Phase Stability, G.M. Stocks and A. Gonis Eds., Nato ASI Series E **163** (1989) 145-175

M. Hillert, Empirical Methods of Predicting and Representing Thermodynamic Properties of Ternary Solution Phases, CALPHAD **4** (1980) 1-12

H.L. Lukas, J. Weiss and E.-Th. Henig, Strategies for the Calculation of Phase Diagrams, CALPHAD **6** (1982) 229-231

C. T. Liu et al. (eds.), Ordered Intermetallics – Physical Metallurgy and Mechanical Behaviour, 89–91.
© 1992 *Kluwer Academic Publishers.*

2.1.2 *Calculation of superalloy phase diagrams*

Part I, L. Kaufman and H. Nesor, Metallurgical Trans. **5** (1974) 1617-1621
Part II, L. Kaufman and H. Nesor, Metallurgical Trans. **5** (1974) 1623-1629
Part III, L. Kaufman and H. Nesor, Metallurgical Trans. **6A** (1975) 2115-2122
Part IV, L. Kaufman and H. Nesor, Metallurgical Trans. **6A** (1975) 2123-2131

2.1.3 *Coupled Phase Diagrams and Thermochemical Data for Transition Metal Systems*

Part I, L. Kaufman and H. Nesor, CALPHAD **2** (1978) 55-80
Part II, L. Kaufman and H. Nesor, CALPHAD **2** (1978) 81-108
Part III, L. Kaufman, CALPHAD **2** (1978) 117-146
Part IV, L. Kaufman, CALPHAD **3** (1979) 45-76
Part V, L. Kaufman and H. Nesor, CALPHAD **2** (1978) 325-348
D. Dew-Hughes and L. Kaufman, Ternary Phase Diagrams of the Mn-Ti-Fe and Al-Ti-Fe systems: a Comparison of Computer Calculations with Experiments, CALPHAD **3** (1979) 175-203
T. Chart, F. Putland and A. Dinsdale, Calculated Phase Equilibria for the Cr-Fe-Ni-Si System - Ternary Equilibria, CALPHAD **4** (1980) 27-46
F. Putland, T. Chart and A. Dinsdale, Thermodynamically Calculated Phase Diagrams for the Co-Cr-Ta and Co-Cr-Nb Systems, CALPHAD **4** (1980) 133-141
C. Allibert, C. Bernard, G. Effenberg, H.-D. Nüssler and P.J. Spencer, A Thermodynamic Evaluation of the Fe-Co-Cr System, CALPHAD **5** (1981) 227-237

2.2. MODELING OF THERMODYNAMIC FUNCTIONS

2.2.1 *Sublattice Model*

B. Sundman and J. Ågren, A Regular Solution Model for Phases with Several Components and Sublattices Suitable for Computer Calculations, J. Phys. Chem. Solids **42** (1981) 297-301
J.O. Andersson and P. Gustafson, A Thermodynamic Evaluation of the Fe-W System, CALPHAD **7** (1983) 317-326
J.-O. Andersson, A. Fernandez-Guillermet, M. Hillert, B. Jansson and B. Sundman, A Compound Energy Model of Ordering in a Phase with Sites of Different Coordination Numbers, Acta Met. **34** (1986) 437-445
J.-O. Andersson and B. Sundman, Thermodynamic Assessment of the Cr-Fe System, CALPHAD **11** (1987) 83-92
K. Frisk and P. Gustafson, An Assessment of the Cr-Mo-W System, CALPHAD **12** (1988) 247-254
A. Fernandez-Guillermet, Thermodynamic Properties of the Fe-Co-Ni-C System, Z. Metallk. **79** (1988) 524-536

2.2.2 *Magnetic effects*

M. Hillert and M. Jarl, A Model for Alloying Effects in Ferromagnetic Metals, CALPHAD **2** (1978) 227-238
T. Nishizawa, M. Hasebe and M. Ko, Thermodynamic Analysis of Solubility and Miscibility Gap in Ferromagnetic Alpha Iron Alloys, Acta Met. **27** (1979) 817-828
G. Inden, The Role of Magnetism in the Calculation of Phase Diagrams, Physica **103B** (1981) 82-100

2.3. STATISTICAL THERMODYNAMIC TREATMENT

2.3.1 *Overview and general aspects*

G. Inden and W. Pitsch, Atomic Ordering, in Materials Science and Technology, Cahn R.W., Haasen P., Kramers E.J. Eds., Vol. 5: Phase Transformations, VCH Verlagsgesellschaft, Weinheim 1991
G. Inden, Thermodynamics of Ordering, Scand. J. Metallurgy **20** (1991) 112-120

2.3.2 *Bragg-Williams Treatments*

G. Inden, Determination of the Interchange Energies $W_{CoSi}^{(1)}$ and $W_{CoSi}^{(2)}$ for BCC Solid solutions from High Temperature Neutron Diffraction on Ternary Fe-Co-Si Crystals, phys. stat. sol. (a) **56** (1979) 177-186

T. Nishizawa, S.M. Hao, M. Hasebe and K. Ishida, Thermodynamic Analysis of Miscibility Gap due to Ordering in Ternary Systems, Acta Met. **31** (1983) 1403-1416

M. Fukaya, T. Miyazaki, P.Z. Zhao and T. Kozakai, A Statistical Evaluation of the Free Energy of Fe-base Ternary Ordering Alloys, J. Mater. Sci. **25** (1990) 522-528

2.3.3 *Cluster Variation Method (CVM)*

R. Kikuchi, Ternary Phase Diagram Calculations - I General Theory, Acta Met. **25** (1977) 195-205

R. Kikuchi, D. de Fontaine, M. Murakami and T. Nakamura, , Ternary Phase Diagram Calculations - II Examples of Clustering and Ordering Systems, Acta Met. **25** (1977) 207-219

J.M. Sanchez, R. Kikuchi, H. Yamauchi and D. de Fontaine, Cluster Approach to Order Disorder, in Theory of Alloy Phase Formation, L.H. Bennett Ed., The Metallurgical Soc. AIME 1980, p. 289-302

J.M. Sanchez and D. de Fontaine, Ising Model Phase Diagram Calculations in the FCC Lattice with First and Second Neighbour Interactions, Phys. Rev. **B 25** (1982) 1759-1765

C. Siggli and J.M. Sanchez, Calculation of Temperature-Concentration Diagrams by the CV Method with Lennard-Jones Pair Interactions, CALPHAD **8** (1984) 223-231

R. Kikuchi and J.L. Murray, Tetrahedron Treatment of the BCC Lattice, CALPHAD **9** (1985) 311-348

T. Mohri, J.M. Sanchez and D. de Fontaine, Binary Ordering Prototype Phase Diagrams in the CVM Approximation, Acta Met. **33** (1985) 1171

H. Ackermann, G. Inden and R. Kikuchi, Tetrahedron Approximation of the Cluster Variation Method for BCC Alloys, Acta Met. **37** (1989) 1-7

M. Sluiter, D. de Fontaine, X.Q. Guo, R. Podloucky and A.J. Freeman, Ab Initio Calculation of Ordered Intermetallic Phase Equilibria, Mat. Res. Symp. Proc. Vol. 133, High-Temperature Ordered Intermetallic Alloys III, C.T.Liu, A.I. Taub, N.S. Stoloff and C.C. Koch Eds., MRS Pitsburgh 1989, p. 3-15

D. de Fontaine, The Cluster Variation Method and the Calculation of Alloy Phase Diagrams, in Alloy Phase Stability, G.M. Stocks and A. Gonis Eds., Kluwer Acad. Press 1989, p. 177-203

ATOMIC ORDERING IN TERNARY AND QUATERNARY COMPOUND SEMICONDUCTORS

S. MAHAJAN AND B.A. PHILIPS
Department of Materials Science
Carnegie Mellon University
Pittsburgh, PA 15213, USA

ABSTRACT. The status of atomic ordering in ternary and quaternary III-V compound semiconductors is reviewed briefly. Evidence favors the occurrence of CuPt-type ordering in these materials. Reconstruction at (001) surfaces is essential for the formation of the ordered structure that is metastable in the bulk. Ordering reduces bandgaps, and degradation resistance of light emitting devices is enhanced.

1. Introduction

One of the remarkable developments of the last decade is that of lightwave communications. The production of extremely low loss fused silica fibers which are used as a transmitting medium has been realized around the world and has become routine. In addition to the transmitting medium, light sources and detectors are required for the systems. Figure 1 shows optical losses in fused silica fibers as a function of wavelength. Two minima are observed: one at 1.33 μm and the other at 1.55 μm. In reference to Fig. 2 that shows bandgaps, corresponding emission wavelengths and lattice parameters of various III-V compound semiconductors, it is evident that ternary and quaternary layers involving different combinations of In, Ga, As and P atoms can emit in the wavelength regime where the fiber losses are minimum. Thus, the emitters based on the InP/InGaAsP system constitute optimal light sources for optical communication systems. Furthermore, a ternary $In_{0.53}Ga_{0.47}As$ is observed to emit at 1.67 μm. Consequently, a layer of this composition can be used in the fabrication of detectors. The elegance of the InP/InGaAsP system is that lattice matched ternary and quaternary layers can be grown on InP substrates that emit in the wavelength regime spanning 1.1 to 1.67 μm, an optimal range for lightwave communication systems.

The ternary and quaternary materials crystallize in the zinc-blende structure that consists of two interpenetrating F.C.C. sub-lattices. One of the sub-lattices is displaced with respect to the other by $\frac{a}{4}$<111>, where a is the lattice parameter of the material. The group III atoms occupy one sub-lattice, whereas group V atoms reside on the second sub-lattice. An obvious question is whether or not atoms within ternary and quaternary compositions are distributed at random within the respective sub-lattices? The answer is a definite 'no'. A number of investigators have addressed this issue in the last decade

C. T. Liu et al. (eds.), Ordered Intermetallics – Physical Metallurgy and Mechanical Behaviour, 93–106.
© 1992 *Kluwer Academic Publishers.*

94

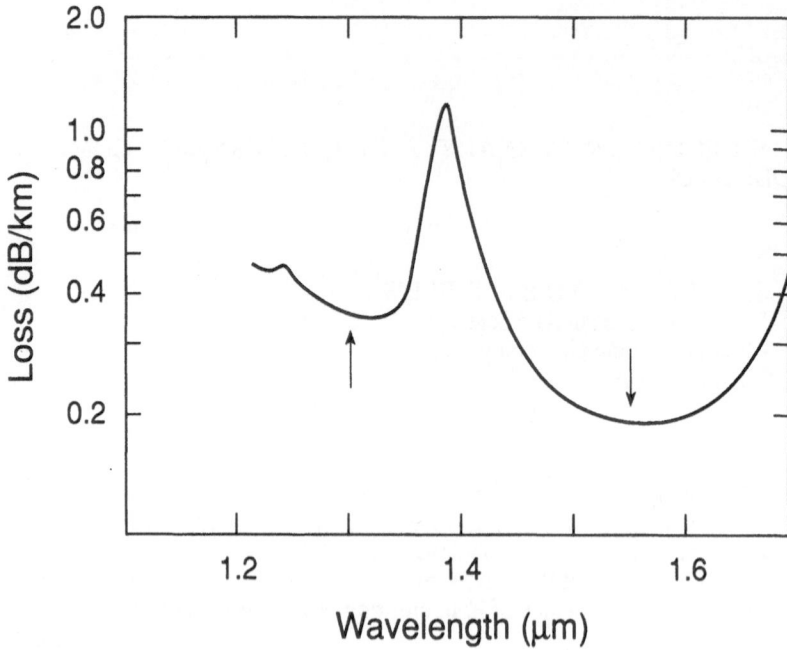

Figure 1. Optical loss in fused silica fibers as a function of wavelength. Note that there are two minimal; one at 1.33 µm and the other one at 1.55 µm.

and find that these materials exhibit two types of deviations from randomness: (i) atomic ordering (Kuan et al. 1985, Jen et al. 1986, Kuan et al. 1987, Shahid et al. 1987, Norman et al. 1987, Ihm et al. 1987, Gomyo et al. 1987, Shahid and Mahajan 1988, Gomyo et al. 1988, McKernan et al. 1988, Kondow et al. 1988a, Bellon et al. 1988, Mahajan and Shahid 1989, Mahajan et al. 1989, Augarde et al. 1989, Kondow et al. 1989, Chen et al. 1990, Murgatroyd et al. 1990, Chen et al. 1991 and Mahajan 1991) and phase separation (Henoc et al. 1982, Launois et al. 1982, Mahajan et al. 1984, Chu et al. 1985, Norman and Booker 1985, Treacy et al. 1985, Shahid and Mahajan 1988, Mahajan et al. 1989, Mahajan and Shahid 1989, McDevitt et al. 1990 and Mahajan 1991). Furthermore, the formation of ordered structures depends on the technique used to grow an epitaxial layer (Norman et al. 1987, Mahajan and Shahid 1989 and Mahajan et al. 1989). For example, with the exception of a single observation where ordering has been observed in a ternary layer grown by liquid phase epitaxy (Nakayama and Fujita 1985), ordering is seen only in layers grown by molecular beam epitaxy (Kuan et al. 1985, Norman et al. 1987, Kuan et al. 1987 and Murgatroyd et al. 1990), organometallic vapor phase epitaxy (Kuan et al. 1985, Jen et al. 1986, Kuan et al. 1987, Ihm et al. 1987, Gomyo et al. 1987, Gomyo et al. 1988, McKernan et al. 1988, Kondow et al. 1988, Bellon et al. 1988, Mahajan and Shahid 1988, Mahajan et al. 1989, Mahajan and Shahid 1989, Augarde et al. 1989, Kondow et al. 1989, Chen et al. 1990 and Chen et al. 1991) and vapor levitation epitaxy (Shahid et al. 1987, Shahid and Mahajan 1988, Mahajan and Shahid 1989, Mahajan et al. 1989 and Mahajan 1991).

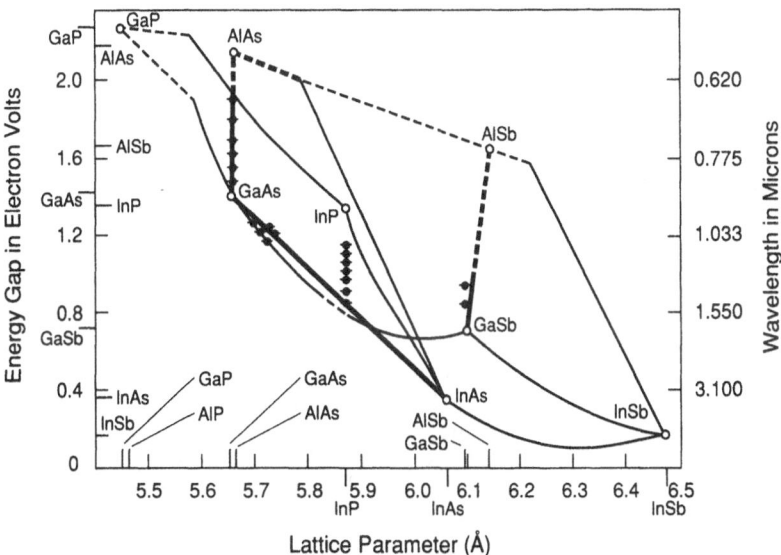

Figure 2. Bandgaps, emission wavelengths and lattice parameters of various binaries, ternaries and quaternaries III-V compound semiconductors. Note that the emission wavelengths of InGaAsP layers can be varied between 1.1 and 1.67 μm and layers are still matched to InP substrates.

The initial observations of Kuan et al. (1985) and Jen et al. (1986) indicated the possibility of CuAu-I and chalcopyrite type ordered structures in GaAlAs and GaAsSb epitaxial layers. Since these investigators examined only the (001) reciprocal section by electron diffraction, it is not possible to discern unequivocally the type of ordered structure based on these observations. Mahajan et al. (1989) have suggested that the observed superlattice spots could arise from the presence of different types of interfaces between two variants of the CuPt-type ordered structure.

Shahid et al. (1987) were the first to observe the occurrence of CuPt-type ordering in InGaAs and InGaAsP layers deposited on (001) InP substrates by vapor levitation epitaxy. Their observation was later on confirmed by a number of investigators (Norman et al. 1987, Ihm et al. 1987, Gomyo et al. 1987, Shahid and Mahajan 1988, Gomyo et al. 1988, McKernan et al. 1987, Kondow et al. 1988, Bellon et al. 1988, Mahajan et al. 1989, Augarde et al. 1989, Kondow et al. 1989, Chen et al. 1990, Murgatroyd et al. 1990, Mahajan 1991 and Chen et al. 1991). These studies included GaInP, (Ga,Al) In P, GaAsSb and InAsSb layers grown by organometallic vapor phase epitaxy and molecular beam epitaxy. The significant points to emerge from these studies are the following. First, as a result of ordering the real space periodicity along the <111> directions is doubled. Second, only two out of the possible four ordered variants are observed. Augarde et al. (1989) have shown experimentally that ordering occurs on the $(\bar{1}11)_B$ and $(1\bar{1}1)_B$ planes in GaInP epitaxial layers. The assignation of these indices is based on the assumption that the origin of the group III sub-lattice is at 0,0,0, whereas that of the group V sub-lattice is at $\frac{1}{4},\frac{1}{4},\frac{1}{4}$. Third, the pair of $\{\bar{1}\bar{1}\bar{1}\}_B$ planes on which ordering takes place is independent of the sub-lattice on which atomic substitutions occur to derive a ternary composition from a parent binary (Chen et al. 1990, Murgatroyd et al. 1990 and Chen et al. 1991).

Since the four <111> directions are equivalent in the bulk, the observation of two ordered variants indicates that ordering may be occurring at the surface while the layer is growing (Shahid and Mahajan 1988, Suzuki et al. 1988, Mahajan and Shahid 1989 and Mahajan et al. 1989). Suzuki et al. (1988) have proposed a model for atomic ordering that involves nucleation of epitaxial growth from the step edges present on a vicinal surface. As discussed later, this model cannot explain in a cogent manner the three observations highlighted above. Recently, Murgatroyd et al. (1990) have invoked the role of surface reconstruction in atomic ordering. While their idea on surface reconstruction is correct, the independence of {111} ordered variants of the sub-lattice on which atomic substitutions occur cannot be rationalized on the basis of their explanation.

The objectives of the present paper are four fold. First, to highlight the salient features of CuPt-type ordering in ternary and quaternary III-V compound semiconductors. Second, to present a model that has been developed to rationalize the above observations pertaining to ordering (Philips et al. 1991). This model is an extension of the surface-stress-induced ordering mechanism proposed by LeGoues et al. (1990) for the evolution of CuPt-type ordering in Si-Ge films. Third, to present results on diffusion- and thermal-induced disordering and to discuss the issue of metastability of ordered structures. Fourth, to discuss the effects of ordering on bandgaps and to evaluate the ramifications of ordering on reliability of light emitting devices.

2. Highlights of CuPt-type Ordering

Figure 3(a), reproduced from the work of Shahid and Mahajan (1988), shows the (110) cross-section of a multi-layer structure consisting of InP, InGaAsP and InP layers; these layers were grown by vapor levitation epitaxy. The electron diffraction pattern observed from the InGaAsP layer is shown in Fig. 3(b). It is clear that, in addition to the diffraction spots which correspond to the zinc-blende structure, superlattice spots are present at $\frac{\bar{1}}{2},\frac{1}{2},\frac{1}{2}$, $\frac{1}{2},\frac{\bar{1}}{2},\frac{1}{2}$ and equivalent positions. This implies that as a result of ordering the real space periodicity along the $[\bar{1}11]$ and $[1\bar{1}1]$ directions is doubled. In addition, superlattice spots were not observed in the $(1\bar{1}0)$ reciprocal section.

Following the suggestion of Shahid et al. (1987), the above observations can be rationalized by referring to Fig. 4 that shows the atomic arrangement in an InGaAs$_2$ layer which has undergone CuPt-type ordering on the $(1\bar{1}1)_B$ planes. The stacking arrangement of the {111} planes in the disordered material is A(III) a(V) B(III) b(V) C(III) c(V) A(III) a(V).... whereas it is A(Ga) a(As) B(In) b(As) C(Ga) c(As) A(In) a(As) B(Ga) b(As) C(In) c(As) A(Ga) a(As).... in the ordered material.

The perfection of the ordered structure is affected by growth conditions (Mahajan and Shahid 1989, Mahajan et al. 1989 and Cao et al. 1991). Figures 5(a) → (e) show diffraction patterns obtained from GaInP$_2$ layers grown at different rates by organometallic vapor phase epitaxy (Cao 1991). The principal effect of the increased

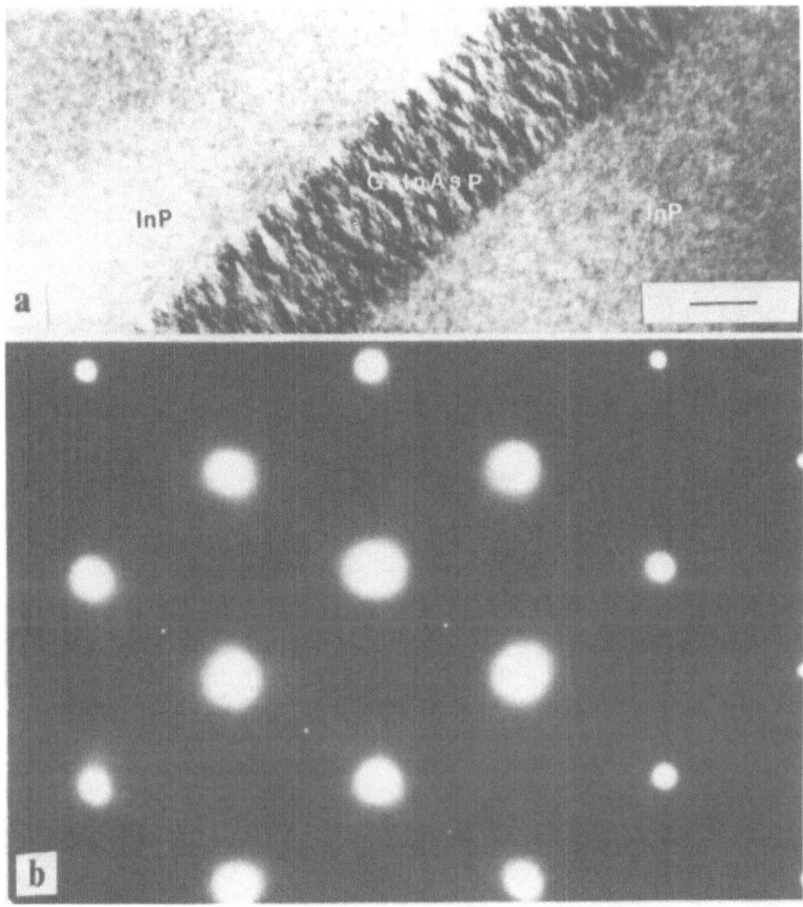

Figure 3. (a) (110) cross-section of a multi-layer structure consisting of InP, InGaAsP and InP that was grown by vapor levitation epitaxy. (b) Electron diffraction pattern observed from the InGaAsP layer shown in (a). After Shahid and Mahajan (1988).

growth rate is to produce intensity spikes which pass through the superlattice spots and are essentially parallel to the growth direction.

The effects of the increased growth rate are also manifested in the microstructures. Figure 6 shows domain structures observed in $(Ga,Al)InP_2$ layers grown under different conditions by low pressure organometallic vapor phase epitaxy. The layer in Fig. 6(a) was grown at a low temperature and high growth rate, whereas a high growth temperature and low growth temperature was used for the layer shown in Fig. 6(b) (Mahajan et al. 1989). A high density of features resembling antiphase domain

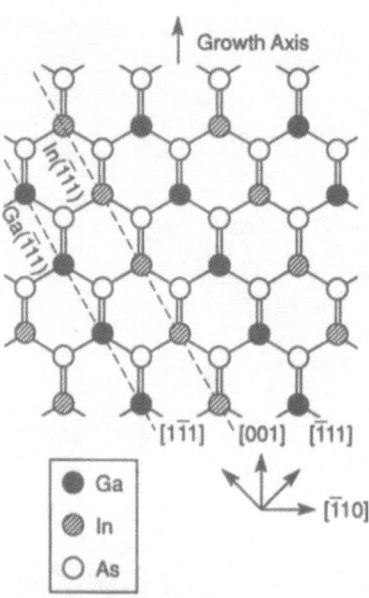

Figure 4. Schematic showing the atomic arrangement in a InGaAs layer that has undergone CuPt-type ordering and can be used to explain diffraction pattern shown in Fig. 3(b).

boundaries observed in ordered metallic alloys is seen in Fig. 6(a). On the other hand, a coarse domain structure is observed in Fig. 6(b).

The origin of the domain boundaries in Fig. 6(a) is not clear. It is possible that they represent interfaces at which switch occurs between the $(\bar{1}11)_B$ and $(1\bar{1}1)_B$ ordered variants. It can be shown that the occurrence of ordering on the $(\bar{1}11)_B$ planes would tilt the growth surface away from the [001] orientation. This however can be restored by ordering on the $(1\bar{1}1)_B$ planes. The alternation between the two variants could lead to a large number of interfaces, which in turn could produce intensity spikes through superlattice spots observed in Fig. 5.

Mahajan et al. (1989) have proposed a model to rationalize the formation of domains shown in Fig. 6(b). For the sake of discussion consider the case of a GaInP$_2$ layer grown on a (001) substrate. To produce CuPt-type ordering in this layer, Ga and In atomic rows parallel to [110] must alternate along the [1$\bar{1}$0] direction, see Fig. 4. It is conceivable that in the presence of steps on vicinal surfaces the arrangement of atomic rows is not commensurate across the steps. They have argued that this would lead to the tubes of binary material within a ternary layer. The interfaces between the binary and ternary materials could produce domain-like contrast.

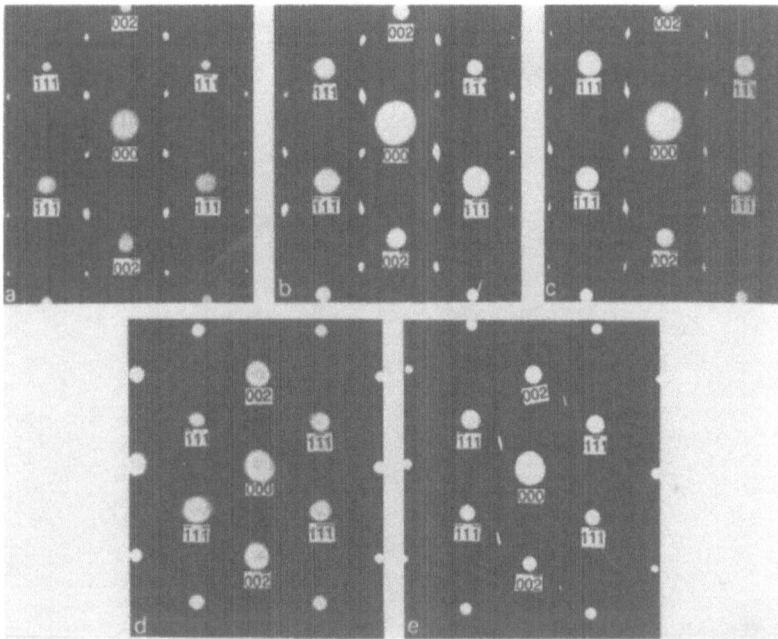

Figure 5. [110] electron diffraction patterns obtained from GaInP layers grown at different growth rates (G_R): (a) G_R=4.1 µm/hr, (b) G_R=6.3 µm/h, (c) G_R=8.3 µm/h, (d) G_R=12 µm/h, and (e) G_R=12 µm/h. After Cao et al. (1991).

The suggestion of Mahajan et al. (1989) is borne out by the observations of Augarde et al. (1989). They have observed that domains appear to nucleate from steps present on the substrate surface. This is illustrated in Fig. 7 which shows domains in a (110) cross-section of a GaInP$_2$ epitaxial layer grown on (001) GaAs substrates by organometallic vapor phase epitaxy.

3. Models for CuPt-type Ordering

Two attempts have been made to explain the evolution of CuPt-type ordering (Suzuki et al. 1988 and Murgatroyd et al. 1990). The model of Suzuki et al. is illustrated schematically in Fig. 8. It assumes the presence of {111} facets on the (001) GaAs surface on which GaInP$_2$ is being grown. They envisage that the corner As atoms would bond preferentially to Ga atoms because the Ga-As bond is stronger than the In-As bond. Subsequently a row of In atoms would form, followed by alternating rows of Ga and In atoms. For the arrangement of Ga and In rows in the next layer to be in phase with that in the underlying layer, the rows must undergo lateral displacement along the [1$\bar{1}$0] direction, see Figs. 4 and 8(b) and (c). It is not clear how is this achieved? Furthermore, the model predicts that if the (001) surface would be subjected to a flux of Ga, As and P atoms, the constituents required to grow a Ga$_2$AsP layer, ordering should be observed on the (111) and (11$\bar{1}$) planes. This is contrary to the observations of Chen et al. (1990, 1991).

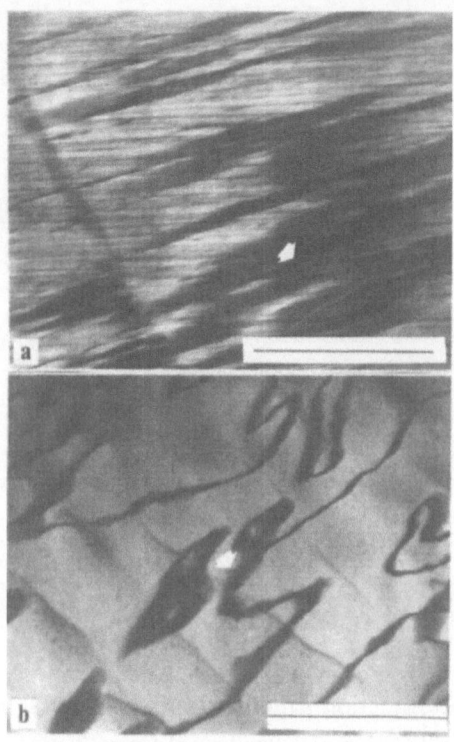

Figure 6. Domain boundaries observed in (Ga,Al)InP layers grown at different rates on (001) GaAs substrates by organometallic vapor phase epitaxy; (a) high, (b) low growth rate. Reflections used for forming images in (a) and (b) are, respectively, 111 and 3$\bar{1}$1 superlattice spots. Markers in (a) and (b) represent 0.2 and 0.5 μm, respectively. After Mahajan et al. (1989).

Recently, Murgatroyd et al. (1990) have proposed a model that invokes the role of surface reconstruction in CuPt-type ordering in Ga_2AsSb layers grown on (001) GaAs substrates. The major drawback of their model is that the driving force for preferential occupation of the [1$\bar{1}$0] rows by As and Sb atoms is not apparent.

Philips et al. (1991) have recently extended the surface-stress-induced ordering mechanism, proposed by LeGoues et al. (1990) for Si-Ge films, to explain the development of CuPt-type ordering in ternary and quaternary III-V semiconductors. The salient feature of their proposal is illustrated in Fig. 9. For the sake of discussion, consider the growth of a $GaInP_2$ layer on a (001) GaAs substrate. Figure 9(a) shows the (110) perspective of the As-terminated GaAs (001) surface, whereas the reconstructed surface is shown in Fig. 9(b). As a result of the formation of As-dimers that accompanies surface reconstruction, tension and compression regions are produced. If the reconstructed surface is subjected to a flux of Ga, In and P atoms, Ga atoms would

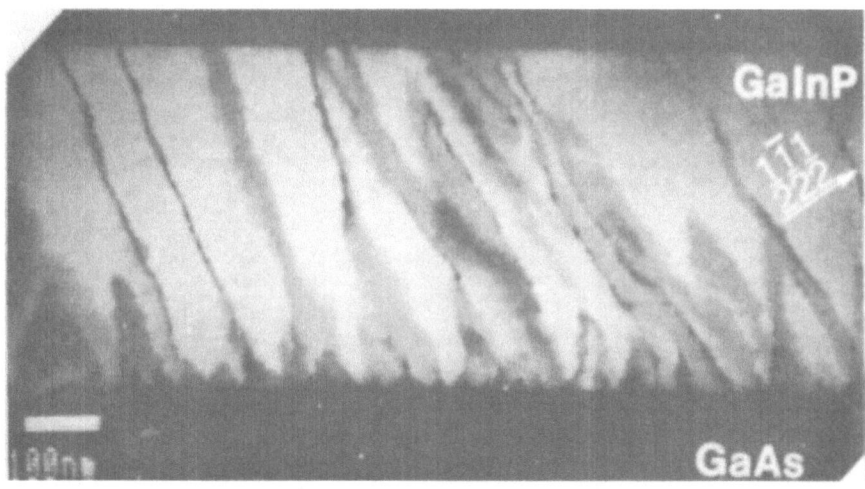

Figure 7. (110) cross-section of a GaInP layer grown on (001) GaAs substrates by organometallic vapor phase epitaxy. Note that domains appear to nucleate from steps present on the surface of the underlying substrate. After Augarde et al. (1989).

occupy the [110] rows in the compression regions while In atoms would cluster along the [110] rows in the tension regions, i.e. the occurrence of surface reconstruction would bias the occupation of sites by certain atomic species. This is inferred because the tetrahedral covalent radius of In atoms is considerably larger than that of Ga atoms. By propagating reconstruction along the [001] growth direction, the lateral displacement of [110] In and Ga rows required for CuPt-type ordering can be achieved.

If the reconstructed surface is subjected to a flux of Ga, As and P atoms, compression and tension regions would be occupied by P and As atoms, respectively. This is again dictated by their size difference, and would lead to ordering on the same pair of $\{1\bar{1}\bar{1}\}_B$ planes as in the case of a GaInP$_2$ layer. It is therefore inferred that the model proposed by Philips et al. (1991) can account for the known features of CuPt-type ordering.

4. Metastability and Thermal- and Diffusion-Induced Disordering

Using first principles total energy calculations, Bernard et al. (1988), Dandrea et al. (1990) and Bernard et al. (1990) have shown that in the bulk chalcopyrite and CuAu-I type ordered structures have lower energy than CuPt-type ordering. However, as discussed in Sections 1, 2 and 3, the latter type of ordering is a common microstructural feature of ternary and quaternary layers grown by vapor phase techniques. One plausible way out of this dichotomy is to invoke that CuPt-type ordering represents an equilibrium situation at the surface, but is metastable in the bulk. This assessment is consistent with the recent calculations of Froyen and Zunger (1991) who have shown that in the presence of surface reconstruction, the energy of CuPt-type structure is lower than that of chalcopyrite and CuAu-I type structures. Presumably the surface equilibrium situation is incorporated into the bulk because bulk diffusion in compound semiconductors is extremely slow.

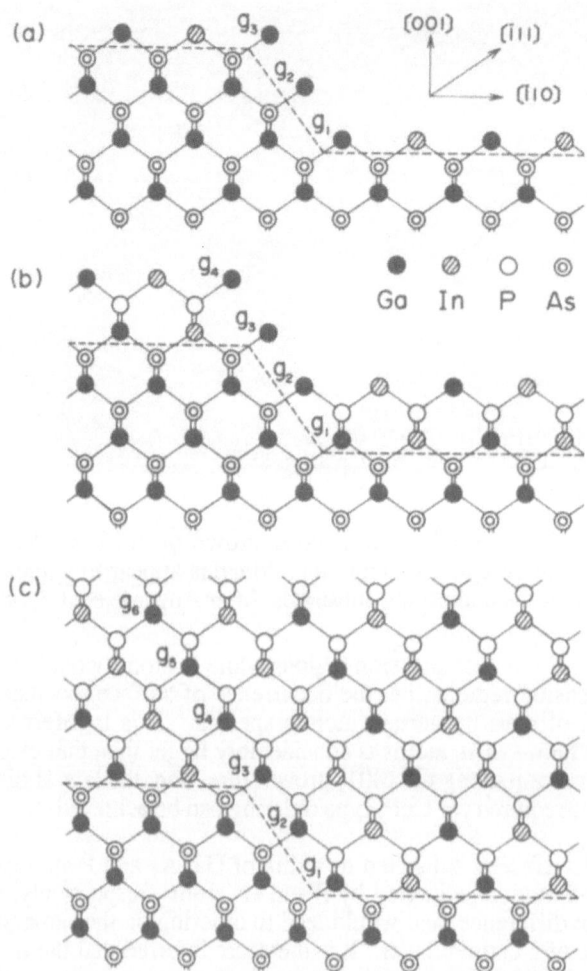

Figure 8. Schematic illustrating the salient features of a mechanism for CuPt-type ordering proposed by Suzuki et al. (1988). G's represent sites where Ga atoms would like to attach.

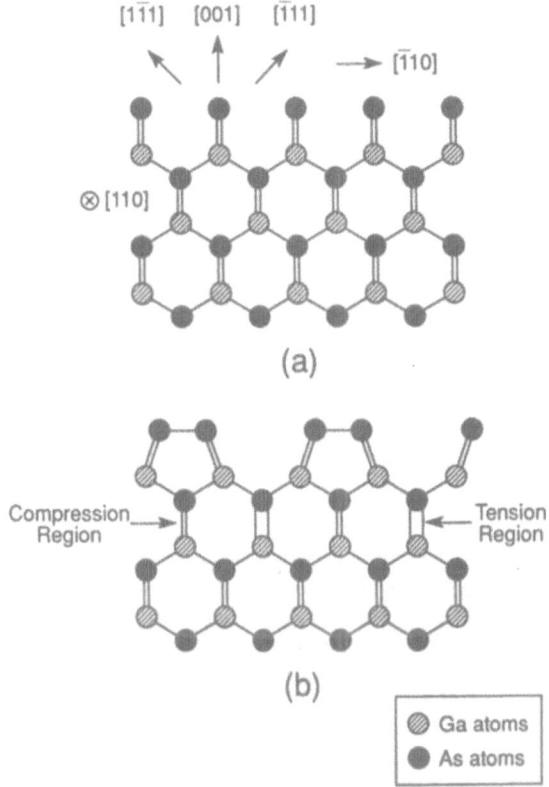

Figure 9. (a) (110) perspective of an unreconstructed GaAs (001) surface, and (b) (110) perspective of a reconstructed GaAs (001) surface. Note that the occurrence of surface reconstruction leads the formation of tension and compression regions in sub-surface layers. After Philips et al. (1991).

That CuPt-type ordering represents a metastable state in the bulk is borne out by the observations of Plano et al. (1988) and Gavrilovic et al. (1988) that GaInP$_2$ and (Ga,Al)InP$_2$ layers can be disordered by annealing above the transition temperature for ordering. Once disordered ordering cannot be restored by further annealing below the transition temperature. Results on diffusion-induced disordering (Plano et al. 1988 and Dabkowski et al. 1988) are also consistent with the issue of metastability.

5. Effects of Ordering on Bandgaps and Reliability of Light Emitting Devices

Theoretical predictions of Zunger and co-workers (Bernard et al. 1988 and Wei and Zunger 1990) that atomic ordering lowers the bandgap have been verified experimentally (Gomyo et al. 1988, Kondow et al. 1988a, Kondow et al. 1988b, Nishino et al. 1988 and McDermott 1990). This appears to be quite a general effect in ordered materials. The magnitude of bandgap narrowing is usually smaller than that predicted by calculations. This discrepancy is most likely related to less than perfect ordering in as-grown layers.

A consensus has emerged that dark line defects observed in degraded GaAs/GaAlAs lasers consist of dislocation networks (Hutchinson and Dobson 1974, Ishida et al. 1977 and Ishida and Kamejima 1979). These networks evolve from existing dislocations by non-radiative recombination enhanced glide and climb. Similar results have been obtained on InGaAsP material which has been degraded optically to simulate the degradation behavior of InP/InGaAsP lasers (Mahajan et al. 1979, Temkin et al. 1982 and Mahajan et al. 1984). Since the presence of atomic ordering should strengthen the InGaAsP lattice, multiplication of dislocations by glide should be difficult. Also, the occurrence of dislocation climb may be retarded because it would entail the formation of antiphase domain boundaries. Thus the presence of ordering in InGaAsP epitaxial layers should enhance the performance of light emitting devices based on the InP/InGaAsP system. This is consistent with the observed degradation resistance of such devices.

6. Summary

The status of current understanding of atomic ordering in ternary and quaternary III-V compound semiconductors is reviewed briefly. The accumulated evidence supports that CuPt-type ordering occurs exclusively in these materials. Models proposed to rationalize the evolution of this type of ordered structure are discussed. It is argued that reconstruction occurring at (001) surfaces plays a very important role in the development of CuPt-type order. The ordered structure is stable at the surface, is metastable in the bulk and is inherited in the bulk because bulk diffusion is extremely slow in these materials. Furthermore, the presence of ordering reduces bandgaps and enhances degradation resistance of light emitting devices.

Acknowledgements

The authors acknowledge fruitful discussions with M. Skowronski, V.G. Keramidas and J.P. Harbison and are grateful to the Department of Energy for the award of a research grant DE-FG02-87ER45329.

References

Augarde, E., Mpaskoutos, M., Bellon, P., Chevalier, J.P., and Martin, G.P. (1989) Inst. Phys. Conf. Ser. #100, 155.

Bellon, P., Chevalier, J.P., Martin, G.P., Dupont-Nivet, E., Thiebault, C., and Andre, J.P. (1988) Appl. Phys. Lett. 52, 567.

Bernard, J.E., Ferreira, L.G., Wei, S.-H., and Zunger, A. (1988) Phys. Rev. B. 38, 6338.

Bernard, J.E., Dandrea, R.G., Ferreira, L.G., Froyen, S., Wei, S.-H., and Zunger, A. (1990) Appl. Phys. Lett., 56, 731.

Cao, D.S., Riehlen, E.H., Chen, G.S., Kimball, A.W., and Stringfellow, G.B. (1991) J. Cryst. Growth 109, 279.

Chen, G.S., Jaw, D.H., and Stringfellow, G.B. (1990) Appl. Phys. Lett. 53, 2475.

Chen, G.S., Jaw, D.H., and Stringfellow, G.B. (1991) J. Appl. Phys. 69, 4263.

Chu, S.N.G., Nakahara, S., Strege, K.E., and Johnston, Jr., W.D. (1985) J. Appl. Phys. 57, 4610.

Dabkowski, F.P., Gavrilovic, P., Meehan, K., Stutius, W., Williams, J.E., Shahid, M.A., and Mahajan, S. (1988) Appl. Phys. Lett. 52, 2142.

Dandrea, R.G., Bernard, J.E., Wei, S.-H., and Zunger, A. (1990) Phys. Rev. Lett. 64, 36.

Froyen, S. and Zunger (1991) submitted for publication to Appl. Phys. Lett.

Gavrilovic, P., Dabkowski, F.P., Meehan, K., Williams, J.E., Stutius, J.E., Hsieh, K.C., Holonyak, Jr., N., Shahid, M.A., and Mahajan, S. (1988) J. Cryst. Growth 93, 426.

Gomyo, A., Suzuki, T., and Iijima, S. (1988) Phys. Rev. Lett. 60, 2645.

Gomyo, A., Suzuki, T., Kobayashi, K., Kawata, S., and Hino, I. (1987) Appl. Phys. Lett. 50, 673.

Henoc, P., Izrael, A., Quillec, M., and Launois, H. (1982) Appl. Phys. Lett. 40, 963.

Hutchinson, P.W. and Dobson, P.S. (1975) Phil. Mag. 32, 745.

Ihm, Y.E., Otsuka, N., Klen, J., and Morkoc, H. (1987) Appl. Phys. Lett. 51, 2013.

Ishida, K., and Kamejima, T. (1979) J. Electron. Mater. 8, 57.

Ishida, K., Kamejima, T., and Matsui, J. (1977) Appl. Phys. Lett. 31, 397.

Jen, H.R., Cherng, M.J., and Stringfellow, G.B. (1986) Appl. Phys. Lett. 48, 1603.

Kondow, M., Kakibayashi, H., and Minagawa, S. (1988a) J. Cryst. Growth 88, 291.

Kondow, M., Kakibayashi, H., Minagawa, S., Inoue, Y., Nishino, T., and Hamakawa, Y. (1988b) Appl.Phys. Lett. 53, 2053.

Kondow, M., Kakibayashi, H., Tanaka, T., and Minagawa, S. (1989) Phys. Rev. Lett. 63, 884.

Kuan, T.S., Kuech, T.F., Wang, W.I., and Wilkie, E.L. (1985) Phys. Rev. Letts. 54, 201.

Kuan, T.S., Wang, W.I., and Wilkie, E.L. (1987) Appl. Phys. Lett. 51, 51.

Launois, H., Quillec M., Glas, F., and Treacy, M.M.J. (1982) Inst. Phys. Conf. Ser. #65, 537.

LeGoues, F.K., Kesan, V.P., Iyer, S.S., Tersoff, J., and Tromp, R. (1990) Phys. Rev. Lett,. 64, 2038.

Mahajan, S. (1991) Proc. of the Fifth Brazilian School on Semiconductor Physics, edited by J.R. Leite, World Science Publishers; Singapore.

Mahajan, S., Dutt, B.V., Temkin, H., Cava, R.J., and Bonner, W.A. (1984) J. Cryst. Growth 68, 589.

Mahajan, S., Johnston, Jr., W.D., Pollack, M.A., and Nahory, R.E. (1979) Appl. Phys. Letts. 34, 717.

Mahajan, S. and Shahid, M.A. (1989) MRS Proc. vol. #144, 169.

Mahajan, S., Shahid, M.A., and Laughlin, D.E. (1989) Inst. Phys. Conf. Ser. #100, 143.

Mahajan, S., Temkin, H., and Logan, R.A. (1984) Appl. Phys. Lett. 44, 119.

McDermott, B.T., Reid, K.G., El-Masry, N.A., Bedair, S.M., Duncan, W.M., Yin, X., and Pollack, F.H. (1990) Appl. Phys. Lett. 56, 1172.

McDevitt, T.L., Mahajan, S., Laughlin, D.E., Bonner, W.A., and Keramidas, V.G. (1990) MRS Proc. Vol. #198, 609.

Mckernan, S., DeCooman, B.C., Carter, C.B., Bour, D.P., and Shealy, J.R. (1988) J. Mats. Res. 3, 406.

Murgatroyd, I.J., Norman, A.G., and Booker, G.R. (1990) J. Appl. Phys. 67, 2310.

Nakayama, H. and Fujita, H. (1985) Inst. Phys. Conf. Ser. #79, 289.

Nishino, T., Inoue, Y., Hamakawa, Y., Kondow, M., and Minagawa, S. (1988) Appl. Phys. Lett. 53, 583.

Norman, A.G. and Booker, G.R. (1985) J. Appl. Phys. 57, 4610.

Norman, A.G., Mallard, R.E., Murgatroyd, I.J., Booker, G.R., Moore, A.H., and Scott, M.D. (1987) Phys. Conf. Ser. #87, 77.

Philips, B.A., Norman, A.G., Tseong, T.Y., Mahajan, S., and Booker, G.R. (1991) submitted for publication to J. Cryst. Growth.

Plano, W.E., Nam, D.W., Major, Jr., J.S., Hsieh, K.C., and Holonyak, Jr., N. (1988) Appl. Phys. Lett. 53, 2537.

Shahid, M.A. and Mahajan, S. (1988) Phys. Rev. B38, 1344.

Shahid, M.A., Mahajan, S., Laughlin, D.E., and Cox, H.M. (1987) Phys. Rev. Lett. 58, 2567.

Suzuki, T., Gomyo, A., and Iijima, S. (1988) J. Cryst. Growth 93, 396.

Temkin, H., Mahajan, S., DiGiuseppe, M.A., and Dentai, A.G. (1982) Appl. Phys. Letts. 40, 562.

Treacy, M.M.J., Gibson, J.M., and Howie, A. (1985) Phil. Mag. A 51, 389.

CRITICAL PHENOMENA AT SURFACES AND INTERFACES

REINHARD LIPOWSKY
Institut für Festkörperforschung
Forschungszentrum Jülich
Postfach 1913
D–5170 Jülich
FRG

ABSTRACT. The multitude of critical phenomena which occur at surfaces and interfaces is briefly reviewed from a theoretical point of view. Three types of critical effects are distinguished related (i) to the 2–dimensional character of the interface, (ii) to its morphology, and (iii) to its structural changes at phase transitions in the bulk. Wetting phenomena in three dimensions belong to the last category (iii). Some recent theoretical results for such phenomena are also discussed: (a) Wetting of a moving interface; (b) Wetting in the 3–dimensional Ising model; and (c) Wetting of an inhomogeneous substrate.

1. Introduction

During the last decade, a lot of effort has been devoted to the study of critical phenomena at surfaces and interfaces. Here and below, the term surface denotes the interface between a condensed phase and an inert vapor or 'vacuum'. Since the latter phase is transparent to most experimental probes, many experimental techniques have been used to study surface critical phenomena. However, from a theoretical point of view, there is no fundamental difference between a surface and any other interface between two thermodynamic phases.

One intriguing aspect of interfaces is their *reduced dimensionality*. Indeed, in many cases, the interface can simply be viewed as a planar 2–dimensional system. However, it can also 'bulge' into the third dimension and then attain nonplanar morphologies. In addition, the interface itself has a third dimension which can become mesoscopic as in wetting phenomena; one then has a system which interpolates between two and three dimensions.

In this paper, I will briefly review the multitude of critical phenomena which can occur at surfaces and interfaces. Three different categories of such phenomena will be distinguished: (i) Critical behavior within the 2–dimensional interface; (ii) Critical effects related to the morphology of interfaces; and (iii) Changes in the interfacial

C. T. Liu et al. (eds.), Ordered Intermetallics – Physical Metallurgy and Mechanical Behaviour, 107–121.
© 1992 *Kluwer Academic Publishers.*

structure as one of the adjacent bulk phases undergoes a phase transition.

Wetting processes such as, e.g., surface melting or surface–induced order and disorder belong to the last category (iii). Some recent theoretical results for these processes will also be discussed: (a) *Wetting of a moving interface* — An example is surface melting of a crystal which slowly evaporates. It is shown that the wetting layer thickens as a result of this motion; (b) *Wetting in the 3–dimensional Ising model* — In this case, thermally excited shape fluctuations of the two interfaces bounding the wetting layer should be important. Until very recently, computer simulations gave no evidence for such fluctuation effects. It is argued here that the interfacial fluctuations observed in the simulations are presumably spikes rather than smooth deformations; and (c) *Wetting of an inhomogeneous substrate* — If one of the phases consists of an inhomogeneous solid, the wetting layer experiences a random substrate potential which leads to an increase of the interfacial roughness.

2. Critical behavior within the 2–dimensional interface

2.1 MULTITUDE OF 2–DIMENSIONAL PHASES AND PHASE TRANSITIONS

Usually, the interface between two 3–dimensional phases has a thickness with is set by the scale of the molecules. It then represents a 2–dimensional subsystem which exhibits a variety of phases and associated phase transitions. Consider, e.g., a mono-layer of adatoms adsorbed onto the surface of a 3–dimensional liquid or solid. Such a layer often exhibits a 2–dimensional vapor, liquid, and solid phase. In general, the molecules possess internal degrees of freedom which can order and then lead to ad-ditional phases. Examples are the liquid crystalline phases of amphiphilic molecules adsorbed at the water–air interface. If the substrate is a crystal, these 2–dimensional layer phases within the adsorbed layer can be commensurate or incommensurate with this substrate.

The multitude of phases leads to a multitude of critical phenomena. A well-known example is the critical point of the 2–dimensional vapor liquid coexistence curve; the associated critical behavior should belong to the universality class of the 2–dimensional Ising model as confirmed in many experiments. Critical effects can also occur at *discontinuous* phase transitions. One example is provided by wetting phenomena such as edge melting which occurs as the 2–dimensional triple point is approached along the solid–vapor coexistence curve, see Fig. 1. In this case, the monolayer contains solid domains surrounded by vapor. As the triple–point temperature is approached, the edges of these domains start to melt and the domain boundaries contain stripes of the 2–dimensional liquid phase.

Another type of transition occurs when a commensurate state of the monolayer is transformed into a (weakly) incommensurate one. The latter state consists of commensurate domains separated by 1–dimensional domain boundaries.

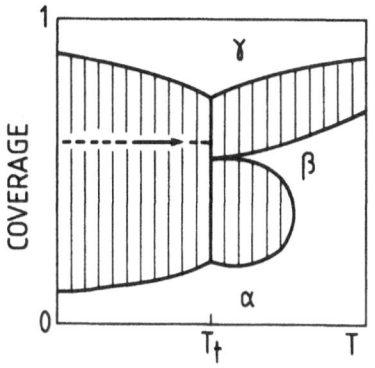
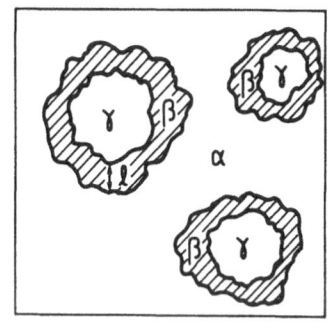

Figure 1: (left) A typical phase diagram for an adsorbed monolayer — The 2–dimensional vapor, liquid, and solid phases are denoted by α, β, and γ, respectively. When the triple point temperature $T = T_t$ is approached along the broken line, the monolayer consists of solid domains, γ, surrounded by vapor, α; (right) Edge melting near the triple point. — The solid domains start to melt along their edges.

The critical effects associated with wetting and commensurate–incommensurate transitions are governed by the behavior of the domain boundaries. This behavior is in itself quite interesting and will be discussed in the next two subsections.

2.2 ROUGHNESS OF DOMAIN BOUNDARIES

First, consider a single domain boundary separating two 2–dimensional domains. In equilibrium, the morphology of this domain boundary reflects the underlying *symmetry* of the system /1/. The following cases have been studied theoretically in some detail: (a) *Periodic systems* — In this case, the domain boundary is flat or smooth at $T = 0$. At $T > 0$, it makes transverse excursions $L_\perp$ which grow with the longitudinal extension, $L_\|$, of the domain boundary. More precisely, one has the scaling law

$$L_\perp \sim L_\|^\zeta \tag{1}$$

with the roughness exponent $\zeta = 1/2$; (b) *Quasi–periodic systems* — In this case, the behavior in non–universal and depends on the model parameters /2-4/. If the domain boundary feels an effective Fibonacci potential, it is rough at all T but the roughness exponent ζ is T–dependent and satisfies $\zeta \leq 1/2$; if it feels an effective Harper's potential, the domain boundary undergoes a morphological transition from a smooth state at low T to a rough state at large T; and (c) *Systems with quenched randomness* — In this case, the domain boundary feels a random potential. It then roughens in order to adapt its shape to the random potential and, thus, to minimize its energy. As a result, these systems are characterized by an increased roughness exponent: for random field and random bond systems, the scaling law (1) holds again but with $\zeta = 1$ (Ref. /5,6/) and $\zeta = 2/3$ (Ref. /7,8/) respectively; and (d) *Kinetic roughening* — Away from equilibrium, growth or shrinkage of the domains implies a moving domain boundary. Various models with local deposition or evaporation rules have recently been studied which lead to a kinetic roughening of the domain boundary /9/.

2.3 FLUCTUATION–INDUCED REPULSION BETWEEN DOMAIN BOUNDARIES

Next, consider two domain boundaries which are, on average, parallel and have separation, ℓ. The roughness of these boundaries leads to an effective fluctuation–induced repulsion, V_{FL}, with

$$V_{FL}(\ell) \sim 1/\ell^\tau \qquad \text{for large } \ell \qquad (2)$$

where V_{FL} represents a free energy per unit length. Two cases must be distinguished: (i) If the shape fluctuations are thermally excited with $\zeta \leq 1/2$, this repulsion arises from a loss of entropy since the configurations of each domain boundary are constrained by the presence of the other one, and

$$\tau = 1/\zeta \geq 2 \qquad (3)$$

as applies to periodic and quasiperiodic systems; /10/ and (ii) For systems with quenched impurities with $\zeta \geq 1/2$, on the other hand, this repulsion arises from an increase in energy since each domain boundary cannot explore the minima of the random potential, which lie beyond the other domain boundary, and /11,1/

$$\tau = 2(1 - \zeta)/\zeta \leq 2 . \qquad (4)$$

The heuristic concept of a fluctuation–induced interaction just described can be used to determine the correct critical exponents for some wetting phenomena such as edge melting and for commensurate–incommensurate transitions.

Edge melting provides an example for complete wetting. In this case, the wetting layer contains a *metastable* phase, denoted by β in Fig. 1. Then, in the absence of shape fluctuations, the free energy per unit length of this β layer has the generic form

$$V(\ell) = H\ell + V_{DI}(\ell) \qquad (5)$$

where the pressure–like variable H is proportional to the difference, $f_\beta - f_{\alpha\gamma}$, of the bulk free energies of the metastable β phase and the stable phases denoted by α and γ in Fig. 1. The direct interaction, $V_{DI}(\ell)$, reflects the underlying molecular forces such as van der Waals, electrostatic, or structural forces and decays to zero for large ℓ. For example, if the adatoms interact via (nonretarded) van der Waals forces, $V_{DI}(\ell) \sim 1/\ell^3$ for large ℓ.

Now, the dependence of the mean separation, $\langle \ell \rangle$, on H can simply be obtained by minimization of the free energy $\Delta f(\ell) = H\ell + V_{DI}(\ell) + V_{FL}(\ell)$. Then, two different scaling regimes or universality classes can be distinguished: (i) If $V_{FL}(\ell) \ll V_{DI}(\ell)$ for large ℓ, the fluctuations do not affect the behavior of $\langle \ell \rangle$ for small H; and (ii) If $V_{FL}(\ell) \gg V_{DI}(\ell)$ for large ℓ, minimization of $\Delta f(\ell) \approx H\ell + V_{FL}(\ell)$ leads to

$$\langle \ell \rangle \sim 1/H^{\psi} \qquad \text{with} \qquad \psi = 1/(1 + \tau) \tag{6}$$

where τ depends on the roughness exponent ζ via (3) and (4).

In the latter case, the surface free energy has a singular contribution $f_s \simeq H\langle \ell \rangle \sim H^{1-\psi}$ which implies the surface specific heat

$$c_s \sim H^{-\alpha} \qquad \text{with} \qquad \alpha = 1 + \psi = (2 + \tau)/(1 + \tau) . \tag{7}$$

If the adatoms interact via van der Waals forces and the domain boundaries have the roughness exponent $\zeta = 1/2$, the relation (7) applies and leads to the specific heat exponent $\alpha = 4/3$ for edge melting. /12/ Experiments on Neon monolayer gave clear evidence for such a melting process. /13/ The data were fitted with $\alpha \simeq 5/4$ which could indicate a crossover effect arising from a relatively large edge stiffness. /1/

A similar line of arguments can be used in order to determine the critical behavior at continuous commensurate–incommensurate transitions. /14/ If $n \sim 1/\ell$ denotes the density of the domain boundaries, the free energy per unit area, $\Delta F(\ell)$, of the incommensurate state is given by

$$\Delta F(\ell) = -n\Delta\mu + n[V_{DI}(\ell) + V_{FL}(\ell)] \tag{8}$$

where $\Delta\mu$ is the effective chemical potential for the domain boundary. Then, if $|V_{DI}(\ell)| \ll V_{FL}(\ell)$ for large ℓ, minimization of (8) with respect to ℓ leads to /11/

$$\langle \ell \rangle \sim 1/\Delta\mu^{\bar{\beta}} \qquad \text{with} \qquad \bar{\beta} = 1/\tau . \tag{9}$$

So far, the critical behavior of the domain boundaries has been determined by simple superposition of the direct interaction, V_{DI}, and the fluctuation–induced interaction, V_{FL}. Such an heuristic approach is, however, *not* valid in general. One important class of critical phenomena for which this approach fails are *wetting transitions* for sufficiently–short ranged interactions which satisfy $V_{DI}(\ell) \approx -W/\ell^{\tau}$ for large ℓ, where W can be negative, zero, or positive. In this case, the interaction is renormalized in a nontrivial way. For periodic systems characterized by $\zeta = 1/2$ and $\tau = 2$, this renormalization can be studied by transfer matrix /15/ and exact functional renormalization group /16/ methods. In this way, a complete classification of wetting transitions in two dimensions has been obtained. /17/ In real systems, such wetting transitions could occur along the steps of vicinal surfaces as has recently been studied in Monte Carlo simulations. /18/

3. Critical effects related to the morphology of interfaces

3.1 ROUGHNESS OF INTERFACES

The morphology of 2–dimensional interfaces in three dimensions is completely analogous to the behavior of 1–dimensional domain boundaries discussed in the previous subsection. Thus, the following cases can again be distinguished: (i) *Periodic systems* — The surface of a periodic crystal is smooth at sufficiently low T but undergoes a morphological transition to a rough state at a characteristic roughening temperature, $T = T_R$. /19/ The roughness is, however, only logarithmic for $T \geq T_R$ with $L_\perp \sim [ln((L_{||}/a)]^{1/2}$; (ii) *Quasi–periodic crystals* — The surface of an ideal quasicrystal is predicted to be smooth for all T /10,4/. This conclusion is based on renormalization group calculations of interface models; (iii) *Systems with quenched disorder* — In this case, the 2–dimensional interface is presumably always rough with roughness exponent $\zeta > 0$. This behavior should also apply to an interface in a random quasicrystal /10/; and (iv) *Kinetic roughening* — Local deposition rules again lead to an increased roughness of the moving interface and typically to an increased value of ζ. In addition, various morphological transitions between two different rough states of the interface have been predicted /9/.

3.2 EDGE OF A CRYSTAL FACET

Another aspect of the morphology which is governed by fluctuations is the singular behavior of the equilibrium crystal shape which occurs near the edge of a facet. At such an edge, the facet often meets a rounded part composed of terraces and steps, see Fig. 2. Let $z(x)$ describe this shape: the facet is given by $z(x) = 0$ for $x \leq 0$ and the edge of this facet is at $x = 0$. Now, the excess free energy of the rounded part arising from the steps can again be estimated as in (8) where $n \sim 1/\ell \sim dz/dx$ and the effective chemical potential $\Delta\mu \sim x$. It then follows from (9) that $dz/dx \sim x^\beta$ or /10/

$$z \sim x^\lambda \qquad \text{with} \qquad \lambda = 1 + 1/\tau \qquad (10)$$

for small x where the exponent τ as given by (3) or (4) arises from the roughness of the 1–d steps. Therefore, the scaling relation $\lambda = 1 + 1/\tau$ connects the $2d$ properties of the interface to its $3d$ morphology. For a periodic crystal, one has $\tau = 2$ and thus $\lambda = 3/2$ as has experimentally been observed, e.g., for small lead crystals. /20/

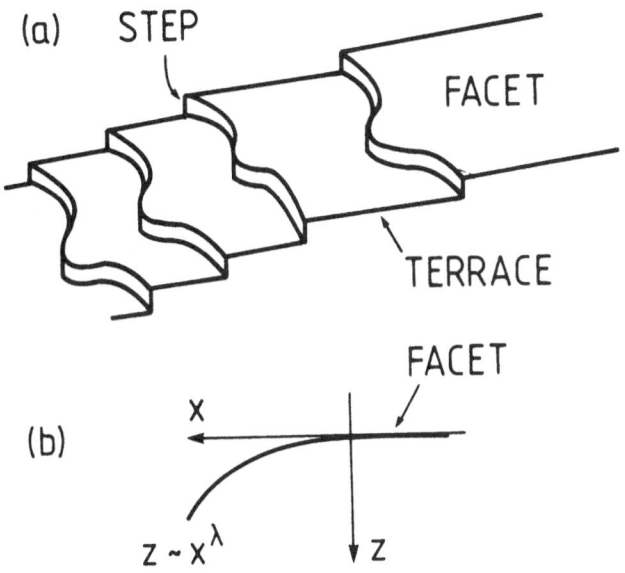

Figure 2: Edge of a crystal facet — (a) On a microscopic scale, the rounded part consists of terraces separated by steps which are, on average parallel to the facet edge; (b) On a macroscopic scale, the rounded part is described by the contour $z(x) \sim x^\lambda$ with λ as given in (10).

4. Interfacial structure at phase transformations in the bulk

When one of the two bulk phases adjacent to the interface undergoes a phase transition, the interface aquires characteristic density profiles. To be more specific, let us consider a crystal–vapor interface, i.e., the surface of a crystal.

4.1 CONTINUOUS PHASE TRANSITIONS

The bulk crystal may undergo a variety of *continuous* phase transitions. If this transition belongs to the universality class of the 3–dimensional Ising model, one can distinguish three types of surface behavior /21-23/: (i) If the coupling constant, J_1, between two "spins" within the surface is smaller than the coupling constant, J, for two "spins" in the bulk, the system undergoes an 'ordinary transition' at the bulk critical temperature $T = T_c$ at which both bulk and surface become ordered simultaneously; (ii) If J_1 is large compared to J, the surface undergoes a 'surface transition' at $T_{c1} > T_c$ while the bulk undergoes an 'extraordinary transition' at

$T = T_c$; and (iii) Finally, for a certain intermediate ratio J_1/J, bulk and surface simultaneously undergo the so–called 'special transition'.

4.2 DISCONTINUOUS PHASE TRANSITIONS

Now, consider again the interface between a crystalline phase, γ, and a vapor phase, α, but assume that the crystal undergoes a *discontinuous* phase transition between a disordered phase at high temperature and several odered phases at low temperature. Then, a third phase, β, appears which may prefer to go into the $(\alpha\gamma)$ interface and then forms a thin wetting layer. Two cases may be distinguished: as the transition temperature is approached (i) from below or (ii) from above, a thin layer (i) of the disordered phase or (ii) of one of the ordered phases can be induced by the surface. /24-28/ Alternatively, if both bulk phases α and γ are in chemical equilibrium, wetting by the β phase may occur as one approaches a triple point at which all three phases α, β and γ coexist. The latter case is schematically shown in Fig. 3.

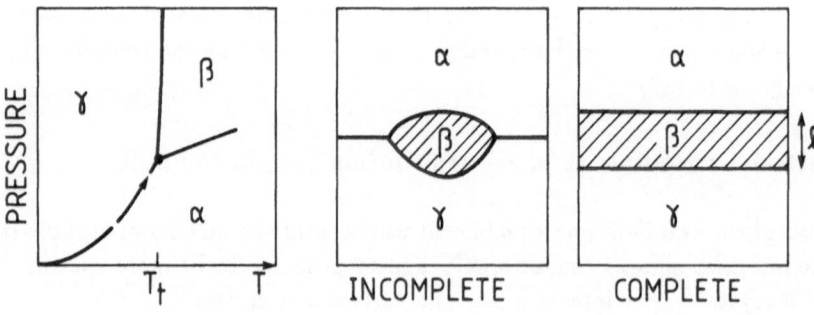

Figure 3: Wetting, surface melting, and related phenomena.— When the $(\alpha\beta\gamma)$ triple point is approached along the $(\alpha\gamma)$ coexistence line (left diagram), the $(\alpha\gamma)$ interface may be (in)completely wet by the β phase. Incomplete and complete wetting corresponds to the formation of droplets and a thin wetting layer, respectively. The layer thickness is denoted by ℓ.

Wetting has to be distinguished from heterogeneous nucleation even though the nucleation barrier is intimately related to the wetting properties. Indeed, heterogeneous nucleation at the interface occurs when the system has passed the triple point and the bulk phases, α and γ, (or at least one of them) have become thermodynamically unstable. The energy barrier for nucleation of β droplets is reduced at the interface. For complete wetting, this barrier is, in fact, reduced to zero, and the transformation into the stable β phase will start at the interface or surface.

When the $(\alpha\gamma)$ interface contains a wetting layer of β phase, it splits up into an $(\alpha\beta)$ and a $(\beta\gamma)$ interface. The separation, $\ell \geq 0$, of these two interfaces is equal to the thickness of the wetting layer. Likewise, the excess free energy, $V(\ell)$, of the layer can be regarded as an effective interaction between these interfaces.

In the absence of interfacial fluctuations, the interaction $V(\ell)$ has the same generic form $V(\ell) = H\ell + V_{DI}(\ell)$ as for domain boundaries, see (5). This interaction is again renormalized by shape fluctuations of the interfaces even though this renormalization is less effective in $d = 3$ than in $d = 2$. The influence of interfacial fluctuations is studied most conveniently within the framework of effective models for the interfacial separation $\ell \geq 0$. For wetting in three dimensions, these models have the generic form /1/

$$\mathcal{H}\{\ell\} = \int d^2x \left\{ \frac{1}{2}\tilde{\Sigma}(\nabla\ell)^2 + H\ell + V_{DI}(\ell) \right\} \tag{11}$$

where $\tilde{\Sigma}$ is an appropriate interfacial stiffness. A hard wall potential at $\ell = 0$ which ensures $\ell \geq 0$ is implicitly assumed. In what follows, I will study three different situations which correspond to three different interactions, $V_{DI}(\ell)$.

4.3 WETTING PHENOMENA IN THREE DIMENSIONS: SOME RECENT THEORETICAL RESULTS

4.3.1 *Wetting away from equilibrium: a moving wetting layer is thicker.* First, consider an $(\alpha\gamma)$ interface which moves with constant velocity. An example is a crystal in ultra–high vacuum which slowly evaporates. Now, assume that this interface contains a thin wetting layer. The question is: does the motion of the interface affect the thickness of the layer?

This question can be addressed in the framework of Landau models or density functional theories which lead to density profiles, $M(z, t)$, where z is the coordinate perpendicular to the $(\alpha\gamma)$ interface and t is the time. Let us assume that the motion of the interface is not limited by diffusion. One may then consider a simple relaxational dynamics for the densities /29-31/. This leads to solutions $M(z, t) = f(z - vt)$ which move with constant velocity v.

The behavior of the densities can be used in order to construct the interaction $V(\ell)$ acting between the $(\alpha\beta)$ and the $(\beta\gamma)$ interface. As a result, one finds a shift of the pressure–like variable H and the direct interaction /30/

$$V_{DI}(\ell) = B \exp(-\ell/\ell_1) + C \exp(-\ell/\ell_2) \tag{12}$$

where the two length scales are given by

$$\ell_1 = \xi_\beta/(-x + \sqrt{1 + x^2}) \quad \text{and} \quad \ell_2 = \xi_\beta/2\sqrt{1 + x^2} \tag{13}$$

with $x \equiv v/v_\beta$. The parameter v_β is a velocity scale which depends on the mobility within the β layer. The length scale ξ_β is the bulk correlation length within the β phase. If the wetting layer does not move, one has $x = v/v_\beta = 0$, and $\ell_1 = \xi_\beta = 2\ell_2$. In general, the scale ℓ_1 increases with increasing velocity. If one ignores the possible effects of shape fluctuations, the expectation value, $\langle \ell \rangle$, is determined by the minimum of $V_{DI}(\ell) + H\ell$. It then follows that the motion of the interface leads to a thickening of the wetting layer /30/.

4.3.2 Wetting in the 3–dimensional Ising model: is it dominated by spikes?

Next, consider the Ising model on a simple cubic lattice with a (100) surface. The "spins" interact with nearest–neighbor couplings, J, and are subject to a short–ranged surface field. This model exhibits critical wetting transitions which have extensively been studied in Monte Carlo simulations /32/. However, the critical behavior of these transitions is not well–understood. Until very recently, the simulations within the Ising model were thought to be consistent with mean–field theory. In contrast, linearized renormalization group calculations of effective interface models with

$$V_{DI}(\ell) = B \exp(-\ell/\ell_0) + C \exp(-2\ell/\ell_0) \tag{14}$$

predicted nonuniversal critical behavior depending on the parameter $\omega = T/4\pi\tilde{\Sigma}\ell_0^2$ where $\tilde{\Sigma}$ is the interfacial stiffness as in (11). /33,34/

In order to address this controversy, we have studied the solid–on–solid (SOS) limit of the 3–dimensional Ising model by Monte Carlo simulations. /35/ A detailed analysis of the MC data revealed that the SOS model belongs to the same universality class as the Gaussian model given by (11) provided (i) one accepts the results of linear renormalization, and (ii) one makes a proper identification of the model parameters. The reduced stiffness of the SOS model has the value $\tilde{\Sigma}/T = c_1(J/Ta)^2$ with $c_1 \simeq$ 10.4, and the direct interaction has the form (14) with $\ell_0 = \xi = c_2 aT/J$ and $c_2 \simeq$ 0.175. As a consequence, the parameter $\omega = T/4\pi\tilde{\Sigma}\ell_0^2$ has the universal value $\omega \simeq 1/4$ for the SOS limit of the 3–dimensional Ising model. /35/

The direct interaction (14) for the Ising model can be derived from mean–field theory for the order parameter density /33,36,37/. In this derivation, the length scale ℓ_0 is identical with the bulk correlation length ξ_β within the β layer as follows from (12) and (13) for $x = v/v_\beta = 0$. The SOS model does not include any bubbles or interfacial overhangs and thus does not contain information about the bulk correlation length. Therefore, the length scale ξ cannot be identical with the bulk correlation length. However, this length scale has another rather direct interpretation in terms of interfacial *spikes*. Thus, consider two planar interfaces at separation ℓ and introduce an interfacial fluctuation which consists simply of a column of reversed spins bridging the gap between the two interfaces. Such a fluctuation has an energy $\triangle E = 8J\ell/a$ and thus a Boltzmann weight $\sim \exp(-8J\ell/aT)$ ignoring entropic effects. Therefore, these spikes are governed by the length scale $\xi_{sp} = 0.125aT/J$ which is comparable to the scale ξ determined in the simulations.

Thus, one may distinguish two different types of shape fluctuations of the interface: (i) smooth deformations of the mean–field profile, and (ii) abrupt spikes in the form of thin fingers. In both cases, the critical behavior resulting from the shape fluctuations can be described by an effective Gaussian model as in (11) with a direct interaction as in (14) *but with different length scales* ℓ_0. The question now is: which type of shape fluctuation is *typical* for critical wetting in the 3–dimensional Ising model?

For $J/T = 0.35$, ξ as found in our simulations is about twice as large as the generally accepted value for the bulk correlation length ξ_β. /35/ In this case, spikes should dominate and $\omega \simeq 1/4$. This conclusion is consistent with a recent analysis of the MC data for the 3–dimensional Ising model /38/. This analysis is based on the amplitude ratio for the interfacial susceptibility χ_1 and leads to the estimate $\omega \simeq 0.3$ which is rather close to the value $\omega \simeq 1/4$ obtained for the SOS model.

4.3.3 *Wetting of an inhomogeneous substrate: lateral disorder translates into interfacial roughness.* Finally, consider wetting of a smooth solid substrate which has an inhomogeneous composition and thus exerts an effective interaction with a random component. /39/ The corresponding direct interaction is taken to be

$$V_{DI}(\ell) = V_R(\ell) - (W + W_r)v(\ell) . \tag{15}$$

The first term, $V_R(\ell)$, and the second term, $-Wv(\ell)$ with $W \geq 0$, describe the repulsive and the attractive part of V_{DI}, respectively. The third term, $-W_r v(\ell)$, corresponds to the random component of V_{DI} where W_r is a quenched random variable with zero mean value, $\overline{W_r} = 0$. The potential $v(\ell)$ is taken to behave as $v(\ell) \sim 1/\ell^s$ for large ℓ; (non-retarded) van der Waals forces are described by $s = 2$.

For complete wetting with $W = 0$, the $(\alpha\beta)$ interface unbinds from the substrate as the pressure–like variable H in (5) goes to zero. Alternatively, this interface undergoes a critical wetting transition at $(\alpha\beta)$ coexistence with $H = 0$ in the limit of zero W. For a relatively thick wetting layer, the interfacial fluctuations are now characterized by $L_\perp \sim L_\parallel^\zeta$ with $\zeta = 1/(2+s)$ which arises from the random component of the substrate potential. /40/ This increased roughness affects the mean separation $< \ell >$ of the $(\alpha\beta)$ interface from the substrate *provided* $V_R(\ell)$ decays faster than $1/\ell^\tau$ with $\tau = 2(1+s)$. In the latter situation, one has $< \ell > \sim 1/H^{\psi_c}$ with $\psi_c = 1/(3+2s)$ for complete wetting and $< \ell > \sim 1/W^\psi$ with $\psi = 1/(2 + s)$ for critical wetting.

5. Summary and outlook

In summary, the multitude of critical phenomena which occur at surfaces and interfaces has briefly been reviewed from the theoretical point of view. These phenomena may be divided up into three large classes: (i) Critical behavior within the interface, i.e., critical phenomena in two dimensions. An example is provided by edge melting, see Fig. 1; (ii) Critical aspects of the interfacial morphology, see, e.g., Fig. 2; and (iii)

Critical changes in the interfacial structure which are induced by phase transitions in the adjacent bulk phases. If this phase transition is discontinuous, the system can exhibit wetting phenomena, see Fig. 3.

This research field still poses many theoretical challenges. For example, surfaces of ideal quasicrystals have been theoretically predicted to be always smooth. Real quasicrystals, on the other hand, usually contain a certain amount of disorder which acts to roughen these surfaces. /10/ Another area with many open problems is the kinetic roughening of surfaces. A variety of theoretical models with local growth rules has recently been introduced and studied. /9/ These studies should be useful in order to understand real growth of solids from the vapor phase as, e.g., in molecular–beam epitaxy.

On the experimental side, new tools have been recently developed such as surface–sensitive x-ray and neutron scattering. /41/ Two different methods have been successfully applied: (i) reflectivity measurements (see, e.g., /42/), and (ii) scattering under total external reflection (see, e.g., /27/). With these methods, one can experimentally study structural changes of the interface on the nm scale such as, e.g., wetting or surface–induced nucleation phenomena.

The surfaces considered in this paper are interfaces between two different bulk phases. Finally, I want to mention another type of surfaces, namely membranes which consist of ultra–thin sheets of molecules. It is interesting to note that these membranes exhibit critical behavior which is rather similar to the critical phenomena described here. /43/

Acknowledgements

I thank Michael E. Fisher for a helpful discussion about interfacial spikes, and Joanna Cook, Willi Fenzl, Stefan Grotehans, Frank Jülicher, Götz Schmidt, Jörn Sonnenburg, Dietrich Wolf and Joachim Wuttke for stimulating interactions. I also thank the organizers of this conference for their invitation.

References

/1/ For reviews, see R. Lipowsky, in *Random fluctuations and patterns growth*, ed. by H.E. Stanley and N. Ostrowsky (Kluwer Academic, Dordrecht 1988) p. 227; G. Forgacs, R. Lipowsky, and T.M. Nieuwenhuizen, to appear in *Phase transitions and Critical Phenomena*, ed. by C. Domb and J. Lebowitz.

/2/ C. Henley and R. Lipowsky, Phys. Rev. Lett. **59**, 1679 (1987).

/3/ A. Garg and D. Levine, Phys. Rev. Lett. **59**, 1683 (1987).

/4/ R. Lipowsky, in *Fundamental problems in statistical mechanics VII*, ed. by H. van Beijeren (North–Holland, Amsterdam 1990) p. 139, and references therein.

/5/ J. Villain, J. Phys. (Paris) Lett. **43**, L551 (1982).

/6/ G. Grinstein and S.-K. Ma, Phys. Rev. **B38**, 2588 (1983).

/7/ D.A. Huse, C.L. Henley, and D.S. Fisher, Phys. Rev. Lett. **55**, 2924 (1985).

/8/ M. Kardar, Phys. Rev. Lett. **55**, 2923 (1985).

/9/ For reviews, see D.E. Wolf in *Kinetics of ordering and growth at surfaces*, ed. by M. Lagally (Plenum Press, New York 1990); and J. Krug and H. Spohn, in *Solids far from equilibrium: Growth, morphology and defeats*, ed. by C. Godreche (Cambridge University Press, 1991).

/10/ R. Lipowsky and C. Henley, Phys. Rev. Lett. **60**, 2394 (1988).

/11/ R. Lipowsky and M.E. Fisher, Phys. Rev. Lett. **56**, 472 (1986).

/12/ R. Lipowsky, Phys. Rev. Lett. **52**, 1429 (1984); and Phys. Rev. **B32**, 1731 (1985).

/13/ D.M. Zhu, D. Pengra, and J.G. Dash, Phys. Rev. **B37**, 5586 (1988).

/14/ For a review, see M.E. Fisher, J. Chem. Soc. Far. Trans. II **82**, 1569 (1986).

/15/ R. Lipowsky and T.M. Nieuwenhuizen, J. Phys. **A21**, L89 (1988); R.K.P. Zia, R. Lipowsky, and D.M. Kroll, Am. J. Phys. **56**, 160 (1988).

/16/ F. Jülicher, R. Lipowsky, and H. Müller–Krumbhaar, Europhys. Lett. **11**, 657 (1990); H. Spohn, Europhys. Lett. **14**, 689 (1991).

/17/ Wetting in 2–dimensional quasiperiodic systems and in 2–dimensional random bond systems has recently been studied by G. Schmidt and R. Lipowsky, to be published, and by J. Wuttke, and R. Lipowsky, submitted to Phys. Rev. B.

/18/ K. Binder, in *Kinetics of ordering and growth of surfaces* ed. by M. Lagally (Plenum, New York 1990).

/19/ A recent review is in H. van Beijeren and I. Nolden, in *Structure and Dynamics of Surfaces II*, ed. by W. Schommers and P. van Blanckenhagen (Springer–Verlag, 1987).

/20/ C. Rottmann, M. Wortis, J.C. Heyraud, and J.J. Metois, Phys. Rev. Lett. **52**, 1009 (1984).

/21/ K. Binder, in *Phase transitions and critical phenomena*, Vol. 8, ed. by C. Domb and J. Lebowitz (Academic Press, London 1983).

/22/ H. Wagner, in *Applications of field theory to statistical mechanics*, ed. by L. Garrido (Springer, Berlin 1985).

/23/ H.W. Diehl, in *Phase transitions and critical phenomena*, Vol. 10, ed. by C. Domb and J. Lebowitz (Academic Press, London 1986).

/24/ R. Lipowsky, Phys. Rev. Lett. **49**, 1575 (1982); R. Lipowsky and W. Speth, Phys. Rev. B**28**, 3983 (1983).

/25/ For reviews, see R. Lipowsky, J. Appl. Phys. **55**, 2485 (1984); and in *Magnetic properties of low–dimensional systems II*, ed. by L.M. Falicov, F. Mejia–Lira, and J.L. Moran–Lopez (Springer–Verlag, Berlin 1990) p. 158.

/26/ Surface melting has recently been studied by density functional methods, see R. Ohnesorge, H. Löwen, and H. Wagner, Phys. Rev. A **43**, 2870 (1991).

/27/ For a recent x-ray study of surface melting of Al(110), see H. Dosch, T. Höfer, J. Peisel, and R.L. Johnson, Europhys. Lett. **15**, 527 (1991).

/28/ Surface–induced order in fcc alloys has recently been studied by W. Schweika, K. Binder, and D.P. Landau, Phys. Rev. Lett. **65**, 3321 (1990).

/29/ T. Meister and H. Müller–Krumbhaar, Z. Physik B**55**, 111 (1984).

/30/ H. Löwen and R. Lipowsky, Phys. Rev. B**43**, 3507 (1991).

/31/ Diffusion–limited growth of wetting layers is studied in R. Lipowsky and D.A. Huse, Phys. Rev. Lett. **57**, 353 (1986).

/32/ K. Binder and D.P. Landau, Phys. Rev. B**37**, 1745 (1988).

/33/ R. Lipowsky, D.M. Kroll, and R.K.P. Zia, Phys. Rev. B**27**, 4499 (1983) used a variational method which is equivalent to linear renormalization. Their parameter τ corresponds to $[4\pi\omega]^{1/2}$.

/34/ E. Brezin, B. Halperin, and S. Leibler, Phys. Rev. Lett. **50**, 1387 (1983); D.S. Fisher and D.A. Huse, Phys. Rev. B**32**, 247 (1985).

/35/ G. Gompper, D.M. Kroll, and R. Lipowsky, Phys. Rev. B**42**, 961 (1990).

/36/ E. Brezin, B. Halperin, and S. Leibler, J. Phys. (Paris) **44**, 775 (1983).

/37/ M.E. Fisher and A.J. Jin, University of Maryland, preprint.

/38/ A.O. Parry, R. Evans, and K. Binder, University of Mainz, preprint.

/39/ This situation has to be distinguished from wetting in 3–d random systems, see R. Lipowsky and M.E. Fisher, Phys. Rev. Lett. **56**, 472 (1986), and from wetting of a rough solid substrate, see M. Kardar and J. Indekeu, Europhys. Lett. **12**, 161 (1990); H. Li and M. Kardar, Phys. Rev. **B42**, 6546 (1990).

/40/ For a $d_{\parallel}$–dimensional interface in $d = d_{\parallel} + 1$, one has $\zeta = (5 - d)/2(2 + s)$ which exceeds the thermally–excited roughness for $s < s^*$ with $s^* = (d - 1)/(3 - d)$ for $1 < d \leq 3$ and $s^* = \infty$ for $3 < d \leq 5$. If $s < s^*$ and $V_R(\ell) \ll 1/\ell^\tau$ with $\tau = 2(d - 1 + 2s)/(5 - d)$, complete and critical wetting are characterized by $\psi = (5 - d)/(d + 3 + 4s)$ and $\psi = (5 - d)/(d - 1)(2 + s)$, respectively.

/41/ See, e.g., *Surface x-ray and neutron scattering*, ed. by H. Zabel and I.K. Robinson (Springer Verlag, to be published)

/42/ C. Cevc, W. Fenzl, and L. Sigl, Science **249**, 1161 (1990)

/43/ R. Lipowsky, Nature **349**, 475 (1991)

GROWN-IN AND SHEAR-PRODUCED APB FAULTS IN ORDERED INTERMETALLICS

D.G. MORRIS
Institute of Structural Metallurgy,
University of Neuchâtel,
Avenue de Bellevaux 51,
2000 NEUCHATEL,
SWITZERLAND

ABSTRACT. Antiphase Domain Boundaries play an important role in phase changes and microstructural stability of ordered alloys and intermetallics as well as affecting mechanical behaviour. The origin of APB faults in annealed material is examined here, and in particular the differences between a sharp boundary, as produced by crystal shear, and a relaxed fault structure are emphasized. The kinetics of relaxation of a shear-produced fault are examined and it is shown that fast relaxation may greatly affect the movement of dislocations by creating locking stresses as well as affecting cross slip behaviour, and hence significantly affecting mechanical properties.

1. Introduction

Antiphase Domain Boundaries (APB's), representing stacking faults in the ordered crystalline arrangement of ordered alloys and intermetallics, are frequently seen in materials prepared by typical procedures such as solidification, forming and recrystallisation, or annealing. The presence and morphology of these defects can be useful for obtaining information on the types and energies of ordering of the material in question. Shear-produced faults are of particular relevance because they are present as part of the structure of dissociated dislocations and hence APB properties, such as energy anisotropy or thermal relaxation, can play a significant role in affecting the mobility of these dislocations. These various factors are examined here, with frequent reference made to recent work on alloys based on Ni_3Al and Al_3Ti.

2. Grown-in APB's and their Origins

2.1. GROWN-IN APB's

The classical idea of the ordering of different atom species within an initially disordered crystal can explain the appearance of domains of ordered regions "out of phase" with each other when it is considered that nucleation of the order can occur randomly within the disordered crystal. For example, within a crystal of ordered structure B2 there are two sub-lattices onto which a given atom species can sit, giving rise to APB displacements of type $1/2<111>$. Within the ordered $L1_2$ structure there are four sub-lattices and the displacement between domain regions is of type $1/2<110>$. For a typical A_3B alloy with this structure

123

C. T. Liu et al. (eds.), Ordered Intermetallics – Physical Metallurgy and Mechanical Behaviour, 123–142.
© 1992 Kluwer Academic Publishers.

the B atom sits on one of the sub-lattices, and each APB is equivalent to all others having the 1/2<110> vector to characterize it. These geometrical aspects of APB faults and their associated displacement vectors have been reviewed many times, for example by Marcinkowski (1).

It is important, however, to realize that the relative orientations of the fault vector (R) and the fault plane will distinguish two categories of APB. In the case when the fault vector lies in the fault plane, which may be defined as Type I APB's (2,3), movement of the boundary can take place by lattice site exchange of the A and B atoms on their respective sub-lattices. When the fault vector is not in the fault plane, however, as for the Type II APB, extra A or B atoms are associated with the fault and movement of this can only occur at sufficiently high temperatures that the excess atom species can be dragged with the boundary.

In the case of the DO_3 structure, which can be regarded as a B2 structure with one further degree of ordering taking place, an alloy of A_3B composition can contain two types of APB's following ordering by random site nucleation, depending on whether the fault vector connects atom sites on the same B2 lattice or not. These two types of APB's are characterized by fault vectors of type 1/4<111> or 1/2<100>. Figure 1 shows domain boundaries within Fe_3Al ordered to the DO_3 structure, where some boundaries are clearly seen (fault vector usually 1/4<111>) and others are visible only in residual contrast (fault vector 1/2<100>), confirming that in this case where random site nucleation has occurred during ordering there exist both types of APB's.

The theory of contrast at APB faults when examined by transmission electron microscopy has been well described (1) in terms of the phase angle α introduced as the electron beam passes through the fault, where the angle α can be written 2π **g.R**, where g is the diffraction vector and **R** the displacement vector characterizing the fault. According to this the APB should be imaged, similar to other stacking faults, as a series of dark and light bands parallel to the intersection line of the fault and the thin foil surfaces for those imaging conditions when α is not zero. In reality the extinction distance associated with these faults is large, about the same as the foil thickness, and the APB is often imaged as a single dark or light band.

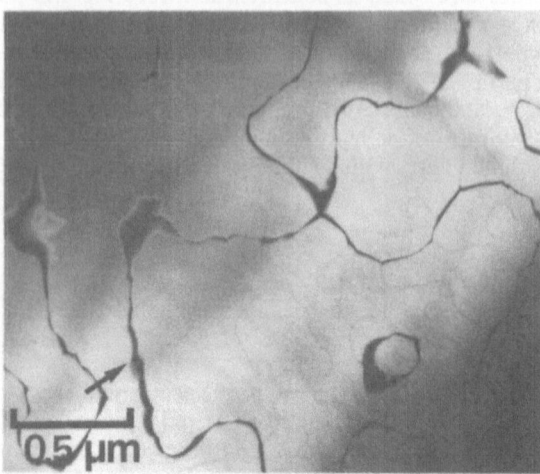

Figure 1.   Micrograph showing 1/4<111> APB's in strong contrast, and 1/2<100> APB's in weak contrast in Fe_3Al ordered for 7 days at 500°C to produce large domains with the DO_3 structure. Foil orientation near (001); g vector 200; dark field micrograph.

The ideas outlined so far are based on the principle that a disordered crystal exists initially which subsequently becomes disordered. As such the observation of the grown-in domain network is proof that the material existed, albeit momentarily, in the disordered crystalline state before becoming ordered (4). In accordance with this, domain networks are commonly seen in weakly ordered alloys, for example Cu_3Au, Ni_3Fe and sometimes FeAl, but not in strongly ordered intermetallics such as Ni_3Al and TiAl.

The existence of only two, or as many as four, sub-lattices within a given ordered crystalline structure has an important consequence as far as the stability of the domain network is concerned (5). Within the B2 structure, where there are only two sub-lattices, it is in theory impossible to have a stable network of interpenetrating domains but only domain islands contained within a primary domain. The observed domain networks exist in a metastable configuration because many domain boundaries take up saddle-shapes with curvature in two opposing directions (6). Observations of domain growth on annealing confirm that domain coarsening in indeed fast, for example within alloys such as β brass or FeAl, and is not even much faster because of these saddle configurations (7). Within the $L1_2$ structure, where four sub-lattices allow a stable three-dimensional domain network to be set up, fine domains are produced after ordering, and domain growth occurs steadily during subsequent annealing by reduction of APB energy, much as during grain growth. Following these comments on the kinetics of domain growth, it is interesting to note that the absence of a domain network in a given ordered material may be a reflection of strong ordering, considered in the preceding paragraphs, or a reflection of rapid domain growth.

According to Flinn (8) the energy of the APB can be approximated by calculating the density of nearest neighbour violations across the fault, which will depend, for a given ordered structure, on the fault vector and on the plane occupied by the fault. Higher order neighbour violations will certainly make an important contribution to the APB energy in many cases, but the results of Flinn nevertheless show many important features. One of the most important conclusions of this analysis is that the APB energy in material with $L1_2$ structure will have an APB energy which is highly anisotropic, indeed zero energy for a fault lying on a cube plane with the shear vector in that plane. For the B2 structure the APB energy is reasonably isotropic. The anisotropy of the APB energy in the $L1_2$ structure is reflected in the morphology of the domain boundary network found after annealing, for example within the Cu_3Au alloy, where a cuboid network is seen. This anisotropy has also been used as an important explanation of the unusual strengthening of many $L1_2$ alloys on testing at higher temperatures, based on the cross slip of superdislocations to cube planes to permit the reduction of their APB energy. The absence of grown-in domain boundaries in intermetallics has meant that there is no visual confirmation of APB energy anisotropy in these cases and the Flinn theory has been accepted as applicable.

An alternative approach has been considered to assess low APB energies on specific planes (9,10). The APB fault represents a disruption in the stacking sequence of the parent phase, but can also be considered as creating a thin slab of new phase of the stacking sequence occurring across the APB. For example, a shear APB fault on the $\{100\}$ plane of the $L1_2$ structure creates the stacking of the DO_{22} tetragonal ordered phase, whilst a shear fault on the $\{111\}$ plane can create a thin slab of material with the DO_{19} structure. If it is known, from thermodynamic considerations or otherwise, that the given $L1_2$ material is fairly unstable relative to either the tetragonal or the hexagonal phase it can be deduced that the APB energy on the $\{100\}$ or $\{111\}$ planes will be especially low.

2.2. APB's PRODUCED BY RAPID SOLIDIFICATION

Rapid solidification techniques offer the possibility to undercool a liquid phase before solidification takes place, to rapidly cool from the initial liquid temperature, through the

126

freezing range, and to rapidly quench the solid. When applied to the Ni₃Al intermetallic, the results obtained vary depending on the precise stoichiometry of the alloy in question and on the ternary alloying additions present. Rapid solidification of an alloy containing 24at% Al produced a fine grained material containing a fine domain network at the grain centres with elongated domain boundaries stretching to the grain boundaries (11), see Fig. 2. The grain interior was poor in Al and it was deduced that solidification took place by the formation of Ni-rich dendrites followed by the final solidification of material richer in Al. Similar results by Cahn (4) are explained by the formation of a disordered crystalline solid, particularly where the Al content is low, which subsequently orders to a fine domain structure. The domains obtained here are somewhat cuboid with faces on the {100} planes, tending to support the Flinn analysis for APB energy anisotropy in this material.

Similar studies on a variety of binary alloys near the Ni₃Al composition and on alloys containing ternary alloying additions (12,13) show essentially similar results but importantly show that the anisotropy of the domain structure is weak and indeed not confirmed for many of the alloys studied. Significantly anisotropic domain networks are clearly seen only for alloys close to the stoichiometric Ni₃Al composition.

An interesting study by Maurer et al (14) illustrates the influence of precise starting composition and segregation during solidification on the phases obtained. When rapidly solidifying by powder quenching a stoichiometric Ni-25at% Al alloy, they obtained the β (NiAl) phase and secondly the Ni₃Al phase, as would in fact be expected from this hyper-eutectic composition.

Another study dealing with fast solidification by atomisation of a hypo-eutectic alloy (15), where Al-poor initially-forming dendrites would be expected, found in fact that the initial material to solidify was rich in Al. This inverse segregation behaviour could be explained by significant undercooling occurring before solidification such that the first phase to form was the Ni₃Al ordered phase. The fairly stable domain boundaries seen in this material, Fig. 3, were explained by the significant segregation at the boundaries which ensured that these remained disordered in some cases and required considerable chemical transport for domain coarsening to occur (16).

These studies emphasize that the domain boundaries observed even after very rapid solidification may have local compositions significantly different from that of the parent alloy, and their stability and orientations will certainly be affected by this.

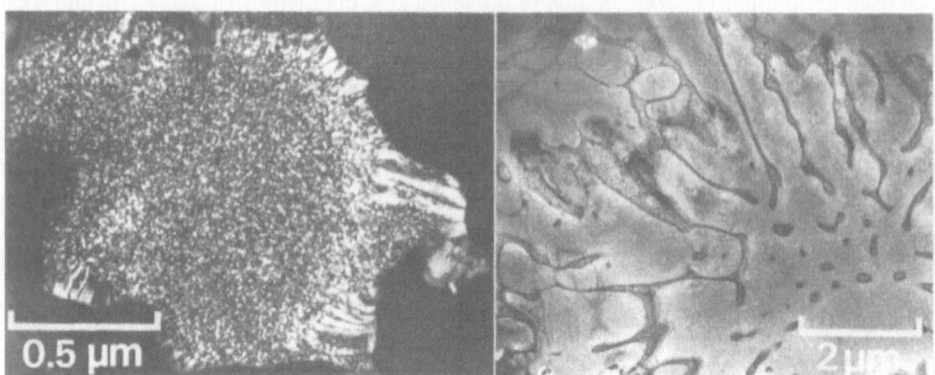

Figure 2.   Fine domain network in grain centres of melt spun Ni₃Al(Fe,Cr) and straight domain boundaries behind grain boundaries.

Figure 3.   Domain network in a rapidly solidified Ni-18%Al-8.7%Cr alloy. Inverse segregation leaves stable domain boundaries which are Al poor.

2.3. APB's PRODUCED BY RECRYSTALLIZATION

Recrystallization involves the change from one crystalline arrangement to another as a grain boundary moves driven by the energy difference given by a high degree of deformation and defects on one side and an essentially defect-free state on the other side. If the grain boundary is sufficiently complex in nature as to represent a disordered region between the ordered volumes on each side it may well be expected that APB defects will be left behind the moving grain boundary. Competitive kinetics of grain boundary movement and reordering and domain growth behind the boundary will determine the final microstructure observed.

Cahn (17) has recently reviewed the interactions of ordering and recrystallization, including the perfection of the ordered regions remaining after grain boundary movement. In the case of Cu_3Au - a weakly ordered $L1_2$ material where a disordered grain boundary film may be expected and showing slow domain growth kinetics - domain boundaries are normally seen stretching behind the grain boundaries, much as illustrated in Fig. 2 (17,18). A recent study in β brass (19) found no such domain boundaries in the just-recrystallized grains - this may be because the ordered structure (B2) permits the much more rapid growth of any domains initially present. On recrystallizing $(Co,Fe)_3V$, no domain network was seen (20) but only a few isolated APB's. This $L1_2$ alloy is disorderable at high temperatures and should be expected to produce a domain network on recrystallizing according to the ideas developed here. In a similar way, recrystallizing Ni_3Al did not leave a network of domain boundaries inside the grains (21,22). While the evidence is not conclusive, it seems likely that in this intermetallic the grain boundaries may also represent disordered zones (23,24)

The interpretation of these results seems to depend more on domain stability and growth kinetics than on the nature of the recrystallization process itself. A comparison of recrystallization kinetics with domain coarsening rates illustrates this point. Recrystallization of heavily deformed Ni_3Al requires typical temperatures of 800°C for times of the order of an hour (21,22). Studies on the growth of domains in rapidly solidified Ni_3Al alloys, as illustrated in Fig. 2 and 3, show that in a highly segregated material domain growth, requiring solute transport for APB motion, may need temperatures above 1100°C, whilst in near stoichiometric Ni-25%Al domains grow rapidly at 800°C and some are even lost during the initial rapid quenching process (13,15).

3. Shear-Produced APB's

Shear-produced APB's occur within the APB-dissociated superdislocations commonly found in many deformed ordered alloys and intermetallics, for example Cu_3Au, Ni_3Fe, Ni_3Al and Al_3Ti. Such APB-linked partial dislocations are not limited to $L1_2$ structures, but are also found in other ordered cubic and non-cubic structures, eg B2, DO_3, DO_{19},.... In $L1_2$ crystals, the dissociation scheme of perfect dislocations has been examined in detail (25) and essentially involves the formation of an APB fault or a complex stacking fault. The present study will consider in detail only the APB dissociation scheme, but many of the arguments presented will apply equally to other faults in intermetallics. The APB faults present within the superdislocations will essentially be rehealed by the trailing partial except at jogs, where APB tubes may be produced (26), or unless the trailing partial dislocation is somehow pinned, as presumably occurs during intense deformation when large numbers of ordered faults are produced and shear disordering will occur.

The fault produced by shear is a sharp discontinuity in the stacking sequence of the ordered crystal. It is known, however, that annealing such a fault will lead to a spread of the discontinuity to produce a fault several atoms in thickness (27,28,29). This can be understood in terms of increasing the low configurational entropy corresponding to the sharp shear-produced discontinuity. In addition, excess atoms in the material, if the composition is not exactly on stoichiometry, will tend to diffuse to the boundary, reducing the overall energy level of the alloy and changing the local composition significantly.

These results were first demonstrated by Brown (27) for an ordered alloy with B2 structure and applied specifically to β brass. The Bragg-Williams approach was used, taking account of nearest neighbour bonds and interactions for the calculation of enthalpy and determining the configurational entropy of the crystal-APB system. In the case of such simple, disordering B2 alloys it appears that this approach is a reasonable approximation of the state of order. At low temperature, where bond enthalpy dominates the fault configuration, a sharp shear-produced interface should be maintained, whilst at progressively higher temperatures as entropic effects reduce both the overall degree of order and particularly that at the fault, the thickness of disordered material at the fault gradually increases. At the critical ordering temperature (T_c) the fault becomes infinitely thick - this may be seen as a mode of heterogeneous transformation - whilst at temperatures about 0.975 T_c the fault thickness is already about 10 atom layers thick. In analogy with Suzuki strengthening found following segregation of solute to the stacking fault of a dissociated dislocation in a normal fcc solid solution, the reduction of order and fault energy at the APB plane of a superdislocation will lead to pinning of that dislocation. It is easy to understand that motion of the superdislocation after relaxation of the APB involves the leading partial dislocation creating fresh, high energy APB, whilst the trailing partial will only incompletely reheal the low energy, relaxed APB. From the calculations of Brown, a maximum extent of dislocation pinning was predicted at a temperature of about 0.7 T_c, which was seen to be reasonably consistent with high temperature strength measurements.

These ideas have developed significantly over many years (eg 28,29,30,31) and been applied to other crystal types and off-stoichiometric alloys. Grinberg and Plishkin (30) developed the results of Brown, using the same nearest neighbour bonding and configurational entropy approach, to consider the influence of non-stoichiometry in a B2 crystal. They deduced that no chemical segregation to the APB would occur for a stoichiometric alloy, but that considerable segregation could occur for different compositions. The maximum degree of segregation occurred when the bulk alloy composition was about 10% off stoichiometry and led to an enrichment of up to 10% further of the excess element, this happening at low temperatures where entropic effects were not important. Such segregation led to a significant reduction in the order parameter at the APB, and a corresponding increase in the extent of dislocation pinning may be expected. Popov et al (28) considered the L1$_2$ crystal structure to demonstrate similar effects taking place here. The order parameter at the APB was reduced by a factor of two at a temperature of about 0.75 T_c, falling much faster than that of the bulk alloy at higher temperatures (illustrated in Fig. 4, taken from reference 28); the APB thickness increased over this range from about 3 planes to 5-10 planes; segregation occurred at the APB, even for a stoichiometric A$_3$B alloy, leading to an enrichment in element A at the APB plane at the intermediate temperatures. These changes led to significant pinning of the superdislocations, as shown in Fig. 5, from reference 31. Dislocation pinning is affected by bulk composition, the extent of chemical segregation and local order change at the APB plane, and the overall degree of order of the alloy at the temperature considered. At low temperatures, the extent of segregation is much smaller at stoichiometry, and the extent of pinning is much less there than for compositions slightly different; at higher temperatures, disordering at the APB is more important, and slight changes in composition play a minor role such that dislocation pinning is greater, but is now independent of alloy composition.

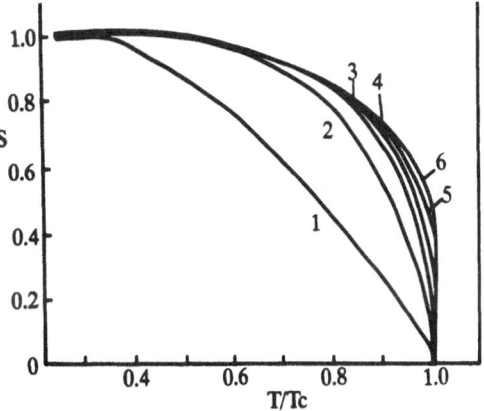

Figure 4.   Reduction in order parameter with temperature for an L1$_2$ crystal - in the bulk alloy, and at various planar distances away from an APB (28).

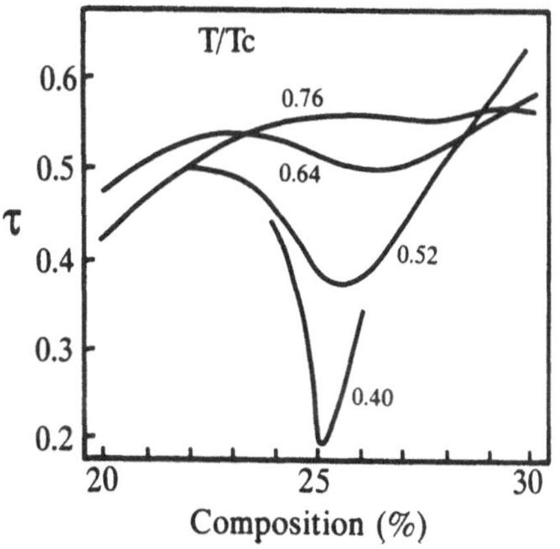

Figure 5.   Dislocation pinning by APB relaxation for a L1$_2$ crystal as the annealing temperature or composition is altered (31).

More recently (29), Sanchez et al have re-examined changes at APB's as the fault relaxes using a more rigorous and sophisticated tetrahedron approximation of the cluster variation method to calculate the minimum energy configuration for a given temperature and composition. The results are very similar to the earlier ones reported here and confirm the essential notions of significant relaxation at the APB on heating, with the extent and thickness of relaxation effects being much more significant for an off-stoichiometric alloy and at higher temperatures. For example, for the L1$_2$ material considered, a stoichiometric A$_3$B composition will be enriched by about 1% in A at the APB fault, with the fault spread over 3-10 atomic planes for temperatures of 0.75T$_c$ to near T$_c$. The long range order parameter is deduced to decrease to about 0.6 at the APB for about 0.75 T$_c$ and to be reduced to about 0.2 at a temperature of about 0.98 T$_c$. A slightly off-stoichiometric alloy (76%A-24%B) showed much greater enrichment at the APB, to about 80% A, at low temperatures (0.5-0.75 T$_c$) but less significant enrichment at higher temperatures. As seen for the earlier calculations, the maximum extent of dislocation pinning was predicted for about 0.8 T$_c$.

In his original work, Brown (27) considered that the relaxation-induced pinning could explain the yield stress at room temperature or at elevated temperatures for the material tested (β brass). It is conceivable that very rapid relaxation at elevated temperatures could produce continuous, steady-state relaxation and hence maintain superdislocation pinning, much as proposed by Flinn, invoking a different superdislocation pinning mechanism (8). It is now not generally accepted that relaxation can be sufficiently rapid as to play a major role in strengthening. Initial, discontinuous yielding or dynamic strain ageing has been observed on several occasions, however, and related to this APB relaxation (32,33).

Before moving on to consider some recent experiments on APB relaxation and the effects on dislocation geometry and behaviour, it is of interest to reconsider in detail some of the characteristics of APB's, in particular their anisotropy, both in terms of the shape of domain networks and in terms of energy of the faults.

4. Anisotropy of APB energy

Attention is given here specifically to $L1_2$ crystals, both because of the importance of materials with this structure (Ni_3Al, Cu_3Au, modified Al_3Ti) and also because of the relationship between APB energy anisotropy and the so-called anomalous strengthening observed at intermediate temperatures. Based on the early work of Flinn (8), which considered that $L1_2$ crystals should show a strong anisotropy of the energy of their APB's, and based on transmission electron microscope observations which suggested that cross slip occurred to place the APB of the dislocations on the cube planes, the anomalous strengthening seen on testing at high temperatures was explained in terms of cross slip to the cube planes pinning dislocations, this cube cross slip driven by the low APB energy there (34). For such cross slip to occur the anisotropy ratio of APB energy (on {111} planes relative to on {100} planes) should be greater than $\sqrt{3}$. Recently Yoo (35) has suggested that it may be the elastic anisotropy that is the driving force for such changes in habit plane of the superdislocation with only minor contribution from anisotropy of the APB energy. In view of this alternative suggestion it is of interest to reconsider the theoretical basis and experimental data in terms of APB energy and its anisotropy.

The theoretical basis (due to Flinn) for considering that the APB energy on the cube plane is very low is not a rigorous one since only nearest neighbour effects are taken into account in the $L1_2$ crystal, and possible effects of higher-order neighbour interactions are neglected. Observations of grown-in domain networks, such as in Cu_3Au (36), nicely confirm these predictions since the domain boundaries lie predominantly on the cube planes. One of the problems in applying such logic to intermetallics, such as Ni_3Al and $L1_2$ modified Al_3Ti, is that there are generally no grown-in domain boundaries. Studies on rapidly solidified materials containing domain networks, as illustrated in Figs. 2 and 3, then take on particular interest. From the shape of the domains it is possible to deduce the anisotropy of fault energy based on the Wulff theorem: for a domain which has essentially adopted an equilibrium shape (a large, slowly growing domain, for example) the shape reflects the APB anisotropy in that the fault planes with low energy will tend to dominate the shape. By way of example, if the APB energy is reasonably isotropic near-spherical domains will be expected, whilst if the APB energy is lowest on the {100} planes then the domains will adopt a cuboid shape with the faces on the {100} planes. Based on this logic, the APB energy anisotropy has been deduced in Ni_3Al-based alloys from the shape of the grown-in or annealed domains in rapidly solidified materials (11,12,13), and some of the results are shown in Table I. The anisotropy factor shown here is the ratio of the APB energy on the {111} plane to that on the {100} plane, whilst the equivalent aluminium content takes account of the presence of elements such as iron and chromium in the alloys which lie partially on Al and Ni sites. These results show clearly that there is a mild anisotropy of the APB energy in these Ni_3Al-based alloys, which increases somewhat as the equivalent aluminium content increases towards the stoichiometric value, but that on its own the APB anisotropy is not sufficient to explain the cube cross slip phenomenon.

TABLE I. Anisotropy of APB energy for Ni_3Al-based intermetallics as a function of the Equivalent Aluminium content.

Aluminium Content (%)	APB Anisotropy
23.5	1
24.0	1.2 - 1.6
24.6	1.3 - 1.5
25.0	1.4

Similar results were obtained by Lasalmonie et al (12), both in terms of the near-isotropy of APB energy as well as the increased tendency to anisotropy as the composition changed towards stoichiometry and hyper-stoichiometry. In an analysis of the displacement vector associated with the APB's in the rapidly solidified materials, it was shown that the total $\mathbf{R}$ vector was not simply $1/2<110>$ as expected for the $L1_2$ crystal. APB's remained visible, albeit in weak contrast, even when the product $\mathbf{g}.\mathbf{R}^1$, where $\mathbf{R}^1$ is taken as being $1/2<110>$, was zero such that no contrast would be expected. An example of this is given in Fig. 6, where it is clearly seen for many of the APB's that incomplete invisibility is not achieved when the diffraction vector selected gives $\mathbf{g}.\mathbf{R}^1 = 0$. A possible explanation of this is that relaxation has occurred at the APB creating effectively a second minor displacement vector across the APB in addition to the major $1/2<110>$ vector expected from the simple crystallographic approach. The major $\mathbf{R}$ vector was seen to be not always in the plane of the APB, that is that many of the grown-in APB's were non-conservative, as can be produced frequently by random site nucleation of order but not by a dislocation shear process. The supplementary vector $(\mathbf{r})$ was generally in the plane of the APB, but not always so, and its magnitude was about $1/10$ that of $\mathbf{R}$. The $\mathbf{r}$ vector was a variable along a given APB, and its function was taken to be to smooth the overall energy anisotropy. According to this analysis, the APB anisotropy seen even after rapid solidification is that of relaxed APB's, and not that of the shear APB's considered to be important in controlling dislocation cross slip.

Another approach to estimating APB energy and anisotropy is to consider the APB as producing a change in free energy of the system since a thin section of different phase is obtained by the change in stacking sequence. For example, a shear of $1/2<110>$ on a $\{001\}$ plane produces a thin slice of material of DO_{22} phase within the $L1_2$ crystal. On the other hand this phase change is not obtained when the APB fault is on a $\{111\}$ plane. Thus, if the $L1_2$ material is intrinsically unstable towards the DO_{22} structure, there is a preference for the APB's to lie on the cube planes where the APB energy will be low. Recent work on Al_3Ti modified by additions of transition elements such as Fe or Mn so that the symmetry changes from DO_{22} to $L1_2$, shows that these may retain some instability towards the DO_{22} structure, such that fault energy can be low there, and dislocations therefore cross slip onto the cube planes. This tendency is seen specifically for Al_3Ti alloys

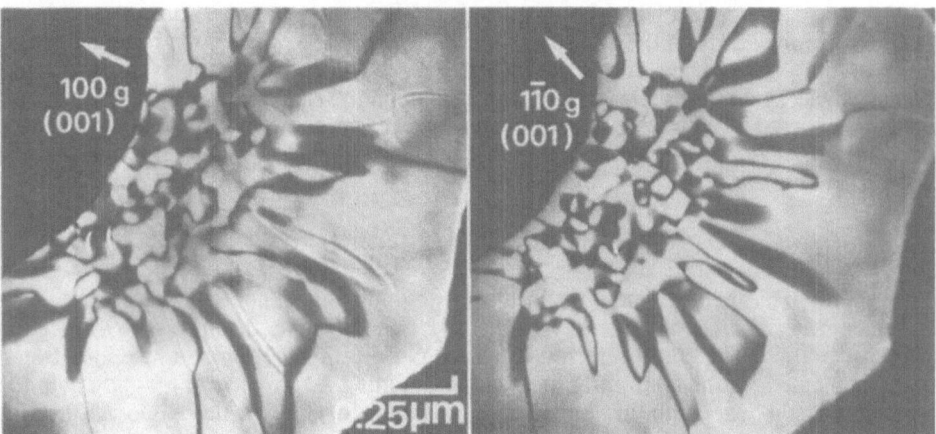

Figure 6.   Transmission electron micrographs showing APB's in rapidly solidified Ni_3Al. Comparison between images shows many cases where the APB's are nominally invisible but in fact a faint contrast remains.

132

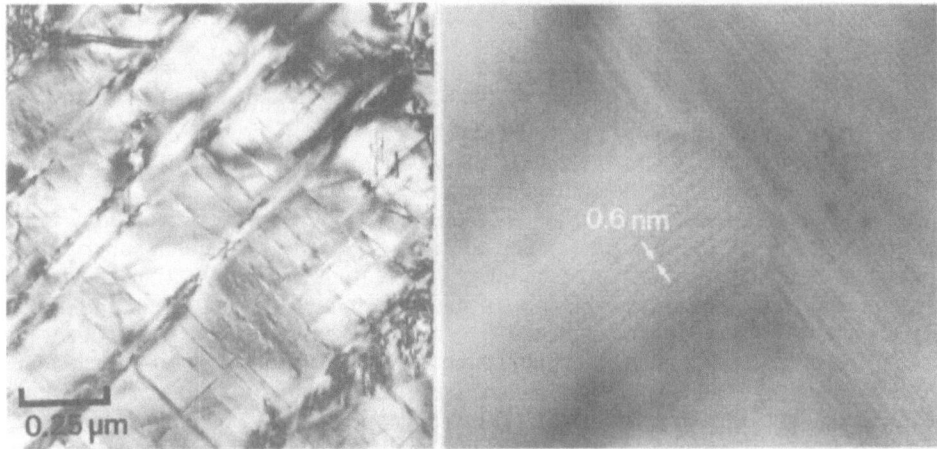

Figure 7.   Transmission electron micrographs showing precipitation on {100} planes in Al$_5$Ti$_2$Fe alloy. High resolution imaging shows that this precipitation corresponds to a local change of stacking sequence to the DO$_{22}$ structure.

with Fe additions and in this case precipitation leading to changes in plane stacking and cross slip of dislocations to cube planes after high temperature deformation is seen. The high temperature strength of the materials modified by Mn and Fe vary in different manners with the temperature (37): while the Fe-modified alloy shows anomalous strengthening at elevated temperature, with dislocations cross slipping to cube planes (38), the Mn-modified alloys show no anomalous strengthening and the dislocations remain on the octahedral planes (37). Using the elastic anisotropy theory of Yoo (35) as a basis for explaining cube cross slip, and inserting into these equations the elastic anisotropy term (A) for Al$_3$Ti with the L1$_2$ structure due to Fu (39), a relatively high APB energy anisotropy of at least 1.46 is required. The anomalous strengthening seen in the Fe-modified Al$_3$Ti alloys can thus be interpreted on the basis of highly anisotropic APB energy, and this due to the inherent instability of this alloy to the DO$_{22}$ structure. Figure 7 shows precipitation occurring in some of the Fe-modified alloys examined, where the precipitation is seen, by high resolution electron microscopy, to lead to a change in the stacking sequence on the {100} planes such that the Fe depletion occurring corresponds also to the creation of fine plates of DO$_{22}$ material. No such changes were seen in the more stable Mn-modified L1$_2$ material.

5.   Annealing effects on dislocations in intermetallics

Little attention has been given to the influence of annealing, either deliberate annealing of deformed material, or simultaneous with the high temperature deformation, on dislocations in intermetallics. The study referred to above concerning the temperature variation of strength and dislocation configuration in an Fe-modified Al$_3$Ti alloy of L1$_2$ structure (38) pointed out the variation in dissociation distance of the partial dislocations of a superdislocation as the deformation temperature changed. At room temperature after about 2% strain the dislocations were perfect, undissociated dislocations, whilst after deforming 2% at 500°C and 700°C the dissociation distance of the APB-dissociated dislocations was about 7nm and 16nm, respectively, in both cases for near-screw dislocations lying on

{111} planes. This variation prompted the study of the influence of annealing alone on the dislocations present in material deformed at low temperatures.

In situ electron microscope annealing of thin foil samples led to almost exactly the same variation in dislocation separations as found in the hot-deformed samples, as shown in Fig. 8, and summarized in Fig. 9 (40). In this case the material was deformed 4% at room temperature before the annealing experiments and many of the dislocations were just visibly dissociated as superdislocations after this strain. The kinetics of the changes seen were very fast at all but the lowest annealing temperatures, and these kinetics will be discussed in detail in the following section. On annealing at temperatures as low as 300°C

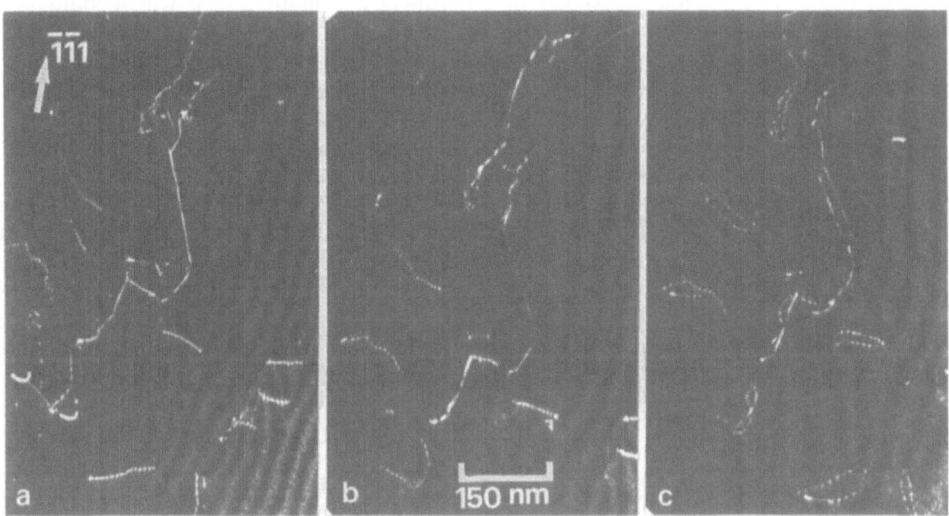

Figure 8. Dislocations in Al_5Ti_2Fe: (a) deformed at room temperature; (b) annealed 5mins at 500°C; (c) annealed 5mins at 700°C.

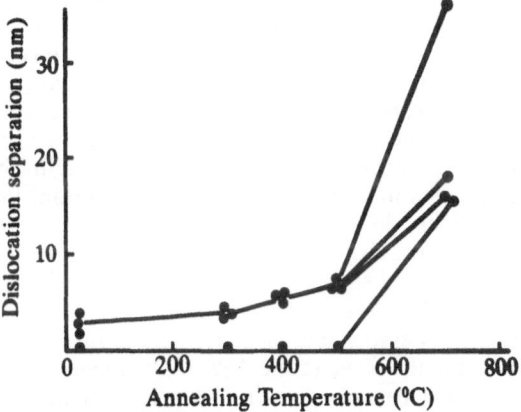

Figure 9. Variation in dissociation of APB-connected superdislocation partials in Al_5Ti_2Fe after annealing at the temperatures shown.

for only a few minutes the dislocation dissociation distance increased visibly from 2-4nm to about 4nm, whilst at progressively higher temperatures the spacing increased steadily to values of near 7nm at 500°C and to 20-30nm at 700°C, again for dislocations on {111} planes. The spacing change seen takes place rapidly to achieve a steady state value and thereafter, over time periods an order of magnitude longer, there is no further observable change. The change seen is not recoverable on cooling down, whether this is done slowly with long annealing pauses or is done quickly. Examination of diffraction patterns taken at suitable zone axes at the annealing temperatures shows no evidence of a change in the intensity of superlattice diffraction spots on heating from room temperature. These observations suggest that the changes seen are due to changes in the local APB structure, as described above, and not due to bulk disordering or to pinning effects which could have maintained a non-equilibrium dissociation distance after the room temperature deformation.

During these annealing treatments edge and screw dislocations behave differently, as illustrated in Figs. 8 and 9. It had previously been noted that the dislocations in this Al_5Ti_2Fe alloy were not dissociated after deforming to 1% strain at room temperature (38). After straining to 4%, however, many of the dislocations are now visibly dissociated into two 1/2<110> partials separated by APB. At this stage most dislocations are in 30°,60° or 90° edge orientations, and are clearly dissociated over 2-4nm. Many of the screw dislocations are not dissociated, however. On annealing the edge dislocations take up more relaxed configurations, already at temperatures as low as 300°C, even though the change in partial separation is small for such low temperatures. On the other hand, the screw dislocations that are undissociated after straining remain so up to temperatures as high as 500°C, and thereafter dissociate into the APB superdislocations with dissociation distance similar to that of the various edge dislocations. It is clear that during normal deformation, dislocation dissociation occurs under the combined action of both stress and temperature, and in the absence of applied stress higher temperatures are required to ensure complete dissociation.

During deformation tests at high temperatures (38) on the same material, it has been seen that dislocation glide takes place on {111} planes for deformation temperatures of 500°C and below, whilst on deforming at 700°C significant dislocation motion also takes place on {100} planes following cross slip from the initial {111} slip systems. It is interesting to note that during the annealing studies on the room temperature deformed samples there is no evidence of cross slip from the initial {111} planes to {100} planes. This is again indication that the cross slip process is activated both by thermal energy as well as by the applied stress. On occasions during annealing some of the dislocations are seen to take up unusually wide dissociation distances - an example is seen in Fig. 8. Such events are often associated with the interaction of the superdislocation with other, nearby dislocations or with free surfaces, but not with such cross slip to a cube plane.

Following relaxation of the APB and the superdislocation configuration, there are signs that the APB fault remaining cannot be described in simple terms by a shear vector of type 1/2<110>. It is clearly seen in some of the imaging conditions used to examine the relaxed dislocations (Fig. 8) that there is a fine fringe contrast associated with the fault. Such fault contrast was generally visible when imaged using "fundamental" 111-type g vectors, and sometimes visible and sometimes invisible when using 200-type g vectors. The use of 202-type g vectors also gave sometimes but not always such fault contrast. The supplementary shear vector (r) associated with the fault can be described as parallel to <010>, a vector which is not in the plane of the APB (habit plane of the superdislocation). This analysis is similar to that described earlier, due to Lasalmonie et al (12), where grown-in domain boundaries are described in terms of a primary 1/2<110> shear vector and a supplementary relaxation vector, non-parallel to the first.

From the spacings of the superdislocation partials it is possible to deduce the relaxed value of the APB energy, and these values are shown in Table II. These values were deduced using reasonable values for shear modulus (38) and for elastic anisotropy of the $L1_2$ crystal (39). The values shown refer in all cases to APB on the $\{111\}$ planes. It is seen that the annealing treatments lead to an order of magnitude decrease in value of the APB energy, and the values deduced are reasonably close to the values deduced after deformation at high temperature. It appears from this table that the APB energy associated with edge dislocations after relaxation may be lower than for screw dislocations. While such an effect may be the result of more complete relaxation around the stronger stress field associated with the edge dislocation, it should also be remembered that the analyses made depend on the values of elastic anisotropy used (39), which are not known precisely for this particular alloy.

TABLE II. APB energy (mJ/m^2) in Al_5Ti_2Fe deduced from dissociation separation of superdislocation partials, as-deformed or after annealing at the temperature.

Temperature (°C)	As-Deformed	After Annealing	
		Screw	Edge
20	385	-	-
200	-	330	290
300	-	250	230
400	-	250	190
500	140	200	145
700	72	40	40

In view of the extensive APB relaxation found in this particular alloy, it is of interest to examine such relaxation phenomena in other ordered alloys and intermetallics. Figure 10 summarizes changes in APB energies deduced from the separations of the partials making up the superdislocations, as a function of the reduced temperature. The temperature is normalised relative to the critical temperature for disordering (T_c), which has a precise value only in the cases of Ni_3Fe and Cu_3Au, whilst for the other, non-disordering intermetallics the melting point has been used instead. The results shown in Fig. 10 are in part new (Al_5Ti_2Fe, Al_5Ti_2Mn, Ni_3Al, and $FeAl$), and in part taken from the literature (41,42,43). Some of the results refer to *in situ* heating experiments in the transmission electron microscope (the new results, as well as for Ni_3Fe(43)), whilst the others (Ni_3Al and Cu_3Au(41,42)) refer to results obtained on samples deformed at the temperatures shown and subsequently examined at room temperature. It should also be emphasized that the APB energies for both Al_5Ti_2X alloys measured during *in situ* experiments as well as from bulk, high temperature tests were essentially identical, thereby confirming the validity of these *in situ* experiments.

In Fig. 10, all the alloys considered show considerable relaxation of APB structure and energy as a result of high temperatures, either the high temperature used for the deformation or for post-deformation annealing. It is clear that the absolute values of the APB energy for a given material depend on the strength of the atomic interactions and there is no reason to expect a correlation or similarity in these values. The variation of APB energy with reduced temperature should, on the other hand, be expected to show similar trends, based on the theories of APB relaxation outlined earlier (27,28,29). Such a similarity in extent of relaxation is not visible in Fig. 10: for example the two Al_5Ti_2X alloys show a rapid and extensive drop in APB energy, from very high values to very low values, at low reduced temperatures, whilst Ni_3Al and Ni_3Fe show only slight reductions

136

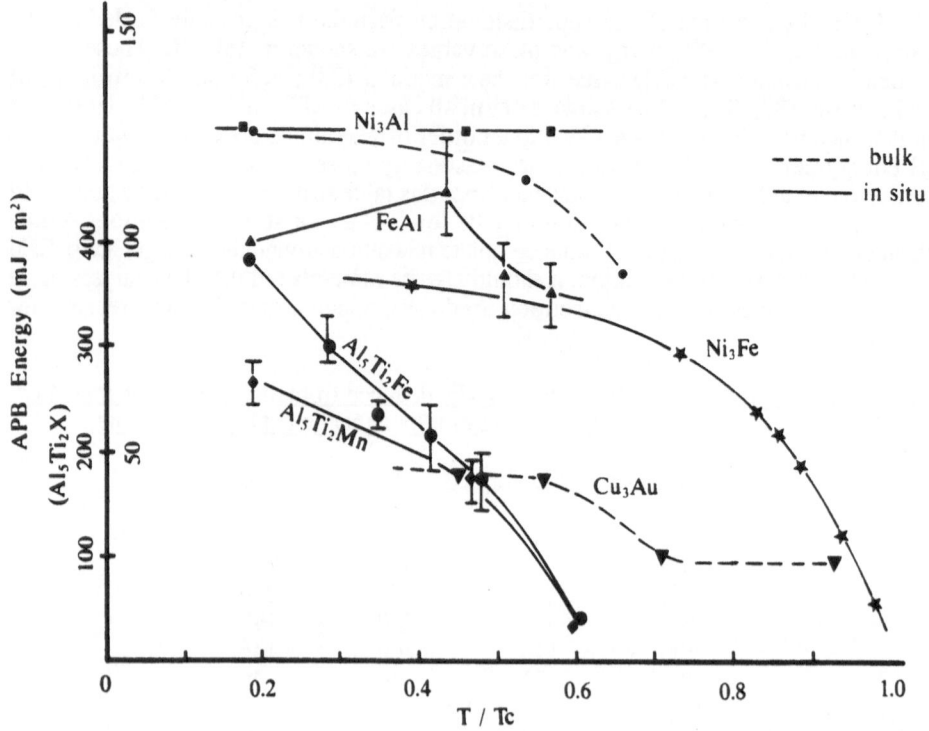

Figure 10.   Variation of APB energy in a variety of ordered alloys and intermetallics as a function of the deformation or post-deformation annealing temperature.

in the APB energy even after relaxation at high reduced temperatures. The reason for such variations is not known, but may relate to the precise composition of each alloy, the presence of ternary element additions and closeness to stoichiometry. The variations of APB energy in Ni_3Al may indicate such effects: the *in situ* experiment described here refers to a binary alloy of 24%at. Al; a similar absence of change in superdislocation dissociation distance has been noted (44) for an alloy containing about 76%at. Ni with about 6%at. Ti and the remainder Al; the results shown in Fig. 10 (42) on bulk material refer to superdislocations on {100} planes in a binary alloy of 24%at. Al to which a small addition of B has been made. The absolute values of APB energy in Ni_3Al, as well as the variations with temperature, vary greatly from case to case. The case of FeAl is a special one, since it appears that additional ordering (to the DO_3) structure may occur during annealing thin foils of this alloy containing 35%at. Al. (The alloy composition is close to the $B2$-DO_3 phase boundary at low temperatures and such ordering on a fine scale may be possible.) During annealing a fine speckling appears in the thin foil material, which leads to a slight reduction in the dissociation distance, and also makes precise measurements of separations difficult since the images obtained are less clear and the dislocation spacings are not constant along the dislocation lines.

The influence of such considerable reduction in APB energy as shown in Fig. 10 is uncertain and will be discussed later. It is certain that relaxation of APB's will lead to the pinning of dislocations present within the material, for example during an annealing treatment, and as such be responsible for discontinuous yielding on loading (32,33). In addition, in cases where the relaxation process is fairly rapid, it can be expected that repeated discontinuous yielding - dynamic strain ageing - will be obtained. Such discontinuities in the stress-strain behaviour have been observed in the Al_5Ti_2X alloys on testing at temperatures in the range about 300-500°C (38). These serrations were not observed on testing at room temperature - no relaxation can take place at this temperature - and were also hardly seen at temperatures from 500 up to 700°C, presumably where relaxation may be very fast. It is interesting to note that this material showed anomalous strengthening at temperatures above 500°C, that is where APB relaxation is fast.

6. Kinetics of APB relaxation

One of the most important factors which will determine the relevance of relaxation effects in affecting dislocation mobility and mechanical properties is the speed of relaxation and the possibility to drag the relaxed zone with the moving dislocation. Earlier studies of APB relaxation, whether theoretical or experimental (27,28,29,41), considered essentially equilibrium configurations of relaxed APB's without consideration of the kinetics of relaxation.

Schoeck and Korner (45) have examined *in situ* the relaxation on annealing of superdislocations in ordered Ni_3Fe. The magnitude of the effect seen on annealing at reduced temperatures above 0.8 is great, with dislocation spacings large enough to be accurately measured by conventional bright field electron microscopy. Over the temperature range studied the rate of change of the dislocation separation - used to determine the change in APB energy - was slow and both temperature and time dependent. Detailed analysis of the rates of change led the authors to deduce (45) that the rate of dislocation separation was determined by the difficulty of dislocation glide, with the actual relaxation process of the APB being much faster. Movement of the dislocation partials leads initially to the creation of a shear APB, as the dislocations glide apart, and this APB must be relaxed by atomic interchange at the APB before the dislocation can move further.

The new examinations of *in situ* relaxation processes in FeAl, Al_5Ti_2Mn and Al_5Ti_2Fe (40) have shown that very fast relaxation occurs, even at low values of reduced temperature. Indeed, for tests performed at temperatures of 300°C and above the relaxation observed took place within the heating period inside the microscope. For example, heating from room temperature to 300°C or to 500°C within a period of about 2-3 mins led to immediate changes in dislocation spacing, with no subsequent changes detected after holding at the temperature for times up to one hour. Heating to the desired temperature followed by immediate cooling and weak beam photography at room temperature, heating to the desired temperature and holding for one hour before weak beam photography at high temperature, or heating to the temperature for one hour before cooling and photographing at room temperature, all gave identical results. It is only at 200°C that some indication of the kinetics of relaxation was obtained. It is also important to note that such annealing at intermediate temperature, eg 300-500°C, followed by subsequent annealing at higher temperature, eg 500-700°C, led to the same results as when heating directly to the high temperature. These observations eliminate a specific quenched-in structure, for example local short range order or vacancy concentration, or frictionally-locked configurations of the original dislocation, as an explanation for the changes seen.

An estimate of the diffusion kinetics responsible for the APB relaxation can be made by noting that at 300°C dislocation dissociation distances of 2-4nm are achieved at relaxed dislocations in a time of 1 min or less, whilst at 200°C a similar scale of relaxation occurs in a time of about 1 hour. Considering that the diffusion distance over which a relaxed structure is set up is in the range about 0.4nm (local relaxation over a few atomic planes) to the dissociation distance (chemical gradient set up because of alloy off-stoichiometry), the diffusion coefficients deduced are about 10^{-23} - 10^{-21}m^2/s at 200°C and 10^{-21} - 10^{-19}m^2/s at 300°C. While diffusion coefficients are not known for these Al$_5$Ti$_2$X intermetallics at these temperatures, a comparison with fcc metals (eg Ni) at similar reduced temperatures (T/T$_m$) suggests that values of 10^{-31} and 10^{-27}m^2/s may be expected at 200 - 300°C. It is seen that relaxation kinetics at the APB are much faster than can be explained on the basis of such bulk diffusivity estimations, and enhanced diffusivity near the dislocation cores may be the explanation of this.

An interesting extension of this study on relaxation kinetics is to consider changes occurring at room temperature. Considering relaxation to occur in 1 h at 200°C and 1 min at 300°C, equivalent relaxation may be expected in 1-2 weeks at room temperature. Indeed, re-examination of thin foil samples after a period of a few months does suggest subtle changes in dislocation configurations.

7. Influence of APB relaxation on mechanical properties

The influence of grown-in APB's on dislocation behaviour has been known for a long time (46) and essentially seen as a weak effect due to ledges created as the dislocations cut through the domain walls. It is also possible that the grown-in domain boundaries may play a more subtle role in dissociating and disconnecting the partial dislocations that make up the superdislocation (13) since the shear APB may be annihilated as the superdislocation passes through the domain wall. For the remainder of this section, we shall restrict the discussion to the analysis of the role of APB relaxation on dislocation behaviour.

It is fairly well established that APB relaxation can lead to dislocation locking, similar to Cottrell locking or Suzuki locking in annealed materials, and lead to the appearance of discontinuous yield points at the initiation of straining as well as dynamically, throughout plastic deformation, if relaxation occurs fast enough (32,33). It is of interest now to consider even faster, continuous relaxation, and its effects on dislocation behaviour.

If relaxation occurs so fast as to be effectively continuous on a moving dislocation, then there will be a steady, viscous drag acting on the dislocation that can lead to significant strengthening. The strengthening will be similar to that analysed by Flinn (8), with the differences that there is no need for change in habit plane of the superdislocations (Flinn considered changes from {111} to {100} planes driven by APB energy anisotropies) and fast, localised relaxation can occur at the dislocation without the need for long range diffusion. By means of example, using the kinetics described before (relaxation in 1 hour at 200°C and in 1 min at 300°C), it can be expected that relaxation in Al$_5$Ti$_2$X can take place at 500-700°C (0.5-0.7 T/T$_c$) in times of the order 10^{-2}-10^{-3}s. When taken in the context of the jerky dislocation motion (47,48) that seems to characterize much of plastic deformation of alloys such as those considered here (*In situ* observations showed dislocations in Ni$_3$Al moving rapidly over time periods less than 1/50sec, and then being locked for periods of seconds in some sessile configuration or against an obstacle.) it is clear that rapid relaxation may affect the structure of even the moving dislocation. It is also interesting to note that deformation testing of Ni$_3$Al alloys at high temperatures (49), when {001}<110> slip systems control deformation, leads to frequent reports of yield points, as would be expected from dislocation locking by some mechanism.

It is relevant to raise the question here of whether the anomalous strengthening observed on testing many intermetallics at high temperature may be caused, or affected by relaxation at the APB. In view of the success of the Paidar-Pope-Vitek model (50) based on the cross slip of superdislocations to cube planes under the influence of thermal and stress assistance, particularly as relates to the orientation dependence of the anomalous strengthening, there seems no reason here to seriously doubt this model, especially since the APB relaxation should not be dependent on the orientation of the crystal and dislocation relative to the applied stress. In addition, it would seem that APB relaxation should also be found for those intermetallics and ordered alloys where anomalous strengthening is not seen - Ni_3Fe is an example - except that the kinetics of relaxation may not be sufficiently fast. It seems more relevant to consider the effect that APB relaxation may have on APB energies, thereby affecting the cross slip models.

The studies reported here have shown that relaxation at the APB can reduce the energy of the fault by at least a factor of 2. Relaxation of the APB when still on the {111} plane can essentially remove the APB energy anisotropy (of the relaxed APB on the {111} plane relative to the new unrelaxed APB yet to form on the {100} plane) and thereby inhibit cross slip to the cube plane unless other factors, elastic anisotropy for example, can provide the driving force. In a similar way, a dislocation already on the {100} plane will suffer APB relaxation on that plane which will essentially stabilise the dislocation in that configuration, that is inhibit reverse cross slip. Such analyses suggest that APB relaxation may greatly influence the frequency or possibility of forward or reverse cube cross slip. For conditions where relaxation is extremely fast, such that the APB behind a moving dislocation is already relaxed, the influence of APB modification will depend on the relative extents of relaxation on the {111} and {100} planes, which will not necessarily be the same. An important principle to be understood is that the true, shear APB and its energy are virtually never seen - unless deformation is carried out at room temperature and the dislocations examined quickly before relaxation occurs, even perhaps at room temperature. An example of this is the study on dislocations in Al_5Ti_2Fe deformed at 500°C and cooled quickly under load - does relaxation occur during the minutes of cooling or were the dislocations really present in the form seen during deformation?

When grown-in APB's are examined in materials such as Ni_3Al (11,12,13), even after rapid quenching or rapid solidification, it is clear that sufficient relaxation has occurred that the APB network seen is essentially a relaxed one and not that of shear APB's. Estimates of the anisotropy of the APB energy in Ni_3Al based on measurements of dislocation dissociation distances show considerable variation (51,52,53) from case to case ($\gamma_{111}/\gamma_{100}$ from 1.0 to 1.6) and it is not possible to establish a clear trend in variation based on alloy composition or test conditions. The rapid solidification studies tend to agree in demonstrating reasonably isotropic domains (and APB energy) for off-stoichiometric alloys (23-24%at Al) and more anisotropic domains for near-stoichiometric alloys. Two explanations for this trend in anisotropy can be proposed:- assuming a certain instability of the $L1_2$ crystal to the DO_{22} structure, this DO_{22} structure will be best created by a <110> fault on the {001} plane in a stoichiometric alloy but in an off-stoichiometric alloy the region of DO_{22} structure will not be so perfectly formed and the reduction in APB energy due to this phase change will not be so significant; segregation of solute to the APB in the off-stoichiometric alloy means that at fairly low reduced temperatures (eg $0.5T/T_c$) the local composition will be greatly changed, ordering largely destroyed, and hence differences due to orientations become minor.

It should also be noted that variations of alloy composition away from stoichiometry may be accommodated by the insertion of selective APB faults. For example, in a hypo-stoichiometric $L1_2$ A_3B alloy, excess component A can be retained at the APB's leading to a modified composition and structure there - these are the Type II boundaries referred to earlier. Considering a domain network of size 10nm, about 1% composition change may be accommodated at the Type II boundaries. In an early, interesting study by Poquette and Mikkola (3), a reduction in the proportion of {100} Type I boundaries was observed as the domain size increased in the alloy, perhaps because Type II boundaries became more frequent. Such preferential segregation of one component to a given domain boundary orientation may lead to an early formation of second phase, as shown in Fig. 7 for the Al_5Ti_2Fe material where DO_{22} Al_2Ti precipitates formed. It is seen that differences of composition away from stoichiometry can lead to the selective segregation of excess elements to specific APB planes, changing the energy there, and this dependent on the preference of the alloy to form specific phases with specific stacking sequences.

Several other unusual observations may also be re-examined in the light of the possible influence of APB relaxation effects. Caillard et al (48) described reversible cross slip between {111} and {100} planes, as well as between {111} planes, in a near-stoichiometric Ni_3Al deformed at room temperature. It should be pointed out that their deductions of changes in habit planes were based on changes in apparent superdislocation dissociation distances and not unambiguously in crystallographic terms. Dislocations were seen to jump quickly from one configuration to another over periods of seconds, while dislocation motion occurred rapidly, in periods less than 1/50 sec. The observations of dislocation relaxation in Ni_3Al (Fig. 10) are variable from one particular alloy to another, but in order to observe a significant change in dislocation separation it seems necessary to raise the temperature to a considerable value (reduced temperature ≈ 0.5), with no effect being observable at room temperature. Other workers (54) have pointed out, however, that irradiation damage, particularly at the dislocations, can be significant at high accelerating electron voltages, even at 200kV as used by Caillard et al (48), and such local increases in atomic mobility may be capable of inducing rapid dislocation relaxation.

The possibility of local radiation damage and enhanced atomic mobility leading to fast, low temperature relaxation may also explain the recent observation (55) of disordering in a slip plane in Ni_3Al by the passage of several superdislocations. This experiment was carried out *in situ* at an operating electron acceleration of 300kV, and so significant local radiation damage may be possible. Relaxation at the APB could thus mean that the passage of the superdislocation leaves behind a diffusely faulted zone of reduced order where easier subsequent dislocation motion can occur. The present proposition is based on an APB having a non-planar configuration because of APB relaxations, rather than a non-planar superdislocation produced by elastic torque effects, as suggested by the authors (55).

8. Conclusions

Antiphase Domain Boundaries may be produced during the ordering of an initially disordered material following solidification or recrystallization. As a result of diffusional relaxation the faults will generally be diffuse, low energy faults. The APB present inside a suitably-dissociated superdislocation will, initially, be a sharp, high energy fault which can rapidly relax on annealing at moderate temperatures. Both the energy and the anisotropy of the relaxed APB will be changed and depend on the precise composition of the alloy. The relaxation may be so fast in many materials that a true shear APB may never be seen, certainly not after an experimental test; the relaxation may in fact be virtually instantaneous and the superdislocation and its behaviour can be significantly modified during deformation of the material, affecting not only the yield or flow stress but also the cross slip frequency and influencing the anomalous strengthening.

141

Acknowledgements

The author would like to thank the Swiss National Science Foundation and the Commission for the Encouragement of Scientific Research for financial support of parts of the work reported here.

References

1. Marcinkowski, M.J. (1963) "Theory and Direct Observation of Antiphase Boundaries and Dislocations in Superlattices", in G. Thomas and J. Washburn (eds.), Electron Microscopy and Strength of Crystals, Interscience Publishers, New York, pp.333 - 440.
2. Fisher, R.M. and Marcinkowski, M.J. (1961), Phil. Mag., **6**, 1385 - 1405.
3. Poquette, G.E. and Mikkola, D.E. (1969), Trans. Met. Soc. A.I.M.E., **245**, 743 - 751.
4. Cahn, R.W. (1987) "Antiphase Domains, Disordered Films and the Ductility of Ordered Alloys based on Ni_3Al", in N.S. Stoloff, C.C. Koch, C.T. Liu and O. Izumi (eds.), High-Temperature Ordered Intermetallic Alloys II, MRS, Pittsburgh, vol. 81, pp27 - 38.
5. Bragg, W.L. (1940), Proc. Roy. Soc., **52**, 105.
6. English, A.T. (1966), Trans. A.I.M.E., **236**, 14.
7. Cupschalk, S.G. and Brown, N. (1968), Acta Met., **16**, 657.
8. Flinn, P.A. (1960), Trans. Met. Soc. A.I.M.E., **218**, 145.
9. Wee, D.M. and Suzuki, T. (1979), Trans. J.I.M., **20**, 634.
10. Yodogawa, M., Wee, D.M., Oya, Y. and Suzuki, T. (1980) Scripta Met., **14**, 849.
11. Horton, J.A. and Liu, C.T. (1985) Acta Met., **12**, 2191.
12. Lasalmonie, A., Chenal, B., Hug, G. and Beauchamps, P. (1988) Phil. Mag., **58**, 543.
13. Morris, D.G. and Morris, M.A. (1990) Phil. Mag., **61**, 469.
14. Maurer, R., Galinski, G., Laag, R. and Kaysser, W.A. (1989) "Structural and Chemical Composition of Ni-Al Powders", in C.T. Liu, A.I. Taub, N.S. Stoloff and C.C. Koch (eds.), High-Temperature Ordered Intermetallic Alloys III, MRS, Pittsburgh, vol. 133, pp293-298.
15. Morris, D.G. and Morris, M.A. (1991) J. Mater. Res., **6**, 361.
16. Carro, G., Bertero, G.A., Wittig, J.E. and Flanagan, W.F. (1989) "The Effect of Anti-Phase Domain Size on the Ductility of A Rapidly Solidified Ni_3Al-Cr Alloy", in C.T. Liu, A.I. Taub, N.S. Stoloff and C.C. Koch (eds.), High-Temperature Ordered Intermetallic Alloys III, MRS, Pittsburgh, vol 133, pp535-541.
17. Cahn, R.W. (1990) "Recovery, Strain-age-hardening and Recrystallization in Deformed Intermetallics", in S.H. Whang, C.T. Liu, D.P. Pope and J.O. Stiegler (eds.), High Temperature Aluminides and Intermetallics, TMS, Warrendale, pp245-270.
18. Hutchinson, W.B., Besag, F.M.C. and Honess, C.V. (1973) Acta Met., **21**, 1685.
19. Morris, D.G. and Morris, M.A. (1991) J. Mater. Sci., **26**, 1734.
20. Cahn, R.W., Takeyama, M., Horton, J.A. and Liu, C.T. (1991) J. Mater. Res., **6**, 57.

142

21. Baker, I., Viens, D.V. and Schulson, E.M. (1984) J. Mat. Sci., **19**, 1799.
22. Gottstein, G., Nagpal, P. amd Kim, W. (1989) Mater. Sci. Eng., **A108**, 165.
23. Mackenzie, R.A.D. and Sass, S.L. (1988) Scripta Met., **22**, 1807.
24. Baker, I. and Schulson, E.M. (1989) Scripta Met., **23**, 345.
25. Yamaguchi, M., Paidar, V., Pope, D.P. and Vitek, V. (1982) Phil. Mag.,**45**, 867.
26. Vidoz, A.E. and Brown, L.M. (1962) Phil. Mag., **7**, 1167.
27. Brown, N. (1959) Phil. Mag., **4**, 693.
28. Popov, L.E., Kozlov, E.V. and Golosov, N.S. (1966) Phys. Stat. Sol., **13**, 569.
29. Sanchez, J.M., Eng, S., Wu, Y.P. and Tien, J.K. (1987) "Modelling of Antiphase Boundaries in L1$_2$ Structures", in N.S. Stoloff, C.C. Koch, C.T. Liu and O. Izumi (eds.), High-Temperature Ordered Intermetallic Alloys II, MRS, Pittsburgh, vol 81, pp57-64.
30. Grinberg, B.A. and Plishkin, Y.M. (1965) Fiz. Metal. Metalloved., **19**, 182.
31. Popov, L.E., Golosov, N.S., Ginzburg, A.E., Kodzemyakin, N.V. and Kozlov, E.V. (1970) "Locking and Impediment of Superdislocations due to Antiphase Boundaries", in B.H. Kear, C.T. Sims, N.S. Stoloff and J.H. Westbrook (eds.), Ordered Alloys - Structural Applications and Physical Metallurgy, Claitors Pub. Div., Baton Rouge, La., pp307-319.
32. Besag, F.M.C. and Smallman, R.E. (1970) Acta Met., **18**, 429.
33. Besag, F.M.C., Morris, D.G. and Smallman, R.E. (1974) Acta Met., **22**, 813.
34. Paidar, V., Pope, D.P. and Vitek, V. (1984) Acta Met., **32**, 435.
35. Yoo, M.H. (1987) Acta Met., **35**, 1559.
36. Marcinkowski, M.J. and Zwell, L. (1963) Acta Met., **11**, 373.
37. Morris, D.G. and Lerf, R., submitted to J. Mater. Res.
38. Lerf, R. and Morris, D.G. (1991) Acta Met., in press.
39. Fu, C.L. (1990) J. Mater. Res., **5**, 971.
40. Morris, D.G. (1991) Scripta Metall. and Mater., **25**, 713.
41. Morris, D.G. and Smallman, R.E. (1975) Acta Met., **23**, 73.
42. Veyssiere, P., Yoo, M.H., Horton, J.A. and Liu, C.T. (1989) Phil. Mag. Letters, **59**, 61.
43. Korner, A. and Schoeck, G. (1990) Phil. Mag., **A61**, 909.
44. Korner, A. (1989) Phil. Mag. Letters, **59**, 1.
45. Schoeck, G. and Korner, A. (1990) Phil Mag., **A61**, 917.
46. Cottrell, A.H. (1955) Seminar on Relation of Properties to Microstructure, American Society for Metals, Cleveland, p151.
47. Pope, D.P. (1990) "Dislocation Mechanisms and Anomalous Flow in L1$_2$ Alloys", in S.H. Whang, C.T. Liu, D.P. Pope and J.O. Stiegler (eds.), High Temperature Aluminides and Intermetallics, TMS, Warrendale, pp51-61.
48. Caillard, D., Clement, N. and Couret, A. (1988) Phil. Mag. Letters, **58**, 263.
49. Umakoshi, Y., Pope, D.P. and Vitek, V. (1984) Acta Met., **32**, 449.
50. Paidar, V., Pope, D.P. and Vitek, V. (1984) Acta Met., **32**, 435.
51. Dimiduk, D.M., Williams, J.C. and Thompson, A.W. (1989) "On APB Dragging and APB Energy Anisotropy in Binary Ni$_3$Al", in C.T. Liu, A.I. Taub, N.S. Stoloff and C.C. Koch (eds.), High-Temperature Ordered Intermetallic Alloys III, MRS, Pittsburgh, vol 133, pp467-473.
52. Douin, J., Veyssiere, P. and Beauchamp, P. (1986) Phil. Mag., **A54**, 375.
53. Korner, A. (1988) Phil. Mag., **A58**, 507.
54. Baluc, N.L. (1990) Doctoral Thesis No 886, Federal Technical School, Lausanne.
55. Horton, J.A., Baker, I. and Yoo, M.H. (1991) Phil. Mag., **A63**, 319.

DEFORMATION AND FRACTURE OF INTERMETALLIC COMPOUNDS HAVING THE L1$_2$ CRYSTAL STRUCTURE

David P. Pope
University of Pennsylvania
Philadelphia, PA   19104 USA

ABSTRACT. Intermetallic compounds having the L1$_2$ structure show many unusual mechanical properties.   For example, many show an anomalous temperature dependence of the flow stress which has been related in various models to the structure of screw dislocation cores.   Others, however, show a bcc-like behavior, which also has been related to core structures.   Apparently unrelated to the flow behavior, L1$_2$ compounds also have a tendency for brittleness at low temperatures, with failure either occurring by transgranular cleavage or by intergranular failure mechanisms.   Cleavage in some L1$_2$ materials, e.g. Fe-modified Al$_3$Ti, tends to occur on a remarkably large number of planes in both poly- and single crystals. The cleavage resistance of this material can be slightly increased only by Mn additions.   Ni$_3$Al, on the other hand, tends to fail intergranularly, unless the grain boundaries contain B, and even then, failure under triaxial stress states tends to occur intergranularly.   Intergranular cracks propagate randomly through the grain boundaries of Ni$_3$Al with little preference for boundaries of any particular type, except that cracks tend to avoid symmetrical twin and low angle tilt boundaries. In this paper our current understanding of the flow and fracture behavior of this class of materials will be reviewed.

1. INTRODUCTION

Most metals having the fcc structure are ductile at low temperatures unless there is some complicating factor, such as S in the grain boundaries of Ni [1], iridium being the one notable exception because of intrinsically brittle grain boundaries [2].   Tensile failure usually occurs by microvoid coalescence, and cleavage failure is uncommon.   L1$_2$ ordered alloys, with a crystal structure based on the fcc have also been commonly thought to be "intrinsically ductile", similar to fcc metals.   As long as Cu$_3$Au was considered the prototypical L1$_2$ material, this analogy seemed reasonable since Cu$_3$Au is so very ductile.   However, as the properties of more L1$_2$ materials have been investigated it has been found that more and more of them

143

C. T. Liu et al. (eds.), Ordered Intermetallics – Physical Metallurgy and Mechanical Behaviour, 143–153.
© 1992 *Kluwer Academic Publishers.*

are intrinsically brittle, either because of cleavage failure, as in $L1_2$ Al_3Ti and Pt_3Al, or because of intergranular failure, as in Ni_3Al. We are discovering that even $L1_2$ intermetallic compounds suffer from the same tendency for brittle failure as do many bcc-based alloys e.g. NiAl, hcp-based alloys, e.g. Ti_3Al, and more complex ordered structures such as A15, Cr_3Si. In the great majority of brittle intermetallics, the brittle failure mode is cleavage fracture; although intergranular failure is also seen, and this leads to the observation that even intermetallics with high symmetry structures may be intrinsically more susceptible to these failure modes than are pure metals and disordered alloys. If this is true, and the evidence certainly seems to indicate that it is, then we need to understand the reasons for this brittle behavior if useful structural materials based on intermetallics are ever to be developed.

In this paper the two brittle fracture modes commonly observed in $L1_2$ intermetallics are reviewed, first intergranular failure in pure (that is, B-free) Ni_3Al and second, cleavage failure in $L1_2$ Al_3Ti. It is shown that in the case of intergranular failure in Ni_3Al the propensity for cracking is not related to grain boundary structure, at least within the confines of a Σ-based model, and in the case of cleavage failure in $L1_2$ Al_3Ti there are many active cleavage planes, even though the flow stress of Al_3Ti is modest, as low as 300 MPa and there are 12 available slip systems. These results suggest that the tendency for brittleness in $L1_2$ materials is not closely related to grain boundary structure in the case of intergranular failure, or to the atomic structure or dislocation mechanisms in the case of cleavage failure, suggesting that the reason for the brittleness is associated with the ordered structure itself.

2. INTERGRANULAR CRACKING IN Ni_3Al

It has been known for a long time that Ni_3Al suffers from an extreme case of grain boundary brittleness, even though the grain boundaries appear to be very clean [3]. The report that the addition of small amounts of B could largely eliminate this problem [4] was received with great enthusiasm, and similar attempts were subsequently made on many other intermetallics, largely without great success. Various explanations for this embrittlement have been proposed, e.g., electronic effects [5,6,7] and dislocation mobility effects in the grain boundary region [8,9], but the fundamental reason is still not clear. If there is an atomic structural basis for the embrittlement, then it would be expected that certain kinds of grain boundaries would be more tough than others, since the atomic arrangement in a grain boundary is certainly expected to change with changes in the relative orientations of the two grains.

The first check of this idea came with the work of Hanada et al. [10], who measured the distribution of grain boundary types in Ni_3Al according to Σ value along an intergranular crack. Although, the statistical significance of the data is open to question, their results suggest that cracks in Ni_3Al do not tend to follow special kinds of boundaries and avoid others. Later, with the availability of improved experimental tools for measuring grain boundary geometry, Lin and Pope

[11] and Lin [12] were able to study a much larger number of boundaries than was previously possible and were then able to draw statistically meaningful conclusions about the intergranular crack path.

The fundamental difficulty of relating grain boundary character to propensity for cracking is the fact that to macroscopically characterize a grain boundary 5 independent parameters must be specified: 3 to orient the two grains relative to each other and 2 to specify the boundary plane. If the goal is to relate, say, grain boundary fracture toughness to structure, one could imagine performing such tests on a large number of bi-crystals of varying grain boundary character, but such an endeavor clearly involves too many experiments to be practical. The alternative that has been used in the past [10,13] is to characterize grain boundaries by Σ value only, thereby considering only the relative orientations of adjacent grains and ignoring the orientation of the boundary plane. Using this method, one characterizes the grain boundaries before performing the fracture test, the crack(s) is introduced, the distribution of cracked boundaries is measured and then compared to the distribution in the bulk of the sample. If the two distributions are statistically the same, then it is concluded that all boundaries crack with equal ease. If they are different, the boundaries absent from the crack are the strong/tough ones.

Our experiments involved several novel ingredients that allowed obtaining statistically significant results: 1. Melt-spun ribbons, approximately 15 μm thick and only one grain thick were used as samples. 2. The electron backscattering pattern technique was used to orient the individual grains. These two features of the experiment are important since because of #1, there are many grains along a crack which crosses a 2 mm wide ribbon and also, there is no complexity due to grains lying below a given grain. Because of #2 we could quickly orient the grains, using the semiautomated method of Dingley et al. [14]. The ribbons used in these experiments were 75 at. % Ni24.8 at. % Al and 0.2 at. % Ta. Details of the specimen preparation technique are given in Lin [12]. Thus the combination of having many grains across the sample width available for cracking and a technique capable of rapid analysis allows obtaining statistically meaningful data. A third novel feature of our experiments is the method of performing the fracture experiments. The ribbon was adhesively bonded to a 100μm thick Al foil and the composite sample was then bent around an edge of known radius of curvature, thereby introducing cracks into the Ni_3Al. The Al foil, being ductile, does not crack and therefore the crack faces of the Ni_3Al can be replaced to their initial relative positions by simply straightening the foil. The orientations of the grains near the crack and far from the crack are then measured and compared.

When assigning a particular Σ value to a grain boundary, the so-called Brandon criterion [15] is used. This criterion says that the structure of a coincidence grain boundary can be preserved if the relative orientation of the grains on either side deviate from the perfect coincidence orientation by a deviation, $\Delta\theta$, that is no larger than $\Delta\theta_{max}$, where

$$\sin\Delta\theta_{max} = \sin 15° \sqrt{N}$$

and N is the value of Σ for the boundary.

Two kinds of cracks were investigated, cracks which completely transversed the width of the ribbon (complete cracks) and cracks that did not traverse the width (incomplete cracks). Incomplete cracks are produced by bending once around a relatively large radius, and complete cracks are produced by bending several times around a smaller radius. The results observed on the two kinds of cracks do not differ significantly. Table 1 and Fig. 1 show the results for incomplete cracks.

Table 1 The distribution in number and proportion of cracked low angle, (Σ=1), Σ=3, low Σ, and high Σ boundaries compared with the 90% confidence intervals of the assumed true proportions taken from those of boundaries measured at random position in the specimen [11].

Grain Boundary Type by Σ value	General Population (A Total of 280 G.B.s)			Cracked (A Total of 155 G.B.s)		
	No.	Percent	90% interval for 155 G.B.s	No.	Percent	Within Interval?
low angle	21	7.5%	4.5-11.0%	3	1.9%	no
Σ3	79	28.2%	22.6-33.5%	3	1.9%	no
$(3 < \Sigma \leq 25)$	(12)	(4.3%)	(1.9-7.1%)	(8)	(5.8%)	yes
$3 < \Sigma \leq 49$	16	5.7%	3.2-8.4%	11	7.1%	yes
$\Sigma > 49$	164	58.6%	52.3-65.2%	138	89.0%	no
Total	280	100%		155	100%	

Considering Table 1 first, the percent of the boundaries in certain ranges of Σ are given, both cracked and in the general population. It is seen that there are many Σ=1 (low angle tilt boundaries) and Σ=3 (twins) in these samples. The boundaries were then characterized as "low sigma", $3<\Sigma<25$ or $3<\Sigma<9$, and "high sigma", $\Sigma>49$. These ranges were chosen arbitrarily. In Table 1 the 90% confidence intervals for the grain boundary fractions are also given. Note that the percentages for Σ=1 and Σ=3 disagree (percentages outside the 90% confidence interval) but for low Σ the percentages agree (the disagreement for $\Sigma>49$ is a result of the lack of cracked Σ=1 and Σ=3 boundaries along the crack). The Σ=3 boundaries were then further analyzed and 90% of them were found to be symmetric twins with either (111) or (112) interfaces. The few broken Σ=3 boundaries were found to be non-symmetrical.

Based on these results we conclude that Σ itself does not control grain boundary cracking in Ni_3Al since all boundaries except Σ=3 and Σ=1 appear to crack with equal facility, independent of whether they are high or low Σ. This agrees quite well with the implications of the structural unit model [16,17]. In this model the atomic structure of a given boundary is a combination of the atomic structure of two delimiting boundaries, the orientations of which lie on either side of the given boundary. Thus a boundary with a very high Σ value can be described in terms of the structures of two delimiting boundaries, which may have low Σ values. The fracture properties of this high Σ boundary is therefore expected to

Fig. 1   The distribution of grain boundary types by Σ values for cracked grain boundaries and boundaries in the general population. The data are taken from table 1.

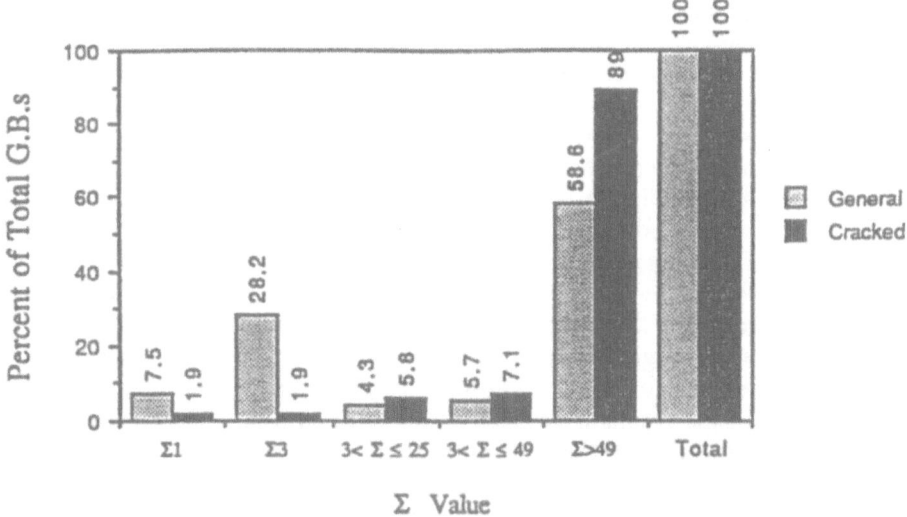

be not very different from those of the component delimiting boundaries. This is observed, except for $\Sigma=1$ and $\Sigma=3$. Why don't they crack? One explanation may lie in the simulations of Vitek et al. [18] in which they found that the atomic structure of a grain boundary in a strongly ordered alloy like Ni_3Al remains ordered right up to the boundary plane. To maintain this order, atomic positions in the boundary plane differ substantially from those in a less strongly ordered, or disordered, material resulting in cylindrical "cavities" in the boundary plane, regions where the atomic density is low. The presence of these cavities seems to be quite general, except in the case of $\Sigma=1$ and $\Sigma=3$, where no such cavities exists.

Thus, we conclude that the relatively high strength at $\Sigma=1$ and $\Sigma=3$ boundaries is not a result of being low Σ, per se, but is actually due to the absence of these "cavities" which are expected to result in reduced interplanar cohesion. Or, stated another way, the propensity for grain boundary failure in Ni_3Al is due to the requirements of the structure being strongly ordered up to the grain boundary plane. This, of course, raises the question of why all strongly ordered $L1_2$ intermetallics are not susceptible to grain boundary failure. Many are [5,6], but not all. Part of the answer lies in the relative ease of cleavage in some of these materials (next section) but the complete answer is not yet known.

3. CLEAVAGE FAILURE IN L1$_2$ Al$_3$Ti

The so-called trialuminides of the form Al$_3$X, where X= Hf, Nb, Ta, Ti, Zr, are of great technological interest because of their low density and possibility high oxidation resistance due to high aluminum contents. They also tend to have melting temperatures considerably higher than Al, e.g. Al$_3$Ti melts at 1350°C compared to 660°C for pure Al. Unfortunately they are all very brittle, but their brittleness was initially ascribed to their tetragonal (DO$_{22}$ or DO$_{23}$) structures. Since these tetragonal structures are expected to have a limited number of slip systems, e.g., see the reviews of Yamaguchi [19] and George et al. [20], this brittleness is to be expected. However some of these tetragonal structures can be converted to L1$_2$ by adding ternary alloying elements, e.g. Fe, Cr, and Mn, to Al$_3$Ti. When Al$_3$Ti is converted to the L1$_2$ form, a decrease in the flow stress is observed, independent of alloying addition, but it still remains brittle due to cleavage failure. Since the L1$_2$ structure has been commonly thought to be "intrinsically ductile", similar to the case of Ni$_3$Al single crystals, this brittleness has generated substantial concern in the metallurgical community. In general the resulting investigations have centered around asking "what is causing this intrinsically ductile material to be brittle?" Most of the results have been obtained to date on L1$_2$ Al$_3$Ti indicate that the material is not at all "intrinsically ductile". It is, in fact, "intrinsically brittle", but the reasons for the brittleness remain obscure. Since so many intermetallic compounds fail in a brittle fashion by cleavage, is it possible that there is a structural reason why all intermetallics should be "intrinsically brittle" and the few ductile compounds are the unusual ones, even for the high symmetry L1$_2$ structure? This is the question addressed in this section of this paper. It will be shown that Al$_3$Ti has the attributes of a normally ductile material - low flow strength, plenty of slip systems - except it is, in fact, brittle. This dichotomy brings into question our understanding of the fundamental differences between ductile and brittle materials, and how these differences relate to dislocation mechanisms. Attention will be focused on the behavior of L1$_2$ Al$_3$Ti+X, where X = Fe, Cr, Mn and many others. For a review of the phase stability of the trialuminides see Kumar [21]. Furthermore, only the properties at 300 K or lower will be considered.

There have been several investigations, using TEM, of dislocation structures in Al$_3$Ti, see [20] for a complete review, and all agree that <110>{111} slip occurs at room temperature. However, some have concluded that the <110> Burgers vector is undissociated [22,23,24]. Others observe dissociated dislocations, but disagree about the nature of the dissociation, e.g. George et al. [25] report dissociation into two 1/2<110> dislocations separated on {111} by APB of energy 270 mJ/m^2 while Inui et al. [26] report dissociation into two 1/3<112> superpartials separated on {111} by SISF of energy 100 mJ/Mm^2, While this disagreement over the scheme of dissociation is important and needs further attention, the most important fact for purposes of the present discussion is that all the slip systems which normally operate in fcc metals can also operate in L1$_2$ Al$_3$Ti, i.e., there is more than an adequate number of available slip systems to allow plastic deformation in a polycrystalline sample.

Confirming the TEM results, Wu et al. [27,28] performed slip trace analyses on single crystalline $L1_2$ Al_3Ti and found that {111} slip, presumably by <110> dislocations, is the predominant slip mode at room temperature. Some samples oriented near [$\bar{1}11$] in the standard unit triangle undergo cube slip, as do many other $L1_2$ materials such as Ni_3Al [29] and Pt_3Al [30,31].

The temperature dependence of the flow stress for poly- and single crystals is similar. The flow stress drops fairly sharply with increasing temperature at low temperatures, reaches a plateau or shows a small peak at intermediate temperatures, then drops again at high temperatures, see Fig. 2 for results on $L1_2$ Al_3Ti+Fe single crystals [28]. The absolute level of the CRSS for single crystals and the flow stress for polycrystals can vary widely depending on composition. For example, for $Al_{65}Fe_8Ti_{27}$ single crystals the CRSS is 150 MPa at the plateau, but for $Al_{64}Fe_8Ti_{29}$ it is 400 MPa, [27]. These differences in CRSS are thought to be the result of differing amounts of a fine dispersion of as-yet unidentified second phase particles in the $L1_2$ material [28]. Nonetheless, independent of strength level, all these materials remain brittle at room temperature.

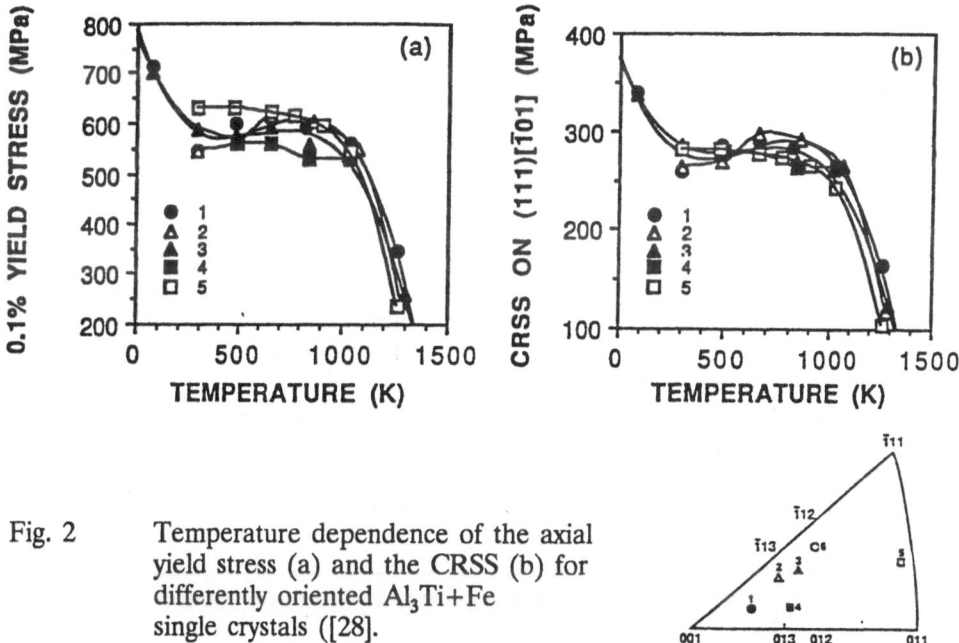

Fig. 2   Temperature dependence of the axial yield stress (a) and the CRSS (b) for differently oriented Al_3Ti+Fe single crystals ([28].

Most measurements of ductility have involved compressive tests, again see [20] for a complete list of previous work. In this section the salient features of these results are discussed. Early experiments on $L1_2$ Al_3Ti suggested the possibility of substantial ductility since compressive ductilities of 14% were observed at room temperature, but at the same time it was reported that final failure occurred by "pulverization" and that extensive internal microcracking is visible after only 2% compressive strain [22]. Subsequent experiments on a number of different $L1_2$ Al_3Ti+X alloys and Al_3Sc (which is $L1_2$ in binary form) have confirmed the generality of this microcracking. For reasons which are not current known, a profusion of stable microcracks develop during compression tests of these materials, and final failure occurs when these cracks link together, causing the sample to crumble into powder. Some plastic flow also occurs during compressive tests, but the relative amounts of deformation due to cracking and to dislocation motion are not known. In general, these cracks are definitely not stable in bending or tensile tests. All known $L1_2$ Al_3Ti+X alloys are totally brittle in bending or tension, with the exception of Al_3Ti+Cr and Al_3Ti+Mn, which have been studied by Zhang, Nic and Mikkola [32] and by Kumar et al. [33]. While there is disagreement about the Cr-containing material, the Mn-containing material shows a very small amount of tensile ductility at 300K in both bending tests and tensile tests. Cleavage failure in $L1_2$ Al_3Ti occurs on a variety of planes, (100), (111), (100) and (013) [28,34]. Al_3Sc, however, cleaves almost entirely on the (110) plane [25].

Since these materials do not have a particularly high strength, nor do they lack the necessary number of slip systems, and if we rule out impurity effects, only an intrinsically low cleavage strength or limited plasticity at crack tips can explain the observed low toughness. Based on the Rice-Thomson criterion, $Gb/\gamma < 10$ for a ductile material (G is the shear modulus, b the Burgers vector and γ the surface energy), Turner, et al. [22] concluded that $L1_2$ Al_3Ti is brittle because it is difficult to emit dislocations from crack tips. In coming to this conclusion they assumed, based on their own TEM results, that the <110> superdislocations are undissociated, thereby leading to a large value of Gb/γ, and therefore a large barrier to dislocation emission from the crack tip. If, on the other hand, it is assumed that the dislocations actually are dissociated, as discussed earlier in this paper, then the brittle trialiminides have a value of Gb/γ not very different from ductile $L1_2$ materials like Ni_3Al (at least in single crystalline form). There is evidence that rather little dislocation activity occurs near cracks since sharp electron channeling patterns are obtained from the cleavage fracture surfaces of both $L1_2$ Al_3Ti and Al_3Sc [25,34,35]. Also the fracture toughness values obtained to date are quite low, 3 MPa$\sqrt{m}$ for Al_3Ti [22] and 3 MPa$\sqrt{m}$ for Al_3Sc [36]. The shear modulus of Al_3Ti [22,25] and Al_3Sc [25,37] are approximately 80 and 70 GPa, respectively, compared to 26 GPa for pure Al. Consequently, Gb/γ for Al_3Ti and Al_3Sc is substantially higher than that for pure Al, but it is comparable to that of Ni_3Al (which does not cleave easily). As discussed by George, et al. [20], a meaningful comparison should involve comparing Gb/γ and the theoretical cleavage strength, see below.

However, it is possible that another mechanism for crack tip plasticity exists, which does not require macroscopic dislocation motion. This mechanism is based on the idea that if the local tensile stresses at the crack tip are sufficiently large to break atomic bonds, then the shear stresses, which are of the same order of magnitude at the tip as the tensile stress, must be large enough to shear the lattice in arbitrary ways on an atomic scale, without the need for long range dislocation motion [38]. These shears, which would be highly localized near the crack tips are expected to reverse once the crack has passed, leaving behind relatively perfect material which has, nonetheless, dissipated substantial plastic work. It is expected that strongly ordered materials, due to their chemical order, would be more resistant to this slip than are pure metals or disordered alloys, and would therefore not benefit as much from such a toughening mechanism. This mechanism has not been studied on the past, but appears to be a useful alternative to the Rice-Thomson criterion.

Finally, the possibility of an intrinsically low cleavage strength is considered. The ideal surface energy of (110) and (100) surfaces in Al_3Sc were calculated by Fu [39] to be 3.7 and 3.4 J/m^2, respectively, not particularly low values compared to, say, pure Al. Also, they calculated the ideal cleavage strength of Al_3Sc to be 19 GPa, substantially lower than the 30 GPa obtained for Ni_3Al, but probably not much lower than what would be obtained for pure Al. Thus, based on considerations of surface energy or the ideal cleavage strength of Al_3Sc, it is difficult to rationalize the low toughness of the trialuminides.

Thus, it appears that none of the normally-used criteria explain why the trialiminides cleave so easily. This suggests that some other mechanism operates which has not yet been carefully investigated, e.g. the crack tip shear mechanism of Jokl et al. [38].

4. SUMMARY

In this paper some possible reasons for the low toughness observed in $L1_2$ intermetallic compounds have been discussed. It was shown that brittle behavior is commonly observed in strongly ordered $L1_2$ intermetallics, either due to intergranular or cleavage failure. Given that the $L1_2$ is the ordered structure most likely to be ductile, these results suggest that there are reasons for brittleness in ordered structures which are intrinsically related to the state of long range order and which are not specifically related to the ease of dislocation generation at a crack tip. Local shears produced at the crack tip which reverse once the crack has passed by is one possible mechanism.

5. ACKNOWLEDGEMENTS

This work was supported by the AFOSR under Grant AFOSR-89-0062 and by the NSF under grant DMR88-22858. Experimental facilities were provided by the LRSM at the University of Pennsylvania supported by the NSF MRL Program under grant DMR88-19885. This paper was originally published in the Proceedings of International Symposium on Intermetallic Compounds - Structure and Mechanical Properties (JIMIS-6) edited by O. Izumi, Japan Institute of Metals (June 1991).

REFERENCES

1. R.A. Mulford: Treatise on Materials Science and Technology, C.L. Briant, and S.K. Banerji eds., 25(1983)1.
2. S.S. Hecker, D.L., Rohr, and D.F. Stein: Metall. Trans., A9(1978)481.
3. T. Takasugi, E.P. George, D.P. Pope and O. Izumi: Scripta Metall., 19(1985)551.
4. K. Aoki and O. Izumi: Trans. JIM 19(1978)203.
5. T. Takasugi, O. Izumi, N. Masahashi: Acta. Metall. 33(1985)1247.
6. T. Takasugi, O. Izumi, N. Masahashi: Acta. Metall. 33(1985)1259.
7. A.I. Taub, C.L. Briant: Acta Metall. 35(1987)1597.
8. E.M. Schulson, T.P. Weihs, D.V. Viens, I. Baker: Acta. Metall. 33(1985)1587.
9. P.S. Khadikar, K. Vedula, B.S. Shale: Met. Trans. 18A(1987)425.
10. S. Hanada, T. Ogura, S. Watanabe, O. Izumi and T. Masumoto: Acta. Metall., 34(1986)13.
11. H. Lin and D.P. Pope: in High Temperature Ordered Intermetallic Alloys IV (Proc. Mat. Soc. Symp.), edited by L. Johnson, D.P. Pope and J.O. Steigler (Materials Research Society, Pittsburgh, PA 1991), Vol. 213, in press.
12. H. Lin: Ph.D. Dissertation, (1991) University of Pennsylvania.
13. T. Watanabe: Met. Trans, 14A(1983)531.
14. D.J. Dingley, M. Longden, J. Weinbrin and J. Alderman: Scanning Elect. Microscopy, 2(1987)451.
15. D.J. Brandon: Acta. Metall., 11(1966)1479.
16. A. Sutton and V. Vitek: Phil. Trans. Roy. Soc., A309(1983)1.
17. M. Khantha, V. Vitek, and M. Goldman: Proc. Mater. Res. Soc. Sym. 193(1990)349.
18. V. Vitek, S.P., Chen, A.F. Voter, J.J., Kruisman, and J. Th. M. DeHosson: Mat. Sci. Forum, 11(1989)237.
19. M. Yamaguchi: Progr. Mat. Sci., 34(1990)1.
20. E.P. George, D.P. Pope, C.L. Fu and J.H. Schneibel: To appear in ISIJ Int., (1991).
21. K.S. Kumar: Int. Mater. Rev., 35(1990)293.
22. C.D. Turner, W.O. Powers, and J.A. Wert: Acta. Metall. 37(1989)2635.
23. V.K. Vausdevan, R. Wheeler, and H.L. Fraser: in High Temperature Ordered Intermetallic Alloys II (Proc. Mat. Res. Soc. Symp.), edited by C.T. Liu, A.I. Taub, N.S. Stoloff, and C.C. Koch (Materials Research Society, Pittsburgh, PA 1989), Vol. 133, Pg. 705.
24. W.O. Powers and J.A. Wert: Metall. Trans., 21A (1990)145.
25. E.P. George, J.A. Horton, W.D. Porter, and J.H. Schneibel, J. Mater. Res. 5(1990)1639.
26. H. Inui, D.E. Luzzi, W.D. Porter, D.P. Pope, V. Vitek and M. Yamaguchi: Philos. Mag. A (1991), in press.
27. Z.L. Wu, D.P. Pope, and V. Vitek: Scripta Metall. 24(1990b)2191.
28. Z.L. Wu, D.P. Pope, and V. Vitek: in High Temperature Ordered Intermetallic Alloys IV (Proc. Mat. Res. Soc. Symp.) edited by L. Johnson, D.P. and J.O. Stiegler (Materials Research Society, Pittsburgh, PA 1991), Vol. 213, in press.

29. C. Lall, S. Chin, and D.P. Pope: Metall. Trans. A., 10A(1979)1323.
30. D.M. Wee, D.P. Pope and V. Vitek: Acta. Metall. 32(1984)829.
31. F.E. Heredia, G. Tichy, D.P. Pope, and V. Vitek, Acta. Metall. 37(1989)2755.
32. S. Zhang, J.P. Nic, and D.E. Mikkola: Scripta Metall. 24(1990)57.
33. K.S. Kumar, S.A. Brown, and J.D. Whittenberger: in High Temperature Ordered Intermetallic Alloys IV (Proc. Mat. Res. Soc. Symp.)., edited by L. Johnson, D.P. Pope, and J.O. Steigler (Materials Research Society, Pittsburgh, PA 1991), Vol. 213, in press.
34. E.P. George, W.D. Porter, and D.C. Joy: in High Temperature Ordered Intermetallic Alloys III (Proc. Mat. Res. Soc. Symp.) edited by C.T. Liu, A.I. Taub, N.S. Stoloff, and C.C. Koch (Materials Research Society, Pittsburgh, PA 1989), Vol. 133, p.311.
35. E.P. George, W.D. Porter, H.M. Henson, W.C. Oliver, and B.F. Oliver: J. Mater. Res. 4(1989)78.
36. J.H. Schneibel and E.P. George: Scripta Metall., 24(1990)1069.
37. R.W. Hyland, Jr., and R.C. Stiffler: Scripta Metall. 25(1991)473.
38. M.L. Jokl, V. Vitek, C.J. McMahon, Jr., P. Burgers: Acta. Metall., 37(1989)87.
39. C.L. Fu: J. Mater. Res., 5(1990)971.

FUNDAMENTALS OF MECHANICAL BEHAVIOR IN INTERMETALLIC COMPOUNDS - A SYNTHESIS OF ATOMISTIC AND CONTINUUM MODELING

C. L. FU and M. H. YOO
Oak Ridge National Laboratory
P.O. Box 2008
Oak Ridge, TN 37831-6114

ABSTRACT. The fundamental aspects of deformation and fracture behavior of ordered intermetallics were investigated on the basis of quantum mechanical total-energy calculations and anisotropic elasticity theory for dislocations and cracks. These first-principles calculations were based on the local-density functional (LDF) theory. The LDF equations are solved either by full-potential linearized augmented plane-wave (FLAPW) method or by mixed-basis pseudopotential method. Our approach represents a major advance in applying LDF to solids, in which the LDF equations are solved without any shape approximation to the potential or charge density and a high degree of variational freedom (and precision) can be achieved. The calculated elastic constants, various shear fault energies, defect self-energies, and cleavage energies were used in conjunction with the continuum modeling of dislocations and cracks to predict the mechanical behavior and to understand the underlying electronic mechanism of observed mechanical properties.

1. Bonding Mechanism - Directionality and Anisotropy

1.1. Ni-ALUMINIDES

1.1.1. *Bonding Charge Denstiy of Ni3Al*. The bonding mechanism of Ni-aluminides can be understood by considering the "bonding charge density" of Ni3Al on the close-packed (111) plane [1]. The bonding charge density, defined as the charge density difference between Ni3Al and the superposition of neutral Ni and Al atomic charge densities at the lattice sites, is the response of the electron distribution (as referred to the atomic density) in the presence of crystal field. In fact, this is the charge density that can be measured experimentally by electron diffraction microscopy. The solid (dashed) lines in Fig.1 show the calculated contours of increased (decreased) electron density as atoms are brought together to form the Ni3Al crystal. We find the depletion of electron density at the Al site accompanied by a significant build-up of directional d-bonding charge along the nearest neighbor Ni-Al direction. It is this directional Ni-Al bond which gives rise to the bonding strength and the anisotropic character of the crystal properties. Thus, the bonding mechanism of aluminides can be best described by the combination of charge transfer and strong p(Al)-d(transition-metals) bonding hybridization effects.

Research sponsored by Div. of Materials Science, Office of BES, U.S. Department of Energy under contract DE-AC05-84OR21400 with Martin Marietta Energy Systems, Inc.

C. T. Liu et al. (eds.), Ordered Intermetallics – Physical Metallurgy and Mechanical Behaviour, 155–164.

Al   Ni   Al

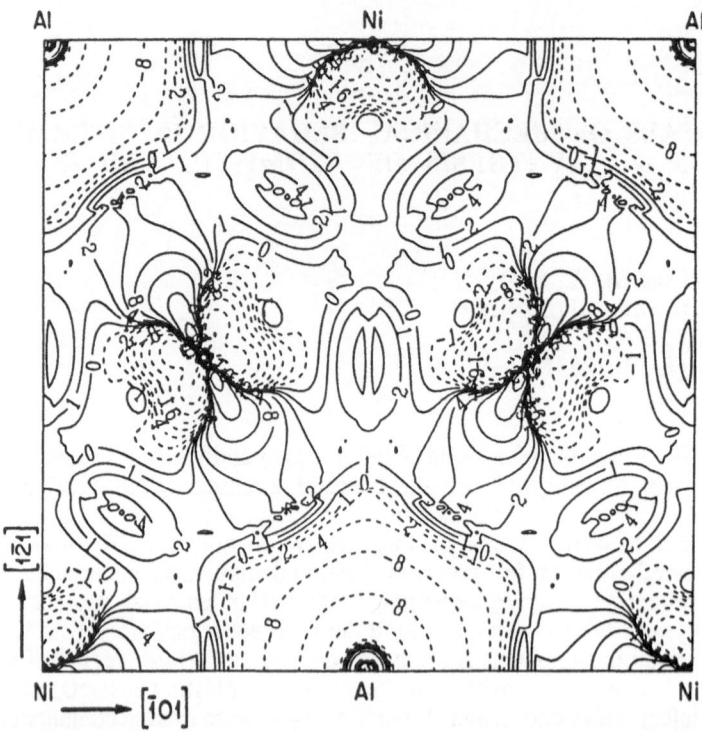

Ni ——→ $[\bar{1}01]$ Al   Ni

Fig. 1. The charge density difference between Ni₃Al and the superposition of neutral Ni and Al charge densities on the (111) plane in units of $10^{-3}e/(a.u.)^3$. The solid (dashed) lines denote contours of increased (decreased) density as atoms are brought together to form Ni₃Al crystal. This charge difference plot clearly shows the formation of directional d-bonding charge on the Ni site.

1.1.2. *Point Defect Structure of NiAl.* Since Al in aluminides acts as an electropositive element, there is a repulsion (or less bonding) of Al atoms as they are brought into close proximity. This explains the mechanism toward ordering (e.g. NiAl) in terms of the size difference and electronegativity difference between Ni and Al [2]. As a further evidence to support this ordering behavior, we have calculated the point defect structure of NiAl. By considering noninteracting defects with the inclusion of configurational entropy, the thermal concentration of point defects can be determined from the calculated defect self-energies and the minimization of the grand potential. The calculated defect concentration of stoichiometric NiAl as a function of temperature is shown in Fig.2. We find that the defect structure is dominated by two types of defects - constitutional vacancy on the Ni site and substitutional antisite defect on the Al site. The defect concentration at stoichiometry is predicted to be 10^{-4} above 1000K. Furthermore, for the cases of slightly off-stoichiometric NiAl, the stable defect configuration is found to be constitutional vacancy on the Ni sublattice in Al-rich NiAl (i.e. in order to minimize Al-Al repulsion) and substitutional antisite defect on the Al site in Ni-rich NiAl (i.e. in favor of metallic bonding among Ni atoms).

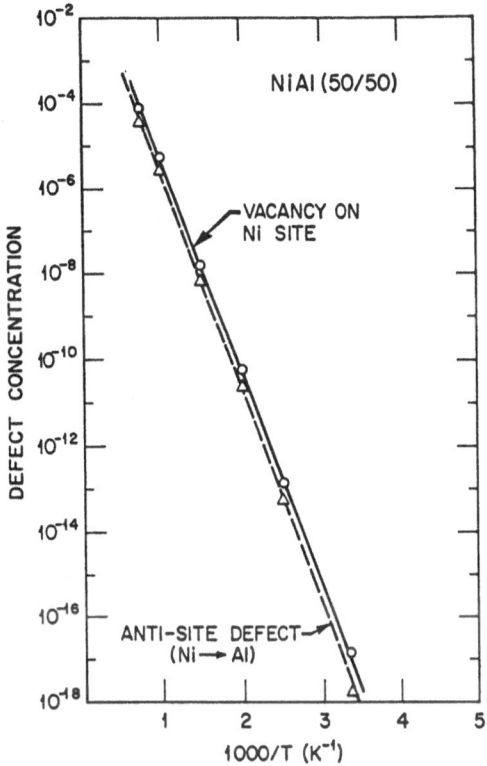

Fig. 2. The temperature dependence of point defect concentration of NiAl at stoichiometry.

1.2 Ti-ALUMINIDES

Although the charge transfer effect is less remarkable for the case of Ti-aluminides, Al atoms still act as electron donor in these systems (e.g., TiAl and TiAl$_3$) [3,4]. The bonding directionality is mainly caused by the polarization of p-electrons at the Al sites as a result of p-d hybridization effect (particularly in the case of trialuminides). In Fig. 3 we present the bonding charge density of TiAl on the (001) and (100) planes. Fig. 3(a) clearly shows the formation of directional d-bond between nearest-neighbor Ti atoms, as expected from the open d-shell of Ti atoms. The most remarkable feature in the bonding charge density, however, is the polarization of p-electrons at the Al sites pointing directly along the [001] direction (c.f. Fig. 3(b)). As a result of this polarization effect, there is a strong cohesion between Ti and Al layers. It is this polarization effect which gives rise to the large bond bending force between Ti and Al layers (i.e., C$_{44}$ value ; see Section 2.1). The relatively large shear modulus of trialuminides (compared to their bulk modulus) is also attributed to the non-spherical charge distribution at the Al sites (i.e., large bond bending force). In other words, the role of transition metals in trialuminides is to induce a p-electron polarization at the Al sites. On the other hand, there is no considerable increase in the bulk modulus (which measure the averaged tensile strength) and surface energies from the corresponding values of pure Al, due to the majority content of Al atoms in trialuminides. The brittleness of trialuminides can be qualitatively understood from these electronic structure factors.

158

pure Ti atomic plane   **mixed Ti-Al plane**

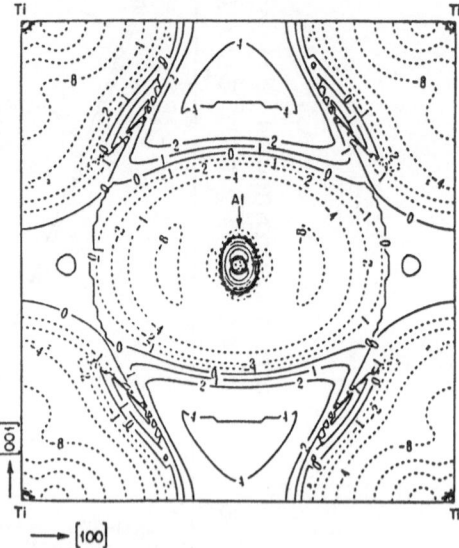

Fig. 3. The bonding charge density of TiAl on the (001) and (100) planes. See Fig. 1 for notations

2. Mechanical Properties

2.1 ELASTIC CONSTANTS

The elastic constants provide valuable information about the bonding strength and its directionality between adjacent atomic planes and, therefore, on the critical shear stress for slip as well as the long-range elastic interaction between dislocations. The calculated values of elastic constants for six transition-metal aluminides of cubic (L1$_2$ and B2) structures are listed in Table 1. It is noticeable that Ni$_3$Al has a relatively high shear anisotropy factor, A=2C$_{44}$/(C$_{11}$-C$_{12}$), as compared to Pt$_3$Al. The anomalous (positive) temperature dependence of yield strength in Ni$_3$Al and its absence in Pt$_3$Al can be explained in terms of the driving force, i.e. the interaction torque which depends on A, acting on screw dislocations to promote the cross-slip pinning process [6]. Elastically, Al$_3$Sc is close to being isotropic, and has a remarkably low Poisson's ratio. This implies that the disparity (edge/screw) of dislocation mobility is small and the resulting shape of dislocation loops is nearly circular. Thus, a low critical stress for dislocation multiplication is suggested (and hence, the soft behavior of Al$_3$Sc) [7].

Table 1. Calculated elastic constants, bulk and shear moduli, Zener's shear anisotropy factor, and Poisson's ratio at 0K

Alloy	$(10^{11}$ N/m$^2)$					A	v
	C_{11}	C_{12}	C_{44}	B	G		
Ni$_3$Al	2.35	1.45	1.32	1.75	0.86	2.93	0.38
Pt$_3$Al	4.36	2.20	1.40	2.92	1.26	1.30	0.34
Al$_3$Ti	1.77	0.77	0.85	1.10	0.69	1.70	0.30
Al$_3$Sc	1.89	0.43	0.66	0.92	0.69	0.90	0.19
NiAl	2.33	1.73	1.15	1.93	0.67	3.83	0.43
FeAl	2.90	1.30	1.65	1.83	1.23	2.06	0.31

For TiAl and TiAl$_3$ of the tetragonal L1$_0$ and DO$_{22}$ structures, there are six independent elastic constants. The calculated elastic constants and the shear anisotropy factors are listed in Table 2. (The shear anisotropy factor is given by $A_1=2C_{66}/(C_{11}-C_{12})$ for the [110] shear direction and by $A_2=4C_{44}/(C_{11}+C_{33}-2C_{13})$ for the [011] direction.) In TiAl, a large A_2 value is indicative of the strong cohesion between pure Ti and pure Al layers. It is this directional Ti-Al bonding which hinders the mobility of superdislocations [3]. For trialuminides, the shear moduli are more than two times that of Al, the primary constituent element. This suggests that the d-electrons of Ti (and Sc) plays a dominant role in the mechanical properties of these trialuminides [4]. Again, TiAl$_3$ is found to have a unusually low Poisson's ratio (v= 0.17), which appears to be the general trend of trialuminides.

Table 2. Calculated Elastic Constants Bulk and Shear Moduli (in units of 10^{11} N/m^2) of TiAl and TiAl$_3$

Alloy	C_{11}	C_{12}	C_{13}	C_{33}	C_{44}	C_{66}	B	G
TiAl	1.90	1.05	0.90	1.85	1.20	0.50	1.25	0.70
TiAl$_3$	2.02	0.88	0.60	2.43	1.00*	1.45	1.18	0.99

*Ref. 5

2.2 APB ENERGIES OF B2 ALLOYS

Using a supercell approach by introducing periodic shear faults, we calculated antiphase boundary (APB), superlattice intrinsic- and extrinsic-stacking fault (SISF and SESF), and twin boundary energies in various transition metal aluminides [2-4,8]. For FeAl, the APB energy is found to be highly anisotropic with respect to different slip planes . For NiAl, we find that the dissociation of <111> superdislocations into partial dislocations is unlikely because of high APB energy and a weak repulsive elastic interaction between the partial dislocations. This result explains why NiAl is observed to deform by <100> slip, which is different from other B2 alloys (e.g. <111> slip in FeAl). Physically, the <100> slip of NiAl is related to the ordering behavior discussed above [2]. The role of these two different types of slip systems in giving rise to crack-tip plasticity is discussed in the following section.

2.3 CLEAVAGE FRACTURE

We can determine the cleavage energy by a supercell geometry. Briefly stated, we calculated the interfacial energy as a function of the separation between adjacent atomic layers displaced from their equilibrium interlayer spacing [4]. The interfacial energy converges to the cleavage energy as the separation between adjacent layers increases. As an example, we present in Fig. 4 the dependence of interfacial energy on the deviation from equilibrium interlayer spacing for the (100) and (110) crystallographic planes for NiAl and FeAl [9]. Our calculation shows that : (1) FeAl has higher cleavage energy and strength than NiAl, as would be expected from the stronger directional d-bonding in FeAl; and (2) the interatomic interaction in NiAl is very much dominated by the short-ranged nearest neighbor interaction, as reflected in the fast convergence of interfacial energy to the cleavage energy for the (110) crystallographic plane. In fact, it is this short-ranged interaction (and/or charge transfer effect) which makes NiAl brittle.

In Table 3 we list the calculated values of Griffith's cleavage energy (G_C), which is the total surface energy of the two cleavage surface planes. By using the calculated elastic constants (Table 1), the critical stress-intensity factor for Mode-I crack (K_{IC}) is obtained for an orthotropic symmetry [10]. The calculated K_{IC} values are also listed in Table 3. As compared to Ni_3Al, Al_3Sc has a relatively low cleavage energy and K_{IC} value, suggesting that the brittle fracture of trialuminides is mainly caused by their intrinsically low cleavage strength (and relatively high shear modulus mentioned above). While a large difference between the (110) and (100) cleavage energies is found for NiAl (about 30%), this difference becomes relatively small for FeAl (about 10%). The low (110) cleavage energy (and low K_{IC} value) of NiAl suggests that the cleavage fracture occurs on that plane, which is consistent with the experimental observation [11]. On the other hand, the small difference between the (110) and (100) cleavage energies of FeAl makes it difficult to identify the cleavage plane solely in terms of cleavage energy. Fig.5 shows the crack-tip shear stress fields for the two cases of Mode-I crack in FeAl. At low temperatures, the driving force for crack-tip plasticity by the primary $(11\bar{2})[111]$ slip (see $\perp$ in Fig. 5) and/or the complementary $(11\bar{2})[11\bar{1}]$ slip (see Π in Fig. 5) in FeAl is much higher for a (110) crack than for a (100) crack, hence the prediction of (100) cleavage plane which has been observed experimentally [11]. (Note that the effect of elastic anisotropy is to reduce the resolved shear stress field by $(11\bar{2})[111]$ slip for the (100) crack.)

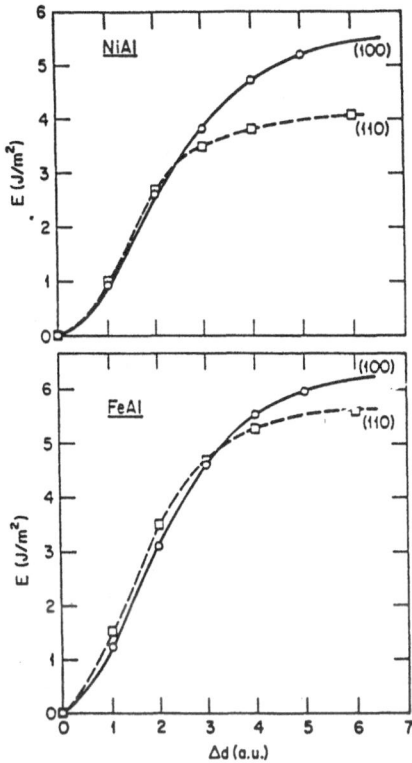

Fig. 4. Cleavage energies of NiAl and FeAl as a function of Δd (the deviation from an equilibrium interlayer spacing of the (100) and (110) crystallographic planes (1 a.u. = 0.529 Å).

Table 3. Calculated cleavage (ideal) energies (in units of J/m^2) and the critical stress-intensity factor for Mode-I crack in the <110> direction (in units of MPa-$m^{1/2}$)

Alloy	G_c			K_{IC}	
	(100)	(110)	(111)	(100)	(110)
Ni$_3$Al	5.8	-	4.6	6.6	-
Al$_3$Sc	3.4	3.7	3.3	4.3	4.5
NiAl	5.5	4.1	-	6.0	4.8
FeAl	6.5	5.8	-	8.0	7.4

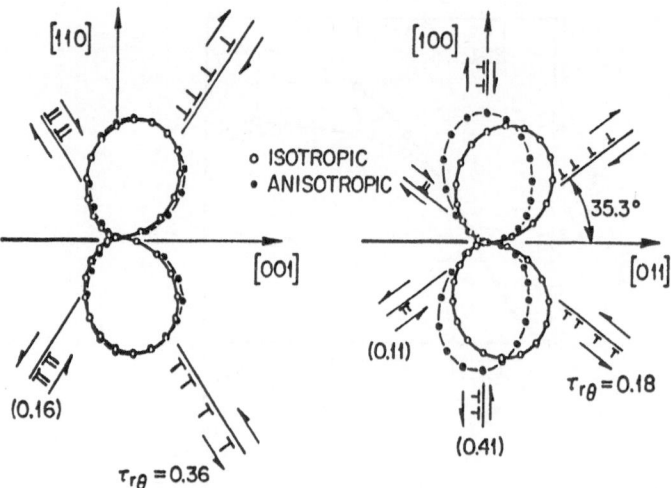

Fig. 5. The resolved shear stress, $\tau_{r\theta}$, for the $(11\bar{2})[111]$ edge dislocations in twinning direction, and those $(\tau_{r\theta})$ for $(11\bar{2})[\bar{1}\bar{1}\bar{1}]$ in the antitwinning direction, at (110) and (100) crack tips in FeAl (in units of $K/(2\pi r)^{1/2}$). The $(11\bar{2})$ slip plane is inclined by $35.3°$ from the <011> direction.

For TiAl, since the cleavage energies are not particularly low {i.e., 4.5, 4.6, 5.3, and 5.6 J/m^2 for the (111), (100), (110), and (001) planes, respectively}, it appears that the brittle fracture of TiAl at low temperatures is not caused by the intrinsic cleavage energy alone. As mentioned already, the lack of plastic work to shield the crack tip is other reason for the low ductility of this alloy. In TiAl and $TiAl_3$, because of low SISF and twin boundary energies, the twin-slip conjugate relationship makes an important contribution to the strain compatility for blunting of a Mode-I type crack [12]. Recently, clear experimental evidence has been reported that deformation twinning in TiAl can lead to an appreciable increase in the fracture resistance [13].

3. Extrinsic Effects

The electronic mechanism underlying the hydrogen-induced embrittlement in FeAl [14] and boron-induced strengthening in Ni_3Al (both effects were observed experimentally in ORNL) were investigated from first-principles calculations Our results show that FeAl is intrinsically resistant to cleavage fracture. The effect of absorbed hydrogen is to reduce the Fe-Al cleavage strength. We find that the bonding in FeAl has strong directional d-bonding character. This bonding charge is, however, considerably weakened in the presence of absorbed hydrogen, due to the charge transfer from Fe to H (i.e., a significant ionic component in the Fe-H bond). In contrast, boron is found to cause an enhancement of the cohesive strength locally about the boron site through the formation of covalent bond between Ni and B (Fig. 6). Furthermore, it is found that, energetically, boron prefers to be absorbed in a defect site with a nearest-neighbor nickel coordination number of four to five (instead of six in the bulk). This indicates that boron tends to segregate preferentially to Ni-rich defect sites and to enhance atomic cohesion there through the formation of localized Ni-B covalent bonds.

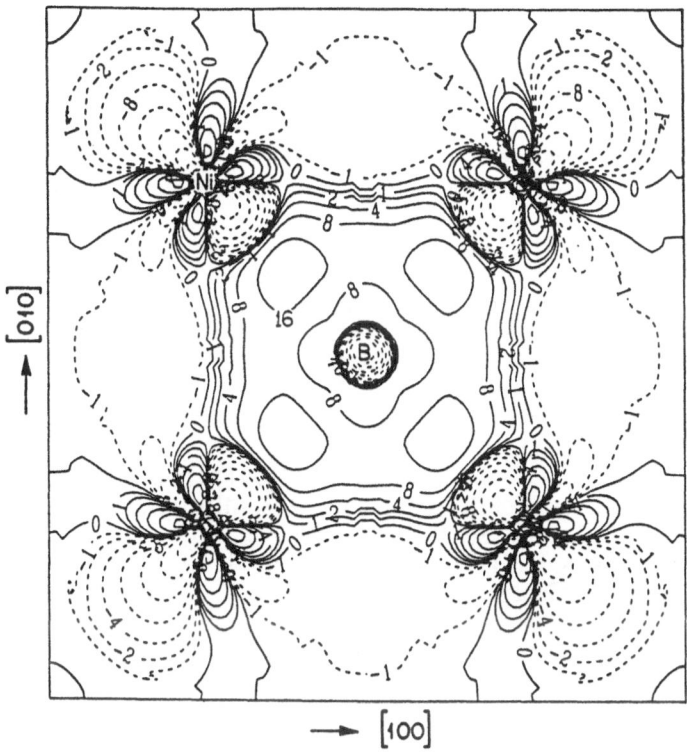

Fig. 6. As in Fig. 1, the plot shows the charge density difference between Ni$_3$Al-B and the superposition of Ni$_3$Al and free-B densities on the (001) plane. It shows the increase of interstitial charge along the Ni-B direction (i.e. covalent bonding).

4. Summary

In summary, we have demonstrated that knowledge of the electronic structure and properties is important for understanding the intrinsic and extrinsic mechanical properties of alloys, and provides a valuable basis for future alloy design strategies. In transition-metal aluminides, the d-band filling through the Al-to-transition metal charge transfer (and p-d hybridization) is the basic bonding mechanism. The degree of directional d-bonding determines the ideal cleavage (and bond) strength.

Nonlinear inelastic aspects of the deformation and fracture behavior can be best treated by atomistic simulation studies dislocation cores and crack propagation based on some form of interatomic potentials including the many-body terms. In transition-metal aluminides, for instance, both the p-d hybridization and charge transfer effects should be properly taken into account in constructing the interatomic potentials.

164

5. References

1. Yoo, M. H. and Fu, C. L. (1991) ISIJ Intl. 31, 1048.
2. Fu, C. L. and Yoo, M. H. (in press) Acta Metall.
3. Fu, C. L. and Yoo, M. H. (1990) Phil. Mag. Lett. 62, 159.
4. Fu, C. L. (1990) J. Mater. Res. 5, 971.
5. Note that an algebraic error in Ref. 4 (involving an $(c/a)^2$ factor) has been corrected here.
6. Fu, C. L. and Yoo, M. H. (1988) Phil. Mag. Lett. 58, 199.
7. Yoo, M. H. (1991) Proc. of JIMIS-6 on Intermetallic Compounds - Structure and Mechanical Properties, JIM, Sendai, Japan, 11.
8. Fu, C. L. and Yoo, M. H. (1989) High-Temperature Ordered Intermetallic Alloys III, MRS Symp. Proc. 133, MRS Pittsburgh, 81.
9. Yoo, M. H. and Fu, C. L. (1991) Scripta Metall. 25, 2345.
10. G. C. Shih and H. Liebowitz (1986) Fracture - An Advanced Treatise, Vol. II, ed. H. Liebowitz, Academic Press, New York, 67.
11. Chang, K. M., Darolia, R. and Lipsitt, H. A. (1991) High-Temperature Ordered Intermetallic Alloys IV, MRS Symp. Proc. 213, MRs Pittsburgh, 529.
12. Yoo, M. H., Fu, C. L. and Lee, J. K. Ibid., 545.
13. Déve, H. and Evans, A. G. (1991) Acta Metall. 39, 1171.
14. Liu, C. T., Fu, C. L., George, E. P. and Painter, G. S. (1991) ISIJ Intl. 31, 1191.

PROPERTIES OF SURFACE DEFECTS IN INTERMETALLICS

P. VEYSSIÈRE
*Laboratoire d'Etude des Microstructures**
CNRS-ONERA
BP 72
92322 Châtillon Cedex
France

ABSTRACT. Surface defects determine the extension of dislocation cores. Information on APBs and stacking faults is now available in a large variety of structures and compositions of intermetallics alloys. It is increasingly evident that real APBs are liable to deviate from geometrical ones.

1. Introduction

More than thirty years of investigation have made clear that, in a large variety of cases, the plasticity of single phase intermetallic alloys is governed by symmetries of dislocation cores (Vitek 1984, Paidar 1984, Veyssière 1988, Duesbery 1990).

Core-controlled plasticity is manifested in a number of specific manners amongst which the most well documented are certainly the various sensitivities of the flow stress to temperature found in the $L1_2$ structure. A better understanding and, possibly, significant improvements of mechanical properties of intermetallics imply that the principal parameters that determine the spreading of a given dislocation in a given crystal structure are studied comprehensively, keeping in mind that whereas the preference for a particular core configuration in a crystal depends upon its energy, the largest mobility is not necessarily associated with the lowest possible energy. Information on the fine structure of dislocations in relation with mobility and/or stability in intermetallics has been gathered at a rate that increased suddenly in the mid-eighties. It then became obvious that, amongst other features, surface defects had to be investigated with the largest possible accuracy and this is the field covered in the present review.

In order to establish links between local information on dislocations and the collective response of these to a stress, comparisons should be conducted between selected alloys within a given ordered structure as well as between a selected variety of ordered structures. In addition, the broadening of defect analysis to a large number of alloys is helpful in order to discriminate between rate-controlling processes and what, in the microstructure, should be regarded as deformation debris. Nevertheless, since the $L1_2$ structure is by far the most documented one, it is unavoidable that most of the body of the present paper concentrates on this family of alloys.

Although this may look a little discouraging to newcomers and somewhat tedious to others, a

---

* *Unité Mixte CNRS-ONERA, UMR104*

C. T. Liu et al. (eds.), Ordered Intermetallics – Physical Metallurgy and Mechanical Behaviour, 165–175.
© 1992 *Kluwer Academic Publishers.*

review on TEM observation of surface defects and dislocation fine structure should include a few words on experimental difficulties currently encountered in the course of these analyses. The principal problems will then be recalled, once more, in the following (§ 3).

One noteworthy result of the recent years of investigation is the increasing proportion of contributions where antiphase boundaries (APBs) are concluded to differ quite significantly from geometrical ones : there is experimental evidence that atoms tend to relax elastically as well as chemically at APBs and that their energy may thus be dramatically influenced by temperature and composition. This property will be considered in a separate section.

2.   Planar defects in intermetallics : definitions

Ordered intermetallics are liable to form two different categories of surface defects depending upon whether lattice site positions are conserved or not. These defects are commonly referred to as antiphase boundaries and stacking faults (SFs), respectively. The former only will disappear when the lattice disorders whereas the latter will not, despite the fact that it may have involved order violations and be thus partly affected upon disordering. Because of generally different surface energies of SFs, a distinction is usually made between those which preserve first-nearest neighbours and those which do not (termed SSFs and CSFs, respectively).

A surface defect is completely defined once its habit plane (hkl) and *a* displacement vector (R_i) are given. All the displacement vectors that differ from one another by a unit translation (T) of the lattice $(R_j - R_i = T)$, give rise to structurally equivalent planar defects (fig. 1). They cannot be discriminated under diffraction contrast since R_j and R_i introduce the same phase shift into the electron beam (g.T is an integer). For instance, in the $L1_2$ structure 3 consecutive shears by $1/6[11\bar{2}]$ in the (111) plane will result in an APB, since then $1/6[11\bar{2}] \rightarrow 1/2[1\bar{1}0] + [01\bar{1}]$ (in this structure, a single and a double $1/6[11\bar{2}]$ translation yield a CSF and a SSF, respectively).

A planar defect may result either from a conservative or from a non-conservative process. When at least one of the equivalent R_is belongs to the defect surface, its non-conservative nature cannot be determined in general, unless it is bordered by a dislocation. In this case of course, the actual displacement vector coincides with the Burgers vector of the partial dislocation $(R_i^* = b)$.

A non-conservative defect results from the addition or from the elimination of a slice of crystal whose thickness is given by $t = n.R_i^*$ where n is the unit normal along the direction [hkl]. The composition of the extra slice (i) depends upon crystal structure and composition through the stacking sequence in the direction perpendicular to the defect, (ii) it may not coincide with the composition of the crystal, when this occurs the defect is termed non-stoichiometric (NS).

Non-conservative APBs are encountered in intermetallic alloys essentially under the form of thermal APBs, that is as APBs formed upon cooling from the disordered phase or from the melt (see the contributions of D.G. Morris and of R.W. Cahn to this confer-

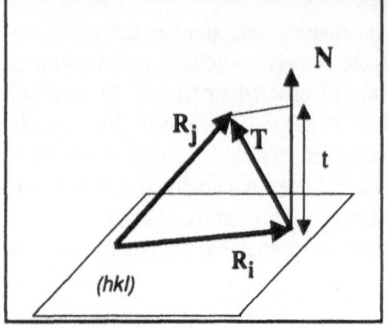
(hkl)

Figure 1. The equivalence between a conservative surface defect (displacement vector R_i in the (hkl) plane) and a non-conservative one (R_j out of the (hkl) plane). In the latter case, the composition of the extra-slice of thickness t determines the stoichiometric/non stoichiometric nature of the defect (Douin *et al.* 1988).

ence). Non-conservative APBs may also result from climb dissociation (Veyssière 1984, Saka and Zhu 1985) of superdislocations or from APB dragging (Brown 1959, Veyssière *et al*. 1989).

In the $L1_2$ structure, the configuration of an APB can be defined entirely by the notation $\{hkl\}^p$ where $p = 2\ N.R = (h - k)$ and $N = [hkl]$: the APB lies in a $\{hkl\}$ plane and the displacement vector is fixed, by convention, to $R = 1/2[1\bar{1}0]$. Stoichiometric and non-stoichiometric APBs correspond with even and odd values of p, respectively (Douin *et al*. 1988). Such a structural diversity is well exemplified in the $L1_2$ structure (Scheunemann-Frerker and Feller-Kniepmeir 1990, Feller-Kniepmeir and Scheunemann-Frerker 1990 and Douin *et al*. 1988). That climb dissociation should occur the most favourably to form non-stoichiometric APBs (Paidar 1985) is not clearly supported experimentally. Reasons for this are discussed by Beauchamp *et al*. (1987) and Douin *et al*. (1988).

The $\{hkl\}^p$ notation applies to tetragonal structures, such as $L1_0$ (Hug 1988) and DO_{22} (François 1991) provided N refers, in the diffraction pattern, to the shortest allowed relection in the [hkl] direction.

In B2 alloys, the above consideration implies that (i) a non-conservative APB with R of the 1/2<111> type in a $\{hkl\}$ plane such that [h + k + l] is even, can be constructed formally by a pure shear in the $\{hkl\}$ plane under consideration, and that (ii) all non-conservative APBs, again with a displacement vector of the 1/21<111> type, are stoichiometric in nature (Dirras 1990).

3. Reliability of observations in transmission electron microscopy

The study of planar defects and the determination of their energies by analysis of dislocation dissociation (see § 4) is nevertheless subjected to a number of difficulties that originate from the method employed to this aim. In principle, atomistic simulations of defect structure are not necessarily more reliable than electron microscopy studies and one expects the latter to describe the most closely defects in real crystals. Both simulators and electron microscopists have to evaluate how close the representation they end up with approximates actual defects.

Serious problems may arise in the course of TEM observations of deformation microstructures; they originate either from the interpretation of contrast or from the analysis of physical properties of dislocations. Experimental artefacts do occur in such studies and several of these are already at the origin of misinterpretations and of significant confusion in the debate about microstructures in intermetallics.

Contrast artefacts constitute one category of problems which, in general, can be elucidated provided the appropriate observations and image simulations are conducted. A frequent artefact is the presence of extra peaks in dislocation images. Other difficulties may arise from the fact that not all the corrections that have to be done in order to determine the *true* separation between companion partials, are established (Baluc 1990); elastic anisotropy is known to effect the distance between image peaks but extensive simulations remain to be worked out.

On the other hand, it is generally impossible to ascertain how faithful observed microstructures are in representing deformation processes typical of those that occur in the bulk : to what extent microstructures have relaxed prior to observation will remain largely obscure in most studies. It is however determined that the slower the cooling after a deformation test, the more extensive the relaxation of microstructures, especially at the level of dislocation cores (Tounsi *et al*. 1990). In addition, more than any other step of a TEM analysis, selecting a representative configuration is not trivial (Veyssière 1991a). Discriminating, within a deformation microstructure, between deformation debris and features that were actually contributing to deformation is also a difficult exercise; furthermore, amongst the latter features, there is large uncertainty in deciding which particular ones are rate controlling (Veyssière 1991b). In this respect, *in situ* straining experiments are not safer than conventional studies conducted on *post mortem* samples (Veyssière 1990a and b, 1991a and b, Dimiduk 1989, 1991).

4. Energies of surface defects and their physical implications in intermetallics

Amongst all possible surface defects, information on APBs dominates in the literature, then most of the following review will cover these defects, unless explicit mention to SFs is made.

With regard to dislocations and plasticity, what actually counts is less the APB energy, γ, than how much this defect contributes to the properties of the superdislocation to which it belongs. In other words APB energy must be compared to other relevant physical properties of the crystal under consideration.

APB energy determines the dissociation distance, d, from a balance between γ and the repulsion elastic forces between partials in the absence of lattice friction. Separation between partials is proportional to the dimensionless quantity $(\mu b^2 / \gamma)$, where μ is the shear modulus; unfortunately γ is not always directly accessible since elastic constants are not yet available in every intermetallic of interest. It is also useful to keep in mind that in a given ordered alloy, irrespective of APB energy, the quantity (γ d) is constant and equal to ($\alpha \mu b^2$), where α depends upon dislocation character.

On the other hand, comparing the spring force exerted by the APB on the bordering superpartials to the mechanical strength of the alloy can be done through the quantity (γ / b) which informs on the minimum stress, τ_{min}, required to unpair companion partials. In the same vein, it is sometimes interesting to consider how much the equilibrium separation between partials can be varied under the effect of a shear stress τ. Assuming for convenience that both partials have collinear Burgers vectors and that the trailing partial is temporarily immobilized at some obstacle, the increase of the separation between partials under an applied stress is given by : $(\Delta d / d_e) = 1 / (\gamma / \tau b - 1)$. This relation is plotted fig. 2, from which it appears that the variation of the separation distance Δd remains smaller than the experimental uncertainty involved in its measurement as long as τb is less than 20% of γ. Difficulties are to be met in ordered materials since :
(i) different from SFs, APBs are not necessarily confined to a single plane : APB energy is more or less the same on crystallographically distinct planes in a number of intermetallics (fig. 4),
(ii) cross-slip of a dissociated superdislocation occurs in a much more straightforward manner than in fcc metals : an APB ribbon bordered by two superpartials may straddle two planes cozonal with the screw direction, with no stair-rod dislocation needed at the intersection between the primary and the cross-slip plane,
(iii) intermetallics are often elastically anisotropic.
This implies essentially that it is necessary but not suf-

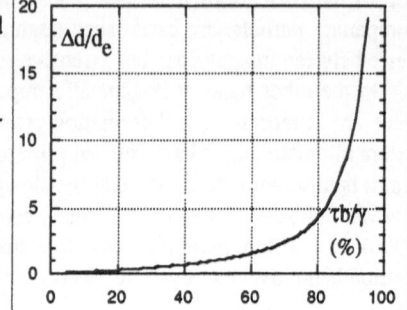

Figure 2. Increase in the dissociation distance Δd between two like companion superpartials separated by a surface defect (energy γ), as a function of the shear stress τ. (d_e is the dissociation distance at equilibrium under no external stress).

ficient to cancel the forces applied to superpartials in the dissociation plane in order to ensure stable equilibrium of the configuration. For the sake of simplicity, illustration of this property may be restricted to screw superpartials. Because of elastic anisotropy, the interaction force between two partials may comrise a tangential component, F_θ, together with the radial component, F_r, and, in a plane inclined at an angle β from the dissociation plane, each superpartial is submitted to an elastic force $F_{hkl} = F_\theta \sin \beta + F_r \cos \beta$. Hence, if F_{hkl} is larger than the spring force exerted by the APB in this plane, γ_{hkl}, and if the mobility of the partial in this plane allows for cross-slip, the configuration will spread on a second crystallographic plane. An equilibrium configuration is reached when forces applied to both partials satisfy simultaneously the following two conditions :

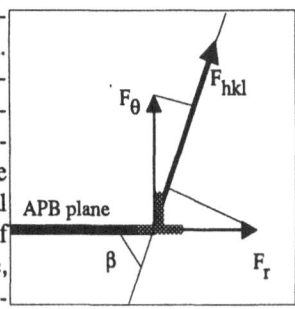

Figure 3. Forces applied to a superpartial under elastic anisotropy.

- $F_{APB} = \gamma_{APB}$ in the plane of the APB which is generated by the partial under study,
- $F_{hkl} \leq \gamma_{hkl}$, in every other plane cozonal with the dislocation line.

Then one expects to observe a variety of dissociated configurations (fig. 4). There is no available example of a complete analytical treatment of this problem in any structure; it has been however partially analyzed in the $L1_2$ structure, firstly by Yoo (1986, 1987), then by Molenat (1991) and by Hirsch (1991).

A similar problem is encountered in the case of a climb dissociated dislocation for which some calculations in the elastically anisotropic case have been carried out (Saka and Zhu 1985, Douin et al. 1988, Feller-Kniepmeir and Scheunemann-Frerker 1990). The implications of these properties are obviously important since mobility can be significantly reduced by the propensity of a dislocation to dissociate out of its primary slip plane. Measurements of APB energy is therefore instrumental to the study of mechanical properties of intermetallics and data is now available for a number of systems some of which are listed in table 1. Experimental uncertainties on such measurements lie between 15% and 20%; they may arise from a number of sources such as calibration of the microscope, conditions of observation (for instance, magnification depends upon the height of the sample in the objective pole-piece, which is subjected to variations during tilting experiments), correction of image separation (Baluc 1990). Furthermore, that the configuration is under equilibrium cannot be assessed in general since lattice friction on superpartial is unknown; moreover, it is never ensured that the dislocation under study has reached either one of the simple configurations of figs. 4(a) or 4(b), since a configuration such as in figs. 4(c) or 4(d) cannot be easily identified in the microscope (Bontemps and Veyssière 1991).

Table 1 demonstrates the role of composition in determining APB energies in $L1_2$ alloys. About all the listed APB energies lie within the 100-250mJ.m^{-2} range and the addition of a few percents of a ternary element in Ni_3Al-based alloys, can effect APB energy dramatically. CSF en-

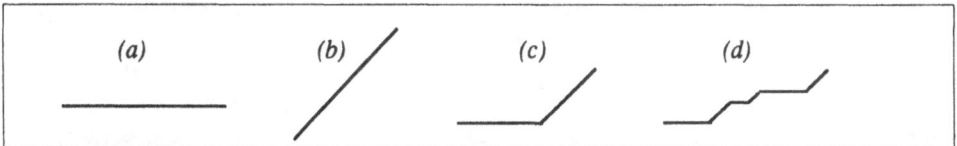

Figure 4. Depending upon the ratio of APB energy on the primary and cross-slip dissociation planes and upon elastic anisotropy, superdislocation may prefer (a) to remain fully dissociated in the primary plane, (b) to form a 'lock' in the cross-slip plane, (c) to lie on both plane as a two-fold or (d) as a manifold configuration.

alloy	remarks	struct.	γ_{001}	γ_{111}	γ_{CSF}	reference
Ni$_3$Al	($\approx$stoi-PX)†	L1$_2$	140	180	-	Veyssière et al. (1985)
Ni$_3$Al	($\approx$stoi-PX)	L1$_2$	90	110	-	Douin et al. (1986)
Ni$_3$Al	(24Al-0.24B-PX)	L1$_2$	126	-	-	Veyssière et al. (1989)
Ni$_3$Al	(24Al-0.24B-PX)#	L1$_2$	72	-	-	"
Ni$_3$Al	(Ni-rich-PX)#	L1$_2$	75 (25#)	-	-	Douin et al. (1991)
Ni$_3$Al	(Al-rich-PX)#	L1$_2$	120 (100#)	-	-	"
Ni$_3$Al	(22.9Al-SX)	L1$_2$	104	170	-	Dimiduk (1991)
Ni$_3$Al	(24.2Al-SX)	L1$_2$	122	163	-	"
Ni$_3$Al	(25.9Al-SX)	L1$_2$	170	190	-	"
Ni$_3$Al	(24.1Al-0.9Sn-SX)	L1$_2$	129	166	-	"
Ni$_3$Al	(21.1Al-3.7Sn-SX)	L1$_2$	146	174	-	"
Ni$_3$Al	(23.4Al-1.0V-SX)	L1$_2$	155	192	-	"
Ni$_3$Al	(20.8Al-4.0Sn-SX)	L1$_2$	201	198	-	"
Ni$_3$Al	(24.7Al-1Ta-SX)	L1$_2$	155	165	-	Baluc et al. (1988)
Ni$_3$Al	(24.7Al-1Ta-SX)	L1$_2$	200	237	250	Baluc (1990)
Ni$_3$Al	(17.4Al-6.2Ti-SX)	L1$_2$	104	170	-	Korner (1988)
Ni$_3$Al	(19.5Al-7.5Ti-SX)	L1$_2$	observed	observed	-	Sun (1990)
Ni$_3$Al	(24Al-0.25Hf, SX)	L1$_2$	120	160	-	Bontemps (1991)
Ni$_3$Al	(24Al-2Hf, SX)	L1$_2$	170	190	-	"
Ni$_3$Al	(24Al-2Hf, SX)	L1$_2$	-	observed	moderate	Liu et al. (1988)
Ni$_3$Ga		L1$_2$	17#	110	-	Suzuki et al. (1979)
Ni$_3$Ga		L1$_2$	observed	observed	-	Sun (1990)
Ni$_3$Si	(23.1S-PXi)*	L1$_2$	220	220	-	Tounsi (1988)
Ni$_3$Si	(24.5S-PXi)*	L1$_2$	250	250	-	"
Ni$_3$Si	(10.9Si-10.7Ti-SX)	L1$_2$	124	-	-	Yoshida and Takasugi (1991)
Ni$_3$Si	(10.9Si-10.7Ti-SX)#	L1$_2$	77	-	-	"
Ni$_3$Fe	(SX)	L1$_2$	55	93	100 *(95)***	Korner et al. (1985)
Cu$_3$Au	(SX/SX)	L1$_2$	-	39	13 *(25)***	Sastry and Ramaswami (1976)
Co$_3$Ti	(23Ti-3Ni-SX)	L1$_2$	210	270		Oliver (1991)
Co$_3$Ti	(22Ti-PX)	L1$_2$	130	155	22 (SISF)	"
Co$_3$Ti	(22Ti-PX)	L1$_2$	observed	observed	SISF-coupled	Liu et al. (1989)
TiAl	●	L1$_0$	100#	120#	60 (SISF)	Hug et al. (1988)
Ni$_3$V	● (RT)	DO$_{22}$	130	220	31(ESF)/26(ISF)	François (1991)
Al$_3$Ti	●	DO$_{22}$	25	>200		Hug et al. (1989)/François (1991)

alloy	remarks	struct.	γ_{110}	γ_{112}		reference
CuZn		B2	50	observed	/	Saka et al. (1984)
CuZn		B2	57	62	/	Dirras (1990)
NiAl		B2	200	-	/	Campany et al. (???)
NiAl		B2	> 500	> 750	/	Veyssière and Noebe (1991)

Table 1. Selected surface defect energies (in mJ.m^{-2}) in various ordered intermetallics.
† assuming elastic isotropy,   # after deformation at high temperature,
* elastic constants of Ni$_3$(Si,Ti),   ** in italic : SF energy in the disordered phase,
● assumed cubic.

ergy in Ni_3Fe should be paid some attention since this is the only documented instance where $\gamma_{CSF} < \gamma_{APB}$.

Composition has important effects on the microstructure for it favours specific configurations relative to others as a consequence of variations of defect energy. For instance the formation of Lomer-Cottrell (LC) type configurations in $L1_2$ alloys (i.e. edge <110> superdislocations undergoing cube slip) seems to be typical of relatively high values of APB energies (or APB energies relative to CSF energy). This point is made clear when superdislocations are compared in Ni-rich and Al-rich Ni_3Al since then, edge segments that have transformed under Lomer-Cottrell type configurations are absent and present, respectively (Douin *et al.* 1991); it is important to recall that still both alloys exhibit a flow stress peak at about the same temperature. From this and from the direct observation of the extremities of Kear-Wilsdorf configurations in many $L1_2$ alloys that show a flow stress anomaly (see for example Tounsi 1988, Bontemps and Veyssière 1991), the formation of LC locks at the terminating segments on either side of KW configurations (fig. 5) is uncertain. This point should be considered carefully since short LC locks are part of a mechanism thought to determine the anomalous flow stress in $L1_2$ alloys (Sun 1990, Hazzledine and Sun 1991, Hirsch 1991; for a complete description of LC-type locks see also Sun *et al.* 1991).

5 Deviations from geometrical APBs

Surface defects are often treated as the result of the shear over **R** of one half-crystal with respect to the other. In general, no further hypothesis is made on the occurrence of a chemical or of an elastic relaxation in the vicinity of the interface, even if the defect had formed non-conservatively. In fact, since by definition the crystal symmetry at the APB differs from that of the host crystal, a geometrical APB is not necessarily stable against local perturbations in its neighbourhood.

Temperature can effect APB structure dramatically. The most obvious - however not always trivial - illustrations of this point are provided in alloys that undergo an order-disorder transition

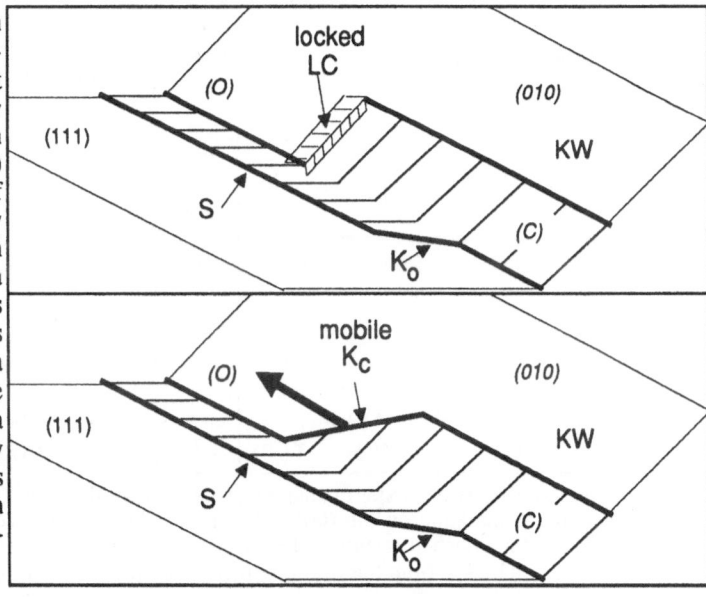

Figure 5. The junction between a Kear-Wilsdorf segment (KW) and a screw segment split in an octahedral plane. (a) the lengthening of the KW segment by lateral expansion in the cube plane of a terminating kink, is impeded when this kink has adopted a LC structure. (b) the lengthening of a KW segment may proceed when this kink is mixed in character as observed experimentally.

172

below their melting point. Second order transitions are expected to give rise to a gradually decreasing APB energy as temperature approaches the order-disorder transition (temperature T_c) and there is ample evidence in FeAl and in FeCoV of a progressive uncoupling of superpartials due to the decrease of APB energy as order degrades. In a first order transition, the domain of temperature in which APB structure is affected is in principle considerably narrower. Nevertheless, reshuffling of atoms is still expectable in the vicinity of APBs : in brief, domain walls are wetted by the disordered phase and the thickness of this disordered film diverges when approaching T_c (Brown 1959, Kikuchi and Cahn 1979, Sanchez et al. 1987, Finel et al. 1990). In Cu-17%Pd ($L1_2$, $T_c = 506.5°C$), the wetting of APBs has been determined experimentally to occur within less than 10°C from the order-disorder transition (Ricolleau et al. 1991).

In ordered alloys that undergo a first order transition, several authors have attempted to measure the dependence of APB energy upon temperature and some effects have been pointed out far below the order-disorder transition (Ni_3Al, Ni_3Fe and Ni_3V). In Ni_3Al, temperature does not effect APB energy both on {001} and on {111} planes up to about 650°C (Veyssière et al. 1989, Korner 1989, Crimp 1989), however, a drop of about 30% of the APB energy on cube planes - significant enough in order to be distinguished from experimental uncertainties - has been detected at 850°C. In Ni_3V (DO_{22}) careful weak-beam experiments have demonstrated that, again, the APB energy could decrease well below T_c (François 1991). Monte-Carlo calculations of the APB structure making use of interaction potentials determined experimentally by neutron diffusion, were then performed in order to check this point (A. Finel, F. Solal and A. François 1991, unpublished results). The agreement, between weak-beam experiments and Monte-Carlo simulations, on the magnitude of the APB energy as well as on its temperature dependence is excellent,and this holds true both on the basal and on {111} planes (figure 6). It is then possible that the dependence of the APB energy upon temperature observed on {001} planes in Ni_3Al, will also be reproduced by simulators provided appropriate approximations (i.e. CVM with a maximum cluster beyond the tetrahedron and Monte-Carlo) are used.

It seems that the role of a deviation from stoichiometry in Ni_3Al alloys discussed in §4 is the largest at elevated temperatures (Douin and Veyssière 1991). Again in Ni_3Al (with 24%Al-

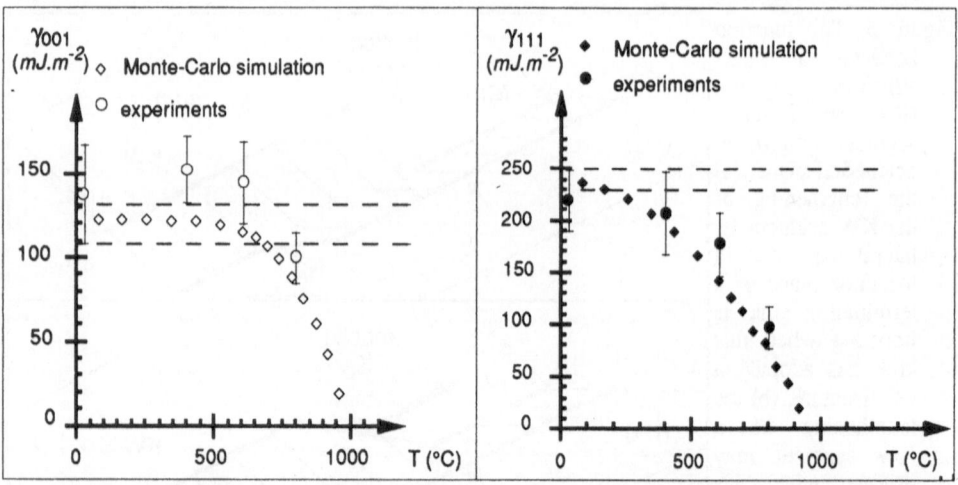

Figure 6. Comparison between experimental measurements of the dependence of the APB energy in Ni_3V upon temperature on the (001) and on the (111) planes (A. Finel, F. Solal and A. François 1991, unpublished results). The dashed lines represent the uncertainty on the APB energy at 0K which arise from the neutron diffusion determination of the atomic interactions.

0,24%B) deformed at elevated temperature, the observation of widely extended superdislocations and of a significant subdissociation of superpartials to form CSFs should also be mentioned (P. Veyssière, J.A. Horton and M.H. Yoo : in preparation). It seems that altogether with dislocations which show a normal separation, other configurations attesting to much lower surface defect energies, tend to exist when the alloy is heated at elevated temperature, as if resting dislocations accepted some chemical relaxation.

In Ni_3Fe ($L1_2$), the equilibrium APB energy varies with temperature because atoms reorder in the vicinity of the geometrical APBs (fig. 7). A study of the kinetics of this process shows that it is not a slowly changing APB structure that is rate-controlling. Rather the equilibrium configuration of the APB is established very fast and the superpartials move apart at a rate controlled by diffusion onto the APB (Schoek and Korner 1990). Ni_3Fe does not exhibit a flow stress peak.

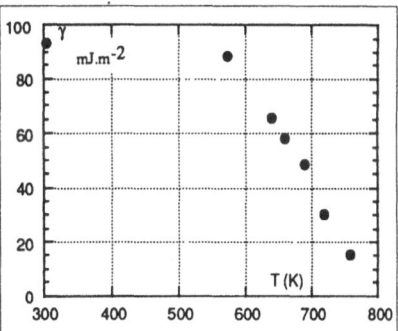

Figure 7. Variation of equilibrium APB energy with temperature in Ni_3Fe Korner and Schoeck 1990)

In β-CuZn, the hypothesis of a structural change at APBs has received a great deal of attention since Brown's original work in 1959. Recently, Beauchamp *et al.* (1991) have shown that there is good consistency between observed sensisivity of the flow stress upon temperature and the prediction from Brown's model when the APB energy is calculated by means of the Cluster Variation Method; nevertheless, the amplitude of the anomaly as calculated by CVM is by far too large. On the other hand, Nohara (1991) has concluded that the kinetics of reordering is sufficiently large in β-brass to allow for full operation of Brown's mechanism (this is also suggested by TEM experiments which show that significant diffusion operates at unexpectedly low temperatures Saka and Zhu 1985, Dirras 1990).

One interesting field of investigation has been open by Horton. *et al.* (1991) in Ni_3Al deformed *in-situ* at room temperature. Under these conditions thin foils exhibit slip localization (which is never observed in bulk deformed samples) that is accompanied by a dramatic modification of dissociation properties with the order of the dislocation in the slip band : correlation between companion superpartials becomes gradually looser, a transition in dissociation mode may occur from APB-coupled to SF-coupled and significant density of cold-work point defects is produced in the wake of the train of dislocations. This suggests that order is affected by the repeated passage of dislocations in a given plane even in alloys with very high ordering energies. This observation is important since it may offer a means to explain the mechanism by which one may disorder Ni_3Al by rolling at room temperature. Further studies are of course required in order to understand this process and in particular a detailed contrast analysis of the effect of a limited number of dislocations on crystal order.

Acknowledgements

A. François and J. Oliver are greatly acknowledged for releasing some of their results prior to publication and for fruitful comments during the preparation of this manuscript. The author has benefited from good interactions with Drs P. Beauchamp, R. Cahn, D. Caillard, D. Dimiduk, J. Douin, A. Finel, K. Hemker, J. Horton, A. Korner, D. Morris, G. Saada, Y. Sun and M. Yoo.

References

Baluc, N. (1990), *Thèse de Doctorat*, Université de Lausanne.

Baluc, N., Karnthaler , P.M. and Mills, M.J. (1988) *Inst. Phys. Conf. Ser., n° 93*, **2**, p. 463-464.

Beauchamp, P., Douin, J. and Veyssière, P. (1987), *Philosophical Magazine A*, **55**, 565-581.

Bontemps-Neveu, C. (1991), *Thèse de l'Université de Paris Sud*.

Bontemps, C. and Veyssière, P. (1990), *Philosophical Magazine Letters*, **61**, 259-267.

Brown, N. (1959), *Philosophical Magazine*, **4** 693-704.

Campany, R.G., Loretto, M.H. and Smallman, R.E. (1973) *J. Micr.*, **98**, 174-179.

Crimp, M.A. (1989) *Philosophical Magazine Letters*, **60** 45-51.

Dimiduk, D. (1989), *PhD Thesis*, Carnegie-Mellon University.

Dimiduk, D. (1991), *J. Phys. III*, **1**, 1025-1053.

Dirras, G.F. (1990), *Thèse de l'Université de Poitiers*.

Douin, J., Veyssière, P. and Beauchamp, P. (1986), *Philosophical Magazine A*, **54**, 375-393.

Douin, J., Beauchamp, P. and Veyssière, P. (1988), *Philosophical Magazine A*, **58**, 923-935.

Douin, J. and Veyssière, P. (1991), *Philosophical Magazine A*, in press.

Duesbery, M. (1989), in F.N.R. Nabarro (ed.), *Dislocations in Solids*, North-Holland, vol. **8**, pp. 67-173.

Feller-Kniepmeir, M. and Scheunemann-Frerker, G. (1990), *Philosophical Magazine A*, **62**, 77-88.

Finel, A., Mazauric , V. and Ducastelle, F. (1990), *Physical Review Letters*, **65** 1016-1019.

François, A. (1991), *Thèse de l'Université de Paris VI.*.

Hazzledine, P.M. and Sun, Y.Q. (1991), in L.A. Johnson, D.P. Pope and J.O. Stiegler (eds.), *High Temperature Intermetallic Alloys IV*, Materials Research Society, Boston, **213**, pp.209-222.

Hirsch, P.B. (1991), *Philosophical Magazine A*, in press.

Horton, J.A., Baker, I. and Yoo, M.H. (1991), *Philosophical MagazineA*, **63**, 319-335.

Hug, G. (1988), *Thèse de l'Université de Paris Sud*.

Hug, G., Douin, J. and Veyssière, P. (1989), in C.T. Liu, A.I. Taub, N.S. Stoloff and C.C. Koch (eds.), *High Temperature Intermetallic Alloys III*, Materials Research Society, Boston, **133**, pp. 125-130.

Hug, G., Loiseau, A. and Veyssière, P. (1988), *Philosophical Magazine A*, **57**, 499-523.

Kikuchi, R. and Cahn, J.W. (1979), *Acta Metallurgica*, **27**, 1337-1353.

Korner, A. (1989), *Philosophical Magazine Letters*, **59**, 1-7.

Korner, A., Karnthaler, H.P. and Hitzenberger, C. (1987), *Philosophical Magazine A*, **56**, 73-88.

Korner, A. and Schoeck, G. (1990), *Philosophical Magazine A*, **61**, 909-915.

Liu, Y., Takasugi, T., Izumi, O. and Ohta, H. (1988), *Philosophical Magazine Letters*, **58**, 81-85.

Liu, Y., Takasugi, T., Izumi, O. and Takahashi, T., (1988), *Philosophical Magazine Letters*, **58**, 437-454.

Molenat, G. (1991), *Thèse de l'Université de Toulouse*.

Nohara, A. (1991), in O. Izumi (ed.), Intermetallic Compounds : Structure and Mechanical Properties', *Proceedings of the International Symposium - JIMIS6'*, The Japan Institute of Metals, Sendai, pp. 561-563.

Oliver, J. (1991), *Thèse de l'Université de Paris Sud*.

Paidar, V. (1984), in V. Paidar and L. Lecjek (eds.), *Materials Science Monographs* **20** : The Structure and Properties of Crystal Defects, Elsevier, Oxford, pp. 19-36.

Paidar, V. (1985), *Acta Met..*, **33**, 1803-1811.

Ricolleau, C., Loiseau, A. and Ducastelle, F. (1991), *Phase Transitions*, **30** 243-254.

Saka, H., Kawase, M., Nohara, A. and Imura, T. (1984), *Philosophical Magazine A*, **50**, 65- 72.

Saka, H. and Zhu, Y.M. (1985), *Philosophical Magazine A*, **51**, 629-637.

Sanchez, J.M., Eng, S., Wu, Y.P. and Tien, J.K. (1987), in N.S. Stoloff, C.C. Koch, C.T. Liu and O. Izumi (eds.), *High Temperature Intermetallic Alloys II*, Materials Research Society, Boston, **133**, pp. 57-64.

Sastry, S.M.L. and Ramaswami, B. (1976) *Philosophical Magazine*, **33** 801-813.

Scheunemann-Frerker, G. and Feller-Kniepmeir, M. (1990), *Scripta Metallurgica et Materialia* **24**, 1381-1386.

Schoeck, G. and Korner, A. (1990), *Philosophical Magazine A*, **61**, 917-928.

Sun, Y.Q. (1990), *Ph.D. Thesis*, University of Oxford.

Sun, Y.Q., Hazzledine, P.M. and Crimp, M.A. (1991), in L.A. Johnson, D.P. Pope and J.O. Stiegler (eds.), *High Temperature Intermetallic Alloys IV*, Materials Research Society, Boston, **213**, pp. 311-316.

Suzuki, K., Ichihara, M. and Takeuchi, S. (1979) *Acta Metallurgica*, **27** 193-203.

Tounsi, B. (1988), *Thèse de l'Université de Poitiers*.

Tounsi, B., Beauchamp, P., Mishima, Y., Suzuki, T. and Veyssière, P. (1990), in C.T. Liu, A.I. Taub, N.S. Stoloff and C.C. Koch (eds.), *High Temperature Intermetallic Alloys III*, Materials Research Society, Boston, **133**, pp.731-736.

Veyssière, P. (1984), *Philosophical Magazine A*, **50**, 189-203.

Veyssière, P., Douin, J. and Beauchamp, P. (1985), *Philosophical Magazine A*, **51**, 469-483.

Veyssière, P. (1988), *Revue de Physique Appliquée* **23**, 431-443.

Veyssière, P., Yoo, M.H., Horton, J.A. and Liu, C.T. (1988), *Philosophical Magazine Letters*, **57**, 17-23.

Veyssière, P., Horton, J.A., Yoo, M.H. and Liu, C.T. (1989), *Philosophical Magazine Letters*, **59**, 61-68.

Veyssière, P. (1990a), in C.T. Liu, A.I. Taub, N.S. Stoloff and C.C. Koch (eds.), *High Temperature Intermetallic Alloys III*, Materials Research Society, Boston, **133**, pp. 175-188.

Veyssière, P. (1990b), in H.E. Exner and V. Schumacher (eds.), *Advanced Materials and Processes*, EUROMAT'89, DGM Informationgesellschaft Verlag, Oberusrsel, pp. 469-474.

Veyssière, P. (1991a), in O. Izumi (ed.), Intermetallic Compounds : Structure and Mechanical Properties', *Proceedings of the International Symposium - JIMIS6'*, The Japan Institute of Metals, Sendai, pp. 745-752.

Veyssière, P. (1991b), *Iron and Steel Institute of Japan International*, Special Issue on Intermetallics, in press.

Veyssière, P. and Hug, G. (1990), in L.D. Peachy and D.B. Williams (eds.), *Electron Microscopy 1990*, San Francisco Press Inc., San Francisco, pp. 450-451.

Vitek, V. (1985), The Institute of Metals (ed.), *Dislocations and the properties of real materials*, Arrowsmith, Bristol, pp. 30-50.

Yoo, M.H. (1986), *Scripta Metallurgicz*, **20** 915-920 ; (1987), *Acta Metallurgica*, **35** 1559-1569.

Yoo, M.H. and Fu, C.L. (1991), *Iron and Steel Institute of Japan International*, Special Issue on Intermetallics, in press.

Yoshida, M. and Takasugi, T. (1991), in O. Izumi (ed.), 'Intermetallic Compounds : Structure and Mechanical Properties', *Proceedings of the International Symposium - JIMIS6'*, The Japan Institute of Metals, Sendai, pp. 403-407.

OBSERVATIONS OF DISLOCATION MECHANISMS GOVERNING YIELD STRENGTH IN L1$_2$ ALLOYS

Y.Q. SUN and P.M. HAZZLEDINE
Department of Materials,
University of Oxford,
Oxford OX1 3PH,
England

ABSTRACT. The structures and properties of dislocations and other defects in deformed L1$_2$ ordered Ni$_3$Al and Ni$_3$Ga are observed with transmission electron microscopy. The principal intention of this paper is to examine the changes in the structure and behaviour of dislocations in a wide temperature range (-196 to 900 C) with a view to understanding the transitions of slip systems and yielding properties. Dislocations with low mobilities in the various temperature regions are first identified with weak-beam microscopy and the core structure of these dislocations are then observed directly by lattice resolution TEM. In order of increasing temperature, dislocations observed to have non-planar structures are: 30° <112>/3 super-Shockley partials (Giamei locks), screw <101> superdislocations (Kear-Wilsdorf locks), edge <101>{001} superdislocations (Lomer-Cottrell type locks), and 45° <100>{001} dislocations (B5 locks). The effects of temperature on the formation and destruction of these locks are discussed and related to the transition of slip systems and the change of yield properties. The operation of a kink mechanism for the unlocking of Kear-Wilsdorf locks is shown to lead to the formation of special kink configurations with switched partials and the formation of antiphase domain boundary tubes which are also observed to characterize the region of the yield stress anomaly.

1. Introduction

The temperature dependence of the flow stress in some L1$_2$ ordered alloys exhibits several interesting transitions, of which the one that is widely known is the transition, with the increase of temperature, from the yield stress anomaly associated with the APB-dissociated <101>{111} slip to the regime of normal temperature dependence associated with <110>{001} slip. Suzuki, Mishima and Miura (1989) have classified the variation of the yield stress with temperature of L1$_2$ ordered Ni$_3$Al alloys into three temperature regions, each region being associated with a separate slip system or dislocation dissociation mode. A fourth region can be recognized when the temperature variation of the work-hardening rate is included; in this region, which in Ni$_3$Al and Ni$_3$Ga occurs around 850 C, the work-hardening rate reaches a second maximum (Staton-Bevan 1983, and Sun 1990) and the operative slip system is <010>{001} (Sun 1990). Another important type of transition occurs in Pt$_3$Al (Wee, Pope and Vitek 1984) and Co$_3$Ti (Liu, et al 1989a) in which the yield stress anomaly is preceded by a normal regime which is believed to be associated with SISF-dissociated <101>{111} slip. With the exception of APB-dissociated <101>{111} slip, which has attracted much attention owing to its relation with the yield stress anomaly, experimental observation of the structure and locking of dislocations in the other three slip modes (<110>{001}, <010>{001} and the SISF-dissociated <101>{111}) has been less well documented. One aim of this paper is to present TEM observations of the structure and core structure of dislocations in the above four slip systems in Ni$_3$Al and Ni$_3$Ga with a special view to

C. T. Liu et al. (eds.), Ordered Intermetallics – Physical Metallurgy and Mechanical Behaviour, 177–196.
© 1992 *Kluwer Academic Publishers.*

elucidating the dislocation mechanisms which controll the transition of slip systems.

Models for the yield stress anomaly associated with the APB-dissociated <101>{111} are based on the formation of Kear-Wilsdorf locks on screw dislocations. In the model by Takeuchi and Kuramoto (1973) the pinning of screw dislocations is caused by small segments on which the leading superpartial has cross-slipped on the {010} cube plane by a small distance. Further development was prompted by the necessity to include the Escaig effect (Escaig 1968, Bonneville and Escaig 1979) in order to explain the orientation dependence of yield stress in full (Lall, Chin and Pope 1978), in particular the tension/compression asymmetry (Ezz, Pope and Paidar 1982). This later development led to a much more elaborate theory of Paidar, Pope and Vitek (1984) in which the effect of locking derives from the resplitting of the leading superpartial after jumping on {010} for a distance of $b/2$, where b is the Burgers vector of the superpartial. Recently, Hirsch (1991), with a view to explaining the very small strain-rate sensitivity of the yield stress in the yield stress anomaly, has proposed a new model in which he made further examination of the processes leading to the formation and destruction of Kear-Wilsdorf locks. One aim of this paper is to analyze some of the geometrical consequences of this new model and to compare them with TEM observations of some of the characteristic features of the deformed structure in the yield stress anomaly, including superkinks with switched superpartials and APB tubes.

2. Dislocations with low mobilities in the four slip systems

The variation of the yield stress of Ni_3Ga and Ni_3Al can be divided into two temperature regions, namely the region of the yield stress anomaly below the strength peak and the region of normal temperature dependence above the strength peak. The variation of the yield stress with temperature in Ni_3Ga is shown in Fig.1. The primary slip system in the region of the yield stress anomaly is $[10\bar{1}](111)$ and above the peak the primary slip system is $[1\bar{1}0](001)$ for most crystal orientations except those very close to [001]. In the region of the yield stress anomaly of $[10\bar{1}](111)$ slip it has been consistently observed since the mid 60's (e.g. Thornton, Davies and Johnston 1967) that the deformed structure consists of a high density of screw dislocations; screws in $[10\bar{1}](111)$ are therefore thought to have relatively low mobilities compared with other dislocations in that system. Figure 2a is a TEM weak-beam picture to show a high density of screw dislocations in a Ni_3Ga specimen deformed at 200 C. More recent weak-beam and atomic resolution observations have shown that the screw dislocations are locked in the Kear-Wilsdorf configuration in which the plane of APB is on (010). In 3.4 we shall show that the deformed

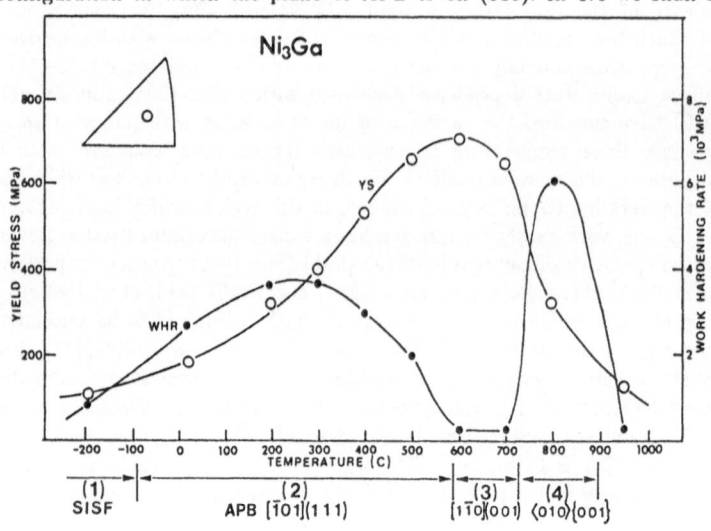

Fig.1 The variation of the yield stress and work-hardening rate with temperature in Ni_3Ga. The dislocation systems in the four temperature regions are indicated.

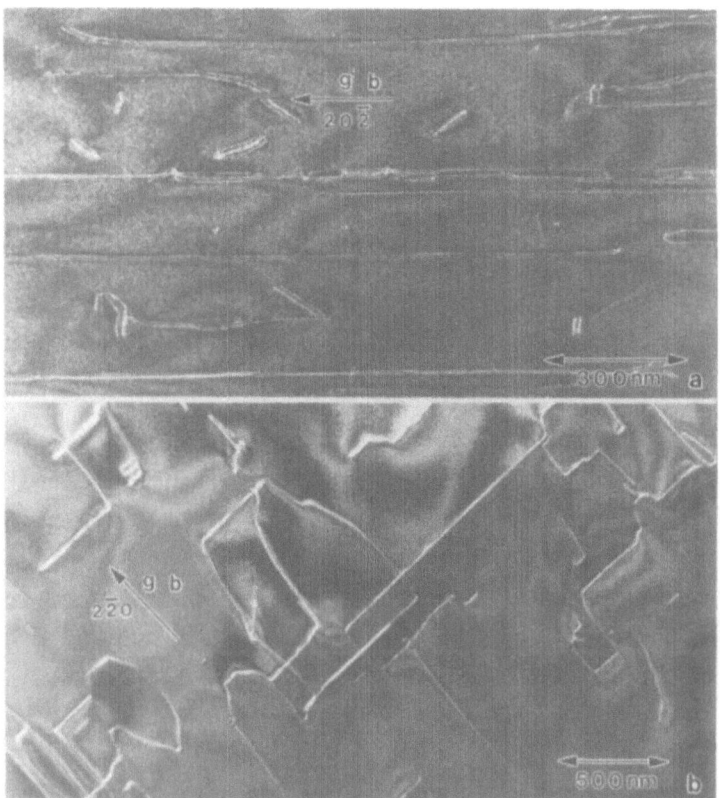

Fig.2.(a)TEM weak-beam image of dislocations in Ni_3Ga deformed at 200 C in the region of the yield stress anomaly. The structure consists of immobile screw dislocations. (b) Weak-beam picture of dislocations in Ni_3Ga deformed at 700 C. Screw and edge dislocations are immobile.

structure in this temperature region is also characterized by superkinks with switched superpartials and by a high density of APB tubes. Above the yield stress peak in the regime of normal temperature dependence, TEM observations have identified the slip system to be $[1\bar{1}0](001)$. In this slip system dislocations are mostly either edges or screws, as shown in Fig.2b for Ni_3Ga deformed at 700 C, suggesting that in this system both screw and edge dislocations are immobile. This feature was also noted in early TEM observations (Thornton et al 1967). In section 3.2 we shall show that in $[1\bar{1}0](001)$ screws are also locked as Kear-Wilsdorf barriers while the edge dislocations are locked as Lomer-Cottrell barriers.

When the work-hardening rate is plotted against temperature, as shown in Fig.1 for Ni_3Ga, a new region emerges at around 850 C in which the work-hardening rate reaches a second peak, a feature that was first noted in Ni_3Al by Staton-Bevan (1983). TEM observations of specimens deformed in this new region show that the operating slip system has changed from $[1\bar{1}0](001)$ to $<010>\{001\}$ since the majority of dislocations have been identified to have $<010>$ Burgers vectors and to lie on $\{001\}$ planes. Figure 2c is a weak-beam image of dislocations in a Ni_3Al specimen deformed at 900 C. It shows that two families of $<010>\{001\}$ dislocations coexist, suggesting multiple slip, and that dislocations in these two systems react to form $<110>$ dislocations. The presence of $<010>\{001\}$ dislocations in the $L1_2$ structure was first noted, but with a very small density, by Veyssiere and Douin (1985) in a Ni_3Al specimen deformed at 650 C. The occurrence of multiple slip is consistent with the special property of $<010>\{001\}$ slip in that for any deformation axis orientation there are at least two slip systems having equal Schmid factors. We shall show in section 4.3 that this property is directly responsible for the high work-hardening rate associated with $<010>\{001\}$ slip. Figure 2c also shows that most of the $<010>$

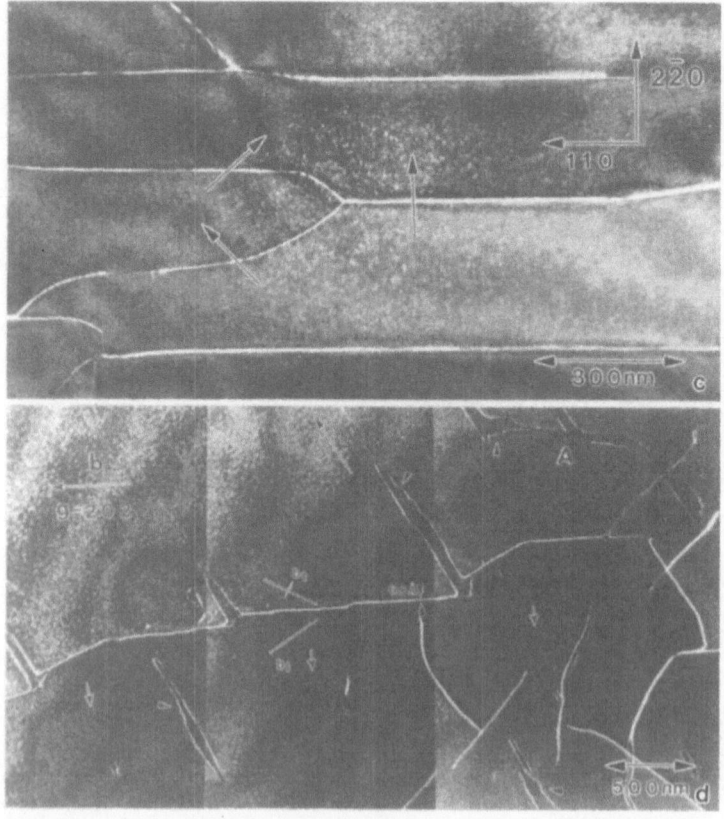

Fig.2 (c) Weak-beam image of <010>{001} dislocations in Ni₃Al de-formed at 950 C. 45° dislocations are immobile. (d) SISF-dissociated dislocations in Ni₃Ga deformed at -196 C. 30° 1/3<112> dislocations are immobile.

dislocations are lying along <110> directions at 45° to their Burgers vectors, suggesting that 45° <010>{001} dislocations are immobile. In 3.3 we shall show that 45° <010>{001} dislocations are locked in non-planar structures known as B5 locks.

In some $L1_2$ ordered alloys the $[10\bar{1}](111)$ octahedral slip is also found to have normal temperature dependence at very low temperatures; these alloys include Pt_3Al (Wee et al 1984, and Heredia et al 1989), Co_3Ti (Liu et al 1989a). The normal temperature dependence of $[10\bar{1}](111)$ slip has been thought to be caused by the splitting of superdislocations into 1/3<112> super-Shockley partials coupled by SISF (superlattice intrinsic stacking faults) (Tichy et al 1986b). In Ni_3Al the yield stress exhibits a continuous increase from the liquid helium temperature (4 K) although the rate of increase of the yield stress with temperature has been found to become very small around and below -196 C (Suzuki et al 1989). Despite the absence of change of yield behaviour towards -196 C, a transition of dislocation dissociation in $[10\bar{1}](111)$ from APB dissociation to SISF dissociation has been observed in Ni_3Al and Ni_3Ga in the present experiments; an example for Ni_3Ga deformed at -196 C is shown in Fig.2d in which the $[10\bar{1}]$ superdislocation, orientated mostly along the screw orientation, is dissociated into two partials which are determined to be 1/3<112> super-Shockleys by standard g.b analysis. Figure 2d also shows that super-Shockley partial loops are created by one super-Shockley being pinched off the propagating superdislocation. Note that the super-Shockley loops are mostly elongated along <110> directions at 30° to the direction of Burgers vector <112>; this observation shows that 30° 1/3<112> super-Shockleys are immobile compared with other characters. The low mobility of 30° 1/3<112> super-Shockley partials was first noted in a cold rolled Ni_3Al by Giamei et al (Giamei et al 1971) and the same observation has since been frequently made by different groups (Pak et al 1976, Baker and Schulson 1985, Veyssiere et al 1985). In the present experiments on Ni_3Ga and Ni_3Al deformed at -196 C the superdislocations in $[10\bar{1}](111)$ slip have been identified to be SISF-dissociated from weak-beam observations, but in most other cases reported in the literature the dislocations are reported to be APB-dissociated and the presence of $1/3<112>$ super-Shockleys, often in the form of loops elongated along $30°$ directions, has been attributed to a mechanism in which $1/3<112>$ loops are pinched off APB-coupled superpartials through a nucleation process. In 3.5 we shall present lattice resolution TEM observations to show that $30°$ $1/3<112>$ super-Shockley partials are locked in a non-planar structure called Giamei lock (Giamei et al 1971).

In the above we have classified the deformation of Ni_3Al and Ni_3Ga into four temperature regions; the regions are distinguished from each other by different dislocation systems as well as by different yield and/or work-hardening properties. The very low temperature region in Ni_3Al seems to be an exception in that in this temperature region a transition from APB-dissociation to SISF-dissociation with the decrease of temperature is not found to correlate with a change of the yield and/or work-hardening behaviour. Here we shall treat SISF-dissociated $[10\bar{1}](111)$ slip as a separate slip mode. In Ni_3Al and Ni_3Ga, in order of increasing temperature, these four regimes are: (1) SISF-dissociated $[10\bar{1}](111)$, (2) APB-dissociated $[10\bar{1}](111)$, (3) $[1\bar{1}0](001)$ and (4) $<010>\{001\}$ (Fig.1). For Ni_3Al samples oriented sufficiently close to [001], $[10\bar{1}](111)$ slip would lead to a falling yield stress caused by the onset of viscous flow (see Suzuki et al 1989); this mechanism will not be dealt with here.

3. Dislocation mechanisms in the four temperature regions

3.1 COMPARISON OF DISLOCATION LINE ENERGIES IN THE FOUR SLIP SYSTEMS

(a) $<110>\{001\}$ vs. $<010>\{001\}$. A comparison of dislocation line energies between APB-dissociated $<110>\{001\}$ dislocations and $<010>\{001\}$ dislocations has been calculated for Ni_3Al by Hazzledine et al (1989) using anisotropic elasticity. The calculation shows that, except for a small angular range in which $<010>\{001\}$ dislocations are elastically unstable, the line energy of $<010>\{001\}$ dislocations is lower than that of APB-dissociated $<110>\{001\}$ dislocations.

(b) APB-dissociated $<101>\{111\}$ vs. $<110>\{001\}$. From isotropic elasticity, the line energy of APB-dissociated $<110>\{001\}$ dislocations is expected to be lower than APB-dissociated $<110>\{111\}$ dislocations over the entire angular range owing to the low APB energy on $\{001\}$. From anisotropic elasticity, screw dislocations have been shown to have the lowest energies in both systems (Yoo 1987, Hazzledine et al 1989). It can be shown that owing to the torque force between screw superpartials (Yoo 1987), the line energy of screw dislocations in $<110>\{001\}$ is lower than that of screws in $<101>\{001\}$ by about 5%.

(c) SISF-dissociated $<101>\{111\}$ vs. APB-dissociated $<101>\{111\}$. In terms of the self energy of the partials, dislocation line energy in the SISF dissociation is expected to be higher than the APB dissociation owing to the longer partial Burgers vector in the SISF dissociation. Line energy in the SISF dissociation may become lower than the APB dissociation only when the SISF energy is exceptionally low. Suzuki et al (1979) found, using isotropic elasticity, that the SISF energy must be lower than about 1/30th of the APB energy to make the SISF dissociation energetically favourable. An estimate of the relative values of the SISF and APB energies from weak-beam observations by Veyssiere et al (1985) showed that this condition is not satisfied in Ni_3Al.

The above comparison shows that in terms of the dislocation line energies, $<010>\{001\}$ slip is the most favoured, followed by $<110>\{001\}$ and APB-dissociated $<101>\{111\}$, and the SISF-dissociated $<101>\{111\}$ slip is the least favoured. In the experimental observations, however, the sequence in which these four slip systems occur with the increase of temperature is just the opposite to the above order (Fig.1). In this paper we shall argue that the occurrence of the slip systems and the transitions between them are determined by the mobilities of dislocations.

3.2 DISLOCATION MECHANISMS CONTROLLING [1$\bar{1}$0](001) SLIP

3.2.1 *The locking of screw dislocations.* In the [1$\bar{1}$0](001) system, a superdislocation dissociates into APB-coupled superpartials according to

$$[1\bar{1}0] \rightarrow 1/2[1\bar{1}0] + APB(001) + 1/2[1\bar{1}0] \qquad (1\text{-}1)$$

Along the [110] screw orientation, the $1/2[1\bar{1}0]$ superpartials may further dissociate into CSF (complex stacking fault) -coupled $1/6<112>$ Shockleys on either (111) or ($\bar{1}\bar{1}$1), both of which intersect (001) along [1$\bar{1}$0]. The further dissociations are

$$1/2[1\bar{1}0] \rightarrow 1/6[2\bar{1}\bar{1}] + CSF(111) + 1/6[1\bar{2}1] \qquad (1\text{-}2)$$
$$1/2[1\bar{1}0] \rightarrow 1/6[2\bar{1}1] + CSF(\bar{1}\bar{1}1) + 1/6[1\bar{2}\bar{1}] \qquad (1\text{-}3)$$

For geometrical reasons, the splitting of a screw $1/2[1\bar{1}0]$ into Shockley partials on (111) and ($\bar{1}\bar{1}$1) does not take place simultaneously but alternates between (111) and ($\bar{1}\bar{1}$1) as the screw superpartial propagates on (001) by steps given by $\sqrt{2}a/4$, or $b/2$, where b is the Burgers vector of the superpartial; this is illustrated in Fig.3a. At intermediate positions the configuration of dissociation is not known and we assume that the screw superpartial is fully constricted. Any splitting at intermediate positions would involve faults other than APB, CSF or SISF and which may not be stable or metastable. A screw dislocation in the above non-planar configuration described by equation (1) is locked since it can neither move on {111} nor on (001). This is the Kear-Wilsdorf configuration first proposed to explain the locking for {111} slip (Kear and Wilsdorf 1962). Once fully transformed into the Kear-Wilsdorf configuration, a screw dislocation formed during <101>{111} slip is essentially in the same state as a screw dislocations in <110>{001}. Yamaguchi, Pope and Vitek (1982) carried out detailed computer simulations of the further splitting of screw superpartials on {001} and the geometry of the simulated core spreading is essentially described by the above crystallography relations.

In experiment, the dissociation of superdislocations into APB-coupled superpartials on {001} has been extensively observed by weak-beam microscopy (for reviews, see Veyssiere 1989 and Dimiduk 1991). The extent of the further splitting of the superpartials into Shockleys is however very small owing to the generally expected very high CSF energy and is usually beyond the resolution limit of the weak-beam technique. The further spreading of the superpartials according to equations (1-2) and (1-3) has been confirmed in Ni_3Al by Baluc et al (1988) and Crimp (1989) using lattice resolution TEM.

3.2.2 *The locking of edge dislocations.* In the [1$\bar{1}$0](001) slip system edge dislocations lie parallel to the [110] direction which, like the screw orientation, is also a line of intersection of (001) with two {111} planes, i.e. ($\bar{1}$11) and (1$\bar{1}$1); the edge dislocation may lower its energy by splitting into Shockleys in these two planes and the final structure would also be non-planar. In the face centered cubic (f.c.c.) structure (from which the $L1_2$ ordered structure is derived), an edge $1/2<110>{001}$ dislocation is known as a Lomer dislocation (Lomer 1951) and its energy is lowered if it dissociates into Shockley partials on the two intersecting {111} planes to form a Lomer-Cottrell lock (Cottrell 1952). In $L1_2$ an edge superdislocation in the APB-dissociated [1$\bar{1}$0](001) system is in the double-Lomer configuration since it consists of two Lomer (partial) dislocations each of which may dissociate according to the scheme proposed by Cottrell, i.e.

$$1/2[1\bar{1}0] \rightarrow 1/6[1\bar{1}\bar{2}] + CSF(1\bar{1}1) + 1/6[1\bar{1}0] \text{ (stair-rod)} + CSF(\bar{1}11) + 1/6[1\bar{1}2] \qquad (2\text{-}1)$$

The structure is illustrated schematically in Fig.3b. The splitting of the edge (Lomer)

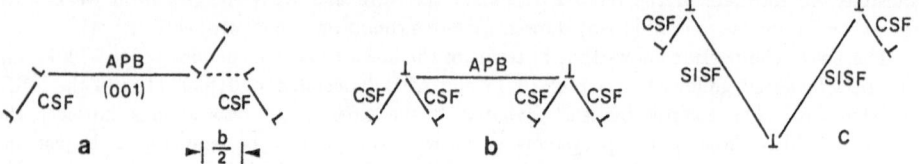

Fig.3. Non-planar structures observed in <110>{001}. (a) Kear-Wilsdorf lock on screw. (b) Double Lomer-Cottrell lock and (c) climb dissociated structure on edge dislocations.

superpartials into Lomer-Cottrell structures is allowed only at positions separated by a periodicity of b. The configuration of dislocations at intermediate positions is not known and we assume that they are fully constricted. Similar to the case for screw dislocations, any additional spreading at intermediate positions would involve other types of fault other than APB, CSF, or SISF. The core structure of Lomer partials in the $L1_2$ structure has not been studied by computer atomistic simulations.

In Ni_3Al and Ni_3Ga, edge dislocations in $[1\bar{1}0](001)$ are observed to be in the double-Lomer configuration above the yield stress peak but below about 700 C. The further splitting of the edge superpartials (Lomers) is not resolved by the weak-beam method; the extent of the further splitting is expected to be very small owing to the high energy of the CSF. Lattice resolution TEM has been used to resolve the further splitting of the Lomer partials by viewing the double Lomer dislocations end-on. Figure 4a shows a lattice image of a double Lomer dislocation in a Ni_3Ga specimen deformed at 600 C. The structure consists of two tent-shaped parts which are aligned along the trace of the (001) plane and are separated by about 50 A, in agreement with the separation observed in the weak-beam mode. The observed structure in Fig.4a is in agreement with the crystallographic relation given by equation (2-1) and illustrated in Fig.3b.

Above 700 C, edge $[1\bar{1}0](001)$ dislocations Ni_3Al and Ni_3Ga are not in the double Lomer configuration; in this temperature region screw dislocations in $[1\bar{1}0](001)$ still dissociate into two APB-coupled superpartials. As a result the dissociation is not continuous but the APB-coupled superpartials along non-edge orientations constrict at junctions with pure edge dislocations (Veyssiere 1984, Douin et al 1986, Sun 1990). In weak-beam observations images of the rectilinear edge dislocations are found to consist of three or more bright lines. An undissociated edge superdislocation in $[1\bar{1}0](001)$ is called a super-Lomer here since its Burgers vector is twice as long

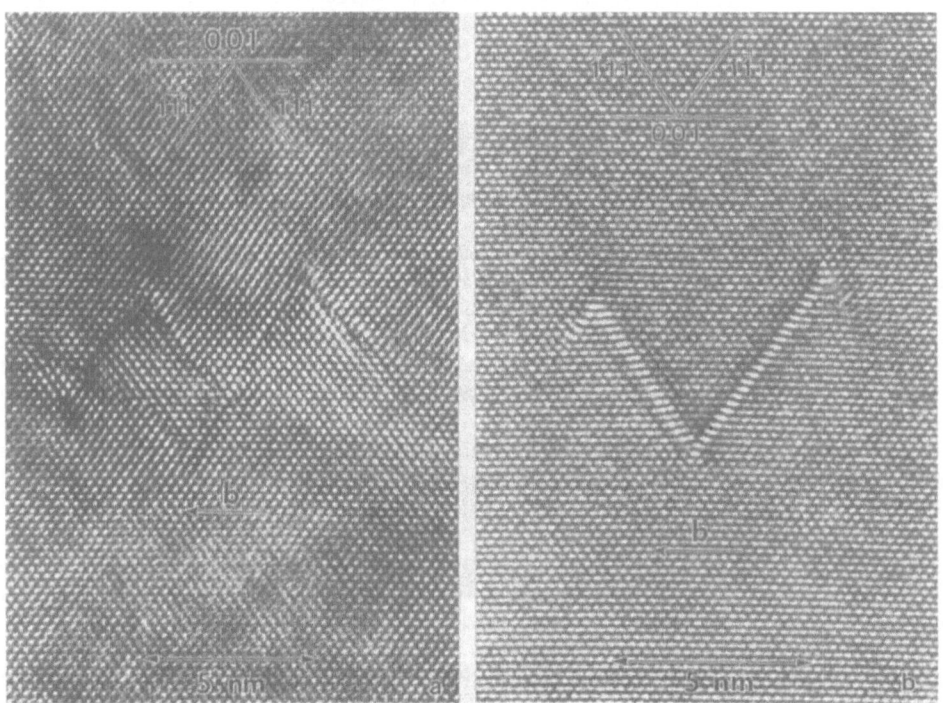

Fig.4 HREM images of edge $<110>\{001\}$ dislocations viewed end-on. (a) Double Lomer-Cottrell lock and (b) climb dissociated structure on super-Lomer.

as that of an ordinary Lomer dislocation. Like a Lomer dislocation, a super-Lomer may also dissociate into partials onto the intersecting {111} planes to form non-planar structures. Owing to a longer Burgers vector a super-Lomer may decompose into more types of partial dislocation, including 1/3<112> super-Shockleys, 1/2<110> superpartials, as well as ordinary 1/6<112> Shockleys. The combination of these partials leads to a large number of possibilities for the dissociation of a super-Lomer. Veyssiere (1984) and Douin et al (1986) attributed the low mobility of super-Lomer dislocations to a climb dissociation in which the plane of the APB is transferred from {001} to {013}, but the proposed structure was not directly confirmed by weak-beam observations because of the small dimension of the structure. Lattice resolution TEM has shown that the dissociation has taken place by climb but the mechanism is different from that proposed by Veyssiere. Figure.4b is an example in which the dislocation is seen to have dissociated into three partials, forming a non-planar structure which spans an acute angle. Two partials lie on the two intersection {111} planes and the third one sits along the line of intersection in the form of a stair-rod dislocation. This lattice resolution image is consistent with the mostly three fold image splitting evidenced on the weak-beam scale. The observed structure in Fig.4b is identified to be formed through the climb of Frank partials on the two intersecting {111} planes according to

$$[1\bar{1}0] \rightarrow 1/3[1\bar{1}1] + SI(E)SF(1\bar{1}1) + 1/3[1\bar{1}0] \text{ (stair-rod)} + SI(E)SF(1\bar{1}\bar{1}) + 1/3[1\bar{1}\bar{1}] \quad (2\text{-}2)$$

whether SESF or SISF is formed depends on the climb direction of the Franks. In lattice resolution TEM experiments, super-Lomers dissociated into both SISF and SESF have been frequently observed. In addition, Fig.4b shows that the Frank partials further spread on another {111} plane. This further dissociation can be understood if the Frank partial dissociates further according to

$$1/3[1\bar{1}1] \rightarrow 1/6[1\bar{1}2] + CSF(1\bar{1}\bar{1}) + 1/6[1\bar{1}0] \text{ (stair-rod)} \quad (2\text{-}3)$$

This further spreading is essentially the same as the mechanism proposed by Silcox and Hirsch (1959) to explain the formation of stacking fault tetrahedra from vacancy discs. The experimental observation of the dissociation of super-Lomers can be illustrated schematically in Fig.3c.

3.2.3 *The difficulty of <110>{001} slip.*

The difficulty of <110>{001} slip lies in that dislocations in this system are locked along two directions, Kear-Wilsdorf locks on screws and Lomer-Cottrell locks on edges. For a given slip system, if the dislocations are locked along only one orientation, slip can still proceed through the propagation of mobile dislocations and the deformed structure would consist of a high density of dislocations locked in the specified orientation. This is the case in some b.c.c. alloys in which screw dislocations are locked in three fold non-planar structures (Duesbery and Hirsch 1968). When the dislocations are locked along more than one direction, two directions in the present case of $[1\bar{1}0](001)$ slip, the slip system is blocked completely. Therefore, to activate the $[1\bar{1}0](001)$ slip, screw and edge dislocations trapped in non-planar structures must unlock through the recombination of the partials which requires the assistance of thermal fluctuations.

A critical piece of information that has not been provided by the present TEM observations of $[1\bar{1}0](001)$ slip is the configuration of dislocations in their unlocked states. If, once unlocked, the screw (or edge) dislocation remains constricted, the next locking event will take place immediately when the dislocation arrives at the next allowed position (in the case of screws in $[1\bar{1}0](001)$, after propagating for $b/2$, where b is the Burgers vector of the superpartial. For edge dislocations this distance is b). In this case, locking would be determined by the geometry and the mechanism is essentially the same as the Peierls mechanism and the critical stress for the operation of $[1\bar{1}0](001)$ slip would be expected to decrease with temperature and to increase with the strain-rate imposed by the test machine. The above conclusion would need reconsideration if the unlocked dislocation does not stay constricted but spreads on the glide plane; in this case there would be a barrier to locking and the gliding distance of the unlocked dislocation would be determined by the probability with which the next locking event takes place. Clement et al

(1988), on the basis of their *in situ* observations, have assumed such an unlocked state for <110>{001} screw dislocations in Ni₃Al but the exact structure was not proposed. In this case the dislocation velocity is determined jointly by the rate of unlocking and the rate of locking. If locking dominates, the rise of temperature would cause an increase of the yield stress. Clement et al (1989) have used this argument to support their conclusion that the <110>{001} slip in the γ' phase of a superalloy is also associated with a yield stress anomaly. Any splitting at intermediate positions would involve the creation of new types of fault (other than APB, CSF of SISF) which might not be stable or metastable. If the new fault is not stable or metastable, the next locking position would be same as if the dislocation remains constricted (itself also an unstable configuration) and the glide mechanism is unchanged. Paidar et al (1982) have simulated the motion of screw superpartials in <110>{001} under an external load, but it is not clear from their work whether the screw superpartials are extended on the glide plane {001} (or any other plane) at positions intermediate between locked positions.

3.3 DISLOCATION MECHANISMS CONTROLLING <010>{001} SLIP

3.3.1 *The locking of 45° <010>{001} dislocations.* In section 2 we have shown with a weak-beam observation that 45° <010>{001} dislocations are relatively immobile compared with other characters. In <010>{001} 45° dislocations are also lying along <110> directions, just like screw and edge dislocations in <110>{001}. A 45° <010>{001} dislocation may therefore also dissociate into partials on the two intersecting {111} planes to form non-planar structures. In the [010](001) system, the [010] dislocation lying along the [1$\bar{1}$0] 45° direction may lower its energy by dissociating according to

$$[010] \rightarrow 1/6[11\bar{2}] + CSF(111) + 1/6[031] \text{ (stair-rod)} + CSF(\bar{1}\bar{1}1) + 1/6[\bar{1}21] \qquad (3\text{-}1)$$

or

$$[010] \rightarrow 1/6[\bar{1}2\bar{1}] + CSF(111) + 1/6[220] \text{ (stair-rod)} + CSF(\bar{1}\bar{1}1) + 1/6[\bar{1}21] \qquad (3\text{-}2)$$

which lead to the formation of obtuse and acute locks respectively, as illustrated in Fig.5a,b. These locks are allowed by crystal geometry to occur only along positions separated by $\sqrt{2}a/2$, where a is the lattice constant. Any additional spreading at intermediate positions would involve potentially unstable faults (other than SISF, APB or CSF) and we assume the dislocations remain fully constricted. In the f.c.c. structures these dislocations may be formed through dislocation reactions and the resultant non-planar structures are known as B5 locks (Hirth and Lothe 1982). So far no computer atomistic simulation has been reported for the core structure of 45° <010>{001} dislocations. In the weak-beam experiments no further dissociation has been resolved on 45° <010>{001} dislocations, attributable to the small dimension of the structures caused by the high energy of the CSF. Figure 6 shows an example of a lattice image of a 45° [010](001) dislocation viewed end-on in a (110) foil. The dislocation has clearly dissociated into a non-planar structure spanning an obtuse angle. The indexing of the individual partials has been found difficult but the image can best be described by equation (3-1).

3.3.2 *<010>{001} slip vs <110>{001} slip.* Dislocations in the <010>{001} system are also locked along two directions in the non-planar structures occurring on 45° <110> orientations. In this respect <010>{001} and <110>{001} systems are very similar. In experiment, <010>{001} slip, despite being energetically more favourable, occurs at a higher temperature than <110>{001}, and this shows that <010>{001} slip requires greater thermal activation for unlocking to take place. The energy increase caused by the recombination of partials in the unlocking can be measured approximately by the difference in the Frank sums between the locked and fully constricted states. For a 45° <010>{001} dislocation dissociated according to equation (3-1) this difference is $7a^2/18$, while for screw and edge superpartials in <110>{001} this difference is $3a^2/18$ and $2a^2/18$ respectively. This estimate shows that the recombination energy required for the unlocking of a B5 lock on 45° <010>{001} dislocations is approximately twice as

Fig.5a,b (above) Non-planar structures on 45° <010>{001} dislocations.

Fig.6 (left) HREM image of a 45° <010>{001} dislocation viewed end-on. The structure is non-planar and can be described by Fig.5b.

large as that of a Kear-Wilsdorf lock on screws in <110>{001}. This explains why <010>{001} slip, despite being energetically more favourable, is more difficult and occurs at higher temperatures than <110>{001} slip.

3.3.3 *The origin of the high work-hardening rate.*

A special property of the <010>{001} slip is that for any given orientation of compression (tension) axis, there are always two <010>{001} slip systems possessing the same Schmid factor; in our notation these two slip systems are [010](001) and [001](010). Dislocations in these two systems are blocking each other like forest dislocations, similar to the stage-II deformation of f.c.c. alloys, and the result is that the flow stress increases rapidly with dislocation density and strain, leading to a high work-hardening rate. Reactions between dislocations in these two systems can be shown to lead to the formation of locked dislocations, an effect that would enhance the work-hardening. Details can be found in Hazzledine et al (1989) and one example is given below. A [011] dislocation lying along [100] is formed through the reaction between [010](001) and [001](010) via

$$[010](001) + [001](010) \rightarrow [011]$$

This dislocation is locked since it may further dissociate into 1/2<110> superpartials on the two {010} planes that intersect along [100] to render the structure non-planar, i.e.

$$[011] \rightarrow 1/2[101] + APB(010) + 1/2[011] \text{ (stair-rod)} + APB(001) + 1/2[\bar{1}10]$$

3.4 DISLOCATION MECHANISM IN THE APB-DISSOCIATED [10Ī](111) SLIP

3.4.1 *The role of superkinks.* 3.1 has shown that in terms of dislocation line energy, APB-dissociated <101>{111} slip is less favoured compared with APB-dissociated <110>{001} slip and <010>{001}. Dislocations in APB-dissociated <101>{111} slip are different from those in <110>{001} in two aspects. First, in the APB-dissociated <101>{111} locking may occur only along the screw orientation while in <110>{001} dislocations are locked in non-planar structures along two directions. Second, in <110>{001} locking occurs along every atomic row similar to the Peierls mechanism while in the APB-dissociated <101>{111} slip screw dislocations also possess a mobile structure, as a result of which there is a barrier to locking and the screw dislocations would glide for many atomic distances before becoming locked. In a slip system containing just one locked orientation, a kink mechanism would provide an easy path for the locked dislocation to overcome the barrier. Let us consider a temperature just below the yield stress peak where we know the macroscopic slip system is [10Ī](111) and where we also know the primary cube slip is just about to start; at this temperature screw dislocations in [10Ī](111) and [1Ī0](001) are most likely in the same configuration, i.e. the Kear-Wilsdorf lock. (In fact, screw dislocations are observed in TEM to be locked in the Kear-Wilsdorf configuration not just near the yield stress peak but also at substantially lower temperatures, e.g. room temperature, see Veyssiere 1989, and Dimiduk 1990). At this temperature, therefore, whether slip takes place on (111) or (001) would be strongly affected by the mobility of dislocations along other orientations. [1Ī0](001) slip can not occur at this temperature because the edge dislocations are locked in the Lomer-Cottrell configuration. [10Ī](111) slip would occur initially through the propagation of mobile edge dislocations. When these mobile edge dislocations are exhausted, or trapped through the formation of dipoles, locked screw dislocations, which are potentially mobile on {010} because of the proximity to the peak, have to be brought back onto (111) for (111) slip to continue. At this temperature, at least, the action of kinks are important in the unlocking of screws and thus in determining slip on (111).

Now if we consider a very low temperature (say near absolute zero) an initially mobile APB-dissociated screw superdislocation would largely remain mobile except very few segments on which the leading superpartial has cross-slipped for $b/2$ onto the cube cross-slip plane (010). At this temperature the configuration of the screw dislocation would be that proposed in the PPV model (Paidar, Pope and Vitek 1984); the configuration of these segments are illustrated in Fig.7a in which the leading superpartial is extended on (1Ī1). Hirsch (1991) shows that as soon as the leading superpartial is able to make another jump of $b/2$ on (010) the leading superpartial becomes mobile on (111) since it spreads into CSF on (111), Fig.7b, and the leading superpartial would move on this (111) layer for a distance equal to the width of the APB, Fig.7c. Both the PPV model and the new model by Hirsch assume locking on small segments; the difference lies in that in the PPV model the leading partial makes a jump of $b/2$ on (010) and would therefore represent a weak lock, while in Hirsch's model the leading partial moves a distance of b on (010) followed by movement on (111) by a large distance equal to the width of the APB. This second movement would make the lock much stronger. Hirsch has also shown that once the screw

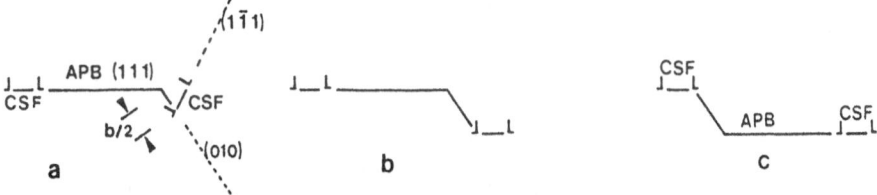

Fig.7 Schematic illustrations of the cross-slip of screw dislocations in APB-dissociated [10Ī](111) slip. (a) The resplitting of the leading superpartial after a jump on (010) for $b/2$. (b) The resplitting after a jump on (010) for b. (c) A locked configuration.

188

dislocation is in the configuration depicted in Fig.7c it can subsequently transform into the full Kear-Wilsdorf configuration with an APB on (010). In this case, as proposed by Hirsch, an effective way to unlock the segments would be through the action of adjacent mobile parts which are essentially acting as kinks.

Detailed processes in the locking and unlocking of Kear-Wilsdorf locks and the critical stress needed have been described by Hirsch (1991). Here we are merely concerned with the geometrical consequences of the kink mechanism in order to make comparisons with experimental TEM observations. Figure 8a illustrates part of an expanding dislocation loop on which the screw dislocation has cross-slipped off the glide plane and may have made several movements on (010) and (111) before eventually becoming locked in the Kear-Wilsdorf configuration. As the mobile edge part continues to propagate new mobile screw dislocations are generated and bow out on (111), as shown in Fig.8b. The junction marked A in Fig.8b is clearly in a high energy state because of the forced high curvature of the trailing superpartial. This high energy configuration can be eliminated through the processes illustrated in Figs.8c,d. In Fig.8c, as the mobile part continues bowing out on (111), a short segment on the leading superpartial is formed which has an opposite line sense to the trailing superpartial. This segment annihilates with the trailing partial and the configuration of the junction becomes as shown in Fig.8d. At this stage the order of the superpartials is clearly switched. Switching of the superpartials is the first stage in the unlocking of Kear-Wilsdorf locks by superkinks.

The switching of superpartials at superkinks has been observed extensively in TEM observations. Figure 9 shows an example of switched superkinks in a Ni$_3$Ga specimen deformed at 400 C. Clearly the two superpartials have changed partner at the superkink. Tilting around the screw dislocation line has shown that the two superpartials are lying on the cube cross-slip plane (010). Bontemp and Veyssiere (1990) have shown that superpartials on a normal kink may appear switched when viewed along special directions. In the tilting experiment for the kink in Fig.9 the partials were seen switched over a range of nearly 90° from [101] to [010], proving that the switching of partials is real and not caused by projection.

After the switching of the partials in Fig.8d, the junction at the kink contains an edge

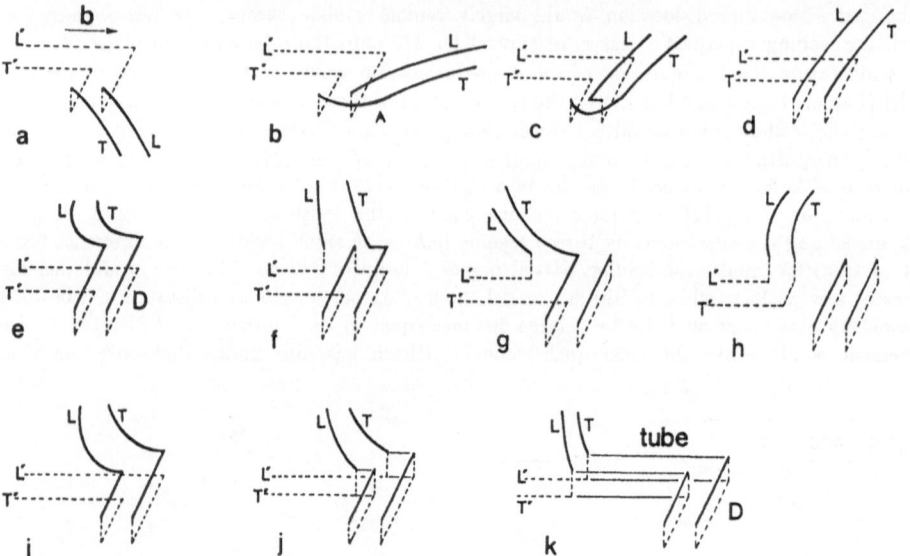

Fig.8 Schematic illustrations of the kink mechanism. (a-d) The switching of the superpartials. (e-h) By-passing the dipole without tube. (i-k) By-passing the dipole with a tube.

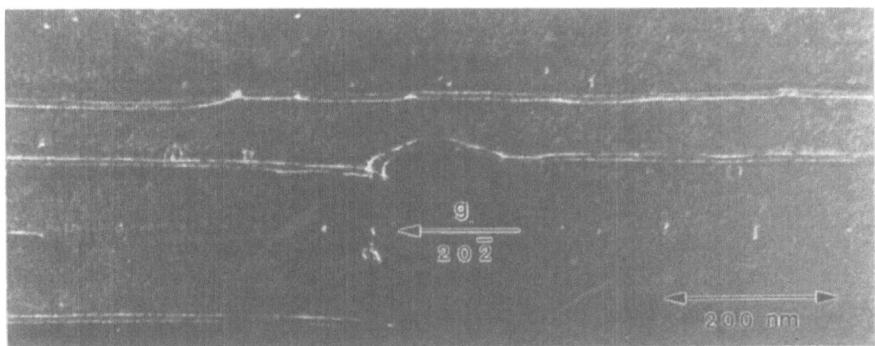

Fig.9 A TEM weak-beam image of switched superpartials at a kink. Ni_3Ga deformed at 400 C.

dipole, marked D in Fig.8e, which is expected to be very strong (owing to its small dimension). The dislocation needs to get rid of the dipole to continue propagation. There are two possible paths for the dislocation to by-pass the dipole. One is for superpartial L to join partial L' through the cross-slip and annihilation of screw segments, Fig.8f, followed by T joining T' through the processes illustrated in Fig.8g,h. In this case the dipole is left behind and the locked screw is brought back onto (111). It is noted that after the above unlocking processes, the order of the superpartials is switched backed. Another path is for partial L to join partial T' and partial T to join partial L', both through the cross-slip and annihilation of the screw segments, as illustrated in Fig.8i,j,k. In this case the dipole D is by-passed but at the same time an APB tube is created. It is noted that in this case the order of the two superpartials is still switched. The identification of APB tubes are explained in the following section.

3.4.2 *Observation of APB tubes.* The formation of APB tubes was originally postulated by Vidoz and Brown (1962) to explain the unusually high work-hardening rate observed in some ordered alloys. In this mechanism, when a APB-coupled superdislocation is intersected by a forest dislocation, two jogs are created, one on each of the two superpartials. If the forest dislocation moves during the cutting, the two jogs on the primary dislocation are not aligned along the direction of Burgers vector, as a result of which the APB created by the leading superpartial is not eliminated by the trailing superpartial and a tube of APB is formed. The Vidoz and Brown mechanism produces Stage-II work-hardening and its operation requires dislocations in a secondary slip system. Chou and Hirsch (1983) proposed that APB tubes could also be formed through the annihilation of APB-dissociated screw dipoles during Stage-I deformation.

APB tubes in Ni_3Al single crystals deformed in the region of the yield stress anomaly were first reported by Chou et al (1982). Despite extensive TEM work on L1$_2$ ordered alloys by several research groups in recent years, APB tubes have only been sporadically mentioned; the structure, the formation and the role of APB tubes in Ni_3Al, Ni_3Ga and other ordered structures are not at all clear. One of the difficulties in the TEM observation of APB tubes is their identification. According to Chou and Hirsch (1983) tube contrast is very weak and can only appear in selected superlattice reflections satisfying the following selection rule

$$\mathbf{g}.\mathbf{R} \neq integer \tag{4}$$

where $\mathbf{g}$ is the diffraction vector and $\mathbf{R}$ is the displacement vector of the APB. In this mechanism of contrast formation, APB tubes are invisible in fundamental reflections. Recent experimental work in the Oxford group has indicated that tubes produce contrast not just in superlattice reflections but also in fundamental reflections, i.e. the contrast selection rule given by equation (4) is violated . A characteristic feature of APB tubes is that they are thin and faint lines lying exactly parallel to the Burgers vector of the primary dislocations. An examination of TEM micrographs in previously published works has indicated that such features can be found, but are

190

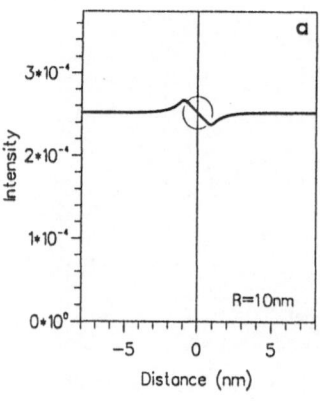 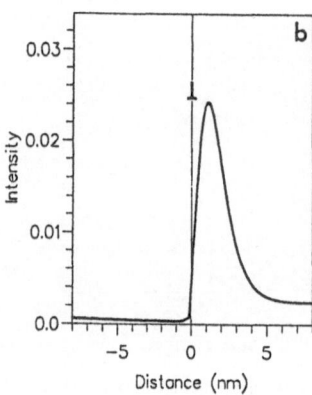

Fig.10 Calculated image profiles of a circular APB tube (a) and a 1/2<101> edge dislocation (b) under identical imaging conditions. The reflection is 020 perpendicular to the tube axis.

not mentioned, in several previous publications, including Pak, Saburi and Nennon (1976) in Ni$_3$Ga, and Korner (1988) and Dimiduk (1989) in Ni$_3$Al; these pictures were also taken in fundamental reflections.

The violation of the selection rule of (4) can be explained if an APB tube also possesses a continuous strain field in addition to the displacement R of the APB. The surface tension of the APB would cause a distortion in the material inside and outside the tube in order to balance the action of the surface tension of the APB. With a view to examining whether such a strain field would cause an observable contrast in TEM, Sun (1991) has considered the strain field of an APB tube with a circular cross-section. For a circular tube, the lattice inside the tube is in a state hydrostatic compression and the lattice outside the tube is in pure shear. Two beam dynamical theory has been used to calculate the contrast of circular APB tubes. Figures 10a,b compare the image profile of a circular APB tube with that of an edge dislocation in a fundamental reflection 020 perpendicular to the tube axis and calculated under identical imaging conditions. The tube contrast, which is substantially weaker than that of the dislocation, consists of a brighter area on one side of the tube and a darker area on the other.

Figures 11a,b are two TEM dark-field images showing the structure of a Ni$_3$Ga specimen deformed at room temperature. Figure 11a is taken in fundamental reflection 20$\bar{2}$ which is parallel to the tube axis and the tubes are invisible. Figure 11b is taken in 020, also a fundamental reflection but perpendicular to the tube axis, and many tubes are observed (dislocations are out of contrast in Fig.11b since 020 is perpendicular to the Burgers vector [10$\bar{1}$]). Note that, in agreement with calculation for circular tubes shown in Fig.10a, the tube image consists of a brighter area on one side and a darker area on the other; a typical place is pointed to by an arrow in Fig.11b. In Ni$_3$Ga, APB tubes, often with very high densities, are observed in the entire temperature region of the yield stress anomaly.

It should be noted that the additional strain field is possible only in the absence of diffusion. When diffusion is able to occur, atoms across the APB may exchange position via short range diffusion to cause the tube to shrink, a process driven by the reduction in surface energy of the APB. To test the effect of temperature on APB tubes, a Ni$_3$Ga specimen deformed at 200 C, which is known to contain APB tubes, was held at 750 C for 10 minutes. An observation after the heat treatment shows that all the tubes have disappeared. The disappearance in deformed and subsequently annealed Ni$_3$Al can be found in the work of Dimiduk (1989) who compared the defect structure of alloyed Ni$_3$Al under deformed and annealed conditions.

The high density of APB tubes observed in the present experiments can not be explained by the Vidoz-Brown mechanism since TEM observations have revealed very few forest dislocations. The density of APB tubes is often higher than that of the screw dislocations which is also very high in the regime of the yield stress anomaly. We propose that the tubes are formed through the kink mechanism, as illustrated in Fig.8i,j,k. An important property of the tubes formed through

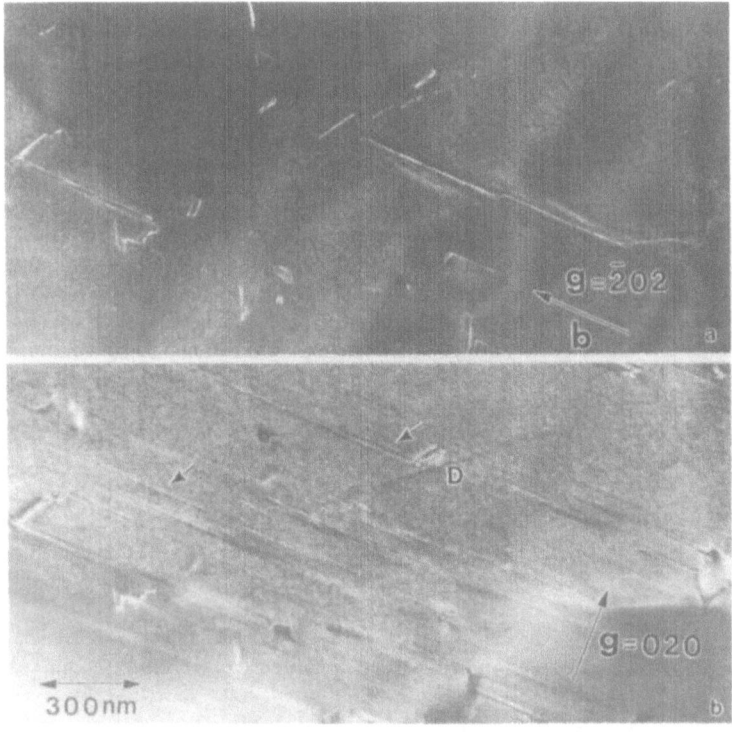

Fig.11 TEM dark-field images of dislocations and APB tubes in Ni₃Ga deformed at room temperature. Tubes are out-of-contrast in (a). (b) Tubes in the fundamental reflection 020 perpendicular to the tube axis.

the kink mechanism is that they are connected with small dipoles, a feature that has been frequently observed. An example is shown in Fig. 11b where a dipole, marked D, is seen joining an APB tube pointed to by an arrow.

In this mechanism the formation of APB tubes is just a by-product of the operation of the kink mechanism and the tube formation itself would not control the deformation. However, if the density of the tubes becomes very high, as is very likely to be the case during cyclic deformation in fatigue tests, the chance for a dislocation to meet APB tubes is larger and APB tubes would be able to affect the propagation of dislocations both through the decoupling of the superpartials, like ordinary APB's, and through their strain fields. This subject is still to be investigated further.

3.5 THE MECHANISM OF SISF-DISSOCIATED [10$\bar{1}$](111) SLIP

3.5.1 *The locking of 30° 1/3<112> super-Shockley partials.* SISF-dissociated [10$\bar{1}$](111) slip operates in Ni₃Al and Ni₃Ga at -196 C. In this regime the yield stress of Ni₃Al and Ni₃Ga is almost independent of temperature. In Pt₃Al and Co₃Ti this slip mode has been associated with the decrease of the yield stress with temperature (Suzuki et al 1989). In Co₃Ti the superdislocations are found to be SISF dissociated when unloaded but APB dissociated when a stress is applied (Liu et al 1989). In Pt₃Al no experimental observation of SISF dissociation has been reported although such a dissociation was predicted from atomistic calculations (Tichy, Pope and Vitek 1986a,b).

In Ni₃Al and Ni₃Ga weak-beam observations have shown that 1/3<112> super-Shockleys are immobile along <110> 30° directions. The first model for the locking of 30° 1/3<112> super-Shockley partials was first proposed by Giamei et al (1971). Let us consider 1/3[11$\bar{2}$], a super-Shockley partial in the primary octahedral system [10$\bar{1}$](111). Along the [10$\bar{1}$] direction, 1/3[11$\bar{2}$] may split into a screw superpartial and an edge Shockley according to

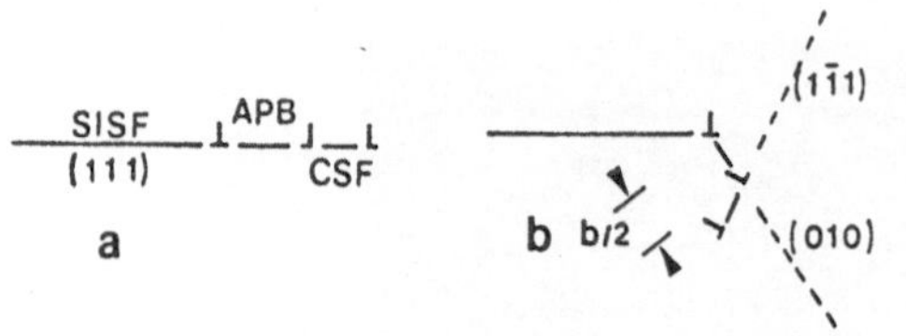

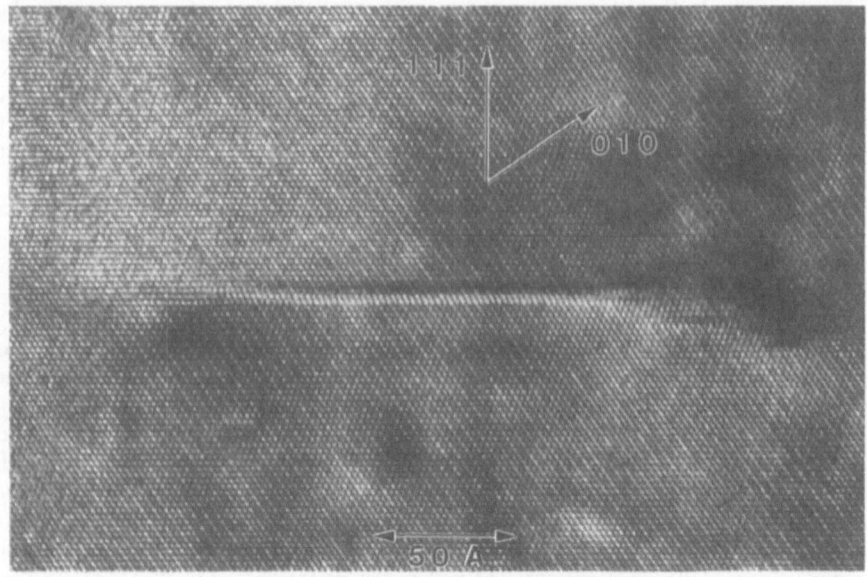

Fig.12 (a) The planar structure of a 30° $1/3<112>$ super-Shockley. (b) The non-planar structure of Giamei lock.

$$1/3[11\bar{2}] \rightarrow 1/6[\bar{1}2\bar{1}] + APB(111) + 1/2[10\bar{1}] \tag{5-1}$$

This structure, which is planar, is illustrated in Fig.12a. If the screw superpartial cross-slips onto the cube plane (010) the structure becomes non-planar, given by

$$1/3[11\bar{2}] \rightarrow 1/6[\bar{1}2\bar{1}] + APB(010) + 1/2[10\bar{1}] \tag{5-2}$$

This non-planar structure, known as a Giamei lock, is schematically shown in Fig.12b, in which the screw superpartial has jumped on (010) for $b/2$ and has further split into CSF-coupled Shockleys on $(1\bar{1}1)$. The edge Shockley is acting as a stair-rod dislocation. Using computer simulations, Tichy et al (1986a) showed that the planar mobile structure given by (5-1) is unstable in alloys with unstable {111} APB and the 30° super-Shockley is therefore always locked in a non-planar structure similar to that given by (5-2). The absence of a mobile configuration, such as that given by (5-1), led Tichy et al (1986b) to propose that the SISF dissociation in alloys with unstable {111} APB should be associated with a normal temperature dependence of the yield stress. The temperature dependence of the SISF-dissociated $<101>${111} slip in alloys with stable or metastable {111} APB has not been clearly proposed. The present observation of SISF dissociation at -196 C in Ni_3Ga and Ni_3Al would suggest that in these alloys SISF-dissociated $<101>${111} slip is associated with a moderate yield stress anomaly.

In the weak-beam observations no further splitting of the 30° $1/3<112>$ super-Shockley partials has been resolved, again because of the resolution limit of the weak-beam technique. Lattice resolution TEM has provided direct evidence that 30° $1/3<112>$ super-Shockleys are locked as Giamei locks. Figure 13 shows a lattice imaging picture of a 30° $1/3<112>$ viewed end-

Fig.13 An HREM image of a 30° $1/3<112>$ super-Shockley partial viewed end-on. Ni_3Al deformed at -196 C. The structure is non-planar with steps on {010}.

on in a thin foil parallel to (10$\bar{1}$). The wider fault on (111) is a SISF which connects two 30° super-Shockleys. It is seen that one of the super-Shockleys has a non-planar structure, appearing in the form of two steps on (010). In most of our observations the steps on (010) (usually just one) are usually smaller than the ones seen in Fig.13.

3.5.2 *Comparison of SISF-dissociated* $<101>\{111\}$ *with* $<110>\{001\}$. If a mobile configuration such as that given by equation (5-1) does not exist, SISF-dissociated [10$\bar{1}$](111) slip would be very similar to APB-dissociated [1$\bar{1}$0](001) for the following two reasons. (1) In both systems dislocations are locked along two orientations. For a super-Shockley partial in [10$\bar{1}$](111) (say 1/3[11$\bar{2}$]), locking occurs on two 30° $<110>$ directions, i.e. [10$\bar{1}$] and [0$\bar{1}$1] (along [1$\bar{1}$0], another $<110>$ direction in (111), the super-Shockley is an edge and mobile). In [1$\bar{1}$0](001) slip, locking occurs on screw and edge orientations. (2) In both systems locking occurs along every atomic row owing to the absence of planar, mobile configurations and dislocation mechanism is essentially the same as the Peierls model; in both slip modes unlocking of dislocations requires the recombination of the partials. As an approximate estimate for the energy increase caused by the recombination of partials, we use the difference in the Frank sums between fully constricted and locked states. The difference in Frank sum for a super-Shockley dissociated according to (5-2) is $3a^2/18$, which is the same as that for a screw superpartial in $<110>\{001\}$ ($3a^2/18$) and bigger than that of the edge (Lomer) superpartial in $<110>\{001\}$ ($2a^2/18$). Based on the above estimate, SISF-dissociated $<101>\{111\}$ slip in alloys with unstable $\{111\}$ APB should be somewhat more difficult than APB-dissociated $<110>\{001\}$. SISF-dissociated $<101>\{111\}$ slip has an added difficulty in that it contains dislocations with high line energies. The fact that SISF-dissociated $<101>\{111\}$ slip, despite being energetically unfavourable and associated with an expected higher energy increase required for unlocking, occurs at temperatures 600 C lower than $<101>\{001\}$ suggests that the planar core structure given by (5-1) must be stable. This is also consistent with the observation that APB on $\{111\}$ is stable or metastable in Ni_3Al and Ni_3Ga. Also, on the basis of the differences in the recombination energies and dislocation line energies, SISF-dissociated $<101>\{111\}$ slip in alloys with unstable $\{111\}$ APB should occur at some higher temperature than $<110>\{001\}$.

3.5.3. *The transition from SISF dissociation to APB dissociation.* The transition between SISF-dissociation and APB-dissociation reported in Pt_3Al and Co_3Ti is quite different from the transition between $<101>\{111\}$ and $<110>\{001\}$. In the $<101>\{111\}\leftrightarrow<110>\{001\}$ transition, the crystal chooses the softer slip while in the SISF$\leftrightarrow$APB transition the crystal chooses to deform with the harder slip mode. If a crystal deforms by several *independent* slip systems of which only one has the yield stress anomaly, common logic suggests that the anomalous regime should occur at the lowest temperatures. The SISF-dissociated and APB-dissociated $<101>\{111\}$ slip modes are not independent of each other but are related through the geometry of the arrangement of the partial dislocations. Suzuki et al (1979) pointed out that the APB dissociation could be changed into the SISF dissociation through the nucleation of a new Shockley partial which takes place under thermal activation. This mechanism, however, would only account for a transition from APB to SISF with the increase of temperature. In Fig.12b the screw superpartial has made a jump of $b/2$ on (010) and its further splitting into CSF-coupled Shockleys can therefore only

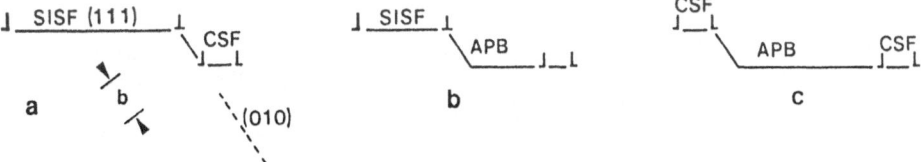

Fig.14a,b,c. Schematic illustrations to show the transition from SISF to APB on screw dislocations in $<101>\{111\}$.

occur on $(1\bar{1}1)$. Once the screw superpartial is able to make another jump of $b/2$, the splitting in CSF is on (111), Fig.14a, and the screw superpartial would move on (111), trailing behind an APB, and the trailing super-Shockley, whose Burgers vector is $1/3[2\bar{1}\bar{1}]$, would close up accordingly, Fig.14b. The trailing super-Shockley combines with the edge Shockley (stair-rod) via

$$1/3[2\bar{1}\bar{1}]+1/6[\bar{1}2\bar{1}]\rightarrow 1/2[10\bar{1}]$$

which has the same Burgers vector as the cross-slipped screw superpartial. The dissociation has therefore become APB dissociated, as shown in Fig.14c. The configuration shown in Fig.14c is the same as that shown in Fig.7c. In this mechanism the transition from SISF to APB occurs with the increase of temperature and the transition is caused by dissociation geometry. The transition from SISF to APB has been observed in *in-situ* TEM observations by Liu et al (1989*b*) who proposed a similar scheme for the transition from SISF to APB on the basis of a planar configuration, also involving the annihilation of the edge Shockley, but their argument was based on the variation of the relative values of SISF and APB energies.

4. Summary

In this paper we have aimed at elucidating the following two aspects concerning the plastic deformation of $L1_2$ ordered alloys. (1) The transitions of slip systems or dissociation modes in relation with transitions in deformation properties. (2) The unlocking of Kear-Wilsdorf locked screws in the regime of the yield stress anomaly and its relation with the formation of kinks with switched superpartials and with the formation of APB tubes.

The discussion of the transition of slip systems is based on the observations of the locking of dislocations. We have shown that in terms of dislocation self energy, the most favoured slip system is probably $<010>\{001\}$, followed by $<110>\{001\}$ and APB-dissociated $<101>\{111\}$. Unless the SISF energy is exceptionally low, SISF-dissociated $<101>\{111\}$ slip is the least favoured in terms of dislocation line energy. In experiment the order in which these slip systems occur is just the opposite: SISF-dissociated $<101>\{111\}$ slip takes place at the lowest temperature while $<010>\{001\}$ slip occurs at the highest temperature investigated. High resolution TEM observations have shown that all four slip systems contain locked dislocations. Both $<010>\{001\}$ and $<110>\{001\}$ contain two locked directions and have been shown possess similar dislocation mechanisms associated with normal temperature dependence of the yield stresses. The partial recombination energy for locks in $<010>\{001\}$ is higher than in $<110>\{001\}$ and this explains the experimental observation that $<010>\{001\}$ slip, despite being energetically more favourable, occurs at higher temperatures than $<110>\{001\}$. $1/3<112>$ super-Shockleys in SISF-dissociated $<101>\{111\}$ are also observed to be locked along two directions in Giamei locks. Its occurrence at very low temperatures is attributed to a mobile, planar core on $30°$ $1/3<113>$ super-Shockleys.

For APB-dissociated $<101>\{111\}$ slip we have concentrated on the role of kinks in causing the switching of superpartials and the formation of APB tubes, two important features that characterize the deformed structure in the regime of the yield stress anomaly. APB tubes are shown to have a continuous strain field based on the observation of their contrast in fundamental reflections.

Acknowledgements

Y.Q. Sun is grateful to the Science and Engineering Research Council and the National Physical Laboratory for financial support during the different periods in which this work was carried out. P.M. Hazzledine would like to thank the Oak Ridge Associated Universities. We would like to thank Prof. Sir Peter Hirsch for many discussions and Dr. M.A. Crimp for providing figures 6,13.

REFERENCES

Baker, I. and Schulson, E.M. (1985) 'The effect of temperature on dislocation structures in Ni_3Al', Phys. Stat. Sol. (a), 89, 163-172.

Baluc, N., Karnthaler, H.P. and Mills, M.J. (1988) Inst. Phys. Conf. Ser. No.93, 2, 463-470.

Bonneville, J., and Escaig, B. (1979) 'Cross-slipping process and the stress-orientation dependence in pure copper', Acta metall., 27, 1477-1486.

Bontemps, C., and Veyssiere, P. (1990) 'The geometrical configuration of kinks on screw super-dislocations in Ni_3Al deformed at room temperature', Phil. Mag. Lett., 61, 259-267.

Chou, C.T. and Hirsch, P.B. (1983) 'Electron microscopy of antiphase domain boundary tubes in deformed ordered alloys', Proc. R. Soc. Lond., A387, 91-104.

Chou, C.T., Hirsch, P.B., McLean, M. and Hondros, E. (1982) 'Anti-phase domain boundary tubes in Ni_3Al', Nature, 300, 621-623.

Clement, N., Caillard, D., Lours, P. and Coujou, A. (1988) '*In situ* deformation of γ' nickel base single crystals: the movement of screw dislocations in (001) planes from -100 C to 880 C', in P.O. Kettunen, T.K. Lepisto and M.E. Lehtonen (eds) Proc. ICSMA8, Pergamon Press, Oxford, pp. 205-210.

Cottrell, A.H. (1952) 'The formation of immobile dislocations during slip', Phil. Mag., 43, 645-647.

Crimp, M.A. (1989) 'HREM examination of [$\bar{1}$01] screw dislocations in Ni_3Al', Phil. Mag. Lett., 60, 45-50.

Dimiduk, D.M. (1989) 'Strengthening by substitutional solutes and the temperature dependence of the flow stress in Ni_3Al', Ph.D. Thesis, Carnegie-Mellon University.

Dimiduk, D.M. (1991) 'Dislocation structures and anomalous flow in $L1_2$ compounds', Rev. Phys. Appl., to be published.

Douin, J., Veyssiere, P. and Beauchamp, P. (1986) 'Dislocation line stability in Ni_3Al', Phil. Mag., 54, 375-393.

Duesbery, M.S. and Hirsch, P.B. (1968) 'Effect of core structure on dislocation mobility with special reference to b.c.c. metals', in A.R. Rosenfield, G.T. Hahn, A.L. Benent and R.I. Jaffee (eds.), Dislocation dynamics, McGraw Hill, New York, pp. 57-85.

Escaig, B. (1968) 'Cross-slip in f.c.c. alloys', in A.R. Rosenfield, G.T. Hahn, A.L. Benent and R.I. Jaffee (eds.), Dislocation dynamics, McGraw Hill, New York, pp. 655-674.

Ezz, S.S., Pope, D.P. and Paidar, V. (1982) 'The tension/compression flow stress asymmetry in $Ni_3(Al,Nb)$ single crystals', Acta Metall., 30, 921-926.

Flinn, P.A. (1960) 'Theory of diformation in superlattices', Trans. AIME, 218, 145-154.

Giamei, A.F., Oblak, J.M., Kear, B.H. and Rand, W.H. (1971) 'The formation of crystallographically aligned a/3<112> dislocations in Ni_3Al', Proc. 29th Ann. Meeting, The Electron Microscopy Soc. of America, Claitor's Pub. Div., Baton Range, pp. 112-113.

Hazzledine, P.M., Yoo, M.H., and Sun, Y.Q. (1989) 'The geometry of glide in Ni_3Al at temperatures above the flow stress peak', Acta metall., 37, 3235-3244.

Heredia, F.E., Tichy, G., Pope, D.P. and Vitek, V. (1989) 'Temperature and orientation dependent plastic flow in Pt_3Al', Acta metall., 37, 2755-2758.

Hirsch, P.B. (1992) 'A new theory of the anomalous yield stress in $L1_2$ alloys', this volume.

Hirth, J.P., and Lothe, J. (1982) Theory of dislocations, John Wiley & Sons, New York.

Kear, B.H. and Wilsdorf, H.G. (1962) 'Dislocation configurations in plastically deformed polycrystalline Cu_3Au alloys', Trans. AIME, 224, 382-386.

Korner, A. (1988) 'Weak-beam study of superlattice dislocations moving on cube planes in $Ni_3(Al,Ti)$ deformed at room temperature', Phil. Mag., 58,507-522.

Lall, C., Chin, S. and Pope, D.P. (1979) 'The orientation and temperature dependence of the yield stress of $Ni_3(Al,Nb)$ single crystals', Metall. Trans., 10A, 1323-1332.

Liu, Y., Takayuki, T., Izumi, O. and Ono, S. (1989a) 'The peculiar temperature and orientation dependence of $L1_2$-type $Co_{74}Ni_3Ti_{23}$ single crystals', Phil. Mag., 59A, 401-421.

Liu, Y., Takasugi, T., Izumi, O. and Takahashi, T. (1989*b*) 'TEM investigation of dislocation dissociation in L1$_2$-type Co$_{74}$Ni$_3$Ti$_{23}$ single crystals I. the effect of applied stress', Phil. Mag. 59A, 423-436.

Lomer, W.M. (1951) 'A dislocation reaction in the face-centered cubic lattice', Phil. Mag., 1951, 1327-1331.

Paidar, V., Pope, D.P. and Vitek, V. (1984) 'A theory of the anomalous yield behaviour in L1$_2$ ordered alloys', Acta Metall., 32, 435-448.

Paidar, V., Yamaguchi, M., Pope, D.P., and Vitek, V. (1982) 'Dissociation and core structure of <110> screw dislocations in L1$_2$ ordered alloys II: effects of an applied shear stress', Phil. Mag., 45A, 883-894.

Pak, H.T., Saburi, T. and Nenno, S. (1976) 'The formation mechanism of superlattice intrinsic stacking faults in Ni$_3$Ga', Script metall., 10, 1081-1085.

Silcox, J., and Hirsch, P.B. (1959) 'Direct observations of defects in quenched gold', Phil. Mag., 4, 72-89.

Staton-Bevan, A. (1983) 'The orientation and temperature dependence of the work-hardening rate of single crystal Ni$_3$(Al,Ti)', Phil. Mag., 47A, 939-949.

Sun, Y.Q. (1990) 'Dislocation processes in γ' and γ/γ' single crystals', D.Phil. Thesis, University of Oxford.

Sun, Y.Q. (1991) 'Diffraction contrast from the displacement field of APB tubes', Phil. Mag., to be published.

Suzuki, K., Ichihara, M. and Takeuchi, S. (1979) 'Dissociated structure of superlattice dislocations in Ni$_3$Ga with the L1$_2$ structure', Acta metall., 27, 193-200.

Suzuki, T., Mishima, Y. and Miura, S. (1989) 'Plastic behaviour in Ni$_3$(Al,X) single crystal: temperature, strain-rate, orientation and composition', ISIJ Inter., 29, 1-23.

Takeuchi, S. and Kuramoto, E. (1973) 'Temperature and orientation dependence of the yield stress in Ni$_3$Ga single crystals', Acta Metall., 21, 415-425.

Thornton, P.H., Davies, R.G. and Johnston, T.L. (1970) 'The temperature dependence of the flow stress of the γ' phase based upon Ni$_3$Al', Met. Trans., 1, 207-218.

Tichy, G., Vitek, V. and Pope, D.P. (1986*a*) 'Core structure and motion of <110> screw dislocations in L1$_2$ alloys with unstable antiphase boundaries on {111} planes', Phil. Mag., 53A, 467-484.

Tichy, G., Vitek, V. and Pope, D.P. (1986*b*) 'Theory of the temperature dependent plastic flow in L1$_2$ alloys with unstable APBs on {111} planes', Phil. Mag., 53A, 485-494.

Veyssiere, P. (1984) 'Weak-beam study of dislocations moving on {100} planes at 800 C in Ni$_3$Al', Phil. Mag., 50A, 189-203.

Veyssiere, P. (1989) 'Transmission electron microscrope observation of dislocations in ordered intermetallic alloys and the flow stress anomaly', in C.T. Liu, A.I. Taub, N.S. Stoloff and C.C. Koch (eds.) High-temperature ordered intermetallic alloys, MRS Proc. vol.133, pp. 175-188.

Veyssiere, P. and Douin, J. (1985) 'Dislocations with <100> Burgers vector in Ni$_3$Al', Phil. Mag. Lett., 51, L1-L4.

Veyssiere, P., Douin, J. and Beauchamp, P. (1985) 'On the presence of superlattice intrinsic stacking faults in plastically deformed Ni$_3$Al', Phil. Mag., 51A, 469-483.

Vidoz, A.E., and Brown, L.M. (1962) 'On work-hardening in ordered alloys', Phil. Mag., 7, 1167-1176.

Wee, D.M., Pope, D.P. and Vitek, V. (1983) 'Plastic flow of Pt$_3$Al single crystals', Acta metall., 32, 829-836.

Yamaguchi, M., Paidar, V., Pope, D.P., and Vitek, V. (1982) 'Dissociation and core structure of <110> screw dislocations in L1$_2$ ordered alloys I: core structure in an unstressed crystal', Phil. Mag., 45A, 867-882.

Yoo, M.H. (1987) 'Stability of superdislocations and shear faults in L1$_2$ ordered alloys', Acta metall., 35, 1559-1569.

LOCKING AND UNLOCKING OF SCREWS AND SUPERKINKS, AND THE YIELD STRESS
ANOMALY IN L1$_2$ ALLOYS

P.B. HIRSCH
University of Oxford
Department of Materials
Parks Road, Oxford OX1 3PH
UK

ABSTRACT.   A new theory is presented to explain the yield stress
anomaly for (111) slip in L1$_2$ alloys. It is shown that strong sessile
dipole barriers are formed by the superkinks at the ends of screws
which have cross-slipped from (111) to (010). These barriers stabilise
the cross-slipped screws, which slip further to form Kear-Wilsdorf
locks.   The yield stress is controlled by the unlocking of the
superkinks, which bypass the screws and Kear-Wilsdorf locks, and which
generate new mobile screws which become locked again.   The unlocking
mechanism is thermally activated with a large athermal component. The
theory accounts for the mechanical properties, including the small
strain-rate sensitivity of the yield stress, and explains many of the
electron microscope observations.   The application of the theory to
deformation of thin foils in the electron microscope is also discussed.

1. Introduction

The anomalous increase in yield stress with increasing temperature in
certain L1$_2$ alloys is well known, and the characteristics have been
reviewed by Pope and Ezz (1984), Suzuki, Mishima and Miura (1989),
Veyssière (1989), Sun (1990), Yamaguchi and Umakoshi (1990), and
Dimiduk (1991). These characteristics are:

(1) The yield stress increases with temperature at a rate equivalent to
     an apparent activation energy of ~0.2ev-0.3ev for binary
     alloys and even less for certain ternary alloys (Takeuchi and
     Kuramoto 1973, Dimiduk 1989);
(2) There is a pronounced orientation dependence (Takeuchi and Kuramoto
     1973), the increase with temperature of the yield stress being
     greatest near [$\bar{1}11$] and smallest near [001] for orientations
     of the deformation axis in the [001][011][$\bar{1}11$] triangle, the
     primary octahedral slip system being (111)[$\bar{1}01$];
(3) The yield stress exhibits a tension/compression asymmetry, the
     magnitude and sign of which is orientation dependent (Ezz,
     Pope and Paidar 1982);

197

C. T. Liu et al. (eds.), Ordered Intermetallics – Physical Metallurgy and Mechanical Behaviour, 197–216.
© 1992 *Kluwer Academic Publishers.*

(4) There is evidence that for certain orientations the yield stress for {111}<101> slip saturates at a maximum value then decreases with increasing temperature (Umakoshi, Pope and Vitek 1984);

(5) The yield stress/temperature variation in stage II is (partially) reversible (Davies and Stoloff 1965, Dimiduk 1989);

(6) The yield stress anomaly largely disappears in the microstrain region, the yield stress being nearly independent of temperature for strains $\leqslant 10^{-5}$ (Thornton, Davies and Johnston 1970);

(7) The strain-rate dependence of the yield stress in stage II is very small; $\Delta\tau/\tau \leqslant 1\%$ for strain-rate changes by a factor of ten (Thornton, Davies and Johnston 1970, Takeuchi and Kuramoto 1973);

(8) Electron microscope observations of the deformed structure show the presence of screw superdislocations, most of which are in the form of Kear Wilsdorf (KW) locks (Kear and Wilsdorf 1962), in which the screw partial dislocations bound an APB on (010). Some KW locks are bowed out in the direction of the applied stress on (010). High resolution electron microscope observations have shown the screw partials are dissociated on the (111) or (1$\bar{1}$1) planes (Baluc, Karnthaler and Mills (1988), Crimp and Hazzledine (1989), and Crimp (1989)).

(9) Electron microscope *in-situ* observations at room temperature have provided direct evidence for the presence of screw superdislocations dissociated on (111), moving in a jerky manner, sometimes with step displacements equal to the separation of screw partials on (111) (Caillard, Clément, Couret, Lours and Coujon 1988, Molenat and Caillard 1991).These workers also report apparently spontaneous transformations of the screw superdislocations between configurations dissociated on (111), (1$\bar{1}$1) and (010).

The theory of Paidar, Pope and Vitek (1984) (PPV theory) accounts for items (1)-(6) in the above list. It does not explain items (7), (8), (9). Fig.1 shows the dynamic locking and break-away model used originally by Takeuchi and Kuramoto (1973), and adapted by PPV. The leading screw partial of the [$\bar{1}$01] superdislocation cross-slips by b/2 onto (010) and pins the screw locally. The driving force for cross-slip arises from the lower antiphase boundary (APB) energy on (010) (Flinn 1960), compared to that on (111), and, as shown more recently by Yoo (1986), from a couple due to elastic anisotropy which acts on the two screw partials. The screws between pinning points advance with free flight velocity (Leibfried 1950) until a critical angle ϕ_p is reached when the barrier collapses by reverse motion of the jogs or is bypassed. The freed dislocation then advances a distance d_c which is inversely proportional to the probability of cross-slip, which is controlled by an activation energy H_{χ}. In the steady state a new pinning point is created for each one which is destroyed. The unpinning is assumed to be athermal, and the mean velocity of the screws is the free flight velocity. Following Takeuchi and Kuramoto (1973) PPV find

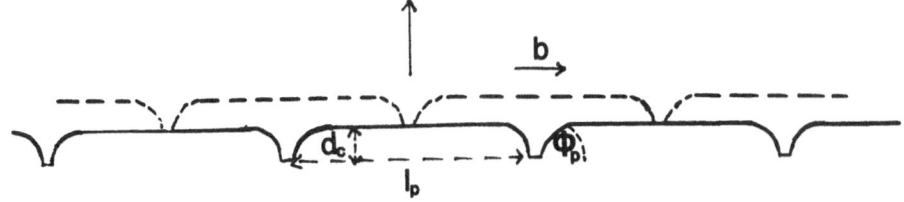

Figure 1.   Dynamic break-away model (Takeuchi and Kuramoto 1973, PPV 1984).

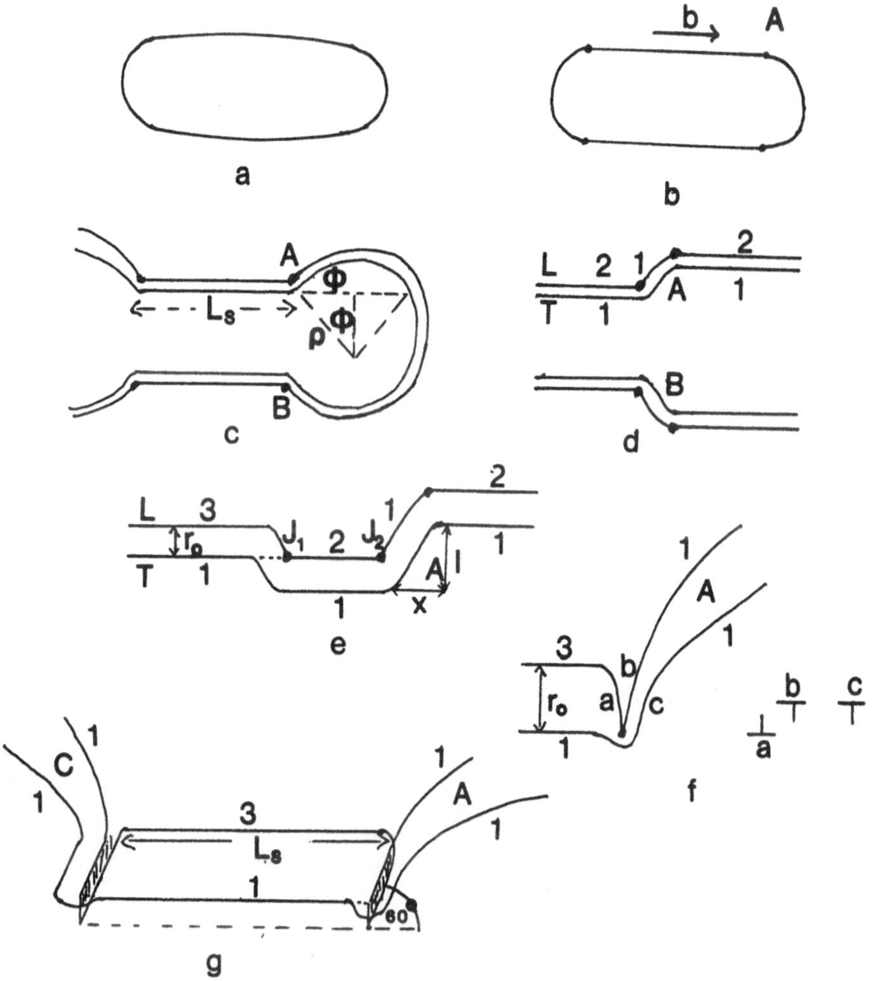

Figure 2.   Evolution of structure and formation of edge dislocation dipole barriers.   Glissile jogs are indicated by dots.   The numbers indicate different levels of planes; the original (111) plane is 1, the next (111) plane b below is 3; 2 is at a level of b/2 below 1.

$$\tau = A' \exp - H_{\ell}/3KT \tag{1}$$

where A' is a constant, H_{ℓ} the activation enthalpy for locking derived by PPV. The strain-rate is given by

$$\dot{\epsilon} = \rho_m bV = \rho_m \tau b^2/B \tag{2}$$

where ρ_m = mobile dislocation density, b the Burgers vector, V the free flight velocity, τ the resolved shear stress and B the viscous damping constant. Using $B \sim 5 \times 10^{-5}$ Nm^{-2} sec, τ=200MPa, for $\dot{\epsilon} \sim 10^{-4}$ sec^{-1} we find $\rho_m \sim 0.04 cm^{-2}$. It is not clear why only such a small fraction of the total dislocation density (typcially 10^8-$10^9 cm^{-2}$) takes part in the deformation at any time, nor why, in order to explain the small strain rate dependence, ρ_m should be inversely proportional to stress. The model must be replaced by one in which unlocking is governed by a thermally activated process, so that the average dislocation velocity is controlled by the waiting time before unlocking can occur, and the mechanism must involve a large athermal component so that the strain-rate sensitivity is small.

Another key assumption in the PPV model is that the pinning is localised. This is inconsistent with the in-situ electron microscope deformation experiments, which show convincingly that the locking mechanism leads to the formation of locks extending over long distances ($\sim 1 \mu m$ or more at room temperature) (Molénat and Caillard 1991). This implies that the jogs formed in the cross-slip process are highly mobile.

Vitek and Sodani (1991) have recently proposed a modification of the original PPV theory, designed to explain the small strain-rate dependence, in which the pinning is still assumed to be localised, but which assumes that unlocking involves a thermally activated mechanism, the nature of which is however unspecified. The theory also assumes that the steady state concentration of pinning points along the dislocation line is controlled by the activation enthalpy for cross-slip. This assumption is highly questionable (Hirsch 1991a), and in any case the implication of local pinning is inconsistent with experiment. For this and other reasons this theory is also unsatisfactory.

The object of the present paper is to suggest a model which is consistent with the above check list (1) to (9).

2. The Locking Mechanism

Fig.2 shows schematically various stages in the motion of the dislocations. Part of an expanding dislocation loop in (a) cross-slips in (b), but contrary to PPV the jogs (e.g. A) are assumed to move rapidly, but not as fast as the edge segments. Thus long lengths L_s of screw cross slip by b/2 at which level the screw partial is extended on ($1\bar{1}1$) (PPV). When ϕ in (c) reaches a critical angle the jog stops and reverses its motion, while the bulge continues to expand, until the screw cross-slips again (d). The superkink A moves to the left until it becomes locked. The locking mechanism is as follows: Suppose the leading screw partial cross-slips a second time, so that it lies again in a (111) plane. Assuming that the trailing partial remains on the original (111) plane, both partials advance by a distance equal to their separation to a new locked position, as shown in the sequence in Fig.3a,b,c. Such motions have been observed in in-situ electron microscope observations (Molénat and Caillard 1991). The leading partial is now extended on a parallel (111) plane, a distance b (projected on (010)) below the original plane. The geometry projected on (111) is shown in Fig.2e, where J_1 is the b/2 jog generated by the second cross-slip step. The different plane levels are indicated by 1,2,3. When the superkink A meets jog J_1, an edge dislocation dipole of height b (projected on ($0\bar{1}0$)) is formed. This is shown in Fig.2f, both parallel to (111) and ($\bar{1}01$). The dipole is effectively a row of vacancies or interstitials, which can lower its energy even further by lying along close packed directions (i.e. at 60° or 120° to the screw), forming low energy jog lines (Fig.2g) (Thompson 1953, Hirsch 1962). Such dipoles constitute strong barriers, stop the motion of the superkink A, and stabilise the length of cross-slipped screw L_s.

Further progress of the screw can now occur either by a bypass mechanism by the superkink, or by unlocking of the trailing dislocation by cross-slip from plane 1 to plane 3 in Fig.2g. By considering the forces acting on the screw partials, it can be shown (Hirsch 1991b) that for orientations in the centre of the triangle and for stresses below a critical value which depends on APB energies and elastic constants, the leading dislocation will more frequently continue to move down on (010) before the trailing dislocation slips down to plane 3 to unlock. This results in the formation of KW locks during the deformation.

3. Formation of Kear-Wilsdorf (KW) Locks

We have calculated the forces on dislocations A and B in Fig.3, taking into account the applied stress, the interaction stress between A and B, including the Yoo (1986) couple arising from elastic anisotropy, and the energies of APBs on (111), (010) and ($1\bar{1}1$) as appropriate (Hirsch 1991b, Chou and Hirsch unpublished). We have assumed that the screws will progress on (010) in steps of b/2, because this mechanism has the lowest activation energy (PPV, Hirsch 1991b). The paths of the dislocations can then be determined by the driving forces on the above three planes. The resistance to slip on (111) and ($1\bar{1}1$) is assumed to be negligibly small, and the dislocations move on these planes until the driving forces are effectively zero. Then they move on (010) if the driving force on that plane is finite. It turns out that for orientations in the centre of the triangle (001)(011)($\bar{1}11$) the path of AB involves one or more intermediate steps on (111) planes displaced from the original (111) plane of the dislocation, before reaching the KW configuration, which is stable. Figs.3d,e illustrate an intermediate and final KW configuration. For crystals with the [001]

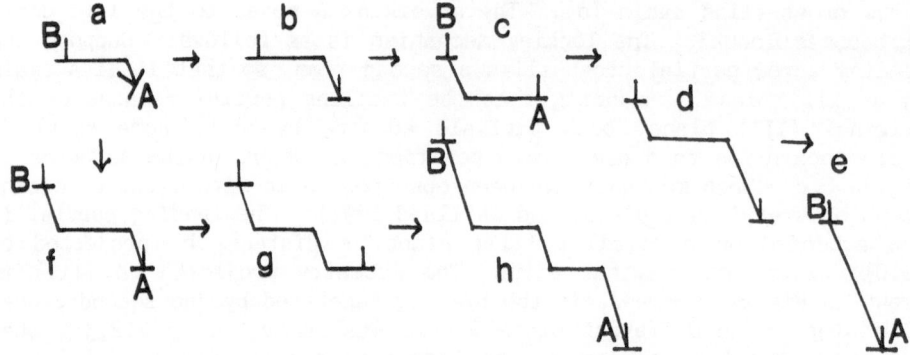

Figure 3. Sequence of locking, unlocking and locking of screw superdislocation, and formation of KW locks (schematic). The distances cross-slipped by A on (010) in a,b are b/2 and b respectively.

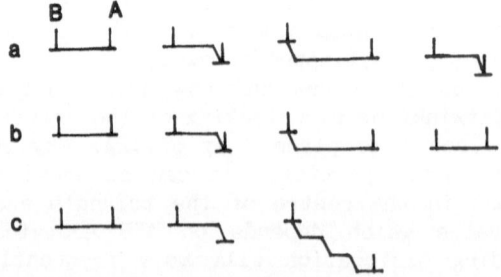

Figure 4a. Jerky motion by screws at critical stress, Eqn(3). b. Glissile motion by screws above critical stress. c. Effect of several cross-slip steps by A on structure.

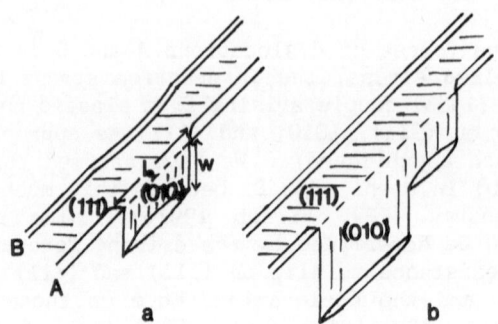

Figure 5a. Cross-slip by nucleation of double jog. b. Nucleation of further cross-slip at existing jog.

orientation, the sequence and final configuration are shown in Figs.3a,f,g,h. Note that B moves up and A moves down on (010). With increasing stress/temperature, a critical stress is reached above which the probability of KW locks being formed is reduced, the trailing partial B in Fig.3c becoming glissile on (111) by slipping down on (010) before A does so. Thus, although the probability of cross-slip from the glissile configuration to Fig.3a increases with increasing temperature, the cross-slipped dislocation will quickly transform into a glissile configuration. Fig.4b shows this sequence. This defines a yield stress of screws, at which they become effectively glissile. At lower stresses the probability of forming locks is greatest when the resolved shear stress is largest along (010), i.e. at [$\bar{1}$11], in agreement with the observed orientation dependence of the rate of increase of yield stress with temperature (Takeuchi and Kuramoto 1973). A lower limit to the yield stress for screws, τ_s, is given by the stress at which the driving force for slip down on (010) is the same for dislocations A and B in Fig.3c, i.e. (Hirsch 1991b)

$$\tau_s \sim (\gamma_1 - \gamma_o/E)/b \qquad (3)$$

where γ_o, γ_1 are the APB energies on (010) and (111) and $E = A\cos\alpha/((A-1)\cos^2\alpha + 1)$, where A is the anisotropy factor equal to 3.3 for Ni$_3$Al, and α the angle between (111) and (010) (i.e. 54°44'). For $\gamma_o = 140$mJm^{-2} and $\gamma_1 = 180$mJm^{-2} this lower limit is ~200MPa. The numbers used are illustrative; in any case τ_s could change with temperature because of a temperature dependence of $\gamma_o, \gamma_{11}E$. At the critical stress the dislocation is expected to advance in jumps equal to the distance between the two partials on the (111) plane. This is shown in Fig.4a. Such repeated jumps have been observed during in-situ experiments at room temperature (Molénat and Caillard 1991).

It should be noted that in Figs.3a,b, the separation between A and B projected onto (111) is smaller than in fig.3c because in a,b B is pressed against the sessile A dislocation, whereas in c the applied stress pulls A away from the sessile B dislocation. This extended configuration is the initial state from which the KW lock forms, because A now moves on (010). At higher stresses and temperatures there is an increasing probability of A cross-slipping again between Figs.3b and 3c, leading to configurations such as that shown in Fig.4c; τ_s under those conditions will be greater than that given by Eqn(3) (Hirsch 1991b).

So far we have assumed that the paths of the dislocations A and B are determined by the relative driving forces, but that otherwise the activation energies for slip on (010) are the same for A and B. The actual mechanism of cross-slip from (111) to (010), or of slip on (010) is considered to be that shown in Fig.5a, where the leading partial nucleates a double jog of height b/2. This has a lower activation energy than other alternatives at low stresses (Hirsch 1991b). The mechanism involves the formation of two constrictions in the leading screw partial, which is expected to be slightly dissociated on (111), bounding a complex stacking fault (PPV). If the next slip step is nucleated at an existing jog, only one new constriction is required (see Fig.5b). Estimates of the activation energies suggest that the constriction energy W_c may account for a major part of it (Hirsch 1991b). Of course this mechanism can only be initiated at an existing jog, whereas homogeneous nucleation can occur anywhere along the screw. Nevertheless, for plausible values of W_c (e.g. ~0.35ev), the jog nucleated mechanism may be favoured over a considerable range of temperatures, and A will continue to slip relatively rapidly on (010) until the driving force is zero, before B becomes unlocked. Furthermore, at higher stresses the activation energies for jumps by multiples of b/2 are close to that for b/2 so that large jogs are formed before B becomes unlocked. Both these mechanisms enhance KW lock formation and increase r_s.

The sequence of formation of KW locks involves sudden jumps; e.g. when the trailing dislocation returns to its original (111) glide plane; this happens for example between Figs.3d and 3e, resulting in sudden changes in the projected width of the dislocation. It is not clear whether such jumps could account for any of those observed by Caillard et al (1988) in in-situ experiments, and which have been interpreted by these authors as instantaneous transformations from (111) to (1$\bar{1}$1) to (010).

When the specimen is unloaded, A and B will move in opposite directions on their respective (010) planes in intermediate configurations (for example Figs.3c,d or 4a,b), forming complex locks with APBs on (111), (1$\bar{1}$1) and (010). If internal stresses are present, e.g. due to other neighbouring screws, the motion will be biased towards forming KW locks. It should be noted therefore that some of the KW locks observed post deformation by electron microscopy may have been formed on unloading. Furthermore, careful electron microscope studies should be carried out to establish whether some of the locks observed have APBs partly on (111), (1$\bar{1}$1) and (010) as expected.

4. Mechanisms of Bypassing Locks

4.1 DIPOLE LOCK

Fig.6 illustrates the mechanism of bypassing the dipole lock. When the angle ϕ in Fig.2 reaches a critical value ϕ_o in Fig.6a, it becomes energetically favourable for the superkink HI (on 111) to bypass the dipole lock. H can move to the left by forming a short screw element at Y which cross-slips from plane 1 to plane 3 annihilating the screw there of opposite sign. (This process will take place via cross-slip on (1$\bar{1}$1), and will leave a step in the APB along XY.) This results in a reduction in energy, ΔE per unit length, where

$$\Delta E = -\frac{Gb^2}{4\pi}\ln\frac{R}{r_c} - \frac{Gb^2}{2\pi}\ln\frac{R}{d_1} \tag{4}$$

where the first term is the self energy and the second the interaction energy; R is the average distance between screws in the crystal, and d_1 is the equilibrium distance of two screws of the same sign on (111); G is the shear modulus, b the Burgers vector of the underlined superpartial, and r_c

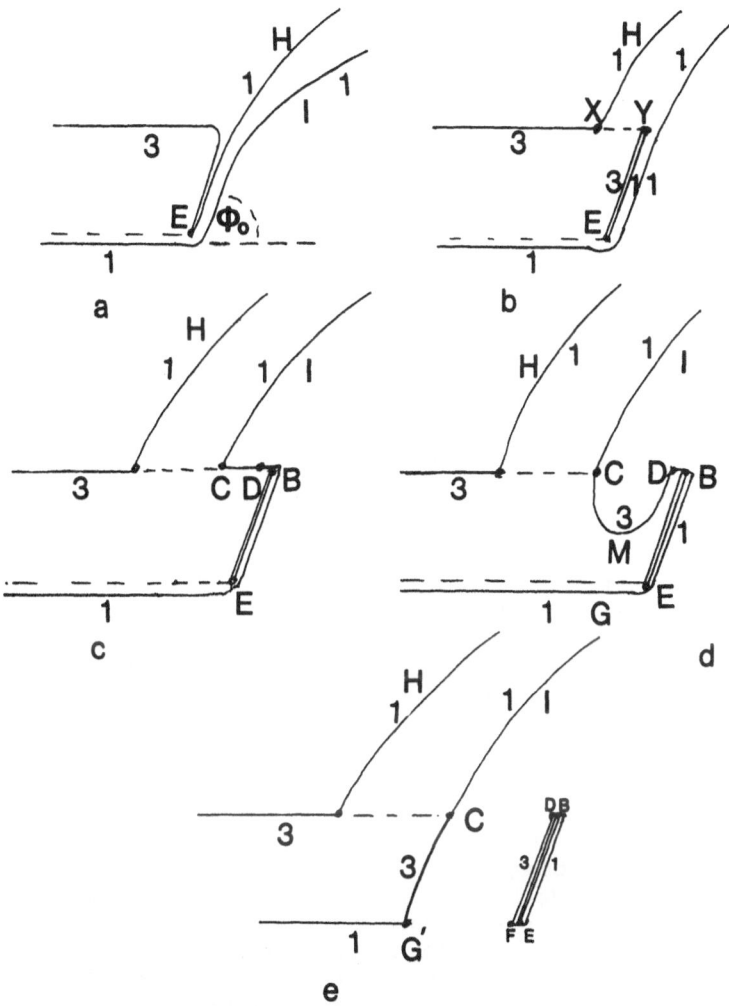

Figure 6. Bypassing of dipole by superkink; numbers indicate different levels of (111) planes.

the core-radius. When H has moved forward by ~d_1, the second dislocation I can follow H and generate a screw segment CD which cross-slips on to plane 3 forming glissile jogs at C and D (Fig.6c). CD is of opposite sign to G and bows out on plane 3 to remove the APB (Fig.6d). G and M annihilate by cross-slip as shown in Fig.6e, leaving a glissile dislocation CG' on 3 and a dipole DF. This completes the unlocking process and HI can now glide freely. (There are other unlocking paths possible, but this one is most likely since the driving force is greatest.) The driving force for the advance of I is greater than that for H, and the critical step is therefore the initial one. The condition for HI to bypass the dipole lock is

$$-\Delta E \geq T_s \cos\phi_o \tag{5}$$

where T_s is the line tension of the superdislocation $\sim 2Gb^2$. For reasonable values of R and d_1 in (1), $\cos \phi_o \sim 0.6$ ($\phi_o \sim 53°$). We have calculated the corresponding athermal stress τ_g for various ratios of x/ℓ for the superkink geometry (Fig.2e); we find

$$\tau_g = p\frac{Gb}{\ell} \tag{6}$$

where p=0.4, 0.36, 0.31 for x/ℓ=2,3,4 respectively (Hirsch 1991b). We have also shown that for $\cos \phi_o$=0.6, the superkink becomes unstable when (6) is satisfied, for x/ℓ>2. This means that new dislocations are generated which can move at low stresses before becoming locked again. This may account for the partial reversibility of the yield stress (Davies and Stoloff 1965, Dimiduk 1989), since there should be some long superkinks left after unloading which could move at low stresses and nucleate in the microstrain region the structure with an average value of ℓ characteristic of the new temperature of deformation. In addition to the athermal stress, the mechanism requires a cross-slip step which is a thermally activated process. There is therefore an additional small temperature and strain-rate dependent stress. While the superkinks wait before bypassing, the screws can cross-slip further forming KW locks.

4.2 KW LOCK

Fig.7a shows the structure of a superkink typical of an orientation in the centre of the stereographic triangle. As explained in Section 3 the formation of the lock under stress involves intermediate steps on (010) and (111) planes before both screw partials finally move on the same (010) plane forming the KW lock. Numbers 1, 2 and 3 in Fig.7a refer to (111) planes at different levels. HI is a superkink lying on the original glide plane of the dislocation (plane 1). The final KW lock is formed by screws L and M on levels labelled 4 and 3 on an (010) plane. There are glissile jogs (height b) at E, F and long jogs in the (010) plane (J_1C and J_2D) which are assumed to be sessile, since such jogs can form Lomer-Cottrell dislocations (Sun 1990). The dislocations joining C and M, and D and J_3 are lying on a (111) plane (labelled 3). As H and I move to the left under the action of the applied stress, HF will first move to the left of J_2F, as shown in Fig.7b. This switch is expected from consideration of the stability of dislocation edge dipoles J_1E, J_2F on plane 2 and HF, IE on plane 1, shown projected normal to the edge dislocations in Fig.7c. The configuration in Fig.7c will change spontaneously to configuration Fig.7d. The kink in dislocation HF near F in Fig.7b attracts and annihilates a short length of dislocation EI near E and F, and the result is a switching of partners at E and F; i.e. the leading dislocation H will now be joined to the originally trailing dislocation EJ_1CM, and the trailing dislocation I will be joined to the originally leading dislocation

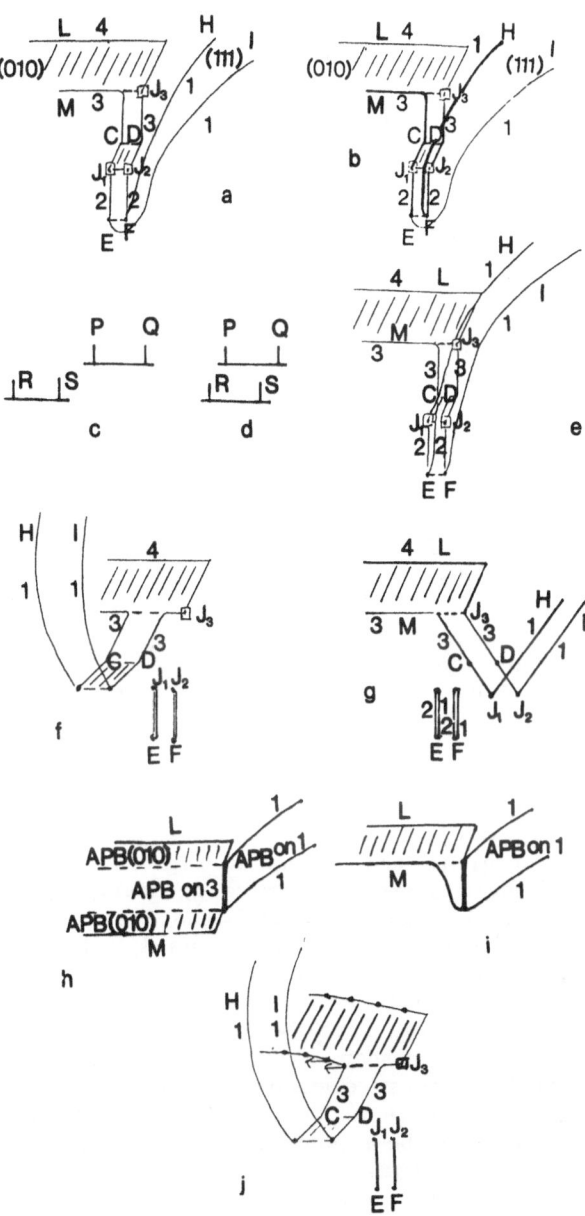

Figure 7a-f. Bypassing of KW lock by superkinks. g. Switched partial configuration. h,i. Locks formed on unloading. j. Bypassing by superkink of KW lock bowed out on (010).

FJ_2DJ_3L (Fig.7e). Suppose now that the sessile jogs J_1C and J_2D are transformed into a series of glissile elementary jogs on (010), by a thermally activated process. Dipoles J_1E and J_2F will be pinched off and HI can then advance, with the screw M on plane 3 being replaced by a screw of opposite sign, as shown in Fig.7f. The driving force for this process is the reduction in energy in replacing the KW lock by a screw dipole which can then annihilate by slip on (010). The athermal stress required for unlocking is given by Eqn(6) with similar values of p, but the activation energy of the unlocking process will be greater than that for unlocking the dipole illustrated in Fig.6.

If the stress is not large enough for the superkink to complete the unlocking process, the process of switching partials can still take place at a lower stress. On unloading line tension forces will then lead to the configuration shown in Fig.7g. Such switched partial configurations are observed by electron microscopy (Sun and Hazzledine 1988, Mills, Baluc and Karnthaler 1989, Sun 1990, Tounsi, Beauchamp, Mishima, Suzuki and Veyssière 1989, Bontemps and Veyssière 1990, Couret, Sun and Hirsch, unpublished work). Many other configurations are possible; Fig.7h is an example of a metastable lock formed on unloading, from the configuration in Fig.6a, and Fig.7i a fully formed KW lock formed from Fig.6a in the presence of an internal stress bias. In neither case are the partials switched at the ends of the lock. Such superkinks are also frequently observed by electron microscopy.

When a superkink C in Fig.8a has been freed to glide by the above mechanism, the dashed dislocations in Fig.8a will be formed. The dashed and full line dislocations between C and L are of opposite sign. The edge kinks at C,G,I,K will annihilate (or form dipoles if they are not on the same glide plane); the dashed and full line screws at F,H,J,K are of opposite sign and will take up the configurations shown schematically normal to the screws in Fig.8b. Such pairs of screws will annihilate by cross-slip forming generally APB tubes (Chou and Hirsch 1981, Chou, Hazzledine, Hirsch and Antis 1987). This leaves the structure shown in Fig.8c, where the APB tubes are labelled T. Such APB tubes are frequently observed by electron microscopy (Sun 1990).

At higher temperatures some KW locks may be bowed out on (010) as observed by electron microscopy (Veyssière 1989). A bowed out screw is likely to consist of screw elements dissociated on octahedral planes plus a series of elementary jogs which should be glissile. Thus a bowed out screw may not be significantly more difficult to bypass than a straight screw (although work will have to be done against the applied stress on (010)). This is shown in Fig.7j where the glissile elementary jogs are indicated schematically.

5. Unlocking by Homogeneous Nucleation

The alternative mechanism to bypassing the locked screws is to unlock them by homogeneous nucleation. In this process the trailing screw on level 1 in Fig.2g cross-slips down onto plane 3 of the leading screw, thus enabling the superdislocations to glide freely on (111) (see also

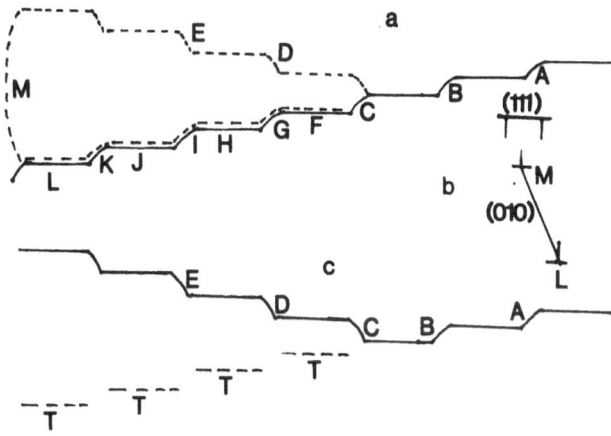

Figure 8.   Dislocation structure formed after superkink C is unlocked, and formation of APB tubes (T).

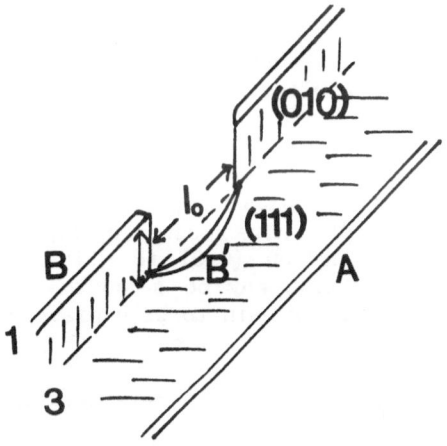

Figure 9.   Unlocking of trailing partial B by homogeneous nucleation.

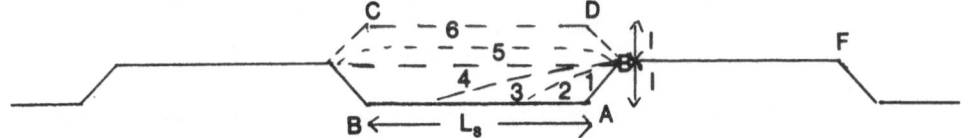

Figure 10. Steady state configuration for unlocking/locking sequence.

Figs.4a,b). The mechanism is illustrated in Fig.9. For small stresses the force on B is upwards, and therefore B has to move from plane 1 to plane 3 in one jump (i.e. the double jog has height b), the driving force arising from the work done by the applied stress and internal forces on the bowed out loop B' on the (111) glide plane. Numerical estimates of the activation energies suggest values of $2W_c + 0.83$ev at a stress of 50MPa, considerably larger than that for locking of $2W_c + 0.17$ev for $\gamma_o = 140$mJm^{-2}, $\gamma_1 = 180$mJm^{-2}, and other parameters appropriate to Ni$_3$Al (Hirsch 1991b). The activation energy decreases rapidly with increasing stress, and above $\tau_p \sim 115$MPa the force on B points downwards and the dislocation moves down in steps of b/2. At about 200MPa the activation energies of locking (which varies little with stress) and unlocking are equal. At the lowest stress (50MPa) the activation volume v is estimated as $v \sim 140b^3$ (b is the Burgers vector of the superpartial); v decreases rapidly with increasing stress to a few b^3 above 115MPa.

There is evidence from the in-situ deformation electron microscope observations on thin foils (Molénat and Caillard 1991), and from post deformation observations of the shapes of the dislocation loops (Couret, Sun and Hirsch, unpublished work) that the mechanism of homogeneous nucleation occurs at room temperature in Ni$_3$Al and Ni$_3$Ga. The question remains whether this mechanism is rate controlling in bulk deformation.

In principle it is also possible to transform a KW lock into a configuration on the (111) plane, by a mechanism somewhat similar to that in Fig.9, except that both screw partials lie on the (010) plane and the leading partial bows out on (111). Estimates suggest that activation energies at low stresses are large (several electron volts), and this mechanism is unlikely to be important at room temperature. Yet Molénat and Caillard (1991) have observed apparently instantaneous transformation from KW locks on (010) to glissile configurations on (111). A possible explanation may be a surface initiated nucleation mechanism; these observations remain to be explained.

6. The Yield Stress/Temperature/Strain-rate Relation

We assume for simplicity the steady state configuration shown in Fig.10. Slip propagates either by unlocking any of the screw segments by homogeneous nucleation, or by a superkink such as AB being unlocked at A and bypassing the screw, successive positions of the free dislocation being indicated by numbers 1 to 6. In either case the mobile screw generated in this way is locked again by cross-slip on (010) along CD. We consider first the case of bypassing by the superkink AB.

If ρ_s is the screw dislocation density, L_s the length of the screw segments, 2ℓ the distance advanced by the dislocation, the strain rate $\dot{\epsilon}$ is given by

$$\dot{\epsilon} = \frac{\rho_s}{L_s} (2\ell\ L_s f)b\ \nu_o\ \exp - H_u/KT \tag{7}$$

where f is a factor which takes into account that when one segment is unlocked a number of others could be unzipped, ν_0 the Debye frequency, and H_u the activation enthalpy for unlocking. Between the locked positions the screw is assumed to move with free flight velocity V, which, following Leibfried (1950), is given by

$$V \sim \frac{10\tau b^3 C_t}{3KT} \sim \frac{10\tau b^4 \nu_0}{3KT} \equiv A\tau \tag{8}$$

where C_t is the transverse sound velocity. The distance ℓ travelled by the screw is given by the condition

$$\nu_0 (\frac{\ell}{V}) (\frac{L_s}{\ell_c}) \exp - H_\ell /KT = 1 \tag{9}$$

where H_ℓ is the activation enthalpy for cross-slip from (111) to (010) given by PPV, ℓ_c is the critical length of nucleus for cross-slip. Hence, from (7),(8) and (9)

$$\dot{\epsilon} = \dot{\epsilon}_0 \exp -(H_u - H_\ell)/KT \tag{10}$$

where

$$\dot{\epsilon}_0 = 2\rho_s fbA\tau(\ell_c/L_s) \tag{11}$$

For the mechanism of unlocking of the screw by homogeneous nucleation, (7) should be multiplied by a factor L_s/ℓ_h where ℓ_h is the critical length of nucleus; $\dot{\epsilon}_0$ should be multiplied by the same factor.

There are now two basically different mechanisms which could control the yield stress. In the first, the strain-rate at the yield stress is maintained by the movement of edge dislocations, in the second by the movement of screws.

6.1 SUPERKINK MECHANISM

In this case the strain-rate at the yield stress is maintained by the continuous generation of freely moving edge dislocations, which trail new mobile screws, which become locked again by cross-slip after moving a certain distance, z. We consider first the statistical distribution of z. If there are N_0 screw elements (length L_s) the fraction dN/N_0 which are locked in length dz is

$$\frac{dN}{N_0} = \frac{N}{N_0} \nu_0 (\frac{L_s}{\ell_c}) \exp(-H_\ell/KT)\frac{dz}{V} \tag{12}$$

where N/N_0 is the fraction remaining at distance z.

Hence $N = N_0 \exp(-z/\ell)$ $\tag{13}$

where ℓ is given by Eqn(9). Eqn(13) is essentially that derived by Couret and Caillard (1989) in their study of the yield stress anomaly for prismatic slip in Be. Couret, Sun and Hirsch (unpublished work) have found that the distribution of superkink lengths in Ni_3Ga deformed at various temperatures follows this law, except that at room temperatures there is superimposed on this distribution a considerable number whose length is equal to the separation of the screw partials on (111). This corresponds to the motion in Fig.4a.

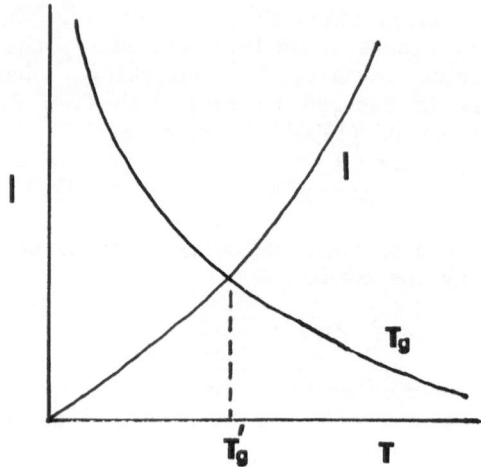

Figure 11. Variation of superkink length ℓ as a function of stress τ, and of τ_g as a function of ℓ.

Consider now the variation of ℓ with stress at a given temperature in the microstrain region. It follows from (9), (8), and assuming $L_s \propto \tau^{-1}$ (Hirsch 1991b), that $\ell \propto \tau^2$; the variation of H_ℓ with (small) stress can be neglected. Fig.11 shows this variation, as well as that of the athermal stress τ_g required to unpin the superkink (Eqn.6). It is clear that above a critical stress, labelled τ_g' on Fig.11, the stress is large enough to move a major fraction of the distribution of superkinks. We have neglected here the small thermal component of the yield stress.

In this model we therefore identify ℓ with that satisfying approximately Eqn(6). There are two mechanisms by which for a given stress ℓ can reach the critical length in Fig.11. In the first the rate controlling processes are the bypass mechanisms described in Section 4. For these we write

$$H_u = H_{uo} - (\tau - \tau_g)v \tag{14}$$

where v is the activation volume. We find that if in eqn.(6), p=0.36, $v \sim 3.1b^2(\ell_c'/b)\ell$, where ℓ_c' is the critical length for cross-slip for the mechanisms in Figs.6,7, and it follows from (6), (10), (14), that

$$\tau = \frac{(H_{uo} + 1.1Gb^3(\frac{\ell_c'}{b})) - H_\ell - KT\ln(\dot{\epsilon}_o/\dot{\epsilon})}{3.1\ b^2 \ell(\frac{\ell_c'}{b})} \tag{15}$$

Substituting from (9) for ℓ, and assuming $L_s = C/\tau$ where C is a constant, we find

$$\tau^3 = \frac{[(H_{uo}+1.1\ Gb^3(\frac{\ell'_c}{b}))-H_\ell-KT\ \ln(\dot{\epsilon}_o/\dot{\epsilon})]}{3.1b^2A\ell_c(\frac{\ell'_c}{b})}\ \nu_o C\ \exp-H_\ell/KT \tag{16}$$

The square bracket in the numerator is dominated by the term $1.1Gb^3(\ell'_c/b)$; using (8), and assuming an experimental value of C~50Pam from room temperature measurements of L_s, ℓ_c~10nm, we estimate the cube root of the pre-exponential factor to be 4400MPa (at room temperature). Dimiduk (1989) has investigated 8 binary and ternary alloys; his results suggest a spread of the corresponding pre-exponential factor from 1100MPa to 3000MPa, with an interpolated value of 2700MPa for Ni_3Al. In view of the approximate nature of the theory, the agreement seems adequate.

Since H_ℓ is the same as that in PPV, the orientation dependence and tension/ compression asymmetry are exactly the same as predicted by PPV, and in agreement with experiment. A typical value of H_ℓ to explain the experimental results is ~0.6ev. The main difference between (16) and PPV is that (16) predicts a small strain-rate dependence of τ. It is easily shown from (10) and (14), at constant structure, i.e. ℓ, and neglecting the small stress dependence of H_ℓ and of $\dot{\epsilon}_o$,

$$\frac{\Delta\tau}{\tau} \sim \frac{KT\ \Delta(\ln\dot{\epsilon})}{3.1b^2\ell(\frac{\ell'_c}{b})\tau_g} = \frac{KT\ \Delta(\ln\dot{\epsilon})}{1.1Gb^3(\frac{\ell'_c}{b})} \tag{17}$$

Assuming ℓ'_c/b~1 for the mechanism described in Fig.6, $\Delta\tau/\tau$~1% at 300°K for a change of strain rate of ten times, which is of the same order as observed experimentally (Thornton et al (1970)). The activation volume v~$3.1b^2\ell=1.1Gb^3/\tau_g$ is predicted to decrease from ~$1450b^3$ for τ_g=50MPa to ~$725b^3$ for τ_g=100MPa. Stress-relaxation experiments for Ni_3Al 1% Ta show v~$1500b^3$ and ~$600b^3$ at these stresses at room temperature and 475°K respectively (Baluc et al (1991)). Above 475° these authors detect a second mechanism for which v decreases from $2700b^3$ at 475°K to small values at the peak (~800°K). We identify this second mechanism as the bypassing of KW locks (Fig.7), which has a larger activation volume, ℓ'_c, and activation enthalpy. We would expect ℓ'_c to be of the order of the lengths J_1C and J_2D of the jogs lying in (010) in Fig.7; good agreement is obtained with experiment if ℓ'_c~4b. The by-pass model therefore accounts for the low strain rate dependence and large activation volumes observed.

In the alternative mechanism the screw BF in Fig.10 is unlocked by homogeneous nucleation, leading to an increase in length of the superkinks above its critical length. Estimates made for the mechanism in Fig.9 suggest activation volumes ~$140b^3$ at low stresses, decreasing with increasing stress. These values are too low explain the experiments. It is therefore unlikely that this mechanism is rate controlling.

The by-pass mechanism also generates edge dislocation segments (see M in Fig.8), some of which should survive on unloading. This explains the partial reversibility of the yield stress (Davies and Stoloff 1965, Dimiduk 1989), since some of these edges should be able to move at low stresses. This is also consistent with the temperature independent yield stress in the small microstrain region (Thornton et al 1970.)

6.2 YIELD STRESS CONTROLLED BY MOTION OF SCREWS

The screws form KW locks at stresses below a critical stress given by (3) at low temperatures. At the critical stress τ_s the screws move in steps of d, the equilibrium distance between the partials, as illustrated in Fig.4a. Above τ_s the screws have an increasing probability of forming glissile configurations as illustrated in Fig.4b, and a decreasing probability of forming KW locks. τ_s is effectively the yield stress for propagation of screws on (111).

Consider now the processes occurring in the microstrain region. For low stresses, edge segments of sufficient length will glide until they get blocked by dislocation interaction; long screws with very few superkinks will be formed, i.e. L_s is large. With increasing stress more loops are formed and L_s decreases. However when $\tau \sim \tau_s$, there is a rapid increase in probability of forming glissile screw segments, and therefore of generating superkinks. The questions therefore arise whether τ_s has to be reached or exceeded by the applied stress τ needed to maintain the given strain-rate, and which of the two mechanisms, unlocking of screws by homogeneous nucleation, or bypassing of locked screws by unlocking superkinks is rate controlling.

Estimates of τ_s can be made from Eqn(3), which should apply at low temperatures. Reliable values of γ_1 and γ_0 have been obtained by Molénat and Caillard (1991) and Molénat (1991) for $Ni_3Al(0.25$ at % Hf); they find $\gamma_1/\gamma_0 \sim 1.14$ at room temperature, and $\gamma_1 = 143mJm^{-2}$. These values give $\tau_s \sim 100MPa$, which is the same order as the yield stress of bulk alloys at this temperature. Couret, Sun and Hirsch (unpublished results) have obtained electron microscope evidence from the length distribution of superkinks that the homogeneous nucleation unlocking mechanism operates at room temperature in Ni_3Ga. This suggests that $\tau_s \sim \tau$ at room temperature in these alloys. The variation of τ_s with temperature is not yet known; it depends on the temperature dependence of γ_0, γ_1 and E in Eqn(3), and on the mechanisms discussed in Section 3. With increasing stress/temperature we would expect the screws to have a higher probability of becoming glissile (see Fig.4b), but also of forming KW locks by large cross-slip steps, see Section 3, particularly from the compressed configuration of Fig.3a for which the driving force for slip on (010) is maximum. This is consistent with the electron microscope observations of KW locks bowed out on (010) and therefore formed during deformation. We suggest that the superkink bypass mechanism is rate controlling at the yield stress up to the peak stress, and that contrary to the suggestion made in Hirsch (1991b), the saturation stress (Umakoshi, Pope and Vitek 1984) is not controlled by the condition of rapid unlocking of the trailing partial, but by the transformation of a KW lock into a glissile configuration on the (111) plane at the level of the leading superpartial. The details remain to be worked out, but that mechanism has qualitatively the correct orientation dependence.

It should be noted that in the case of deformation of thin foils in the electron microscope (Molénat and Caillard 1991), yielding must be controlled by the motion of screws, since superkinks, even if they are formed, run out to the surface, and cannot act as sources for new dislocations. The electron microscope observations have in fact shown that some dislocations move in jumps on (111) equal to the superpartial separation, and occasionally the screws jump by ~500Å; this behaviour is as expected at the yield stress for screws. In the bulk such jumps would generate superkinks which according to Eqn(6) could move at a stress ~120MPa. Unfortunately Molénat and Caillard do not give any estimate of stress.

7. Conclusions

The bypass model presented here accounts for most of the known mechanical property characteristics of the yield stress anomaly, items (1) to (3) and (5) to (7) in the list in Section 1, including the small strain-rate dependence of the yield stress, and the large activation volumes observed in stress relaxation experiments, provided it is assumed that the rate controlling process is that of unlocking the superkinks by the mechanisms in Figs.6,7. The mechanisms of locking and unlocking superkinks also account for several of the electron microscope observations, such as the formation of KW locks during the deformation, the structure of superkinks, (e.g. switching of dislocations), and the generation of APB tubes, i.e. item (8) in Section 1. There are several aspects of the theory which require further development, e.g. the formation of KW locks at high stresses; an understanding of the saturation stress (item (4)) is still lacking, and the assumption that $L_s \propto r^{-1}$ needs checking by computer simulation.

With regard to the in-situ electron microscope deformations (item 9), the jerky motion and the occasional large jump by screws on (111) observed by Molénat and Caillard (1991), are exactly as predicted by the theory to occur for stresses close to the yield stress for screws, which must be rate controlling for the deformation in thin foils. Unfortunately the yield stress in the thin foils is not known.

Acknowledgements

I would like to express my gratitude to the many colleagues with whom I have discussed the problems in this paper on many occasions, and from whose advice and ideas I have benefitted considerably - notably Drs. P.M. Hazzledine, Y.Q. Sun, M.A. Crimp, S. Ezz, T.C. Chou, A. Couret, D. Caillard, D. Dimiduk, A. Korner, H.P. Karnthaler, P. Veyssière and Professors V. Vitek and F.R.N. Nabarro.

References

Baluc, N., Karnthaler, H.P. and Mills, M.J., (1988) *Instit. Phys. Conf. Ser. No.93, Vol.2, Ch.13 (Instit. of Phys. Publishing Ltd.)*, 463.

216

Baluc, N., Mills, M.J., Bonneville, J. and Stoiber, J., (1991) *Rev. Phys. Appl.* in press.

Bontemps, C. and Veyssière, P., (1990) *Phil. Mag. Lettres* **61**, 259.

Caillard, D., Clément, N., Couret, A., Lours, P. and Coujon, A., (1988) *Phil. Mag. Lettres* **58**, 263.

Chou, C.T., Hazzledine, P.M., Hirsch, P.B. and Anstis, G.R., (1987) *Phil. Mag.* **56**, 799.

Chou, C.T. and Hirsch, P.B., (1981) *Phil. Mag.* **44**, 1415.

Couret, A. and Caillard, D., (1989) *Phil. Mag.* **59**, 801.

Couret, A. and Sun, Y.Q., (1991) unpublished work.

Crimp, M.A., (1989) *Phil. Mag. Lettres* **60**, 45.

Crimp, M.A. and Hazzledine, P.M., (1989) in High Temperature Ordered Intermetallic Alloys III (Eds. Kock, C.C., Liu, C.T., Stoloff, N.S. and Taub, A.I.), *MRS Sympos. Proceedings* **133**, 131.

Davies, R.G. and Stoloff, N.S., (1965) *Trans. TMS AIME*, **233**, 714.

Dimiduk, D.M., (1989) Ph.D. Thesis, Carnegie-Mellon University.

Dimiduk, D.M., (1991) *J.de Phys. III*, **1** No.6, 1025.

Ezz, S.S., Pope, D.P. and Paidar, V., (1982) *Acta Metall* **30**, 921.

Flinn, P., (1960) *Trans. TMS-AIME* **218**, 145.

Hirsch, P.B., (1962) *Phil. Mag.* **7**, 67.

Hirsch, P.B., (1991a) *J.de Phys. III*, **1** No.6, 989.

Hirsch, P.B., (1991b) *Phil. Mag.*, in press.

Kear, B.H. and Wilsdorf, H.G., (1962) *Trans. TMS-AIME* **224**, 382.

Leibfried, G., (1950) *Z. Phys.* **127**, 344.

Mills, M.J., Baluc, N. and Karnthaler, (1989) in High Temperature Ordered Intermetallic Alloys III (Eds. Kock, C.C., Liu, C.T., Stoloff, N.S. and Taub, A.I.), *MRS Sympos. Proceedings* **133**, 203.

Molénat, G., (1991) Ph.D. Thesis.

Molenat, G. and Caillard, D., (1991) *Rev. Phys. Appl.*, in press.

Paidar, V., Pope, D.P. and Vitek, V., (1984) *Acta Metall.* **32**, 435.

Pope, D.P. and Ezz, S.S., (1984) *Intern. Metals Reviews* **29**, 136.

Sun, Y.Q., (1990) D.Phil. Thesis, University of Oxford.

Sun, Y.Q. and Hazzledine, P.M., (1988) *Phil. Mag.* A **58**, 603.

Suzuki, T., Mishima, Y. and Miura, S., (1989) *Trans. Iron Steel Inst. Jpn.* **29**, 1.

Takeuchi, S. and Kuramoto, E., (1973) *Acta Metall.* **21**, 415.

Thompson, N., (1953) *Proc. Phys. Soc. Lond.* B **66**, 481.

Thornton, P.H., Davies, R.G. and Johnston, T.L., (1970) *Metall. Trans.* **1**, 207.

Tounsi, B., Beauchamp, P., Mishima, Y., Suzuki, T. and Veyssière, P., (1989) in High Temperature Ordered Intermetallic Alloys III (Eds. Kock, C.C., Liu, C.T., Stoloff, N.S. and Taub, A.I.), *MRS Sympos. Proceedings* **133**, 731.

Umakoshi, Y., Pope, D.P. and Vitek, V., (1984) *Acta Metall.* **32**, 449.

Veyssière, P., (1989) in High Temperature Ordered Intermetallic Alloys III (Eds. Kock, C.C., Liu, C.T., Stoloff, N.S. and Taub, A.I.), *MRS Sympos. Proceedings* **133**, 175.

Veyssière, P., Douin, J. and Beauchamp, P., (1985) *Phil.Mag.* A **51**, 469.

Vitek, V. and Sodani, Y., (1991) *Scripta Met.*, in press.

Yamaguchi, M. and Umakoshi, Y. (1990) *Progr. in Mat. Science* **34**, 1.

Yoo, M.H., (1986) *Scripta Metall.* **20**, 915.

DEFORMATION BEHAVIOR OF TiAl COMPOUNDS WITH THE TiAl/Ti$_3$Al LAMELLAR MICROSTRUCTURE

M. YAMAGUCHI and H. INUI
Department of Metal Science and Technology
Kyoto University
Sakyo-ku, Kyoto 606, Japan

ABSTRACT. In the last few years, substantial progress has been made in our understanding of the microstructure and the deformation and fracture behavior of TiAl compounds. We have made uniaxial tension tests of specimens of the TiAl compounds whose lamellar orientation is controlled so that shear deformation occurs parallel to the lamellar boundaries only in the TiAl lamellae. Then, the TiAl phase itself, which is in equilibrium with the Ti$_3$Al phase, has been found to be easily deformable, and yet brittle fracture occurs in the TiAl phase when the specimens fail. This paper reviews such a deformation and fracture behavior of the TiAl phase together with some key aspects of the progress that was made on TiAl compounds with the TiAl/Ti$_3$Al lamellar microstructure.

1. INTRODUCTION

TiAl possesses a wide composition range. However, it extends primarily on the Al-rich side and the TiAl compounds, with nearly stoichiometric or Ti-rich compositions, exhibit a two-phase micro-structure composed of the TiAl phase and a small volume fraction of the Ti$_3$Al phase. The so-called TiAl-based alloys, on which there has been a recent enormous increase in the research and development activity, are such two-phase compounds rather than Al-rich single-phase compounds (for recent reviews, see [1]~[5]). This is because the two-phase compounds are somewhat more ductile than the single-phase compounds [6]. However, the room-temperature tensile ductility of the two-phase compounds is limited to 2-3%. Why is the room-temperature ductility limited to 2-3%? Is it an inherent property of the TiAl phase which is the major constituent phase of the two-phase TiAl

C. T. Liu et al. (eds.), Ordered Intermetallics – Physical Metallurgy and Mechanical Behaviour, 217–235.
© 1992 *Kluwer Academic Publishers.*

compounds? How does the lamellar structure affect the deformation behavior of the TiAl phase itself? For further improvement in their ductility, studying these questions should be given the highest priority.

When the two-phase TiAl compounds are prepared by usual ingot-metallurgy methods, TiAl and Ti_3Al phases constitute a lamellar structure. Ingots are usually composed of randomly oriented grains with such a lamellar structure. However, when such ingots are remelted and unidirectionally solidified at an appropriate rate, single-crystal-like ingots (PST crystals[†]) composed of only a single grain with the lamellar structure can be obtained. We have made a systematic study on the deformation behavior of such PST crystals in order to find answers to the above mentioned questions.

The results obtained so far can be summarized as follows :

(1) Yield stress depends strikingly on the angle between the lamellar boundaries and loading axis. It is high when the lamellar boundaries are parallel or perpendicular to the loading axis and it is very low for intermediate orientations [1,7].

(2) The above result indicates that two deformation modes, i.e. hard and easy modes, can occur in PST crystals of TiAl. The hard and the easy modes of deformation were found to correspond to shear deformation across the lamellar boundaries and shear deformation parallel to the lamellar boundaries, respectively [7]. The easy mode of deformation is found to be the deformation of the TiAl phase itself in equilibrium with the Ti_3Al phase.

(3) Tensile elongation at room temperature can be as large as 20% for specimens which deform in the easy mode. Furthermore, when PST crystals of TiAl are oriented such that the easy mode of deformation is operative and produces length strain during rolling, they can be rolled to about 50% reduction in thickness at room temperature. Thus, the TiAl phase coexisting with the Ti_3Al phase is seen to be easily deformable.

Then, why is the room-temperature ductility of polycrystalline compounds with the $TiAl/Ti_3Al$ two-phase structure limited to 2-3%? We still continue our systematic study on PST crystals of TiAl to find the answer to this question. The principal part of this paper is devoted to the review of recent results from our systematic study. We do not consider the Al-rich single-phase TiAl compounds whose deform-

[†]Since numerous thin twin related TiAl lamellae are contained in the single-crystal-like ingots, we call these ingots <u>polysynthetically twinned (PST) crystals</u> from analogy with the phenomenon, <u>polysynthetic twinning</u> which is often observed in mineral crystals.

ation behavior is reviewed in detail by Whang in this proceedings [8].
The readers are also recommended to refer to recent general reviews on
the deformation behavior of intermetallic compounds [1,9].

2. PST CRYSTALS OF TWO-PHASE TiAl COMPOUNDS

2.1. CRYSTAL GROWTH

The master ingot was produced by melting high-purity titanium and
aluminum in a plasma arc-furnace, yielding an ingot of composition Ti-
49.3at%Al. Rods, 10 mm in diameter and 100 mm long, were cut from
the master ingot. These rods were remelted and solidified unidirectionally using an ASGAL FZ-SS35W floating zone furnace at a solidification rate of 5 mm h^{-1} under an Ar gas flow [7]. PST crystals have
been found to grow at solidification rates slower than 5 mm h^{-1}. The
average lamellar spacing depends on the solidification rate. At
solidification rates slower than 5 mm h^{-1}, the faster the solidification rate is, the wider the lamellar spacing is [10]. Decreasing the
solidification rate from 5 mm h^{-1} to 2.5 mm h^{-1} was reported to reduce
the lamellar spacing by about 40% [10]. Orientations of as-grown PST
crystals were determined using both X-ray and electron diffraction
methods. When the X-ray beam is directed perpendicularly to the
lamellar plane, ⟨110⟩ directions common to both the matrix and twin in
the twinned TiAl lamellae can be easily determined. Then, oriented
compression specimens, approximately 3 mm x 3 mm in cross-section and
6 mm long and tension specimens, approximately 2 mm x 0.5 mm in cross-
section and 5 mm in gauge length were cut from as-grown crystals.

2.2. MICROSTRUCTURE

Recent high temperature X-ray diffraction and solidification
experiments for the Ti-Al system, have shown that the phase in equilibrium with the liquid phase at a nearly stoichiometric composition is
the h.c.p. solid solution (α), which is based on the structure of
αTi [11]. When our PST crystals are held at temperatures where only
the α phase is stable, the TiAl phase disappears and a single crystal
of the α-phase is obtained. Then, when such a single crystal of the
α phase is held at temperatures in the range of 1200-1350℃
corresponding to the $\alpha + \gamma$ (TiAl) two-phase field, the lamellar
structure is restored [12]. Figure 1 shows the microstructure of a
PST crystal quenched into ice water from a temperature (1300℃) in the
$\alpha + \gamma$ two-phase field. It is seen that the lamellar structure is
formed and it is composed of lamellae of the γ phase and the α_2

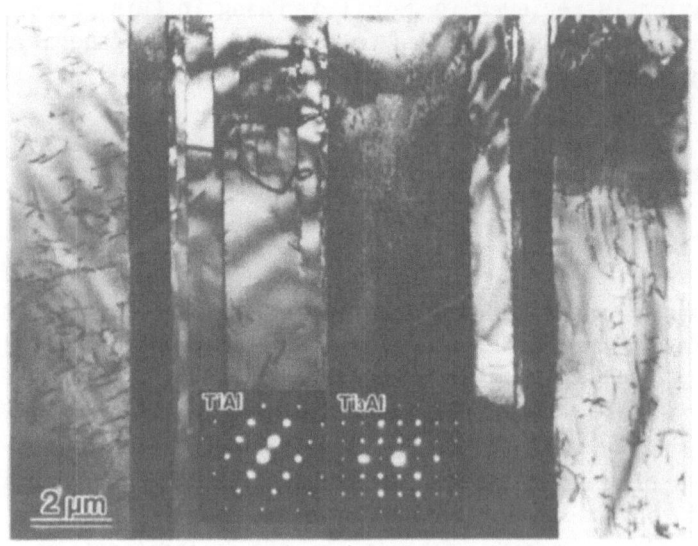

Figure 1. Microstructure of a TiAl PST crystal quenched from
1300℃ to ice–water and diffraction patterns from TiAl and Ti₃Al
(α_2) lamellae.

phase containing numerous antiphase domains [12]. The existence of
such antiphase domains in the lamellae of the α_2 phase indicates that
the α_2 phase is disordered at 1300℃, otherwise antiphase domains may
not be formed. This leads us to a conclusion that the lamellar
structure in the present PST crystals is formed during the reaction,
$\alpha \rightarrow \alpha + \gamma$ and therefore the ordering of the α phase to the
α_2(Ti₃Al) phase with the DO_{19} structure occurs after the formation of
lamellar structure, which has been already observed in the previous
reports [2,13]. Microstructures in PST crystals of TiAl quenched from
temperatures in the range 1000–1500℃ and their implications for the
formation mechanism of the lamellar structure will be reported
elsewhere.

TiAl and Ti₃Al phases constituting the lamellar structure in the
two-phase TiAl compounds have been shown to have the orientation
relationship of

$$\{111\}_{TiAl} /\!/ (0001)_{Ti3Al} \quad \text{and} \quad \langle \bar{1}10 \rangle_{TiAl} /\!/ \langle 11\bar{2}0 \rangle_{Ti3Al} \,.$$

Close-packed $\langle 11\bar{2}0 \rangle$ directions on the basal plane in both α_2 and α

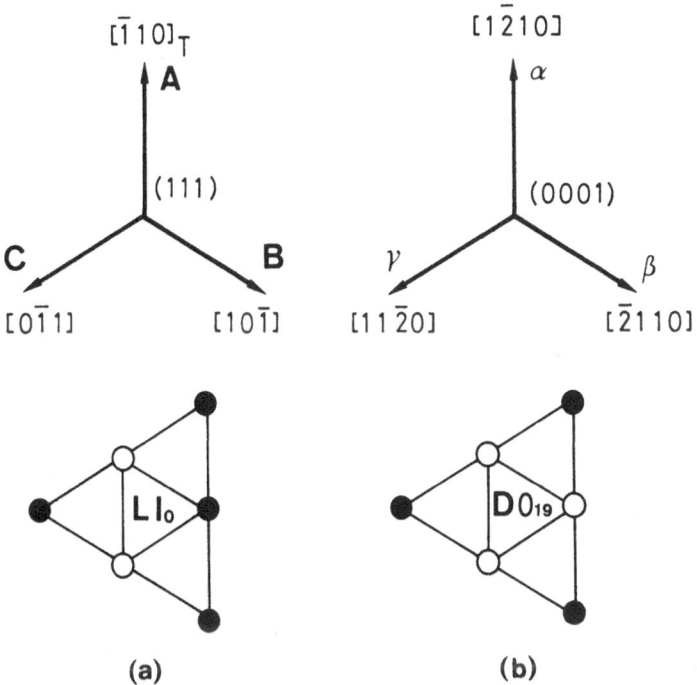

Figure 2. (a) ⟨ī10⟩ directions and the atomic arrangement on a (111) plane in TiAl and (b) ⟨11$\bar{2}$0⟩ directions and the atomic arrangement on the basal plane in Ti_3Al.

phases are all equivalent. However, ⟨ī10]† and ⟨ī01] directions on {111} in the TiAl phase with the tetragonal Ll_0 structure are not equivalent to each other (Figs. 2(a) and (b)). This suggests the following two points. Firstly, the $\alpha \rightarrow \alpha_2$ transformation does not affect the orientation relationship. For nearly stoichiometric compositions, the lamellar structure is formed during the reaction, $\alpha \rightarrow \alpha + \gamma$ and maintained through the reaction, $\alpha + \gamma \rightarrow \alpha_2 + \gamma$ [2,13]. Secondly, the TiAl phase can be formed in six orientation variants corresponding to the six possible orientations of ⟨ī10] on {111} in the TiAl phase with respect to ⟨11$\bar{2}$0⟩ on (0001) in the α phase, i.e.

---

†The mixed notations {hkl} and ⟨uvw] are used to differentiate the first two indices from the third one which does not play the same role as the first two because of the tetragonality of the Ll_0 structure. Directions and planes in superlattices with the DO_{19} structure are referred to the underlying h.c.p. lattice for the sake of simplicity.

and

$$A \mathbin{/\!\!/} \alpha, \quad A \mathbin{/\!\!/} \beta, \quad A \mathbin{/\!\!/} \gamma$$

$$A \mathbin{/\!\!/} \alpha, \quad A \mathbin{/\!\!/} \beta, \quad A \mathbin{/\!\!/} \gamma,$$

where A is, for example, [$\bar{1}$10] on (111) in the TiAl phase as shown in
Fig. 2(a) and α, β, γ are three ⟨11$\bar{2}$0⟩ directions on the (0001)
plane in the α phase (Fig. 2(b)) [1]. $\mathbin{/\!\!/}$ and $\mathbin{/\!\!/}$ mean parallel and
antiparallel. This suggests that a domain structure composed of six
domains corresponding to these six orientation variants will result in
the TiAl phase coexisting with the Ti$_3$Al phase [1,14,15]. Orientation
relationships between the two neighboring domains can be described as
follows :

$$A \mathbin{/\!\!/} A', \quad A \mathbin{/\!\!/} B', \quad A \mathbin{/\!\!/} C'$$
$$A \mathbin{/\!\!/} A', \quad A \mathbin{/\!\!/} B', \quad A \mathbin{/\!\!/} C', \tag{1}$$

where A, B, C and A', B', C' are three ⟨$\bar{1}$10⟩ directions (Fig. 2(a)) in
each of the two domains, respectively.

When $A \mathbin{/\!\!/} A'$, a translational order-fault-type boundary is formed
between the two neighboring domains, otherwise no interface is produced. For the other two parallel cases, which are equivalent to each
other, a 120°-rotational order-fault-type domain boundary is formed.
In this case, the c-axes of the two neighboring domains are perpendicular to each other, but the stacking sequence of close-packed planes
abcabc⋯ is preserved across the domain boundary. When $A \mathbin{/\!\!/} A'$, two
adjacent domains have the (111)[11$\bar{2}$] true twin relation. For the other
two antiparallel cases, which are equivalent to each other, pseudo-
twins are formed. For these antiparallel cases, the stacking sequence
of close-packed planes is reversed across the domain boundary.

The growth of TiAl lamellae in the direction perpendicular to the
basal plane of the α phase gives rise to the occurrence of domain
boundaries parallel to the lamellar boundaries. In other words, the
lamellar structure of PST crystals of TiAl is composed of TiAl lamellae separated by such domain boundaries and Ti$_3$Al lamellae. Each TiAl
lamella also has a domain structure as shown in Figs. 3(a) and (b).
Figure 3(a) shows domains observed on the surface parallel to the
lamellar boundaries. Figure 3(b) shows an electron microscope image
of such domains in a thin foil whose foil plane is parallel to the
lamellar boundaries. The figure is viewed along [111]. The domains
A, B and C are indeed related by 120°-rotation about the foil normal.
It has been found that the orientation relationship between any two of
the domains within any TiAl lamellae is described by one of the three
parallel orientation relationships in eq. 1 [15,16]. Although the
reason for this has not been made clear yet, this may have important

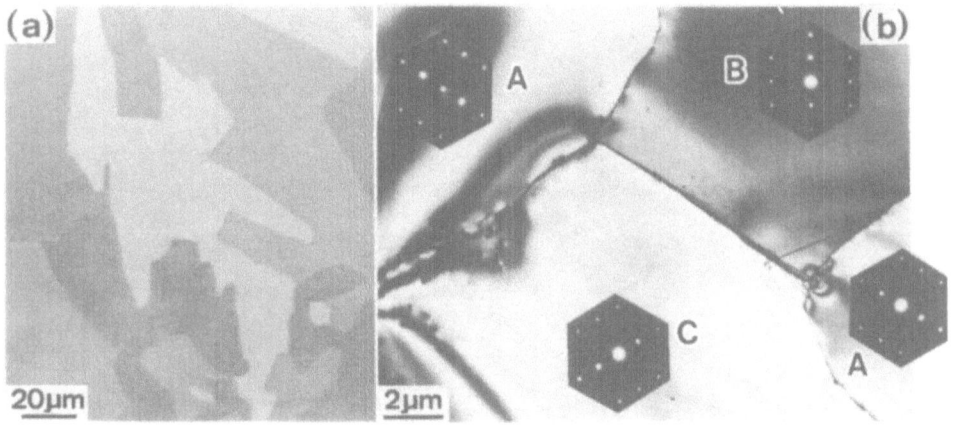

Figure 3.   (a) Optical and (b) transmission electron micrographs of
domains observed along the direction perpendicular to the lamellar
boundaries in an as-grown TiAl PST crystal [1,18].

implications for the formation mechanism of the lamellar structure.

        While the domain boundaries in TiAl lamellae do not show any
crystallographic preference as seen in Figs. 3(a) and 3(b), the
boundary plane of TiAl lamellae is always {111} and atomistically flat
except for the sporadic presence of ledges of two $(111)_{TiAl}$ planes
high [16].   This has been found to be the case for all the TiAl/TiAl
lamellar boundaries corresponding to six orientation relationships
given in eq. 1 [16].   Figure 4 shows a result of HREM observations of
domain boundaries between two neighboring TiAl lamellae which are
related by A ∥B′ or A ∥C′.   In the figure, the boundary is located
horizontally at the center of the figure and the upper and the lower
halves are TiAl lamellae viewed along [$\bar{1}$10] and [0$\bar{1}$1] directions,
respectively [16].   The boundary is seen to be, in general, atomistically very flat, although a ledge of two (111) planes high exists
between arrows A and B forming a pair at both the left- and right-hand
sides in the figure.   The dark contrast associated with the ledge can
be seen between arrows A and B, suggesting the presence of defects
responsible for the ledge.   An enlargement of the framed part in Fig.
4 is shown in Fig. 5.   It should be noted in the figure that the Ti_3Al
phase is not present at the boundary between the two TiAl lamellae.
Our recent HREM observations of domain boundaries between TiAl
lamellae show that the Ti_3Al phase generally does not exist even
between two neighboring TiAl lamellae with pseudotwin relationships.
One of the authors used to believe that the energy of the TiAl/TiAl

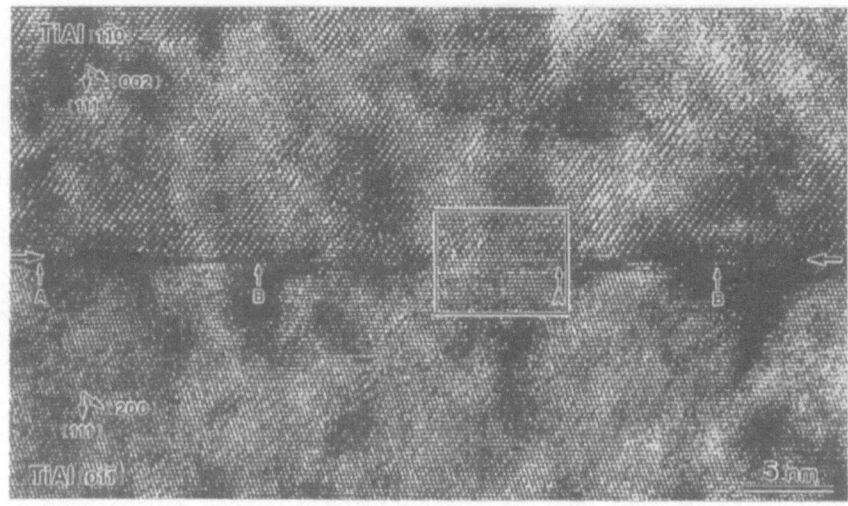

Figure 4. HREM image of a TiAl$_{[1\bar{1}0]}$/TiAl$_{[011]}$ interface parallel to the lamellar boundaries in an as-grown TiAl PST crystal [16].

Figure 5. An enlargement of the framed part in Fig. 4 [16].

lamellar boundaries corresponding to pseudotwin relationships is high because of wrong bondings across the boundaries and therefore the TiAl/Ti$_3$Al/TiAl sandwich structure is formed to lower the energy of such interfaces [1]. However, our recent HREM observations do not support this idea. The energy of the pseudotwin-type lamellar boundaries should be at least lower than twice the energy of the TiAl/Ti$_3$Al lamellar boundaries. Energies of TiAl/TiAl lamellar domain boundaries have been estimated in [18] on the basis of hard sphere model.

Formation of ordered domains is thought to play a roll for accommodations of strains created by the hexagonal-to-tetragonal transformation. In particular, the c-axes of domains related by 120°-rotation are perpendicular to each other. Such a perpendicular arrangement of the c-axes of ordered domains has been found also in the L1$_0$ CuAu and it has been believed to accommodate the lattice strains associated with the cubic-to-tetragonal transformaton [17]. More detailed observations of the domain structure in the TiAl phase will be reported elsewhere [18].

2.3. DEFORMATION BEHAVIOR AT ROOM TEMPERATURE

Figure 6 shows the average values of yield stress for specimens with various lamellar orientations as a function of the angle ϕ between the loading axis and the lamellar boundaries. The specimens with $\phi = 31°$, 51° and 68° deformed in the easy mode and those with $\phi = 0°$ and 90° in the hard mode. No significant difference in yield stress between tension and compression was observed. Thus, we may conclude that no tension-compression asymmetry in yield stress exists in the two-phase TiAl PST crystals. At room temperature, both the true twinning of the {111}⟨11$\bar{2}$]-type and slip on {111}⟨$\bar{1}$10] are operative for both the hard and easy modes of deformation. However, deformation twinning is the principal mechanism of deformation of the TiAl phase in the present PST crystals and slip along ⟨110] is the complementary one to the deformation twinning. Both deformation twinning and slip along ⟨110] are often operative in same TiAl domains, however, in such domains the largest value of Schmid factors for the {111}⟨110] slip is always much higher than that of Schmid factors for the {111}⟨11$\bar{2}$] twinning [19]. In specimens with $\phi = 68°$, the hard mode of deformation has been observed to be mixed with the easy mode of deformation, although the latter mode dominates over the former mode. The transition from the easy mode to the hard mode may be due to the Schmid factor ratios between primary and secondary slip and twinning systems.

The existence of lamellar boundaries is believed to constitute effective barriers to the propagation of deformation twinning and slip

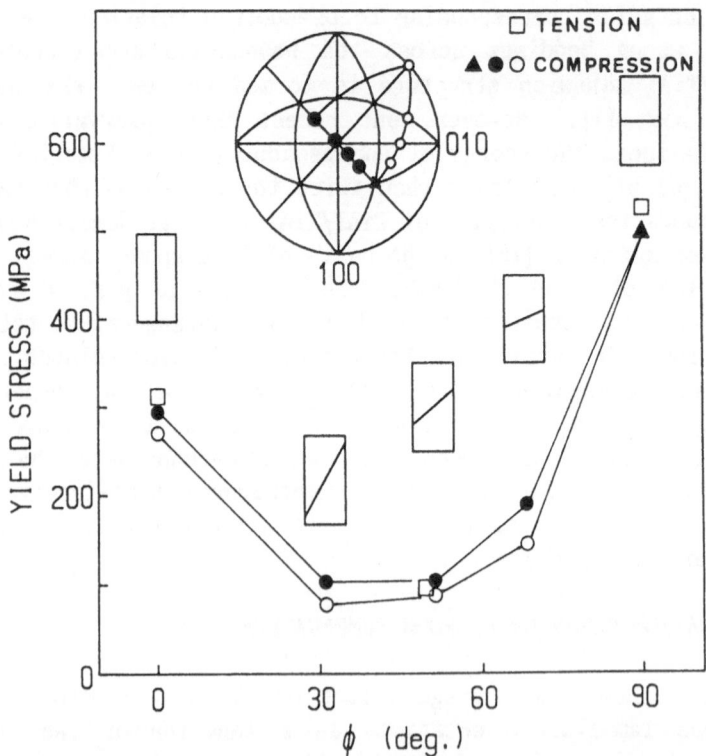

Figure 6. Orientation dependence of yield stress of TiAl PST crystals.

across them. At the same time, there is a large difference in the domain size between the two directions perpendicular and parallel to the lamellar boundaries. They are about 2μm and 50μm, respectively for the present TiAl PST crystals [18]. Therefore, the large difference in yield stress between the hard and easy modes of deformation appears to be explained mainly in terms of the Hall-Petch relationship [19].

Figure 7 shows a microstructure observed in a specimen deformed in compression along an easy direction inclined at 31° to the lamellar boundaries. Compression strain was about 1%. Deformation in the region indicated by T occurs by true twinning whose twinning plane is parallel to the lamellar boundaries. Such a twinning system can be operative only when it produces compressive strain in compression or tensile strain in tension. For the orientation of the specimen of Fig. 7, twinning parallel to the lamellar boundaries can occur in domains of two types of the six different types. In domains of four other types where twinning can not occur, deformation is thought to have occurred by slip along ⟨$\bar{1}$10] on {111} since the Burgers vector of

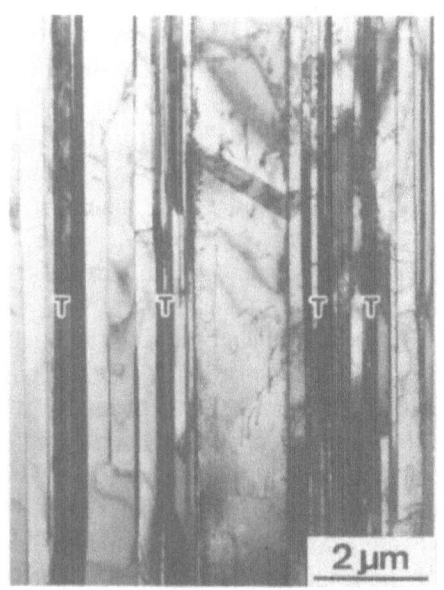 

Fig. 7   Fig. 8

Figure 7. Microstructure of a TiAl PST crystal deformed in compression ($\phi = 31°$) at room temperature.

Figure 8. Microstructure of a TiAl PST crystal deformed in compression ($\phi = 90°$) at room temperature.

the majority of dislocations are of the $\frac{1}{2}\langle\bar{1}10]$-type. Such slip occurs, of course, on the primary and often on the secondary {111}⟨110] systems in each domain.

In hard orientations, true twinning of the {111}⟨11$\bar{2}$]-type and slip along ⟨$\bar{1}$10] occur on {$\bar{1}$11} planes inclined to the lamellar boundaries. Figure 8 shows a microstructure observed in a specimen deformed in compression perpendicularly to the lamellar boundaries. Shear deformation is seen to propagate across the lamellar boundaries. As mentioned before, such a propagation of shear deformation across the lamellar boundaries must be closely associated with the high yield stress for the hard mode of deformation. Since the TiAl/TiAl lamellar boundaries are in fact domain boundaries, some of them are of the parallel-type (eq. 1), that is the two neighboring TiAl lamellae separated by such lamellar boundaries are related by 120°-rotation about the boundary plane normal. The interaction between such domain boundaries and the twinning and slip can occur also for the easy mode of deformation since each of TiAl lamellae contains 120°-rotational

domains. Figure 9 shows a deformation structure in a specimen rolled
to 15% at room temperature by fully utilizing the easy mode of
deformation [20,21]. The figure is viewed along the $\langle\bar{1}10\rangle$ direction
which is parallel to the lamellar boundaries. In the domains A, B, D
and E in the figure, only dislocations are observed, while in domain
C, a number of thin deformation twin lamellae are observed. Shear
strain is thought to propagate across the C/D domain boundary by
changing the deformation mode from twinning to slip. In order to
clarify the mechanism of the propagation of shear strain across the
domain boundaries, more detailed microscopic observations on the
interaction between the domain boundaries and the twinning and slip
are required. However the interaction between the 120°-rotational
domain boundaries and the twinning and slip may not be relevant to the
high yield stress for the hard mode of deformation because such interactions can frequently occur even for the easy mode of deformation.
The TiAl/TiAl lamellar boundaries corresponding to the antiparallel
cases in eq. 1 and Ti_3Al lamellae, which play a role only when shear
deformation occurs across the lamellar boundaries, are presumed to be
associated with the high yield stress for the hard mode of
deformation.

No tension-compression asymmetry in yield stress exists in the
two-phase TiAl PST crystals. The principal deformation mechanism of
the TiAl phase in the TiAl PST crystals is deformation twinning which

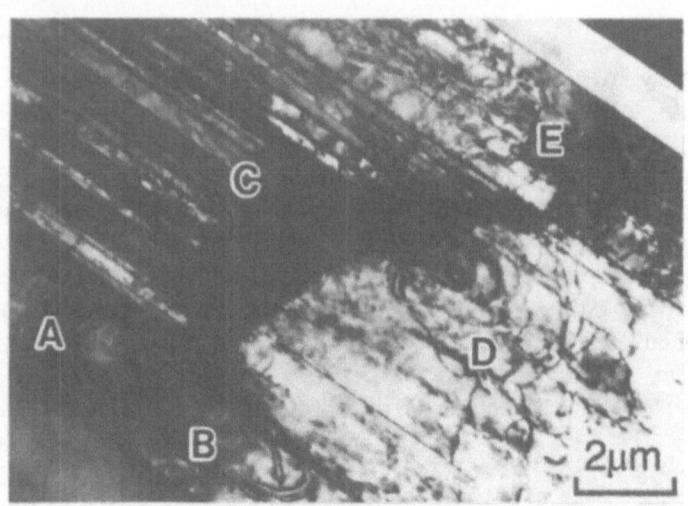

Figure 9. Microstructure of a TiAl PST crystal cold-rolled to 15%
at room temperature utilizing the easy mode of deformation [20].

can only go in one direction. However, since six different domains exist in the TiAl phase, the domains which deform by twinning in compression deform by slip in tension and vice versa [21]. When deformation occurs in the easy mode, the Schmid factor for the twinning system whose twinning plane is parallel to the lamellar boundaries can be zero in some domains. Such domains are expected to deform by slip along ⟨110] in both compression and tension, however, no tension–compression asymmetry should exist for the ⟨110] slip. This is the reason why no tension–compression asymmetry in yield stress is observed in the TiAl PST crystals. As would be expected, there is no significant difference in deformation structure between tension and compression. The tensile deformation behavior of PST crystals of TiAl will be reported elsewhere in detail.

2.4. FRACTURE AT ROOM TEMPERATURE

Tensile ductility of PST crystals of TiAl depends on the lamellar orientation with respect to the tensile axis. A room–temperature elongation of 10~20% can be generally achieved for specimens whose tensile axis is inclined at $30° \sim 60°$ to the lamellar boundaries [22]. On the other hand, the stress–strain curves for specimens whose tensile axis is perpendicular to the lamellar boundaries ($\phi = 90°$) are terminated by fracture soon after yielding. Specimens deformed in tension parallel to the lamellar boundaries ($\phi = 0°$) show a plastic elongation which is often comparable to that for specimens with intermediate lamellar orientations.

In contrast to the yield stress and elongation, the fracture behavior does not depend on the lamellar orientation. In uniaxial tension tests, all of the specimens we investigated fail suddenly without showing and local contraction. Even in specimens showing a tensile elongation as large as 20%, fracture occurs in such a brittle manner. In particular, specimens with $\phi = 30° \sim 60°$ and $\phi = 90°$ have been found to fail by cleavage–like mode on a macroscopic habit plane parallel to the lamellar boundaries. Figure 10 shows a side view of a broken specimen with $\phi = 90°$ (a) and its fracture surface (b). From the electron channelling pattern attached to Fig. 10(b) which shows a three–fold symmetry, the failure plane is thought to be {111} in the TiAl phase or {0001} in the Ti$_3$Al phase. If cleavage fracture occurs along TiAl/Ti$_3$Al phase boundaries, the average height of steps observed on the fracture surface should coincide with the average spacing of Ti$_3$Al lamellae which is about 10μm. However, this does not seem to be the case. Thus, the PST crystals of TiAl are thought to fail by cleavage–like mode on TiAl/TiAl lamellar boundaries such as that shown in Fig. 4 or on {111} planes in the TiAl lamellae which are

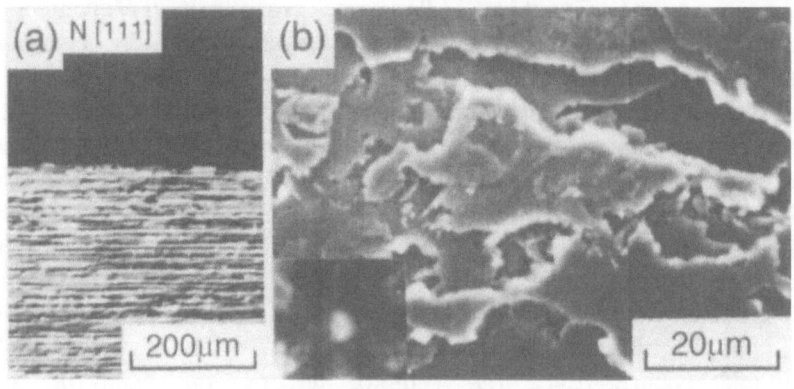

Figure 10. (a) A side view of a broken specimen with $\phi = 90°$ tested in tension at room temperature and (b) its fracture surface.

parallel to the lamellar boundaries.

Recent calculations of ideal surface energies in TiAl [23] suggests that the surface energy would be the lowest on the (111) plane. This is in agreement with the present observaton that TiAl PST crystals with $\phi = 30\sim90°$ fail by cleavage-like mode on (111) in the TiAl phase which is parallel to the lamellar boundaries. However, the theoretical cleavage energy of TiAl itself is not particulary low [24,25]. The lack of plastic work due to the low mobility of dislocations to shield the crack tip has been suggested to be the main cause for the brittle fracture of TiAl at low temperatures [24,25]. The fracture stress of specimens with $\phi >60°$ is practically the same as their yield stress. However, the fracture stress of specimens with $\phi <60°$ depends on the angle ϕ. When the measured fracture stresses are plotted as a function of ϕ, a concave curve similar to that shown in Fig. 6 is obtained. Uniaxial tension tests of PST crystals of TiAl always reveal yielding even if fracture occurs soon after yielding. This indicates that some plastic deformation plays a role as a precursor to cleavage-crack initiation, and this is a reason for the ϕ-dependence of the fracture stress. However, nothing has been known about the plastic precursor. As mentioned above, specimens with $\phi = 30°\sim60°$ show smooth stress-strain curves and deform to $10\sim20\%$ strain like ordinary metals and alloys, but their failure behavior is very brittle. In order to clarify the mechanism of such a deformation and fracture behavior, more work is needed not only on the plastic precursor but also on whether the fracture stress depends on the normal stress on the fracture plane, which would tell us whether the failure mode is true cleavage or not. At the same time, intensive experimental work to clarify the environmental effects on the deformation and fracture behavior of TiAl is needed. The preliminary results of our recent investigation on the environmental effects on ductility of TiAl PST crystals reveal that specimens with $\phi = 30 \sim 60°$ show much higher room-temperature ductilities when tested in vacuum than in air. Therefore, TiAl is also likely to be susceptible to environmental embrittlement like many other intermetallic compounds (for reviews on environmental embrittlement in intermetallic compounds, see, for example, [26,27]).

2.5. DEFORMATION AT HIGH TEMPERATURES

The results of our preliminary study on the high-temperature deformation of PST crystals of TiAl show that the yield stress for specimens with $\phi = 90°$ deformed in compression exhibits an anomalous temperature dependence, while that for the easy mode of deformation does not seem to show any anomaly, similarly to the case of polycrystalline Ti-rich TiAl compounds (for example, see [28]). However, whether the yield stress anomaly found for specimens with $\phi = 90°$ is an intrinsic behavior of the TiAl phase in our PST crystals is not clear yet since the anomaly may be a reflection of the yield stress anomaly in the Ti_3Al phase coexisting with the TiAl phase. Recently, Ti_3Al has been found to exhibit an anomalous temperature dependence of yield stress when the compound is deformed along the c-axis [29,30].

A transition in the operative slip modes and a change in the propensity for twinning have been found to occur at $600 \sim 800°C$ for both the hard and easy modes of deformation. At low temperatures, deformation of PST crystals occurs by true twinning of the $\{111\}\langle11\bar{2}]-$type and slip on $\{111\}\langle\bar{1}10]$. These twinning and slip systems are still operative at high temperatures. However, the propensity for the true twinning was found to decrease at high temperatures. Figures 11(a) and (b) show microstructures of specimens with $\phi = 31°$ deformed at room temperature (a) and at 800°C (b). Specimens of Figs. 11(a) and (b) were cut from the same PST crystal and deformed under exactly the same conditions except for the deformation temperature. At room temperature, numerous thin deformation-twin lamellae are observed at regions indicated by T, while at 800°C such deformation twin lamellae are not found.

Another important point concerns the activity of $\langle\bar{1}01]$ superlattice dislocations. No $\langle\bar{1}01]$ dislocations are observed in PST crystals of TiAl deformed at room temperature, while at 800°C $\langle\bar{1}01]$ dislocations are frequently observed. This is indicative of an increased activity of $\langle\bar{1}01]$ slip at high temperatures. $\frac{1}{2}\langle11\bar{2}]$ dislocations are

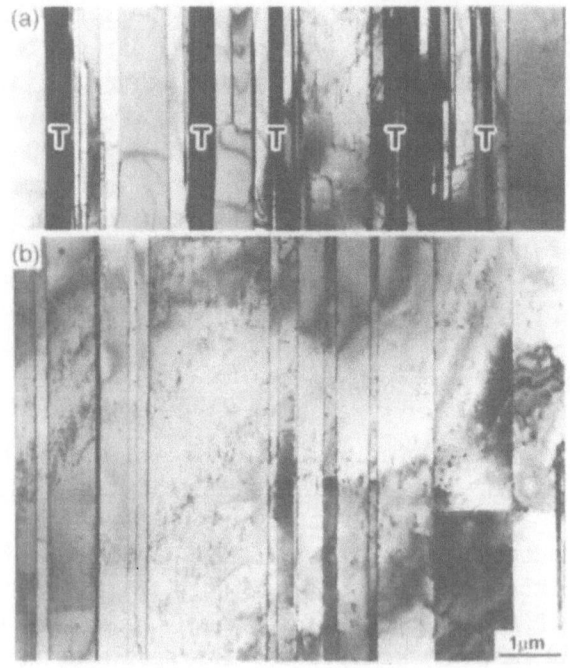

Figure 11. Microstructures of TiAl PST crystals deformed in
compression ($\phi = 31°$) at room temperature (a) and at 800℃ (b).

found in specimens deformed at both room temperature and 800℃.
However, they are not expected to play a major role in the plastic
deformation of PST crystals of TiAl since their density is generally
small.

Thus, outstanding aspects characterizing the deformation behavior
of PST crystals of TiAl at high temperatures are a decreased propensity for twinning and an increased activity of slip along $\langle \bar{1}01]$. The
occurrence of slip along $\langle \bar{1}01]$ as well as twinning and slip along
$\langle \bar{1}10]$ at high temperatures is common to both the TiAl phase in PST
crystals and the Al-rich single-phase compounds [31-34]. Since the
TiAl phase in equilibrium with the Ti_3Al phase is supposed to be Ti-
rich, this seems to suggest that both Ti-rich and Al-rich TiAl phases
show the same sort of deformation behavior at high temperatures.
However further work is needed to obtain more conclusive evidence to
demonstrate this. In particular, deformation experiments using
specimens with a controlled amount of oxygen are needed since the
deformation structures of TiAl compounds have been found to be significantly affected by the amount of oxygen [35].

3. CONCLUDING REMARKS

In the last few years, there has been an enormous increase in the research and development activity on titanium aluminides, in particular, TiAl with the TiAl/Ti$_3$Al two-phase structure, however there are still many remaining road-blocks to the commencement of the structural applications. One of them, which is, we believe, the most important one, is the problem of low ductility at room temperature. As has been shown in this paper, the TiAl phase itself constituting the TiAl/Ti$_3$Al two-phase structure is easily deformable and its yield stress is low. Whereas brittleness of materials with high yield stresses is plausible, the brittleness of TiAl is difficult to understand. Further work on the deformation and fracture behavior of TiAl is need. In particular, we believe, we should make intensive and systematic studies of the "easily deformable but brittle" problem from the point of view of environmental effects on the mechanical properties of TiAl since TiAl is also likely to be susceptible to test environment at ambient temperatures. In our group, tensile tests of PST crystals of TiAl under various atmospheres including high vacuum are in progress.

REFERENCES

1. M. Yamaguchi and Y. Umakoshi, Prog. Mater. Sci., 34, 1 (1990).
2. Y.W. Kim, High Temperature Ordered Intermetallic Alloys Ⅳ (ed. by L.A.Johnson, D.P.Pope and J.O.Stiegler) MRS Sympsia Proceedings, Vol.213, p.777 (1991).
3. Y.W. Kim and D.M. Dimiduk, JOM, 43, 40 (1991).
4. F.E. Froes, C. Suryanarayana and d. Eliezer, ISIJ International, 31, 1235 (1991).
5. O.M. Dimiduk, D.B. Miracle, Y.W. Kim and M.G. Mendiratta, ISIJ International, 31, 1223 (1991).
6. E.L. Hall and S.C. Huang, J. Mater. Res., 4, 595 (1989).
7. T. Fujiwara, A. Nakamura, M. Hosomi, S.R. Nishitani, Y. Shirai and M. Yamaguchi, Philos. Mag., A 61, 591 (1990).
8. S.H. Whang, in this proceedings.
9. C.T. Liu and J.O. Stiegler and F.H. Froes, Metals Handbook Tenth Edition, Vol.2, p.913, ASM, Metals Park, Ohio (1991).
10. Y. Umakoshi, T. Nakano and T. Yamane, Scripta Metall. Mater., 25, 1525 (1991).
11. C. McCullough, J.J. Valencia, C.G. Levi and R. Mehrabian, Acta Metall., 37, 1321 (1989).
12. Y. Toda, M.H. Oh, H. Inui and M. Yamaguchi, Proc. 1991 Hiroshima Meeting of JIM, Japan Institute of Metals, Sendai, p.311 (1991).

13. P.A. MoQuary, P.A. Dimiduk and S.L. Semiatin, Scripta Metall. Mater., 25, 1689 (1991).
14. C.R. Feng, D.J. Michel and C.R. Crowe, Scripta Metall., 22, 1481 (1988); 23, 241 (1989); 23, 1135 (1989).
15. Y.S. Yang and S.K. Wu, Scripta Metall. Mater., 24, 1801 (1990) ; 25, 255 (1991).
16. H. Inui, A. Nakamura, M.H. Oh and M. Yamaguchi, Ultramicroscopy, in press.
17. M. Hirabayashi and S. Weissman, Acta Metall., 10, 25 (1962).
18. H. Inui, M.H. Oh, A. Nakamura and M. Yamaguchi, submitted to Philos. Mag. A in Nov. 1991.
19. H. Inui, A. Nakamura, M.H. Oh and M. Yamaguchi, submitted to Philos. Mag. A in Nov. 1991.
20. M.H. Oh, H. Inui, A. Nakamura and M. Yamaguchi, Acta Metall. Mater., in press.
21. S.R. Nishitani, M.H. Oh, A. Nakamura, T. Fujiwara and M. Yamaguchi, J. Mater. Res., 5, 484 (1990).
22. H. Inui, A. Nakamura and M. Yamaguchi, High Temperature Ordered Intermetallic Alloys Ⅳ (ed. by L.A.Johnson, D.P.Pope and J.O. Stiegler) MRS Symposia Proceedings, Vol.213, p.569 (1991).
23. M.H. Yoo, in this proceedings.
24. C.L. Fu and M.H. Yoo, Alloy Phase Stability and Design (ed. by G.M.Stocks, D.P.Pope and A.F.Giamei) MRS Symposia Proceedings, Vol.186, p.265 (1991).
25. M.H. Yoo and C.L. Fu, ISIJ International, 31, 1049 (1991).
26. C.T. Liu, in this proceedings.
27. C.T. Liu, High Temperature Aluminides and Intermetallics (ed. by S.H.Whang, C.T.Liu, D.P.Pope and J.O.Stiegler) p.133, TMS Warrendale, PA (1990).
28. S.C. Huang and E.L. Hall, High Temperature Ordered Intermetallic Alloys Ⅲ (ed. by C.T.Liu, A.I.Taub, N.S.Stoloff and C.C.Koch) MRS Symposia Proceedings, Vol.133, p.373 (1989).
29. Y. Minonishi and M.H. Yoo, Philos. Mag. Lett., 61, 203 (1990).
30. Y. Minonishi, Philos. Mag., A 63, 1085 (1991).
31. D. Shechtman, M.J. Blackburn and H.A. Lipsitt, Metall. Trans. 5, 1373 (1974).
32. H.A. Lipsitt, D. Shechtman and R.E. Schafrik, Metall. Trans., A 6, 1991 (1975).
33. G. Hug, A. Loiseau and P. Veyssiére, Philos. Mag., A 57, 499 (1988).
34. S.H. Whang and Y.D. Hahn, High Temperature Aluminides and Inter-metallics (ed. by S.H.Whang, C.T.Liu, D.P.Pope and J.O.Stiegler) p.91, TMS Warrendale, PA (1990).

35. S. Sriram, V.K. Vasudevan and D.M. Dimiduk, High Temperature Ordered Intermetallic Alloys IV (ed. by L.A.Johnson, D.P.Pope and J.O.Stiegler) MRS Symposia Proceedings, Vol.213, p.375 (1991).

SOLUTE-DISLOCATION INTERACTIONS AND SOLID-SOLUTION STRENGTHENING MECHANISMS IN ORDERED ALLOYS

D. M. DIMIDUK
Wright Laboratory
WL/MLLM
Wright-Patterson AFB, OH 45433-6533
USA

SATISH RAO
National Research Council
WL/MLLM
Wright-Patterson AFB, OH 45433-6533
USA

T. A. PARTHASARATHY
Universal Energy Systems, Inc.
4401 Dayton-Xenia Road
Dayton, OH 45432-1894
USA

C. WOODWARD
Universal Energy Systems, Inc.
4401 Dayton-Xenia Road
Dayton, OH 45432-1894
USA

ABSTRACT. The brief development history of intermetallic alloys as structural materials has led to several examples which illustrate the important effects of alloy chemistry on the flow and fracture properties of these alloys. Several recent investigations have begun to investigate the theoretical basis underlying such phenomena. However, such treatments have typically not addressed the aspects of these property variations which may be driven by solute-dislocation interactions. Within dilute substitutional solid solutions of intermetallic alloys, several distinct glide systems can often be identified over a range of deformation temperatures. For each of these the distinct solute-dislocation interactions must be quantitatively described, related to the mobility of dislocation loop segments, and then combined in a general way to describe the observed alloy behavior. Within the present manuscript, several examples of chemically-dependent deformation phenomena are reviewed. Following this, solute hardening effects in Ni_3Al alloys are reviewed to illustrate specific mechanisms which must be investigated and described by solute-strengthening theory. Finally, some results from anisotropic elasticity calculations and atomistic simulations within the embedded atom formalism, of solute-dislocation interactions in Ni_3Al alloys, are presented and discussed.

1.0 INTRODUCTION

As a class of structural materials, intermetallic alloys present many intriguing deformation and fracture phenomena. A great many of the close-packed and body-centered cubic derivative compounds are known to exhibit slip and/or twinning deformation accompanied by ductile-rupture failure at temperatures below one-half the melting point. Other intermetallic compounds having more complex crystal structures exhibit remarkable strength at high temperature, often while exhibiting no low temperature crystal plasticity. Still others from each of these structural groups exhibit properties intermediate to these extremes, for example intense localized dislocation slip followed by brittle cleavage-like

C. T. Liu et al. (eds.), Ordered Intermetallics – Physical Metallurgy and Mechanical Behaviour, 237–256.
© 1992 *Kluwer Academic Publishers.*

failure at small macroscopic strains. Unfortunately for the design and processing engineer, these properties are too often heavily influenced by the chemistry of the gaseous environment in which the alloys are tested, with ambient conditions having a strong influence on the aluminide based alloys.

Alloys which exhibit slip may often exhibit dramatic changes in their strain to failure with changes in alloy chemistry. In polyphase alloys, many of these changes are due to variations in the microstructure which result from composition selection, synthesis methods and processing. However, even single-phase single crystals may exhibit changes in ductility with small alloying additions, as recently observed in studies of NiAl [1]. Changes in the constitutive behavior or micromechanical response of interacting phases cannot explain these observations. They must be understood through the atomic aspects of plasticity. Some compound alloys, such as those based on Al_3Ti, exhibit significant slip at low stresses in compression and fail in a brittle fashion under small tensile strains [2,3]. Even this unusual property may be partially altered by choice of alloy chemistry within the single-phase region [4,5]. These phenomena suggest a need to understand the role of chemistry on the dislocation mobility, other plasticity mechanisms, and fracture of compounds. Understanding the relationship between bonding in the perfect lattice, bonding in the vicinity of solute atoms and defects, and the aggregate behavior of a distribution of defects in the crystal, will play a role in understanding the mechanisms which control ductility, fracture properties, and strength of alloys.

Modern investigations of intermetallic alloys have often employed computational approaches for examining the chemical dependence of structural stability using techniques which are well established in the scientific communities of quantum physics and chemistry [6-8]. For example, three separate methods have been applied independently in order to explain the observed stabilization of the $L1_2$ structure in ternary trialuminide compounds based on Al_3Ti [9-11]. Recently, electronic-structure methods have included calculations of the single-crystal elastic properties, including shear constants, and planar defect energies [12,13]. These in turn were employed within anisotropic elasticity theory in order to predict the behavior of families of intermetallic compounds of various structure types. Such studies show that choice of chemistry dictates the elastic behavior, and thereby, broad aspects of the mechanical behavior of compounds. However, the details of dislocation motion and the role of solutes in changing the dislocation motion cannot be sufficiently addressed by elasticity methods or bulk electronic structure methods alone. The flow properties of some materials (including Ni_3Al and NiAl alloys) change significantly with small changes in composition which are either known or not expected to have significant, or even measurable, influences on the elastic or bulk physical properties of the compound.

Atomistic computational approaches have focused on the properties of individual dislocations and their role in controlling the deformation characteristics of the compound. Static core structures of dislocations have been investigated by atomistic-simulations using pair potential and newer theoretical techniques such as the embedded atom method (EAM) [14,15]. However, these approaches are incomplete, and like the electronic structure methods, they do not capture the statistical and dynamical nature of deformation processes in real materials. Also, the chemical nature of directional bonding is lost in these methods; consequently, "design" of structural alloys by computational methods cannot be performed.

In spite of the inadequacies of current theoretical and computational methods, there are many possibilities for extending theories of flow in compounds through computational techniques. Here we use atomistic simulation methods to investigate the possible role of antisite defects on the intrinsic flow response found in Ni_3Al alloys. An understanding of the elements of structural alloy design is being built through a hierarchy of computational methods, and this is expected to provide considerable insight into the currently empirical process of alloy design. Building a sound, generalized solute-strengthening theory is one step in a coherent set of models required for establishing the chemical dependence of flow and fracture of compounds. The classical approaches to solution hardening are incomplete even for disordered alloys and some of these shortcomings have been recently reviewed [16]. Modern computational tools appear to provide a means for extending these models and addressing some of the long-standing shortcomings therein. In this manuscript we review the investigations of solute effects in Ni₃Al and present some initial atomistic simulation results dealing with solute-dislocation interactions. This is just one aspect of a theory for solute strengthening in ordered alloys. We also discuss other aspects of the problem which must be included in any approach toward extending solute-hardening theory in general. Alloys of Ni₃Al are used as an example to illustrate the inadequacies within the classical approaches as well as the details involved in our more recent analysis. The problem is a complex one and is by no means complete in the present form. However, some of the limiting aspects which diminish the predictive capability of classical empirical approaches have been addressed conceptually and a new approach is being developed around these concepts for a few ordered alloys.

2.0 BACKGROUND

In order to build a solute-strengthening model, the dislocation mechanisms which contribute to glide at various temperatures must be established. A few of these aspects are reviewed, first in general, then specifically for Ni₃Al alloys.

2.1 FLOW PHENOMENA

In a previous paper we identified two classes of temperature-dependent flow-stress behavior for each of two structure classes shown below in Fig. 1 [16]. Aspects of these flow curves have been attributed to the behavior of dislocation cores subjected to thermal activation and the interaction of solute atoms with the extended cores. A comprehensive model of solute strengthening in ordered alloys must explain the influence of varying solute type and concentration on these classes of yield curves throughout the temperature range where glide processes dominate. Such a model must include the influence of dislocation core structures on the range and character of the solute-dislocation interaction forces (or vis versa), and the thermally-activated behavior of the alloy through appropriate statistical models. Finally, modeling the full range of yield behavior depicted in Fig.1, requires that a combination of different types of dislocations be statistically introduced with the appropriate coupling equations to capture the transitions in slip systems and flow regimes. For compounds of the Class II or anomalous type within the close-packed derivative (CPD) systems, dislocation-core transformations have an important role in controlling the yield stress in the temperature regime of rising flow stress. This relationship has been well studied, though intense controversy remains. One point which has emerged, however, is the possible role of the planar fault energies and in particular, that of the complex stacking fault (CSF) in controlling the thermally-activated processes which govern flow. We illustrate this point in further detail by summarizing the experimental investigations of solute strengthening in binary and ternary alloys of Ni₃Al.

2.2 SOLUTES IN Ni₃Al AND THE ACTIVATION MODEL

Prior to 1984, solute strengthening studies of the compound Ni₃Al fell into two categories (see [17,18] for reviews). Many studies were performed using both polycrystalline and single crystalline samples and the results of such studies were analyzed within classical

240

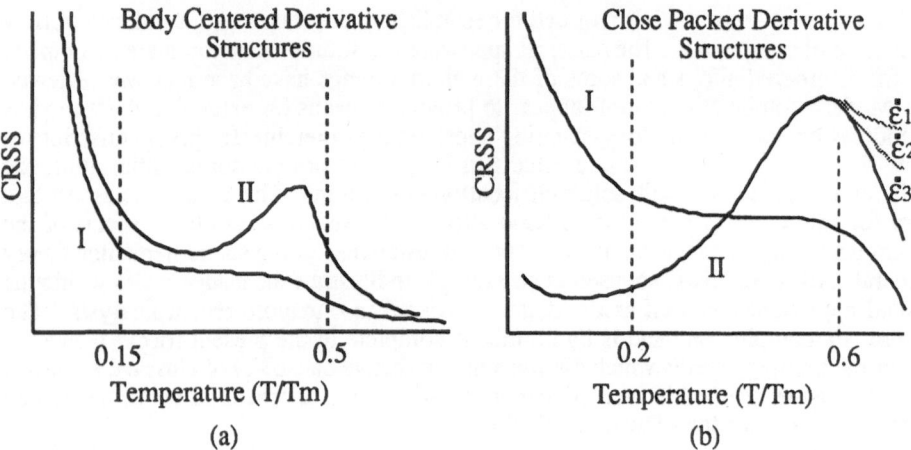

Figure 1. Schematic descriptions of the critical resolved shear stress versus homologous temperature for ordered alloys of two structure types.

elastic theories of solute-strengthening mechanisms. These studies were typically limited to examination of the low temperature data (room temperature and below) without much attention given to the thermally-activated processes governing high-temperature yielding. A second group of studies attributed the results entirely to a form of chemical strengthening characterized by the role of solutes on the antiphase boundary (APB) energies and the crystallographic anisotropy of these energies. Neither of these treatments have proven adequate, even after more detailed study [19,20]. By way of example, Fig. 2 shows the incremental change in the resolved shear yield stress of single crystals (ds/dc) as function of a size misfit parameter (da/dc), for Ni_3Al solid-solution alloys deformed at room temperature, 475 K and 675 K. The data show that simple elastic models do not adequately describe the behavior. Variations in the Ni/Al ratio in binary alloys, or even modest thermal activation introduces a different mechanism(s), possibly including solute effects on fault energies and variations in the kinetics of dislocation motion. In 1984, the Paidar, Pope and Vitek (PPV) model [21] for the thermally-activated behavior was introduced and a new basis was formed for interpreting the behavior of these alloys.

Within the PPV model for $L1_2$ compounds, when the effects of elastic anisotropy are included, the activation enthalpy for closs-slip has been written as:

$$H = W(\gamma_{CSF}, \tau_{111}, \tau_{1\bar{1}1}) + b\left\{ \frac{\mu b^2}{8\pi} - \left[\left(\frac{\gamma_{111}}{\sqrt{3}}(1 + f_{111}\sqrt{2}) - \gamma_{010} + \tau_{010}b \right) \frac{\mu b^3}{8\pi} \right]^{\frac{1}{2}} \right\} \quad (1)$$

here W is a function which gives the energy to form a constriction of the superpartial dislocation core and spread it in the saddle-point geometry in the octahedral cross-slip plane; γ_{CSF}, γ_{111} and γ_{010} are the complex-stacking-fault energy, the APB energy on the (111) plane and the APB energy on the (010) plane, respectively; f_{111} is the anisotropic elastic interaction force factor; and τ_{111}, $\tau_{1\bar{1}1}$ and τ_{010} are the resolved-stress components on the primary octahedral, octahedral cross-slip and cubic cross-slip planes, respectively. The function W reflects the fact that the saddle-point configuration formation energy is influenced by the general stress tensor applied to the crystal and specifically by glide

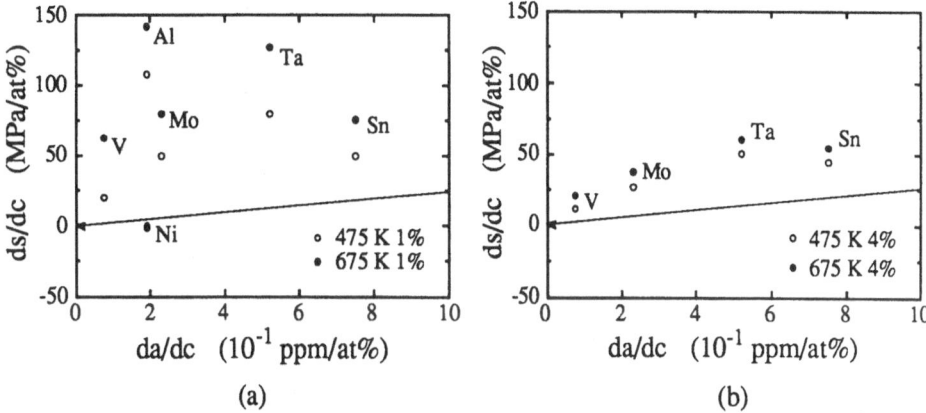

Figure 2. Incremental variation in Ni₃Al alloy single-crystal strength with "size" misfit for various solutes; a) 1% solute at 475 and 675K, and b) saturation of solute strengthening for 4% solute at 475 and 675K. Line is for ternary alloys at room temperature. Flow stress data from references [20] and [23], for detailed discussion see references [16,18,38].

stresses on the primary octahedral and the octahedral cross-slip systems. The flow stress for a single crystal is written as:

$$\tau = \tau_0 + A \exp\left(-\frac{H}{kT}\right) \tag{2}$$

where τ_0 is the lattice friction stress which may be varied within this approximation by changes in alloy chemistry (solute hardening).

Other models, have relied upon the above analysis to describe the activation of the initial core-transformation events. However, the specific details used in developing the morphological character of dislocations and the details of annihilation of the stress-controlling barriers remains controversial. The thermally-activated event which controls unlocking of dislocation segments has never been fully analyzed. Consequently, the view presented below adopts this activation analysis for pinning dislocations. More complete analysis of the unpinning events and their relationship to the flow stress is required in order to complete a model.

According to relations (1) and (2), specific experiments in alloyed crystals having different orientations may identify the role of individual parameters on the flow stress and illuminate the primary mechanism(s) of solute strengthening. Additionally, relations (1) and (2) suggest that specific experimental trends can be anticipated, including the dissociation distances observed by electron microscopy. Through a series of investigations by different groups, many of these experiments have been performed and the role of individual terms may be illustrated by a process of elimination.

2.3 SOLUTES IN Ni₃Al—CURRENT UNDERSTANDING

Heredia and Pope [19,22] have examined the flow stress versus temperature behavior of a number of alloyed Ni₃Al crystals for the crystal orientation where the tension-compression asymmetry is zero. This orientation captures the stress dependence of the function W in (1), entirely through the dependence upon the complex fault energy [21]. This allows the stress dependence of the driving force (influence τ_0 on H) to be examined by seeking correlations between changes in τ_0, the anisotropy factor, and changes in the temperature dependence of the flow stress for various solutes. This assumes that the ratio, $\gamma_{111}/\gamma_{010}$, is essentially a constant. With these experimental conditions, and assuming (1) to be the correct form of the activation enthalpy, only compositionally-dependent changes in τ_0 or the CSF energy can influence the flow stress. After examining the variation of resolved shear stress with temperature for different alloys, these investigators concluded that either the fault energies or the anisotropy factor are varying with solute content to give solute strengthening. The chemical and elastic effects could not be isolated from each other.

Curwick studied the role of specific solutes on the flow properties of single crystals of Ni3Al [23]. As a part of these investigations, the single-crystal elastic constants were examined under the influence of solute additions, including deviations from the stoichiometric composition. These measurements revealed that the individual constants had a minor variation (less than two percent) with solute content, but that the elastic anisotropy was invariant with alloy chemistry.

Separate studies have shown that the APB energy ratio does not vary strongly with solute content, and when changes are observed, they tend to be in the opposite direction from that expected from relations (1) and (2) [20,24-27]. Alloys showing a strong enhancement of the anomalous flow behavior show a decrease in the ratio of octahedral-plane to cubic-plane APB energy, suggesting that the driving force for the rising flow stress is diminished. Distinct morphological changes could be observed in the dislocation lines as a function of solute content, particularly for Al-rich alloys [20,28]. Further, studies of alloys having similar APB energies, tested in orientations near {001}, suggest that the influence of solutes on the function W in (1) are expected to dominate the temperature dependence of yielding [20]. Judging from the above results, and the PPV activation enthalpy, the dominant role for solutes is in changing the activation barrier for the thermally-activated process controlling yield. This may occur by changes in the complex-fault energy and/or changes in the ability to form constrictions or core transformation events on dislocations.

Given these results for Ni3Al alloys and expecting related behavior in CPD systems, we have begun to investigate the chemical and elastic components of the solute-dislocation interactions and the kink/constriction formation events as two parts of a generalized model for compounds. These components must be analyzed for all active dislocation types within the compound. The results to date presented below are for Ni3Al specifically but should be qualitatively similar to possible interactions in other CPD structures, provided that the individual elements of glide can be appropriately isolated. This may be deduced simply from the geometrical aspects of glide which are common to these structures. As the symmetry of the structure is reduced, the number of possible configurations of dislocation loops tends to increase, but these are restricted to combinations of Shockley partial dislocation reactions dictated by the geometrical aspects of the structure type. Similar examinations must be made for BCC derivative structures but these are more complicated since dislocation cores are generally not easily parameterized by such concepts as stacking faults, and screw dislocations are generally not contained in one plane. Further, some of the more interesting compounds within this class (such as MoSi2, NiAl, and FeAl) exhibit transitions in the primary Burgers vector as temperature is increased. While the core structures of these dislocations have not been completely defined, an approach similar to the one introduced here should be applicable.

3. SOLUTE-DISLOCATION INTERACTIONS—THE METHODS

Historically, it is well established that solid-solution hardening (softening) results from the interaction of solute atoms with the core structure of a dislocation. These are conveniently described as the "size", "modulus", and "chemical" interactions [16]. The size and modulus interactions are short-range interactions which have commonly been described using continuum elasticity theory. However, it is widely recognized that the role of the dislocation core is not adequately treated by this approach [16]. Recent electronic-structure calculations, have shown that the elastic calculations could be in substantial error, even in simple metals [29,30]. In ordered alloys, the extended nature of the dislocation core introduces the possibility of multiple "chemical" interactions between solute atoms and planar defects. Additionally, the effects of asymmetries in the strain field around a substitutional atom must be incorporated [16]. Therefore, solute-dislocation interactions must be calculated on an atomic level, using electronic-structure methods or semi-empirical techniques, such as the EAM [14,15]. We separate these interaction simulations into four parts: i) the structure of a single point defect in the "stoichiometric" infinite crystal, ii) the possible core structures of dislocations in the stoichiometric lattice, iii) the interaction between an isolated point defect and an infinite planar interface, and iv) the interaction of point defects with the various core configurations in the nearly stoichiometric lattice. Using i) and iii) above, the same interaction energies may be calculated from anisotropic elasticity theory.

Within the EAM, we have only investigated deviations from the stoichiometric composition in a binary $L1_2$ structured alloy since a limited set of interatomic potentials is currently available. The potentials which have been used and the reliability with which they represent the behavior of Ni_3Al have been described previously [31-33]. Given the complexities of representing a specific compound within this semiempirical model, one must always be careful when interpreting the results of these computer simulations. For the case presented here, the results represent effects which are within the range of those possibly occurring in the compound Ni_3Al. Specific magnitudes of interactions found through simulation, for example, must not be taken as quantitatively representative of Ni_3Al.

3.1 ELASTICITY CALCULATIONS

The displacement fields around single antisite defects of Ni and Al were calculated within a perfect lattice of Ni_3Al, using periodic boundary conditions. Within anisotropic elasticity theory, the first-order size interaction between a point defect and dislocation line is given by

$$E_{int} = \sum_{ij} P_{ij}\varepsilon_{ij}^{D} \tag{3}$$

where P_{ij} is the dipole tensor of the point defect in the bulk crystal and ε_{ij}^{D} is the strain field of the dislocation at the defect position [34,35]. The dipole tensor completely describes the elastic strength of the defect and includes the symmetry of the defect site. The dipole tensors for the point defects described below were calculated from the interatomic potentials using Schober's formalism [36]. The strain field of the dislocation at the point defect position may be calculated using Stroh's sextic formalism assuming liner superposition of the strain fields for individual dislocation components [37]. The gradient of the interaction energy perpendicular to the dislocation line in the direction of motion is the force resisting dislocation motion. Equation (3) above was used to describe the "symmetric" size interaction between Ni antisites (excess Ni) and edge dislocations, and the "tetragonal" size interaction between Al antisite defects (excess Al) and both edge and screw dislocations. The size interactions were examined for two possible core structures, one planar and one transformed to nonplanar (see Fig. 3), for given values of fault energies within anisotropic elasticity theory. The tetragonal interaction must be considered for three unique cases since there are three orthogonal variants of the tetragonal environment for the Al antisite. Figure 4 illustrates the stacking sequence for two of these sites near a {111} APB. The modulus interaction was not considered at this time since experiment has suggested that this is relatively unimportant in Ni₃Al [38].

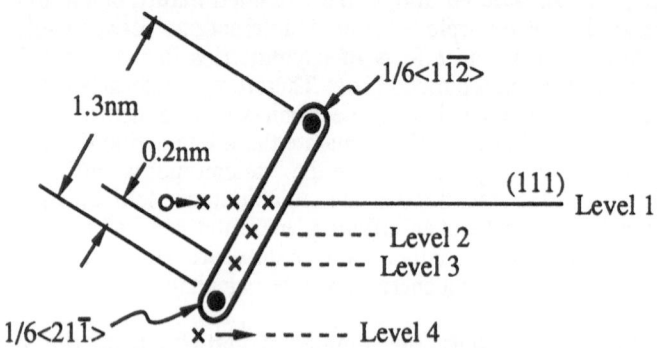

Figure 3. Schematic representation of nonplanar dislocation core considered in anisotropic elasticity calculations. Open circle represents antisite defect, "x's" represent positions sampled by defect. Levels indicate defect planes sampled at ~0.2nm intervals perpendicular to APB plane.

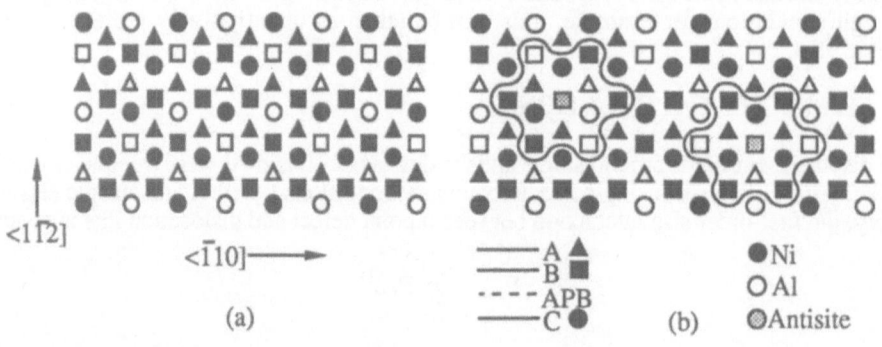

Figure 4. Stacking geometry for L1₂ structure; a) perfect lattice, and b) two tetragonal antisites outline in vicinity of octahedral plane APB.

3.2 CALCULATIONS USING THE EMBEDDED ATOM METHOD

The elasticity calculations were compared with EAM simulations of the interaction energies for the same defect geometries such that the energetic contribution of core level relaxations could be examined. The interaction energies as a function of the position of the solute were computed for solutes both in the glide plane, and residing just above or below the glide plane. Beyond the size and modulus interactions, solute atoms will exhibit an interaction with the planar faults formed by dislocation dissociation; in this case the CSF and the APB. These have been studied, first independently of the dislocation as infinite interfaces, and then in the presence of the dislocation while computing the total interaction between the solute atom and the dislocation. For the case of the infinite planar defect, the results were interpreted as a global change in the fault energy for a specified concentration of solute atoms randomly distributed in the lattice.

A block of the compound crystal (1.7 x 2.0nm) having periodic boundary conditions in the "x-y" plane, and 4.8nm in the "z" dimension having free boundary conditions, was sheared in the "x-y" plane by the translation vector for the perfect shear CSF or APB interface; see Fig. 5a. The potential energy of this block was then minimized. Several simulations were performed, each with random distributions of solute, in order to simultaneously sample the effects of a varying solute concentration and the influence of a volume strain introduced by the solute. Also, an individual antisite atom was moved through the interface in a "stoichiometric" crystal to examine the range of the interaction between the defect and the fault while minimizing the effect of volume strain, as described in Fig. 5b. Both excess Ni and Al antisite defects were examined. Similarly, a dissociated screw dislocation was introduced into a volume of the crystal in the presence of a point defect and the energy of the block was minimized for differing relative positions between the point defect and dislocation line. The diagram of Fig. 6 depicts the geometry used for these simulations and they have been described in greater detail elsewhere [33].

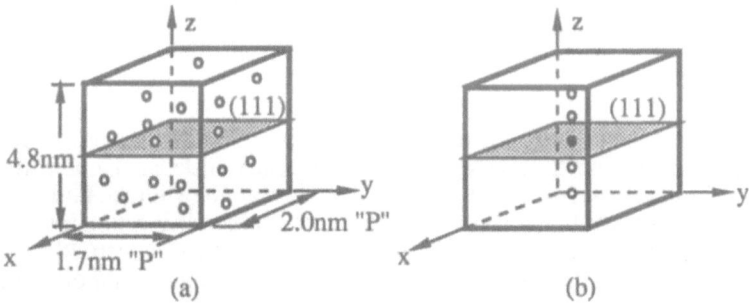

Figure 5. Geometries used to simulate interactions between antisite defects and planar faults; a) random distribution of antisite defects, and b) isolated antisite moved relative to fault plane. Shaded region represents fault plane, either CSF or APB type. Case b) represents concentration of ~1.1% solute.

These results, are useful when interpreting the interaction of solutes with planar defects. However, in real crystals the solutes would have some opportunity to segregate toward or away from the fault. This may lead to more pronounced changes in the fault energies and to the "APB dragging" effect described by Brown, [39] Yoo, [40] and Wu et al. [41]. One influence of the antisite interaction with the faults within a dissociated dislocation core, is a possible enhancement in the ability to form constrictions to transform the core. This interaction may be examined independently from the elastic interactions and the elastic forces influencing cross slip. Current theory suggests that constriction formation and core transformation are functions of the fault energies, stress state, and elastic properties of the

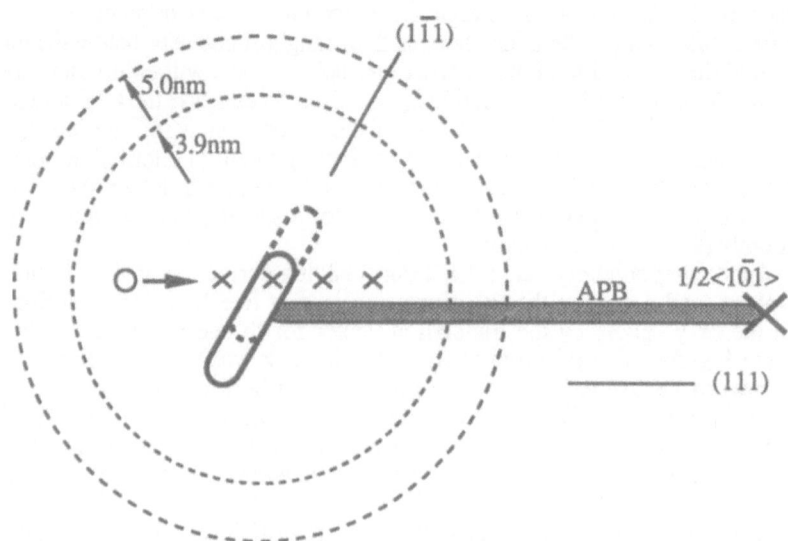

Figure 6. Geometry used for EAM simulations of antisite-dislocation interactions. Open circle represents antisite defect, "x's" are positions sampled. Atoms in region ranging from 3.9-5.0nm radius established fixed boundary condition computed from anisotropic elasticity theory. Atoms at smaller radii, around $a_0/2<110>$ dislocation on cross-slip plane, were relaxed. The adjoining superlattice partial dislocation was included by anisotropic elasticity alone. The aperiodic-unit length varied from 0.5-2.5nm along direction perpendicular to the page.

crystal. For a given stress state, chemical effects on the global elastic properties may be ignored and the local elastic effects are captured in the solute-dislocation interaction forces. Consequently, it may be possible to treat the antisite interaction with the faults as a global effect on the thermal-activation response, separate from solute-dislocation interactions, by varying the statistical functions which describe the frequency of constriction formation (core transformation). In the present study, we examine whether or not other nonelastic contributions may be identified.

In order to extend understanding of solute strengthening to ternary ordered systems and systems for which a large set of solute potentials does not exist, a more generalized parametric method has been developed, the details of which cannot be reported in this manuscript and will be introduced independently [42]. Within the parametric method, an interatomic potential for solute-atom lattice-atom interactions may be fit to simulate the "elastic strength" of the defect (dipole tensor), and the "modulus strength" of the defect (local force constant changes around the solute) as well as the other required lattice properties. Such development of the interaction potentials was needed for the solute-atom lattice-atom interactions and solute-dislocation interactions to converge to the size and modulus effects at large separation distances. Since only empirical potentials are used in these calculations, sets of potentials are desirable for describing a certain structure type so that the range of possible core structures may be simulated for a range of possible solute strengths. The complete parametric method allows for generalized simulations of the size, modulus and "chemical" interactions, each independently. This method is under develop-ment and only now being applied in our studies.

4. SIMULATION RESULTS

4.1 SIZE INTERACTION, ANISTROPIC ELASTICITY THEORY

The Ni antisite defects and ternary solutes which substitute for Al in Ni$_3$Al create defects having a dipole tensor with cubic symmetry. The interaction energies, computed using anisotropic elasticity, for the Ni antisite defect interacting with a dissociated a$_o$/2<110> screw superlattice partial dislocation, are shown by the open triangles in Fig. 7a The individual Shockley partial dislocations are at the zero position and positive 1.7nm. Defects having cubic site symmetry (symmetric defects) are not expected to have a first order interaction with screw dislocations. However, this interaction is symmetric about the center of the Shockley partial dislocations, covers a range of about 3.0nm, and has a maximum energy of about one-half of the asymmetric defect interactions shown in Fig. 7b. This is interpreted as the interaction between the symmetric antisite defect and the edge components of the Shockley partial dislocations. As expected, such an interaction changes sign with the sign of the edge component of the Shockley partial dislocation.

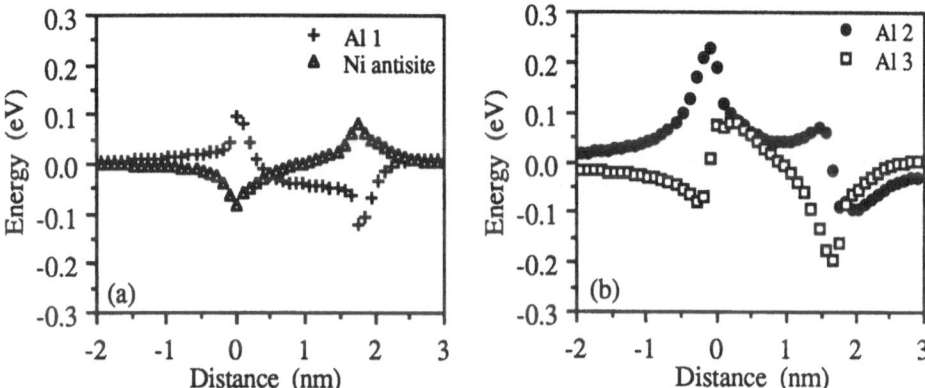

Figure 7. Elastic antisite-dislocation interaction energies for CSF-coupled Shockley partials of screw dislocation dissociated in octahedral primary plane. Elastic center for leading partial at distance zero, trailing partial at +1.7nm.

The asymmetric-defect interaction energies are also shown in Fig. 7. The asymmetric defect having a tetragonal axis oriented perpendicular to the dislocation line ("+"'s in Fig. 7a) exhibits an interaction-energy profile which is similar to that for the symmetric defects in that the interaction changes sign with changing sign of the edge component of the Shockley partial. The interaction is nonzero at the center point between the Shockley partials since this interaction is asymmetric with the Shockley partials individually. This leads to a nonzero sum of two interactions with opposite sign. Although not shown, the interactions between tetragonal antisites and a total a$_o$/2<110> edge dislocation were shown to yield interaction profiles which were asymmetric about the dislocation line. As expected the screw components are non-interacting with this antisite defect. Examining the interaction profile as a function of position, from negative to positive position, the first Shockley dislocation is repelled by the antisite while second one experiences an attraction which is of comparable magnitude, due to the asymmetry of the edge-component interactions. For the antisite defects considered, if the crystal is sheared on the next adjoining plane, then the sequence of the partial dislocations is reversed and the interaction profiles exhibit a change in sequence and a sign reversal (positive to negative and vis versa) from those shown.

The remaining two Al antisites have tetragonal axes oriented at acute angles (±45°) to the superpartial dislocation line and their interaction profiles are given in Fig. 7b. These are much more complicated and difficult to interpret directly since they depict the sum of asymmetric interactions between the point defect and both the edge and screw components of the individual Shockley partial dislocations. These latter two interactions are longer ranged than the previous two, covering a distance greater than 4.0nm along the glide plane. They are considered further by examining the interaction forces rather than the energies, and these are shown below.

The negative gradient of the interaction energy, is the interaction force for an individual antisite, and is defined such that a positive interaction force produces an attraction between the antisite defect and the dislocation in the glide plane. These forces, calculated for the planar dissociation of the screw dislocation, are shown in Fig. 8. Forces out of the glide plane and along the dislocation line have not yet been considered. These data reveal more clearly the asymmetry in the interactions between a particular antisite defect and the individual Shockley partial dislocations. The force due to the Al antisite defect labeled "Al 1" suggests that a nonzero interaction energy for an antisite positioned half way between the Shockley partials, can give a zero net force acting on the Shockley partials. This is due to the asymmetry of interaction with the edge components of the Shockley partials. For the tetragonal antisites oriented at ±45° to the screw line ("Al 2" and "Al 3"), the interaction forces are rather unusual exhibiting a highly asymmetric interaction about an individual Shockley dislocation, while the pair of Shockley partial dislocations is either generally attracted to one antisite and generally repelled by the other. While similar in kind, the two antisites exhibit opposite interaction forces to each other, suggesting that a net intrinsic kinking of the dislocation line in the glide plane may occur for a random distribution of tetragonal antisite defects. However, the details of these interactions are sufficiently complex and long range that computer simulation of the dislocation line interacting with a distribution of defects is required to see the complete effects.

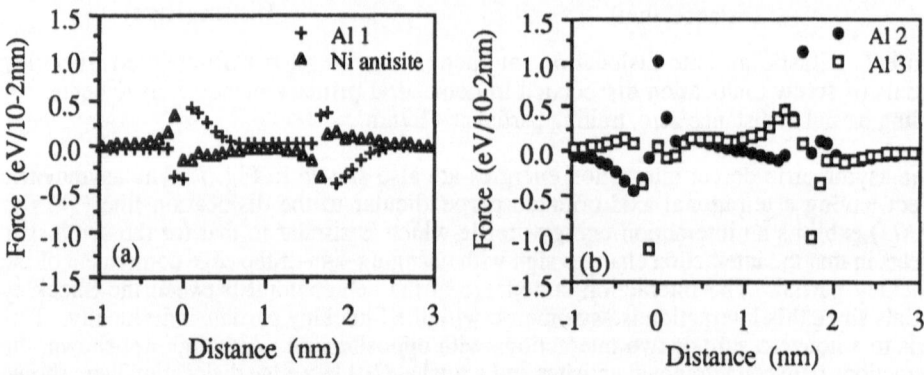

Figure 8.   Elastic force profiles for interaction energies shown in Fig. 7.

Similar solute-dislocation interaction profiles were computed for one of the transformed cores existing in the $L1_2$ structure, as schematically illustrated in Fig. 3. These are shown in Fig. 9. For this configuration the interaction energies are slightly weaker when the solute atom is positioned sequentially along the plane adjoining the APB. In this case the individual Shockley partials are not distinguishable in the interaction profiles. The symmetric Ni antisite has a much weaker interaction than that for the planar-core case and the three unique Al antisites are clearly distinguishable. For the asymmetric antisite having its tetragonal axis oriented perpendicular to the dislocation line, a symmetric attractive elastic interaction is observed when the antisite is located at the center position between the Shockley partials. For the other two tetragonal antisite defects, the dislocation line is initially repelled by one defect while being slightly attracted by the other, followed by a reversal of this situation as the solute is moved past the center point of the Shockley partials. Such an asymmetry in the interaction energies should, again, lead to an intrinsic kinking of the dislocation line in the presence of these solutes.

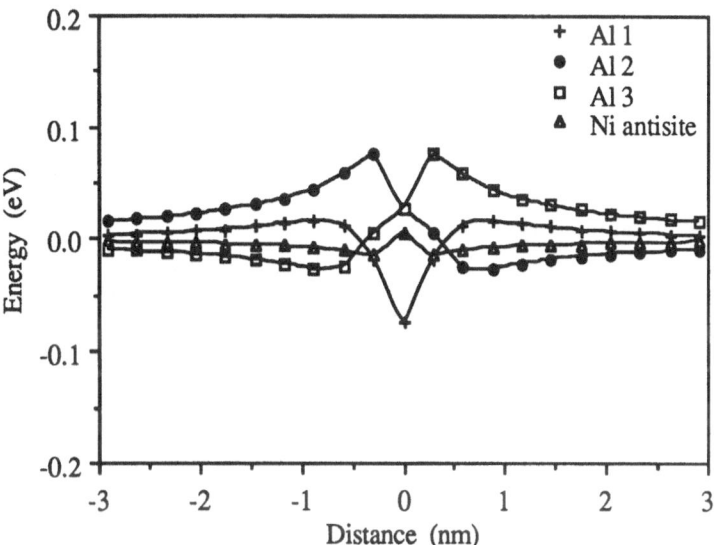

Figure 9. Elastic antisite-dislocation interaction energies for CSF-coupled Shockley partial dislocations dissociated in cross-slip plane as schematically shown in Fig. 3. Intersection of CSF and APB occurs at distance zero.

4.2 ANTISITE-FAULT INTERACTIONS

As discussed above, the EAM simulation technique was employed in order to compute the interaction between antisite defects and an infinite planar shear interface. The interactions were computed in two ways as described earlier and illustrated in Fig. 5. The resultant fault energies for the configurations illustrated in Fig. 5b are summarized in Table I. The Al antisites "1" and "2" have a tetragonal axis contained in the "x-y" plane of the fault; antisite "3" is oriented perpendicular to the fault plane. The values shown suggest that only one of the three Al antisite variants interacts with the faults within the assumptions of the EAM. The possibility exists that many body, angular-dependent potentials could lead to antisite-fault interactions for each of the antisite variants. In the present case, the interacting variant is the same variant which exhibits a symmetric or cubic elastic interaction with the dislocations. The tetragonal antisite defect having the weakest elastic interaction with a dislocation also exhibits the strongest "chemical" interaction with that dislocation. The Ni antisites or symmetric antisites, also show an interaction with the faults. Both "chemical" interactions are short ranged, dominated by first near neighbor atomic interactions. Contrary to recent experiments [20], both antisite defects were found to lower the

Table I Planar Fault Energies in the Presence of Antisite Defects in Ni$_3$Al*

Configuration	{111} APB (mJm^{-2})	{111} CSF (mJm^{-2})
stoichiometric crystal	141.4	120.6
Ni antisite at fault	133.4	113.4
Ni antisite one plane away	141.3	120.7
Al antisites 1&2 at fault	141.4	120.6
Al antisite 3 at fault	132.8	112.4
Al antisite 3 one plane away	141.7	120.2

* Values correspond to a solute concentration of approximately 1.1% and are computed within the EAM formalism using Voter's potential [31].

fault energy when an interaction occurred. A single antisite defect interacting on the fault plane leads to a fault energy which is approximately 7% lower than that in the stoichiometric crystal. The computed fault energies were converted to an effective interaction energy per unit length of dislocation and used to modify the elasticity calculations as described in the next section.

4.3 COMBINED INTERACTIONS

The elastic antisite-dislocation interaction energies shown in Fig. 7 and those shown in Fig. 9 were modified to include the effects of interaction between the antisite solute and the planar fault. One set of these new interaction profiles is shown in Fig. 10 for the nonplanar core configuration. In Fig. 10 the CSF has been introduced on the cross-slip plane and the APB on the primary plane for positive values of distance. This figure may be directly compared with Fig. 9. As shown in Fig. 10, the symmetric Ni antisite is always attracted to the dislocation core, although with a slight fluctuation near the Shockley partials, when both "chemical" and elastic effects are summed. Likewise, the "symmetric" Al antisite ("Al 1") is attracted to the dislocation line. The minima in energy at zero distance for the "Al 1" antisite in Fig. 10 is a refection of one of the difficulties encountered in summing the elastic and fault contributions. Some uncertainty exists over selection of the proper position to begin making the correction, suggesting that the minima may be an artifact. The other two tetragonal antisite interactions are relatively unchanged in the presence of faults.

Finally, the antisite-dislocation interaction energies were computed entirely within the EAM, wherein the dislocation core could continuously relax to the minimum energy as a function of antisite defect position. The results for the nonplanar dislocation core configuration schematically illustrated in Fig. 6 are shown in Fig. 11. This figure may be compared directly to Fig. 10. The general features of the interaction profiles are nearly identical for the two methods of calculation with the exception of the profile for the "Al 3" antisite. In this case, the antisite exhibits an entirely attractive elastic interaction with the transformed core which is not observed in the elasticity calculations. Closer examination of the EAM results revealed that the superpartial core spontaneously readjusted its position along the cross-slip plane as this antisite was introduced into the core region, yielding a totally attractive, rather than an attractive followed by repulsive interaction. Examination of atom positions along the dislocation line gave no indications of fault width fluctuations expected from the varying fault energy in the vicinity of "chemically" interacting solutes.

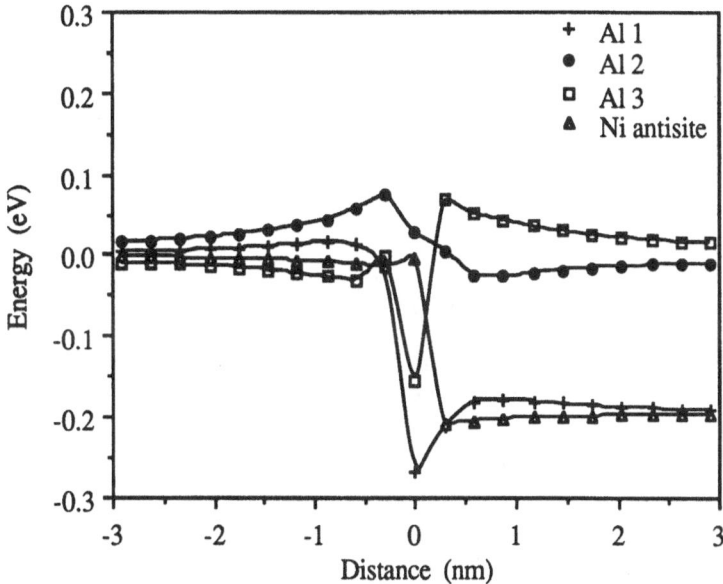

Figure 10. Combined elastic and fault, antisite-dislocation interactions for one superpartial screw dislocation core dissociated in octahedral cross-slip plane.

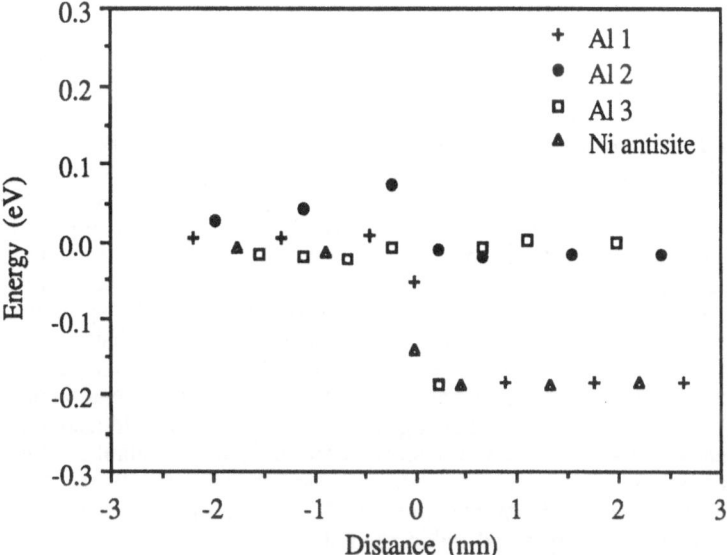

Figure 11. EAM simulation antisite-dislocation interaction energies for dislocation configuration depicted in Fig. 6.

Apparently, any such effects are masked by line tension forces acting to keep the line straight. The possibility of the intrinsic kinking phenomenon described previously has not been investigated by EAM simulations to date. In addition to the data provided herein, we have investigated several other dislocation components looking at both interaction energies and force profiles within anisotropic elasticity theory. The EAM simulations of the planar dissociated screw dislocation will be treated at a later date.

5.0 DISCUSSION

5.1 INTERACTION ENERGIES AND FORCES

Through the elasticity calculations and EAM simulations described above, we have examined several aspects of classical solute-hardening theory and their relevance to binary Ni$_3$Al alloys. Additionally, we have begun to investigate the apparent hardening effects of solute-fault interactions which are suggested from the phenomenological behavior of single-crystal alloys. Considering first the Ni antisites, and thereby the anticipated effects of symmetric (or cubic) hardeners, the following aspects are noteworthy.

First, the magnitude of interaction between symmetric antisite defects and the dislocations examined, is about one third of that found for the tetragonal antisites. However, there is only one type of symmetric site and therefore all of the antisites or solutes of this type should contribute equally to the forces experienced by a dislocation loop expanding in the material. The incremental change in flow stress should be linearly proportional to the concentration of solute or antisite defects for substitutional solutes which form symmetric defects, and these give rise to ordinary solute hardening effects. However, this is before considering the statistics of interaction and the possible influence of thermal activation. The influence of forces acting out of the glide plane on screw dislocations (not yet examined), may lead to significant variations in the thermal-activation response and hence variation from the classical view of symmetric hardeners.

Additionally, some solutes such as V (data shown in Fig. 2), have a very small size mismatch in the Ni$_3$Al, and there appears to be an additional hardening effect possibly due to solute-fault interactions for these symmetric hardening agents. This is supported by a measured change in the APB energies for V-containing alloys [20]. A strong solute-fault component to the interaction energies was observed for the case of Ni antisites in the EAM simulations and such interactions are as large in magnitude as any of the individual elastic effects. The "chemical" interaction as revealed by this measure leads to a net force encountered by any component of a dislocation loop as the faults approach the solute position. However, this interaction is essentially independent of the dislocation strain field. The influence of such forces on the steady-state glide of dislocations has not been previously considered. As mentioned earlier, another possible interpretation of the effect of solute-fault interactions is that a net change in the CSF energy should lead to a net change in the frequency of dislocation core transformations and a concomitant change in the flow stress at a given temperature. However, the observed changes in fault energies, within the approximations of the EAM potentials are small, leaving the macroscopically apparent "chemical" interaction unexplained and in need of further theoretical treatment. Electronic structure methods are currently being employed to investigate this interaction [43].

Three distinct tetragonal interactions were identified for the Al antisite defects. For any selected line direction, two-of-the-three antisites will have a strong elastic interaction with the dislocation while the third site will appear as a weaker symmetric hardener. The tetragonal antisites exhibit a longer range interaction than the symmetric ones. These tetragonal defects lead to interactions which are unequal (asymmetric) about the dislocation line. Such an interaction profile has not been considered in the classical statistical treatments and may lead to additional constraints on the dislocation line morphologies adopted during flow. Such aspects should be established through simulations of dislocation motion following the Labusch formalism so that the importance of these features may be established. Interestingly, from all of the studies of single-crystal alloys of Ni₃Al, Al antisites are the only tetragonal hardener which has been evaluated. The importance of the details of these defects cannot be properly evaluated until additional experimental data are produced for tetragonal hardeners having a wide range of size misfit are examined. Solute additions such as Cu and Co should be evaluated for this purpose.

For the tetragonal antisite which has a weak elastic interaction with a particular dislocation line, a chemical interaction will occur which is comparable in magnitude to that observed for the symmetric antisite interactions (see Fig. 11). However, within the range of the approximations of our EAM simulations, antisite-fault interactions lead to changes of only six to seven percent in the CSF and APB energies for dilute solutions (Table I). Experimentally, Ni antisites are found to yield an immeasurable change in the octahedral plane APB energy for a one percent solute addition—suggesting that the change is less than ten percent [20]. However, Al antisites lead to as much as a twenty percent increase in the APB energy for a one percent solute addition which is contrary to the prediction from the interatomic potential used for this study. Clearly, better techniques are required for investigating the "chemical" component of solute hardening.

5.2 STATISTICS AND THERMAL ACTIVATION

Labusch has treated the statistics of solute dislocation interactions through computer simulation using various empirically developed defect-dislocation force profiles. The method involves solving the equation of motion for a nearly straight dislocation under the application of an applied stress. The defects are distributed randomly in the lattice as per the concentration of the solute. Finite temperatures are treated by adjusting the ranges and magnitudes of the interaction forces as a function of position alone. A major drawback of this treatment is that the extended and non-planar nature of the core structure around dislocations, and interactions out of the glide plane are not included, but both must be considered for treatments of ordered alloys. Also, the interaction of solutes with kinks and the movement of dislocations via kink-pair formation is not explicitly included. A treatment of the Labusch type, with extended cores and kink motion should be developed.

In addition to examining the interaction statistics for a given dislocation line and a solute atom, the statistics for a group of moving dislocations and for multiple slip systems operating in parallel must be considered in order to capture the proper transitions in yield behavior suggesting by the flow curves in Fig. 1. A good example of this requirement is contained in the data by Curwick for deformation of single-crystalline Ni₃Al alloys [23]. His data show that the critical resolved shear stress for primary octahedral slip as a function of temperature, for two orientations away from the {010} orientation, exhibit a double peak suggesting that either more than one slip system or more than one pinning mechanism is operative under these conditions. The curves and peaks are shifted slightly by variation in solute content. This point has been discussed within the phenomenological theory by Greenberg et al. [44]. One other experimental study has reported similar behavior [45]; however, the subject has not received adequate treatment to date.

The view of deformation and solute-dislocation interactions reviewed above suggests an important role of thermally-activated kinking of dislocation lines which may be intrinsically enhanced by solute interactions. However, current flow models for intermetallics introduce an additional added complexity in the treatment of yielding by comparison to their disordered counterparts. Classically, disordered alloys may be classified into simple groups based on the behavior of individual dislocations and their motion [46]. One group of alloys is characterized by a very low (~µ/10,000) lattice-friction stress and thermally-activated glide, wherein the rate and stress controlling process of dislocation motion is that of dislocations bowing around and breaking away from local obstacles in the lattice, such as solute atoms. The yield stress at a given temperature is dictated by the magnitude and statistics of solute-dislocation interactions and these are varied by thermal activation and the mobility of the solute atoms. Such "bowing mode" behavior is characteristic of single-phase FCC alloys, NaCl and MgO. A second group of alloys is characterized by a higher lattice-friction stress (~µ/100 or higher) and the need for kink nucleation and propagation in order for dislocations to move. In this group solutes may influence both the statistics of kink nucleation and those of kink motion. These "kink mode" characteristics are typical of BCC single-phase alloys and some diamond-cubic materials.

On the other hand, within the two structure classes of intermetallics discussed above, the kink-mode versus bowing-mode distinction is diffuse. This is particularly true within CPD structures where thermally-activated kinks (jogs) may be viewed as either the source of obstacles or the source of mobile dislocations, depending on the following: i) the crystallographic plane in which they are contained, ii) the temperature of deformation, iii) the strain range being examined, and iv) the particular model of the deformation process being considered. At strains large enough to require dislocation multiplication and complete loop mobility (strains approaching 0.2 percent), thermally-activated double kinks out of the primary glide plane are believed to limit mobility of the screw segments. These may be localized double kinks which are defeated in a classical Orowan-Friedel "bowing mode" as treated in the model by Takeuchi and Kuramoto [47] or they may be lengths of dislocations having a transformed nonplanar core (cross-slipped) bounded by a kink on each end as considered by several other groups [48-51]. Several different geometries for such dislocations have been proposed, including the Kear-Wilsdorf lock. These dislocations may be mobilized by the reversal of the glide direction of the kinks at the ends in a "bowing mode", or by thermally-activated core transformations, away from the ends, back to a mobile configuration in a kink mode. Several athermal release mechanisms have also been considered for some of the proposed loop geometries. The specific structure of any thermally-activated kinks, which should be closely related to their mobility, has not been examined either theoretically or experimentally (for example through dislocation velocity determinations) as a function of the degree of directional bonding in the pure compounds or as a function of solute content. Furthermore, as discussed above, both structural families include compounds which show transitions in the active slip system (of the glide plane or direction) in the glide regime.

SUMMARY

In this manuscript we have suggested that a major aspect of understanding the role of composition on the flow and fracture behavior intermetallic alloys, lies in understanding the mechanisms of solute strengthening in these alloys. Solute-strengthening studies of alloys based on Ni_3Al were summarized, indicating that the effects of solutes on the process of transforming dislocation cores are the most important mechanisms. Some of the mechanisms for altering the activation barrier have been considered in further detail by examining solute-dislocation interactions. The dipole tensors for antisite defects in binary Ni_3Al were computed within the EAM and used to calculate antisite-dislocation interactions using anisotropic elasticity theory. Ni antisite defects having cubic symmetry, and three orthogonal variants of Al antisite defects having tetragonal symmetry were identified. Interaction energies were determined and used to compute the interaction forces in the glide plane acting perpendicular to straight dislocations. EAM simulation was also used to evaluate the fault energies as a function of composition and to directly evaluate the solute-dislocation interaction energies.

Screw dislocations were found to exhibit a significant interaction with both symmetric and asymmetric antisite defects. For a given defect position and dislocation line direction, the symmetric Ni antisite and the Al antisite variant oriented perpendicular to the dislocation line, yield interactions which are asymmetric about the individual Shockley partials and of opposite sign for the two partials. These two defects also exhibit a strong interaction with the CSF and octahedral plane APB. The magnitudes of these interactions may be as large as the tetragonal interactions when the antisite is at the fault. Such "chemical" interactions give rise to an antisite-fault interaction force which has not been previously considered in solute-strengthening theory. The other two variants of the tetragonal Al antisites lead to complex interactions of opposite sign to each other for both edge and screw dislocations. Unlike the symmetric antisite interactions, each asymmetric defect shows the same sign of interaction for both Shockley partial dislocations. The effects of these interactions on the thermal-activation process will be considered in further detail in future studies.

ACKNOWLEDGEMENT

The authors gratefully acknowledge several helpful discussions with Prof. E. Nadgorny.

REFERENCES

1. R. Darolia, JOM, **43**, 44 (1991).
2. W.O. Powers and J.A. Wert:, Metal. Trans. A, **21A**, 145 (1990).
3. Z. Wu, D.P. Pope and V. Vitek, in *High-Temperature Ordered Intermetallic Alloys IV*, J.O. Stiegler, L.A. Johnson and D.P. Pope, eds. Materials Research Society, Pittsburgh, PA, **213** (1991), 487.
4. H. Mabuchi, K-I. Hirukawa and Y. Nakayama, Scripta Metall., **23**, 1761 (1989).
5. J.P. Nic, S. Zhang and D.E. Mikkola, Scripta Metall., **24**, 1099 (1990).
6. D.G. Pettifor, Mat. Sci. and Tech. **4**, 675 (1988).
7. A.J. Freeman, T. Hong, W. Lin, and J-h. Xu, in *High-Temperature Ordered Intermetallic Alloys IV*, J.O. Stiegler, L.A. Johnson and D.P. Pope, eds. Materials Research Society, Pittsburgh, PA, **213** (1991), 3.
8. A.E. Carlsson and P.J. Meschter, J. Mater. Res., **6**, 1512 (1991).
9. A.E. Carlsson and P.J. Meschter, J. Mater. Res., **4**, 1060 (1989).
10. M.E. Eberhart, K.S. Kumar and J.M. Maclaren, Phil. Mag. B, **61**, 943 (1990).
11. T. Hong and A.J. Freeman, J. Mater. Res., **6**, 330 (1991).
12. C.L. Fu and M.H. Yoo, in *High-Temperature Ordered Intermetallic Alloys IV*, J.O. Stiegler, L.A. Johnson and D.P. Pope, eds. Materials Research Society, Pittsburgh, PA, **213** (1991), 667.
13. M.H. Yoo and C.L. Fu, ISIJ International, Special Issue on Advanced High Temperature Intermetallics, (1991) in press.
14. M.S. Daw and M.I. Baskes, Phys. Rev. Ltrs., **50**, 1285 (1983).
15. S.M. Foiles and M.S. Daw, J. Mater. Res., **2**, 5 (1987).
16. D.M. Dimiduk and S. Rao, in *High-Temperature Ordered Intermetallic Alloys IV*, J.O. Stiegler, L.A. Johnson and D.P. Pope, eds. Materials Research Society, Pittsburgh, PA, **213** (1991), 499.
17. D.P Pope and S.S. Ezz, Int. Mat. Rev., **29**, 136 (1984).

256

18. T. Suzuki, Y. Mishima, and S Miura, Trans. Iron Steel Inst. Jpn., **29**, 1 (1989).
19. F. Heredia and D.P. Pope, Jol. de Phys. III, **1**, 1055 (1991).
20. D.M. Dimiduk, Ph.D. Dissertation, Carnegie Mellon University (1989).
21. V. Paider, D.P. Pope and V. Vitek, Acta Metall. **32**, 435 (1984).
22. F. Heredia, Ph.D. Dissertation, University of Pennsylvania (1990).
23. L.R. Curwick, Ph.D. Dissertation, University of Minnnesota (1972).
24. J. Douin, P. Veyssière and P. Beauchamp, Phil. Mag. A, **54**, 375 (1986).
25. N.Baluc, H.P. Karnthaler, and M.J. Mills, Inst. Phys. Conf. Ser. No. 93, **2**, 463 (1988).
26. A. Korner, Phil. Mag. A, **58**, 507 (1988).
27. C. Bontemps and P. Veyssière, Phil. Mag. Lett., **61**, 259 (1990).
28. P. Veyssière and C. Bontemps, Private communication.
29. D.M. Esterling, D.K. Som, and A. K. Chatterjee, J. Phys. F: Met. Phys., **17**, 109 (1987).
30. T. Shinoda, K. Masuda-Jindo, Y. Mishima, and T. Suzuki, Phys. Rev. B, **35**, 2155 (1987).
31. A.F. Voter, and S.P. Chen, in *Characterization of Defects in Materials*, R.W. Siegel, J.R. Weertman and R. Sinclair, eds. Materials Research Society, Pittsburgh, PA, **82** (1987), 175.
32. D. Farkas and E.J. Savino, Scripta Metall., **22**, 557, (1988).
33. T.A. Parthasarathy, D.M. Dimiduk, C. Woodward, and D. Diller, in *High-Temperature Ordered Intermetallic Alloys IV*, J.O. Stiegler, L.A. Johnson and D.P. Pope, eds. Materials Research Society, Pittsburgh, PA, **213** (1991), 337.
34. G. Leibfried and N. Breuer, in *Point Defects in Metals I: Introduction to the Theory*, G. Höhler, ed. Springer Tracts in Modern Physics, New York, **81** (1978).
35. K. Schroeder, in *Point Defects in Metals II: Dynamical Properties and Diffusion Controlled Reactions*, G. Höhler, ed. Springer Tracts in Modern Physics, New York, **87** (1980), 171.
36. H.R. Schober and K.W. Ingle, J. Phys. F Metal. Phys., **10**, 575 (1980).
37. A.N. Stroh, Phil. Mag., **3**, 625 (1958).
38. Y. Mishima, S. Ochiai, N. Hamao, M. Yodogawa and T. Suzuki, Trans. Jap. Inst. Met., **27**, 656 (1986).
39. N. Brown, Phil. Mag., **4**, 693 (1959).
40. M.H. Yoo, J.A. Horton and C.T. Liu, Acta Metall., **36**, 2935 (1988).
41. Y.P. Wu, J.M. Sanchez, and J.K. Tien, in *High-Temperature Ordered Intermetallic Alloys IV*, J.O. Stiegler, L.A. Johnson and D.P. Pope, eds. Materials Research Society, Pittsburgh, PA, **213** (1991), 87.
42. S. Rao and C. Woodward, in preparation.
43. C. Woodward, J. M. MacLaren and S. Rao, in preparation.
44. B.A. Grinberg, V.N. Indenbaum and YU.N. Gornostyrev, Phys. Met. Metall., **63**, 60 (1987).
45. A.E. Staton-Bevan, Scripta Metall., **17**, 209 (1983).
46. E. Nadgorny, in *Dislocation Dynamics*, Progress in Materials Science, Pergamon Press, Oxford, **31** (1988).
47. S. Takeuchi and E. Kuramoto, Acta Metall., **21**, 415 (1973).
48. Y. Sun and P.M. Hazzledine, Phil. Mag. A, **58**, 603 (1988).
49. M.J. Mills, N. Baluc, and H.P. Karnthaler, in *High Temperature Ordered Intermetallic Alloys III*, C.C. Kock, C.T. Liu, N.S. Stoloff and A.I. Taub, eds. Materials Research Society, Pittsburgh, PA, **133** (1989), 203.
50. D.M. Dimiduk, Jol. de Phys. III, **1**, 1025 (1990).
51. P.B. Hirsch, Scripta Metall., **25**, 1725 (1991).

CYCLIC DEFORMATION OF INTERMETALLIC ALLOYS

N.S. STOLOFF
Department of Materials Engineering
Rensselaer Polytechnic Institute
Troy, New York 12180-3590

ABSTRACT. Recent studies of fatigue behavior of intermetallic alloys under both stress and strain control conditions are reviewed. Effects of composition, temperature, frequency and environment on high cycle fatigue, crack growth and low cycle fatigue are discussed. In addition, crack initiation phenomena in Ni_3Al single crystals are described and a new model of fatigue crack initiation is outlined.

1. Introduction

Improvements in low temperature ductility or high temperature creep resistance remain as major goals for most research studies on intermetallic compounds. Nevertheless, use of intermetallics in aerospace or other structural applications will require an adequate knowledge of their behavior under cyclic loading conditions. It already has been established that several intermetallics, including Ni_3Al+B, display very good resistance to stress controlled fatigue (high cycle or crack growth) at room temperature. Other intermetallics, such as Fe_3Al and Ti_3Al, display very rapid crack growth, probably due to high notch sensitivity. Moreover, at elevated temperatures, interactions with the test environment or with creep components arising from the load cycle can adversely influence fatigue behavior. With respect to strain controlled fatigue, detailed studies of crack initiation have been carried out on crystals of Ni_3Al+B. Low cycle fatigue behavior of polycrystalline Ni_3Al alloys also have been reported recently.

The purpose of this paper is to review progress in understanding of fatigue behavior since the last general review of the subject published in 1987[1]. The emphasis will be upon microstructural, temperature and environmental effects.

C. T. Liu et al. (eds.), Ordered Intermetallics – Physical Metallurgy and Mechanical Behaviour, 257–277.
© 1992 Kluwer Academic Publishers.

2.   High Cycle Fatigue

2.1.   Ni₃Al ALLOYS

2.1.1. Effects of Temperature. High cycle fatigue studies on three Ni$_3$Al alloys, compositions of which are listed in Table 1, have been carried out at room and elevated temperatures. All of the alloys, IC-218, IC-221 and IC-396M, contain chromium and boron as the principal alloying additions; chromium is added to improve the resistance of these alloys to oxygen in the test atmosphere at elevated temperatures, but does not suppress embrittlement by hydrogen in tensile tests at room temperature[2]. IC-396M contains 3% Mo. The importance of test environment when Ni$_3$Al alloys are

Table 1
Ni$_3$Al Alloys Tested in High Cycle Fatigue
Composition, wt%

	IC-218	IC-221	IC-396M
Al	8.50	8.04	9.0
Ni	81.2	81.6	79.45
Cr	7.56	7.55	7.7
Zr	0.83	1.75	0.85
Mo	-----	-----	3.0
O	0.001	0.0008	-----
B	0.02	0.02	0.005

cyclically loaded is readily seen in Figs. 1 and 2, depicting behavior at 25°C and 750°C respectively[3]. At low temperatures precharged hydrogen decreases the fatigue life of IC-218, especially at low test frequencies, see Fig. 1. At 750°C there is a substantial improvement in HCF life of IC-218 when tests are run in 10^{-5} torr vacuum, see Fig. 2. This presumably results from the adverse effects of oxygen at elevated temperatures, as has been well documented for Ni$_3$Al alloys tested under monotonic loading[4]. Note also in Fig. 2 that powder processed IC-396M displays much lower HCF lives than IC-218 at 750°C when both are tested in vacuum. At a stress range of 534 MPa ($0.9\sigma_{ys}$) the fatigue life of IC-396M is reduced by nearly a factor of three by testing in air (2.6×10^4 cycles) vs. vacuum (7.5×10^4 cycles), a considerably larger effect than for IC-218 at the same ratio of stress range to yield stress. This striking environmental effect is probably due to the presence of Mo in 396M; environmental effects are discussed further in the section on creep-fatigue-environment interactions. Note, however, that substituting cast for P/M IC-396M increases the grain size and produces a response to cyclic loading at 850°C that is only slightly lower than that of P/M material at 750°C.

Finally, it is worth noting that IC-396M is comparable in yield strength to IN-713C, a cast nickel base superalloy, to about 800°C, but that IC-396M reportedly displays considerably better HCF resistance at 650°C in air[5]. This result is consistent with crack growth data, showing that Ni$_3$Al alloys are superior to wrought superalloys in FCP resistance at both 25° and 650°C[6,7].

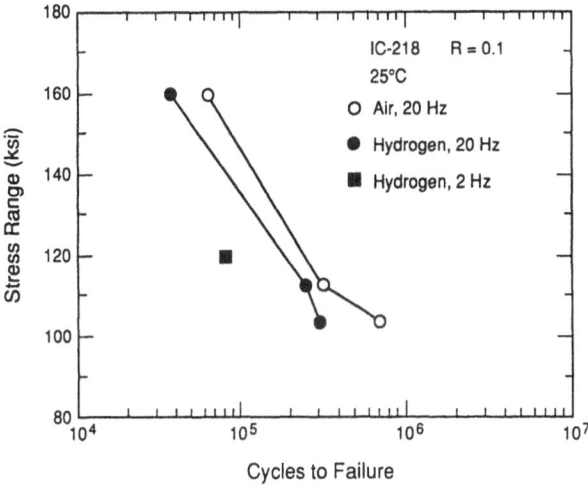

Fig. 1.   Effect of hydrogen on HCF of IC-218.

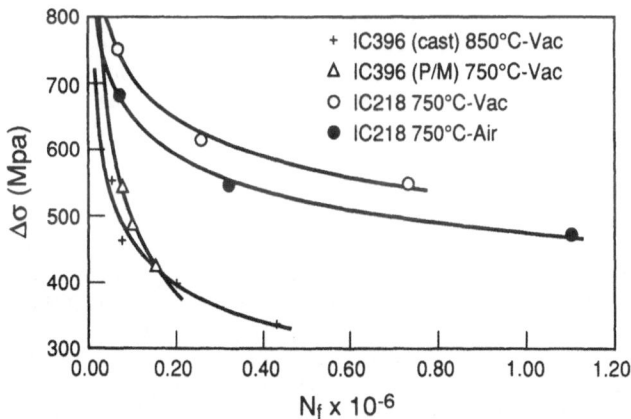

Fig. 2.   HCF of IC-218 and IC-396M at elevated temperatures.

2.1.2.   Frequency Effects. An important aspect of high temperature operation of structural materials under cyclic loading is the possibility of interaction among the fatigue cycle, creep processes in the material and test environment.   These are most likely to occur at low test frequencies. The only intermetallic alloys for which such studies have been carried out are based upon Ni_3Al. Bellows et al[8] suggested that there is very good intrinsic creep/fatigue resistance of a directionally solidified Ni_3Al alloy containing B and Hf at 450°C and 760°C.   Tests were carried out under stress controlled conditions in vacuum.   A more recent study[9] has shown that two fine grained Ni_3Al alloys, IC-218 and IC-221, are affected by test frequency in stress-controlled tests in vacuum or air at 600°C or 800°C, as shown in Fig. 3. Note that the number of cycles to failure is reduced by about 100x at both 600°C and 800°C as frequency decreases from 20 to 0.2 Hz, Fig. 3a). Although there is little effect of environment on number of cycles to failure, cyclic ductility is sharply lower in air, Fig. 3b).

Samples tested at 800°C under vacuum exhibit a fracture initiated internally by microvoid growth and coalescence at each of the frequencies employed. At 600°C the fatigue zones decreased in size as the frequency was lowered. The large increase in ductility observed with decreasing frequency at both temperatures is particularly strong evidence for the role of creep in the fatigue process. Additional evidence for a creep-fatigue interaction was provided by experiments conducted in air[9]. Although fatigue lives in air were slightly shorter compared to the same tests run in vacuum, the general trend of the results in these experiments parallels those obtained in vacuum (i.e. the slopes of N_f vs. ν are identical), indicating that environment plays a minor role in the observed frequency dependence of the stress-controlled fatigue life.

Although creep-fatigue interactions largely account for the reduction in high temperature fatigue life of IC 218 at low frequencies in vacuum, tests performed in air also reveal some environmental effects. Ductilities, in terms of reductions of area as well as total elongation, were considerably reduced when compared to the tests performed in vacuum, the trend being enhanced by increased temperature, see Fig. 3b)[9]. Moreover, major changes in the fracture morphology were observed when comparing air to vacuum. At 600°C fatigue zones were consistently transgranular when the tests were performed in vacuum, but were partially or totally intergranular for tests in air, the degree of intergranular failure increasing with decreasing frequency. At 800°C, in air, dimples were still present in the center, although intergranular failure zones around the outside of the sample became more numerous as the frequency was decreased. The results indicate that there is little or no environmental effect on crack nucleation, while a strong interaction must exist once the crack in initiated, i.e. on crack propagation and total elongation. It is concluded that most of the fatigue life is spent in initiation. These results also confirm previous studies performed on the same class of chromium-containing Ni_3Al alloys[7]. By minimizing high temperature oxygen-induced embrittlement, chromium additions to Ni_3Al improve high temperature tensile properties,[4] as well as high cycle fatigue properties in air[10]. On the other hand, the decreased ductilities and intergranular fatigue zones in air show that, once a crack is initiated, "dynamic" oxygen embrittlement occurs.

2.1.3. Ti$_3$Al Alloys. Considerable effort has been devoted recently to studying the influence of microstructure, environment and fiber reinforcement on the HCF behavior of α_2 (Ti$_3$Al) alloys. Unidirectional composites of Ti-24Al-11Nb, reinforced by continuous SCS-6 fibers and tested in the longitudinal and transverse orientations, are compared to the monolithic alloy, another α_2 alloy (Ti-24Al-17Nb-1Mo) and Ni-base IN-100 in Fig. 4[11]. Note that the longitudinal composite displays the best fatigue resistance of the aluminides and the transverse composite has the lowest resistance. These data scale with tensile strength, as expected for HCF behavior. On a normalized (by weight) basis, the relative

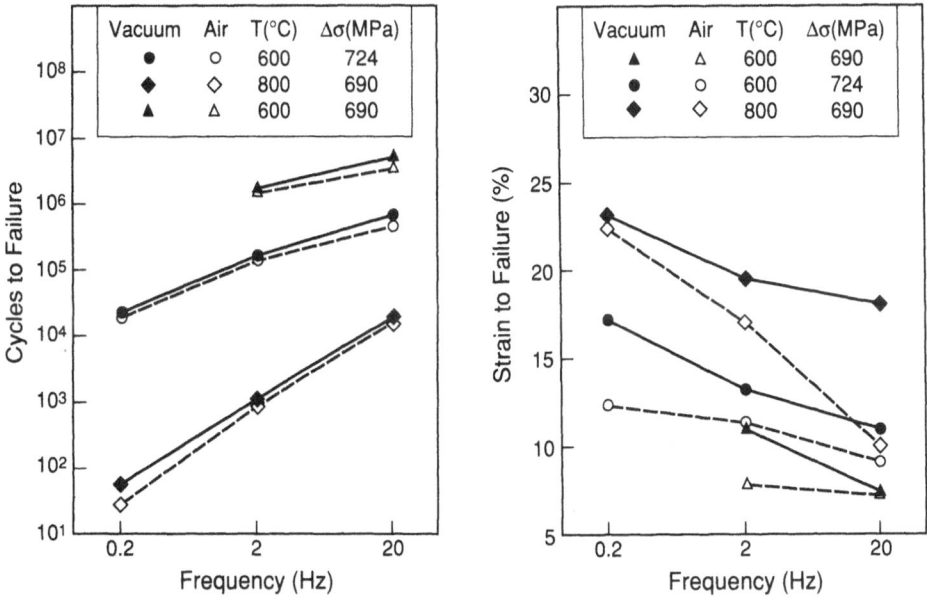

Fig. 3.   Effects of frequency and environment on fatigue behavior of
IC-218[9]. a) cycles to failure, b) cyclic ductility

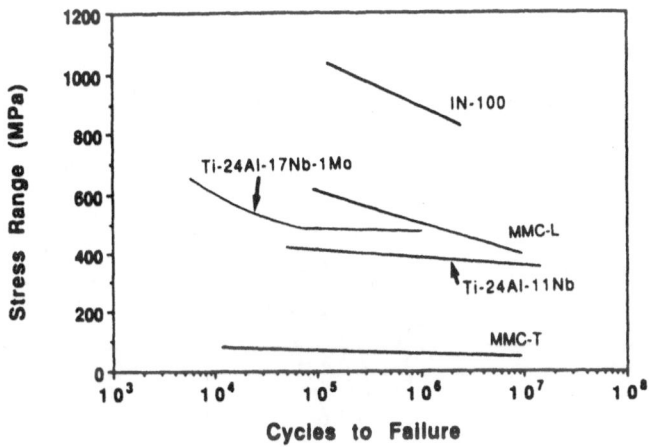

Fig. 4.   High cycle fatigue behavior of α_2 Ti$_3$Al and its composites,
compared to IN 100 at 650°C, air $\nu \approx 0.2$Hz, $R \approx 0$[11].

positions of IN 100 and the longitudinal composite are reversed, while the
other alloys retain their relative positions.   The low density of the
aluminide clearly outweighs its inferior absolute fatigue resistance.

3.   Crack Propagation

3.1.   Ll$_2$ ALLOYS

Crack growth rates for Ll$_2$ alloys (FeNi)$_3$V, Ni$_3$Al+B and IC-221 tend to be lower than for other structural alloys, especially at low ΔK[7]. Crack growth rates in these alloys increase with increasing temperature, in spite of increased yield stresses over the same temperature range, see Fig. 5. Plastic zone sizes are reduced at high temperatures, indicating that a portion of the temperature effect noted in Fig. 5 may be due to a shift from plane stress to plane strain conditions.

In many models of stress controlled crack propagation, the constant, C, in the Paris-Erdogan[12] equation:

$$\frac{da}{dN} = C\Delta K^m$$

is postulated to vary inversely with yield stress, modulus and/or tensile ductility[1].   This clearly is not the case with the LRO alloys and Ni$_3$Al, suggesting that one or more atomic species in the gaseous environment of moderate vacuum or argon causes an increase in growth rates at elevated temperature. Chang, et al[13] suggested, in fact, that air is embrittling to cyclically loaded Ni$_3$Al+B at temperatures as low as 400°C, resulting in Stage II crack growth rates more rapid than those of superalloys.   The very low crack growth rates near threshold, on the other hand, were confirmed in their work.   In vacuum, stage II crack growth rates for Ni$_3$Al remain below those of superalloys, even though the vacuum of 10^{-5} torr is not sufficient to eliminate environmental attack completely.

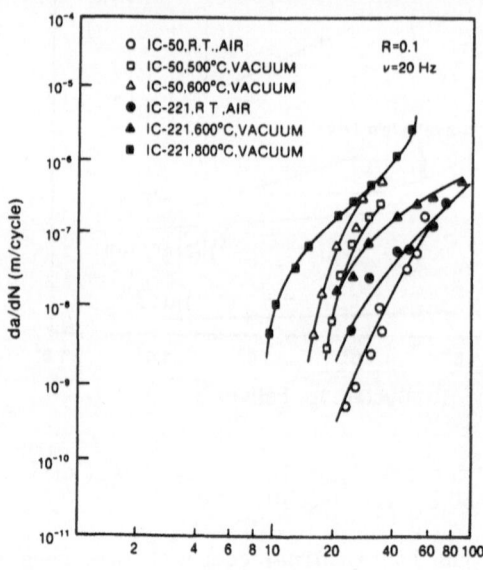

Fig. 5.   Effect of temperature on crack growth in Ni$_3$Al-base alloy IC-221[3].

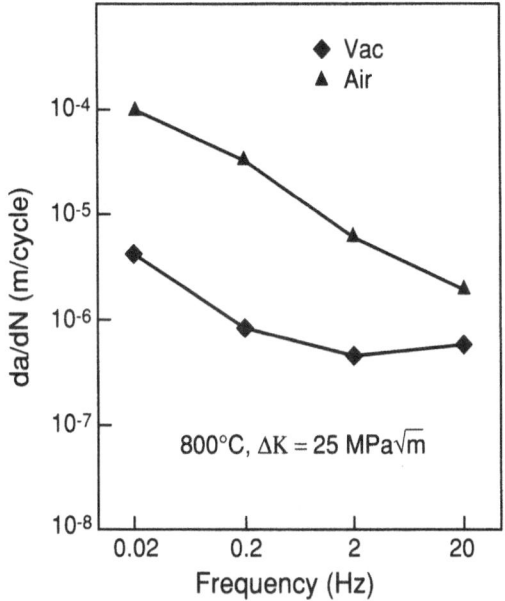

Fig. 6.   Crack growth as a function of frequency and environment in Ni₃Al alloy IC-221[7].

At elevated temperatures both creep and environmental effects (due to oxygen) play a significant role in crack growth behavior. For example, Fig. 6 shows that crack growth rates, da/dN, increase with decreasing frequency in both vacuum and air for IC-221 at 800°C[7]. The increase in growth rate in vacuum at low frequencies arises primarily from a creep effect, while in air there is an interaction among fatigue, creep <u>and</u> environmental processes. Intergranular fracture was noted at all test frequencies in air, and transgranular fracture was noted in vacuum. These results are consistent with those described earlier for HCF of IC-218.

3.2   Fe₃Al Alloys

Fe₃Al alloys generally are intermediate in growth rates between superalloys and Ll₂ intermetallics, e.g. Ni₃Al or (FeNi)₃V, at low ΔK levels; at high ΔK, crack growth is more rapid in Fe₃Al[14].

Recent work has shown that hydrogen or moisture in air has very significant effects on ductility of Fe₃Al[15,16]. Similarly, crack growth behavior of Fe₃Al is adversely affected by both hydrogen gas and precharged hydrogen, as shown in Fig. 7, for an alloy containing 28%Al and 8%Cr (FA-129)[17].   Fractographic   examination   revealed   distinct   striations superimposed on cleavage facets in samples free of hydrogen or precharged, both tested in air. Striations were less visible in a sample tested in 1atm. H₂ gas. Tests are underway to determine whether high vacuum can significantly improve crack growth resistance.

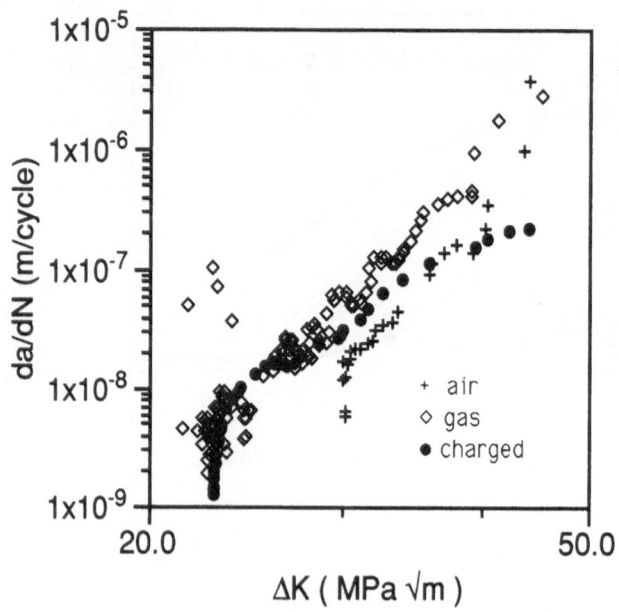

Fig. 7.   Influence of environment on fatigue crack growth of Fe₃Al alloy
FA-129 at 25°C[17].

3.3   Ti₃Al

Although crack growth phenomena in Ti₃Al alloys have been studied
extensively, little had been published until recently. Williams et al[18]
have shown that crack growth rates of Ti-24-11 and Ti-25-10-3-1 (super α₂)
at 25°C are lower than that for the commercial alloy Ti-6246 in the near
threshold region, but above a ΔK of about 10 MPa√m growth rates are more
rapid in the intermetallics, see Fig 8[18]. Note that the super α₂, which
is stronger than the Ti-24-11 alloy, has a much higher growth rate than
either of the others. Davidson et al[19] report about the same threshold
for super α₂, but show that the range of thresholds in α+β alloys is
higher, and crack growth rates are lower than for super α₂ at 20°C, see
Fig. 9. As temperature increases to 650°C relative crack growth rates of
Ti-24-11 change in a complex manner. However, at 650°C, the growth rate
of Ti-24-11 is higher than that of either IMI 834 or the nickel-base
superalloy IN 100[18]. The influence of microstructure on fatigue life is
as great or greater than compositional or temperature effects[18].   For
example, super α₂ that has been solutioned in the β range has a higher
growth rate at 650°C than the same alloy with a 20% equiaxed α, transformed
β microstructure. There is a further improvement in fatigue crack growth
resistance when the alloy is cycled in vacuum.

Further evidence for an environmental effect on crack growth of α₂
at both room and elevated temperature has been provided by Aswath and
Suresh[20] and Balsone et al[21]. As shown in Fig. 10, there is an effect
of frequency on crack growth rates in laboratory air and in vacuum

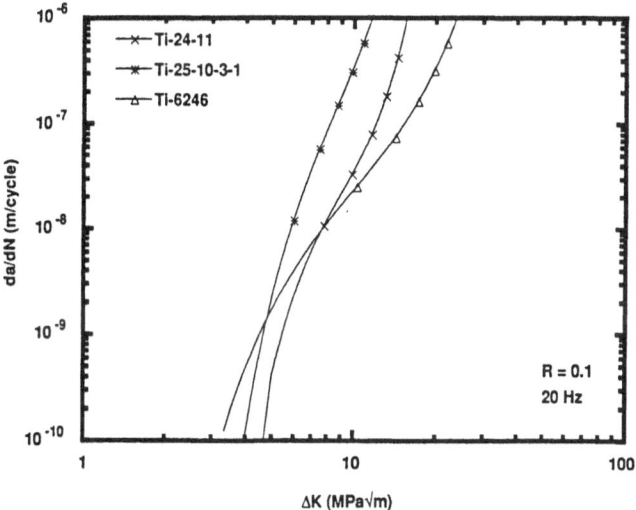

Fig. 8.   Crack growth of α_2 alloys compared to commercial Ti alloy 6-2-4-6[18].

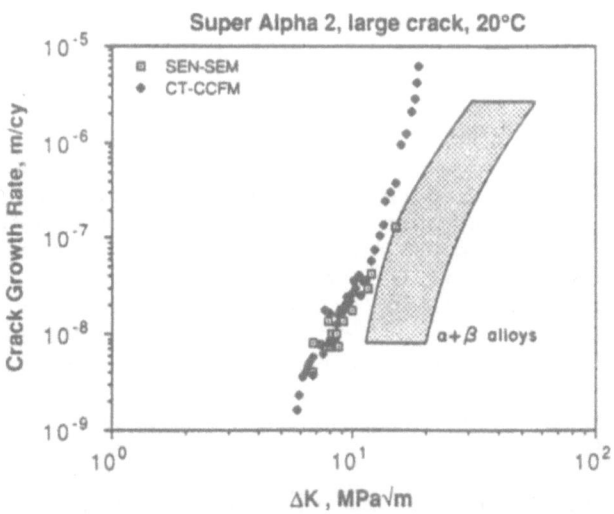

Fig. 9.   Effects of microstructure and environment on crack growth in Ti-25-10-3-1 (super α_2) at 25°C[20,21].

(10^{-5}torr) for Ti-24-11 with coarse Widmanstatten α_2 surrounded continuously by transformed β (this is a microstructure with relatively high FCG resistance). Note that crack growth is much more rapid in air than in vacuum. Similar observations have been made at temperatures in the range 650-800°C[20,21].

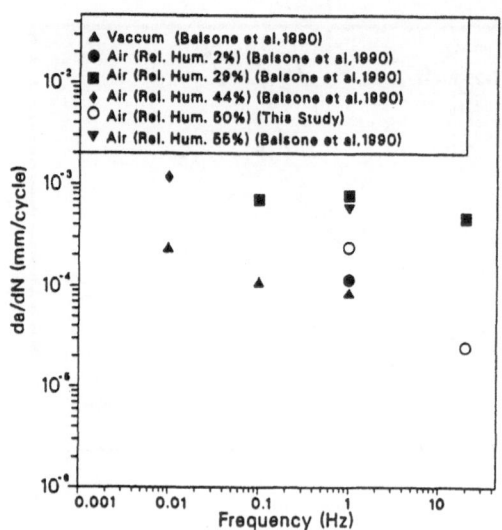

Fig. 10.   Influence of microstructure and environment on crack growth rates of Ti-25-10-3 (super α_2)[18].

Wessels et al[22] also have reported strong microstructural effects on crack growth at 425°C of a Ti-24.5Al-8-Nb-2Mo-2Ta alloy. A β heat treated sample with a coarse aligned lath or colony microstructure produced by slow cooling provided the lowest growth rates and highest threshold (about 6 MPa√m). The lowest threshold, 3 MPa√m, and highest growth rate was exhibited by a salt quenched $\alpha_2+\beta$ sample. Vekaturaman et al[23] report that microstructure has little effect on crack growth rates of Ti-16w%Al-10%Nb in air, but that testing in vacuum lowered the crack growth rate. Therefore, it may be concluded that hydrogen embrittles α_2 alloys at low temperatures, while oxygen is detrimental at elevated temperatures, similar to the behavior of Ni₃Al alloys. In general, fatigue crack growth rates in α_2 increase with moist environments (probably due to the release of hydrogen) lowered test frequency and increased test temperature.

3.4   Ti₃Al COMPOSITES

Perhaps the single factor most likely to influence the FCG resistance of titanium aluminides is the use of fibrous reinforcements, as was suggested by the HCF data in Fig. 4[11]. Fig. 11 shows that although the crack growth rates of α_2 and γ are higher than for the P/M Ni base superalloy IN 100, the longitudinally oriented composite of α_2 (Ti-24Al-11Nb) with continuous SCS6 SiC fibers is far more crack growth resistant than is IN 100[11]. Unfortunately, the transversely oriented α_2 composite displays the highest growth rate of all. (Note: the specimen geometry differed in that thin tubes of Ti-48Al-1V (γ) were utilized[24]; also, the frequency of tests of the α_2 composite was 3.33 Hz, which would be expected

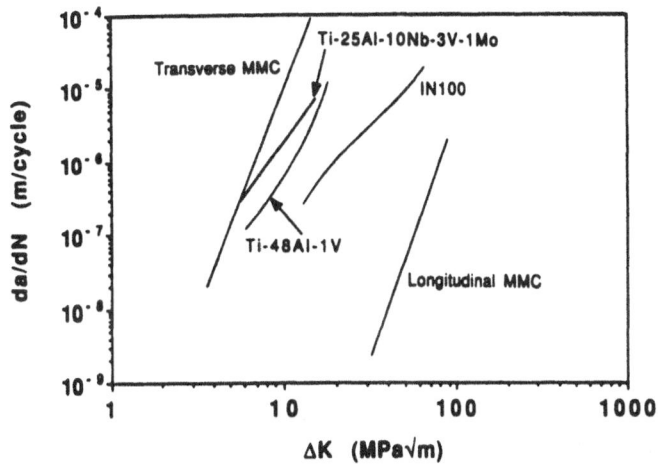

Fig. 11.   Comparison of crack growth rates of titanium aluminides, α_2 composites and IN 100; tests at 650°C, R=0.1, ν=0.2Hz except for the composite, for which ν=3.33Hz[11].

to result in a somewhat lower growth rate than at 0.2 Hz).   It is noteworthy that the fracture toughness of the longitudinally tested α_2/SCS-6 composite is about 70 MPa√m at 650°C, while IN 100 displays a K_I of about 100 MPa√m.   Transverse specimens of the α_2 composite   as well ^{C}as the γ alloy, on the other hand, display about the same level of K_I approximately 15 MPa√m.

3.5   TiAl

Data for TiAl(γ) alloys are even more limited than for Ti_3Al. Note in Fig. 11 that the growth rate of Ti-48Al-1V is comparable to that of super α_2[11].   Soboyejo et al[25] have reported the influence of microstructure and temperature on FCG behavior of a P/M Ti-48%Al alloy tested in air.   Crack growth at room temperature was slightly slower than for a Ti-6Al-4V alloy, see Fig. 12.   At 700°C, crack growth was even slower, probably due to oxide-induced crack closure[25].

4.   Low Cycle Fatigue

4.1   Ni₃Al ALLOYS

Studies of low cycle fatigue (LCF) of intermetallic compounds remain sparse.   In a previous review[1], results of LCF studies on Cu_3Au[26] and Ti_3Al[27] were described.   Recently the LCF behavior of the Ni_3Al alloy IC-218 was reported[28].   The results of that study are in general agreement with LCF behavior of an FeCo-V alloy (B2 structure) as described below. Hot extruded powders of IC-218 were tested under total strain

268

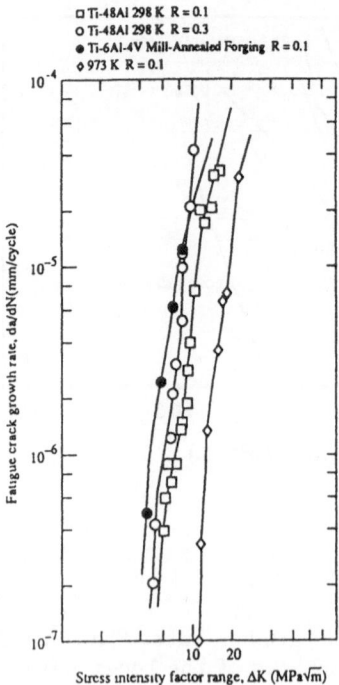

Fig. 12.   Effect of temperature on crack growth of Ti-48Al[25].

control (fully reversed, R=-1) in laboratory air, utilizing a triangular
wave form and frequency of 10 cpm. The tests showed that LCF performance
of IC-218 is superior to that of other structural alloys, especially at
high strain amplitudes, as shown in Fig. 13[28].   These results were
explained in terms of the high tensile ductility and good crack growth
resistance of alloys of this type.

        The IC-218 alloy displayed considerable cyclic hardening at 25°C,
followed by rapid softening, see Fig. 14[28]. The magnitude of hardening
increased and the number of cycles to maximum strength decreased as strain
amplitude decreased. Similar behavior has been noted in ordered FeCo-V,
also tested at room temperature in air with R=-1[29].   The rapid cyclic
hardening observed in FeCo-V was directly attributable to the presence of
long range order, as the magnitude of hardening was much less in
disordered FeCo-V[29]. Therefore, the high cyclic hardening rate of IC-218
is assumed to arise from the long range ordered state of the alloy.

        Gordon and Unni[30] have reported that cyclic hardening/softening
behavior under R=-1 conditions correlates well with HCF results, see Table
2.   Large ratios of $\Delta\sigma_{10}{}^6/\sigma_{ys}$ are observed when rapid hardening occurs,
while cyclic softening usually leads to lower HCF resistance.   This
behavior is readily accounted for by the rapid decrease in plastic strain
per cycle when rapid hardening occurs in a stress-controlled HCF test.
However, the reverse situation applies under strain controlled conditions;
at high strain amplitudes the LCF behavior of IC-218 is apparently

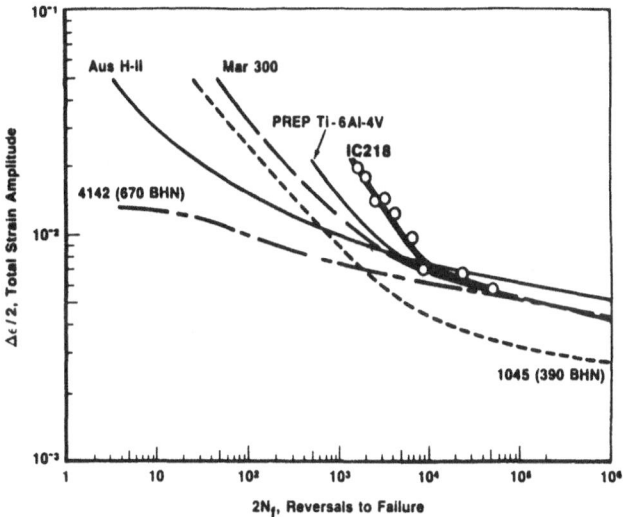

Fig. 13.   Low cycle fatigue of Ni₃Al alloy IC-218 compared to commercial alloys at 25°C[28].

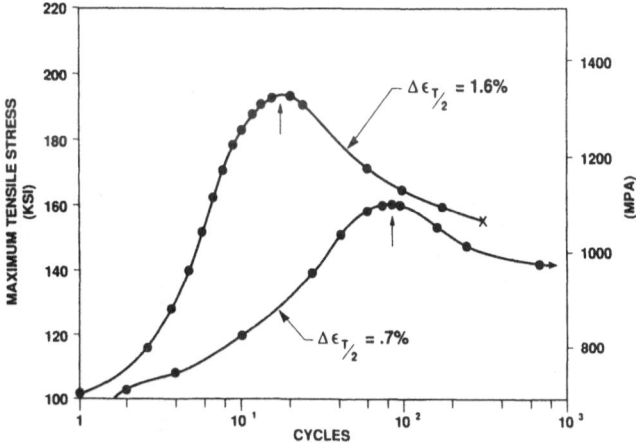

Fig. 14.   Cyclic hardening and softening of Ni₃Al alloy IC-218 at 25°C[28].

improved due to the cyclic softening that occurs after 10-100 cycles. Again there is good correlation with the results of LCF tests on FeCo-V which showed that disordered specimens (low cyclic hardening) displayed longer strain controlled fatigue lives than ordered specimens (high cyclic hardening)[29].

270

4.2 Ti₃Al Alloys

The LCF behavior of Ti-24Al-11Nb has been studied between 25°C and 650°C, see Fig. 15[18]. Note that there is a crossover in properties between 25°C and 425°C, but at 650°C the LCF life is considerably reduced, especially at low strain levels. This is again suggestive of an environmental effect. As in the case of HCF and crack growth, microstructure and composition play significant roles in the LCF behavior of α_2 alloys. For example, β processed super α_2 has superior LCF resistance to Ti-24-11 at both 25°C and 650°C[18].

4.3 SURFACE FILM EFFECTS

Thin nickel films applied to extruded Fe-40Al (which deforms by <111> slip) and Ni-30Al-20Fe (which deforms by <100> slip) have been shown to enhance ductility in both monotonic and cyclic tests. Fig. 16 shows that small amounts of compressive prestrain increase the accumulated strain to failure for the ternary alloy from 18% to 40%, while a thin nickel film increased the accumulated strain to failure to 34%[31]. Combining prestrain and coating increased the strain to failure to almost 90%. These results have been explained in terms of mobile <100> dislocations induced by the two techniques. In the binary Fe-40Al alloy, on the other hand, as-extruded material already contains an adequate dislocation density to accomodate deformation. Introducing new dislocations by prestrain has little effect on cyclic deformation. However, surface films in bcc-like materials (such as Fe-40%Al) may generate mobile non-screw dislocations which compensate for the reduced mobility of screw dislocations at room temperature and below.

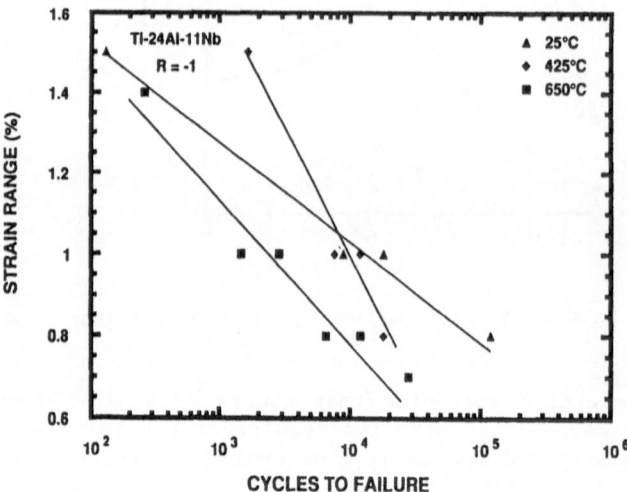

Fig. 15.   Effect of temperature on LCF of Ti-24Al-11Nb (α_2)[18].

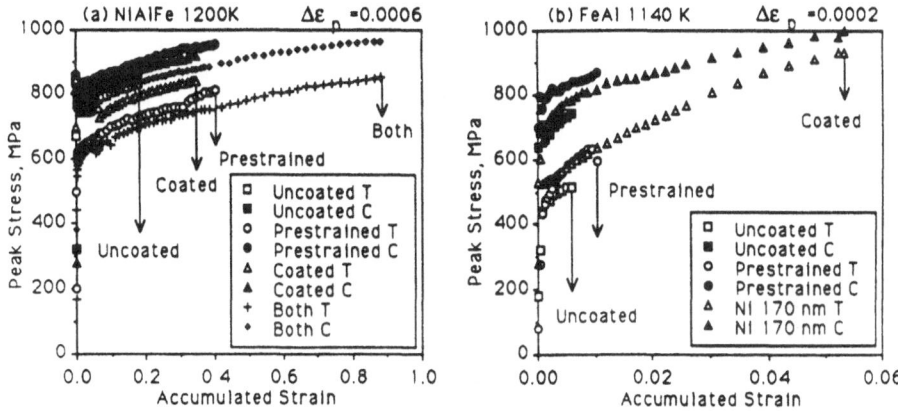

Fig. 16. Influence of surface films on cyclic ductility[31].
a) Fe-40a%Al; b) Ni-30Al-20Fe.

Table 2. Cyclic Hardening and HCF
Behavior of Intermetallics[30]

HIGH CYCLE FATIGUE (HCF) DATA, T = 25C, R ≈ 0.

ALLOY	$\frac{\Delta\sigma \text{ (At } 10^{6} \text{ Cycles)}}{\sigma \text{ (ys)}}$	CYCLIC HARDENING / SOFTENING BEHAVIOR UNDER R = -1 LOADING
ALUMINIDES		
Ni–24Al HIP	1.79 (1)	Hardens (Ni₃Al+B Single Crystals Harden - 300%) (2)
Fe–24Al–DO₃	.84 (3)	–
Fe–24Al–B2	.76 (3)	–
Ti–25Al–10Nb–3V–1Mo	.83 (4)	Hardens (~ 30%) (5)
Ti–25Al–11Nb	.85 (4)	Hardens (~ 30%) (5)
TiAl	.80 (6) * (R = -1)	Hardens (~ 20%) (4)
SUPERALLOYS		
Waspaloy	.57 (7)	Hardening followed by softening observed in most
Udimet 700	.60 (8)	precipitation hardened alloys
Mar M 200	.55 (9)	
TITANIUM ALLOYS		
Ti–24%V	.71 (10)	Hardens (~ 60%)
Ti–32%V	.36 (10)	Softens (~ 10%)
Ti–6Al–4V (β-Annealed)	.45 (11)	Softens (~ 10%)
ALUMINUM ALLOYS		
2024–T351	.47 (12)	Hardens (~ 16% at $\Delta\varepsilon_T$ / 2 = .93%)
2124–T851	.28 (12)	Softens (~ 11% at $\Delta\varepsilon_T$ / 2 = .93%)
7075–T7651	.35 (11)	Cyclically Stable

5. Crack Initiation

Most of the recent studies of fatigue crack initiation have been
carried out on Ni₃Al, FeCo and Fe₃Al alloys. We will review the results
for Ni₃Al only, as these are the most comprehensive.

The earliest study of fatigue behavior of Ni₃Al single crystals
showed that Stage I fatigue cracks initiated at defects or inclusions, but
propagated along slip planes[32]. Stage I cracks along [111] slip planes
also were noted by Bonda et al[33] in strain controlled fatigue of Ni₃Al

crystals. A recent detailed investigation of fatigue crack initiation in
Ni₃Al+B single crystals by Hsiung and Stoloff[34,35] suggested that the
initiation of Stage I cracks is intimately connected with the
formation of persistent slip bands (PSBs) and point defects.   We will
describe here the principal findings of that research.

5.1   DEVELOPMENT OF PERSISTENT SLIP BANDS

     Single crystals with nominal chemical composition of Ni-24at%Al-
0.25%B were homogenized at 1150°C for 178 hours in vacuum. The orientation
of the stress axis in all crystals was parallel to [2 13 17].   This
orientation is favorable for the primary octahdedral slip system (111)
[101].
     Strain-controlled tension-compression cyclic deformation tests, all
beginning in tension, were performed on a closed loop machine. Several
tests were interrupted periodically to allow observation of surface slip
traces by a Noymarski-type optical microscope or by scanning electron
microscopy.   Dislocation structures were observed in a transmission
electron microscope; the weak-beam imaging method was applied to study the
dissociation of superdislocations and formation of point defects.
     Successive surface observations on a specimen cycled at a total-
strain amplitude of 0.1% were made on the same area of gage section.   At
an early stage of cycling, the observed (111) primary slip bands were
dense and somewhat uniformly distributed.   As cycling proceeded, some of
the bands become coarser in appearance, indicating that strain
localization was occurring.   Some short and fine slip lines along the
traces of secondary slip planes also were observed. The density of these
slip lines did not increase with increasing number of cycles, and their
presence was attributed to local stress concentrations.   The coarse
primary slip bands eventually developed into PSB-like bands associated
with shallow cracks, as shown in Fig. 17[35].
     Detailed observation of a sample after 1,000 cycles at a
total-strain amplitude of 0.15% showed a morphology of well-developed
extrusion/intrusion pairs, similar to those observed on copper[36,37].
Eventually, distinct stage I cracks on (111), with occasional small
segments along the (111) cross slip plane, were observed.

5.2   OBSERVATION OF MICROVOIDS

     An example of persistent slip bands accompanied by microvoids is
shown in Fig. 18[35].   Microvoids also were found on a Stage 1 fracture
surface.   The observations of microvoids on both PSBs and a fracture
surface indicated that condensation of point defects plays an important
role in fatigue crack initiation.   This conclusion was supported also by
transmission election microscopic observations of point defect clusters.

     The observations of an apparent excess of vacancies formed in Ni₃Al
crystals under cyclic deformation are in general agreement with mechanisms
proposed by others to explain fatigue crack initiation in fcc metals.
Although point defects in fatigued copper have been identified by using

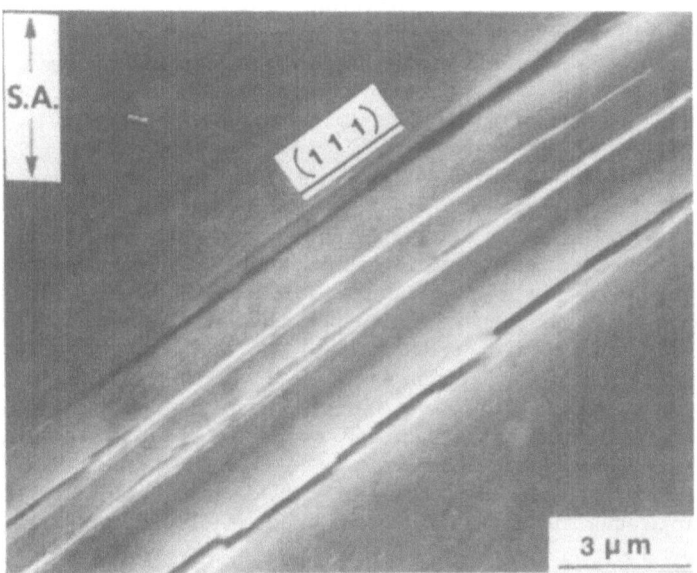

Fig. 17.   Persistent slip bands in Ni₃Al+B single crystals[35].

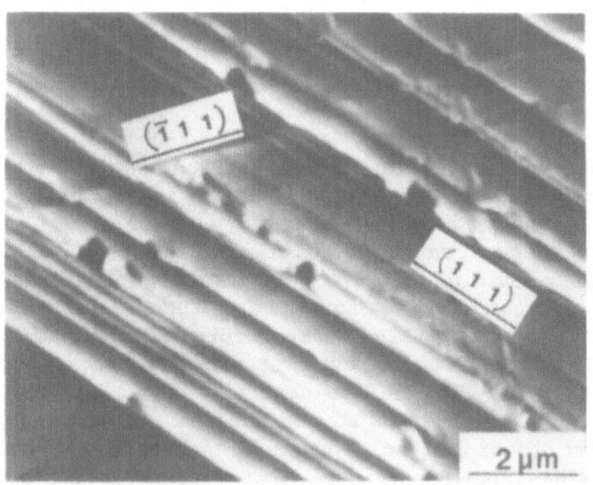

Fig. 18.   Microvoids in persistent slip bands formed in Ni₃Al+B single crystals[35].

weak-beam TEM techniques[36] and the indirect method of electrical resistivity[38] measurement, direct observations of large point defect agglomerates, e.g. voids and vacancy tetrahedra, in copper fatigued at room temperature have not been reported.

The models proposed by Antonopolous et al[36], Essmann et al[37] and Polak[38] to explain the formation of PSBs and fatigue crack initiation in copper were based on a typical dislocation structure, edge dislocation walls, formed in PSBs. Since no such structure is observed in fatigued Ni₃Al, the above models could not be applied to explain fatigue crack initiation in Ni₃Al. With this situation in mind, a new model was proposed for fatigue crack initiation in Ni₃Al single crystals[35].

The principal features of the proposed model can be summarized as follows:
1. Fatigue cracks form by a nucleation process with a critical crack size that is dependent on environment and sense of applied stress.
2. The model does not require prior surface roughening. However, the presence of surface roughening increases the local stress, resulting in the promotion of crack initiation.
3. The model implies that the morphology of extrusion/intrusion pairs on PSBs is a product of short range vacancy migration and mass transfer caused by lattice misfit strain accumulated at the PSB/matrix interfaces. The condensation of vacancies at the interfaces forms intrusions, which break down the coherency between PSB and matrix and relieve the accumulated misfit strain. The counter-flow of atoms may form extrusions at the PSBs.
4. Fatigue resistance can be improved by decreasing the misfit strain at PSB/matrix interfaces, i.e. decreasing the excess-vacancy generation rate in slip bands.
5. The critical radius of a vacancy cluster is dependent on test environment.

6. Discussion

Studies of fatigue behavior of intermetallics continue to lag behind other mechanical property evaluations. Most of the published fatigue literature of the past five years pertains to Ni₃Al and Ti₃Al, with isolated reports of fatigue data for Fe₃Al and TiAl. It is clear that fatigue behavior of aluminides is very much affected by the external environment, just as is the case for other properties. Hydrogen or moisture adversely affect the low temperature HCF and crack growth behavior of Ni₃Al, Fe₃Al and Ti₃Al alloys. At elevated temperatures oxygen is very detrimental to the fatigue resistance of Ni₃Al and Ti₃Al alloys. Chromium is highly beneficial in improving the tensile ductility of Ni₃Al in air; under cyclic loading conditions some protection against oxygen penetration is provided, but this is partly overcome by the much longer times necessary to carry out fatigue tests. Compositional and microstructural effects are particularly strong in Ti₃Al alloys.

One of the least understood aspects of cyclic behavior of intermetallic compounds is the sequence of events leading to crack initiation. Studies on polycrystals show that grain boundaries, pores, inclusions and/or slip bands can be preferred sites for initiation, depending upon the alloy system and processing technique[1]. Although not explicitly reviewed here, slip bands are preferred sites in (Fe,Ni)₃V, FeCo-V and wrought Ni₃Al+B at room temperature, while grain boundaries are the dominant sites in unnotched Fe₃Al at room and elevated temperatures[14].

Studies on Ni₃Al single crystals oriented for single slip show clearly that persistent slip bands form readily and are sites for early crack initiation[35]. Although a few other studies of fatigue in single crystals of Ni₃Al have been carried out[32,33], none have included detailed surface damage observations.

There is need for work on single crystals of other ordered alloys to establish the relations among ordering energy, crystal orientation and crack initiation sites. Also, the role of cyclic hardening (or softening) in crack initiation remains unclear. For example, while a flow stress asymmetry has been noted in single crystals of Cu₃Au[26], Ni₃Al[35], and Ni₃Ge[39], the effect is very small in Cu₃Au. For Cu₃Au and Ni₃Ge the flow stress is higher in compression than in tension, while in Ni₃Al the reverse is true. Differences in ordering energy among the various intermetallics, as well as different orientations of crystals in each investigation, could be responsible for a portion of the discrepancies, but clearly much more work is needed in this field.

7. Acknowledgements

This research was supported by the National Science Foundation under Grants No. DMR-8409593 and DMR-8911975. Discussions with Professor K. Rajan have been of great benefit. The author is grateful to Dr. C. Liaw of Pratt and Whitney Aircraft Div. of United Technologies Corporation for supplying single crystals of Ni₃Al+B, and to Dr. David Gordon of General Dynamics - Fort Worth Div. for providing unpublished data.

8. References

1. Stoloff, N.S., Fuchs, G.E., Kuruvilla, A.K. and Choe, S.J. (1987) 'Fatigue of Intermetallic Compounds' in High Temperature Ordered Intermetallic Alloys II, Mat. Res. Soc., v. 81, Pittsburgh, PA., 247-261.

2. Camus, G., Duquette, D.J. and Stoloff, N.S., (1990), J. Mater. Res., 5, 950-954.

3. Shea, M., Stoloff, N.S. and Gordon, D.E. (1991) Rensselaer Polytechnic Institute, Troy, NY; unpublished.

4. Liu, C.T. and Sikka, V.K., (1986) J. Met., 38, (5), 19-21.

5. Oak Ridge National Laboratory (1991), unpublished.

6. Kuruvilla, A.K. and Stoloff, N.S. (1987), Scripta Met., 21, 873-877.

7. Matuszyk, W., Camus, G., Duquette, D.J. and Stoloff, N.S. (1990) Met. Trans. A, 21A, 2967-2976.

8. Bellows, R.S., Schwarzkopf, E.A. and Tien, J.K. (1988) Met. Trans. A, 19A, 479-486.

276

9. Camus, G., Duquette, D.J. and Stoloff, N.S. (1991), submitted to J. Mater. Res.

10. Matuszyk, W. (1988) M.S. Thesis, Rensselaer Polytechnic Institute, Troy, NY.

11. Larsen, J.M., Williams, K.A., Balsone, S.J. and Stucke, M.A. (1990) 'Titanium Aluminides for Aerospace Applications' in High Temperature Aluminides and Intermetallics, S.H. Whang et al (eds), TMS-AIME, Warrendale, PA, pp. 521-556.

12. Paris, P. and Erdogan, F. (1963), J. Basic Eng., Trans ASME Series D, 85, 528.

13. Chang, K.M., Huang, S.C. and Taub, A.I. (1986) General Electric Company, Report 86CRD202.

14. Fuchs, G.E. and Stoloff, N.S. (1988) Acta Met., 36, 1381-1388.

15. Liu, C.T., McKamey, C.G. and Lee, E.H. (1990) Scripta Met., 24, 385-390.

16. Castagna, A., Shea, M. and Stoloff, N.S. (1991) 'Hydrogen Embrittlement of FeAl and Fe$_3$Al' in High Temperature Ordered Intermetallic Alloys IV, Mat. Res. Soc., v. 213, Pittsburgh, PA, pp. 609-616.

17. Castagna, A. and Stoloff, N.S. (1991), Rensselaer Polytechnic Institute, unpublished.

18. Williams, K.A., Balsone, S.J., Stucke, M.A. and Larson, J.M. (1990) 'An Assessment of the Fatigue and Fracture Capability of Titanium Aluminides for Aerospace Applications', presented at Aeromat 90, Long Beach, CA, 21-24 May, 1990.

19. Davidson, D.L. Campbell, J.B. and Page, R.A. (1991) Met Trans. A, 22A, 377-391.

20. Aswath, P.B. and Suresh, S. (1989) Mat. Sci. and Eng., A114, L5-L10.

21. Balsone, S.J., Maxwell, D.C., Khobaib, M. and Nicholas, T. in Fatigue 90, (1990) (H. Kitagawa and T. Tanaka, eds.), Materials and Components Eng. Publ. Ltd., Warley, UK, v. III, pp. 1905-1910.

22. Wessels, J.F., Marquardt, B.J. and Krueger, D.D. (1989) 'The Influence of Microstructure and Temperature on the HCF Behavior of Alpha-2 Titanium Aluminides', presented at TMS-AIME Symposium on Creep and Fracture of Titanium Aluminides, October, Indianapolis, IN.

23. Vekaturaman, S. (1987) 'Elevated Temperature Crack Growth Studies of Advanced Titanium Aluminides', AFWAL-TR-87-4103, Air Force Materials Lab, Wright Patterson Air Force Base, Ohio.

24. Blackburn, M.J., Hill, J.T. and Smith, M.P. (1984) 'R & D on Composition and Processing of Titanium Aluminide Alloys for Turbine Engines, AFWAL-TR-84-4078, Wright Patterson Air Force Base, Ohio.

25. Soboyejo, W.O., Aswath, P.B. and Deffeyes, J.E. (1991), submitted to Mat. Sci. and Eng. A.

26. Chien, K.H. and Starke, E.A. Jr. (1975) Acta Met., 23, 1173-1180.

27. Sastry, S.M.L. and Lipsitt, H.A. (1977) Met Trans. A, 8A, 299-308.

28. Gordon, D.E. and Unni, C.K. (1991) J. Mater. Soc., (1991) in press.

29. Choe, S.J. and Stoloff, N.S. (1987) Rensselaer Polytechnic Institute, Troy, NY, unpublished.

30. Gordon, D.E. and Unni, C.K. (1989) presented at TMS-AIME Fall Meeting, October 2-5, 1989, Indianapolis, IN.

31. Hartfield-Wunsch, S.E. and Gibala, R. (1991) 'Cyclic Deformation of B2 Aluminides' in Proc. High Temperature Ordered Intermetallic Alloys IV, Mat. Res. Soc., v. 213, Pittsburgh, PA, pp. 575-580.

32. Doherty, J.E., Giamei, A.F. and Kear, B.H. (1975) Met. Trans. A, 6A, pp. 2195-2199.

33. Bonda, N.R., Pope, D.P. and Laird, C. (1987) Acta Met., 35, 2371-2384.

34. Hsiung, L.M. and Stoloff, N.S. (1989), 'Electron Microscopy Study of Fatigue Crack Initiation in Ni_3Al Single Crystals', in High Temperature Ordered Intermetallic Alloys III, Mat. Res. Soc., v. 133, Pittsburgh, PA, pp. 261-267.

35. Hsiung, L.M. and Stoloff, N.S. (1990), Acta Met., 38, 1191-1200.

36. Antonopolous, J.G., Brown, L.M. and Winter, A.T. (1976) Phil. Mag., 34, 549.

37. Essmann, U. Gosele, U. and Mughrubi, H. (1981) Phil. Mag. 44, 401.

38. Polak, J. (1986) Mat. Sci. and Eng., 92, 71.

39. Hsiung, L. (1986) M.S. Thesis, New Mexico Institute of Technology, Socorro, New Mexico.

TEMPERATURE AND COMPOSITION DEPENDENT DEFORMATION IN γ-TITANIUM ALUMINIDES

S. H. WHANG
Polytechnic University
Six Metrotech Center
Brooklyn, N.Y., NY 11201, USA

ABSTRACT.

γ-titanium aluminides exhibit a deformation behavior like normal fcc metals from 4K to RT, but an anomalous strain hardening from RT up to 1073 K similar to L1$_2$ Ni$_3$Al. Unlike Ni$_3$Al, γ-titanium aluminides have two different slip systems: 1/2⟨110] {111} and ⟨101] {111} slip, both of which are responsible for a strong hardening behavior at high temperatures. Although little has been known about the mechanisms causing the hardening behavior, preliminary results indicate that the non-planar structures of the dislocations in the TiAl appear to be a key to the hardening at the low temperature regime (lower than 800 K) while cross-slip of the APBs of the superdislocations onto cube planes and the cross slip or climb of the ordinary dislocations onto non-glide planes are responsible for the hardening above 800K. Other potential mechanisms for the hardening are discussed. The composition dependence of TiAl on mechanical properties at RT in various ternary γ-titanium aluminides is reviewed.

1. INTRODUCTION

Detailed experiments on deformation behavior of γ-titanium aluminides as a function of temperature in a wide temperature range was reported by Kawabata (1985 and 1990). The yield stress curves over two different temperature ranges: 1) 4K to RT, and 2) RT to 1273K shows a striking contrast in which the curve from 4K to RT may be characterized as a fcc-like deformation while the curve from RT to 1273K is abnormal, which means that the yield stress increases with increasing temperature, showing a characteristic positive temperature dependence. This positive temperature dependence is very similar to those observed in some L1$_2$ compounds such as Ni$_3$Al, Ni$_3$Ga, etc.

Such positive temperature dependence of yield stress was also reported in polycrystalline γ-titanium aluminides (Hahn and Whang 1990; Whang and Hahn 1990). Although the yield stress curves have a similarity with those of Ni$_3$Al, the slip systems and dislocation structures responsible for the anomaly are significantly different between the two compounds as shown in Table 1. The two slip systems of TiAl exhibit anomalous hardening independently. One of the systems, 1/2⟨110] {111}, is called "ordinary dislocations" slip and the other, ⟨101]{111}, called "Superdislocation" slip. Each ordinary dislocation may dissociate into two partial dislocations with a complex stacking fault which, however, has not been observed. The superdislocation ⟨101] has two superpartials with an APB ribbon between them. Each superpartial may dissociate into

C. T. Liu et al. (eds.), Ordered Intermetallics – Physical Metallurgy and Mechanical Behaviour, 279–298.
© 1992 *Kluwer Academic Publishers.*

partial dislocations, but they are not identical in that the leading superpartial contain a SF and the trailing superpartial, a CSF. On the other hand, the superpartials in Ni$_3$Al are symmetrical in their structure. Although the yield stress of single crystal TiAl has been determined over a wide temperature range; 4 K to 1173 K, neither the orientation dependence nor the compression-tension asymmetry has not yet been studied.

Composition dependent deformation of single phase γ titanium aluminides on mechanical properties has been found in the ternary systems. The study has been mainly focused on alloying effect on ductility and fracture toughness. These aspects will be reviewed.

TABLE 1. Comparison of Properties between γ-TiAl and L1$_2$ Ni$_3$Al

	L1$_0$ TiAl	L1$_2$ Ni$_3$Al
Crystal Structure	FCT Layered Structure P4/mmm	Ordered FCC Pm3m
Elastic Anisotropy (Fu & Yoo 1990)	$A_1 = 1.18$ $A_2 = 2.46$	$A_1 = 3.3$
Slip System	1. 1/2⟨110] {111} 2. ⟨101] {111}	L. T. ⟨101] {111} H. T. ⟨101] {100}
Yield Stress Peak	873 K - 1073 K	674 K - 873 K
Deformation Anomaly	Orientation Dependence	Orientation Dependence Tensile-Compression Asymmetry
Fracture Mode	Cleavage	Transgranular

where $A_1 = 2C_{44} / (C_{11} - C_{12})$; $A_2 = 4C_{44} / (C_{11} + C_{33} - 2C_{13})$.

2. TEMPERATURE DEPENDENCE

2.1 DEFORMATION

Mechanisms for the strain hardening at high temperatures in Ni$_3$Al(L1$_2$) have been extensively studied based on the cross-slip models in the past (Takeuchi & Kuramoto 1973; Paidar, Pope & Vitek 1984; Yoo 1986; Hirsch in these proceedings). Nevertheless, these models may not be directly applicable to the tetragonal structure TiAl. Primary difficulties in employing these models for the TiAl arise from different characteristics of dislocation structures and deformation behavior as described as follows:

1.   In TiAl, the ordinary dislocations (1/2⟨110]) do not have a dissociation form containing APB. Instead, each of them may dissociate into two partials with a complex stacking

fault ribbon. Neither the identity itself in a {111} plane nor the cross-slip of the CSF has been yet observed by electron microscopy.

2. The plot of ln CRSS vs. 1/T (dτ/dT) does not yield a single straight line, indicating that two or even three different activation energies may be obtained from the curve between room temperature (RT) and 1073K as shown in Fig. 1.

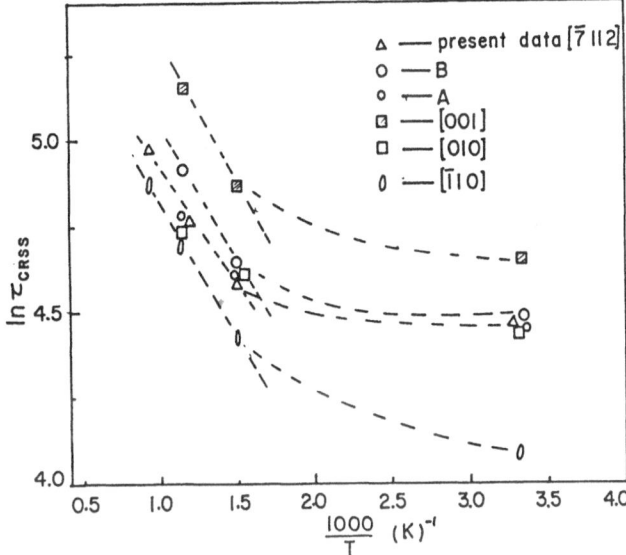

Fig. 1 Plot of ln CRSS vs. 1/T in orientations for single crystals TiAl based on data reported by Kawabata et al 1985.

3. The strain hardening rate (dτ/dT) increases with temperature in the $1/2\langle110\rangle$ {111} slip whereas in the $\langle101\rangle$ {111} slip, the rate decreases with increasing temperature. Consequently, the hardening mechanism by the superdislocation slip in TiAl must be somewhat different from that in Ni_3Al.

4. The orientation dependence and tensile-compression asymmetry which can be explained by the cross-slip of APBs have not been established, though the cross-slip of APBs in the $\langle101\rangle${111} slip was observed in the alloy deformed at 873 K and 1073 K, respectively. There was no correlation established between the orientation dependence and the Schmid factor ratio in TiAl (Kawabata et al 1985; and Whang et al JIM-6 1991) as was demonstrated in Ni_3Al (Liang et al 1977; and Lall 1979). An absence of tension-compression asymmetry in yield stress was reported in polysynthetically processed TiAl(Fujiwara, T, et al 1990). The TiAl variants which are deformed by twinning in compression can be deformed by slip in tension with the same magnitude of Schmid factor for the twinning, i.e., the tension-compression deformation is symmetry. Therefore, this is not a tension-compression asymmetry in the conventional sense, where the twinning is not involved.

5. The existence of the different activation energies at two different temperature ranges in TiAl adds a more complexity in the interpretation of the anomalous hardening.

If the CRSS of TiAl consists of thermal and athermal terms, the thermal term may be expressed using an Arrhenius equation. The activation energy determined from this slope between 673 K and 1073K (see Fig. 1) has a range of 0.05-0.057 eV which is close to the upper end value

282

of the activation energy 0.05 eV reported in Ni₃Al. However, below 673K, the curve has a slow slope compared with the one at the high temperature range. Practically, two linear lines can reasonably be drawn to represent the curve. Thus, two different activation energies representing the low and high temperature kinetics, respectively must describe the deformation in TiAl in contrast with a single activation energy for the deformation in Ni₃Al. Furthermore, the TiAl also exhibits a very low strain rate sensitivity near 700 K (Kawabata 1985, Hahn 1990) similar to 770 K in polycrystalline Ni₃Al (Thornton, Davies, & Johnston 1970).

A) Deformation by $\langle 101]$ {111} Slip

In single crystal TiAl, slip in the $\langle 110]$ and $\langle 101]$ directions introduce ordinary dislocations and superdislocations, respectively. Consequently, the deformation orientations for single slip $\langle 101]${111} in single crystal TiAl should be chosen inside the triangle I and II (see Fig. 2) based on Schmid factor consideration alone.

Single slip $\langle 101]${111} was obtained in single crystal TiAl at 4K, 77K, 196K, 293K, 673K, 873K, 1073K and 1273K (Kawabata et al 1985 & 1990). The yield stress decreased slowly from 4K to 293K as do many fcc metals. Nevertheless, it increases slowly and later rapidly from RT up to 873K. The peak value of the yield stress occurs at 873K. Also, the strain rate sensitivity parameter $(S=1/2(\partial\log\sigma/\partial\log\dot{\varepsilon})_T)$ approaches zero with its minimum near 700K. This behavior is very similar to that of L1₂ Ni₃Al.

On the contrary, the work hardening rate defined as the difference in the CRSS for given strains, $\Delta\tau = \tau_{1.5\%} - \tau_{0.2\%}$, decreases with increasing temperature from RT up to 873K, which is one of the major discrepancies between the deformation in TiAl and that in Ni₃Al. In other words, the dislocation structure for the 0.2 % deformation is responsible for the high temperature hardening while the structure for the 1.5% deformation, for high temperature softening. Hence, the structural characteristics of the TiAl should be different from that of the Ni₃Al for 1.5% strain, which may be an important subject for the future investigation.

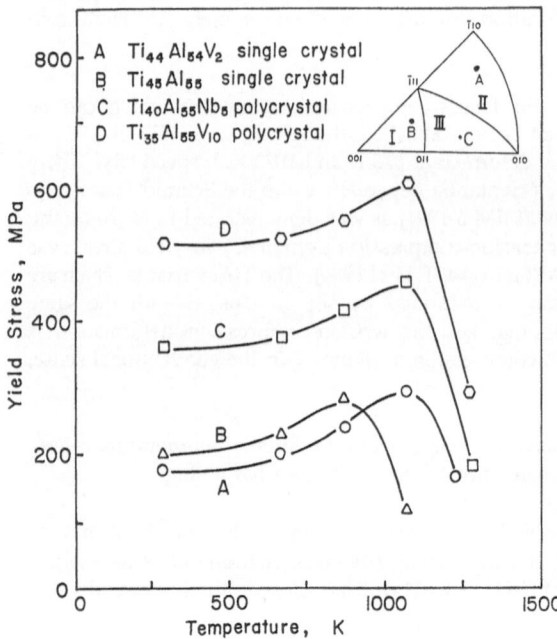

Fig. 2 Plot of yield stress vs. temperature in single and polycrystalline γ-titanium aluminides.

In a single crystal $Ti_{44}Al_{54}V_2$ alloy, deformation orientation "B" in the stereographic projection (Fig. 2) introduced single slip $\langle101]\{111\}$ with a yield stress maximum at 1073K(Whang, Hahn, Li & Li 1991) instead of 873K. This orientation dependence in TiAl can't be explained by the Schmid factor ratio (in $\{111\}$ plane vs. in $\{100\}$ plane) of superpartials, which was successfully applied to Ni_3Al (Copley and Kear 1967; Lall, Chin and Pope 1979). The understanding of the orientation dependence of the yield stress in TiAl requires a further systematic study on deformation as well as deformation microstructure.

B) Deformation By $1/2\langle110]\{111\}$ Slip

The probability of having $1/2\langle110]\{111\}$ slip vs. $\langle101]\{111\}$ slip is $1/3$ in the TiAl compounds. Nevertheless, the mobility of these dislocations is considered to be higher than that of the superdislocations. For this reason, these dislocations are important in the deformation of TiAl. Single slip $1/2\langle110]\{111\}$ in single crystal TiAl was obtained by Kawabata et al (1985) from 4K to 1273K using the orientations in triangle III.   Conspicuously, in the crystal deformed at 196K, the ordinary dislocations were absent. Instead, $1/2\langle112]$ dislocations were present in replacing the ordinary dislocations. The reasons for such absence are not known. Hence, the deformation below 196K in this orientation can't be considered as $1/2\langle110]\{111\}$ slip.

On the other hand, the deformation in an orientation within the triangle III between RT and 1273K results in $1/2\langle110]\{111\}$ slip. The Kawabata's results on single crystal TiAl in this temperature range show that the maximum yield stress occurs at 873K. In this alloy, the value of the strain rate sensitivity parameter approaches zero between 673K and 873K. Also, the deformation orientation C resulted in a positive dependence of yield stress and a positive work hardening rate, accompanied by a maximum yield stress at 873K in contrast with the negative rate for single slip superdislocations. So far, the deformation behavior of $1/2\langle110]\{111\}$ slip in TiAl is much similar to that of superdislocation slip in Ni_3Al.

C) Deformation in Polycrystalline TiAl Compound Alloys

The deformation in polycrystalline TiAl alloys may be characterized as a multiple slip which will result in a high frequency of dislocation interactions. In the earlier work, the deformation of polycrystalline γ-titanium aluminides didn't exhibit anomalous hardening for some reasons (Lipsitt et al 1975), though the positive temperature dependence of yield stress has later been identified in Ti-Al-Nb alloy (Hahn and Whang 1989) and also in Ti-Al-V alloy (Whang and Hahn 1990). Hahn has shown that the yield stress peaks in these alloys are centered near 1273 K instead of 1073K. Further, the strain rate sensitivity parameter approaches zero near 750K(Hahn 1991), similar to that of single crystal TiAl. It is anticipated that the deformation behavior of polycrystalline TiAl or the ternary alloys largely depends on dislocation interactions. In this regard, the hardening mechanism in the polycrystalline alloys would be complex and the analysis of the deformation would be very difficult.

Recently, Huang(1991) has shown that the anomalous hardening becomes diminishing as the composition changes from the Al rich composition to the stoichiometric TiAl. Furthermore, the two phase alloy ($\gamma + \alpha_2$) did not show the anomalous hardening at all.

2.2 DISLOCATION STRUCTURE

 I. General Morphology

A) Superdislocations

The dislocation structures in deformed, single phase γ-titanium aluminides have been widely studied by numerous authors. However, the investigations on the dislocation structures were limited to the room temperature deformation structure, hence the results are not sufficient to explain the hardening behavior at high temperatures. The information available up to now is reviewed in this paper. In polycrystalline TiAl deformed at RT, the superdislocations, b=⟨101], were observed, but neither their dissociation nor cross-slip of their APBs was observed (Shechtman, Blackburn and Lipsitt 1974). Later, in $Ti_{46}Al_{54}$ alloy deformed at RT, faulted dipoles with b=1/6⟨112] lying in (111) plane and with the dislocation lines in [01T] direction were observed (Hug, Loiseau and Lasalmonie 1986; Hall & Huang 1989). The faulted dipoles appear to form when the superdislocations pinned by jog split into 1/6⟨112] sessile dislocations which in turn trail the glissile part of the superdislocations (Hug et al 1980). No such faulted dipoles were reported in the single crystal material.

At high temperatures, popular configurations of the superdislocations includes dipoles and cross-slip. The superdislocation dipoles (b=⟨101]) were observed in single crystal $Ti_{44}Al_{54}V_2$ deformed to 1 % at 873 K and 1073 K (Hahn, Li, Whang and Kawabata 1990; Whang, Hahn, Li and Li 1991). The density of the dipoles is about 30 % of total superdislocations in the 873 K alloy (Fig. 3a), and remains constant in the 1073 K alloy (Fig. 3a). The distance between the dipole dislocations ranges from several tens of nm to few hundreds nm. The elastic interaction of these dipoles will be weakening at high temperatures. Consequently, the contribution of the dipoles to the yield stress from 873 K to 1073 K in the single crystal Ti-Al-V alloy appears constant.

At 873 K, another prevailing configuration of superdislocations (screw) is cross-slip of the APBs and the superpartials onto a {010} plane. The cross-slip of the APBs was reported by Hug, Loiseau & Veyssiere (1988) in single crystal $Ti_{46}Al_{54}$ deformed at 873 K. The cross-slip of the APBs and superpartials takes place from octahedral planes onto cubic planes. Furthermore, in some cases, the superdislocations themselves can cross-slip onto a {101} plane (Hahn and Whang 1990). The planar fault energies determined from the measurements of the planar defect width by TEM were rather small, specially, in the case of APB energy, γ(APB) in a {111} plane is 120 mJm^{-2}, and γ(APB) in (010), 100 mJm^{-2}. Hence, the energy reduction due to the cross-slip is rather small in this case.

Nevertheless, the cross-slip mechanism becomes more favorable when the torque that arises from the elastic anisotropy in Ni_3Al or TiAl is considered (Yoo 1986; Fu and Yoo 1990). The alloy deformed at 1073K shows predominant cross-slip of the APBs from a {111} plane onto a cube plane (Li and Whang 1991). The superdislocations (b=[101]) still remained straight in shape in $Ti_{35}Al_{55}V_{10}$ alloy(L1$_0$) deformed at 1073K (Hahn 1991) where the alloy became very soft. No kinks or jogs were observed in the alloy deformed at high temperatures.

Other superdislocations with b=1/2⟨112], in which the Burgers vector connects the third nearest neighbors, do not play a significant role in the deformation of this material. These dislocations were observed sporadically in the matrix regardless of the deformation temperatures. Also, It was not successful to initiate single slip of these superdislocations in single crystal TiAl (Kawabata et al 1985) or TiAl-V alloys (Hahn et al 1991) from RT to high temperatures from various deformation orientations. However, strange enough the 1/2⟨112] dislocations were observed in single crystal TiAl deformed below 196K in [010] orientation (Kawabata et al 1990). In fact, in this orientation, the slip of the ordinary dislocations was anticipated instead of the 1/2⟨112] dislocations. This means that 1/2⟨112] dislocations can operate when the ordinary dislocations become sessile. The sessile nature of these dislocations may be an important subject of future investigation.

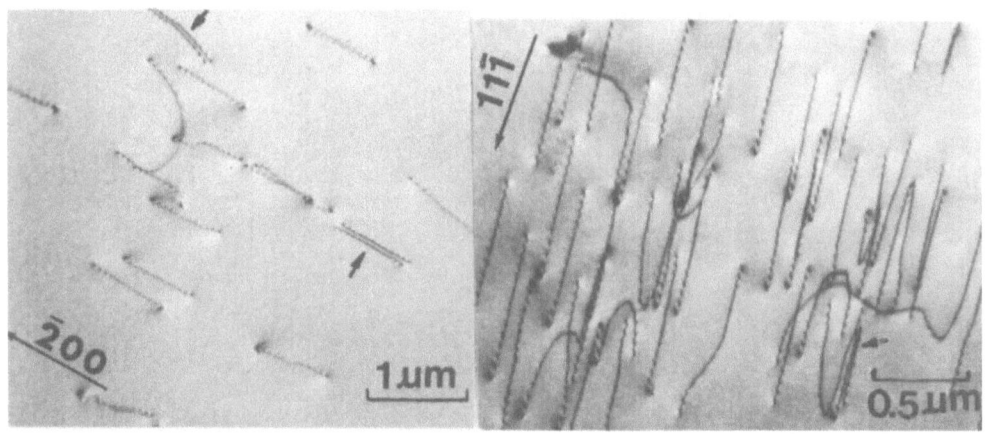

Fig. 3a Superdislocations (b=[10T]) in single crystal Ti$_{44}$Al$_{54}$V$_2$ after 1 % plastic deform-ation under compression at 873 K.

Fig. 3b Superdislocations (b=[10T) in the same alloy after 1 % deformation at 1073 K.

B) Ordinary Dislocations

The term "Ordinary Dislocations" mean the dislocations with b=1/2⟨110⟩, which are typical dislocations in FCC metals. However, an important distinction exists between those ordinary dislocations in FCC metals and those in L1$_0$ TiAl compounds. The dislocations in FCC metals can easily decompose into partial dislocations and stacking faults whereas the dislocations in TiAl may decompose into partial dislocations with complex stacking faults (CSF), though no observation of such decomposition has been made so far (Shechtman 1974; Hahn 1991). The CSF contains APB character and therefore, is expected to have a higher defect energy than that of SISF. The energy of the CSF predicted by the theory is lower than the APB in a {111} plane(Table 2).

Contrary to the prediction, the APB has been observed in TiAl deformed at 873 K(Hug et al 1986) and in Ti$_{44}$Al$_{54}$V$_2$ at 1073 K (Li et al 1991) whereas the CSF has not been observed at these temperatures. Another difference in deformation behavior between the two groups of ordinary dislocations is that in TiAl the ordinary dislocations are absent in the single crystal TiAl deformed below 196 K whereas in FCC metals this is not the case. The ordinary dislocations in the TiAl appear approximately straight in shape, but consist of curved segments as shown in Fig. 4. In addition, in the TiAl-V alloy deformed at RT, the ordinary dislocations contain superjogs along the dislocation lines (Hahn 1991). In the middle group of the dislocations in Fig. 4a, the dislocations are mingled, indicating possible cross-slip or dipoles formed at this temperature.

In the same alloy deformed at 1073K, the dislocation morphology is very different from that of the alloy deformed at 873K. The dislocations no longer appear straight, but have pinning points (indicated by arrow 1 & 2). When the edge segment of the ordinary dislocation climbs onto a non-slip plane such as {113} plane by thermally activated process, it may become a pinning point. The dislocation loops shown in Fig. 4b (arrow 3) lie in a {113} plane (Whang, Hahn, Li & Li 1991). They are not prismatic in nature. These pinnings and the loops may contribute to the hardening at this temperature.

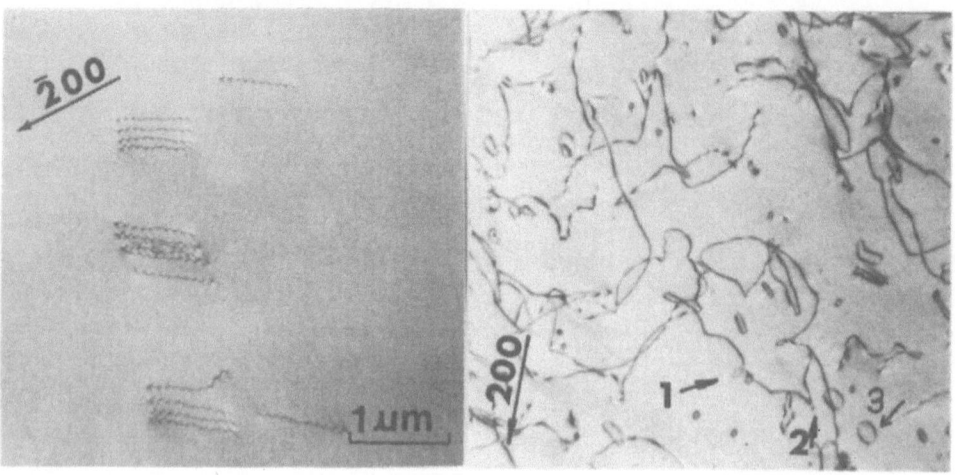

Fig.4a Ordinary dislocations(b=1/2[1T0]) Fig. 4b Ordinary dislocations in the same alloy
of screw character in Ti$_{44}$Al$_{54}$V$_2$ crystal   deformed to 1 % at 1073K in orientation [7,11,15]
deformed to 1 % at 873K.

II. Other Fine Structures

A) Dislocation Decomposition

Superdislocations in TiAl are not stable under applied stress, thus they decompose into perfect dislocations during deformation. We must understand the decomposition behavior of superdislocations in this material since the mobility of the decomposed dislocations with smaller Burgers vectors may be greater than that of the superdislocations. This information will promote to a better understanding on the deformation of these compounds. Such decomposition of superdislocations (b=⟨101]) into two different perfect dislocations was predicted by Greenberg[4] and expressed as

$$⟨101] = 1/2 ⟨110] + 1/2 ⟨1\bar{1}2]$$ (1

The Burgers vectors of these two decomposed dislocations are orthogonal, but they are non-orthogonal with the associated superdislocation. Consequently, the decomposed dislocations from the screw superdislocation are not screw, but are mixed. From this logic, it is likely that superdislocations of screw character can decompose itself into the perfect dislocations of mixed character as suggested by eq. 1 when the interaction energy between the cores and the surrounding dislocations is favorable for the decomposition.   To date, three different decomposition modes were observed. These three modes include 1) self-decomposition, 2) super-ordinary dislocation intersection and 3) super-super dislocation intersection.

First mode: Hug et al(1988) observed that a segment of the superdislocation with mixed character in single crystal Ti$_{46}$Al$_{54}$ deformed at 873 K decomposed into two perfect dislocations with b=1/2⟨110] and b=1/2⟨112]. Also, the self-decomposition was observed in γ-phase of the lamellar structure in Ti$_{55}$Al$_{45}$ alloy deformed at RT (Kim, Hahn and Whang 1991) . Fig. 5a and b show the decomposition of superdislocations [10$\bar{1}$] in the γ-phase of Ti$_{55}$Al$_{45}$ alloy where the

"A" dislocation(b=[10$\overline{1}$]) is invisible due to the g•b=0 condition. It is interesting to note that the same reaction can occur only at high temperatures in Al rich γ-TiAl. The decomposition reaction(eq. 1) was also found in the γ-phase grains containing Al less than 50 at % in TiAl-2 wt% Mn(Hanamura et al 1989) and in Ti$_{55}$Al$_{45}$ alloy (Kim et al 1991). Both of the alloys were deformed at RT instead of high temperatures. This is another example of self-decomposition of the superdislocations.

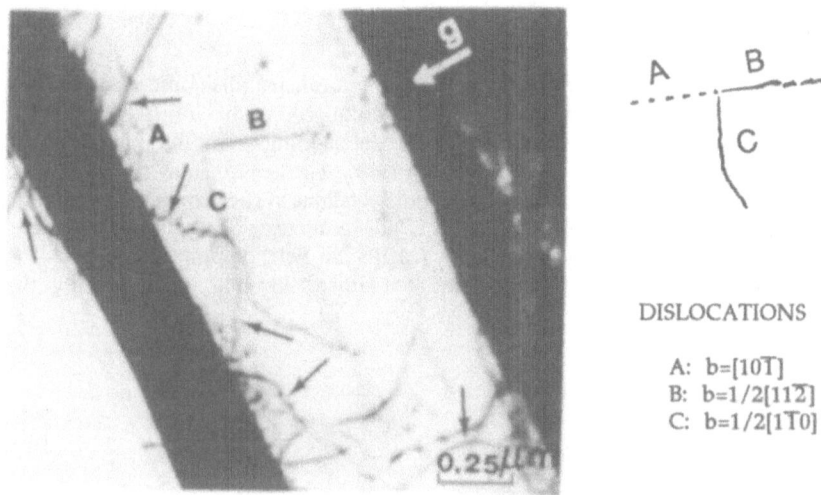

DISLOCATIONS

A: b=[10$\overline{1}$]
B: b=1/2[11$\overline{2}$]
C: b=1/2[1$\overline{1}$0]

Fig. 5 Decomposition of Superdislocations (b=[10$\overline{1}$]) into perfect dislocations ("B" & "C") in γ-phase of the lamellar structure in Ti$_{55}$Al$_{45}$ alloy.

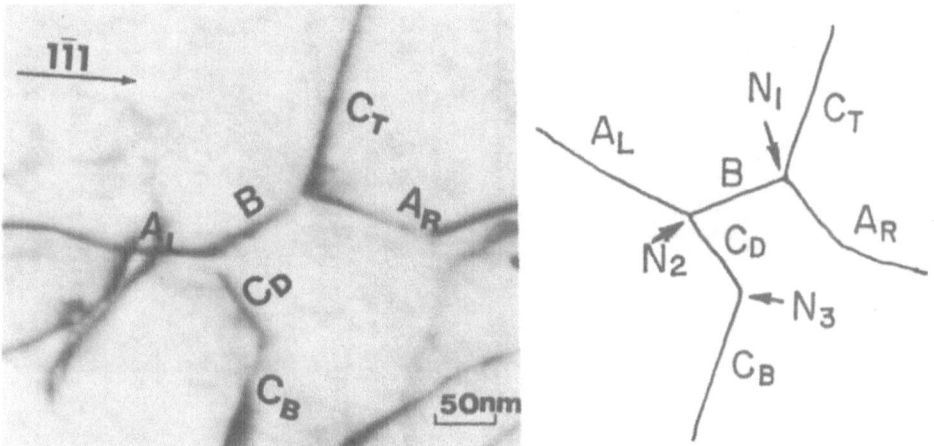

Fig. 6 Decomposition of superdislocations (b=[101]) in γ-grains of Ti$_{55}$Al$_{45}$ deformed to 2% at RT, g=(1$\overline{1}$1), B=[110]; a) bright field b) schematic drawing of the decomposition.

Second mode: When the superdislocation intersects an ordinary dislocations in TiAl, the decomposition occurs at the intersecting point. Otherwise, the intersection makes a strong pinning, which will be discussed in the next section. The morphology of the decomposed dislocations and associated dislocations were observed by Li et al (1991) shown in Fig. 6a & b. Dislocation A_L & A_R with b=[101] were the same dislocations before separation and as were dislocations C_T & C_B with b=1/2[112]. Dislocation "B"(b=1/2[112]) and "C_D"(b=1/2[1$\bar{1}$0]) are the decomposed dislocations. Hence, node N_1 & N_3 were the identical points on the dislocation C before separation. It appears that the interaction energy may contribute to the decomposition and rearrangement of the dislocation cores.

Third mode: The superdislocations aided by the ordinary dislocations can easily decompose into perfect dislocations. When the superdislocation and an ordinary dislocation intersect each other, the decomposition occurs at the intersecting point in polycrystalline $Ti_{40}Al_{55}Nb_5$ alloy deformed at 1073K (Hahn and Whang 1990). Fig. 7a & b show square shaped dislocations which were connected to the original superdislocations at each corner of the square. The square consists of two sets of 1/2[$\bar{1}$10] and 1/2[11$\bar{2}$] dislocations . The square dislocations were also observed in polycrystalline $Ti_{35}Al_{55}V_{10}$ alloy (L1$_o$)(Hahn 1991) deformed at 1073 K. The square dislocations are not co-planar with the associated superdislocation. Consequently, the square dislocations are sessile.

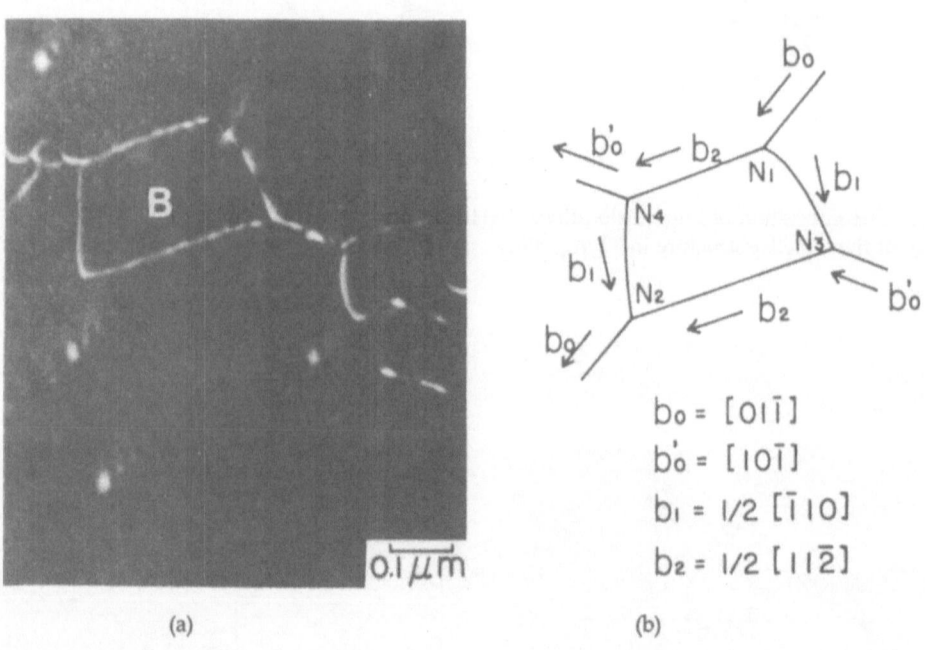

$b_o = [01\bar{1}]$

$b'_o = [10\bar{1}]$

$b_1 = 1/2 [\bar{1}10]$

$b_2 = 1/2 [11\bar{2}]$

(a)   (b)

Fig. 7 Decomposition of two superdislocations by forming two sets of perfect dislocations of "square shape" in γ-phase $Ti_{40}Al_{55}Nb_5$: a) dark field, b) schematic drawing.

B) Dislocation Interaction

Without decomposition, the intersecting dislocations form strong pinning points.

Consequently, the work hardening due to such pinnings likely occurs in a multiple slip system in the single crystals or the polycrystalline alloys in this material. To date, three different types of interactions were observed. Each case will be discussed in the following.

i) Super-superdislocation Pinning

Two superdislocations on two conjugate planes form a pinning point upon intersecting. A superdislocation (b=[10$\bar{1}$]) on (1$\bar{1}$1) plane was pinned by another superdislocation (b=[$\bar{1}$01]) on the (111) conjugate plane in single crystal TiAl deformed at 873K (Hug et al 1988). However, it is difficult to predict when the superdislocations are pinned instead of being decomposed into perfect dislocations. The calculations of the core energy to predict the outcome of the interaction must be complex.

ii) Ordinary-Superdislocation Pinning

The ordinary dislocations in $Ti_{35}Al_{55}V_{10}$ deformed at 1073K can be pinned by an array of superdislocations. Fig. 8 shows an ordinary dislocations ("A") with b=1/2[110] pinned by an array of superdislocations, labelled "B" with b=[10$\bar{1}$], in polycrystalline $Ti_{35}Al_{55}V_{10}$ deformed to 2 % at 1073K.

iii) Ordinary-Ordinary Dislocation Pinning

Unlike the superdislocations, the ordinary dislocations show neither decomposition or dissociation, nor mutual pinning at RT. The anomalous hardening by 1/2⟨110] dislocations may be attributed to the dislocation core structure at this temperature.

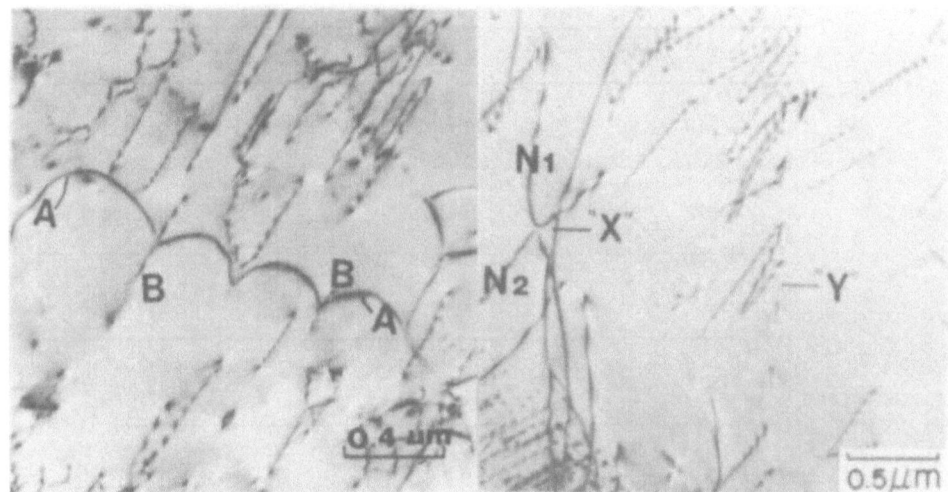

Fig. 8   1/2[110] dislocations are pinned   Fig. 9 Ordinary dislocations N_1 & N_2 are intersecting
by superdislocations [10$\bar{1}$]; B=[$\bar{1}$10],   at "X" and form a stair rod dislocation(b= [010]) at X.

However, at very high temperatures (i.e, 1273K), two ordinary dislocations can form stair rod dislocations through reaction (Whang and Hahn 1990). The two intersecting ordinary dislocations in polycrystalline $Ti_{40}Al_{55}Nb_5$ alloy($L1_o$) deformed at 1273K (Fig. 9) formed a stair rod dislocation [010] which is not glissile on a {111} plane, i.e.,

$$1/2[\bar{1}10] + 1/2[110] = [010]$$

Conversely, the cube dislocation (b=⟨100⟩) of edge character can dissociate into two ordinary dislocations. At this temperature, formation of the cube dislocations by deformation becomes possible. No stair rod dislocations of this kind was observed at the low temperature deformation.

2. 3. PLANAR DEFECT ENERGY

The cross-slip behavior of the superdislocations in TiAl undoubtedly raises questions as to possible existence of lower planar defect energy in other planes other than the {111} slip planes. The larger the energy difference is, the stronger the driving force for the cross-slip can be or the stronger the pinning force will be.

TABLE 2. Shear Planar Defect Energies (mJm^{-2})

	APB energy (100)	APB energy (111)	SISF	CSF	SESF	Twins	References
		470					Shechtman et al 1974
	520	435	60	485			Yamaguchi et al 1985
	430	510	90		80	60	Fu et al 1990
Theory	347	670	174	337	165		Woodward et al 1991
	66	306	71	119			Rao et al 1991
	291	522	158	212	219		Whang et al 1991
	131	322	67	153	76		Li, Z.C et al 1991
Experi-ments RT		130-160	70-80				
873 K	100	120	60				Hug et al 1988

Therefore, the determination of the APB energy in potential cross-slip planes has been the main interest among the researchers. Some experimental measurements are given in Table 2. However, the majority of data on the defect energies tabulated in Table 2 were generated by the theoretical calculations. The data shown in Table 2 may be summarized as follows.

1. The APB energy is smaller in {100} type planes than in {111} type planes. Both experiments and theory agree on this.

2. The ISF energy is the lowest among planar defect energies in {111} planes.

3. The temperature dependence of the APB width is small, evidenced by experimental measurements at RT and 873 K. This means the temperature dependence of the APB energy is also small Hug et al 1988).

The major discrepancy between the experiments and theoretical calculations is in that the theory predicts a large energy reduction in the APB energy by cross-slip whereas the experiments show a small energy reduction in the TiAl.

2.4 ANOMALOUS HARDENING MECHANISM

The hardening mechanism in TiAl has been discussed in recent years. Several mechanisms for the hardening were suggested. Nonetheless, no consensus on the mechanisms has been reached due to partly contradicting evidence presented by experiments and partly insufficient experimental evidence to support the models. Thus, in this paper, we introduce all these models and discuss them in a proper context. Most experiments quoted here have been derived from the single crystal work since each type of the slip in TiAl can be isolated in single crystal deformation. The models dealt with in this section include 1) cross-slip of APB(or K-W model), 2) sessile structure of dislocation cores, 3) superdislocation dipoles, 4) Lomer-Cottrell-Hirth barrier etc.

A) Cross-Slip of APB (K-W Model)

The concept of the mechanism was borrowed from those developed for L1$_2$ Ni$_3$Al. Initial discovery of the cross-slip of APBs onto {010} planes in the Ni$_3$Al was made Kear et al (1962), and further development of this model for the Ni$_3$Al was made by numerous authors in recent years. Recently, this model was further refined to explain the strain rate dependence of yield stress, bowed-out of APBs in (010) plane and jerky motion of screw superdislocations in Ni$_3$Al by Hirsch (1991).

However, the K-W type models can not apply directly to the TiAl without a proper modification. The reasons for this is as follows.

First, the APBs of the superdislocations were observed by Hug et al(1988) in the single crystal TiAl deformed at 873K. Predominant cross-slip of the APBs in this material was observed in the crystal deformed at 1073K by Li and Whang (1991). In addition, the superdislocation spilt on {111} planes in a triplet containing an APB & a SISF, not a CSF(Hug et al 1988). In other case, the majority of the superdislocations in TiAl deformed at 873K and 1073K were dissociated into superpartials and their APBs on {111} planes (Hahn et al 1991; Li et al 1991). The difference between TiAl and Ni$_3$Al, however, is that TiAl has a negative strain hardening rate opposed to the positive strain hardening rate in Ni$_3$Al. This means that for a small strain (0.2%), the TiAl behaves like Ni$_3$Al, but for a large strain (1.5%) it behaves like fcc metals. Therefore, it is suspected that the sessile or pinning of the cross-slipped APBs in TiAl is strain dependent, different from those of Ni₃Al. Thus, the cross-slipped APBs in TiAl have a good mobility at high temperatures. It is important to study the dislocation structure as a function of strain in TiAl to elucidate this discrepancy.

Another difficulty in applying the K-W model directly to TiAl is that the orientation dependence of the yield stress existing in TiAl can't be correlated with the ratio of the Schmid factors between the superpartials on a {111} planes and those on {010} planes. Attempts to find a similar correlation between the yield stress and the ratio of the Schmid factors in TiAl were futile (Kawabata 1985, and Whang 1991). An assumption that the cross-slipped APBs have a higher mobility than those in Ni₃Al makes sense in this case. As a result, it is clear that the K-W type models for the Ni₃Al can't be applicable in this case while a modified model, if any, has a better chance to be adopted.

B) Dislocation Core Structure

In some intermetallics, the anisotropy in elastic properties makes a dissociation of the superdislocation core onto another plane possible. For example, the APBs of superdislocations in Ni₃Al are more stable in {010} planes. Likewise, both experiments and calculations show that the APBs of the superdislocations in TiAl are more stable in {010} planes than in {111} (see Table 2). Furthermore, the experimental evidence shows that the ordinary dislocations in TiAl do not split into partials with CSFs in {111} planes while they show positive temperature dependence of yield stress. The TEM morphology of the ordinary dislocations in $Ti_{44}Al_{54}V_2$ alloy deformed at 873K reveals a sessile-like structure while no cross-slip plays a role in the mobility. Thus, it is worth investigating possible configurations of the non-planar core structure of the superdislocations and the ordinary dislocations. The important aspects of this study on the core structure include 1) to determine the stability of various core configurations on various planes, 2) to investigate the mechanisms for the glissile-to-sessile transformation and sessile-to-glissile transformation in the core structures.

In the past, numerous sessile configurations (configurations I, II-a, IV-a, IV-c in Fig. 10) of the superdislocation cores were proposed by Indenbaum et al (1985), and Greenberg et al (1991).

Fig. 10 shows possible different configurations of the superdislocation cores in TiAl (Li, Z.C., and Whang, S.H. 1991). In Fig. 10, Conf. I represents a dissociation of (101] dislocation in a {111} plane. This configuration is not stable than others in Fig. 10 since the energy of the APB and CSF in {111} planes is large (see Table 2). The arguments on the stability of these configurations will be discussed below.

Conf. II-b has a lower energy than that of Conf. I. The width of the APB in Conf. II-b is narrower than that of Conf. I. However, Conf. II-b has a lower energy than that of Conf. II-a since the two octahedral planes in Conf. II-a has an acute angle while those in Conf. II-b has an obtuse one.

Conf. III-a, -b, -c, -d have each one APB on (T01) plane, but Conf. III-b with an obtuse angle between the two octahedral planes has lower energy than the counterpart Conf. III-a.

Similarly, Conf. IV-b has the lowest energy among Conf. IVs. The calculations of the total energies of all these configurations in Fig. 10 show that Conf. II-a, IV-a and IV-c don't have low energies contrary to the suggestion by Indenbaum et al (1985).

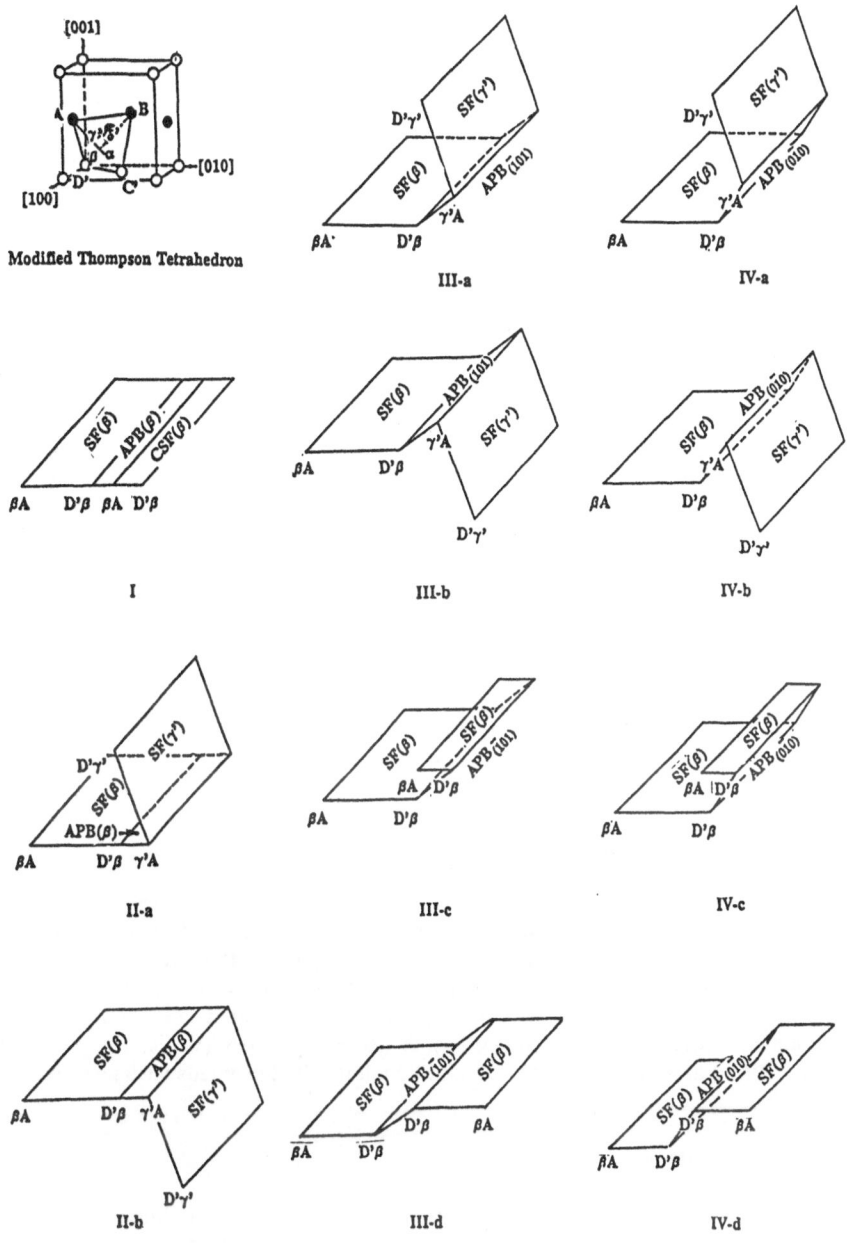

Fig. 10 Possible Configurations of Dissociated Superdislocations: on the top, a modified Thompson Tetrahedron where A and B are occupied by the same type of atoms, and so are D' and C', β on ($\bar{1}$11) plane; γ' on (11$\bar{1}$) (Li, Z.C. and S.H. Whang 1991).

C) Dipoles

The initial mention of 1/2[011] dislocation pairs was made by Shechtman et al (1974). In fact, the superdislocation dipoles (b=⟨011]) are a popular structure in the alloys deformed at 873K and 1073K (Hahn et al 1990; Whang et al 1991). Kawabata et al (1990) proposed that the dipoles could be a mechanism for the anomalous hardening in this material. In this mechanism, the leading superpartials of screw character in the dipoles could cross-slip onto a {101} plane due to the applied force coupled with the interaction force between the superpartials.

At high temperatures, this process may be facilitated due to the thermally activated process. Although it is clear that these dipoles make a contribution to the hardening, to what extent they contribute to the hardening is not yet clear.

D) Lomer-Cottrell-Hirth Barriers

Some dissociated dislocations from the superdislocations can combine to form a sessile dislocation, which is well known form of stable barrier in FCC metals. If these barriers occur in $L1_0$ TiAl, they may be considered as potential mechanisms for the anomalous hardening in this material (Greenberg et al 1991). In order for these barriers to be effective mechanisms, it has to be shown that the formation of the barriers is temperature dependent and it is active near the yield stress peak temperature. Experimentally, no observation on the Lomer locks has been made except that the ordinary dislocations form stair rod dislocations at 1273K in the $Ti_{35}Al_{55}V_{10}$ alloy (previous section).

Meanwhile, it was reported that the weak beam micrographs of Ni_3Ga deformed at 873K and 973K show two Lomer partials separated by APB on (001) plane (Hazzledine et al 1990). Both single and double Lomer locks appeared at 973K and 873K were formed by climb and by glide, respectively.

3. COMPOSITION DEPENDENCE

Composition dependence deformation in TiAl and its alloys has been studied primarily in conjunction with the fracture toughness and ductility. Little work has been done in these areas for the Al rich γ-TiAl. Instead, more attention has been paid to the two phase, γ-phase rich compositions. Meanwhile, the effort to investigate alloying effect on mechanical and high temperature properties by numerous third elements has continued. The ternary elements employed for these investigations include group IV (Zr, Hf), group V (V, Nb, Ta), group VI (Cr, Mo), group VII(Mn) and group III (Ga). In addition, interstitial elements such as C and N, and Be were studied.

A) Binary System

Earlier work on the composition dependence was performed by Bumps et al (1952) by hardness testing. The analysis of the hardness which is functions of composition and microstructure is difficult in relating to mechanical or physical parameters. Thus, little can be gained from the results. However, the tendency shown in the hardness often agrees with those of other mechanical measurements. The hardness in Ti-Al system shows a sharp drop near $Ti_{50}Al_{50}$, beyond which it increases with increasing Al content. The uniaxial compression tests on single phase γ-TiAl show that with increasing Al content the yield stress increases moderately while the fracture strain decrease sharply (Hahn 1991).

B) Ternary Systems

Investigation on the ternary γ-titanium aluminides has been focused on alloying effect on mechanical and high temperature properties. The review in this paper is limited to low temperature mechanical properties, i.e., strength and ductility in Al rich single γ-phase TiAl.

Zr and Hf addition (group IV) to TiAl was investigated by Kawabata et al (1989). The ternary phase diagram Ti-Al-Zr basically show an extensive solubility of Zr in the TiAl with constant Al content indicating that Zr replacing Ti at the Ti sublattice sites. Zr addition in $(Ti_{458}Al_{542})_{100-x}Zr_x$ alloys and $(Ti_{479}Al_{521})_{100-x}Zr_x$ alloys, where x=1-5 at %, increases the hardness and the fracture stress, but moderately decreases the fracture strain. Similarly, in $(Ti_{1-x}Hf_x)_{50}Al_{50}$ alloys, where x=0.01-0.04, the fracture stress increases and the fracture strain decreases as Hf content increases (Kawabata, Tamura and Izumi 1989). Hf is less effective in increasing the fracture stress than Zr, but more effective in decreasing the fracture strain than Zr.

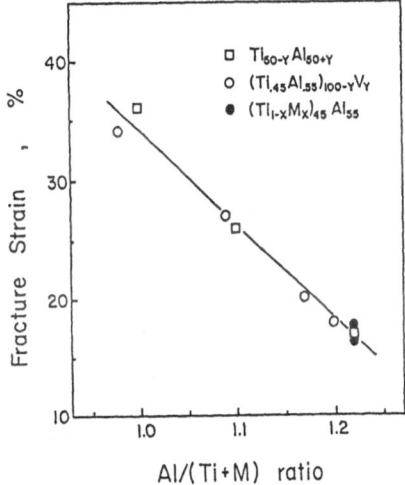 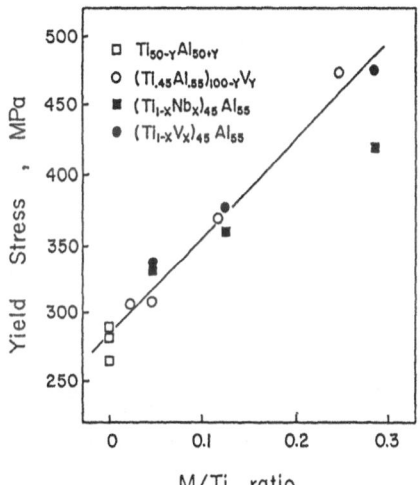

Fig. 11a Plot of Yield Stress vs. M/Ti ratio in Binary and Ternary γ-titanium Aluminides , where M is alloying element.

Fig. 11b Plot of Fracture Strain vs. Al/ (Ti+M) Ratio in Binary and Ternary γ-titanium Aluminides, where M is alloying element.

Effect of group V elements, V and Nb, as alloying elements, on yield stress and fracture strain under uniaxial compression were studied (Hahn 1991; Whang & Hahn 1989; Hashimoto et al 1988). For a fixed Al content, $(Ti_{1-x}V_x)_{45}Al_{55}$, x=0-0.2, the yield stress increases strongly with V content while the fracture strain decreases moderately with V content. Nb addition to TiAl follows a similar path in terms of yield stress and fracture strain (Hahn & Whang MRS Proc. 1990; Kawabata et al MRS Proc. 1990). On the contrary, Hashimoto et al (1988) reported an improvement in the fracture strain of $(Ti_{458}Al_{542})_{100-x}Nb_x$. The results on the yield stress and the fracture strain in the Ti-Al-V and the Ti-Al-Nb alloys were plotted as shown in Fig. 11a and b. The yield stress increases with V and Nb content approximately following a linear path, but the fracture strain (Fig. 10b) sharply decreases with increasing Al in these systems.

Little work has been done on alloying TiAl with group VI elements (Cr, Mo, W). Almost no effect of Cr addition on plastic fracture strain in rapidly quenched single γ phase $Ti_{100-x}Al_xCr_2$ alloys (x=52 - 55 at %) while a great improvement in the fracture strain in the alloys with 47-50 at %Al, the duplex alloys (Huang et al JIM-6 1991). The large improvement in the fracture strain has not yet been examined. No Mo addition to single γ phase TiAl has been reported. However, the two phase alloy $Ti_{51}Al_{48.5}Mo_5$ exhibits an improvement in elongation indicating that the ductility is apparently linked to the grain refinement by Mo addition (Maeda et al 1991). It is interesting to investigate whether or not the improvement in fracture strain in the alloy containing Cr is attributed to the grain refinement by Cr.

Ternary γ-phase Ti-Al-Mn alloys have received much attention because of their improved ductility. According to Hashimoto et al (1990), in single phase γ-phase Ti-xAl-yMn alloys, x=36-38 wt %; and y=0.5-2.5 wt %, the failure strain by bending increases with Mn content. The failure stress increases in the 36 Al series alloys and in the 37 Al series alloys, but slightly decreases in the 38 Al series alloys. The estimated elongation is 3% based on the bend tests. The improved deformation of these alloys are believed to be attributed to twinning. The ease of twin formation in the Ti-Al-Mn alloys was discussed in terms of the effect of Mn addition on the stacking fault energy and the stabilization of twin dislocations (Hanamura 1988). Nonetheless, some of these alloys have a single phase, Al lean γ-phase. Thus, the dislocation structure in these alloys likely different from those in the Al rich γ-titanium aluminides, which should be taken into account.

Effect of beryllium on mechanical properties of TiAl was also investigated (Kawabata JIM-6 1991). Since Be has a small atomic size, it may be treated as an interstitial element. In $Ti_{50}Al_{50-x}Be_x$ alloys with x=0-3 at%, a mild increase in the proof stress with Be concentration is seen. However, the failure strain decreases with Be content in these alloys. The proof stress and failure strain strongly depend on microstructure such as the amount of α_2 phase, and Ti_2Be_{17} precipitates, etc.

4. CONCLUDING REMARKS

The information on deformation and the related dislocation structures in the γ-titanium aluminides is less extensive than that for $L1_2$ Ni_3Al. Furthermore, current trends of the research on γ-titanium aluminides has an emphasis on the two phase ($\gamma + \alpha_2$) titanium aluminides for practical reasons. Nevertheless, the scientific value of the research on the single phase titanium aluminides has not diminished. Furthermore, the potential for developing into high temperature materials remains in both single and two-phase forms.

1) The crux of the temperature dependence of deformation is the anomalous hardening in the temperature range of RT to 1073K. The hardening is known to be caused by two different slip systems: $\langle 101]\{111\}$ and $1/2\langle 110]\{111\}$. The former, superdislocation slip, has many similarities in deformation as well as dislocation structures with those found in $L1_2$ Ni_3Al. The details of the deformation and the dislocation structure in TiAl, however, are different from those of Ni_3Al. For a better understanding, these discrepancies in deformation and dislocation structure between the two compounds should be understood in the context of crystal structure characteristics, elastic moduli and its anisotropy, bond character, and stability of planar defects.

2) Composition dependence on γ-titanium aluminides has been limited to Al rich γ-titanium aluminides due to the unavailability of single phase Al lean γ-titanium aluminides. It is, therefore, important to study the composition dependent deformation in Al lean single phase

γ-titanium aluminides if developed in the future.

ACKNOWLEDGEMENT

I would like thank the organizing committee of the NATO Advanced Materials Work Shop for their financial support.

REFERENCE

Bumps, E.S., Kessler, H.D. and Hansen, M. (1952) AIME Trans., Vol. 194, p. 609.

Farkas, D., Pasianot, R., Savino, E.J. and Miracle, D.B. (1991)
MRS Sym. Proc., Vol. 213, pp. 223-228.

Farenc, S., Caillard, D. and Comet, A. (1991) JIM-6 Sym., pp. 791-96.

Fu, C.L. and Yoo, M.H. (1990) Mat. Res. Sym. Proc. Vol. 186.

Fu, C.L. and Yoo, M.H. (1990) Phil. Mag. Lett., Vol. 62, 159-165.

Fujiwara, T., Nakamura, A., Hosomi, M., Nishitani, S.R., Shirai, Y., Yamaguchi, M. (1990) Phil. Mag. A, Vol. 61, 591.

Greenberg, B.A. (1970) Phys. Stat. Sol. 42, 459.

Greenberg, B.A. (1973) Phys. Stat. Sol. 55, 59.

Greenberg, B.A. (1991) Acta Metall. Mater. Vol. 39, 233-242.

Greenberg, B.A. (1989) Sripta Met. Vol. 23, 631-636.

Hahn, Y-D. & Whang, S.H. (1990) Scripta Met. Vol. 24, 139-144.

Hahn, Y-D.(1991) Ph.D Thesis, Polytechnic University.

Hahn, Y-D. and Whang, S.H. (1990) Mat. Res. Soc. Symp. Proc. Vol. 133, 385-390.

Hahn, Y-D., Li, Z.X., Whang, S.H. and Kawabata, T. (1991) Mat. Res. Sym., Johnson, L.A., Pope, D.P. and Stiegler, J.O. (eds) Vol. 213, 291-296.

Hahn, Y.-D. and Whang, S.H. (1990) Scripta Met et Mat., Vol. 24, 139-144.

Hall, E.L. & Huang, S-C.(1989) J. Mater. Res., 4, 595-602.

Hanamura, T. and Tanino, M. (1989) J. Mat. Sci. Lett. 8, 24-28.

Hanamura, T., Uemori, R. and Tanino, M. (1988) J. Mater. Res.

Hashimoto, K., Doi, H., Kasahara, K., Tsujimoto, T. and Suzuki, T (1990) J. Japan Inst. Metals, Vol. 54, 539-548.

Hashimoto, K., Haruo, D., Kasahara, K., Nakano, O., Tsujimoto, T. and Suzuki, T. (1988) J. Japan Inst. Metals, Vol. 52, No. 11, pp. 1159-1166.

Hazzledine, P.M. and Sun, Y.Q. (1990) Mat. Res. Soc. Sym. Proc. Vol. 213, 209-222.

Hazzledine, P.M., Yoo, M.H. and Sun, Y.Q. (1989) Acta Metall, Vol. 37, 3235-3244.

Hirsch, P.B. (1991) Scripta Met. et Mat. Vol. 25, 1725-1730; and Proceedings of this workshop.

Huang, S.C. and Hall, E.C. (1991) to appear in Scripta Met.

Huang, S.C. and Hall, E.L. (1991) Mat. Res. Soc. Sym. Vol. 213, pp. 827-832.

Huang, S.C., McKee, D.W., Shih, D.S. and Chesnutt, J.C. (1991) Japan Inst. Metals (JIM) Sym. Proc. on Intermetallic compounds, June 17-20, pp. 363-370.

Hug, G., Loiseau, A. and Lasalmonie, A (1986) Phil. Mag. A, 54, 47-65.

Hug, G., Loiseau, A. and Veyssiere, P (1988) Phil. Mag. A, 57, 499-523.

Indenbaum, V.N., Grinberg, B.A., Gornostyrev, Y.N. and Karkina, L.Y. (1985) Phys. Met. Metall. Vol. 59, No. 2, 52-63.

Kasahava, K., Hashimoto, K., Doi, H. and Tsujimoto, T. (1987) J. Japan Inst. Metals, Vol. 5, pp. 278-284.

Kawabata, T., Kanai, T. and Izumi, O., (1985) Acta Metall, Vol. 33, 1355-66.

Kawabata, T., Tamura, T. and Izumi, O. (1989) Mat. Res. Soc. Proc. Vol. 133, pp. 329-334.

298

Kawabata, T., Abumiya, T., Kanai, T. & Izumi, O(1990) Acta Metall. Mater. 38, 1381-1393.

Kear, B.H. and Wilsdorf, H.G.E. (1962) Trans Metall. Soc. AIME, 224, 382.

Kim, J.Y., Hahn, Y.D. and Whang, S.H. (1991) Vol. 25, 543-548.

Lall, C., Chin, S. and Pope, D.P. (1979) Met. Trans A, Vol. 10A, 1323-32.

Li, Z.C., and Whang, S.H.(1991) Proc. Sym. High Temperature Aluminides & Intermetallics, September 16-19, 1991, San Diego, CA., to be published as a special issue of J Materials Science and Engineering, 1992.

Li, Z.X., and Whang, S.H., ibid.

Li, Z.X., Kim, J.Y. and Whang, S.H. (1991) Scripta Met., Vol. 25, No. 11.

Liang, S.J. and Pope, D.P. (1977) Acta Metall. Vol. 25, 485-493.

Lipsitt, H.A., Shechtman, D. and Schafrik, R.E. (1975) Met. Trans A, Vol. 6A, 1991-1996.

Maeda, T., Okada, M. and Shida, Y. (1991) Mat. Res. Soc. Sym. Proc., Vol. 213, pp. 827-832.

Maeda, T., Anada, H., Okada, M. and Shida Y. (1991) JIM Sym.(6) on Intermetallic Compounds pp. 463-468.

Marcinkowski, M.J., Brown, N. and Fisher, R.M. (1961) Acta Metall. Vol. 9, 129-137.

McAndrew, J.B. aand Kessler, H.D. (1956) J. Metals, Trans. AIME, pp. 1348-53.

Nonaka, K., Tanosaka, K., Fujita, M., Kawabata, T. and Izumi, O. (1991) JIM-6 Sym. pp. 489-494.

Paidar, V., Pope, D.P. and Vitek, V. (1984) Acta Metall. Vol. 32, 435.

Pasianot, R., Farkas, D. and Savino, E.J. (1990) Vol. 24, 1669-1674.

Rao, S.I., Woodward, D. and Parthasarathy, T.A. (1991) MRS Sym Proc. Vol. 213, pp. 125-130.

Shechtman, D., Blackburn, M. J., Lipsitt, H.A. (1974) Met. Trans. A vol. 5, 1373-81.

Takeuchi, S. and Kuramoto, E. (1973) 21, 415.

Thornton, P.H., Davies, R.G. and Johnston, T.L. (1970) Met. Trans., Vol. 1, pp. 207-218.

Tsujimoto, T. and Hashimoto, K. (1989) Mat. Res. Soc. Sym. Proc. Vol. 133, pp. 391-96.

Umakoshi, Y., Pope, D.P. and Vitek, V (1984) Acta Metall. Vol. 32, 449-456.

Vitek, V. (1985) in Dislocations and Properties of Real Materials, edited by M.H. Loretto (The Institute of Metals, London, 1985) P. 30.

Vitek, V., Sodani, Y., and Cserti, J. (1991) Mat. Res. Sym. Proc. Vol. 213, 195-208.

Yoo, M.H., (1986) Scripta Met. et Mat., 20, 915.

Whang, S.H. and Hahn, Y.D. (1989) Mat. Res. Soc. Sym. Proc. Vol. 133, pp. 687-92.

Whang, S.H. and Hahn, Y.D. (1990) TMS-AIME Sym. Proc. on High Temperature Aluminides & Intermetallics, pp. 91-110.

Whang, S.H. and Hahn, Y.D. (1990) Scripta Met. et Mat., Vol. 24, 485-490.

Whang, S.H. and Hahn, Y.D. (1990) Scripta Met. et Mat. Vol. 24, 1679-1684.

Whang, S.H., Hahn, Y.D., Li, Z.X. and Li, Z.C.(1991) Proc. JIM Int. Sym. on Intermetallics, Izumi, O. (ed) 763-770.

Woodward, C., Maclaren, J.M. and Rao, S. (1991) MRS Sym. Proc. Vol. 213, pp. 715-720.

FRACTURE MECHANISMS IN INTERMETALLICS

H. VEHOFF
Max Planck Institut für Eisenforschung
Max Planck Str. 1
4000 Düsseldorf 1
Germany

Abstract. Models on cleavage fracture, interfacial fracture, and on the brittle-to-ductile transition are overviewed for intermetallics and compared with recent experimental results on the rate and temperature dependences of the fracture toughness of inter-metallics. The influence of grain size, phase distribution, temperature and environment on the fracture toughness of NiAl and TiAl based alloys was measured and discussed. It was found that the brittle-to-ductile transition temperature in single phase NiAl alloys cannot be lowered below the transition temperature of suitably oriented single crystals by grain refinement. But the toughness of two-phase alloys is improved when the ductile phase completely surrounds the brittle phase in sufficient thickness. The fracture toughness of many intermetallic alloys was found to be extremely rate sensitive. For NiAl (B2) single crystals the influence of orientation, strain rate and temperature was measured and discussed in view of recently developed dynamic models of the brittle-to-ductile transition.

1. Introduction

New intermetallic materials should combine the advantages in material behavior of ceramics (high strength at high temperature) with a high fracture toughness and sufficient ductility at room temperature. Therefore in many countries the development of new and improved intermetallic alloys is currently under way as is documented in several conference proceedings on intermetallic alloys [1].

The most challenging problem is still the brittleness at room temperature. The purpose of this overview is to review briefly our current understanding of fracture. This knowledge, mainly obtained from the examinations of conventional metals and ceramics, is then applied to intermetallic alloys. Differences and similarities to the behavior of conventional metals are discussed. Models for the brittle/ductile transition and interfacial fracture will be reviewed and compared with experimental results on the strain rate and temperature dependences of the fracture toughness in intermetallic alloys.

The fracture behavior of common and less common intermetallic alloys was examined. The examples which will be presented were obtained within a larger German research activity. This effort had as a primary intention the screening of different intermetallic phases for the application of intermetallics as new high temperature materials [2].

C. T. Liu et al. (eds.), Ordered Intermetallics – Physical Metallurgy and Mechanical Behaviour, 299–320.
© 1992 *Kluwer Academic Publishers.*

A thorough understanding of the mechanisms which lead to fracture must be gained in order to develop intermetallics which meet the demands for a material not only with great strength at high temperature but with sufficient ductility and fracture toughness at room temperature. Materials with high strength at high temperatures are usually brittle at room temperature. For intermetallics with their large variety of crystal structures and compositions, first principle calculations of cleavage strength and surface energies which become currently more and more available, are important. These calculations yield an understanding of the effects of order and composition on binding, crystal and interface structure. In combination with thermodynamic data on phase equilibria, guidelines for the development of new alloy systems can be obtained.

In the first section, some results from first principle calculations of surface energies and elastic constants are given and compared with experimental results of the fracture toughness of NiAl single crystals. In addition, the effect of segregation and order on interfacial strength is discussed.

The main advantage of intermetallics compared to ceramics is the possibility of plastic deformation even at low temperature. Some intermetallics show even a rising yield strength with increasing temperature as is well known from the classical example Ni_3Al with the $L1_2$-structure [3]. But this phenomenon is neither confined to A_3B aluminides nor to the $L1_2$ structure, rather, it depends on details of the interatomic binding. Therefore, in the second section the brittle/ductile transition in intermetallic alloys will be discussed in view of recent experimental results. Models which describe the kinetic processes at crack tips, dislocation emission, kinetics and crack stability are still outside the scope of computer simulations. However, models which combine dislocation theory with self consistent computer calculations currently become available. These models describe the role of loading rate and dislocation dynamics on the brittle/ductile transition correctly. But they are not able to make quantitative predictions.

In the third section, the influence of microstructure and environment on fracture toughness will be discussed. For multiphase alloys our knowledge of fracture is still based mainly on empirical results.

2. Mechanisms of Brittle Fracture

2.1. Brittle Cracks

The concepts of the stress intensity factor, K_M, and the energy release rate, G, are introduced in the following, since they are used permanently throughout the paper. The general form of the stress field in front of a general elastic crack can be described by:

$$\sigma_{ij}(r,\Theta) = \frac{K_M}{\sqrt{2\pi r}} f_{ij}^M(\Theta)$$

$$K_M = \sigma_M^\infty \sqrt{\pi a}\, Y(a/w)$$

where K_M is the stress intensity factor with M = I,II,III for the different loading modes, f^M_{ij} are angle dependent geometric functions which do not depend on the specimen geometry, σ_M^∞ is the stress applied externally, a is the crack length, w the specimen width, and Y contains the part of the solution which depends on the specimen geometry.

The change in elastic energy, W_{el}, when the crack propagates by an increment Δa is denoted by G:

$$G = \frac{1}{B} \frac{\partial W_{el}}{\partial(a)}$$

B is the specimen thickness, G is called energy release rate and relates to K by the following formula:

$$K_I^2 = EG$$

where E is the Young's modulus. The crack propagates when G exceeds twice the surface energy:

$$G \geq 2\gamma$$

This concept can be even applied to slow crack growth, for example due to environmental effects. In this case the bonds attract each other over a finite distance till final separation. This gives [4]:

$$G = \int_0^\infty \sigma(\delta)d\delta = 2\gamma_{int}$$

where $\sigma(\delta)$ is the local force elongation curve, γ_{int} is the interfacial energy and $2\gamma_{int}$ is the work for reversibly separating an interface; using the universal bonding concept by Rose and co-workers [5] it can be shown that reducing the surface/interface energy by segregation or lowering the bond strength by trapping are identical concepts.

2.1.1. Cleavage in Single Phase Single Crystals Recently, the cleavage energies of the B2-type aluminides, FeAl and NiAl, were calculated by Yoo and Fu by a first-principles energy approach [6]. From the cleavage energy they calculated the critical stress intensity factor for a mode-I crack, K_I, according to an equation given by Sih and Liebowitz [7] for anisotropic perfectly brittle materials:

$$K_{IC} = \left(\frac{G_c}{\pi}\right)^{1/2} \left\{\frac{S_{11}S_{22}}{2}\left[\left(\frac{S_{22}}{S_{11}}\right)^{1/2} + \frac{2S_{12} + S_{66}}{2S_{11}}\right]\right\}^{-1/4}$$

where the S_{ij}'s are the reduced elastic constants for the chosen crack geometry and material given in a review paper by Yoo and Fu [8] for intermetallics with L1$_2$ and B2 structure. The fracture toughness of NiAl single crystals with different orientations as a function of temperature and strain rate was measured recently by Reuss and Vehoff [9]. In table 1, their room temperature results are summarized together with the calculations of Yoo and Fu [6]:

Alloy	G_C		{100} K_{IC}		{110} K_{IC}	
NiAl	(100)	(110)	<010>	<011>	<110>	<001>
Theory [6]	5.5	4.1	5.5 (4.7)	6.0 (5.2)	5.6 (5.4)	4.8 (3.7)
[12]	1.8	3.2				
Exp. [9]			10	12	5.2	4.8
[11]			8			4

Table 1 Calculated cleavage (ideal) energies, G_C, (in units of J/m^2) and fracture toughness, K_{IC} ($MPam^{1/2}$) together with the measured fracture toughness of corresponding orientations.

The nomenclature used in table 1 deviates from Yoo's nomenclature since we used the crack line and the direction perpendicular to the crack plane as coordinate system. The orientations of the single crystals tested are given in Fig. 1 and in Fig. 2. In the text, the orientation is given by $<hkl>\{h'k'l'\}$ where $<hkl>$ is the direction of the crack line, and $\{h'k'l'\}$ is the plane perpendicular to the specimen axis. The tests were done in four point bending on notched specimens ($4 \times 6 \times 50 mm^3$). The K_{IC}- values were evaluated according to the relationship

$$K_{IC} = F_{crit} \cdot \frac{3(l-e)}{2Bw^2} \cdot \sqrt{a} \cdot Y\left(\frac{a}{w}\right)$$

where the geometry function $Y(a/w)$ is given by Gross and Srawley [10], l and e are the distances between the lower and higher support roller, respectively. The other constants have the usual meaning. A typical curve of K_{IC} versus temperature for a NiAl single crystal with $<110>\{110\}$-orientation is given in Fig. 3. In addition the slope of the force, F, elongation curve, dF/ds, at the point of unstable fracture is plotted into the same diagram (open circles). At the onset of pronounced plasticity, the slope of this curve decreases drastically. The temperature for which this change in slope occurs is taken as the brittle/ductile temperature. Since all $K_{IC}(T)$ curves presented in this work are obtained with the same testing and specimen geometry, the brittle-to-ductile transition temperatures of different alloy systems can be directly compared.

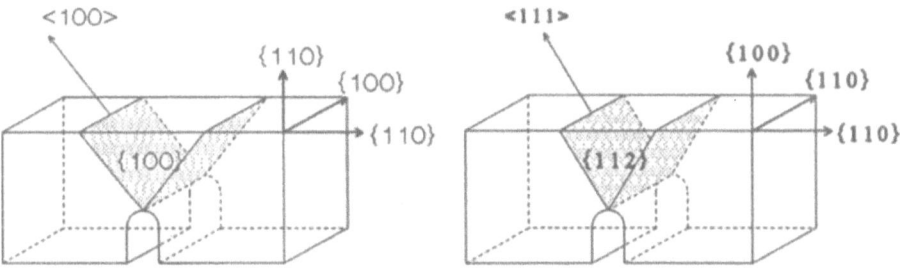

Fig. 1 Single crystal orientations which were chosen for maximum normal stress on {110} (cleavage plane for NiAl) and for easy stress relaxation by alternate slip at medium and high temperatures, respectively.

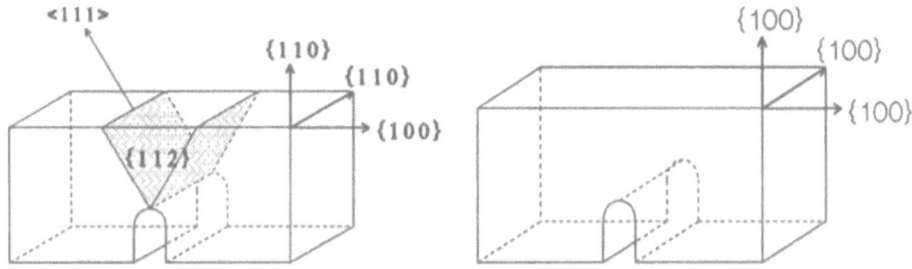

Fig. 2 Single crystal orientations with maximal normal stress on {100} and with high or low stresses for dislocation emission, respectively.

Even the crystals which were oriented favorably for slip showed unstable brittle fracture at room temperature. Pronounced slip was detected at around 200°C at a deformation rate of 0.2 μm/s (Fig. 3). For the specimens which were unfavorably oriented for slip, this transition occurs at 250°C (Fig. 10). Hence, the K_{IC}-values obtained at room temperature can be considered to represent values for cleavage, however, the TEM results showed that some plasticity at this temperature cannot be excluded (Fig. 5). In addition, the crystals are not dislocation free. Therefore, shielding due to dislocations in the vicinity of the crack tip can result in K_{IC}-values which are higher than the ideal cleavage strength of the lattice.

The calculations yield the lowest cleavage energy for the (110)-plane in correspondence with the experimental findings that at room temperature all specimens examined failed along {110}-planes even if they were oriented unfavorably. The TEM foils showed cleavage along {110} at low temperatures, as well. However, recent results given by Darolia [11] showed evidence for cleavage along {115}, hence a careful re-analysis of cleavage planes by an electron channeling technic is necessary.

304

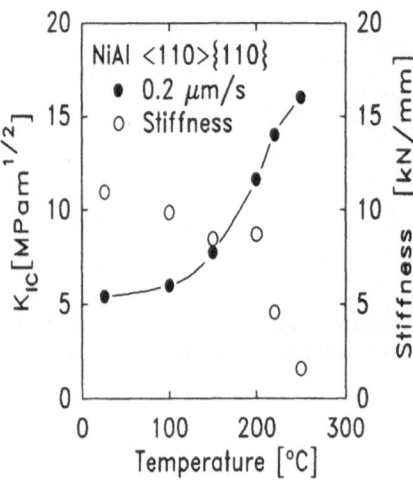

Fig. 3 K$_{IC}$ as a function of temperature for NiAl single crystals with <110>{110} orientations

Fig. 4 K$_{IC}$(T) in NiAl polycrystals with different Ni compositions

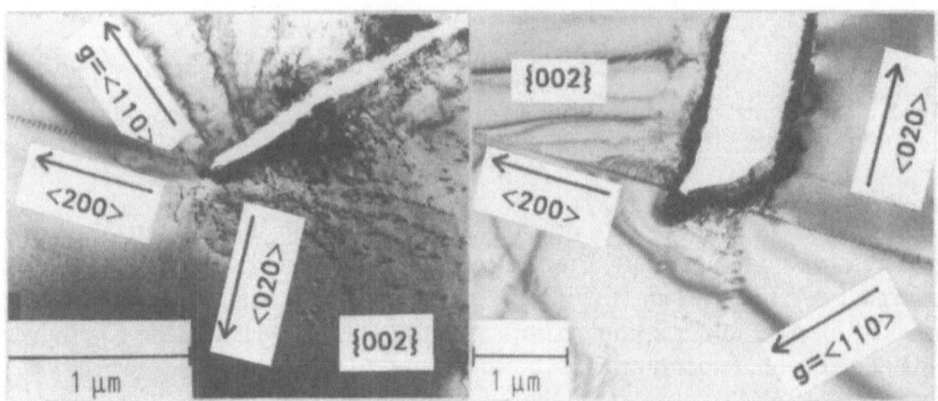

Fig. 5 a) Preparation crack in a TEM foil (NiAl) fractured at room temperature, foil normal <100>, crack flanks <110>
b) Crack produced by ex-situ straining of a TEM foil at 200°C, foil normal <100>, crack flanks <100>

The correspondence between the measured and predicted values is astonishingly good. But this correspondence should not be taken too seriously, since, as mentioned above, even at room temperature some dislocation activity by the activation of near crack tip sources or by dislocation emission cannot be excluded. Therefore the measured fracture strength can still be much higher than the ideal fracture strength of NiAl. Furthermore, calculations with other technics yielded much lower surface energies of around $1\,J/m^2$ for the (100)-plane and of $1.6\,J/m^2$ for the (110)-plane which would predict the (100)-plane as the plane with the lowest energy and therefore as the cleavage plane [12]. Atomistic simulation studies depend on interatomic potentials, which are obtained by fitting to known physical properties. For most materials these potentials are lacking or inadequate. For intermetallics, much research is still needed to obtain reliable potentials. Calculations which predict if a material behaves brittle or ductile are still outside the current scope of computer models. The available continuum models, however, which will be reviewed below increase our understanding of near crack tip processes but are not capable for quantitative predictions.

2.2. Interfacial Fracture

In general, the nucleation and growth of intergranular cracks depend on the cohesive strength of the boundary and on the stress state at the boundary, which is affected by dislocations which can cross the boundary or pile up at the boundary depending on the boundary orientation. The cohesive strength of a boundary depends on the grain boundary structure, the degree of disorder and on the chemical composition. In addition, for an interfacial crack the stress state at the crack tip is altered by the different elastic properties of the adjacent phases or grains resulting in a multiaxial stress state even for uniaxial loading [13].

Intermetallics which are still difficult to process often fail along the grain boundary. The cohesion of the boundary can be altered by segregation which can be easily shown by thermodynamic arguments. Following Rice and Wang [14] the work of separation for a segregated boundary can be estimated when the path in configurational space is specified. The Helmholtz excess free energy at fixed T is given by

$$df = \sigma d\delta + \sum_i \mu^i d\Theta^i$$

where Θ^i is the interfacial coverage of the segregant i.

For separation at constant Γ (non-mobile segregants) this yields [15]

$$(2\gamma_{int})_{\Theta = const} = (2\gamma_{int})_0 - \int_0^\Theta \left\{ \mu_b(\Theta') - \mu_s\left(\frac{\Theta'}{2}\right) \right\} d\Theta'$$

where $(2\gamma_{int})_0$ is the work to separate a clean interface, $\mu_b(\theta)$ the chemical potential of an unstressed interface with coverage θ, and $\mu_s(\theta/2)$ the chemical potential for a single free surface with coverage $\theta/2$. In cases where adsorption can be described by a Langmuir/McLean isotherm this gives directly

$$2\gamma_{int} = (2\gamma_{int})_0 - (\Delta g_b - \Delta g_s)\Gamma$$

In other words, the reduction of the surface energy due to segregation is directly given by the difference in the free enthalpies of segregation for the boundary (Δg_b) and the fracture surface (Δg_s). Based on these arguments, γ_{int} can be either increased or decreased by segregation, and the effect of boron can be rationalized when it is assumed that boron is more strongly bonded to the grain boundary than to the free surface.

The arguments given above are not confined to intermetallics. But in intermetallics grain boundary fracture is often observed on boundaries which are completely clean. A thorough examination of the factors which influence grain boundary weakness in A_3B alloys was undertaken by Takasugi and Izumi [16]. They showed for Ni_3Al that the substitution of Al by third elements with similar chemical bonding nature as Ni prohibited grain boundary fracture when they substitute for the Al site [17]. Recent observations in our group on stoichiometric NiAl polycrystals which were produced by HIP with a grain size of 10 to 20 μm showed intergranular fracture at room temperature (20% of the grains failed by intergranular fracture). With increasing temperature the portion of intergranular fracture increased up to 50%. If, however, the nickel content was increased to 55 at%, only transgranular fracture was observed. This did not increase the fracture toughness but shifted the brittle/ductile transition to higher temperatures (Fig. 4). Again, as recent observations by George and co-workers have shown [18], the addition of boron did suppress intergranular fracture for the stoichiometric alloy but did not improve the fracture behavior since the addition of boron had no influence on the active slip systems but a pronounced influence on the strength at room temperature.

These observations are in agreement with recent calculations by Vitek and co-workers [19], which showed that for strongly ordered $L1_2$ alloys the ordering behavior can persist up to the grain boundary forming atomic size cavities in the boundary. If, however, the alloy deviates from the stoichiometric composition the grain boundary becomes similar to fcc metals [20]. Ni_3Al is found to be ordered up to the boundary plane, with surplus of Ni the boundaries become increasingly similar to those found in pure fcc metals as the deviation from stoichiometry increases. When we assume a similar behavior for the strongly ordered B2 alloy NiAl, the transition from intergranular to transgranular fracture by Ni addition can be understood. However, preservation of order up to the boundary is not necessarily synonymous to grain boundary weakness since calculations of Baskes and co-workers [21] have shown that the grain boundary energies for pure Ni and for Ni_3Al are nearly identical.

2.3. Ductile vs. Brittle Behavior

According to the Rice-Thomson [22] approach for the ductile vs. brittle behavior of metals the energy U necessary to nucleate a dislocation from a loaded crack tip is given by three terms. They are: (1) the self energy of the dislocation including image terms produced by the presence of the crack surfaces, plus (2) the energy of the ledge created at the crack tip, minus (3) the work of the dislocation loop through the crack tip stress field.

$$U = U_{self} + U_{ledge} - U_{\sigma}$$

For different materials, U can be calculated, and estimates were obtained for the emission of dislocations at absolute zero.

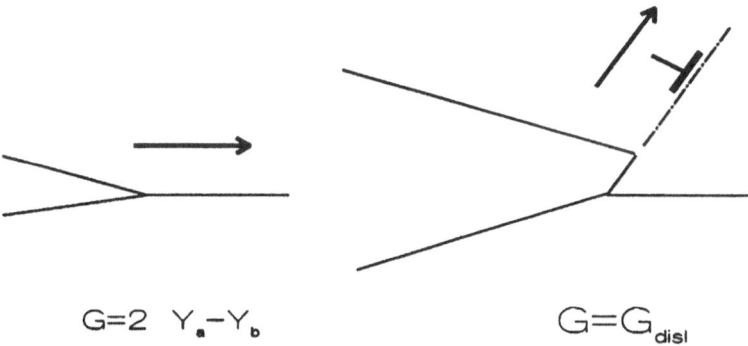

$$G = 2 \ Y_a - Y_b \qquad\qquad G = G_{disl}$$

Fig. 6 Schematic drawing of a crack at a grain boundary

The critical combination of K_e at which dislocation emission is spontaneous must be compared with the critical load for cleavage, which is calculated by equating the elastic energy release rate G to 2 γ. The situation is shown schematically in Fig. 6. The line $G_{disl} = G_{cleav}$ denotes the boundary between ductile and brittle behavior for an exact model. Anderson and Rice [23], using a refined description of the Rice-Thomson model, have calculated estimates for the critical stress intensity factor K_e for several metals. They found that most fcc metals beside Ir should be intrinsically ductile, most bcc metals, however, intrinsically brittle. In addition, they found that the results of the calculation depended strongly on the loading condition and crystal orientation which indicates that ductile vs brittle behavior depends on the microstructure at the crack tip. Similar calculations which take into account the special dislocations in ordered alloys are not done yet, but they could serve as first estimates for the fracture behavior of intermetallics.

The above interpretation of ductile versus brittle crack response suggests that either dislocation emission or atomistic brittle cleavage occurs at a tip of a micro crack. However, the brittle/ductile transition of a metal is a dynamic process which depends on loading rate, dislocation kinetics, and temperature. This will be discussed below.

3. Dynamic Models of the Brittle-To-Ductile Transition

So far only the competition between dislocation emission and bond breaking was considered. In reality, the response of a crack tip to the external loading rate is a dynamic process. Possible dislocation sources along the crack tip emit dislocations, these dislocations shield the source until the emitted dislocation has propagated far enough to raise the stress at the source above the value for the emission of the next dislocation. This situation is drawn schematically in Fig. 7.

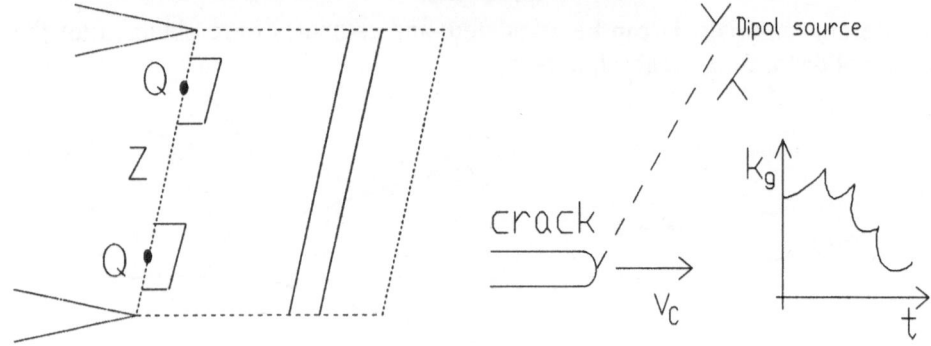

Fig. 7 a) Dislocations emitted from sources, Q, at the crack tip [24]
b) Crack stopped by the emission of dislocations from a dipole near the crack tip [28]

The brittle/ductile transition depends on the emission criterion, the dislocation rate and on the number and distribution of dislocation sources. The dynamic three-dimensional problem is not solved, yet. But, based on experimental results obtained with Si single crystals, several one- and two-dimensional models were developed by Hirsch and Roberts [24] and by Brede and Haasen [25] which will be reviewed briefly below and discussed together with measurements of the brittle/ductile transition in intermetallics obtained in our group.

Lin and Thomson [26] have considered the different solutions of the near crack tip stress field for different loading conditions, slip geometries and dislocation distributions. For the simple case of a Mode III crack they obtained for the force on a screw dislocation in the distance x_j from the crack tip in the force field of other dislocations:

$$f_d(x_j) = \frac{K_{III}b}{\sqrt{2\pi x_j}} - \frac{\alpha\mu b^2}{4\pi x_j} - \sum_{\substack{i=1 \\ i\neq j}}^{N} \frac{\mu b^2}{2\pi}\sqrt{\frac{x_i}{x_j}}\cdot\frac{1}{x_i-x_j}$$

where the first term is the force due to the applied load, the second is the image term and the third is the force due to the other dislocations. If this force exceeds a prescribed load (for example due to impurity pinning) the dislocation starts to move by thermally activated glide with a velocity according to

$$v_d = A(f_d)^m \cdot e^{-\frac{U}{kT}}$$

When now the crack is loaded by a given stress rate, dislocations will be emitted continuously. These dislocations shield the crack from the applied load according to

$$k_{III} = K_{III} - \sum_{i=1}^{N} \frac{\mu b}{\sqrt{2\pi x_i}} = K_{III} - \Sigma K_D$$

where K and k are the applied and local stress intensity factors, respectively, and K_D is the shielding K of an emitted dislocation. In the simple one-dimensional picture of mode III loading, the crack will always behave ductile when the force to operate the source is below the Griffith stress. Therefore Hirsch and coworkers [24] have assumed that the dislocation sources are distributed along the crack front. In their still one-dimensional computer simulation they have calculated the stress intensity at a point Z (Fig. 7) between the sources. If this stress intensity exceeds the Griffith value, the crack will propagate. The calculations yield $K_{IC}(T)$-curves in which K_{IC} increases smoothly with increasing temperature. The parameters which enter into the model are the distance of the dislocation sources, the operational stress of the dislocation sources and the rate law for the dislocation velocity. Brede and Haasen [27] have recently simulated the same problem. In their calculations they used the real rate law obtained from experiments and a more realistic picture of the slip geometry in Si, and they considered all three components of the stress intensity factor for mode I loading of single crystals. In contrast to the mode III calculations they found that k_I at the crack tip increases with the number of emitted dislocations even at the dislocation source. No assumptions for a discrete distribution of dislocation sources were necessary.

So far, only the case of cracks which emit dislocations is considered. For metals, however, the case of external preexisting dislocation sources which emit dislocations which might blunt or shield the crack from the external load can be equally important. The situation is shown schematically in Fig. 7b together with a schematic representation of the results of a preliminary calculation by R. Thomson [[28]]. A crack which approaches a dislocation source will be shielded or anti-shielded depending on the sign of the Burgers vector and the relative position between crack and dislocation when the source starts to emit dislocations. Depending on the relative velocities of the crack and the dislocations, the crack will be either stopped or only slowed down. Calculations of this type are currently in progress. These models describe correctly the processes at a crack tip, but are far from being quantitative.

3.1 Brittle-to-Ductile Transition in Intermetallics

How are these models related to the fracture behavior of intermetallics? It was shown above that the brittle/ductile response of an alloy depends on details of the dislocation dynamics and on the availability of dislocation sources either at the crack tip or in the vicinity of the crack.

The temperature dependence of the yield strength and of the dislocation dynamics are critical parameters which enter into the theory. Order not only can alter the crystal structure of the lattice, but also has a drastic influence on the core structure of the dislocations [29], which in turn influences the temperature dependence of the yield

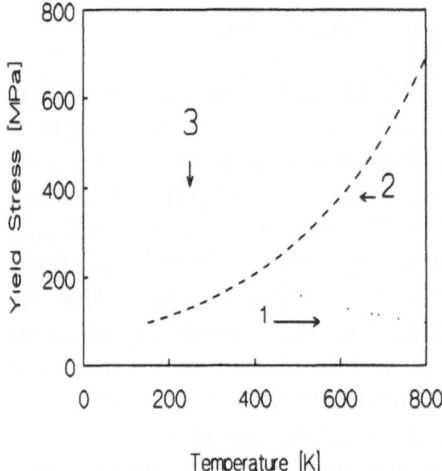

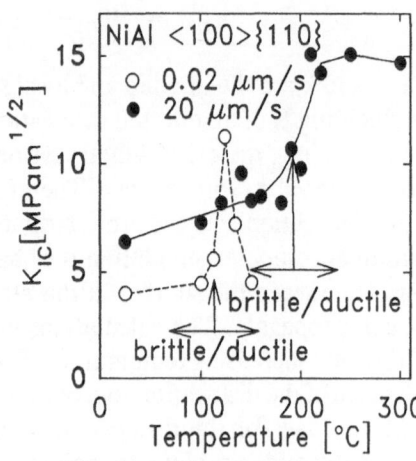

Fig. 8 Dependence of the yield stress on temperature in intermetallics (schematically)

Fig. 9 Brittle/ductile transition for two different displacement rates in NiAl single crystals with the orientation <100>{110}

stress and the rate equations for thermally activated glide. Typical yield stress vs temperature curves are given schematically in Fig. 8. In intermetallics, increasing yield stress with temperature (curve 2 in Fig. 8), decreasing yield stress (curve 1) and a minimum of the yield stress at intermediate temperatures (curve 3-2) [30] was observed.

3.1.1 Single Crystals Recently, Takasugi and co-workers [30] examined the temperature and rate dependence of the yield stress in B2 CoTi single crystals. They observed a pronounced minimum in the yield stress at 400 K. At lower temperatures, σ_y was found to be independent of strain rate, at intermediate and higher temperatures, however, a pronounced rate dependence of σ_y was found.

In order to understand the influence of the deformation behavior on the brittle/ductile transition in intermetallics, in our group $K_{IC}(T)$ was measured for several intermetallic alloys. In addition, the influence of strain rate and orientation was examined systematically in NiAl single crystals. The orientations examined are given in Figures 1a-2b.

In NiAl, the critical resolved shear stress on <100>{110} slip systems increases sharply below 200°C, and is nearly independent of temperature up to 700°C [31]. This sharp increase in yield stress corresponds to the brittle/ductile transition observed in the NiAl single crystals (Figs. 3 and 9) with the symmetric alternate slip orientation (Figs. 1a,b). Specimens tested at room temperature showed yield strengths nearly independent of strain rate, however, above 200°C marked rate sensitivity was observed [31]. Again, this corresponds to the observed shift in the brittle/ductile transition shown in Fig. 9.

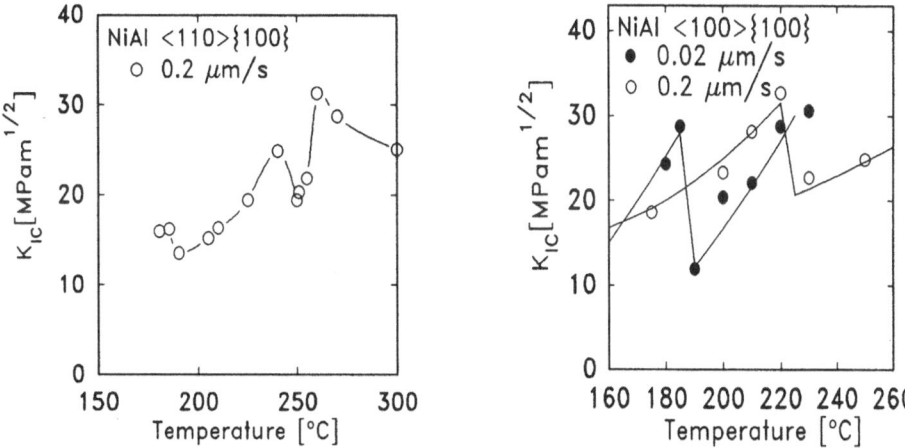

Fig. 10   a) K_{IC} as a function of the temperature for NiAl single crystals with <110>{100} orientation

b) Maximum of the first brittle/ductile transition in NiAl single crystals with <100>{100} orientation for two rates.

The other two orientations (Figs. 2a,b) showed a marked difference in the brittle/ductile transition behavior. The cleavage stress was more than a factor of two higher (cleavage plane {110}). For the same rate, the ductile/brittle transition temperature was shifted by more than 100°C towards higher temperature, and the transition occurred in two steps, i.e. K_{IC} increases, decreases and increases again with temperature (Fig. 10a). This was proven on several specimens and is therefore not an artefact due to scatter in the results. The minimum shifted systematically with displacement rate (Fig. 10b).

Fig. 11 shows the brittle-to-ductile transition for the symmetrical orientation (Fig. 1a) as a function of the elongation rate and temperature (filled circles). From the slope an apparent activation energy of 1.1 eV for this transition was estimated. The shift of the first maximum in Fig. 10a shows a similar slope (filled triangles, Fig. 11), however, for the hard <100>-orientation (Fig. 10b) a definitely steeper slope was found [32] (open squares, Fig. 11).

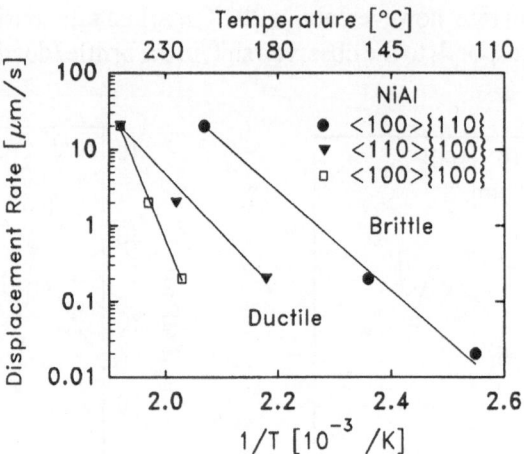

Fig. 11 Dependence of the ductile/brittle transition temperature on strain rate for NiAl single crystals of different orientations

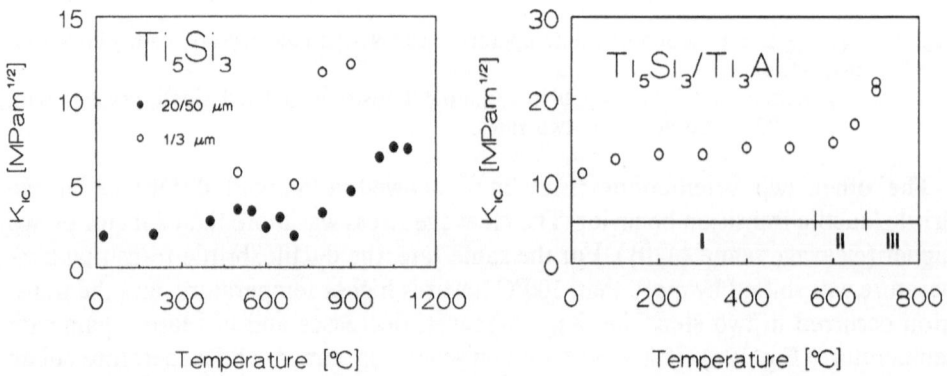

Fig. 12 $K_{IC}(T)$ for Ti$_5$Si$_3$

Fig. 13 $K_{IC}(T)$ for Ti$_5$Si$_3$/Ti$_3$Al

Fig. 5 shows a TEM photograph of a crack tip near the transition temperature. Slip occurred perpendicular to the crack flanks and in the direction of crack growth. The crack was produced by ex-situ straining the foil at the transition temperature. Hence, only cracks which were stopped by extensive dislocation emission could be observed. For comparison in Fig. 5a a typical crack produced at room temperature with the same technic is shown. In this case the crack had propagated along the {110}-plane and had stopped by the emission of dislocations. Examinations of many high temperature cracks showed that the plastic zone at the crack was very small and that much dislocation interaction had occurred. Therefore it seems plausible to assume that near the maxima shown in Fig. 10b, additional sources at the crack tip start to operate with Burgers vectors perpendicular to the crack flanks (Fig. 5) which inhibit the free emission of dislocations from the tip. Hence, the stress is raised at the tip which then resulted in a decrease of K_{IC} with increasing temperature. Further work, especially in-situ straining experiments with different foil orientations as a function of temperature, is needed to clarify the details of the dislocation processes at the crack tip in this alloy. In addition, the dislocation velocity as a function of temperature and stress must be examined in more detail to allow the modeling of the brittle/ductile transition in the same way as in Si.

3.1.2 Polycrystals NiAl polycrystals showed a $K_{IC}(T)$ dependence similar to the single crystals (Fig. 4). Furthermore, the brittle/ductile transition of two phase alloys was examined, in which one phase behaved more ductile. Fig. 12 shows $K_{IC}(T)$ for the single phase alloy Ti_5Si_3 with the complicated D_{88} structure in comparison to the two phase alloy Ti_5Si_3/Ti_3Al (Fig. 13). The silicides tested were developed by Frommeyer and Rosenkranz [33]. They contained two classes of grains, 20 and 50μm. Examinations of the fracture surfaces revealed that the small grains failed along the grain boundary, whereas for the larger grains transgranular fracture was observed. The eutectic two phase alloy Ti_5Si_3/Ti_3Al showed stable crack growth (range I in Fig. 13) at low temperature and unstable crack growth in the range where the first Ti_3Al grains start to deform plastically (II). In range I, only transgranular fracture was observed whereas in range II only interfacial fracture was found.

In this alloy, the brittle/ductile transition was very steep and characterizes the onset of interfacial fracture due to the yielding of the weaker phase. An apparent activation energy of 2.8 eV was measured for this process (Fig. 14). However, the alloy still behaved brittle, but with a higher K_{IC}.

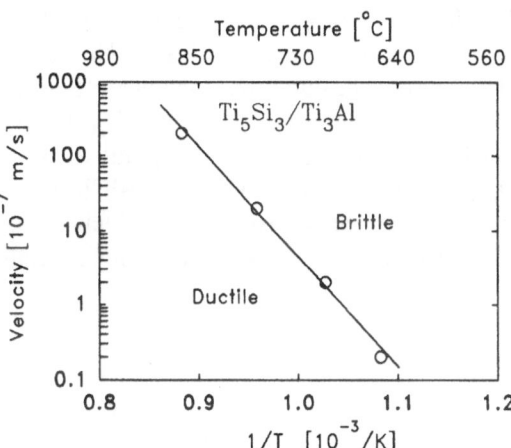

Fig. 14 Dependence of the ductile/brittle transition on temperature and strain rate for a Ti_5Si_3/Ti_3Al alloy

4. Role of Microstructure in Brittle Fracture

The discussion above has shown that intermetallics with good high temperature toughness are usually brittle, since beside for the cubic systems the number of slip systems in the complicated structures are not sufficient to fulfil the von Mises criterion. In addition, the nucleation and movement of dislocations is difficult and often strongly rate dependent. Even cubic systems like in the case of NiAl, often have only three independent slip systems and should be ductile as single crystals only. However, Schulson and Barker [34] have shown that the tensile ductility in NiAl could be dramatically improved by refining the grains to sizes smaller than a critical value. Guided by the Stroh mechanism [35]

$$\sigma_f > \frac{2\mu\gamma}{c_o\sqrt{d}}$$

they assumed that in fine-grained structures, cracks are not able to propagate when the grain size falls below a critical value. However, they conducted their experiments at 400°C. Our experiments showed that at this temperature the single crystals, at least the orientations tested, all behaved ductile. Additional slip systems are able to operate at this temperature. Therefore enough slip systems are available, and the conventional toughening mechanism due to grain refinement can operate. Within a joined research effort [36], we examined the fracture toughness of many alloy systems. It was never possible to improve the fracture toughness by grain refinement in a temperature range where the alloys showed no ductility (see for example Figs. 4,12).

The fine mixture of a ductile phase with a brittle phase is another approach to improve the fracture toughness of an alloy. This approach was very successful for NiAl with a small percentage of γ' [37] and for NiAl/NiAl$_3$ alloys, when the NiAl grains were completely surrounded by the ductile phase and the ductile phase was thick enough to stop cracks initiated in NiAl. Careful experiments by Pank and co-workers [38] have shown that a $>5\ \mu m$ skin around the grains of a NiAl based alloy was sufficient to stop the cracks and to improve the ductility. Machon and Sauthoff [39] tried a similar approach in the NiAlNb/NiAl system. Fig. 15 showed K_{IC} values obtained with their alloys [40]. The filled triangles (tip down) correspond to pure NiAlNb, the filled triangles (tip up) to the eutectic composition NiAl/NiAlNb. The eutectic composition had the best mechanical data but no improvement in ductility compared to NiAl was obtained. Hence, at least one phase must have enough ductility to inhibit crack propagation.

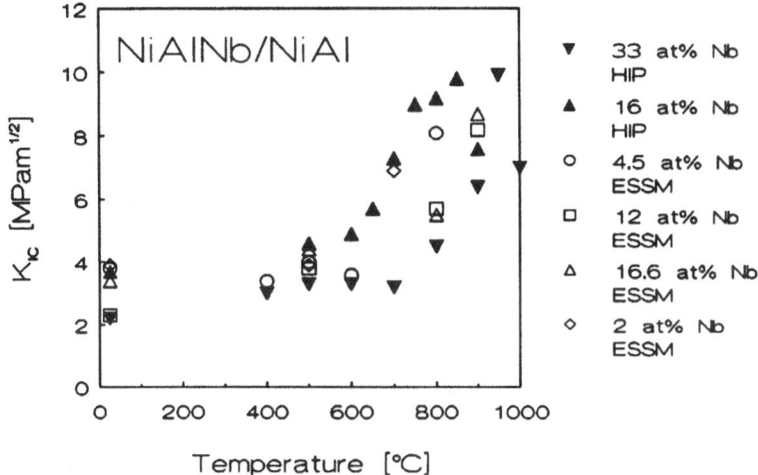

Fig. 15 $K_{IC}(T)$ for NiAlNb/NiAl alloys with different compositions

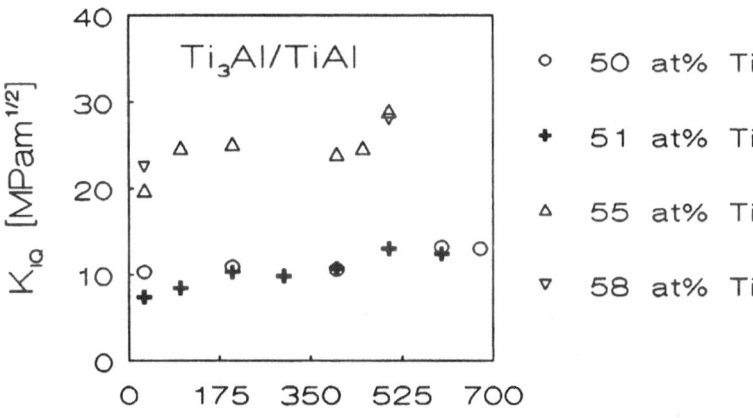

Fig. 16 $K_{IC}(T)$ for Ti₃Al/TiAl alloys with different compositions

A special case are TiAl/TiAl₃ alloys. The alloy contained two intermetallic phases, the hexagonal DO₁₉ Ti₃Al (α_2) and the tetragonal L1₀ TiAl (γ)-phase which interface along $\{0001\}_{\alpha 2}//\{111\}_\gamma$ planes [41]. Quasistatic crack growth in this alloy occurs by the nucleation and coalescence of microcracks in front of the main crack. In this system the fracture toughness could be increased dramatically by micro-alloying and appropriate heat treatments. Table 2 shows fracture toughness values of TiAl alloys for room temperature and 500°C obtained after different heat treatments. The alloy with the Cr and Si addition showed the best toughness values.

Material	Heat treatment	K_{IQ},RT	K_{IQ} 500°C
Ti51-Al	-	7	13
	1h/1000°C/Al_2O_3	7	12
	1h/1100°C/Vacuum	7	14
	electron beam	10	12
Ti55-Al	-	19	29
	1h/1000°C/Al_2O_3	16	28
	1h/1100°C/Vacuum	10	12
	168h/600°C/air	17	23
	168h/700°C/air	**23**	27
	336h/650°C/air	21	25
	168h/1100°C/Ar	11	19
	electron beam	18	21
	HIP	13	23
	168h/1100°C/Ar HIP	14	20
+ Cr3	-	23	21, 32
	168h/1100°C/Ar	15	21
+ Cr1/Si0.1	-	34	37
+ Cr1/Si0.1/Nb2		20	35
Ti58-Al	-	18	27
	168h/1100°C (Ar)	21	28

Table 2 : Fracture toughness of different TiAl alloys after different heat treatments [42,43]

This alloy had a stable layered structure with very fine Ti_3Al and TiAl lamellae. Fig. 16 shows the temperature dependence of the fracture toughness for alloys with different Ti compositions [42]. The alloy with 58 at% of Ti had the highest toughness. Wunderlich and Kremser [43] have shown that this increase in toughness corresponds to an increase in the volume fraction of Ti_3Al and a refinement of the lamella size down to 200 nm.

4.1 Environmental effects.

The fracture toughness of the $TiAl/Ti_3Al$ alloy was observed to be independent of temperature. However, when the alloy was tested in vacuum (VL Fig. 17), the fracture toughness was found to be markedly higher for all temperatures (VH = vacuum, 700°C) compared to air (AL,AH = air at room temperature and 700°C, respectively). Tests in water vapor (W), hydrogen (H_2), and oxygen (O_2) revealed that the normal humidity embrittles the alloy by hydrogen. With increasing temperature the effect of hydrogen diminishes but the alloy starts to corrode by stress corrosion cracking due to local oxidation. The weak dependence of K_{IC} on temperature (Fig. 16) could therefore be due to different environmental effects which operate at lower and higher temperatures and not due to an intrinsic quality of the alloy. The pure γ-alloy showed no dependence on the fracture toughness of the environment (Fig. 17), therefore only the α_2 phase embrittles by hydrogen, possibly due to hydride formation. Many intermetallics showed strong embrittlement due to hydrogen. Liu and co-workers [44] examined the effects of hydrogen on the brittleness of CoTi alloys. In tests with Co_3Ti single crystals they showed, that hydrogen alters the glide processes at the crack tip due to the formation of stacking faults which were not observed in a vacuum environment. These faults hinder the emission and motion of dislocations which results in brittle fracture. Details of the different effects of the environment will be given in the following paper in this book.

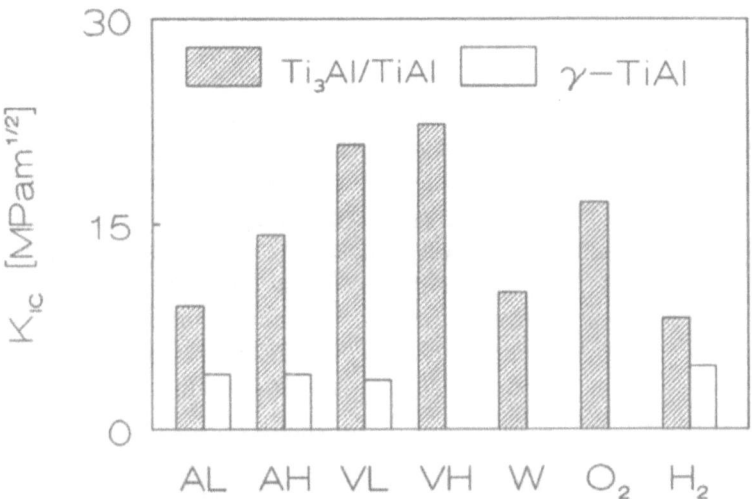

Fig. 17 Dependence of K_{IC} on the testing environment for the two and single phase TiAl alloy

5 Summary and Conclusions

Models on fracture and on the brittle/ductile transition were reviewed. These models were then used to explain the fracture behavior of single-phase and multiphase intermetallic alloys. Special intention was given to the temperature and strain rate dependence of the fracture toughness. For NiAl single and polycrystals, the brittle/ductile transition was examined in detail. It was found that the brittle-to-ductile transition temperature and the fracture toughness depended on the orientation of the single crystals. NiAl single crystals with <100>-specimen axis showed the largest toughness at room temperature and the highest transition temperature. For crystals with <110>-specimens axis, an apparent activation energy of 1.1 eV for the

318

brittle/ductile transition was measured whereas for crystals with <100>-orientation two transition where observed, one with an apparent activation energy of 2 eV. Evidence was given that the activation of additional slip systems at higher temperatures can reduce the toughness by blocking the further emission of dislocations which then results in a second transition.

Dual-phase fine-grain materials usual showed higher toughness values as their single phase components. Alloys which contained not enough slip systems could not be made ductile at room temperature even when the grain boundary strength was increased by appropriate alloy design. Alloys with strong covalent bonds showed a steep brittle/ductile transition whereas alloys like NiAl showed a smooth transition with temperature. The transition could be made much steeper when the grain size was reduced. Titanium based aluminides and silicides were found to be extremely environment sensitive. In TiAl/Ti$_3$Al alloys, even humid air was sufficient for a pronounced reduction in toughness due to hydrogen.

The effects of dislocation kinetics, impurities, and of micro alloying on fracture must be studied in detail to improve our understanding in the mechanical behavior of intermetallics. The experimental studies should be supported by theoretical studies of the ideal strength, of the dislocation core structure and of the segregation effect on the core structure. In addition, classical dislocation theory should be further developed to obtain an increased understanding of the dislocation kinetics and how the kinetics affect the brittle/ductile transition. Crack nucleation at interfaces, and the role of interfaces in blocking the further emission of dislocations from crack tips, must be studied in detail to obtain a deeper understanding of the influence of the microstructure on the brittle/ductile transition.

Acknowledgements

The authors would like to thank G.Frommeyer and G. Sauthoff for providing alloys, and W.Vogt for his help with the experiments. A special thank is extended to B. Schaff for her successful efforts in growing NiAl single crystals. We are grateful for the financial support by the BMFT.

References

1. High-Temperature Ordered Intermetallic Alloys, I, II, III and IV, Mat. Res. Soc., Pittsburg (1985, 1987, 1989 and 1991)
2. Sauthoff, G., (1989) Z. Metallkde, **80**, 337
3. Copley, S.M., B.H. Kear (1967) Trans. Metall. AIME **339**, 977
4. Rice, J.R.: in "Chemistry and Physics of Fracture" R.M. Latanision, J. Pickens, eds. Martinus Nijhoff, Dordrecht, 22, (1987)
5. Rose, J.H., Smith, J.R., Guinea, F., Ferrante, J. (1984) Phys. Rev. B **29**, 2963
6. Yoo, M.H., C.L. Fu (1991), Scripta Metall. Mater. **25**, 2345
7. Sih, G.C., Liebowitz, H.(1968) 'Mathematical Theories of Brittle Fracture' in: Liebowitz (ed), Fracture, Academic Press, London, pp. 68-190
8. Yoo, M.H., Fu, C.L. (1991) in: "Advanced High Temperature Intermetallics", ISIJ Intl., **31**, 1048
9. Reuss, S., Vehoff, H., (1991) Brittle to Ductile Transition in Intermetallic Alloys, in: Proc. of the 2nd European Conference on Advanced Materials and Processes, EUROMAT 91, Cambridge, UK, in Press
10. Gross, B., J.E. Srawley (1965) NASA TN-D, 2603
11. Darolia, R., (1991), JOM, **43**, March, 44
12. Clapp, P.C., Rubins, M.J., Charpenay, S., Riffkins, J.A, Yu, Z.Z., Voter, A.F. (1989) Mat. Res. Soc. Symp., Boston, Proc., Vol. **133**, 29
13. Rice, J.R. (1988) Transactions ASME, **55**, 98
14. Rice, J.R., Wang, J.S., (1989) Mater. Sci. Eng. A **107**, 23
15. Hirth, J.P., Rice, J.R. (1980) Metall. Trans. A **11**, 1502
16. Takasugi, T., O. Izumi (1985) Acta Metall., **33**, 1247
17. Takasugi, T., Izumi, O., Masahashi, N. (1985), Acta Metall. **33**, 1259
18. George, E.P., Liu, C.T., (1990) J. Mater. Res. **5**, 754
19. Ackland, G.J., Vitek, V. (1989) MRS, Vol. **133**, 105
20. Kruisman, J.J., Vitek, V., De Hosson, J.Th. M. (1988), Acta Metall. **36**, 2729
21. Baskes, M.I., Foiles, S.M., Daw, M.S. (1988), Journal de Physique, C5, Tome **49**, 483
22. Rice, J.R., Thomson, R. (1974) Phil. Mag. **29**, 73
23. Anderson, P.M., Rice, J.R. (1986) Scritpa Metall., **20**, 1467
24. Hirsch, P.B., Roberts, S.G., Samuels, J. (1989) Proc. R. Soc. Lond. A **421**, 25
25. Brede, M., P. Haasen (1988) Acta Metall. **36**, 2003
26. Lin, I. H., Thomson, R. (1986), Acta Metall. **34**, 187
27. Brede, M., Haasen, P., (1991) ICSMA 9, eds. D.G. Brandon, R. Chaim, A. Rosen, Haifa, Freund Publishing House, London, 813
28. Thomson, R., To be published
29. Paidar, V., Pope, D.P., Vitek V. (1984) Acta Metall. **32**, 435
30. Takasugi, T., Tsurisaki, K., Izumi, O., Ono, S. (1990), Phil. Mag. A **61**, 785
31. Wasilewski, J.R., Butler, S.R., Hanlon J.E. (1967) TMS of AIME, **239**, 1357
32. Reuss, S., Vehoff, H. (1991) Fracture toughness of intermetallics in: Intermetallische Phasen als Strukturwerkstoffe, Konferenzen des Forschungszentrums Jülich, **6**, 65
33. Frommeyer, G., Rosenkranz, R., Lüdecke, C. (1990) Z. Metallkde. **81**, 307
34. Schulson, E.M., Barker D.R. (1983), Scripta Metall. **17**, 519
35. Stroh, A.N. (1954) Proc. R. Soc., A **223**, 404
36. Sauthoff, G. (1990) Z. Metallkde **81**, 855
37. Ishida, K., Kainuma, R., Ueno, N., Nishizawa, T. (1991),Metall. Trans. A, **22**, 441

38. Pank, D. R., Nathal, M. V., Koss, D. A. (1990) J. Mater. Res. **5**, 942
39. Machon, L., G.Sauthoff, to be published
40. Reuss, S., Vehoff, H. (1990) Scripta Metall. **24**, 1021
41. Yang, Y.S., Wu, S.K. (1990) Script. Metall. **24**, 1801
42. Reuss, S., Dissertation, TH Aachen, 1991
43. Wunderlich, W., Kremser, Th., Frommeyer, G. (1990) Z. Metallkde. **81**, 802
44. Liu, Takasugi, Y. T., Izumi, O., Yamada, T. (1989) Acta Metall. **37**, 507

MOISTURE-INDUCED ENVIRONMENTAL EMBRITTLEMENT OF ORDERED INTERMETALLIC ALLOYS AT AMBIENT TEMPERATURES

C. T. Liu
Metals and Ceramics Division
Oak Ridge National Laboratory
P. O. Box 2008
Oak Ridge, TN 37831-6115
U.S.A.

ABSTRACT. Recent studies have demonstrated that moisture-induced environmental embrittlement is a major cause of low ductility and brittle fracture in ordered intermetallics with high crystal symmetries (e.g., L1$_2$ and B2). The embrittlement involves the reaction of reactive elements in intermetallics with moisture in air and the generation of atomic hydrogen at crack tips. The loss in ductility at ambient temperatures is generally accompanied by a change in fracture mode from ductile appearance to brittle grain-boundary fracture in many L1$_2$ intermetallics, and to brittle cleavage in body-centered cubic (bcc)-ordered intermetallics. In a number of cases, the embrittlement was alleviated by alloy design through control of microstructure and alloy composition.

1. Introduction

Aluminides and silicides generally possess attractive high-temperature properties; however, brittle fracture and poor ductility at ambient temperatures have limited their use as structural materials for engineering applications [1-7]. For the past ten years, substantial efforts have been devoted to understanding brittleness in ordered intermetallics. As a result, significant progress has been made in understanding both intrinsic and extrinsic factors contributing to the low ductility and brittle fracture in these intermetallics.

This paper briefly reviews recent research on environmental embrittlement of ordered intermetallics in air and other atmospheres at ambient temperatures. The studies have demonstrated that environmental embrittlement — an extrinsic factor — is a major cause of brittle fracture and low ductility in many intermetallic alloys when tested in moist air at ambient temperatures [8-19]. The embrittlement involves decomposition of moisture in air and generation of atomic hydrogen at crack tips [9]. The loss in ductility is caused by hydrogen embrittlement. In many cases, the ductility reduction is accompanied by a change in fracture mode from ductile appearance to brittle grain-boundary fracture or cleavage fracture [8-19].

This review paper contains five sections. The introduction section is followed by a description of environmental effects on room-temperature ductility and fracture in bcc-type intermetallics. The third section summarizes the environmental effects in face-centered cubic (fcc)-type intermetallic alloys. The fourth briefly mentions alleviation of environmental embrittlement by alloy design. The last section provides a brief summary and remarks.

C. T. Liu et al. (eds.), Ordered Intermetallics – Physical Metallurgy and Mechanical Behaviour, 321–334.

2. Environmental Embrittlement of bcc-Type Ordered Intermetallics

The FeAl aluminide with the B2 structure is a good model material for study of environmental embrittlement in ordered intermetallics because of its high sensitivity to test environment at ambient temperatures [20]. Figure 1 and Table 1 show the tensile properties of the aluminide containing 36.5 at. % Al as a function of test environment at room temperature [9]. The yield stress is not sensitive to environment, whereas the ultimate tensile strength scales with tensile ductility, which, in turn, depends strongly on test environment. The aluminide with a coarse grain structure showed a ductility of 2.2% in air. The ductility of the aluminide increased to 5 to 6% when tested in vacuum or Ar + 4% H_2 environment. The specimens showed, surprisingly, a high ductility of 17.6% when tested in dry oxygen. The water-vapor test confirmed the low ductility found in the air tests, indicating that water vapor is the embrittling agent.

The FeAl aluminide shows a change in fracture mode with increasing ductility (Fig. 1) [9]. The aluminide exhibited mainly transgranular cleavage fracture when tested in air, whereas it showed mainly grain-boundary separation when tested in an oxygen environment. A mixed fracture mode was observed for the vacuum test. The occurrence of cleavage fracture, presumably along {100} planes, indicates that cleavage planes are more susceptible to embrittlement than are the grain boundaries. When the embrittling effect of moisture is removed by testing in dry oxygen, the fracture path follows the grain boundaries, which are the (intrinsic) weakest links in the aluminide.

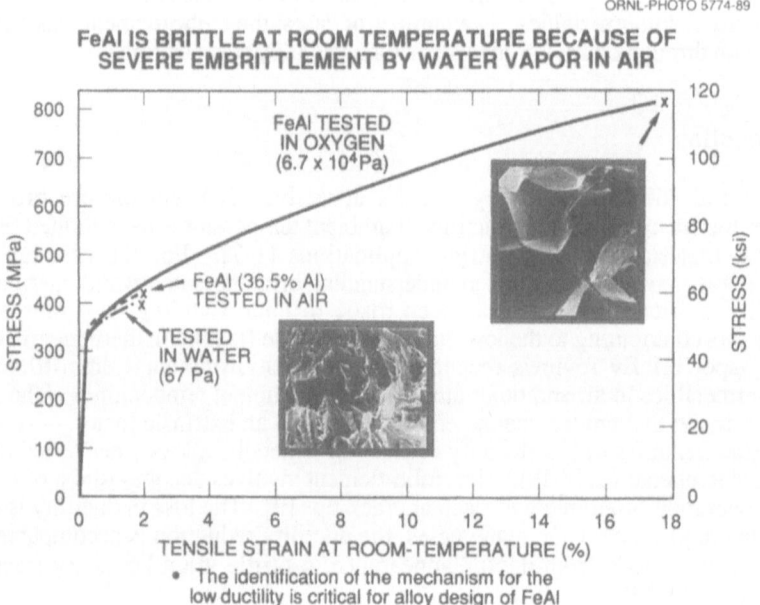

Figure 1. Tensile stress-strain curves comparing the effects of different environments on FeAl fracture behavior.

TABLE 1. Effect of test environment on room-temperature tensile properties[a] of FeAl (36.5% Al) [9,10]

Test environment (gas pressure)	Elongation (%)	Yield strength (MPa)	Ultimate tensile strength (MPa)
Air	2.2	360	412
Vacuum ($< 1 \times 10^{-4}$ Pa)	5.4	352	501
Oxygen (6.7×10^4 Pa)	17.6	360	805
Ar + 4% H_2 (6.7×10^4 Pa)	6.2	379	579
H_2O Vapor (67 Pa)	2.4	368	430

[a]All specimens were annealed 1 h/900°C + 2 h/700°C in vacuum.

The mechanism that has been postulated [9] for this type of environmental embrittlement is similar to what is observed in aluminum and its alloys [21,22]. In general, aluminum alloys are not embrittled by dry hydrogen; sometimes, however, they can be severely embrittled by moist air. The chemical reaction that is thought to cause this embrittlement is shown below [21]:

$$2Al + 3H_2O \rightarrow Al_2O_3 + 6H. \tag{1}$$

It is believed that the high-fugacity atomic hydrogen, produced in this reaction between the aluminum atoms and H_2O in moist air, enters the metal at crack tips and causes hydrogen embrittlement. Since FeAl alloys contain relatively large concentrations of aluminum, it is postulated that a reaction similar to the one described above for aluminum alloys also takes place in FeAl [9]. Preliminary experiments using nuclear reaction analysis have shown that FeAl does, in fact, react with heavy water (presumably in a reaction analogous to the H_2O dissociation shown above), producing deuterium which is then absorbed into FeAl [23]. The underlying mechanism of environmental embrittlement in FeAl may be similar to hydrogen embrittlement in other ordered intermetallics like Co_3Ti, $Ni_3(Al_{0.4}Mn_{0.6})$, $(Fe,Ni)_3V$, and B-doped Ni_3Al [13-19], the principal difference being the manner in which atomic hydrogen is generated and absorbed in the different studies (e.g., reaction with moist air [9] versus cathodic hydrogen charging [17]). Consistent with this, Table 1 shows that the yield strength of FeAl is insensitive to test environment, a common observation also in other studies of hydrogen embrittlement [13-19]. The best ductility (17.6%) is obtained in a dry oxygen environment, indicating that the reaction of aluminum atoms with oxygen competes with the moisture reaction in Eq. 1, thereby reducing the generation of atomic hydrogen from the moisture in air. It is surprising that even the residual moisture in the vacuum test is enough to cause significant embrittlement compared with the oxygen test.

The environmental embrittlement in FeAl occurs along cleavage planes rather than grain boundaries, suggesting a possible reduction of the cohesive strength by absorbed hydrogen [9-12]. This is supported by recent first-principles total-energy calculations, which indicate that interstitial hydrogen draws electrons to itself and reduces the cleavage strength and energy of FeAl by as much as 20 to 70% (depending on the hydrogen concentration) [24]. Recent studies of the prestrain effect also support the decohesion mechanism. In this study, FeAl specimens containing 35 at. % Al were first strained in dry oxygen to different levels, followed by further straining to fracture in moist air. As

324

shown in Table 2, all the specimens fractured almost at the same fracture strength, independent of prestrains or total strains. Note that the fracture strength in air is lower than the maximum strength obtained by prestraining in dry oxygen. This result can be interpreted as follows: hydrogen released from moisture reduces atomic bonding and causes crack propagation at a relatively lower strength at crack tips.

For FeAl containing more than 38% Al, its low ductility is essentially caused by two mechanisms: (1) intrinsic grain-boundary weakness and (2) environmental embrittlement [11,12]. The first mechanism dominates the ductility and fracture when Fe-40% Al is tested at room temperature in air or in dry oxygen (see Table 3). To eliminate the intrinsic grain-boundary problem, boron at a level of 300 wt ppm was added to the 40% Al alloy. Boron tends to segregate strongly to grain boundaries and suppresses intergranular fracture in FeAl [11]. As shown in Table 3, the boron-doped 40% Al alloy tested in air showed mainly transgranular fracture rather than grain-boundary fracture, as was seen in the boron-free alloy. The increase in ductility resulting from boron doping is around 3%, which generally agrees with the data reported by Crimp et al. [25,26] and Gaydosh et al. [27]. The boron-doped alloy showed a distinctly higher ductility (16.8%) when tested in the oxygen environment. This result clearly demonstrates that boron-doped FeAl containing 40% Al, just like the lower aluminum alloys, exhibits environmental embrittlement in air, and that this embrittlement can be reduced by testing in a dry oxygen atmosphere.

Environmental embrittlement is also the major cause of low ductility in Fe_3Al [10] at ambient temperatures. Table 4 shows the tensile properties of Fe_3Al (28% Al) heat treated to produce either the DO_3 or B2 structure and then tested in various environments at room temperature. Both crystal structures showed essentially the same dependence on test environment. The air tests produced a ductility of about 4%. When tested in vacuum, the tensile elongation increased three-fold to about 12%. Similarly high values of ductility (~ 12%) were obtained in dry oxygen and somewhat lower values (~ 8%) in a mixture of Ar + 4% H_2. The water vapor tests, on the other hand, resulted in ductilities even lower than those obtained in the air tests, although the pressure in the water vapor tests was comparable to the vapor pressure of moisture in the air tests.

TABLE 2. Effect of prestrain in dry oxygen on room-temperature tensile properties of FA-317 (35% Al) tested in air

Pre-strain in oxygen			Follow-on test in air	
Yield strength (MPa)	Maximum stress (MPa)	Strain (%)	Fracture strength (MPa)	Total elongation[b] (%)
a	--	0	442	0 + 3.4 = 3.4
349	412	2.6	425	2.6 + 0.5 = 3.1
351	487	4.3	427	4.3 + 0 = 4.3
342	578	7.8	442	7.8 + 0 = 7.8

[a]For FA-317, yield strength = 347 MPa.
[b]Total elongation due to pre-strain and follow-on straining at room temperature.

TABLE 3. Effect of test environment on room-temperature tensile properties of FeAl (40 at. % Al) with and without boron [11]

Test environment	Strength (MPa)		Ductility (%)	Fracture mode[a]
	Yield	Ultimate		
No boron				
Air	390	405	1.2	GBF
Oxygen[b]	402	537	3.2	GBF
Doped with 300 wt ppm boron				
Air	391	577	4.3	TF
Oxygen[b]	392	923	16.8	TF

[a]GBF = grain-boundary fracture; TF = mainly transgranular fracture (cleavage).

[b]Oxygen pressure: 6.7×10^{-4} Pa.

TABLE 4. Room-temperature tensile properties of Fe_3Al (28 at. % Al) tested in various environments [10]

Test environment (gas pressure)	Elongation (%)	Yield strength (MPa)	Ultimate strength (MPa)
(A) B2 Structure Produced by Annealing 900°C/1 h + 700°C/2 h			
Air	4.1	387	559
Vacuum ($\sim 10^{-4}$ Pa)	12.8	387	851
Ar + 4% H_2 (6.7×10^4 Pa)	8.4	385	731
Oxygen (6.7×10^4 Pa)	12.0	392	867
H_2O vapor (1.3×10^3 Pa)	2.1	387	475
(B) DO$_3$ Structure Produced by Annealing 850°C/1 h + 500°C/5 d			
Air	3.7	279	514
Vacuum ($\sim 10^{-4}$ Pa)	12.4	316	813
Oxygen (6.7×10^4 Pa)	11.7	298	888
H_2O vapor[a]	2.1	322	439

[a]Air saturated with water vapor was leaked into the test chamber.

These results clearly indicate that Fe₃Al alloys are also susceptible to moisture-induced environmental embrittlement, just like the FeAl alloys. It is interesting to note that there is no difference between the ductilities obtained in the oxygen and vacuum tests (Table 4). This is quite different from the case of FeAl alloys where the best ductility was obtained in dry oxygen [9]. Since a dry oxygen environment is not needed to completely eliminate environmental embrittlement in Fe₃Al, it appears that the Fe₃Al alloys are less susceptible to moisture-induced embrittlement than are the FeAl alloys. Presumably, this is a result of their lower aluminum contents. The environmental effect has also been studied recently in several Fe₃Al and FeAl alloys tested in different atmospheres and at different strain rates [28].

In addition to the iron aluminides, β-brass (CuZn) also showed some degree of environmental embrittlement [29,30]. For instance, CuZn typically exhibits a 55% elongation to failure in air and about 6% elongation in distilled water. The embrittlement here is due to stress-corrosion cracking in distilled water.

3. Environmental Embrittlement of fcc-Type Ordered Intermetallics

A number of Ll₂-ordered intermetallic alloys also show environmental embrittlement in moist air at room temperature. Izumi and his group at Tohoku University have done extensive work on characterization of deformation and fracture in Co₃Ti and its alloys [13,14,18,31]. They observed environmental embrittlement in Co₃Ti both with and without boron additions (Fig. 2) [13,31]. The alloy specimens with and without hydrogen charging showed a substantial reduction in room-temperature ductility when tested in air. Co₃Ti alloys also exhibited severe embrittlement when tested in hydrogen-gas environments [13,15,18]. With decreasing ductility, their fracture mode changed

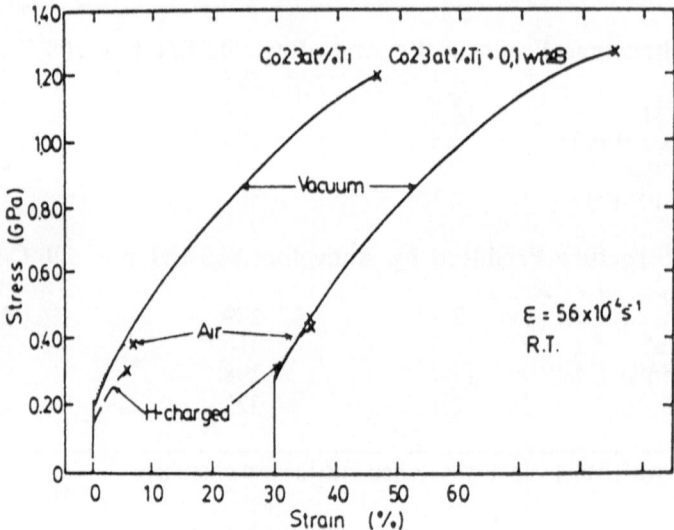

Figure 2. Effect of test environment on room-temperature stress-strain curves of Co₃Ti (23 at. % Ti) with and without 0.2 wt % B [13].

from transgranular to intergranular. The embrittlement was attributed to hydrogen, which promotes brittle intergranular fracture. The environment embrittlement is not affected by boron doping but is influenced by alloy stoichiometry of Co_3Ti. The alloy is much less susceptible to environmental embrittlement when titanium is reduced from 23 to 21 at. % [13].

Similar to the Co_3Ti alloys, a $(Co,Fe)_3V$ alloy has been found recently to be susceptible to environmental embrittlement at room temperature [32-33]. As indicated in Fig. 3, $(Co_{78}Fe_{22})_3V$ showed ductile transgranular fracture, with a tensile ductility of 35.8% in vacuum. The room-temperature ductility reduced to 20% in air and 15.3% in distilled water. The reduction in ductilities is accompanied by a change in fracture mode from ductile transgranular to mixed transgranular and intergranular fracture. The grain-boundary fracture was observed mainly at corners of the fracture surface, indicating that moisture-induced hydrogen diffuses from surface to interior, mainly through the grain boundaries, and causes intergranular fracture. Note that the yield strength and work hardening behavior are insensitive to test environment. The environmental embrittlement in $(Co,Fe)_3V$ was completely eliminated by increasing strain rate from 3.3×10^{-5} to $3.3 \times 10^{-1}s^{-1}$ [33].

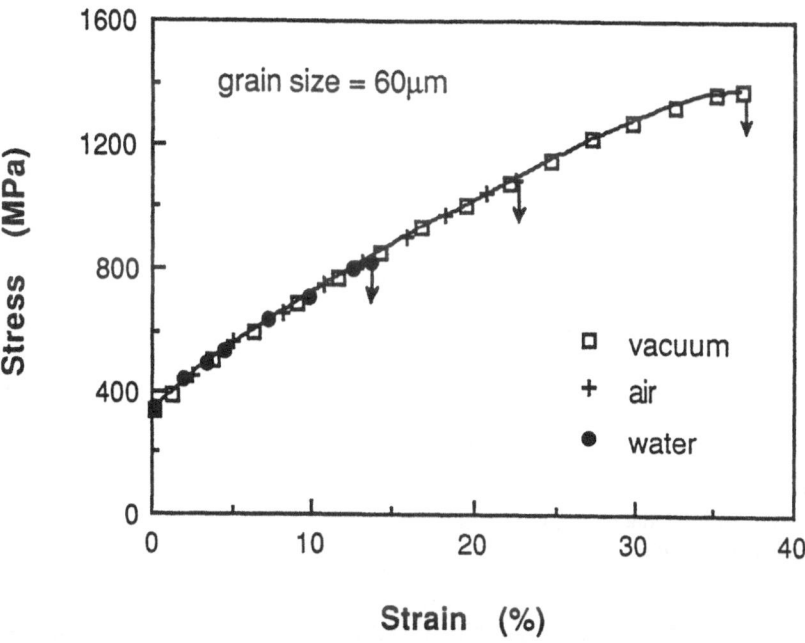

Figure 3. Effect of test environment on stress-strain curves of $(Co, Fe_3)V$ tested at room temperature [33].

Recently, Takasugi et al. [34,35] and Liu et al. [36] have studied the effect of test environment on room-temperature tensile properties of $Ni_3(Si,Ti)$ and Ni_3Si, respectively. Ni_3Si showed no appreciable plastic deformation when tested in air but an elongation of 7.5% when tested in dry oxygen (Table 5). This result demonstrates that Ni_3Si is prone to environmental embrittlement at ambient temperatures. Since the elimination of the environmental effect by testing in dry oxygen does not lead to extensive ductility (e.g., 30% or above) and complete suppression of intergranular fracture in the silicide, the environmental effect appears not to be the sole source of grain-boundary brittleness in Ni_3Si. This is further indicated by comparison of the tensile data of Ni_3Si and $Ni_3(Si,Ti)$. As shown in Table 5, $Ni_3(Si,Ti)$, like Ni_3Si, is also susceptible to environmental embrittlement. The $Ni_3(Si,Ti)$ alloy, however, exhibited a considerable ductility (7%) when tested in moist air at room temperature. It exhibited excellent ductility (29%) when tested in vacuum, indicating that the titanium addition enhances intrinsic grain-boundary properties (such as grain-boundary cohesion) of Ni_3Si, as suggested by Takasugi and Izumi (37,38). The effect of boron additions on environmental embrittlement will be discussed in the next section.

TABLE 5. Effect of Test Environment on Room-Temperature Tensile Properties of Ni_3Si (22.5 at. % Si) and $Ni_3(Si,Ti)$ Alloys[a]

Alloy	Test environment	Tensile ductility (%)	Yield strength, MPa (ksi)	Ultimate strength, MPa (ksi)
Ni_3Si	Air	~ 0[b]	b	627 (91.0)
Ni_3Si	Vacuum	4.7	677 (98.3)	853 (123.8)
Ni_3Si	Oxygen	7.5	685 (99.4)	1040 (151.0)
$Ni_3(Si,Ti)$	Air	7	606 (88)	813 (118)
$Ni_3(Si,Ti)$	Vacuum	29	586 (85)	1323 (192)
B-doped $Ni_3(Si,Ti)$[c]	Air	36	599 (87)	1323 (192)
B-doped $Ni_3(Si,Ti)$[c]	Vacuum	34	599 (87)	1323 (192)

aAlloys containing 9.5 at. % Ti [34].
bFracture prior to macroscopic yielding.
cDoped with 50 ppm boron [34].

In addition to Co_3Ti, $(Co,Fe)_3V$, Ni_3Si, and $Ni_3(Si,Ti)$, the environmental effect was observed in other $L1_2$-ordered intermetallics, such as $Ni_3(Al,Mn)$ [14] and Be-doped Ni_3Al [39]. The titanium aluminide, TiAl, has the $L1_0$-ordered structure, which is also a derivative of the fcc structure. Recently, Nakamura [40] has found that the bend ductility of TiAl (50 at. % Al) is highly susceptible to environmental embrittlement at room temperature as shown in Fig. 4. The aluminide fractured almost immediately after macroscopic yielding when bend testing in air. It showed, on the other hand, a significant plastic strain when testing in vacuum. The bend strain fell between the two limits when tested in hydrogen atmosphere. All these results are essentially consistent with the moisture-induced environmental embrittlement observed in FeAl.

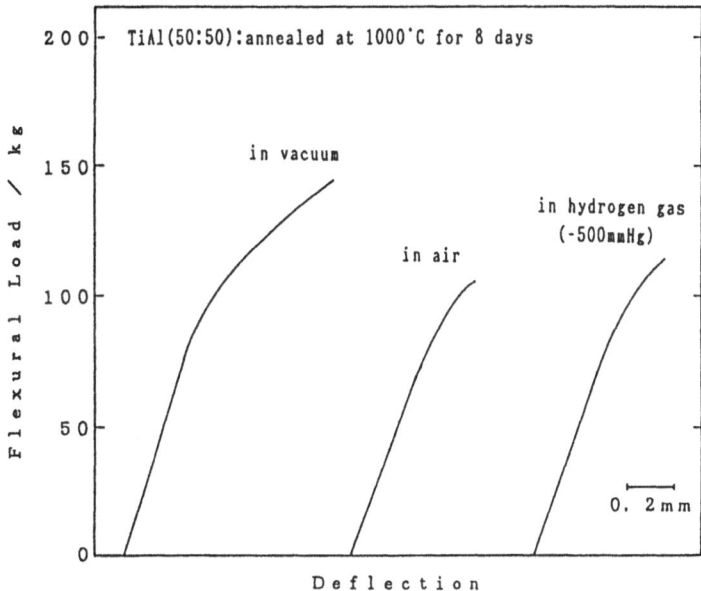

Figure 4. Effect of test environment on load-deflection curves of TiAl (50% Al) bend tested at room temperature [40].

4. Alloy design of ordered intermetallics

As shown in the foregoing sections, environmental embrittlement is a major cause of the brittle fracture and low ductility of many ordered intermetallics at ambient temperatures. The environmental problem has to be solved satisfactorily in order to use intermetallic alloys as engineering materials. There are possible ways to alleviate this problem:

(1)   to control surface composition to reduce generation of hydrogen from the moisture/intermetallic reaction [Eq. (1)],
(2)   to reduce bulk and grain-boundary diffusion of hydrogen by alloying additions,
(3)   to reduce the solubility limit of hydrogen in the material,
(4)   to refine microstructure (such as grain size) to reduce stress concentration, and
(5)   to lower yield strength.

All these are metallurgical means possible to reduce generation of hydrogen at alloy surfaces, to slow down penetration of hydrogen into the material, and to reduce stress concentration at microcrack tips.

Recently, considerable efforts have been devoted to alleviating environmental embrittlement in intermetallic alloys. The most interesting one is to solve the environmental embrittlement in Co_3Ti by alloying. In this study, Izumi and Takasugi [15,18] added individually a variety of alloying elements to Co_3Ti containing 23 at. % Ti. As indicated in Fig. 5, the elements Y, Ta, Cr, Mo, W, and Ge at a level of 3 at. % appear not to significantly affect environmental embrittlement, whereas the elements Fe and Al completely suppress the embrittlement. This result has demonstrated that the environmental embrittlement in intermetallics can be effectively reduced or completely

330

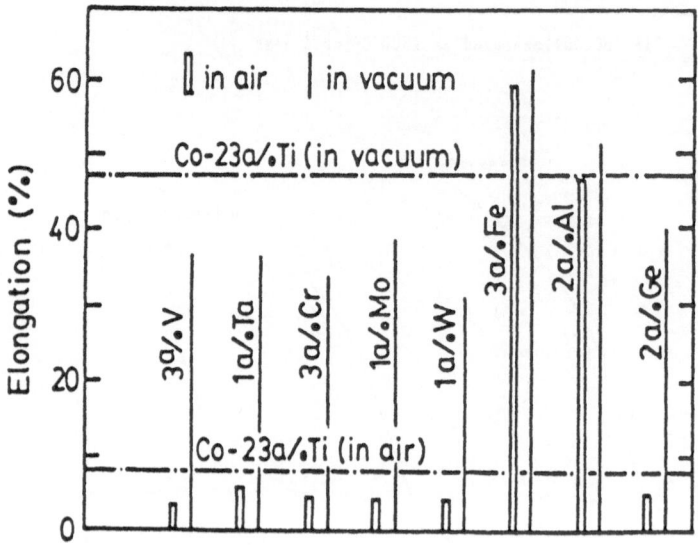

Figure 5. Effect of alloy addition on room-temperature tensile elongation of Co₃Ti (23 at. % Ti) tested in air and vacuum [15,18].

eliminated by alloy additions. The beneficial effect of iron has been attributed to creation of a more homogeneous electronic distribution at grain boundaries where some of the Co-Ti bonds are replaced by Co-Fe bonds. This suggestion appears to be reasonable; however, it is difficult to explain why aluminum is as effective as iron in reducing the embrittlement.

Boron has been found to be effective in reducing environmental embrittlement of grain boundaries in certain Ll₂ intermetallics. As shown in Table 5, undoped Ni₃Si and Ni₃(Si,Ti) are prone to environmental embrittlement, whereas boron-doped Ni₃(Si,Ti) is insensitive to test environment at room temperature [34,35]. As a matter of fact, the boron-doped Ni₃(Si,Ti) showed excellent ductility (34 to 36%) when tested in both air and vacuum. This result clearly indicates that boron is very effective in alleviating environmental embrittlement in Ni₃(Si,Ti). Carbon-doped Ni₃(Si,Ti) also exhibited an excellent ductility, independent of test environment at room temperature. Boron and carbon are known to segregate strongly to grain boundaries in Ni₃(Si,Ti), and their beneficial effect may come from reducing diffusion through reduction in site occupation by hydrogen at the boundaries [35].

Ni₃(Al,Mn) [14] and beryllium-doped Ni₃Al [39] showed environmental embrittlement in air, whereas boron-doped Ni₃Al is not sensitive to test environment and strain rate (Fig. 6) [41]. The difference between beryllium and boron doping is that beryllium occupies substitutional sites while boron takes interstitial sites in the aluminide. Also, beryllium is not known to segregate to Ni₃Al grain boundaries. Boron, on the other hand, does not alleviate environmental embrittlement in Co₃Ti, as reported by Takasuki and Izumi [13]. In this case, it is possible that boron does not segregate to grain boundaries in Co₃Ti. Auger study is required to verify this point.

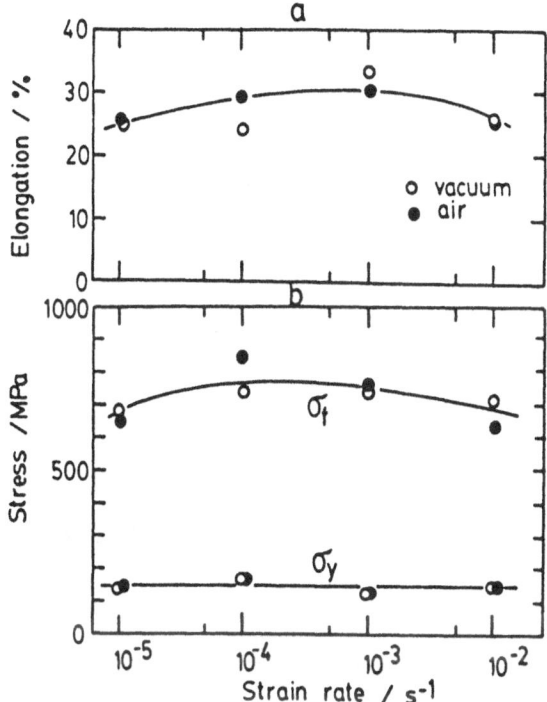

Figure 6. Effect of strain rate on tensile properties of boron-doped Ni₃Al (24 at. % Al) tested at room temperature in air and vacuum [41].

Fe₃Al and FeAl exhibited severe embrittlement when tested in moist air at ambient temperatures [9-12,28,42,43]. Recent effort on alloy design of Fe₃Al by McKamey et al. [42,44,45] showed that the ductility of the aluminide can be substantially improved by increasing aluminum content from 25 to 28% and by adding chromium at a level of 2 to 6%. The increase in aluminum concentration sharply decreases the yield strength of the aluminide. The chromium additions double the tensile ductility when there are oxide scales formed on specimen surfaces during hot rolling or subsequently heat treatment in air [46]. As the oxide scales are removed by electropolishing, there is no apparent difference in tensile ductilities of Fe₃Al with and without the chromium additions. Refinement of grain structure and control of degree of recrystallization have been proven to be effective in improving the room-temperature ductility of Fe₃Al and FeAl alloys tested in moist air [27,42,43,47,48]. Preoxidation and formation of protective oxide scales on surface also reduce environmental embrittlement of iron and nickel aluminides in air or hydrogen environments [17,47].

5. Summary and Concluding Remarks

Recent studies have demonstrated that environmental embrittlement is a major cause of low ductility and brittle fracture in many ordered intermetallics. Surprisingly, the embrittlement occurs when intermetallic alloys are tensile tested in air at conventional strain rates at ambient temperatures. It involves the reaction of reactive elements (such as Al, Si, Ti, V, etc.) in intermetallics with moisture in air, and the generation of atomic hydrogen. The embrittlement is a dynamic process involving continuous hydrogen generation and crack propagation during tensile tests.

Both bcc- and fcc-ordered intermetallics show severe environmental embrittlement at ambient temperatures. In the case of iron aluminides with bcc-ordered crystal structures, the embrittlement occurs along cleavage planes, resulting in brittle cleavage fracture. The brittle cleavage was suppressed when FeAl alloys were tested in dry oxygen instead of moist environments. In the case of Ll_2 intermetallics, the loss in ductility is accompanied by a change in fracture mode from ductile appearance to brittle grain-boundary separation. The fcc-ordered intermetallics prone to environmental embrittlement include Co_3Ti, $(Co,Fe)_3V$, Ni_3Si, $Ni_3(Si,Ti)$, $Ni_3(Al,Mn)$, Be-doped Ni_3Al, and TiAl.

Environmental embrittlement as discussed here is similar to traditional hydrogen embrittlement [49], except that, in this case, hydrogen is released from moisture. The underlying mechanisms suggested for hydrogen embrittlement in ordered intermetallics can be grouped into three categories: (1) reduction of atomic bonding across cleavage planes, (2) reduction of cohesive strength across grain boundaries, and (3) influence of dislocation activities and crack-tip plasticity. Environmental embrittlement in the bcc-ordered iron aluminides occurs along cleavage planes rather than grain boundaries, suggesting the reduction of cleavage strength by absorbed hydrogen [9-12]. This is supported recently by first-principles total-energy calculations, which indicate that absorbed hydrogen significantly reduces the cleavage strength and energy of FeAl (by as much as 20 to 70%, depending on the hydrogen concentration) [24]. Superdislocations have been suggested to be the carrier for enhanced diffusion of hydrogen to crack tips [50]. In the case of fcc-ordered intermetallics [13-19,34,35] such as $Ni_3(Al_{0.4}Mn_{0.6})$, Co_3Ti, $(Fe,Ni)_3V$, $(Co,Fe)_3V$, and $Ni_3(Si,Ti)$ alloys, hydrogen embrittlement takes place mainly along grain boundaries, suggesting the reduction of cohesive strength along the boundaries. Fast diffusion [51,52] and segregation [53] of hydrogen to grain boundaries have been detected experimentally in conjunction with hydrogen embrittlement in conventional metals and alloys, but not yet in ordered intermetallics. In situ transmission electron microscope (TEM) studies of crack-tip plasticity indicate that hydrogen causes the dislocation velocity and crack growth rate to increase significantly in Ni_3Al and Fe_3Al alloys [54,55]. This observation basically supports the mechanism of decohesion in the presence of hydrogen at crack tips.

Limited effort has been devoted to alleviating environmental embrittlement in ordered intermetallics. Izumi and Takasugi [15] found that the embrittlement in Co_3Ti can be completely eliminated by alloying with 3% Fe or Al. Boron is effective in reducing environmental embrittlement of grain boundaries in Ni_3Si and $Ni_3(Si,Ti)$, but not in Co_3Ti. The beneficial effect of boron is believed to be due to the fact that boron tends to segregate to grain boundaries and thus reduces the site occupation by hydrogen at the boundaries. In other words, the segregation of boron may slow down the penetration of hydrogen along the boundaries. The environmental embrittlement in iron aluminides was reduced by control of alloy composition, surface condition, and microstructure [42-48]. A tensile ductility as high as 20% has been achieved for Fe_3Al alloys tested in moist air at room temperature [42,43,48].

It should be noted that the environment affects not only ductility, but also alloy preparation and fabrication. For instance, iron aluminides showed hair-line cracks on cutting surfaces when water was used as a coolant during cutting processes [54]. The cracking is caused by hydrogen generated from the reaction of iron aluminides at the cutting edge with water coolant. It can be eliminated by either control of cutting parameters or change of coolant (e.g. use of an oil coolant). Iron aluminides also absorb moisture from environments, resulting in the formation of porosities (like the formation of pin holes in aluminum alloys) during melting and casting. The porosity density can be dramatically reduced either by casting in an inert-gas environment or by adding alloying additions [47].

6. Acknowledgments

The author is grateful to E. P. George and J. A. Horton for reviewing the manuscript. Thanks are also due to Fay Christie, Connie Dowker and Shirin Badlani for manuscript preparation. Research sponsored jointly by the Division of Materials Sciences, U.S. Department of Energy, and Assistant Secretary for Conservation and Renewable Energy, Office of Industrial Technologies, Advanced Industrial Concepts Division, Advanced Industrial Concepts Materials Program, under contract DE-AC05-84OR21400 with Martin Marietta Energy Systems, Inc.

7. References

1. Liu, C. T. and Stiegler, J. O. (1984) *Science*, **226**, 636.
2. High-Temperature Ordered Intermetallic Alloys, Proc. Mat. Res. Soc. Symp., ed. C. C. Koch, C. T. Liu, and N. S. Stoloff, p. 39 in Materials Research Society, Pittsburgh, PA (1985).
3. High-Temperature Ordered Intermetallic Alloys II, Proc. Mat. Res. Soc. Symp., ed. N. S. Stoloff, C. C. Koch, C. T. Liu, and O. Izumi, Materials Research Society, Pittsburgh, PA (1987) 81.
4. High-Temperature Ordered Intermetallic Alloys III, Proc. Mat. Res. Soc. Symp., ed. C. T. Liu, A. I. Taub, N. S. Stoloff, and C. C. Koch, Materials Research Society, Pittsburgh, PA (1989), 133.
5. High-Temperature Aluminides and Intermetallics, Proc. TMS/ASM Symp., ed. S. H. Whang, C. T. Liu, D. P. Pope, and J. O. Stiegler, TMS Publication, Warrendale, PA (1990).
6. High-Temperature Ordered Intermetallic Alloys IV, Proc. Mat. Res. Soc. Symp., ed. L. Johnson, D. P. Pope, and J. O. Stiegler, Materials Research Society, Pittsburgh, PA (1991) 213.
7. C. T. Liu, J. O. Stiegler, and F. H. Froes: Ordered Intermetallics, pp. 913-42 in ASM Handbook, Vol. 2, 10th Edition, ASM International Publication, Metals Park, Ohio (1990).
8. Liu, C. T. and McKamey, C. G., pp. 133-51 in ref. 5.
9. Liu, C. T., Lee, E. H., and McKamey, C. G. (1989) *Scr. Metall.* **23**, 875.
10. Liu, C. T., McKamey, C. G., and Lee, E. H. (1990) *Scr. Metall.* **24**, 385-90.
11. Liu, C. T. and George, E. P. (1990) *Scr. Metall.* **24**, 1285-90.
12. Liu, C. T. and George, E. P. (1991) pp. 527-32 in ref. 6.
13. Takasugi, T. and Izumi:, O. (1986) *Acta Metall.* **34**, 607.
14. Masahashi, N., Takasugi, T., and Izumi, O. (1988) *Metall. Trans.* **19A**, 353.
15. Izumi, O.and Takasugi, T. (1988) *J. Mater. Res.* **3**, 426.
16. Kuruvilla, A. K., Ashok, S., and Stoloff, N. S. (1982) p. 629 in *Proc. Third Intl. Congress on Hydrogen in Metals*, Pergamon, Paris, Vol. 2.
17. Kuruvilla, A. K. and Stoloff, N. S. (1985) *Scr. Metall.* **19**, 83.
18. Liu, Y., Takasugi, T., Izumi, O., and Yamada, T. (1989) *Acta Metall.* **37**, 507-17.
19. Stoloff, N. S. (1988) *J. Met.* **40**, 18.
20. Liu, C. T., Fu, C. L., George, E. P., and Painter, G. S. (1991) *International Journal of Iron and Steel Institute, Japan (ISIJ)*, Special Issue on Advanced High-Temperature Intermetallics, October 1991, ISIJ.

334

21. Speidel, M. P. (1977) p. 329 in *Hydrogen Damage*, C. D. Beachem (ed.), American Society for Metals, Metals Park, Ohio.
22. Gest, R. J. and Troiano, A. R. (1974) *Corros.* **30**(8), 274.
23. George, E. P., Lewis, M. B., and Liu, C. T. (1989) unpublished research, Oak Ridge Natl. Lab., Oak Ridge, Tenn.
24. Fu, C. L. and Painter, G. S. (1991) *J. Mater. Res.* **6**, 719-723.
25. Crimp, M. A., Vedula, K. M., and Gaydosh (1987) pp. 499-504 in ref. 3.
26. Crimp, M. A. and Vedula, K. M.(1986) *Mater. Sci. Eng.* **78**, 193.
27. Gaydosh, D. J. and Nathal, M. V. (1990) *Scr. Metall.* **24**, 1281-84.
28. Shea, M., Castagna, A., and Stoloff, N. S. (1991) submitted to *J. Mat. Res.*
29. Kramer, I. R., Wu, B., and Feng, C. R. (1986) *Mater. Sci. and Eng.* **82**, 141.
30. Kasul, D. B., White, C. L., and Heldt, L. A. (1990) in R. Gangloff (ed.), *Proc. Int. Conf. Environment-Induced Cracking of Metals*, Kohler, Wisconsin, National Association of Corrosion Engineers, Houston [in press].
31. Takasugi, T. and Izumi, O. (1985) *Scr. Metall.* **19**, 903-07.
32. Nishimura, C. and Liu, C. T. (1991) *Scr. Metall.* **25**, 791-94.
33. Nishimura, C. and Liu, C. T. (1991) submitted to *Acta Metall.*
34. Takasugi, T., Suenaga, H., Izumi, O. (1991) *J. Mater. Sci.* **25** [in press].
35. Takasugi, T. and Izumi, O. pp. 403-416 in ref. 6.
36. Liu, C. T. and Oliver, W. C. (1991) *Scripta Metall.* (in print).
37. Takasugi, T. and Izumi, O. (1985) *Acta Metall.* **33**, 1247-58.
38. Takasugi, T. and Izumi, O., and Masashashi, N. (1985) *Acta Metall.* **33**, 1259.
39. Takasugi, T., Masahashi, N., and Izumi, O. (1986) *Scr. Metall.* **20**, 1317-21.
40. Nakamura, M., National Research Institute for Metals, Tokyo, Japan, private communication, March 1991.
41. Masahashi, N., Takasugi, T., and Izumi, O. (1988) *Acta Metall.* **36**, 1823-36.
42. McKamey, C. G., Devan, J. H., Tortorelli, P. F., and Sikka, V. K. (1991) *J. Mat. Res.* (in print).
43. McKamey, C. G. and Liu, C. T. (1991) *Proc. Environmental Effects on Advanced Materials*, National Association of Corrosion Engineers, Houston.
44. McKamey, C. G., Horton, J. A., and Liu, C. T. (1988) *Scr. Metall.* **22**, 1679.
45. McKamey, C. G., Horton, J. A., and Liu, C. T. (1989) *J. Mater. Res.* **4**, 1156-63.
46. McKamey, C. G. and Liu, C. T. (1990) *Scr. Metall.* **24**, 219-22.
47. Liu, C. T. (1990) unpublished results, Oak Ridge Natl. Lab., Oak Ridge, Tenn.
48. Sanders, P. G., Sikka, V. K., Howell, C. R., and Baldwin, R. H. (1991) *Scr. Metall.* [in print].
49. Hirth, J. P. (1980) *Metall. Trans. A*, **11A**, 861.
50. Camus, G. M., Stoloff, N. S., and Duquette, D. J. (1989) *Acta Metall.* **37**, 1497-501.
51. Yao, J. and Cahoon, J. R. (1991) *Acta Metall.* **39**, 111-18.
52. Palumbo, G. et al. (1991) *Scr. Metall.* **25**, 679.
53. Birnbaum, H., Sirois, E., and Ladna, B. (October 1988) *Hydrogen Segregation to Grain Boundaries and External Surfaces*, USN 00014-83-K-0468, Office of Naval Research, Arlington, Va.
54. Bond, G. M., Robertson, I. M., and Birnbaum, H. K. (1989) *Acta Metall.* **37**, 1407-13.
55. Robertson, I. M., University of Illinois, Urbana, Illinois, private communication, April 19, 1991.
56. Ferguson, P. and Liu, C. T. (1991) submitted to *Scripta Metall.*

ATOMIC STRUCTURE AND CHEMICAL COMPOSITION OF GRAIN BOUNDARIES IN L1₂ INTERMETALLIC COMPOUNDS: RELATION TO INTERGRANULAR BRITTLENESS

MIN YAN and V. VITEK
Department of Materials Science and Engineering, University of Pennsylvania, Philadelphia, PA 19104-6272, U.S.A.

G. J. ACKLAND
Department of Physics, University of Edinburgh, Mayfield Road, Edinburgh EH9 3JZ, U. K.

ABSTRACT. Intergranular fracture in L1₂ compounds such as Ni₃Al poses two basic questions. First, whether the grain boundary brittleness is intrinsic and if yes what is its origin and why it is not found L1₂ alloys such as Cu₃Au. Second, how and why is the brittleness affected by alloying and deviations from stoichiometry. These problems are discussed here in the light of the results of the Monte Carlo atomistic studies of the structure and composition of grain boundaries. For stoichiometric Ni₃Al virtually no compositional disorder occurs even at very high temperatures, while in Cu₃Au a significant disordering takes place already at room temperature. As suggested earlier [27, 30, 31], preservation of the compositional order in grain boundaries may be the principal reason for their intrinsic brittleness and, therefore, these results may explain why grain boundaries are brittle in Ni₃Al but not in Cu₃Au. In non-stoichiometric Ni₃Al the compositions of grain boundaries are very different in nickel rich and aluminum rich alloys, respectively. Whereas both components segregate to the boundaries when in surplus, nickel invokes disordering while aluminum does not. This may be the reason why the intergranular brittleness can be alleviated by additional alloying, for example by boron, in nickel rich alloys only.

1.   Introduction

1.1.   INTRINSIC GRAIN BOUNDARY BRITTLENESS

Intergranular brittle fracture is a common failure mode in both disordered alloys and intermetallic compounds.   In the former case this fracture mode is always associated with segregation of embrittling impurities to grain boundaries (see e.g. [1,2]).   In contrast, intermetallic alloys of high purity fracture intergranularly. This implies an intrinsic nature of grain

C. T. Liu et al. (eds.), Ordered Intermetallics – Physical Metallurgy and Mechanical Behaviour, 335–353.
© 1992 *Kluwer Academic*

boundary brittleness in these materials (for recent reviews see [3]). Whereas grain boundary brittleness has been observed, for example, in NiAl ([4,5]) which crystalizes in a b.c.c. based B2 structure, this phenomenon is particularly intriguing in the case of f.c.c. based $L1_2$ structure, for alloys, such as Ni_3Al, Ni_3Ga, Ni_3Ge etc., which are quite ductile in the single crystalline form (see e.g. review [6]). From these materials Ni_3Al has been studied most extensively and a number of authors concluded that grain boundaries in this compound are intrinsically brittle [3, 7-14]. At the same time this is not a general property of $L1_2$ ordered alloys. For example, Cu_3Au, which crystalizes in the same structure, does not exhibit intergranular brittleness. Hence, the challenging question is the explanation of the physical nature of the inherent boundary brittleness in some $L1_2$ alloys and lack of it in others.

A natural approach to look for the source of the intrinsic brittleness, and also the alloying effects discussed below, is to consider the cohesive properties of grain boundaries. These are governed by the local electronic structure [8, 10, 14-17] and a significant correlation between the grain boundary brittleness and the valency and/or Pauling electronegativity difference of the A and B elements forming the A_3B compounds has, indeed, been established [14, 16, 17]. However, extrapolations of the results of the quantum mechanical cluster calculations [18, 19] to metallic solids, which may provide a physical justification for such correlations, are questionable [20, 21]. Moreover, calculations performed using the local volume potentials [22], analogous to the embedded atom method [23], indicate that the ideal work to fracture (equal to $2\gamma_s - \gamma_b$, where $2\gamma_s$ is the energy of the surfaces formed during cracking and γ_b the grain boundary energy) is not significantly different in Ni_3Al and pure Ni [24, 25]. On the other hand, computer simulations of the application of the tensile stress to bicrystals of Ni_3Al showed that breaking did occur in the boundary region rather than in the bulk [26].

An alternative though related explanation of the intrinsic brittleness of grain boundaries in alloys like Ni_3Al emerges from the results of the calculations of the atomic structures of grain boundaries. These calculations have been made using pair-potentials representing a stable $L1_2$ structure [27], local volume potentials [22, 24-26] and N-body potentials of the Finnis-Sinclair type [28, 29]. (The latter are described in more detail in the following section). The most interesting result, common to all these calculations, follows from the comparison between structures of crystallographically identical boundaries in $L1_2$ alloys with high and low ordering energies [30, 31]. In the case of the low ordering energy (Cu_3Au) the relaxation in the boundary region is substantial and the atoms in the boundary region cannot be uniquely assigned to either the upper or the lower grain. On the other hand, in the case of a high ordering energy (Ni_3Al) all the atoms can be regarded as uniquely attached to either the upper or the lower grain so that the ideal $L1_2$ structure is practically undisturbed on either side of the boundary up to the boundary plane. This is illustrated in Figs. 1a and b which show the structures of the $\Sigma = 29$ (520)/[001] symmetrical tilt boundaries in Cu_3Au and Ni_3Al, respectively. The features described above are clearly visible when following the distributions of atoms within the repeat cells of these boundaries, marked by the solid lines in these figures.

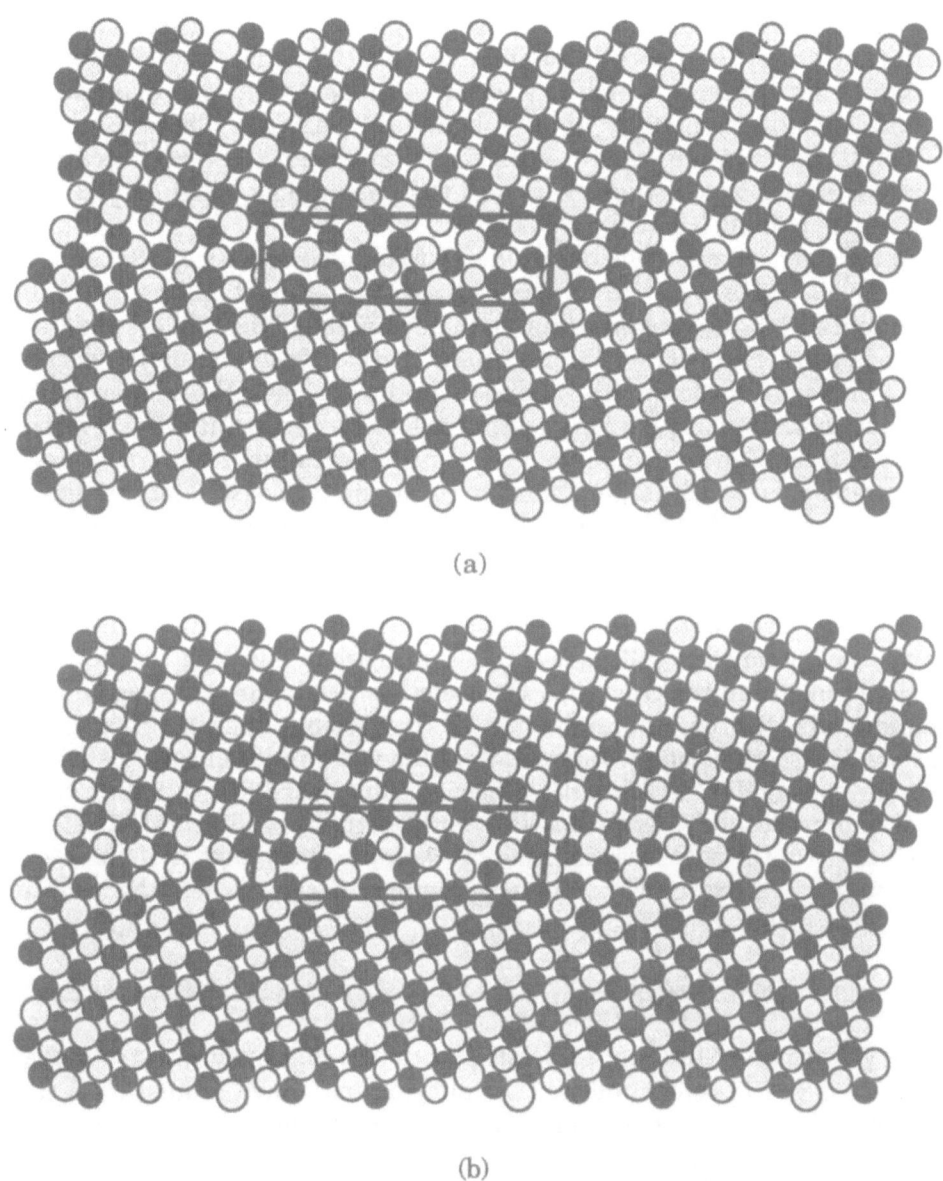

(a)

(b)

Figure 1. Calculated structures of the $\Sigma = 29$ (520)/[001] symmetrical tilt boundaries in Cu_3Au (a) and Ni_3Al (b) projected onto the (001) plane. In these and the following figures atoms are depicted as shaded circles of two different sizes. The larger circles correspond to Au or Al and smaller circles to Cu or Ni, respectively. The shading distinguishes two different (002) layers in the [001] period.

The reason for these structural differences is obviously related to the differences in ordering energies. In the case of a strongly ordered alloy (Ni_3Al), the chemical order is the principal factor controlling the energy of the system. Its preservation dominates the grain boundary structure. On the other hand, in a weakly ordered alloy (Cu_3Au), more relaxed grain boundary structures, with a possible chemical disorder, are energetically favored over those in which the order is preserved at the expense of a significant inhomogeneity in the boundary region. The recent high resolution electron microscopy (HREM) observations of several short period boundaries in Ni_3Al, indeed, reveal that an almost ideal $L1_2$ order is preserved up to the boundary [32-36].

In strongly ordered compounds distinct columnar cavities occur in grain boundaries owing to the mismatch of the almost undistorted but misoriented $L1_2$ lattices meeting at the boundary plane. These may then serve as suitable nuclei for intergranular cracks. Such a situation does not arise in weakly ordered $L1_2$ alloys or pure f.c.c. metals. Moreover, and perhaps more importantly, in metallic materials a substantial irreversible shear deformation accompanies even brittle cracking. The true work to fracture is then much larger than the ideal one (cohesive energy) since its major component is the energy dissipated by the local irreversible shear deformation at the crack tip [37, 38]. This work is strongly dependent on both the cohesion and the ability to produce such shears [37]. When the ideally ordered structure extends up to the boundary, high energy antiphase boundaries will have to be formed when emitting plastic shears (dislocations) from the crack tip. On the other hand, when the region of the boundary is disordered, production of irreversible shears will be much easier. Hence, in strongly ordered alloys the irreversible shearing at the tip of a grain boundary crack will be considerably more difficult than in the weakly ordered alloys and thus the work to fracture substantially lower. Furthermore, the yield stress in Ni_3Al is appreciably higher than in pure Ni and this favors cracking in the compound even when the cohesive energies are very similar [39].

1.2. ALLOYING EFFECTS

A very important finding associated with the intrinsic grain boundary brittleness of $L1_2$ compounds is that alloying may have dramatic effect on the ductility of these materials. The most pronounced is the influence of boron which segregates to the grain boundaries in an interstitial manner and may completely suppress the intergranular fracture in Ni_3Al and other nickel based alloys [3, 7-15, 40]. A very important related observation is that addition of boron inhibits intergranular fracture only in *non-stoichiometric, Ni rich,* alloys [7-14, 40, 41].

Similarly as in the case of the binary alloys, it can be speculated that boron directly affects the cohesion at grain boundaries. This view is supported by the fact that while boron segregates readily to the grain boundaries it does not segregate strongly to free surfaces [8, 42] which implies, following the arguments based on the equilibrium thermodynamics [43], that it should increase the boundary cohesion. The possible strengthening of bonding across grain boundaries has also been indicated in recent first principle calculations [44], studies using a tight-binding method [45, 46] as well as in calculations employing empirical local volume potentials [25].

Notwithstanding, the above rationalization does not explain why the beneficial influence of boron on intergranular cohesion is limited to Ni rich Ni_3Al alloys while it has no effect in stoichiometric or aluminum rich alloys. An obvious suggestion is then that a surplus of Ni in the boundary region is an essential precursor for the decrease of the grain boundary brittleness and an important role of boron might be to attract additional Ni via co-segregation. If this is the case chemical disordering will occur in the grain boundary region which may then affect both the cohesive strength and plasticity in this region. Frost [47] was the first to suggest that disordering at the grain boundary may enhance plasticity by relieving the geometric constraints upon dislocation interactions with grain boundaries during the transmission of slip through them. This idea was further developed by King and Yoo [48, 49]. Moreover, it was inferred from the measurements of the Hall-Petch relation [50-57] that transfer of slip through the grain boundaries is, indeed, easier in boron doped, Ni rich, Ni_3Al. Schulson and co-workers proposed this as the main reason for the dramatic effect of boron on ductility improvement of this alloy [54-57]. On the other hand, Lee et al. [58] conclude from their in-situ observations of the slip transfer through grain boundaries that segregation of boron does not affect the slip transmission and that the slip transfer follows the same criteria as in f.c.c. structures. Similarly, recent atomistic studies of the interaction between dislocations and grain boundaries do not indicate any pronounced differences between pure f.c.c. metals and $L1_2$ alloys [59]. Nevertheless, the local disordering may play an important role in ductilizing the intergranularly brittle $L1_2$ compounds. However, the major impact may be on the ability to produce localized irreversible shears near the tips of intergranular cracks, as discussed above, rather than on slip transfer through the polycrystal.

Nevertheless, the extent of disordering at grain boundaries in Ni_3Al is at present a controversial subject. Recently Schulson and Baker [54-57] observed in a boron doped Ni-24%Al alloy disordered regions spreading up to 20nm away from the boundaries. In contrast, disordered grain boundary phase of such an extent has not been observed by most other authors although Mackenzie and Sass [60] reported disordered region up to 2nm in their HREM studies. On the other hand, in more recent observations Sass and co-workers [34, 35] report compositional order up to the boundary in most boundaries studied and a relatively narrow disordered region near some general high angle boundaries. A similar conclusion has been reached by Mills and co-workers [33, 36]. At the same time, combined HREM and chemical analysis [61], as well as FIM/atom probe analysis [62] of grain boundaries in boron doped Ni_3Al, suggest an enrichment of boundaries by nickel. Hence, while nickel segregation is probably an important factor in decreasing the brittleness of grain boundaries in Ni$_3$Al, the region of disorder is likely to exist only very near the boundary.

In this paper we present results of atomistic Monte Carlo calculations of the structure and composition of grain boundaries in Cu$_3$Au and Ni$_3$Al at finite temperatures. The goal of these calculations was to address the problems discussed above. They were carried out for a number of symmetrical and asymmetrical tilt boundaries as well as for several [001] twist boundaries. We only present here the results for the $\Sigma = 29$ (520) symmetrical tilt boundary with the [001] rotation axis but since no qualitative differences were found for different boundaries they exemplify the principal features common to all the boundaries studied.

This atomistic study consist of two parts. First are the calculations of the boundaries in stoichiometric L1$_2$ alloys. The aim is to investigate the possibility of grain boundary segregation and analyze the differences between strongly and weakly ordered alloys. The second are the calculations of the grain boundaries in both Ni rich and Al rich Ni$_3$Al. These calculations have been made for binary alloys only and thus the effect of boron has not been studied directly. However, understanding of the boundary structures in non-stoichiometric ordered alloys is essential for our comprehension of the boron effects. In particular, we attempt to clarify the difference between the off-stoichiometric alloys with the surplus of nickel and aluminum, respectively, since they appear to have a very different behavior when considering the ductilizing effect of boron. Some calculations of this type have been made in the past but only to Ni$_3$Al and a few short period boundaries [63, 64].

2. Method of calculation

2.1. TOTAL ENERGY OF THE SYSTEM AND INTERATOMIC FORCES

The energy of the systems studied and the corresponding interatomic forces have been described in terms of the empirical N-body potentials constructed within the Finnis-Sinclair scheme [65]. In this approach the energy associated with an atom **i** in a system of N atoms is written as

$$E_i = \sum_{j=1}^{N} V_{ij}(R_{ij}) - \sqrt{\sum_{j=1}^{N} \Phi_{ij}(R_{ij})}$$

where the first term is a sum of pair potential interactions and the second term the N-body attractive part of the cohesive energy. Both the potentials V and Φ are fitted empirically; the suffices ij of these potentials refer to the species of the atoms involved and for the binary systems they are denoted V_{AA}, V_{AB}, V_{BB}, Φ_{AA}, Φ_{AB}, and Φ_{BB}. Functions V_{AA}, V_{BB}, Φ_{AA}, and Φ_{BB} were identified with those for pure metals and the function Φ_{AB} was taken as a geometrical mean of Φ_{AA} and Φ_{BB}. which is consistent with its interpretation in terms of hopping integrals [66]. Therefore, only the pair potential V_{AB} needs to be fitted to alloy properties. Details of the empirical fitting and functional forms of these potentials can be found in Refs. [28, 29].

For the Cu-Au system the potential V_{AB} was fitted to reproduce the alloying energies

$$E^{alloy} = E^{dis} - c_{Cu}E^{Cu} - c_{Au}E^{Au}$$

in disordered substitutional alloys for various concentrations c_{Cu} and c_{Au}, tabulated in [67]; E^{Cu} and E^{Au} are energies per atom in pure Cu and Au, respectively, and E^{dis} is the energy per atom in the disordered alloy. In the fitting we assume that the alloying energy is independent of temperature below the melting point but take into account the relaxation of the lattice around the substitutional atom. Furthermore, the lattice parameter for the $L1_2$ ordered Cu_3Au alloy was fitted. The details of the corresponding fitting procedures are described in [28].

The alloying energies of $L1_2$ Cu_3Au and Au_3Cu alloys, as well as of the $L1_0$ and B2 CuAu alloys, were evaluated [28]. These calculations show that the $L1_2$ and $L1_0$ structures are favored over the disordered phase and in the case of CuAu over the B2 structure. Hence, the potentials reproduce the basic features of the phase diagram of this alloy system [67] without any further adjustment. The ordering energy of the $L1_2$ Cu_3Au alloy, deduced from these calculations, is 0.02eV which is close to the experimental value of 0.012eV [67]. For the disordered alloys the potentials also reproduce very well the concentration dependence of the lattice parameter [28] and for the ordered Cu_3Au the calculated energy of the $1/2<110>$ antiphase boundary

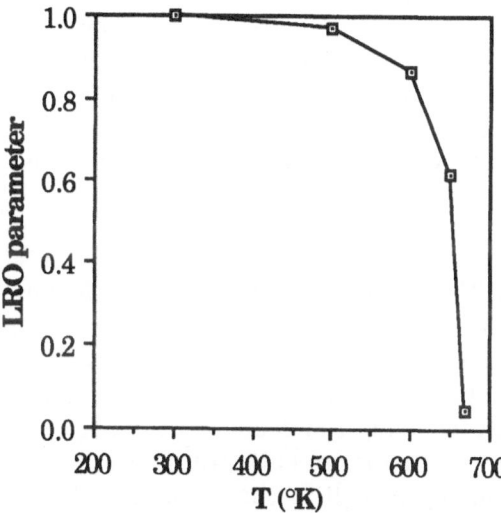

on {111} planes is 54mJm^{-2} which compares well with the available experimental estimates of 40-60 mJm^{-2} [68]. This energy is, of course, controlled by the ordering energy.

Using the same Monte Carlo technique as in the studies of grain boundaries we have also calculated the temperature dependence of the long range order parameter (LRO). This calculation was made at constant volume since the potentials have not been fitted to reproduce the thermal expansion and this

Fig. 2. Calculated temperature dependence of the long range order parameter in the Cu-Au system.

would be significantly overestimated in constant pressure calculations. The result is shown in Fig. 2. Both the shape of this curve and the calculated order-disorder temperature (~660°K) agree well with experimental data [69].

The potentials for pure Ni and Al were constructed using the same method as described in Ref. [70]. The physical quantities fitted are the cohesive energy, the equilibrium lattice parameter for the f.c.c. structure, the elastic constants and the vacancy formation energy; the potential for Ni differs somewhat from that given in [70] and does not reproduce the stacking fault energy. The potential for Al is of the same type as those for the noble metals and Ni and while it is appropriate in alloys containing a lower percentage of Al (e.g. Ni₃Al), it should not be used for pure aluminum since the electron density per atom is then much higher than in the alloys and the potential will have a different form [71]. Similarly, this potential will not be applicable for alloys with high concentration of aluminum.

In the case of Ni-Al system the potential V_{AB} was fitted such as to reproduce for $L1_2$ (Ni₃Al) and B2 (NiAl) structures the alloying energies (0.39eV and 1.68eV, respectively) and lattice parameters (3.567Å and 2.95Å, respectively) as well as, approximately, the elastic constants for the $L1_2$ structure. Furthermore, the energy of the 1/2<110> APB on {111} planes was fitted to 226mJm^{-2} which is close to the value estimated experimentally [72]. Finally, the stability of $L1_2$ and/or B2 structures relative to other possible ordered structures (DO_{19}, DO_{22}, A15, $L1_0$) as well as with respect to the disordered structure was tested using the methods described in [28]. Comparison of the energies of the ordered and disordered states gives the ordering energy for the $L1_2$ Ni₃Al about 0.1eV which is five times higher than for Cu₃Au. Unfortunately, no direct comparison with experiments can be made since this alloy remains ordered until melting. The temperature dependence of the LRO was calculated similarly as for Cu₃Au and the order-disorder temperature evaluated in this way is 1600°K which is, indeed, very close to the melting temperature of this compound (1658°K).

2.2. RELAXATION TECHNIQUES

The structure of a grain boundary at 0°K in an $L1_2$ alloy has been calculated using the similar technique as in a number of previous studies of boundaries in pure metals and alloys (e.g. [27, 73]). First a block consisting of the atomic coordinates of an unrelaxed bicrystal containing the chosen boundary is constructed in the computer using the geometrical rules of the coincidence site lattice (CSL) theory. At this stage both grains possess the stoichiometric $L1_2$ structure and the repeat cell in the boundary plane is that imposed by the CSL. For the Σ = 29 (520)/[001] symmetrical tilt boundary the unit cell imposed by the CSL is bounded by the vectors [$\bar{2}50$] and [001].

A relaxed structure is then found using a molecular statics method in which the total internal energy is minimized with respect to both the local atomic displacements and relative rigid body translations of the adjoining grains. During the relaxation the periodic boundary conditions are applied in the directions parallel to the boundary. In the direction normal to the boundary the block is divided into two regions. Region I contains the boundary plane and the atoms inside it are allowed to move relative to each other. Region II serves as the border in which atoms do not move relative to each other but are allowed to move as a slab in any direction when a relative translation of the grains occurs. This procedure allows for relative displacement of the grains parallel to the boundary plane and for expansion and/or contraction perpendicular to the boundary plane. Hence, the calculation is carried out at constant pressure.

The number of atoms in the region I was always chosen such that any further increase of this region did not lead to any significant change of the final result of the calculation. The relaxation was regarded as complete when the force acting on any atom of the relaxed block did not exceed 10^{-3} eV/Å. This is a somewhat arbitrary criterion but it was found in a number of previous calculations that a further decrease of the force did not lead to any significant change in the boundary structure and energy.

When calculating the structure and composition of grain boundaries at a non-zero temperature a Monte Carlo (MC) method was utilized. The starting configurations were always the corresponding relaxed 0°K stoichiometric structures obtained as described above. However, a new repeat cell is then chosen as a multiple of the CSL cell and the overall composition in the block set to be the desired bulk composition, with the corresponding excess atoms randomly distributed on the lattice sites in the block. For the $\Sigma = 29$ (520)/[001] symmetrical tilt boundary the unit cell used is bounded by the vectors 3[$\bar{2}$50] and 3[100]. This cell is nine times larger than the CSL cell and contains eighteen atoms per each (10 4 0) plane parallel to the boundary. The region I was then chosen as containing 152 (10 4 0) layers. It extends, therefore, to the distance of about $7a$ on each side of the boundary and the total number of movable atoms is about 2740.

The MC calculations were carried out at constant volume rather than constant pressure since in polycrystals the grains are constrained and the segregation cannot invoke a relative movement of the grains with respect to each other while the orientation of the boundary plane is fixed. However, the expansions present in the starting configurations were retained.

The MC method employed here is similar to that originally used by Foiles [74]. A modified grand canonical ensemble is utilized, in which the total number of atoms, temperature (T), volume and chemical potential difference between the two species ($\Delta\mu$), corresponding to a given bulk composition, are fixed; $\Delta\mu = \mu_A - \mu_B$, where μ_A and μ_B are the chemical potentials of the species A and B in the A_3B alloy, respectively. The last condition assures that while the ratio of the two species in the block can be altered, the bulk composition, far away from the grain boundary, remains fixed. Clearly, the dependence of $\Delta\mu$ on composition need to be known prior to the grain boundary calculations. This dependence was determined from a series of MC simulations made for the block without any boundary. Each of these simulations is performed for a given value of $\Delta\mu$ and the corresponding equilibrium composition is found as a result.

Two types of configurational modifications represent the MC steps. The first corresponds to a small random displacement of a randomly chosen atom away from its current position and the second to the change of a randomly chosen atom into the alternate species. The change of the total internal energy, ΔE, associated with each MC step is then evaluated and the decision whether to retain this configurational modification is made in dependence on the quantity $\lambda = \Delta E \pm \Delta\mu$, where the plus sign applies when A was replaced by B and the minus sign applies in the opposite case. If λ is smaller than zero the modification is retained. If it is larger than zero the modification is retained with the probability $\exp[-\lambda/kT]$, where k is the Boltzmann's constant. Several millions of such MC steps were always performed until the equilibrium state had been attained.

3.   Grain boundaries in stoichiometric L1$_2$ alloys

The calculated structure of the $\Sigma = 29$ (520) boundary in Cu$_3$Au at T = 300°K is shown in Fig. 3. Comparison with Fig. 1a reveals that the relaxation in the boundary region is even more extensive and many atoms in the boundary region cannot be unambiguously assigned to either the upper or the lower grain. In fact, if neglecting the chemical nature of individual atoms, the structure is rather similar to that found in pure f.c.c. metals.

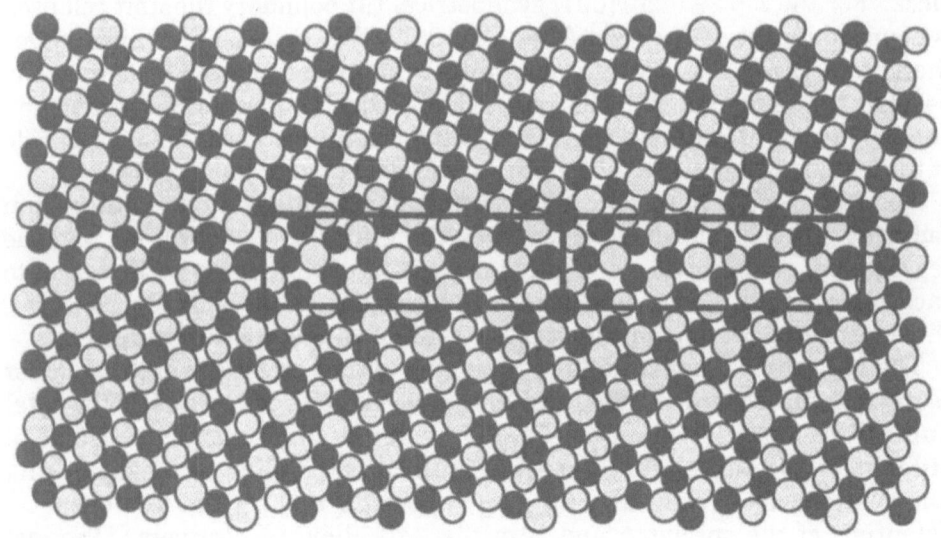

Figure 3.   Calculated structure of the $\Sigma = 29$ (520)/[001] symmetrical tilt boundary in the stoichiometric Cu$_3$Au at T = 300°K.

However, the compositional disorder at the boundary is quite significant. This is seen in Fig 4 where we plot the excess amount of Cu (measured in atomic %) for individual (10 4 0) layers parallel to the boundary. Since the layers are numbered in such a way that the layer number zero is always near the center of the boundary, this plot demonstrates the dependence of the excess amount of Cu on the distance from the boundary. (It should be noted that the distribution of the excess Cu is not symmetrical owing to the loss of mirror symmetry in this structure invoked by the relative rigid body displacements of the grains). Since the (10 4 0) layers are separated by $0.0925a$, where a is the lattice spacing[1], this picture reveals that the compositional disorder spreads into a ribbon of width of about $2a$ around the boundary. As the temperature increases the structure of the boundary does not change significantly but the compositional disorder increases and spreads gradually further into the bulk.

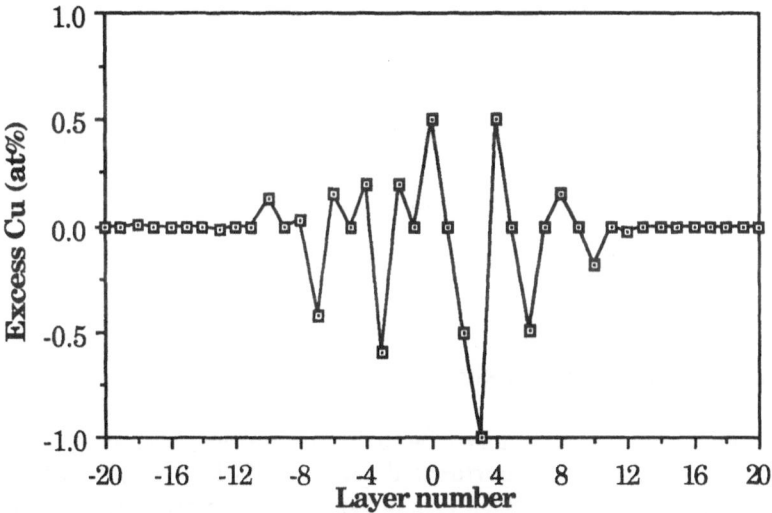

Figure 4. Excess amount of Cu for individual (10 4 0) layers parallel to the the $\Sigma = 29$ (520)/[001] symmetrical tilt boundary in the stoichiometric Cu_3Au at $T = 300°K$.

The situation is, however, very different in Ni_3Al. No significant structural change is observed even at very high temperatures and a noticeable disorder occurs only at temperatures in excess of 1200°K. For this reason the high temperature structure of the $\Sigma = 29$ (520) boundary in Ni_3Al is not shown. The excess amount of Ni for individual (10 4 0) layers parallel to the boundary at $T = 1200°K$ is shown Fig. 5. It is seen that the local deviations from stoichiometry are much smaller than in the case of Cu_3Au and a limited compositional disorder occurs only within the ribbon of the width of about $1.5a$ around the boundary. These results show that the differences between grain boundaries in weakly and strongly ordered $L1_2$

---

[1] $a = 3.748Å$ for Cu_3Au and $3.567Å$ for Ni_3Al.

alloys, established on the basis of the 0°K calculations (Figs. 1a and b), are even more pronounced at high temperatures.

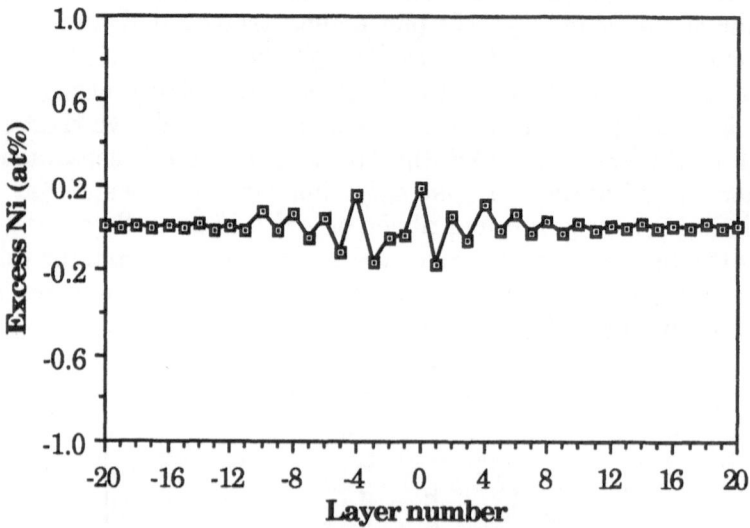

Figure 5. Excess amount of Ni for individual (10 4 0) layers parallel to the the $\Sigma = 29$ (520)/[001] symmetrical tilt boundary in the stoichiometric Ni_3Al at T = 1200°K.

4. Grain boundaries in non-stoichiometric Ni_3Al

Two nickel rich compositions have been investigated, Ni-24%Al and Ni-20%Al, and calculations were performed for T = 600K. In both cases the overall structure of the boundaries is not very different from that shown in Fig. 1b although in the latter case it starts to resemble more closely the structure found in Cu_3Au, shown in Fig. 1a. For this reason the calculated atomic structures are not presented here. However, an appreciable compositional disorder occurs in the case of Ni-24%Al and it is very extensive in the case of Ni-20%Al. This is seen in Figs. 6a and b which show the excess amount of Ni (in atomic %) for individual (10 4 0) layers parallel to the boundary for the two cases studied. It should be noted that interchanges of Ni and Al atoms occur in the boundary region in addition to the replacement of Al by segregating Ni. This is manifested by the negative values of the Ni excess at some (10 4 0) layers, seen in Fig. 6b. Hence, as nickel segregates to the grain boundary this becomes compositionally disordered and the extent of the disordered region increases with the surplus of nickel. In the case of Ni-24%Al the disordered region spans the ribbon of the width of about $0.75a$ around the boundary and in the case of Ni-20%Al it spans the ribbon of the width of about $4.3a$ around the boundary.

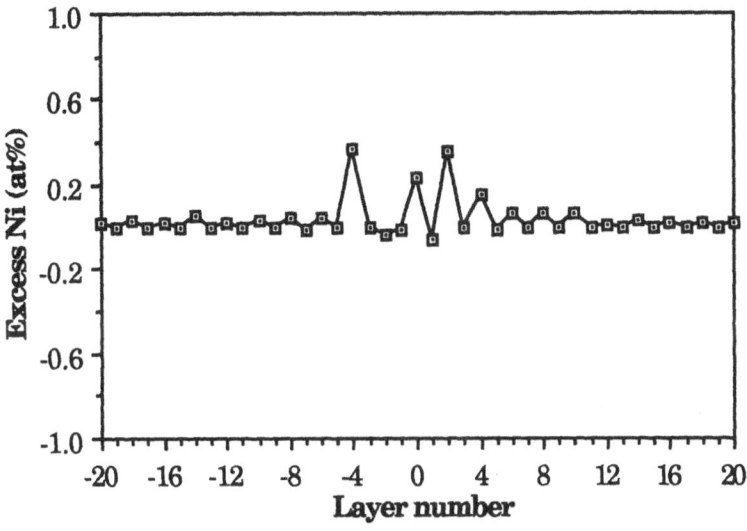

(a)

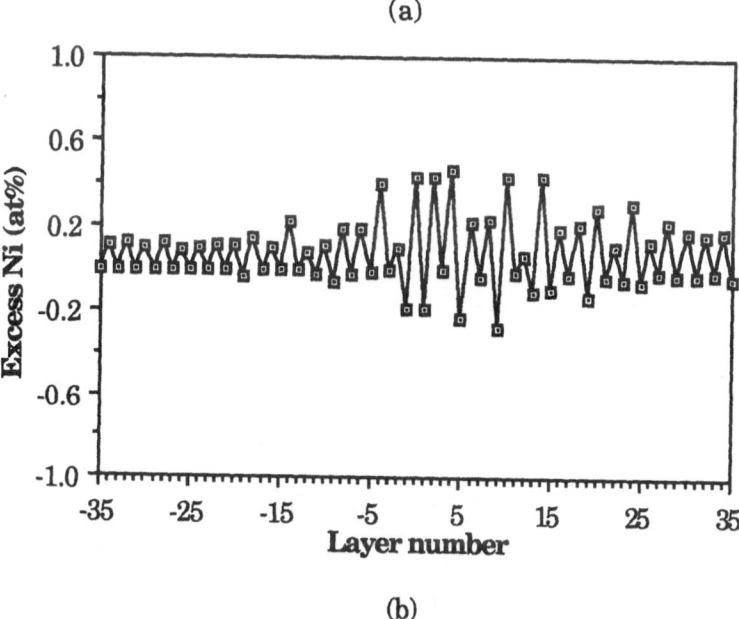

(b)

Figure 6. Excess amount of Ni for individual (10 4 0) layers parallel to the the $\Sigma = 29$ (520)/[001] symmetrical tilt boundary at T = 600°K for Ni-24%Al (a) and Ni-20%Al (b).

Analogously, two aluminum rich compositions have been studied for the same temperature, Ni-26%Al and Ni-30%Al. The atomic structure is in both cases practically the same as that shown in Fig. 1b and it is not, therefore, presented here. Nevertheless, aluminum also segregates to the grain boundary but in a very different fashion than nickel. This is seen from Figs. 7a and b which show the excess amount of Al (in atomic %) for

individual (10 4 0) layers parallel to the boundary for the two cases studied.
It is seen that in both cases aluminum segregates to certain sites in layers
3, -4 and -7 which are all within the ribbon of the width smaller than a
around the boundary. Unlike in the case of Ni, no interchanges of Ni and
Al take place in addition to the segregation. Hence, no significant
compositional disorder develops in the vicinity of the boundary.

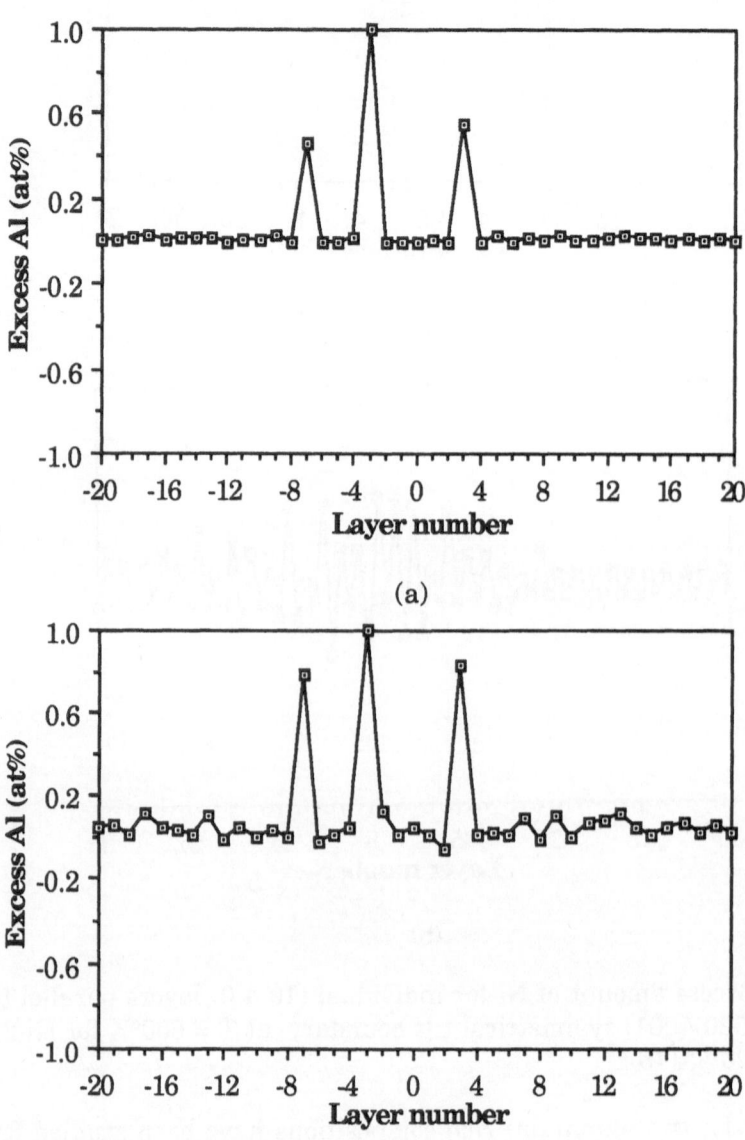

(a)

(b)

Figure 7. Excess amount of Al for individual (10 4 0) layers parallel to the
the $\Sigma = 29$ (520)/[001] symmetrical tilt boundary at T = 600°K for Ni-26%Al
(a) and Ni-30%Al (b).

The most probable reason why nickel segregation in Ni rich and aluminum segregation in Al rich Ni$_3$Al, respectively, are very different, is the different atomic size of these two elements. Aluminum, being a large atom, can only segregate very selectively, principally to the sites associated with a large free volume. Once these sites have been filled the boundary becomes saturated. At this point a two-dimensional ordered Ni-Al structure has formed in the boundary region rather than a disordered structure. On the other hand, nickel is a smaller atom and thus there are many more grain boundary sites to which it can segregate and, moreover, new ones can be formed if disordering occurs in the boundary region.

5. Discussion

The MC computer simulations of the structure and composition of grain boundaries in L1$_2$ ordered alloys, presented in this paper, had two key motivations. The first was to explain why in some materials with this crystal structure, such as Ni$_3$Al, the grain boundaries are intrinsically brittle, while in others, such as Cu$_3$Au, no intergranular brittleness occurs. The second was to elucidate the effects of compositional deviations away from the stoichiometry upon the structure and fracture related properties of grain boundaries in Ni$_3$Al. In particular, an outstanding question is why the surplus of nickel has a very different influence on the intergranular brittleness than the surplus of aluminum although both elements may segregate to the grain boundaries. This problem relates directly to other alloying phenomena, in particular to the ductilizing effect of boron which takes place only in nickel rich alloys.

Investigations pertinent to the intrinsic brittleness are studies of the crystallographically identical boundaries in stoichiometric Ni$_3$Al and Cu$_3$Au, two L1$_2$ alloys with distinctly different propensities to ordering. The former remains ordered until melting while the latter undergoes an order-disorder transition well before melting. In the framework of N-body potentials employed in this study the ordering energy of Ni$_3$Al is five times higher than that of Cu$_3$Au. Analogous molecular statics studies had been made before and alluded to the fact that in strongly ordered alloys the preservation of the compositional order dominates the grain boundary structure so that the ideal L1$_2$ structure is practically undisturbed on either side of the boundary up to the boundary plane. In contrast, in the weakly ordered alloys a chemical disorder occurs owing to the substantial structural relaxation in the boundary region [27, 30, 31, 75]. The present MC calculations which correspond to non-zero temperatures, strongly support this proposition. In Cu$_3$Au an extensive compositional disorder occurs in the boundary region already at room temperature, while in Ni$_3$Al the structure remains compositionally ordered even at very high temperatures and a significant disordering can be expected only at temperatures close to the melting temperature.

As discussed in §1.1, and in more detail in our previous papers [27, 30, 31, 38], the boundaries in which the compositional order is preserved are likely to be more susceptible to cracking. The reason is that crack nucleation may be easier owing to the presence of columnar cavities in these boundaries and the work to fracture, controlling the crack propagation, will be low since the irreversible, energy dissipating shearing at the crack tip, will be difficult. Hence, the brittleness of grain boundaries in compounds like Ni_3Al follows from the distinct features of the atomic structure of grain boundaries in these materials.

The intrinsic brittleness of grain boundaries can, therefore, be expected to be a general property of strongly ordered $L1_2$ compounds while the weakly ordered alloys will not be intergranularly brittle. Nonetheless, it does not mean that all strongly ordered compounds must possess intrinsically brittle boundaries. For example, if segregation of one of the components of the alloy occurs even when the bulk composition is stoichiometric, the boundaries will disorder compositionally and a situation analogous to that found in weakly ordered alloys will ensue. However, the present calculations show that this is not the case in Ni_3Al and thus this compound will have intrinsically brittle grain boundaries even at high temperatures.

Alloying effects were studied by MC simulations of nickel rich and aluminum rich Ni_3Al alloys, respectively. Both nickel and aluminum segregate to the grain boundaries when in surplus in the bulk. However, the impact of the nickel segregation on the boundary structure is very different from that of aluminum segregation. When nickel segregates the grain boundary region disorders compositionally since additional interchanges of nickel and aluminum atoms accompany the segregation process. In the case of aluminum segregation no significant compositional disorder occurs. On the contrary, since the segregation of aluminum is very selective, a two-dimensional ordered Ni-Al compound with a high aluminum content is formed at the grain boundary. This is in fact similar to what occurs, for example, in the disordered copper-bismuth alloy. After segregation of bismuth to grain boundaries two-dimensional ordered Cu-Bi compounds are formed in them [76-78] which may be the reason for the strong embrittlement of copper by bismuth.

The compositionally disordered region formed in nickel rich Ni_3Al is not of macroscopic dimensions but spans a few lattice spacings. This is, however, sufficient to affect the crack nucleation and propagation along grain boundaries. The grain boundary region becomes more akin to that found in weakly ordered alloys and, as discussed above, both the crack nucleation and propagation will be more difficult. In contrast, in aluminum rich Ni_3Al the two dimensional compositionally ordered structure formed due to the segregation of aluminum may render the material even more brittle. This considerable difference between the structural changes invoked by segregation of nickel and aluminum, respectively, may be the principal reason why only in nickel rich alloys the intergranular brittleness can be inhibited by alloying with boron.

Acknowledgements

This research was supported by the U.S. Department of Energy, Office of Basic Energy Sciences, Grant. No. DE-FG02-87ER45295. The computations were performed at the computing facility of the Laboratory for Research on the Structure of Matter supported by the National Science Foundation, MRL Program DMR88-19885 and at the Pittsburgh Supercomputing Center.

References

1. W. C. Johnson and J. M. Blakely, editors, *Interfacial Segregation*, ASM, Metals Park, Ohio (1977).
2. C. J. McMahon, Jr., *Grain Boundary Chemistry and Intergranular Fracture* , edited by G. S. Was and S. M. Bruemmer, *Mat. Sci. Forum* **46**, 61 (1989).
3. C. T. Liu, editor, *Intergranular Fracture and Boron Effects in Ni_3Al and other Intermetallics*, *Scripta Metall. et Mater.* **25**, 1231 (1991).
4. K. H. Hahn and K. Vedula, *Scripta Metall.* **23**, 7 (1989).
5. E. P. George and C. T. Liu, *J. Mat. Res.* **5**, 754 (1990).
6. D. P. Pope and S. S. Ezz, *Int. Met. Rev.* **25**, 233 (1984).
7. K. Aoki and O. Izumi, *J. Japan Inst. Metals* **43**, 1190 (1979).
8. C. T. Liu, C. L. White and J. A. Horton, *Acta Metall.* **33**, 213 (1985).
9. T. Takasugi, N. Masahashi and O. Izumi, *Acta Metall.* **33**, 1247, 1259 (1985).
10. O. Izumi and T. Takasugi, *J. Mater. Res.* **3**, 426 (1988).
11. C. T. Liu, *Interfacial Structure, Properties and Design* (edited by M. H. Yoo, W. A. T. Clark and C. L. Briant), Mater. Res. Soc. Symp. Vol. 122, p. 429 (1988).
12. T. Takasugi, E. P. George, D. P. Pope and O. Izumi, *Scripta Metall.* **19**, 551 (1985).
13. T. Takasugi, M. Nagashima and O. Izumi, *Acta Metall.* **38**, 747 (1990).
14. T. Takasugi and O. Izumi, *Scripta Metall. et Mater.* **25**, 1243 (1991).
15. E. P. George, C. L. White and J. A. Horton, *Scripta Metall. et Mater.* **25**, 1259 (1991).
16. T. Takasugi and O. Izumi, *Acta Metall.* **33**, 1247 (1985).
17. A. I. Taub and C. L. Briant, *Acta Metall.* **35**, 1597 (1987).
18. R. P. Messmer and C. L. Briant, *Acta Metall.* **30**, 487, 1811 (1982).
19. M. E. Eberhard and D. D. Vvedenski, *Phys. Rev. Lett.* 58, 61 (1986).
20. M. A. Saqi and D. G. Pettifor, *Phil. Mag. Lett.* **56**, 245 (1987).
21. L. Goodwin, R. J. Needs and V. Heine, *Phys. Rev. Lett.* **60**, 2050 (1988).
22. A. F. Voter and S. P. Chen, *Mater. Res. Soc. Symp.*, Vol. 82, p.175 (1987).
23. M.S.Daw and M.I.Baskes, *Phys.Rev.B* **29** 6443 (1984).
24. S. P. Chen, A. F. Voter and D. J. Srolovitz, *Scripta Metall.* **20**, 1389 (1986).

352

25. S. P. Chen, A. F. Voter, R. C. Albers, A. M. Boring and P. J. Hay, *J. Mater. Res.* **5**, 955 (1990).

26. S. P. Chen, D. J. Srolovitz and A. F. Voter, *J. Mater. Res.* **4**, 62 (1989).

27. J. J. Kruisman, V. Vitek and J. Th. M. DeHosson, *Acta Metall.* **36**, 2729 (1988).

28. G. J. Ackland and V. Vitek, *Phys. Rev. B* **41**, 10324 (1990).

29. V. Vitek, J. Cserti and G. J. Ackland, *Alloy Phase Stability and Design*, (edited by G. M. Stocks, A. P. Giamei and D. P. Pope), Mat. Res. Soc. Symp., Vol. 186 (1991).

30. V. Vitek, S. P. Chen, A. F. Voter, J. J. Kruisman and J. Th. M. DeHosson, *Grain Boundary Chemistry and Intergranular Fracture* (edited by G. S. Was and S. M. Bruemmer) *Mat. Sci. Forum* **46**, 237 (1989).

31. V. Vitek and S. P. Chen, *Scripta Metall. et Mater.* **25**, 1237 (1991).

32. G. Sasaki, D. Shindo, K. Hiraga, M. Hirabayashi and T. Takasugi, *Acta Metall.* **38**, 1417 (1990).

33. M. J. Mills, *Scripta Metall.* **23**, 2061 (1989).

34. H. Kung, D. R. Rasmunsen and S. L. Sass, *Proc. Int. Symp. on Intermetallic Compounds (JIMIS-6)*, The Japan. Inst. Metals, Sendai, p. 347 (1991).

35. H. Kung, D. R. Rasmunsen and S. L. Sass, *Scripta Metall. et Mater.* **25**, 1277 (1991).

36. M. J. Mills, S. H. Goods, S. M. Foiles and J. R. Whetstone, *Scripta Metall. et Mater.* **25**, 1283 (1991).

37. M. L. Jokl, V. Vitek, C. J. McMahon , Jr., and P. Burgers, *Acta Metall.* **37**, 87 (1989).

38. V. Vitek, *J. Phys. Paris III* **1**, 1085 (1991).

39. J. E. Hack, S. P. Chen and D. J. Srolovitz, *Acta Metall.* **37**, 1957 (1989).

40. A. I. Taub, S. C. Huang and K. M. Chang, *Metall. Trans. A* **15**, 399 (1984).

41. D. L. Lin, D. Chen and H. Lin, *Acta Metall. et Mater.* **39**, 523 (1991).

42. C. L. White, C. T. Liu and R. A. Padgett, *Acta Metall.* **36**, 2229 (1988).

43. J. P. Hirth and J. R. Rice, *Metall. Trans. A* **11**, 1501 (1980).

44. G. S. Painter and F. W. Averill, *Phys. Rev. Lett.* **58**, 234 (1987).

45. K. Masuda-Jindo, *J. Phys. Paris* **49**, C5-557 (1988).

46. K. Masuda-Jindo, *Proc. Int. Symp. on Intermetallic Compounds (JIMIS-6)*, The Japan. Inst. Metals, Sendai, p. 111 (1991).

47. H. J. Frost, *Acta Metall.* **35**, 519 (1987).

48. A. H. King and M. H. Yoo, *Scripta Metall.* **21**, 1115 (1987).

49. A. H. King, H. J. Frost and M. H. Yoo, *Scripta Metall. et Mater.* **25**, 1249 (1991).

50. E. M. Schulson, T. P. Wiehs, D. V. Viens and I. Baker, *Acta Metall.* **33**, 1587 (1985).

51. E. M. Schulson, T. P. Wiehs, I. Baker, H. J. Frost and J. A. Horton, *Acta Metall.* **34**, 1395 (1986).

52. P. S. Khadkikar, K. Vedula and B. S. Shabel, *Met. Trans. A* **18**, 427 (1987).

53. I. Baker, E. M. Schulson and J. A. Horton, *Acta Metall.* **33**, 1533 (1987).

54. I. Baker and E. M. Schulson, *Scripta Metall.* **23**, 1883 (1989).

55. I. Baker, E. M. Schulson, J. P. Michael and S. J. Pennycook, *Phil. Mag.* B **62**, 659 (1990).

56. E. M. Schulson, *Proc. Int. Symp. on Intermetallic Compounds (JIMIS-6)*, The Japan. Inst. Metals, Sendai, p. 339 (1991).

57. E. M. Schulson and I. Baker, *Scripta Metall. et Mater.* **25**, 1253 (1991).

58. T. C. Lee, R. Subramanian, I. M. Robertson and H. K. Birnbaum, *Scripta Metall. et Mater.* **25**, 1265 (1991).

59. B.J. Pestman, J.Th.M. de Hosson, V. Vitek and F.W. Schapink, *J. Phys. Paris*, **51**, C1-311 (1990); *Phil. Mag.* A **64**, to be published (1991).

60. R. A. D. Mckenzie and S. L. Sass, *Scripta Metall.* **22**, 1807 (1988).

61. J. E. Krzanowski, *Scripta Metall.* **23**, 1219 (1989).

62. S. S. Brenner and Hua Ming-Jian, *Scripta Metall. et Mater.* **25**, 1271 (1991).

63. S. M. Foiles, *High Temperature Ordered Intermetallic Alloys II* (edited by N. S. Stoloff, C. C. Koch, C. T. Liu and O. Izumi), Mat. Res. Soc. Symp., Vol. 81, p. 51 (1987).

64. R. Najafabadi, H. Y. Wang, D. J. Srolovitz and R. LeSar, *High - Temperature Ordered Intermetallic Alloys IV* (edited by L. A. Johnson, D. P. Pope and J. O. Stiegler), Mater. Res. Soc. Symp. Vol. 213, p. 51 (1991).

65. M. W. Finnis and J .E. Sinclair, *Phil. Mag.*A **50**, 45 (1984).

66. G. J. Ackland, M. W. Finnis and V. Vitek, *J.Phys.F* **18**, L153 (1988).

67. R. Hultgren, R. L. Orr, P. D. Anderson and K. K. Kelley, *Selected values of the Thermodynamic Properties of.Metals and Binary Alloys*, Wiley, New York (1963).

68. S. M. L. Sastry and B. Ramaswami, *Phil. Mag.* **33**, 375 (1976).

69. P. Haasen, *Physical Metallurgy*, Cambridge University Press, London, p. 148 (1978).

70. G. J. Ackland, G. Tichy, V. Vitek and M. W. Finnis, *Phil.Mag.*A **56**, 735 (1987).

71. Y. Minonishi and V. Vitek, *Surf. Science* **199**, 196 (1984).

72. J. Douin, P. Veyssiere and P. Beauchamp, *Phil. Mag.* A **55**, 565 (1987).

73. A. P. Sutton, and V. Vitek, *Phil. Trans. R. Soc. London* A **309**, 1 (1983).

74. S. M. Foiles, *Phys. Rev.* B **32**, 7685 (1985).

75. G. J. Ackland and V. Vitek, *High-Temperature Ordered Intermetallic Alloys III* (edited by C.T.Liu, A.I. Taub, N.S.Stoloff and C. C. Koch), Mater. Res. Soc. Proc., Vol. 133, p. 105 (1988).

76. B. Blum, M. Menyhard, D. E. Luzzi and C. J. McMahon, Jr., *Scripta Metall.* **24**, 2169 (1990).

77. D. E. Luzzi, *Proc. XII Int. Cong. for Electron Microscopy* (edited by L. D. Peachey and D. B. Williams), San Francisco Press, Inc., p. 318 (1990).

78. D.E. Luzzi, Min Yan, M. Sob and V. Vitek, *Phys. Rev. Lett.*, to be published (1991).

THE LOCAL COMPOSITIONAL ORDER AND DISLOCATION STRUCTURE OF GRAIN BOUNDARIES IN Ni₃Al

H. Kung, D. R. Rasmussen and S. L. Sass
Department of Materials Science and Engineering
Cornell University, Ithaca, NY 14853-1501
U.S.A.

ABSTRACT. The local compositional order and dislocation structure of grain boundaries in Ni₃Al, with and without boron, were examined using electron microscopy techniques. Lattice imaging studies showed that small angle twist, tilt and mixed boundaries and large angle (near $\Sigma = 5$) [001] twist boundaries are ordered up to very close to the interface plane. A compositionally disordered region ~1.5 nm thick is present in the vicinity of a large angle general boundary in boron-doped Ni₃Al, while similar boundaries in boron-free material are ordered. Image simulations were performed and it was shown that the experimental observations of a locally disordered region cannot be explained as being an artifact. Dislocations with Burgers vectors that correspond to anti-phase boundary (APB)-coupled superpartials were found in small angle [001] twist boundaries in both boron-free and boron-doped Ni₃Al and a small angle [011] tilt boundary in boron-doped Ni₃Al. The APB energies determined from the dissociation of the boundary dislocations were smaller than reported for bulk Ni₃Al. For small angle twist boundaries the presence of boron reduced the APB energy at the interface until it approached zero.

1. Introduction

The ordered compound, Ni₃Al, has received considerable attention during the past few years due to its potential high temperature applications. The intrinsic brittleness of polycrystalline Ni₃Al, however, is a major drawback to its use as a structural material. Adding small amounts of boron to a slightly Ni-rich alloy was found to increase its ductility [1,2]. The boron segregated to the grain boundaries [2,3,4].

One explanation for the ductilizing effect of boron is that it increases the cohesive strength of the boundaries [5]. Another suggestion is that boron eases slip transmission across interfaces by increasing the mobility of grain boundary dislocations [6,7]. It is known that the easier the propagation of slip, the smaller is the stress concentration that can build up at the grain boundary, which reduces the tendency for cracking to occur. A theoretical model by King and Yoo [8] indicated that slip transmittal is easier when crossing compositionally disordered (face centered cubic structure) rather than ordered (L1₂ structure) grain

C. T. Liu et al. (eds.), Ordered Intermetallics – Physical Metallurgy and Mechanical Behaviour, 355–370.
© 1992 *Kluwer Academic Publishers.*

boundaries, since the number of possible dislocation reactions at the interface is larger for the disordered case.

A number of experimental studies have attempted to check these hypotheses. Sieloff, Brenner and Burke [9] and Baker, Schulson and Michael [10], using atom probe and high resolution energy dispersive X-ray spectroscopy (EDS), respectively, showed that the boundary regions were Ni-rich in boron-doped Ni₃Al, whereas in boron-free Ni₃Al the boundaries had the bulk composition. Baker et al. [10] suggested that the excess Ni at the boundary made it effectively partially disordered. Recently, however, George, Liu and Padgett [11] used Auger Electron Spectroscopy to show that grain boundaries in both boron-doped and boron-free Ni₃Al had similar Ni-enrichment. Local compositional order at grain boundaries can be studied directly using high resolution transmission electron microscopy (HRTEM). Mackenzie and Sass [12] reported evidence for an ~2.0 nm thick compositionally disordered region in the vicinity of a large angle general boundary in Ni₃Al doped with 300 wt. ppm boron. Krzanowski [13], using the same technique, found that the ordered Ni₃Al structure existed up to the interface plane in Ni-rich Ni₃Al, both with and without boron. Mills [14] also reported that the Ni₃Al was ordered up to the interface in boron-doped material, for a small angle boundary and another with mainly tilt character. Finally, Baker, Schulson, Michael and Pennycook [15] showed that a disordered phase, up to 20 nm wide, exists along portions of grain boundaries in boron-free Ni₃Al and that this phase was found to cover all the boundaries examined in boron-doped Ni₃Al.

In summary, Ni-enrichment was reported in the vicinity of grain boundaries in boron-doped Ni₃Al, which lends support to the idea of local disordering at grain boundaries, though similar enrichment was reported in one study in the absence of boron. There was disagreement among the results of the HRTEM studies. The character and composition of the boundaries examined were frequently not known. In order to understand the role of boron in influencing the mechanical properties of Ni₃Al, the present study examined its effect on the local compositional order and dislocation structure of grain boundaries with similar character, in Ni₃Al with and without boron. The complete results of this study will be published elsewhere [16].

2. Experimental Procedure

The results reported here were obtained from bicrystal and polycrystalline specimens. Single crystal slices of Ni₃Al with composition of Ni-24 at. pct. Al, both doped with 500 wt. ppm boron and boron-free, were first oriented using the X-ray Laue back reflection technique and then mechanically polished to obtain flat {001} surfaces. Small squares (5 mm x 5 mm x 3 mm thick), with edges along <100> directions, were hot pressed together to form a bicrystal using conditions of 1000°C and 2 MPa for 24 hr, with a vacuum of 1x10⁻⁵ Pa. Polycrystalline alloys with composition of Ni-24 at. pct. Al, both doped with 500 wt. ppm boron and boron-free, were supplied by C. T. Liu of Oak Ridge National Laboratory. Materials grown by two different methods were studied. One was obtained by directional solidification (DS) using a levitation zone melting technique. In order to maximize boron segregation, these specimens were given a final heat treatment of 72 hr at 1000°C in an Ar-4%H$_2$ atmosphere, followed by furnace cooling. Another was in the form of a boron-free as-cast alloy, which was cold-worked and annealed to give a fine grain structure, with a final heat treatment of 30 min at 1000°C for the purposes of recrystallization. Thin slices were obtained from these bulk samples. Then 3 mm disks were cut using an electrodischarge machining method, followed by mechanically thinning to ~200 μm.

The final step to prepare electron microscopy specimens was to electropolish the 3 mm disks using a 10% sulfuric acid-methanol solution. The local compositional order at grain boundaries was examined using the lattice imaging mode in a JEOL 4000EX electron microscope operating at 400 KV. In order to identify possible imaging artifacts which may lead to erroneous conclusions, simulations of the image contrast were performed. The dislocation structure of these boundaries was studied using a JEOL 1200EX electron microscope operating at 120 KV.

3. High Resolution Image Simulation

Mills [14] showed that under certain imaging conditions a wedge-shaped crystal could exhibit disordered contrast even though an ordered structure was present. In order to account for the effects of the inclination of the interface plane on the lattice images of grain boundaries, an effort was made in the present work to establish methods to distinguish a truly disordered boundary from an inclined ordered one. A series of image simulations were performed to investigate the change in contrast for inclined ordered and disordered boundaries as a function of objective lens defocus, using the TEMPAS program [17]. In order to explain the 1.5 - 2.0 nm wide region with disordered contrast observed both in the present work and by Mackenzie and Sass [12], in terms of a boundary inclination, a misalignment of the interface plane away from the edge-on orientation of at least 4° is needed for a 20 nm thick sample. The instrumental parameters used in the image simulation were: spherical aberration coefficient, Cs = 1.0 mm; semi-angular beam divergence, 0.6 mrad; and aperture size, 0.7 Å^{-1}. The foil thickness was 20.3 nm and the boundary inclination 4°.

Calculated images for an inclined boundary are shown in Fig. 1 (a-f). Rectangular patterns of spots, which correspond to the {110} orientation for the L1$_2$ structure, are seen in the region with uniform thickness. As the defocus condition changed from (a) to (f), the contrast also changed but the rectangular pattern remained. In the wedge-shaped part of the crystal, the image contrast changed rapidly with defocus and thickness. When moving across the specimen in the direction normal to the trace of the boundary plane, occasionally small areas of a diamond-shaped spot pattern are present, which look like the compositionally disordered (f.c.c.) structure. At the defocus Δf = -30 nm (Fig. 1 (a)), the image contrast varies as the thickness changes in the wedge-shaped part of the sample; a disorder-like contrast appears in a small region as the thickness starts to decrease, but the rectangular pattern corresponding to the ordered structure resumes up to the trace of the boundary plane. At Δf = -40 nm and -60 nm a small region of apparent disorder appears

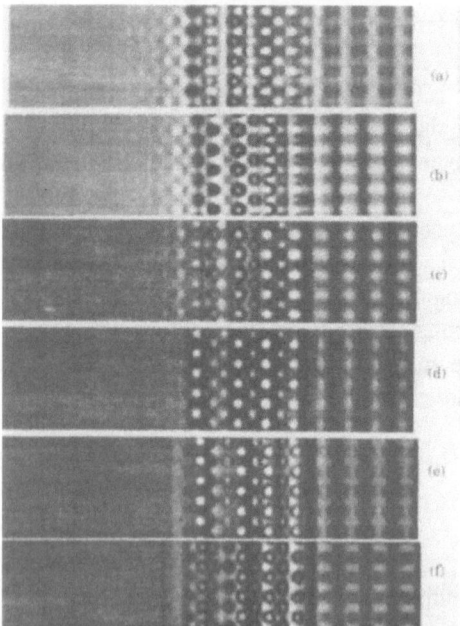

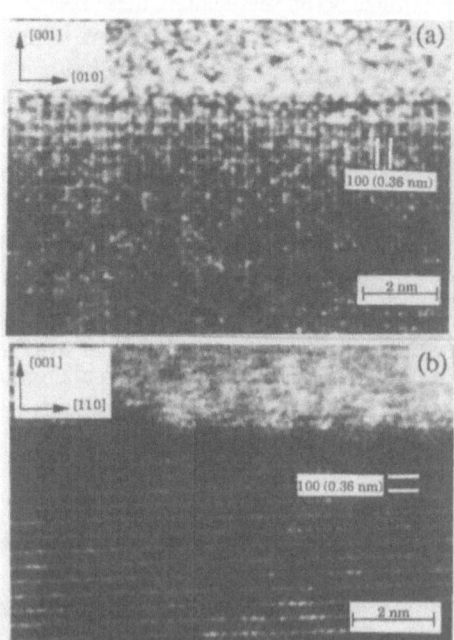

Fig. 1 Computed images for an inclined boundary in ordered Ni$_3$Al with objective lens defocus values: (a) -30 nm, (b) -40 nm, (c) -50 nm, (d) -60 nm, (e) -70 nm, (f) -80 nm.

Fig. 2 Lattice images of small angle [001] twist boundaries: (a) without, and (b) with boron.

close to the boundary trace. The disordered diamond-shaped pattern reappears again at $\Delta f = -80$ nm (Fig. 1 (f)) in from the trace.

The simulations in Fig. 1 and ref. [16] indicate that, for an inclined ordered boundary, disordered contrast may appear in small areas close to the interface due to boundary misalignment but only at particular defocus conditions, while ordered contrast is present in the remaining regions. The position of the disordered contrast changes with defocus. In general, for a particular defocus value, the image contrast changes rapidly across the wedge-shaped crystal (i.e. with thickness). For the model used in the image simulations the mean inner potential was allowed to vary across the boundary, because one crystal was removed. This could lead to additional contrast that was not present in the real case. To test for this effect, simulations were carried out for a 20 nm thickness, but with a single step of height 4.0 nm. For a few values of defocus Fresnel fringes were visible, but in all cases they were weak compared to the lattice image contrast.

The image simulations lead to the conclusion that the appearance of disordered contrast due to boundary inclination occurs only at special defocus values in small areas of the wedge-shaped crystal. In addition, it was noted that the image contrast changed rapidly with thickness across the wedge-shaped crystal, so that if there is disordered contrast present, it was only in a small area with neighboring regions of ordered contrast.

4. Experimental Results

4.1. LOCAL COMPOSITIONAL ORDER

The lattice image of an [001] small angle twist boundary, with misorientation angle $\theta = 7°$, is shown in Fig. 2 (a) for boron-free Ni_3Al. The image was taken with the beam parallel to the [100] direction for one grain, so the boundary is in an edge-on orientation. Lattice fringes with 0.36 nm spacing, which come from the 100-type superlattice reflections, are seen extending from the bulk up to very close to the boundary plane. No change in contrast due to disordering was observed along the plane of the grain boundary. The lattice image of an [001] twist boundary with θ of 4° is shown in Fig. 2 (b) for boron-doped Ni_3Al. A rectangular array of lattice fringes corresponding to the {100} and {110} planes are seen inside one of the crystals, as well as close to the boundary. The observations in Fig. 2 indicate that the atomic structure is compositionally ordered up to very close to the interface plane in both boron-free and boron-doped Ni_3Al. For this type of imaging compositional disorder which is highly localized to within one or two planes of the interface might not be detected.

Additional observations were made on tilt and mixed-type boundaries. Fig. 3(a,b) shows small angle mixed boundaries in boron-doped and boron-free Ni_3Al, respectively. In both cases the {100} fringes with 0.36 nm spacing are seen extended to close to the boundary plane without changes in intensity, which indicates that the Ni_3Al is ordered up to very close to the interface. A similar observation was made on a faceted small angle tilt boundary in B-doped Ni_3Al, as shown in Fig. 4. The boundary has an average plane normal ~12° from [001] and a facet period of 8 nm; (010) fringes were seen continuously from inside the crystal up to the interface. Faceting was also commonly observed in small angle tilt boundaries in B-free Ni_3Al. Fig. 5 shows a lattice image for such a boundary and an ordered structure can be seen extending up to the interface in both crystals. For all the small angle boundaries examined, with twist, tilt and mixed character, the Ni_3Al was ordered up to the vicinity of the interface, both in the absence and presence of boron.

Two types of large angle boundaries were studied, [001] twist boundaries with misorientation angle close to the Σ=5 value of 36.9° and general boundaries with a high index rotation axis. Fig. 6 (a,b) shows lattice images of large angle [001] twist boundaries in boron-doped and boron-free Ni_3Al, respectively. The electron beam was oriented along the [001] direction of the lower crystal so the boundary could be viewed edge-on. A square array of 0.36 nm lattice fringes, corresponding to the (100) and (010) planes, is present up to close to the interface plane in both cases. No differences in the lattice images with respect to fringe periodicity, variation in intensity and spacing can be detected between Fig. 6 (a) and Fig. 6 (b). It is interesting to note that the contrast does change close to the interface plane in the lower crystal in Fig. 6 (b). This may be due to a slight inclination of the interface plane away from the edge-on orientation.

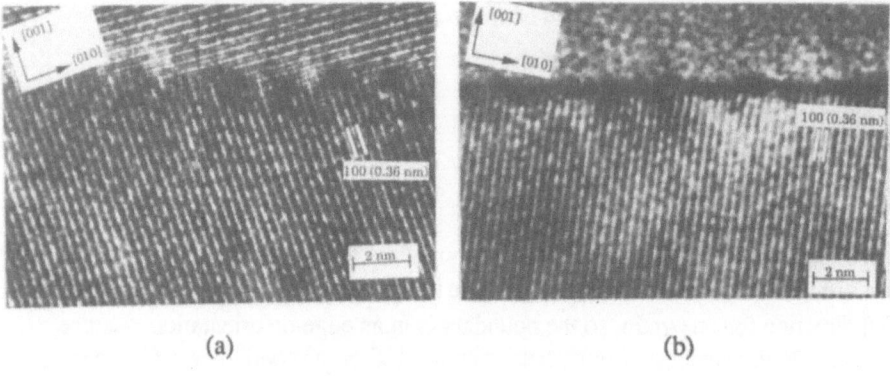

(a)   (b)

Fig. 3 HRTEM images of small angle mixed boundaries in directionally solidified polycrystalline Ni₃Al. (a) Boron-doped. (b) Boron-free.

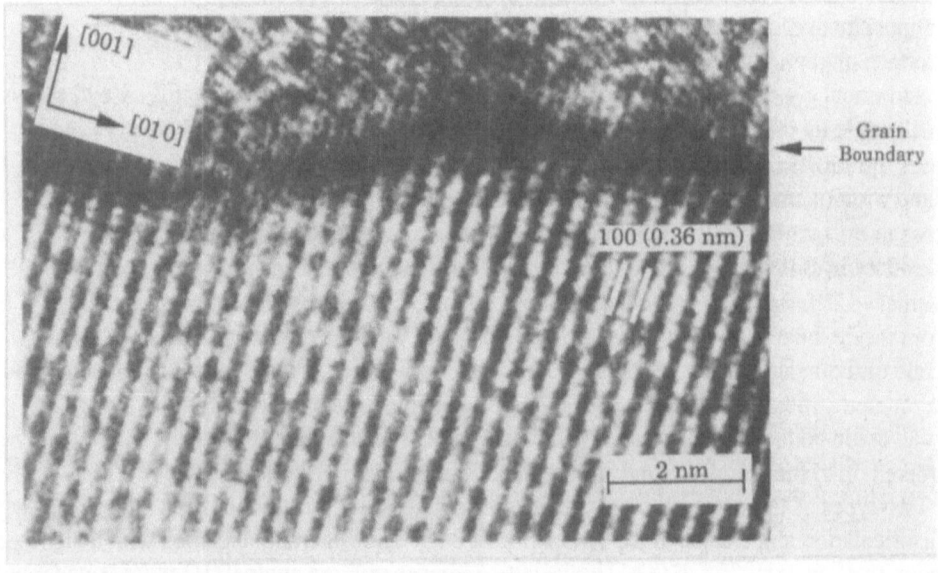

Fig. 4 Lattice image of a small angle tilt boundary ($\theta = 4°$) in boron-doped directionally solidified polycrystalline Ni₃Al.

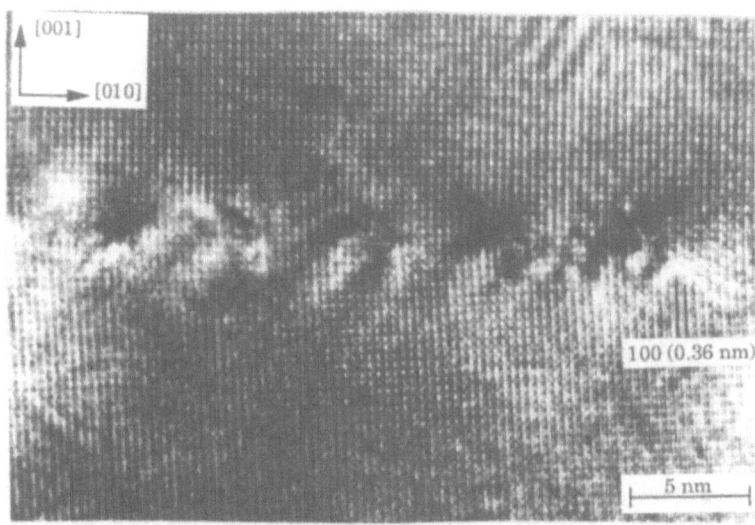

Fig. 5 Lattice image of an [001] small angle tilt boundary ($\theta = 4°$) in boron-free directionally solidified polycrystalline Ni$_3$Al.

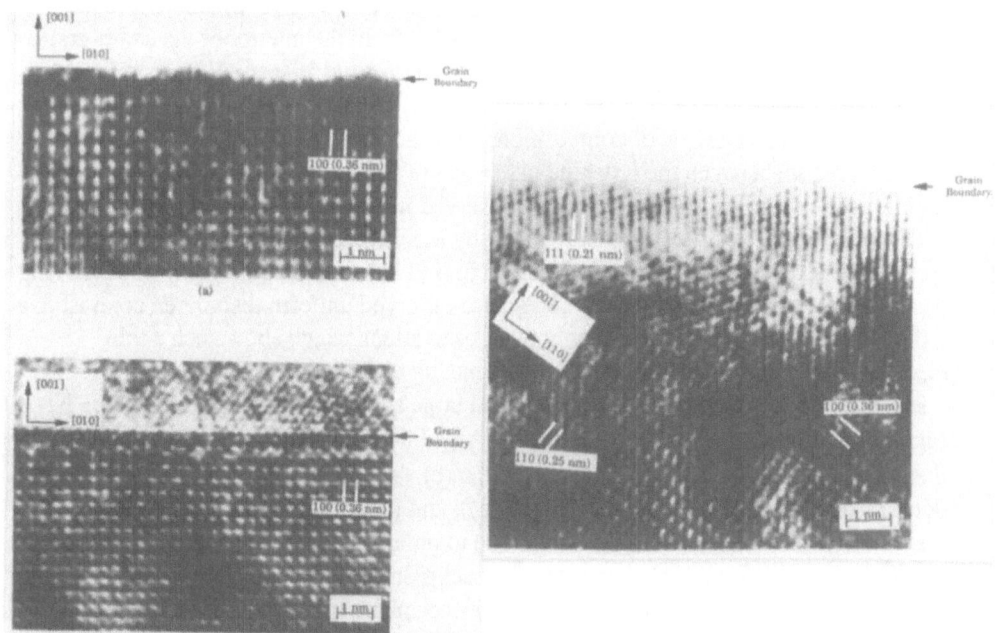

Fig. 6 Lattice images of large angle [001] twist boundaries: (a) with boron and (b) without boron.

Fig. 7 Lattice image of large angle general grain boundary in Ni$_3$Al with boron.

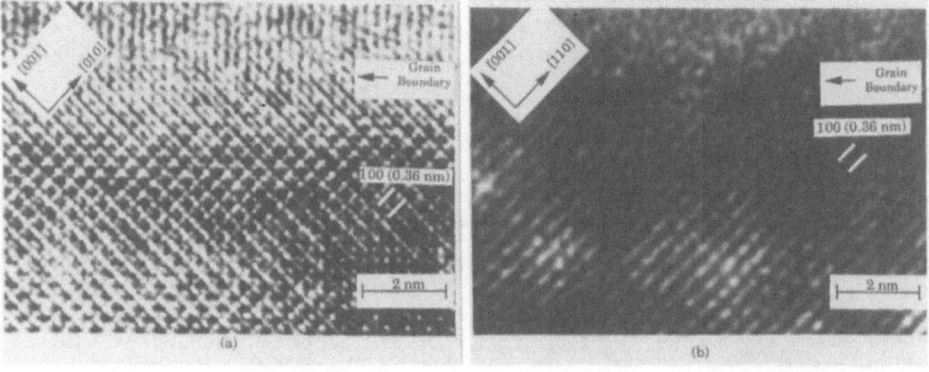

Fig. 8 Lattice images of large angle general grain boundaries in Ni₃Al without boron.

Fig. 7 shows a more general type large angle boundary which was found in the same boron-doped Ni₃Al specimen as the large angle twist boundary in Fig. 6 (a). In one crystal the boundary normal and rotation axis were determined to be ~ [112] and [2, 9, -43], respectively, while the misorientation angle was ~26°. In the interior of the lower grain, a rectangular array of spots due to the 100 and 110 reflections is seen. When approaching the boundary plane, the rectangular pattern gradually disappears and diamond shaped {111} fringes characteristic of compositional disorder are seen. There appears to be a transition region between the ordered bulk structure and the disordered grain boundary region, with uniform disordered contrast observed within 1.5 - 2.0 nm of the interface.

This observation is similar to that reported by Mackenzie and Sass [12], where a rectangular array of spots gradually changed into {111} disordered fringes when moving from the bulk toward the boundary. They also observed uniform disordered contrast in a 2.0 nm thick layer. These are the only two observations which show such a thin disordered boundary structure using lattice imaging techniques.

Fig. 8 (a,b) shows images from two general large angle boundaries in boron-free Ni₃Al. Fig. 8 (a) is from as-cast polycrystalline Ni₃Al, while Fig. 8 (b) is from the same sample as the large angle twist boundary shown in Fig. 6 (b). The rotation axes and angles were determined to be: [0, 2, 5] and 32° (Fig. 8 (a)), and [48, 5, 4] and 50° (Fig. 8 (b)), respectively. In both cases lattice fringes due to ordering persist up to very close to the interface plane. In Fig. 8 (a) changes in the background and fringe contrast near the grain boundary region are an indication that the interface plane is inclined. Nevertheless, ordered contrast was observed along the entire length of the boundary. It can be concluded that compositional order remains up to the vicinity of the interface for this boundary.

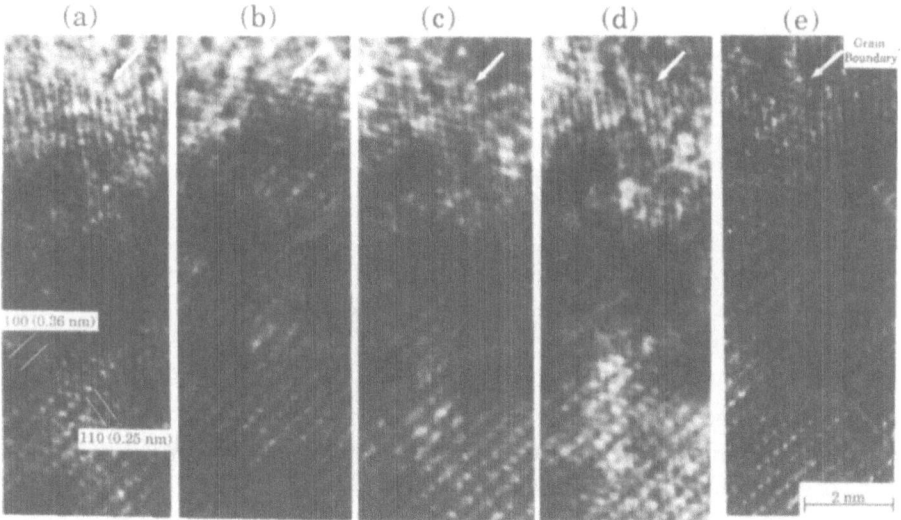

Fig. 9 Through focus series, in steps of $\Delta f = -16$ nm, of the large angle general boundary in Fig. 7 .

Comparisons between near $\Sigma=5$ twist boundaries in boron-doped and boron-free Ni_3Al indicate that ordering occurs up to the interface in both cases. In the general large angle boundaries studied, for boron-doped Ni_3Al, contrast corresponding to compositional disorder is present in the vicinity of the interface, while in the boron-free Ni_3Al, contrast due to ordering exists up to the interface. As pointed out by Mills [14], images that exhibit disordering contrast may be the result of the boundary being inclined away from the edge-on orientation. The image simulation results in this study suggest that one way to identify this kind of artifact is to examine the change of the image contrast in a through focus series. The contrast from the large angle general boundary in Fig. 7 was examined in this manner in Fig. 9. Consistent disordered contrast was present in the vicinity of the interface throughout the range of defocus values studied. According to the simulation results, only a disordered boundary would show such consistent disordered contrast for different defocus values [16]. It is also important to note that the observations in Fig. 7 and by Mackenzie and Sass [12] both show uniform contrast across the entire 1.5-2.0 nm disordered region, which is not expected if the boundary was inclined. The observation in Fig. 7 together with the through focus series results in Fig. 9 are evidence that the disordered contrast is not an artifact due to inclination of the interface plane.

4.2. DISLOCATION STRUCTURE

Fig. 10 (a) shows the lattice image of an edge-on [001] tilt boundary in boron-doped

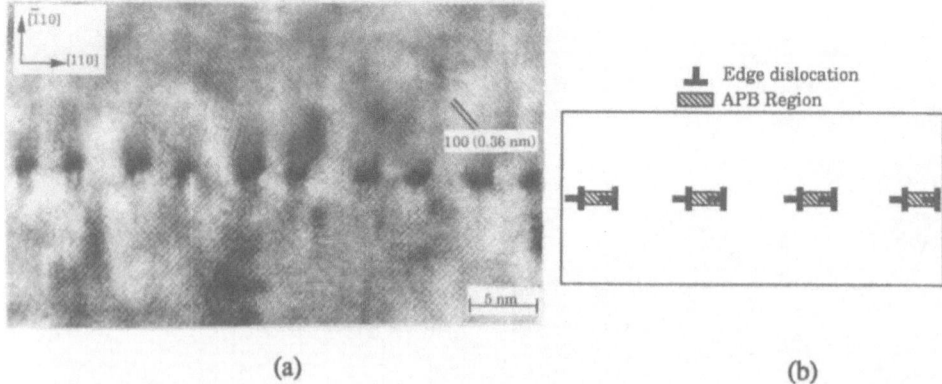

Fig. 10 (a) Lattice image of an edge-on small angle tilt boundary in boron-doped DS Ni₃Al. (b) Schematic diagram showing a possible grain boundary dislocation configuration consisting of APB coupled superpartial dislocations.

Ni$_3$Al. The dislocations are seen to be paired, with the spacing of successive dislocations alternating between two values. From Frank's rule, $d_D = |b|/\theta$, with the spacing of the pair of dislocations, $d_D = 9.3$ nm and the measured $\theta = 3°$, the magnitude of the total Burgers vector, $|b|$, corresponds to a<110>. The individual dislocations could then have a Burgers vector of the type a/2<110>, if they are coupled by an APB as shown in Fig. 10 (b). This geometry is very similar to that proposed by Marcinkowski [18] for a small angle tilt boundary in the L1$_2$ structure.

In order to detect the APB contrast and check the model in Fig. 10 (b), the boundary was tilted ~15° from the edge-on orientation. The bright field image in Fig. 11(a) shows the inclined boundary and confirms the dislocation pairing configuration. Fig. 11(b) is a dark field image made using the 100-type reflection, from a different region with a slightly different foil orientation than Fig. 11(a). Due to the low intensity of this type of image, the resolution is poor. The distinct dark-bright alternation in contrast associated with adjacent dislocation pairs is consistent with the existence of an APB between the pairs of closely spaced dislocations.

Fig. 12 (a,b) shows BF and DF images, respectively, of the dislocation structure of an [001] twist boundary in boron-doped Ni$_3$Al. A square array of dislocations is observed with an average spacing of 8.3 nm. In Fig. 12 (c), a weak beam image with 220 reflection operating, one set of dislocations is out of contrast. The twist angle θ was measured to be 1.6° from the separation of the two (200) matrix spots in the electron diffraction pattern, and from this value and the dislocation spacing, the Burgers vector of the dislocations in the network was determined to be a/2<110>-type. Fig. 12 (c) shows that the Burgers vector is in the same plane as the dislocation line, which is consistent with these dislocations being pure screw-type.

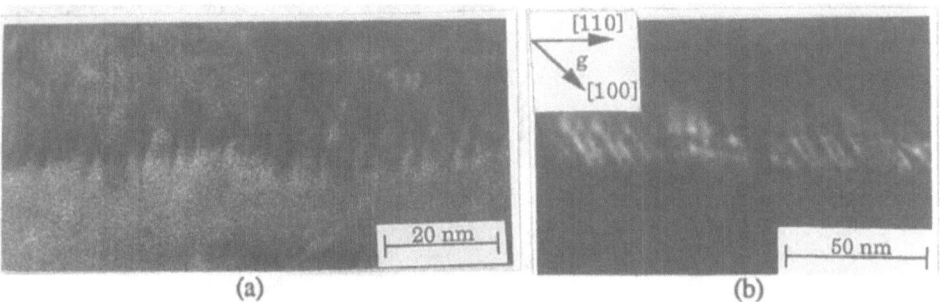

Fig. 11 (a) Bright field image showing the same boundary as in Fig. 10 but tilted away from the edge-on orientation. (b) Dark field image, using the 100 superlattice reflection, of a different region of the boundary in (a).

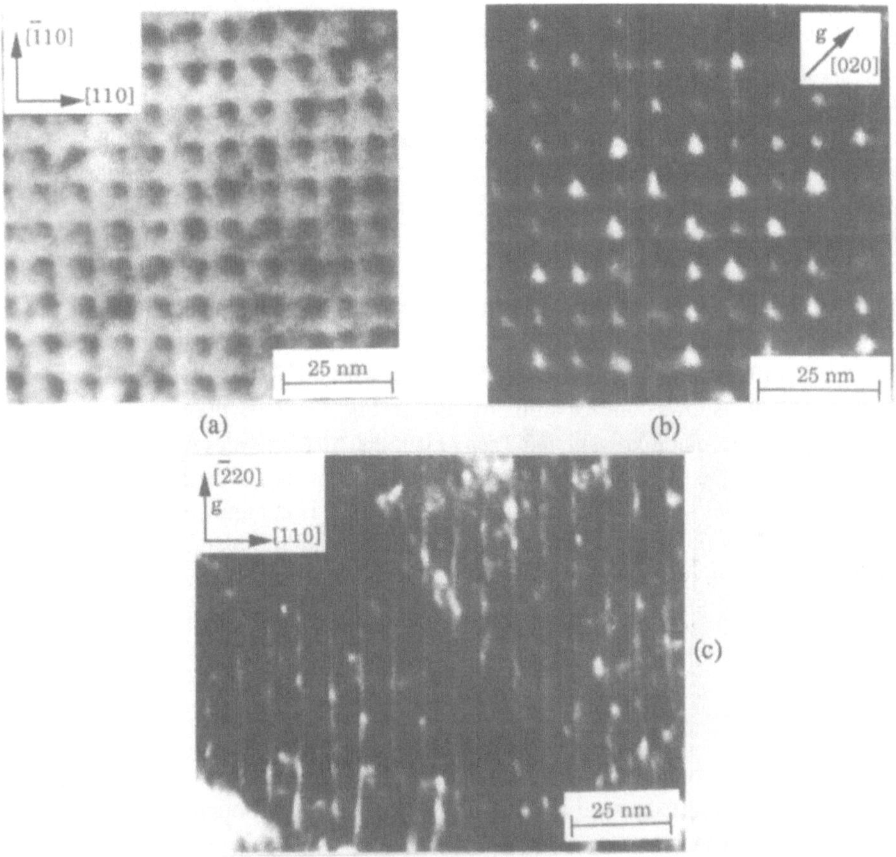

Fig. 12 Micrographs of a square dislocation network in an [001] small angle ($\theta = 1.6°$) twist boundary in boron-doped Ni_3Al bicrystal: (a) BF, (b) DF (g = 020) and (c) Weak beam DF (g = 220). Deviation parameter, $s_{220} = 0.23$ nm^{-1}.

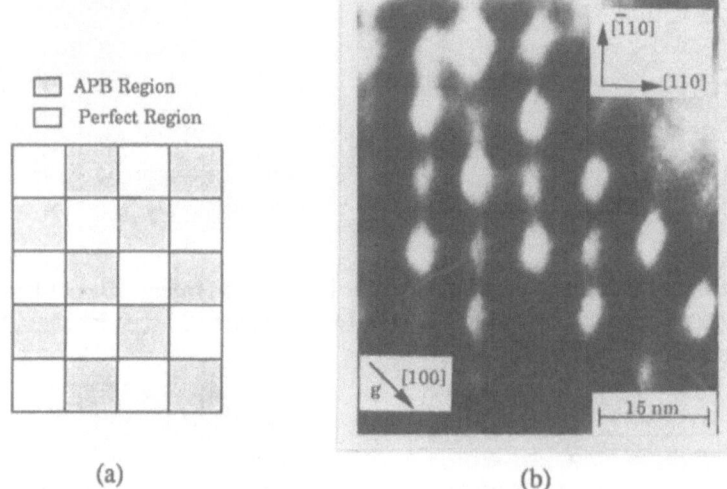

(a) (b)

Fig. 13 (a) Schematic diagram showing the expected contrast from a APB-coupled superpartial grain boundary dislocation array in a small angle [001] twist boundary. (b) Dark field image of small angle [001] twist boundary in boron-doped Ni$_3$Al bicrystal using a 100-type superlattice reflection.

Fig. 13(a) illustrates the checkerboard pattern of contrast which is expected from a small angle [001] twist boundary, if the dislocations with **b** of a/2<110>, are separated by APB's. Fig. 13(b), which is a dark field image taken with a 100 superlattice reflection, in boron-doped Ni$_3$Al, shows contrast similar to that in Fig. 13(a). Using the treatment of Czernichow, Gudas, Marcinkowski and Tseng [19], the APB energies were determined for a variety of small angle twist boundaries, from measurements of the partial dislocation spacing, with the results summarized in the Table. Since it is not known at what temperature the grain boundary dislocations reached their equilibrium dissociation separation, to cover all possibilities the APB energy was determined using elastic constants for both 1000°C and 25°C.

5. Discussion

The suggestion that a compositionally disordered region is present in the vicinity of grain boundaries in Ni$_3$Al was examined in this study. All small angle boundaries and large angle twist boundaries were observed to be ordered up to very close to the interface plane, both in the absence and presence of boron. Only in the case of large angle general boundaries was there a difference in the ordering close to the interface, between Ni$_3$Al with and without boron. In the presence of boron one large angle boundary was found to be

Table

The APB energies determined for small angle twist boundaries (γ_{gb})

and the bulk APB energy (γ_{bulk}), in B-free and B-doped Ni$_3$Al.

Ni$_3$Al	Angle(°)	γ_{gb} (mJ/m^2) 1000°C	γ_{gb} (mJ/m^2) 25°C	γ_{bulk} (mJ/m^2)	Reference for γ_{bulk}
B-free	1.1	50±10	70±11	90±5	[20]
B-doped	1.6	5±3	7±4	92	[21]
B-doped	2.0	7±2	10±3	92	[21]

compositionally disordered over a region 1.5 to 2.0 nm thick in one crystal next to the interface. This boundary and that examined by Mackenzie and Sass show similar behavior. In Ni$_3$Al without boron two large angle boundaries were ordered up to very close to the interface. The image of the boundary showing disordered contrast was compared to simulation results. It was concluded that the contrast is due to compositionally disordering in the vicinity of the interface. Since, thus far, only two boundaries have been observed with a disordered region 1.5 - 2.0 nm thick, many more must be examined before it can be concluded that all large angle general boundaries are compositionally disordered.

The study by Mackenzie, Vaudin and Sass [22] showed that at least 70% of the boundaries present in Ni$_3$Al with and without boron were the large angle general type. Thus the observations on these boundaries are probably most relevant to understanding the influence of boron on the mechanical properties of Ni$_3$Al.

The observations on the boundary dislocations were analyzed using the approach of Marcinkowski et al to obtain values for the APB energy in small angle twist boundaries. In general the APB energy was lower in the grain boundaries than in the bulk. Most strikingly, in the presence of boron the APB energy approached zero. This indicates that in Ni$_3$Al containing boron the grain boundary is able to accommodate the potentially "bad" Al-Al bonds with very little energy increase, thus allowing partial dislocations to form to take advantage of the decrease in strain energy associated with their presence. In Fe$_3$Al, which has the ordered B$_2$ structure, Buis, Tichelaar and Schapink [23] obtained a smaller APB energy in small angle tilt boundaries than in the bulk. They suggested that segregation of Fe to the APB was responsible for this difference. It can be speculated, therefore, that the low value of the APB energy in B-doped small angle twist boundaries is related to the Ni-rich nature of boundaries in Ni$_3$Al, where Ni segregates to the APB, and thereby, decreases the number of Al-Al bonds. The very low APB energy raises a question about the degree of compositional order in the atomic plane just adjacent to the interface. It seems reasonable to suggest that by reducing its APB energy the boundary approaches the situation of a compositionally disordered structure which is localized in small regions in the first atomic plane next to the interface. The lattice imaging study was

368

not capable of providing information on just this situation.

The large disordered region (~20 nm thick) reported by Baker and Schulson [15] was not seen in any of the observations in the present study. Even though such a region may only appear on one side of the boundary and not be continuous, it is still surprising that a microstructural feature of such dimension was not observed.

Finally, as to the role of boron in ductilizing Ni_3Al, the results of the present study lend support to the suggestion that slip transmittal is facilitated by the presence of a disordered region in the vicinity of a boron-doped large angle general boundary. However, as discussed by Kruisman, Vitek and De Hosson [24], it is also possible for compositional disorder, through its effect on boundary structure, to influence grain boundary cohesion. These two explanations are probably interrelated and both may play a role in the ductilizing effect of boron.

6. Conclusions

For twist, tilt and mixed-type small angle ($\theta \leq 7°$) boundaries, the compositionally ordered structure is present up to very close to the interface plane in the absence and presence of boron. Similar results were obtained for large angle (near $\Sigma=5$) twist boundaries in the absence and presence of boron. A large angle general boundary in boron-doped Ni-rich Ni_3Al exhibited a compositionally disordered region ~1.5 nm thick in the vicinity of the interface. Similar general-type boundaries in boron-free Ni_3Al were compositionally ordered up to the interface plane. The observation of a disordered region in this study cannot be explained as being an imaging artifact. The APB energy determined from the dissociation of grain boundary dislocations was smaller than values measured in the bulk. For small angle twist boundaries the presence of boron reduced the APB energy at the interface until it approached zero.

7. Acknowledgements

This research was supported by U.S. Department of Energy, Basic Sciences-Materials Sciences, under grant DE-FG02-85ER45211. We are especially grateful to Dr. C. T. Liu of Oak Ridge National Laboratory for providing us with the polycrystalline Ni_3Al alloys. The use of the Electron Microscope and Material Preparation Facility of the Materials Science Center at Cornell University, which is supported by the National Science Foundation, is acknowledged. The assistance of B. F. Addis and R. Coles was greatly appreciated.

8. References

1. Aoki, K. and Izumi, O. (1979) 'Improvement in Room Temperature Ductility of the L1$_2$-Type Intermetallic Compound Ni$_3$Al by Boron Addition' Trans. J. Japan Inst. Metals 43, 1190-1196

2. Liu, C. T., White, C. L. and Horton, J. A. (1985) 'Effect of Boron on Grain Boundaries in Ni$_3$Al', Acta met. 33, 213-229

3. Horton, J. A. and Miller, M. K. (1987) 'Atom Probe Analysis of Grain Boundaries in Rapidly Solidified Ni$_3$Al', Acta met. 35, 133-141

4. White, C. L., Padgett, R. A., Liu, C. T. and Yalisove, S. M. (1984) Scripta met. 18, 1417-1420

5. Messmer, R. P. and Briant, C. L. (1982) 'The Role of Chemical Bonding in Grain Boundary Embrittlement', Acta met. 30, 457-467

6. Schulson, E. M., Weihs, T. P., Veins, P. V. and Baker, I. (1985) 'The Effect of Grain Size on the Yield Strength of Ni$_3$Al', Acta met. 33, 1587-1591

7. Schulson, E. M., Weihs, T. P., Baker, I., Frost, H. J. and Horton, J. A. (1986) 'Grain Boundary Accommodation of Slip in Ni$_3$Al Containing Boron', Acta met. 34, 1395-1399

8. King, A. H. and Yoo, M. H., (1987) ' On the Availability of Dislocation Reactions at Grain Boundaries in Cubic Ordered Alloys', Scripta met. 21, 1115-1119

9. Sieloff, D. D., Brenner, S. S. and Burke, M. G., (1987) ' FIM/Atom Probe Studies of B-Doped and Alloyed Ni$_3$Al', MRS Symposium 81, 87-97

10. Baker, I., Schulson, E. M. and Michael, J. R. (1988) 'The Effect of Boron on the Chemistry of Grain Boundaries in Stoichiometric Ni$_3$Al', Phil. Mag. B57, 379-385

11. George, E. P., Liu, C. T. and Padgett, R. A. (1989) 'Comparison of Grain Boundary Compositions in B-Doped and B-Free Ni$_3$Al', Scripta met. 23, 979-982

12. Mackenzie, R. A. D. and Sass, S. L. (1988) 'Direct Observation of the Compositional Disordering of Ni$_3$Al in the Vicinity of Grain Boundaries Using High Resolution Electron Microscopy Techniques', Scripta met. 22, 1807-1812

13. Krzanowski, J. E. (1989) 'Nickel Enrichment Near Grain Boundaries in Chill Cast Ni$_3$Al Alloys and Its Relation to Grain Boundary Cohesive Strength', Scripta met. 23, 1219-1224

14. Mills, M. (1989) 'Determination of the Compositional Ordering at Grain Boundaries in Boron-Doped Ni$_3$Al', Scripta met., 23, 2061-2066

15. Baker, I., Schulson, E. M., Michael, J. R. and Pennycook, S. J., (1990) 'The Effects of Both Deviation From Stoichiometry and Boron on Grain Boundaries in Ni$_3$Al', Phil. Mag. B62, 659-676

16. Kung, H., Rasmussen, D. R. and Sass, S. L. (1991) 'Grain Boundaries in Ni$_3$Al, Part I: The Local Compositional Order, and Part II: The Dislocation Structure of Small Angle Boundaries', Acta met. To be published (1991)

17. Kilaas, R. (1987) 'Interactive Simulation of High Resolution Electron Micrographs', Proc. 45th Annual Meeting EMSA, 66-67

18. Marcinkowski, M. J. (1968) 'Theory of Grain Boundaries in Ordered Alloys', Phil. Mag. 17, 159-168

19. Czernichow, J., Gudas, J. P., Marcinkowski, M. J. and Tseng, Wen Feng (1971) 'Twist Boundaries in Ordered Alloys', Met. Trans., 2, 2185-2188

20. Douin, J., Veyssiere, P. and Beauchamp, P. (1986) 'Dislocation Line Stability in Ni_3Al', Phil. Mag. A, 54, 375-393

21. Veyssiere, P., Yoo, M. H., Horton, J. A. and Liu, C. T., (1989) 'Temperature Effect on Superdislocation Dissociation on a Cube Plane in Ni_3Al', Phil. Mag. Let. 59, 61-68

22. Mackenzie, R. A. D., Vaudin, M. D. and Sass, S. L. (1988) 'Grain Boundary Structure in Ni_3Al', Proc. 46th Annual Meeting EMSA, 602-603

23. Buis, A., Tichelaar, F. D. and Schapink, F. W. (1989) 'Transmission Electron Microscopy Observations on a Small Angle Symmetrical Tilt Boundary in B_2-Ordered Fe_3Al', Phil. Mag. A, 59, 861-871

24. Kruisman, J. J., Vitek, V. and De Hosson, J. Th. M. (1988) 'Atomic Structure of Stoichiometric and Non-Stoichiometric A_3B Compounds with $L1_2$ Structure', Acta met. 36, 2729-2741

THE BRITTLE TO DUCTILE TRANSITION AND THE TRANSMISSION OF SLIP ACROSS GRAIN BOUNDARIES IN L1₂ INTERMETALLIC COMPOUNDS

E.M. SCHULSON AND I. BAKER
Thayer School of Engineering
Dartmouth College
Hanover, New Hampshire 03755
U.S.A.

ABSTRACT. It is argued that the improvement in the ductility of Ni₃Al, Ni₃Ga, Ni₃Si and Ni₃Ge through the addition of boron (in the ppm range) and of Zr₃Al through fast neutron irradiation is related to an increase in the transmission of slip across the grain boundaries. The transmittal mechanism is dislocation nucleation at the heads of pile-ups, and this is made easier through the disordering of the boundaries. The evidence supporting this view is reviewed. It is suggested that grain boundary disorder may be a general requirement for ductility.

1. Introduction

Strongly ordered intermetallic compounds are generally brittle at room temperature. This characteristic is a serious shortcoming to their utilization, particularly in components vulnerable to impact. Attempts over the past decade or so to modify this behavior have led to a number of successes, resulting in remedies such as grain refinement, macroalloying, and microalloying. The appropriate one depends upon the nature of the brittleness (Baker and Munroe 1988).

A solution that has attracted extraordinary attention is microalloying with boron. Aoki and Izumi (1979) first reported that the addition of the element (in the ppm range) imparts extensive ductility to Ni₃Al, which is otherwise brittle. Liu et al. (1985) confirmed the finding, correlated it with the segregation of boron to grain boundaries, and optimized the chemistry. This approach subsequently has been pursued with vigor, and success has been achieved to varying degrees in several other systems; e.g., Ni₃Ga (Taub and Briant 1987), Ni₃Si (Taub and Briant 1987, Oliver and White 1987, Schulson et al. 1990) and FeAl (Crimp et al. 1987; Liu and George 1990).

As is often the case, the physical processes underlying materials phenomena are not always immediately apparent. Nor are they necessarily the same in different alloy systems. This is particularly true for the effect of boron on the ductility of Ni₃Al where, more than ten years after the event, the debate continues. The authors have contributed to this process and have come to the view that the brittle-to-ductile transition reflects an increase in the ease with which slip is transmitted across grain boundaries. This paper reviews the evidence which supports this view and describes the mechanism.

C. T. Liu et al. (eds.), Ordered Intermetallics – Physical Metallurgy and Mechanical Behaviour, 371–389.
© 1992 *Kluwer Academic Publishers.*

The connection between the ductility of polycrystals and the transmission of slip across grain boundaries is not a new idea; nor is it restricted to $L1_2$ alloys. It is reflected in Gilman's (1966) paper in which he noted that the suppression of localized plastic flow can embrittle polycrystals. Also, the accommodation of plasticity near grain boundaries has been discussed in connection with a similar brittle-to-ductile transition in Fe-Mn steels (Strum et al. 1988). It was suggested (Schulson et al. 1986) in the present context for the reasons which will become apparent.

2. The Characteristics of the Brittle-to-Ductile Transition

In presenting the evidence it is worth summarizing some of the salient characteristics as revealed through the earlier work. (Ni_3Al is the focus at this point, unless otherwise noted.) They are:

•Single crystals of undoped, Ni-rich alloys are ductile in tension at room temperature and slip on systems of the form $\{111\}<110>$ (Copley and Kear 1967).

•Undoped polycrystals fracture intergranularly (Grala 1960).

•Deleterious segregants have not been detected at grain boundaries in material of high purity, which also fails intergranularly (Liu and Stiegler 1984, Ogura et al. 1985, Takasugi et al. 1985a). This is true also of Ni_3Si.

•Boron segregates to grain boundaries, as already noted. This occurs in alloys of all compositions, but possibly to a lesser degree in Al-rich material (Liu et al. 1985, Briant and Taub 1988, Lin et al.1991).

•Boron (100 - 1000w ppm) imparts extensive ductility (>50% elongation) to Ni-rich (24 Al) alloys (Liu et al. 1985), less to stoichiometric material (around 10 - 30%, depending upon grain size, Schulson et al. 1985), and none to Al-rich (26 Al) alloys (Liu et al. 1985).

•The doped (and ductile) Ni-rich alloys fracture transgranularly (Liu et al. 1985); the doped (and ductile) stoichiometric alloy fails through a mixture of intergranular and transgranular fracture (Schulson et al.1986); the doped (and brittle) Al-rich alloys fracture intergranularly (Liu et al. 1985).

Several inferences can be drawn. Dislocations in Ni_3Al are mobile, and sufficient independent deformation modes (five; Von Mises 1928) exist to allow polycrystals extensive deformation without cracking. That they are brittle implicates the grain boundaries. The absence of segregants suggests that the boundaries are intrinsically the "weak links" within the aggregate. That boron ameliorates the brittleness of the Ni-rich and stoichiometric alloys through segregation to the boundaries implies further that the transition is the result of a fundamental change in the structure and properties of the grain boundaries. And, the fact the boron does not impart ductility to Al-rich alloys, even though it still segregates to their grain boundaries, implies that it does not modify the boundary structure on the Al-rich side of the stoichiometric composition in the way that it does on the other side of stoichiometry.

The important implication, therefore, is that the brittleness and its selective amelioration through the addition of boron reflect primarily the behavior of the grain boundary region. The behavior of the lattice may be a factor as well, for boron-doped single crystals appear to be somewhat more ductile than undoped crystals (Heredia and Pope 1989). This, however, cannot be the primary factor in the face of the evidence implicating the boundaries.

3. The Grain Boundary Cohesion Model

The first explanation invoking the grain boundaries was that boron raises their cohesive strength (Liu et al. 1985, Takasugi et al. 1985b, and Taub and Briant 1987). The idea here is that the ability of the boundaries to support a tensile stress is increased. As a result, stresses above the yield strength may be supported and yielding and subsequent plastic flow may be spread throughout the aggregate before it cracks. Increased slip transmission across boundaries follows, but is thought to occur not because the localized shear stresses to activate the process are reduced, but because the boundary can support higher tensile stresses.

Supporting this view are calculations by Chen et al.(1989,1990). They found that both the surface energy, γ_s, and the grain boundary energy, γ_{gb}, decrease in the presence of boron and that the decrease in γ_{gb} is about twice the decrease in γ_s. Also, they found that these changes are greater for Ni-rich boundaries than for stoichiometric and for Al-rich boundaries. By implication the calculations suggest that the grain boundary cohesive energy, γ_c, is increased and that the increase is greatest for the most ductile alloys. (γ_c is defined as $\gamma_{s1} + \gamma_s -2\gamma_{gb}$ where 1 and 2 denote the two grains.) Unfortunately, there are no experimental measurements for comparison.

Additional support, it could be argued, is the fact that the boundaries in the ductile alloys support a higher far-field or applied tensile stress. The problem with this argument is that it ignores stress concentrations. These may be lower in the ductile alloys, as is argued below, in which case their boundaries may actually be required to support lower localized tensile stresses.

The main problem with the cohesion model is that even the cohesive energy of the Al-rich boundaries is calculated to increase upon the presence there of boron, yet these alloys remain brittle.

4. The Slip Transmission Model

The alternative explanation is based upon the slip transmission model (Schulson et al. 1986). The idea here is that boron, through a mechanism described below, leads to an easing of slip transmission across grain boundaries by lowering the localized shear stresses which are required to activate the process. In other words, unlike the cohesion argument in which boron leads to an increase in the localized tensile stresses needed to nucleate cracks without affecting the shear stresses needed to transmit slip, the argument here is that boron leads to a reduction in the stresses required for slip transmission without significantly affecting the boundary cracking stresses. Figure 1 sketches the distinction. It is possible, of course, that boron may both ease slip transmission and raise the cohesive strength of the boundaries, but it is probable that one effect is more important than the other.

How to make the distinction unambiguously is not easy. Yet, in the interests of alloy design, it is important to distinguish. The approach taken here is that if the ductile-to-brittle transition is truly related to an easing of the slip transmittal process, then the grain boundaries within the ductile alloys should impede slip less. As will become apparent, the evidence suggests that this is the case. This characteristic cannot easily be explained in terms of the cohesion model, and thus serves as the principal support for the alternative explanation.

In reviewing the evidence it seems appropriate to first describe briefly some of the technical details of our experimental work. Further information may be obtained from the references cited.

374

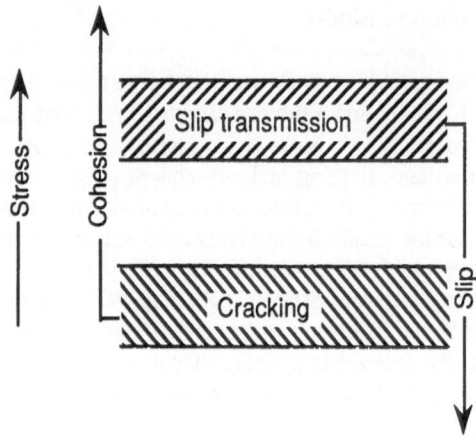

Figure 1.   Schematic illustration of the distinction between the grain boundary cohesion and the slip transmission models. In the former boron raises the tensile stress for cracking. In the latter model boron leads to a reduction in the shear stress required to transmit slip across grain boundaries.

4.1. THE MATERIALS

Three Ni-based systems were examined (all compositions are given in at. %): Ni-rich (24 Al), stoichiometric (25 Al), and Al-rich (26 Al) Ni_3Al with (0.3) and without boron; Ni-rich Ni_3Si (23 Si) with (0.2) and without boron; and Ni-rich Ni_3Ge (23.5 Ge) with (0.1) and without boron. In each case the alloy was prepared as an ingot, transformed to powder, canned in stainless steel and then hot-extruded to rod. Detailed procedures and chemical analyses are given elsewhere (Schulson et al. 1985, 1986, 1990; Baker et al. 1990b). This processing produced fully recrystallized, chemically homogeneous material with an equiaxed microstructure. A small number density of sub-micron sized second phase particles was present at the grain boundaries in both the boron-doped silicide and the doped germanide; although not unambiguously identified, the particles are probably Ni-rich borides given their enrichment in Ni and B. Otherwise, the boundaries were very clean, at least as far as Auger electron spectroscopy could detect. X-ray diffraction and pole figure analyses of the undoped, stoichiometric Ni_3Al showed no texture, implying that the grains in this alloy were randomly oriented. Such analyses were not performed on the other alloys, but transmission electron diffraction patterns from several grains revealed no evidence of a strong texture. Both tensile and compression specimens were prepared and, prior to testing at room temperature at an initial strain rate of $10^{-4}s^{-1}$, were annealed at an elevated temperature(1000 - 1200°C) for different lengths of time to vary the grain size. Subsequently, the specimens were annealed at an intermediate temperature (600 - 700°C) for 0.5 h and then air cooled. Before annealing the tensile specimens were electrochemically polished. Transmission electron microscopy revealed that prior to deformation all alloys were remarkably free from dislocations (i.e., the dislocation density was less than about $10^8/m^2$).

4.2 THE OBSERVATIONS

4.2.1 *Ni₃Al* The first important point is that yielding always precedes fracture (Schulson et al. 1985, 1986,1991). This feature was observed from tensile tests on each brittle alloy and is illustrated in Figure 2. The yielding occurs discontinuously, as it does for both weakly ordered (Ni₃Mn, Arko and Liu 1971; Ni₃Fe, Johnston et al. 1965; and Cu₃Au, Sastry 1976) and other strongly ordered L1₂ alloys (Zr₃Al, Schulson and Roy 1978; Ni₃Si, Schulson et al. 1990; Ni₃Ge, Fang and Schulson 1991), and corresponds to the propagation of a Lüders band along the gauge section. In the brittle alloys failure often occurs before the band propagates all the way along the gauge, as it did for the specimen from which Figure 2 was obtained. The point is important because it signifies that dislocation slip is a prerequisite to brittle intergranular fracture. In other words, cracks must first nucleate and the mechanism of nucleation involves dislocations.

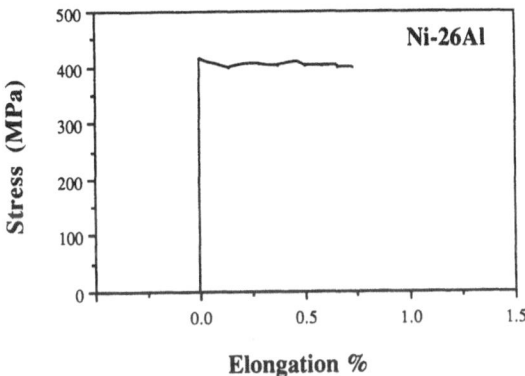

Figure 2.   Stress-strain curve for Al-rich Ni₃Al illustrating discontinuous yielding. The yield strength is defined as the stress at the plateau.

The coupling of slip and fracture is also evident from another observation. For each brittle alloy selected area electron channelling patterns from the fracture surfaces of specimens broken in tension are blurred (Schulson, Davidson, Viens 1983; Hanada et al. 1986a, Baker et al. 1990a), Figure 3. In comparison, patterns from grains along the gauge section of unbroken specimens are as sharp as from unloaded material. This is consistent with the fact that localized yielding precedes fracture. Furthermore, because the electron channelling patterns originate from within around 50nm of the surface when generated using conventional scanning electron microscopes, this observation suggests that the deformation was concentrated within the grain boundary region.

The concentration is probably in the form of dislocation pile-ups at grain boundaries. While it is impossible to reveal such features within the interior of bulk specimens, they have been seen within thin foils which were deformed in-situ in a transmission electron microscope (Baker et al. 1987). Figure 4 shows an example in a foil of undoped, stoichiometric Ni₃Al. It seems reasonable to assume that they also develop within bulk material. (Incidentally, from careful in-situ TEM deformation studies one cannot easily distinguish one alloy from another, brittle or ductile. In other words, foils from different alloys display similarities which are much greater than the differences. This contrasts with the mechanical behavior of bulk specimens, and reflects presumably the difference in stress states.)

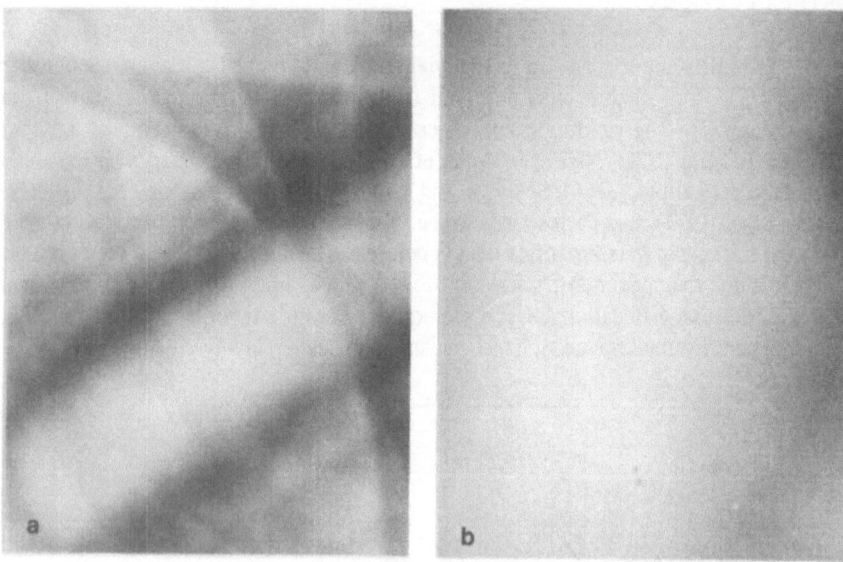

Figure 3.   Selected area electron the channelling patterns from stoichiometric Ni3Al: a) from a grain along the undeformed gauge section of a tensile specimen; b) from an intergranular facet on the fracture surface. Plasticity localized near the grain boundary distorts the pattern. (Schulson, Davidson and Viens 1983).

From these observations a picture begins to emerge. Under increasing tensile stress undoped Ni3Al polycrystals of any composition first yield locally and work harden. Once the hardening has become sufficient the material then begins to transmit slip into the unyielded, adjacent regions. Throughout this process the grain boundaries act as barriers against which dislocations pile up. Stress is thus concentrated. Eventually, but generally before the whole aggregate has yielded, at some site the local stress becomes high enough to nucleate cracks. Either immediately or shortly thereafter the cracks propagate and brittle fracture ensues.

What boron does in the cases where it is effective, according to the slip transmission model, is to lead to an easing of the transmission of slip from grain to grain. In other words, boron reduces the grain boundary blockage of slip. As a result, the stress concentrated on the boundaries is reduced, thereby reducing the likelihood of crack nucleation.

The primary experimental support for this view (Schulson et al. 1986, 1991) is that in the alloys where boron is effective the grain boundaries contribute less to the yield strength when they contain boron. For the Ni-rich and the stoichiometric alloys this is evident from the slopes of plots of yield strength vs (grain size)$^{-n}$, Figures 5a and 5b. When the boundaries contain boron the slopes are lower by about 40% (i.e., for Ni-rich from 2360 to 1370 MPa.μm$^{0.8}$ and for stoichiometric from 2080 to 1200 MPa.μm$^{0.8}$). For the Al-rich alloy, which remains brittle upon doping, the slope does not change significantly when boron is present, Figure 5c (i.e., from 1650 to 1600 MPa.μm$^{0.8}$). Of course, the intercept for the doped alloys is higher, owing to a solid solution strengthening of around 400 MPa/ at. % B (Baker et al. 1988a) which is largely independent of the Ni:Al ratio (Schulson et al. 1991). (Incidentally that n = 0.8 in Fig.5, and not

0.5, is irrelevant to the argument. The reason for this dependence is explained in Schulson et al. 1985, 1986, 1991.)

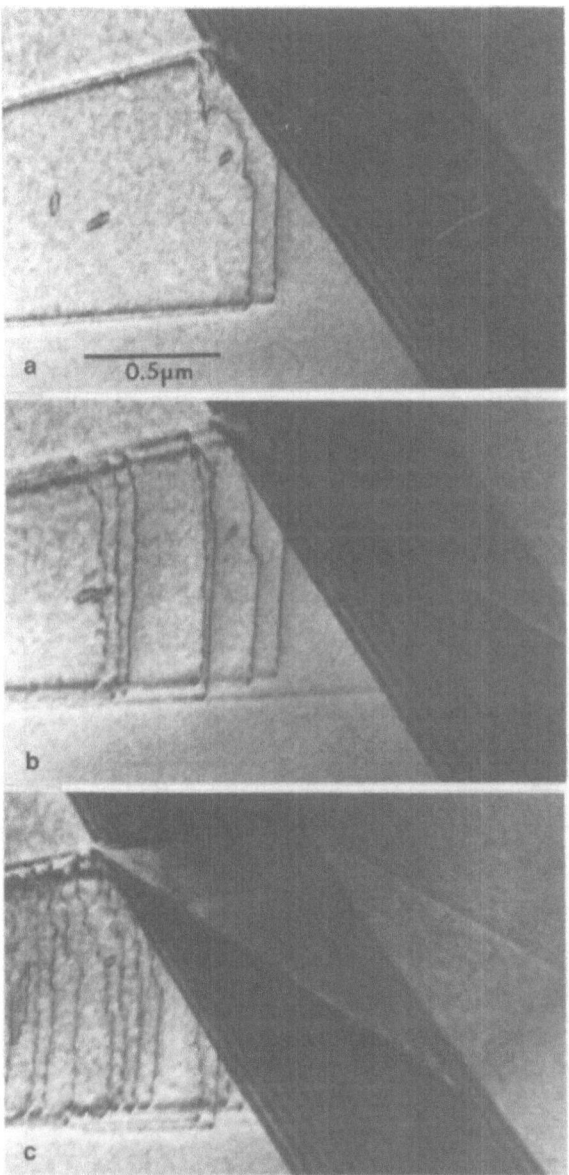

Figure 4.   Transmission electron micrographs of stoichiometric Ni₃Al deformed in-situ in a TEM: a) dislocations begin to pile up at a grain boundary; b) with further applied strain more dislocations pile up; c) eventually, slip is transmitted across the boundary ahead of the pile-up. (Baker, Schulson and Horton 1987).

378

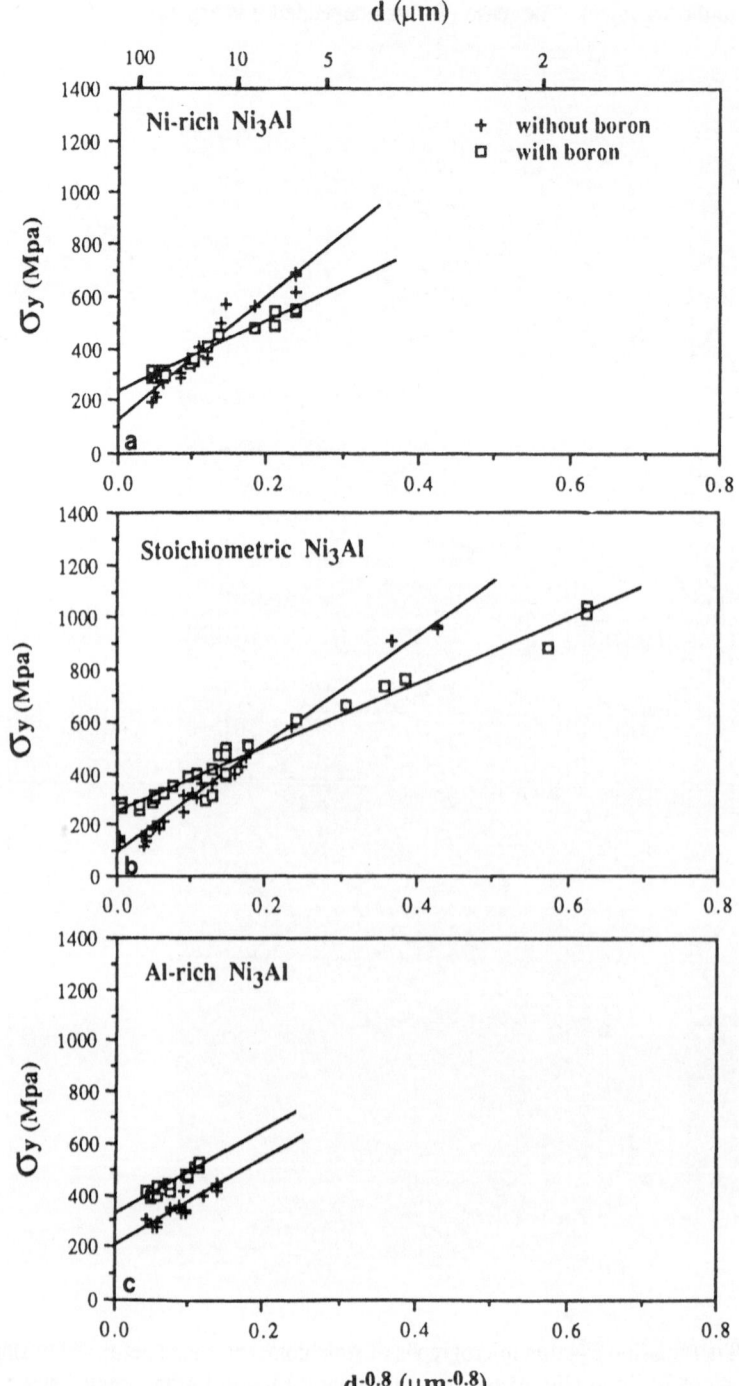

Figure 5.   Graphs of yield strength (defined in Fig.2) versus (grain size)$^{-0.8}$ for Ni$_3$Al of different compositions with and without boron: a) Ni-rich; b) Stoichiometric; c) Al-rich. Boron lowers the slope of the curves for the Ni-rich and the stoichiometric alloys, but has no significant effect for the Al-rich alloy. (Schulson et al. 1986, 1991).

4.2.2 *Ni3Si and Ni3Ge* Although to a lesser degree, brittle-to-ductile transitions are also seen in Ni-rich compositions of Ni3Ga and Ni3Si upon microalloying with boron, as noted above, and in Ni3Ge (Fang and Schulson 1990). For instance, the ductility of Ni3Si with 0.2 B increases from essentially nothing to around 5%, and the ductility of Ni3Ge with 0.1 B from nothing to only about 1.5 to 2%. In these cases, too, the element segregates to the grain boundaries. Single crystals of Ni3Ga (Takeuchi and Kuramoto 1973) and Ni3Ge (Aoki and Izumi 1978) are ductile, and polycrystals of undoped Ni3Si can be compressed plastically without fracturing. Again, therefore, the transition is attributed primarily to a change in the structure of the grain boundaries.

As in Ni3Al, discontinuous yielding precedes fracture, blurred selected area electron channelling patterns are obtained from intergranular fracture facets, and dislocations pile up at grain boundaries. The chain of events leading to failure appears to be similar in each L1$_2$ case, implying that the same mechanism accounts for the transition.

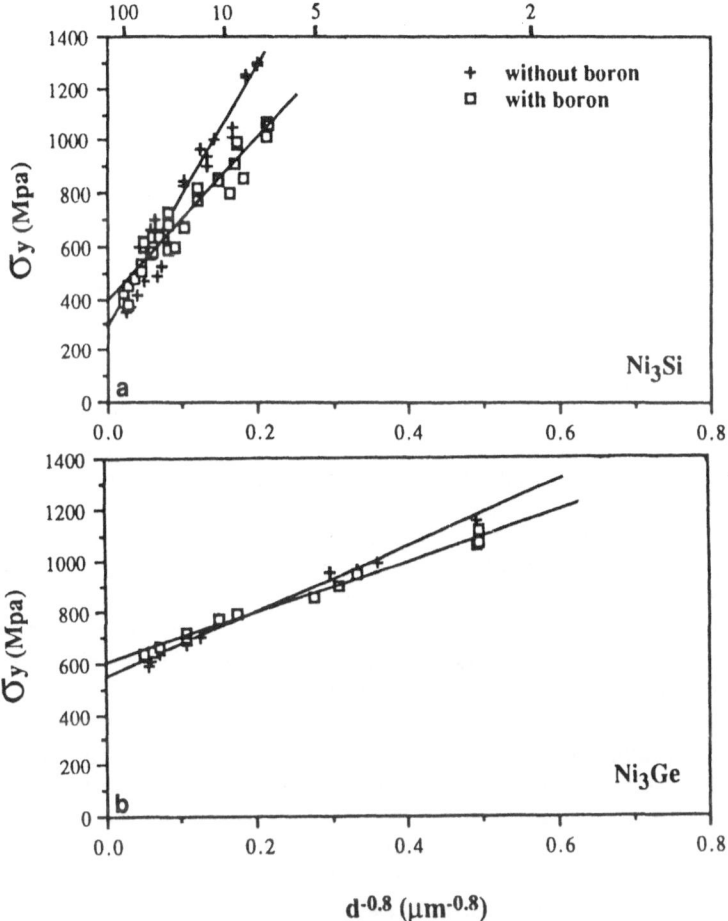

Figure 6. Graphs of yield strength (defined in Fig. 2) versus (grain size)$^{-0.8}$ for a) Ni-rich Ni3Si and for b) Ni-rich Ni3Ge, with and without boron. Boron lowers the slope in both cases. (Schulson, Briggs and Baker 1990; Fang and Schulson 1991).

Figures 6a and b, respectively, show that the slopes of graphs of yield strength vs (grain size)$^{-0.8}$ for Ni$_3$Si (Schulson et al. 1990) and for Ni$_3$Ge (Fang and Schulson 1991) again decrease upon the addition of boron (i.e., for Ni$_3$Si from 5010 to 3120 MPa.μm$^{0.8}$, and for Ni$_3$Ge from 1270 to 980 MPa.μm$^{0.8}$). Curiously, the relative reduction for Ni$_3$Si is about the same as for both Ni-rich and stoichiometric Ni$_3$Al (i.e., about 40%); for Ni$_3$Ge, it is closer to 20%. It thus appears that in these cases, too, boron eases the transmission of slip across grain boundaries. It is expected that Ni$_3$Ga will exhibit similar behavior.

At this juncture it is seen that in the four cases where the addition of boron increases the ductility of an L1$_2$ intermetallic, it also eases the transmission of slip across grain boundaries. In the one case where the element has no effect on ductility, it has none on slip transmission.

It is noted that implicit in the foregoing discussion is the assumption that the boron-induced change (or the lack of it) in the slope of the strength-(grain size)$^{-0.8}$ plots is truly reflective of a change in dislocation-grain boundary interactions and not in the propensity for cross slip of dislocations within the lattice. This assumption rests on two other observations; viz., that slip is highly planar in these alloys, and that the dislocation structure in the doped alloys (Ni-rich, stoichiometric and Al-rich Ni$_3$Al; Ni$_3$Si and Ni$_3$Ge) is indistinguishable from the structure of the undoped alloys (Baker and Schulson 1985; Schulson et al. 1991), at least at the level of weak-beam transmission electron microscopy. Implicit, too, is the assumption that dislocation-solute interactions are not important, because they would probably lead to an increase in slope. The other assumption is that boron does not affect the distribution of special boundaries, in keeping with the observations of Mackenzie et al. (1988).

4.2.3 *Zr$_3$Al* Before the ductile-to-brittle transition was reported in Ni$_3$Al one was reported in Zr$_3$Al, an experimental nuclear alloy and another strongly ordered L1$_2$ intermetallic (Schulson and Roy 1977). In this case the transition related to notch brittleness, for Zr$_3$Al in the form of smooth tensile specimens is ductile. Microalloying (36±ppmw) with boron had no effect. On the other hand, irradiation with energetic neutrons (E>1 Mev; 1.4 - 5 × 10^{24} n/m^2 or 0.1 - 0.4 displacements per atom) suppressed notch brittleness and changed the mode of fracture from intergranular to transgranular. Correspondingly, the irradiation reduced the contribution of the grain boundaries to the yield strength (Schulson 1978) by about 35%, Figure 7 -- curiously, by about the same amount as boron affects the same property in Ni$_3$Al and Ni$_3$Si. Little change in slip character was detected, implying again that the effect is related to the grain boundaries.

Thus, once more a correspondence is seen between the suppression of brittle behavior (notch brittleness in this case) and an improvement in the ease of slip transmission across grain boundaries of strongly ordered L1$_2$ polycrystals. In all these instances the correspondence, it is suggested, is not fortuitous, but indicative of a fundamental link.

4.2.4 *Other Cases* In all the cases in which the correspondence is seen, the transition is from essentially no ductility to some ductility. In such cases it seems that an easing of an early step in the deformation process should encourage ductility. In cases where alloying leads to more ductility where significant ductility already exists, such as Ni-23Al-2Pd (Chiba et al. 1991b) microalloyed with boron which increases the ductility from around 12% to around 40%, slip transmission appears to be adequate in the first place. In such cases the improvement in ductility is probably related to another process.

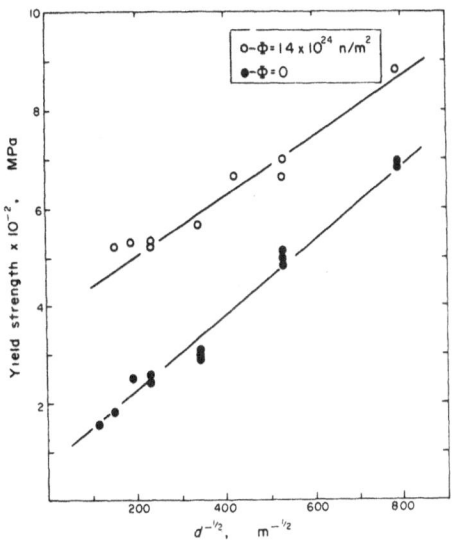

Figure 7.   Graph of yield strength versus (grain size)$^{-0.5}$ for Zr$_3$Al before and after neutron irradiation. The irradiation lowers the slope. (Schulson 1978).

4.3 THE GRAIN BOUNDARY CHEMISTRY AND STRUCTURE

The question now is: how is the transfer of slip made easier? As discussed below, it appears to be related to the constitutional disordering of the grain boundary region and to the attendant reduction or elimination of the anti-phase boundary barrier to the nucleation of superlattice partial dislocations. The evidence was obtained primarily from detailed examinations of grain boundaries in Ni$_3$Al using a combination of electron optical methods.

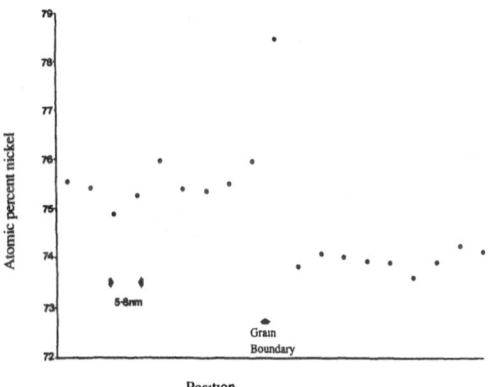

Figure 8.   Plot of Ni concentration across a grain boundary in stoichiometric Ni$_3$Al doped with boron. Note the enrichment at the boundary. (Baker, Schulson and Michael 1988).

Figure 8 shows a plot of the Ni concentration across a grain boundary in stoichiometric Ni₃Al with boron (Baker et al. 1988b). The data were obtained from a thin foil which was examined using a high resolution scanning transmission electron microscope and a beam of 1nm diameter. The plot shows a boundary region about 1nm wide in which the Ni concentration is higher than in the matrix. A similar examination of boundaries in undoped material revealed no Ni enrichment. The implication is that boron led to co-segregation of Ni, as had been predicted (Frost 1987) and observed earlier (Seiloff et al. 1987). The further implication is that the co-segregation led to a reduction in the localized degree of long range order.

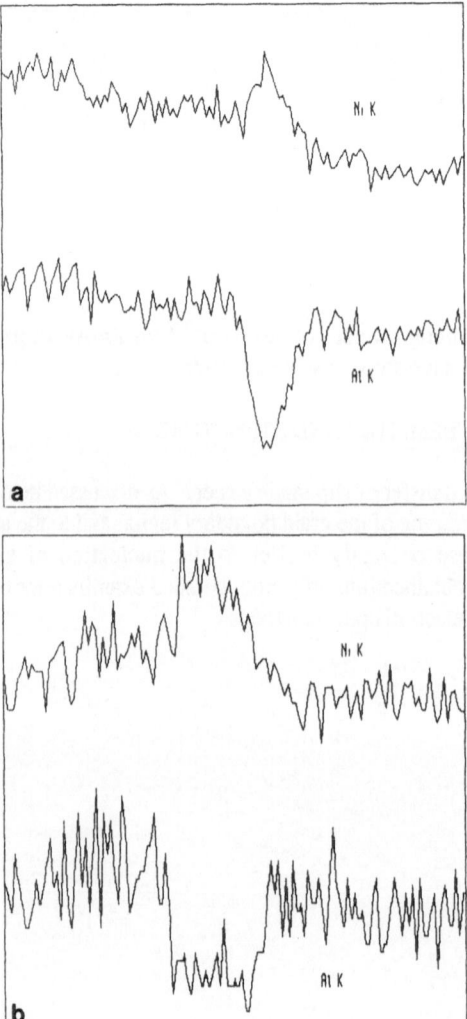

Figure 9.   Plots of Ni and Al concentration across grain boundaries in a) Ni-rich Ni₃Al and in b) Ni-rich Ni₃Al doped with boron. The boundaries are enriched in Ni. Boron increases the enrichment. (Baker, Schulson, Michael and Pennycook 1990b).

Figure 9 shows X-ray line scans across grain boundaries (which were oriented normal to the foil surface) in Ni-rich Ni$_3$Al (Baker et al. 1990). Without boron the boundaries are enriched in Ni relative to the matrix, in some cases highly enriched. With boron every boundary is highly enriched in Ni, to the point that that little Al remains. Figure 10 is a Z-contrast STEM image from a grain boundary in the same doped alloy and confirms the Ni enrichment. Extensive constitutional disordering is implied.

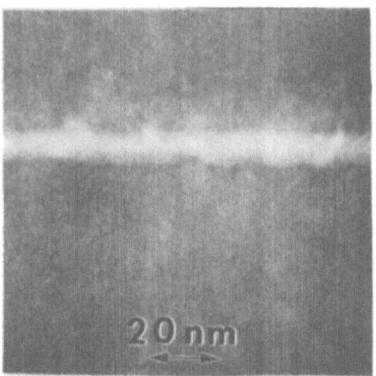

Figure 10.   Z-contrast STEM image of a grain boundary in Ni-rich Ni$_3$Al doped with boron. The bright band indicates Ni-enrichment within the region of the boundary and confirms the observation from Fig. 9b. (Baker, Schulson, Michael and Pennycook 1990b).

Figure 11 shows an X-ray line scan across a grain boundary in Al-rich Ni$_3$Al with boron (Baker et al. 1990b). In this case the boundary is indistinguishable from the matrix, even when boron is present. No disordering is implied.

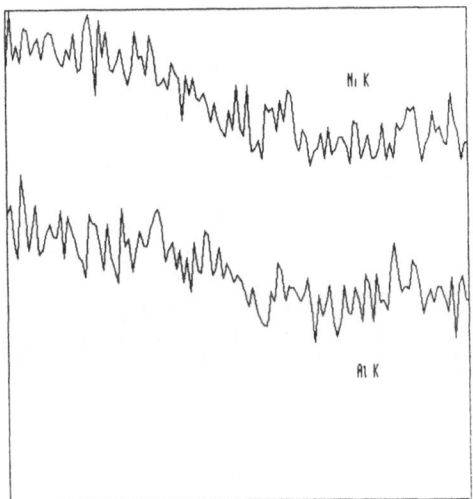

Figure 11.   Plots of Ni and Al concentration across a grain boundary in Al-rich Ni$_3$Al doped with boron. The boundary cannot be distinguished from the matrix. (Baker, Schulson, Michael and Pennycook 1990b).

384

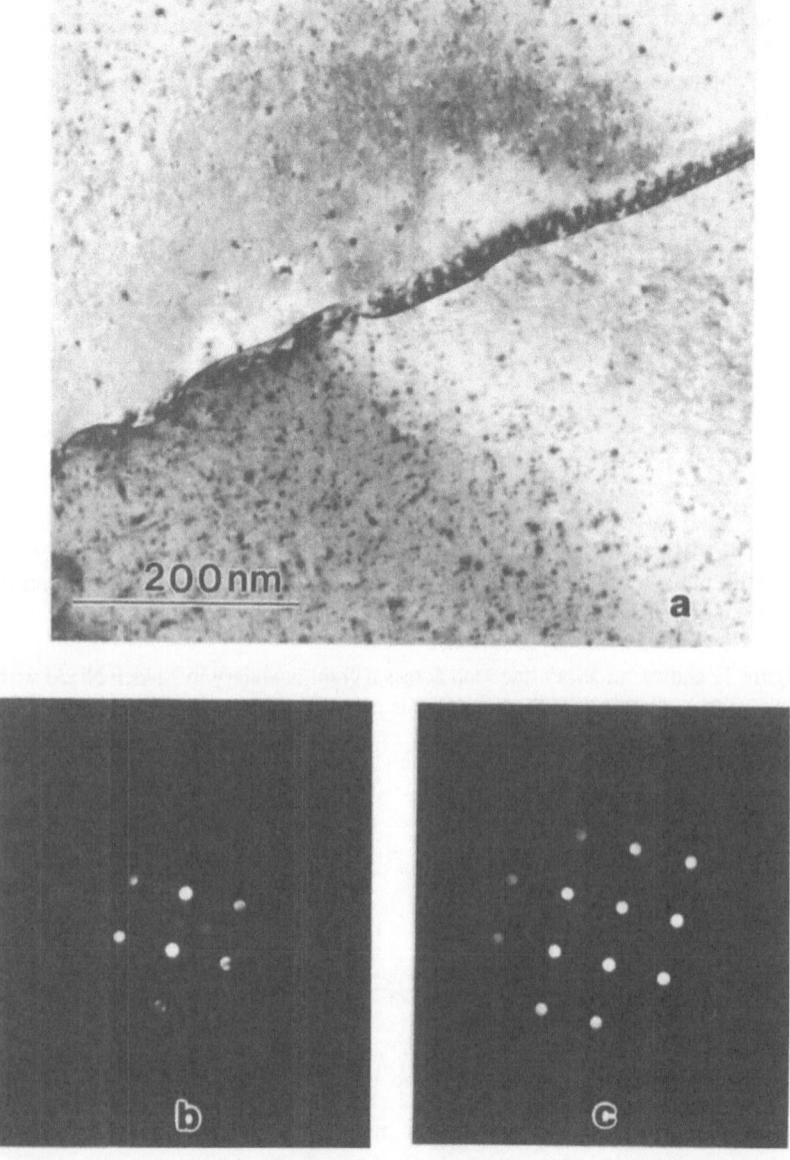

Figure 12.   a) Transmission electron micrograph of Ni-rich Ni3Al doped with boron. Note the rope-like feature along the grain boundary, bonded to one of the grains. Electron diffraction patterns b) from the matrix and c) from the grain boundary phase. Note the absence of superlattice reflections from the boundary feature, implying that it is a disordered region. (Baker and Schulson 1989a).

Figure 12a shows a transmission electron micrograph of the doped, Ni-rich alloy from which the X-ray line scan in Fig.9b was obtained (Baker and Schulson 1989a). The feature decorating the boundary was seen on all boundaries studied, either on one side or alternating from side to side along the boundary (Baker and Schulson 1989a). It corresponds to the region highly enriched in Ni and, although generally only a few nanometers wide, was as wide as 20nm in places. This feature was also seen on some boundaries in the undoped, Ni-rich alloy and on some segments of some boundaries in this alloy. Figures 12b and 12c show the corresponding electron diffraction patterns. The pattern from the matrix (12b) shows both fundamental and superlattice diffraction spots, corresponding to the $L1_2$ lattice. The pattern from the boundary phase (12c) shows the same fundamental pattern, but not the superlattice spots, and corresponds to the face-centered cubic lattice. The boundary phase is thus a constitutionally disordered region coherently or semi-coherently bonded with the matrix of one of the grains. This feature was not detected in doped, stoichiometric Ni_3Al, probably because it was too narrow to be resolved. Nor was it seen in the Al-rich alloys, probably because it did not exist.

A disordered grain boundary region in Ni_3Al has also been reported by Mackenzie and Sass (1988) and by Kung and Sass (1991), but has not been found by Mills (1989), Krzanowski (1989) or Horton and Liu (1990). Sass et al. used high-resolution transmission electron microscopy /lattice imaging and revealed the feature on two, high-angle boundaries in Ni-rich Ni_3Al doped with boron. They reported its width to be about 2nm; i.e., narrower than the largest feature reported by Baker et al. Also, they found no evidence of the feature on low-angle or special boundaries, in agreement with the observation of Baker et al. (1990b) who did not detect it on twin boundaries. Why Mills did not see it may be related to his examination of special boundaries. Why others did not see it is not clear, but may be related to a number of factors, including examining the "wrong" side of the boundary, examining "special" boundaries, and using electron probes of insufficient spatial resolution. The other explanation is that the feature was not there, owing possibly to differences in heat treatment, in the Ni:Al ratio, or in the interstitial concentration.

The disordered phase appears to be somewhat elusive. This character is further evident from recent work by Xu (1991) who performed conventional transmission electron microscopy on a series of five specimens of boron-doped, Ni-rich Ni_3Al heat treated in different ways. She could detect the disordered region (about 20nm wide) on the grain boundaries in one specimen only, suggesting that the development of the feature is rather sensitive to the thermal history of the material. These observations do not preclude the possibility that a thin disordered region was present, but they underline the need for further investigation.

Elusive though it is, the constitutionally disordered region on high-angle grain boundaries, we suggest, is the primary microstructural feature which aids the transfer of slip.

4.4 THE MECHANISM OF SLIP TRANSMISSION

How the transmittal mechanism works is not clear from the studies to date, but may operate as follows. Generally, slip is probably not transmitted by dislocations from one grain penetrating the boundary and entering the adjacent grain, because the relative orientations of the grains are such that the boundaries block dislocations. Slip must then be re-nucleated, either on the other side or on a different system within the same grain. Presumably, the process occurs through the nucleation of $a/2 <110>$ dislocations (or partials thereof) and is localized to within a few dislocation core spacings from the boundary. The structure of the boundary and its immediate surroundings is then important.

When the boundary region is fully ordered, the dislocation is a partial superlattice dislocation and its nucleation requires the creation of an antiphase boundary (APB). This consumes energy, γ_{apb}, and raises the shear stress for the process by γ_{apb}/b, where b is the Burgers vector of the dislocation. In Ni3Al the increase is about $0.11/(0.25 \times 10^{-9}) = 440$ MPa (where γ_{apb} on (111) is about 110 mJ/m^2 (Douin et al. 1986) and b = 0.25nm) which, when converted to a normal stress, is of the order of the macroscopic yield strength. The removal of this barrier through localized constitutional disordering thus lowers significantly the stress for nucleation. This reduction, we suggest, is the key to the enhanced transmission. In other words, the disordered, high-angle boundaries appear to be better sources of dislocations than the ordered boundaries.

The width of the disordered zone may not be important, provided it is at least as wide as a few dislocation core sizes.

4.5 FURTHER DISCUSSION

That the issue of the boron-induced, brittle-to-ductile transition reduces to one of a disordered grain boundary region is consistent with earlier observations. Hanada et al.(1986b) noted such a region decorating the boundaries in recrystallized Ni-rich Ni3Al (22.5 Al) without boron. The material exhibited 15% elongation which is less than the 52% reported by Liu et al. (1985) owing probably to the nonuniform distribution of the boundary phase. Takasugi et al. (1985b) and Dimiduk et al. (1987) macroalloyed Ni3Al with Mn or Fe and noted an association between ductility and the presence of a grain boundary phase. And Chiba et al. (1991a) noted a marked correlation between ductility and the tendency for a disordered phase to form within Ni3Al upon macroalloying, and an equally marked correlation between brittleness and the tendency to form an ordered phase upon macroalloying. (Incidentally, Chiba's later work (1991b) on Ni-23Al-2Pd indicated no reduction in the contribution of the grain boundaries to the yield strength when boron is present, even though the element raises the ductility of the alloy from around 12% to around 40%. This is not necessarily surprising because Pd is one of the elements which disorders the alloy, and so it is possible that the grain boundaries were disordered before boron was added.)

What evidence is there that a disordered region encases the grains of the other ductile L1$_2$ alloys? Auger electron spectroscopy (Baker et al. 1989) of grain boundaries in the Ni-rich Ni3Si alloy described above indicates an enrichment of Ni in the presence of boron (i.e., an increase of about 30% in the Ni:Si ratio), as does a similar analysis of grain boundaries in the ternary L1$_2$ alloy Ni-18.9Si-3.2Ti doped with boron (Oliver and White 1987). Direct evidence, however, does not exist.

The Zr3Al result, perhaps, deserves further discussion. In this case, localized disordering is still thought to be important, but not constitutional disordering. Instead, irradiation disordering of the grain boundaries has been suggested (Schulson 1984). The idea here is as follows. Under energetic particle irradiation both vacancies and interstitials are created in equal number. Disordering then occurs through a variety of mechanisms. At the same time reordering occurs through atom-vacancy exchanges. A dynamic equilibrium then sets in. Vacancies created near grain boundaries participate in fewer reordering exchanges before they are lost to the boundary sinks. As a result, the degree of long-range order near the grain boundaries may be lower than within the lattice. Again, although the details differ, the structure of the boundaries changes in a way which allows slip to be more easily transmitted through a reduction in the energy barrier to dislocation nucleation.

Another point to note is that in earlier discussions (Schulson et al. 1986) it was suggested that the slip transmission was aided by an enhancement through boron in the movement of residual grain boundary dislocations. Indeed, theory (Frost 1987; King and Yoo 1987) shows that the number of dislocation-grain boundary reactions increases significantly upon disordering the boundary plane, and so such an enhancement might be expected. However, Swiatnicki and Grabski's (1989) in-situ TEM annealing experiments on the effect of boron on the mobility of grain boundary dislocations could be taken to suggest the opposite response. Barring the possibility that this result is inapplicable because mechanical behavior at elevated temperatures is different from that at room temperature, or that the result is questionable owing to limitations in resolution, this development does not negate the view taken. Easing the nucleation of dislocations is the important point, and that process is achieved by reducing the degree of long-range atomic order around the grain boundaries.

The other point concerns the application of the slip transmission model to boron-induced, brittle-to-ductile transitions in strongly ordered polycrystals having different superlattices. If dislocations pile up at grain boundaries, then the model may apply. But if pile-ups do not form, then the model does not apply. An example of the latter case is the B_2 compound FeAl. In-situ TEM experiments (Baker et. al 1991) did not reveal pile-ups in Fe-40 Al. In this case, therefore, the fact that the transition (Liu and George 1990) is accompanied by an increase (and not a decrease) in the grain boundary contribution to the yield strength (Liu 1991) does not contradict any of the ideas presented here. Presumably, the Hall-Petch type slope in Fe-40 Al is not directly related to slip transmission, but to an intragranular process.

5. Grain Boundary Disorder: A Requirement for Ductility?

Finally, one wonders whether grain boundary disorder of any origin - constitutional, thermal or irradiation- may be a requirement for extensive ductility, not just of the $L1_2$ intermetallics but of intermetallics in general, or at least those with sufficiently simple crystal structures (B2, DO_3, DO_{19}, DO_{22}). This suggestion is discussed elsewhere (Baker and Schulson 1989b).

6. Conclusion

It is argued that the improvement in the ductility of Ni_3Al and of other strongly ordered $L1_2$ intermetallics upon microalloying with boron is related to a decrease in the degree to which grain boundaries impede slip and, correspondingly, to an increase in the ease with which slip is transmitted across grain boundaries. It is proposed that the transmittal mechanism is dislocation nucleation at the heads of dislocation pile-ups, and that this is made easier by the constitutional disordering of the grain boundaries. Owing to the elimination of the antiphase boundary barrier, the boundaries may then be viewed as more effective sources for dislocations.

7. Acknowledgements

We would like to acknowledge the many people who through their work contributed to the research reviewed here-D.V. Viens, T.P. Weihs, V. Zinoviev, B. Huang, L.J. Briggs, Y. Xu, S. Guha, J.Fang, H.J. Frost, J.R. Michael, J.A. Horton, S.J. Pennycook, R.A. Padgett, P.R. Munroe and L. Nazé.

388

The work was supported by U.S. Dept. of Energy, Basic Energy Sciences Program, grant numbers DE-FG02-86-ER45260 and DE-FG02-87ER45311.

8. References

Aoki, K. and Izumi, O. (1978), J. Mater. Sci., **13**, 2313.
Aoki, K. and Izumi, O. (1979), Trans. J. Japan. Inst. Met., **43**, 1190.
Arko, A.C. and Liu, Y.H., (1971), Metall. Trans., **2**, 1875.
Baker, I. and Munroe, P.R. (1988), J. Metals, **40**, 28.
Baker, I. and Schulson, E.M. (1985), Phys. Stat. Sol. (a), **89**, 163.
Baker, I. and Schulson, E.M. (1989a), Scripta Metall., **23**, 1883.
Baker, I. and Schulson, E.M. (1989b), Scripta Metall., **23**, 345
Baker, I., Huang, B. and Schulson, E.M. (1988a), Acta Metall., **36**, 493.
Baker, I., Padgett, R.A. and Schulson, E.M. (1989), Scripta Metall., **23**, 1969.
Baker, I., Schulson, E.M. and Horton, J.A. (1987), Acta Metall., **35**, 1533.
Baker, I., Schulson, E.M. and Michael, J.R. (1988b), Phil. Mag. B, **57**, 379.
Baker, I., Schulson, E.M., Michael, J.R. and Padgett, R.A. (1990a), J. de Physique, **51**, C1-77.
Baker, I., Schulson, E.M., Michael, J.R. and Pennycook, S.J. (1990b), Phil. Mag. B, **62**, 659.
Baker, I., Nagpal, P. Guha, S. and Horton, J.A. (1991), Proc. Inter. Symp. on Intermet.
    Compds., JIMIS-6, Japan Inst. of Metals, ed. by O. Izumi, 603.
Briant, C.L. and Taub, A.I. (1988), Acta Metall., **36**, 2761.
Chen, S.P., Strolovitz, D.J. and Voter, A.F. (1989), J. Mater. Res., **4**, 62.
Chen, S.P., Voter, A.F., Albers, P.C., Boring, A.M. and Hay, P.J. (1990), J. Mater. Res., **5**,
    955.
Chiba, A. , Hanada, S. and Watanabe, S. (1991a), Scripta Metall. et Mater. **25**, 303.
Chiba, A., Hanada, S. and Watanabe, S. (1991b) Scripta Metall. et Mater., **25**, 1053.
Copely, S.M. and Kear, B.H. (1967), TMS-AIME, **239**, 977.
Crimp, M.A., Vedula, K.M. and Gaydosh, D.J. (1987), Proc. MRS, **81**, 499.
Dimiduk, Weddington, V.L. and Lippsitt, H.A. (1987), Proc. MRS., **81**, 221.
Douin, J., Veyssiere, P. and Beauchamp, P. (1986), Phil. Mag. A, **54**, 375.
Fang, J. and Schulson, E.M. (1991) unpublished work.
Fang. J. and Schulson, E.M., (1990), Proc. MRS., **213**, 751.
Frost, H.J. (1987), Acta Metall., **35**, 519.
Gilman, J.J. (1966), T.A.S.M., **59**, 597.
Grala, E.M. (1960), Proc. Symp. on Mech. Prop. of Intermet. Cmpds., Electrochem Soc., ed. by
    J.H. Westbrook, Johy Wiley & Sons, New York, p. 358.
Hanada, S., Ogura, T., Watanabe, S., Izumi, O. and Masumoto, T. (1986a), Acta Metall., **34**,
    13.
Hanada, S., Watanabe, S. and Izumi, O. (1986b), J. Mater. Sci.,**21**, 203.
Heredia, F.E. and Pope, D.P. (1989), MRS, **133**, 287.
Horton, J.A. and Liu, C.T. (1990), Scripta Metall., **24**, 1251.
Johnston, T.L., Davies, R.G. and Stoloff, N.S. (1965), Phil. Mag., **12**, 305.
King, A.H. and Yoo, M.H. (1987) Scripta Metall., **21**, 1115.
Krzanowski (1989), Scripta Metall., **23**, 1219.
Kung, H. and Sass, S.L. (1991). Acta Metall. et Mater., (in press).
Lin, T.L., Chen, D. and Lin H. (1991), Acta Metall. et Mater., **39**, 523.

Liu, C.T. (1991) Scripta Metall. **25**, 1231.

Liu, C.T. and George, E.P. (1990), Scripta Metall. et Mater., **24**, 1285.

Liu, C.T. and Steigler, J.O. (1984), Science, **226**, 636.

Liu, C.T., White, C.L. and Horton, J.A. (1985), Acta Metall., **33**, 213.

Mackenzie, R.A.D. and Sass, S.L. (1988), Scripta Metall., **22**, 1807.

Mackenzie, R.A.D., Vaudin, M.D. and Sass, S.L. (1988), Proc. MRS, **122**, 461.

Mills, M.J. (1989), Scripta Metall., **23**, 2061.

Ogura, T., Hanada, S., Masumoto, T. and Izumi, O. (1985), Metall. Trans. A, **16A**, 441.

Oliver, W.C. and White, C.L. (1987), Proc. MRS, **83**, 241.

Sastry, S.M.L. (1976), Mater. Sci. Engg., **22**, 237.

Schulson, E.M. (1978), Acta Metall., **26**, 1189.

Schulson, E.M. and Roy, J.A. (1977), J. Nucl. Mater., **71**, 124.

Schulson, E.M. and Roy, J.A. (1978), Acta Metall., **26**, 29.

Schulson, E.M.(1984), Intermet. Met. Rev., **29**, 195.

Schulson, E.M., Briggs, L.J. and Baker, I. (1990), Acta Metall., **38**, 207.

Schulson, E.M., Davidson, D.L. and Viens, D. (1983), Met. Trans., **14A**, 1523.

Schulson, E.M., Weihs, T.P., Baker, I., Frost, H.J., and Horton, J.A. (1986), Acta Metall., **34**, 1395.

Schulson, E.M., Weihs, T.P., Viens, D.V. and Baker, I. (1985), Acta Metall., **33**, 1587.

Schulson, E.M., Xu, Y., Munroe, P.R., Guha, S. and Baker, I., (1991) Acta Metall. et Mater., (in press).

Sieloff, D.D., Brenner, S.S. and Burke, M.G. (1987), Proc. MRS, **81**, 87.

Strum, J.M., Hwang, S.K. and Morris, J.W., Jr. (1988), Proc. MRS, **122**, 467.

Swiatnicki, W.A. and Grabski, M.W. (1989), Acta Metall., **37**, 1307.

Takasugi, T., George, E.P., Pope, D.P. and Izumi, O. (1985a), Scripta Metall., **19**, 155.

Takasugi, T., Izumi, O. and Masahashi, N. (1985b), Acta Metall., **33**, 1259.

Takeucki, S. and Kuramoto, E. (1973), Acta Metall., **21**, 415.

Taub, A.I. and Briant, C.L. (1987), Acta Metall., **35**, 1597.

Von Mises, R. (1928), Z. Agnew Math. Mech., **8**, 161.

Xu, Y. (1991) unpublished results, Dartmouth College.

ALLOYING EFFECTS AND GRAIN-BOUNDARY FRACTURE IN L1$_2$ ORDERED INTERMETALLICS

T. TAKASUGI and O. IZUMI[*]
Institute for Materials Research
Tohoku University
Katahira 2-1-1, Sendai 980
Japan
* Professor Emeritus of Tohoku University

ABSTRACT. This article involves a number of studies on the alloying effects on intergranular fracture of L1$_2$ ordered intermetallics, and demonstrates that their ductility strongly depends upon atomistic composition and associated crystal and electronical structures at grain boundaries. Component atoms with large size, which occupy anti-structure site and disordered sites, and their preferential sites as third (or quaternary atoms), are shown to influence the grain boundary strength and fracture behavior in moderate levels of additions. Addition of interstitial atoms such as boron, carbon and beryllium with small size is extremely effective to enhance or to reduce the grain boundary cohesive strength in trace amount of levels in matrix but in highly enriched levels at grain boundaries. Also, gaseous hydrogen atoms with small size, which is mobile and penetrated from environment at room temperature, are shown to dynamically reduce the grain boundary strength, resulting in the environmental embrittlement. It is suggested that the grain boundary strength and therefore ductility of L1$_2$ ordered intermetallics can be controlled by compositions of components and interstitials, state of ordering and prohibition of penetration of hydrogen.

1. INTRODUCTION

Ordered intermetallics with L1$_2$ structure are very attractive with potential for applications as high temperature structural materials and chemical plant parts. This is because some of these intermetallics contain sufficient amounts of light elements such as aluminum and silicon to form protective thin film in corrosive environment (and solution), and also have low densities and high melting points to introduce the high performance structural materials. In addition, some of these intermetallics involve specific dynamic property of dislocations to display abnormally increasing strength at an elevated temperature. However, most ordered intermetallics with L1$_2$ structure have a very serious shortage of brittle fracture and poor ductility at ambient temperatures, associated with grain boundary

C. T. Liu et al. (eds.), Ordered Intermetallics – Physical Metallurgy and Mechanical Behaviour, 391–411.

properties.

L1$_2$ ordered intermetallics consist of various kinds of components, permitting off-stoichiometric compositions, disordering, certain amount of and various kinds of ternary additions, small amounts of impurities (or artificially doped) atoms, and gaseous atoms from environment or as residue. This implies that various kinds of properties, including the grain boundary, of L1$_2$ ordered intermetallics can be modified by alloying method which have been widely applied to develop a large number of metals and alloys. During the last decade, a great deal of understanding on the grain boundary fracture of L1$_2$ ordered intermetallics has been gained and significant progress has been done on improving the ductility and toughness of L1$_2$ ordered intermetallics particularly through the alloying technique, with helps of many micro-analytical instruments, the alloy design and sophisticated calculations. This article summarizes the observation and understanding for the intergranular fracture/ductilization of L1$_2$ ordered intermetallics in terms of the selection of constituent elements, the addition of substitutional elements, the control of alloy stoichiometry, the variation of ordering and the doping of the interstitial elements. Also, the gaseous atoms hydrogen which can penetrate from environment (or solution) are shown to strongly affect the grain boundary fracture at ambient temperatures. The high temperature grain boundary fracture of L1$_2$ ordered intermetallics, the subject of which involves the diffusion processes of component atoms, is not included in this article.

2. THE EFFECT OF COMPONENT ELEMENTS ON INTERGRANULAR FRACTURE

2.1. The effect of constituent elements

Among many L1$_2$ ordered intermetallics, a systematic study of the grain boundary strength has been done on A$_3$B alloys consisting of VIII elements (i.e. Ni and Co) as a major component, A atom [1]. The alloys consisting of B-sub group elements as a minor component, B atom failed through intergranular fracture without measurable ductility whereas the alloys consisting of other transition elements as a minor component failed through transgranular fracture, or mixed fracture with intergranular fracture, with measurable ductility. Based on this observation, it was found that valency difference between two components A and B atoms is, among some possible factors, the dominant factor controlling the grain boundary strength, and that the tendency toward grain boundary fracture becomes stronger with increasing valency difference. Furthermore, when a criterion of the valency difference is combined with atomic size difference between two component atoms, it was correctly predicted that the grain boundary strength of Ni-based L1$_2$ ordered intermetallics ranks in the sequence Ni$_3$Fe > Ni$_3$Mn > Ni$_3$Al > Ni$_3$Ga > Ni$_3$Si > Ni$_3$Ge as shown in Table 1. This rank was also in an agreement well with experimental data observed by Taub et al. [2,3]. It was introduced by this observation that Co$_3$Ti alloy, which exhibited an anomalous positive temperature dependence of strength [4], consisting of two transition elements was extremely ductile over a wide range of temperature [5].

The electronic parameter owing to electronegativity extended from the valency criterion was shown to be more successful in predicting the grain boundary fracture in boron-doped and undoped pseudobinary intermetallics based on Ni_3X (X = Al, Ga, Si or Ge) [6,7] (Fig. 1 and also Table 1). Thus, it was recognized that the transition of ductile (i.e. transgranular) fracture to brittle (i.e. intergranular) fracture occurred on a certain value of electronegativity [6].

TABLE 1. Valency difference, size difference and Pauling's electronegativity correlation with the brittleness in $L1_2$ ordered A_3B alloys

Compound	Valency difference (Ref. 1)	Lattice dilation (Ref. 3)	Pauling's electronegativity difference (Ref.3)
Ni_3Ge	4.0	+1.5 %	+0.10
Ni_3Si	4.0	-0.04 %	-0.01
Ni_3Ga	3.0	+1.6 %	-0.10
Ni_3Al	3.0	+1.5 %	-0.30
Co_3Ti	3.4	-	-
Ni_3Mn	0.9	+2.2 %	-0.36
Ni_3Fe	0.2	+1.0 %	-0.08

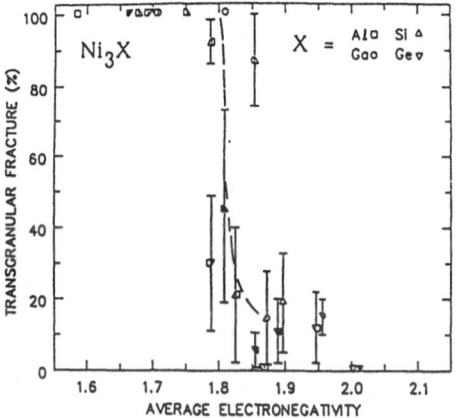

Figure 1. The fracture mode of the Ni_3X (X = Al, Ga, Si and Ge) alloys as a function of average electronegativity [6].

2.2. The effect of ternary additions

$L1_2$ ordered intermetallics generally have certain amounts of solubility limits for various solute atoms. Therefore, ternary additions appear to be most useful modification method for mechanical properties of $L1_2$ ordered intermetallics. This alloying effect on the grain boundary fracture has been investigated for Ni_3Al [8,9], Ni_3Si [10-13] and Co_3Ti alloys [14]. The ternary elements were added to Ni_3Al of stoichiometric composition at the expense of Ni or Al content, keeping their preferential site occupancies [8]. The replacement of Al element with Fe and Mn elements in Ni_3Al enhanced the grain boundary strength and thereby resulted in improved ductility [8]. This alloying reduces the valency difference between Ni and Al elements. The ductilization by the addition of Fe element was more effectively observed in boron-doped Ni_3Al alloy [9] and in Co_3Ti alloy [14]. Also, Oliver and White found that the addition of Ti into Ni_3Si alloy suppressed the grain boundary fracture at concentration of about 3 at% and thereby improved room temperature ductility [10]. This alloying also has an effect to reduce the valency difference between Ni and Si elements. A systematic observation on the mechanical behavior in $L1_2$ Ni-Si-Ti alloys was performed as functions of Ti content and alloy stoichiometry [11]; ductilization was found in higher levels of Ti concentrations than the value reported by Oliver and White [10] and became obvious with further increasing Ti content and in more Ni-excess compositions [11]. Also, the additions of many quaternary elements X into Si or Ti sites of $Ni_3(Si,Ti)$ alloys were performed. Among these elements, Cr and Mn were effective to furthermore enhance the elongation values at ambient temperatures, as shown in Fig. 2 [13].

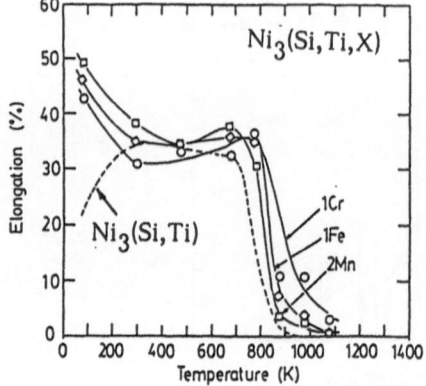

Figure 2.   Changes of elongation with temperature for $Ni_3(Si,Ti,X)$ alloys [13].

2.3. The effect of alloy stoichiometry

In $L1_2$ structure, excess atoms from stoichiometry generally occupy sites of deficient atoms, introducing the anti-structure in both sides of off-stoichiometric compositions. Alloy stoichiometry was found to have a strong effect also on the ductility and fracture mode of many $L1_2$ A_3B ordered intermetallics [15]. Figure 3 indicates that considerable variations of the tensile elongation occur at Ni-excess concentrations and then the elongations always increase and also the transgranular fracture mode becomes dominant as concentration of Ni increases from stoichiometry. Also, the compositional dependence (i.e. the slope of the curves) on the elongation was more remarkable in the order $Ni_3Al > Ni_3Ga > Co_3Ti > Ni_3(Al,Mn)$ although the former two alloys were doped with a small amount of boron (the correlation between boron doping and alloy stoichiometry will be discussed later). This rank is again rationalized by the valency difference (or electronegativity difference) between two constituent atoms; the larger the valency difference between two component atoms is the steeper the slope of the elongation becomes. Similar effect of alloy stoichiometry on the grain boundary brittleness was also reported in boron-doped Ni_3X (X = Al, Ga and Si) alloys [2,16], and boron-doped and -undoped $Ni_3(Si,Ti)$ alloys [11,12]. For these alloys, Ni-excess alloys were more ductile than Ni-poor alloys.

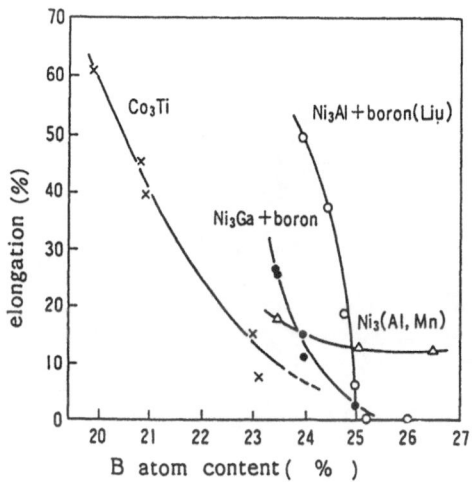

Figure 3. Summary of elongation of $L1_2$ ordered A_3B alloys as a function of B atom content [15].

It was very recently reported that when transition elements, e.g. Pd, Pt, Co and Cu were added into Ni_3Al of Ni-excess composition (i.e. Ni-23Al-2X) some amounts of tensile elongation was observed at room temperature [17]. It was claimed by the authors that the ordering energy of Ni_3Al was lowered by the alloying of these elements and thereby ductilized, and also this lowering of the ordering energy was more effective at Ni-excess concentrations [17]. Although the lowering of the ordering energy by the alloying of these elements may be partially responsible for the ductility improvement of Ni_3Al, as will be actually described in the latter section, another explanation is possible; since this kind of additions have a strong tendency to occupy Ni sites and therefore Ni atoms are enforced to occupy Al site. As a result, rejected Ni atoms and inherently excess Ni atoms must occupy Al site of Ni_3Al. This situation has a greatest effect to enhance the grain boundary strength of Ni_3Al, as discussed in the previous sections.

2.4. The effect of degree of ordering

Not only for ordered alloys which undergo the transition from disordered to ordered structure below their melting points but also for intermetallics which remain ordered structures up to their melting points, the state of ordering can be generally affected by temperature, alloy stoichiometry and plastic deformation. Here, the degree of ordering can be considered as a sort of alloying parameter since wrongly occupied constituent atom behaves as a solute and affect the mechanical properties. The description here stands on the situation in which the state of ordering in the grain inte-

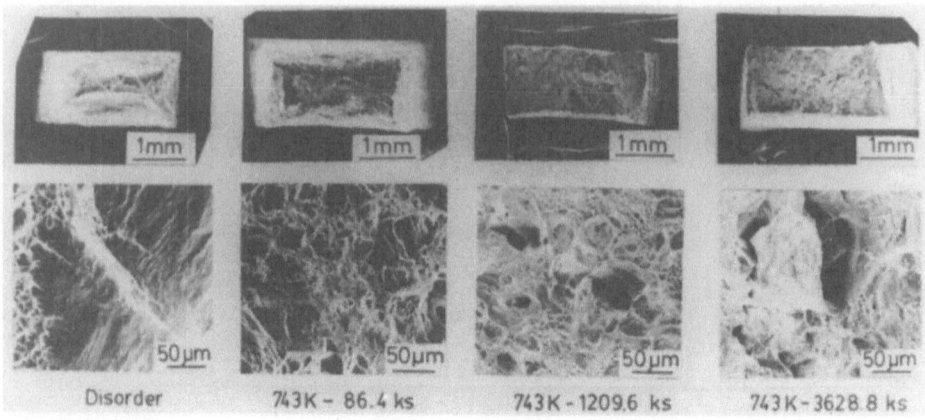

Figure 4. Variation of SEM fractographs in Ni_3Fe alloy with annealing time at 743 K [18].

rior remains the same also in the grain boundary region. On relatively ductile Ni_3Fe alloys, the effect of long range ordering parameter S on the grain boundary fracture was investigated through annealing at temperatures below a transition temperature (776 K) [18]. As ordering proceeds (i.e. as annealing proceeds), the elongation decreased to a constant value, the level of which was lower than that of an initial (disordered) state although the elongation first made a faint plateau [18]. Corresponding to this variation, the reduction in area of alloys decreased and the intergranular fracturing occurred in some areas (approximately 5 % of total area)(Fig. 4). For intermetallics with relatively high ordering energy, the lowering of degree of ordering could be obtained by methods of a rapid quenching from liquid phase or of a radiation damage. Indeed, Ni_3Al based alloys were ductilized by the former method [19,20].

Lowering of the ordering is caused also by the deviation from stoichiometry. It was estimated by a few metallurgical techniques that the order-disorder transition temperature in Ni_3Al decreases with deviating toward Ni-excess compositions, causing the disordering at this composition range [21,22]. It was postulated, based on these experimental results, that the ductilization by boron addition in the Ni-excess Ni_3Al was closely associated with this compositional disordering [21,22]. The compositional disordering occurs also at Ni-poor Ni_3Al and also many Ni-poor Ni_3X alloys. However, the ductilization of Ni_3Al was not established at Ni-poor compositions. Therefore, it is unlikely that the compositional disordering is a whole mechanism responsible for this phenomenon. More detailed experiment and discussion are required.

3. THE DOPING EFFECT OF INTERSTITIAL ELEMENTS ON INTERGRANULAR FRACTURE

In contrast to the component atoms with large diameter which affected the grain boundary keeping the same composition as in the matrix, the interstitials with small diameter affected the grain boundary enriching their compositions, i.e. segregating at grain boundaries. Dramatic ductility improvement on $L1_2$ ordered intermetallics was accomplished by beneficial dopants such as boron, carbon and beryllium as summarized in Table 2. This effect was first found in Ni_3Al alloyed with a trace amount of boron through suppressing their brittle intergranular fracture [23], as shown in Fig. 5. This ductilization was, by Auger studies, observed to occur over a wide range of boron concentration where boron is in solid solution and then attributed to boron segregation to grain boundaries [16]. The beneficial effect of boron has been observed also in Ni_3Si [2], $Ni_3(Si,Ti)$ (Fig. 6) [10-12] and Ni_3Ga alloys [2]. On the other hand, the doping of boron in Co_3Ti alloy reduced the elongation through enhancing solid solution hardening of matrix [24]. Next, carbon also showed the tendency of the segregation to grain boundaries in Ni_3Al alloy [25,26] but did not improve its ductility [25,27]. However, carbon considerably improved, at extremely low levels of their dopings, the ductility of Ni_3Si [7,28] and $Ni_3(Si,Ti)$ (Fig. 6) [29,30] alloys. On the other hand, the doping of carbon did not affect the elongation value of Co_3Ti alloy [24]. Finally, a small amount of beryllium, which was determined to occupy the substitutional sites in Ni_3Al

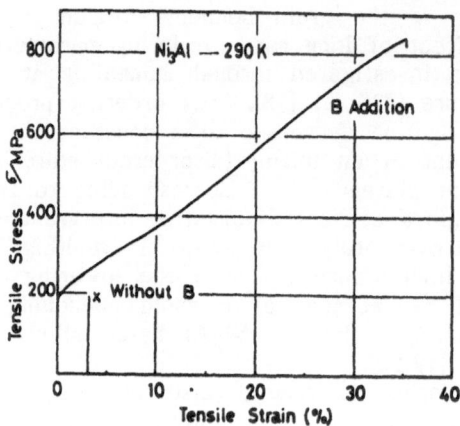

Figure 5. Tensile stress-strain curves at room temperature, showing the effect of boron addition on the elongation of Ni_3Al [23].

[31] and Co_3Ti [32] alloys, has been shown to have a beneficial effect on ductility of Ni_3Al [33] and Co_3Ti [24] alloys, but a harmful effect on ductility of $Ni_3(Si,Ti)$ alloy (Fig. 6) [29,30].

TABLE 2. Doping effect of boron, carbon and beryllium on intergranular fracture of $L1_2$ ordered intermetallics

	Ni_3Al based alloys	Ni_3Si based alloys	Co_3Ti based alloys
boron	oo	oo	-
carbon	-	oo	o
beryllium	o	x	o

* oo;extremely beneficial, o;beneficial, -;neutral, x;harmful

Thus, different behavior of boron, carbon and beryllium for the grain boundary fracture of $L1_2$ ordered intermetallics could be attributed to difference of electronic nature of these doping elements and also difference of electronic interaction between constituent element and doping element.

A large number of studies have been conducted to know the mechanism of the boron ductilization since this dramatic effect was discovered in Ni_3Al. As described already, the doping effect of boron (or carbon) on the ductility and fracture behavior of $L1_2$ ordered intermetallics was shown to be closely related to the alloy stoichiometry [2,16]. Boron addition was shown to be effective only in Ni-excess Ni_3Al alloys. Auger studies [16,34] indicated that deviation from alloy stoichiometry influenced the grain boundary chemistry; in Ni-rich alloys boron segregation was sig-

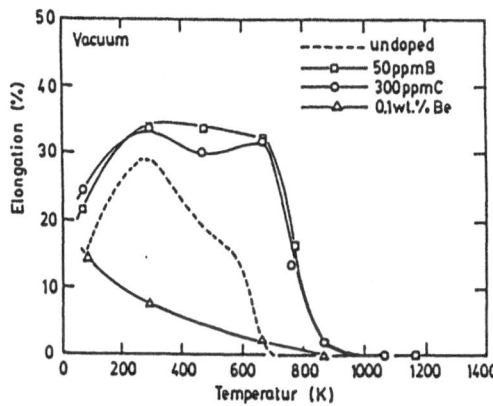

Figure 6.   Variation of elongation with test temperature for undoped, boron-doped, carbon-doped and beryllium-doped $Ni_3(Si,Ti)$ alloys [12,30].

nificant and nickel concentration at grain boundary was higher than that in bulk while in Ni-poor alloys boron segregation still occurs but not so strong and aluminium concentration at grain boundary was higher than that in bulk.   Thus, ductility improvement in Ni-excess Ni_3Al (or Ni_3Si based) alloys doped with boron (or carbon) may be established by cooperation of boron (or carbon) effect and alloy stoichiometric effect, and therefore a number of studies are dealt with this problem [35].

4. THE EFFECT OF HYDROGEN ON INTERGRANULAR FRACTURE

A number of component atoms and dopant atoms mentioned in the foregoing section affected the grain boundary by a static way.   That is, these atoms are not mobile at grain boundaries as well as in matrix and therefore neither enriched by stressing nor injected from environment. However, as many fabricatable $L1_2$ ordered intermetallics are discovered, it has been widely documented that hydrogen dynamically affects the grain boundary enriching by stressing and penetrating from environment.   This is due to the fact that hydrogen is quite mobile at an ambient temperature. That is, hydrogen introduced from environment or contained as residue in the material has been shown to severely embrittle a large number of $L1_2$ ordered intermetallics.

The electro-charging of hydrogen or hydrogen gas exposure led the brittleness in many $L1_2$ ordered intermetallics of $(Fe,Ni)_3V$ [36], Ni_3Fe [37], boron-doped Ni_3Al [38] and Co_3Ti [39] alloys.   Also, recent studies on Co_3Ti [39], beryllium-doped Ni_3Al [33], $Ni_3(Al,Mn)$ [40], $Ni_3(Si,Ti)$ [11, 41], Co_3Ti alloyed with various additions [14] and doped with boron, carbon and beryllium [42], and $(Fe,Co)_3V$ [43] indicated that hydrogen embrittlement was more serious; that is, their tensile ductilities strongly depended upon testing atmosphere whether air, vacuum or hydrogen.   The tensile ductilities were generally lower in samples deformed in air and also at a low strain rate while they were higher in samples deformed in vacuum and at a rapid strain rate, as shown in Figs. 7 and 8. It can be considered that as strain rate decreases sufficient amount of hydrogen atoms are provided to the associated places, resulting in severe embrittlement. Loss of the tensile ductility was recovered after a degassing treatment; at room temperature in the case of Co_3Ti [39], 473 K in the case of $(Fe,Ni)_3V$ [36], at 673 K in the case of boron-doped Ni_3Al [38] and at 573 K in the case of $Ni_3(Al,Mn)$ [40] (see Fig. 7). These results suggest that the reduced ductility could not be attributed to the introduction into the sample interior of permanent damage as a hydride or swelling void. Ductility loss in Co_3Ti was shown to be limited at ambient temperatures [44,45], as shown in Fig. 8. In other words, the hydrogen embrittlement in samples deformed in air was little operative at temperatures below 77 K and also above 473 K. Thus, ductility recovering at cryogenic temperatures and at high temperatures can be attributed to the lowerings of the mobility of hydrogen and of the condensation of hydrogen into the region of propagating microcrack along a grain boundaries, respectively [45]. Also, it is noted that the yield strength of hydogen-embrittled alloys were little affected (Fig. 7). This result implies that the influence of hydrogen is operative at the stage of plastic deformation, i.e. under generation and propagations of dislocations.

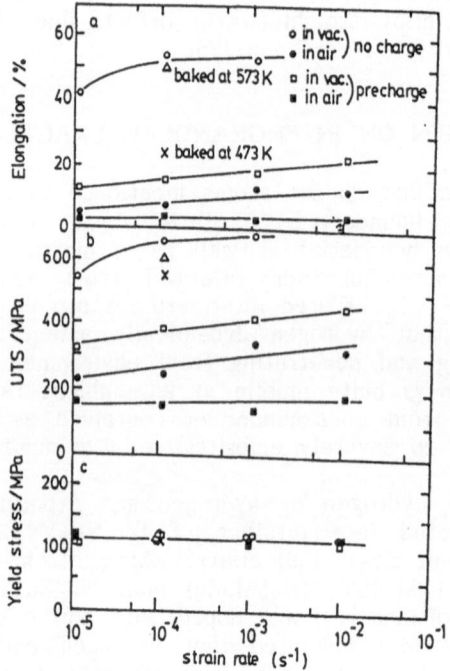

Figure 7. Strain rate dependence of (a) elongation, (b) UTS and (c) yield strength of $Ni_3(Al,Mn)$ alloys tested under various atmospheres [40].

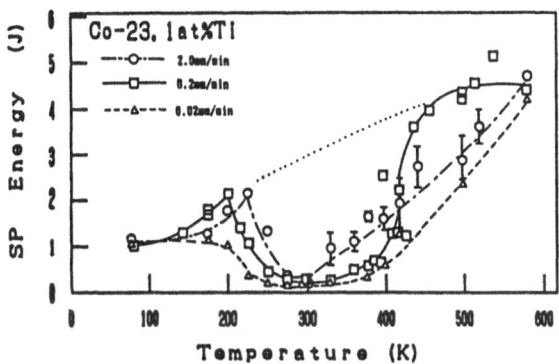

Figure 8. Variations of small punch fracture energy (SPEE) with temperature for Co$_3$Ti alloys which were deformed in air [44].

Hydrogen embrittlement observed in L1$_2$ ordered intermetallics involves diffusion kinetics and the bond-breaking mechanism. Regarding the diffusion kinetics, the decomposition into hydrogen atoms (or ions) from water, hydrogen molecule in air and hydrogen gas (H$_2$), and then the penetration into the specimen must be considered. Probably, slip steps or free surfaces with cracks play an important role in such a decomposition as a sort of catalysis. Similarly, generated dislocations or free surfaces of cracks could provide path of permeation (penetration) of hydrogen atoms. The bond-breaking at grain boundaries will be described in the latter section, in relation to the grain boundary cohesive strength.

There are some observations showing the interaction between hydrogen atoms and additive atoms on hydrogen embrittlement. Most striking result was found on Ni$_3$Al alloys doped with boron [27] and on Ni$_3$(Si,Ti) alloys doped with boron and carbon [29,30,41]. The ductility of boron-doped Ni$_3$Al alloys was basically insensitive to the test environment and test strain rate [27] although this alloy has been embrittled by compulsory injected hydrogen [38]. Similarly, room temperature elongations of boron-doped and carbon-doped Ni$_3$(Si,Ti) alloys were insensitive to test atmosphere and were then higher than those of undoped alloys, as shown in Fig. 9. Whereas, room temperature elongations of undoped Ni$_3$(Si,Ti) alloys were sensitive to test atmosphere and were then lower in air than in vacuum [29,30,41]. Thus, doping of boron or carbon was shown to have the effect prohibiting the hydrogen embrittlement in these alloys. This result implies that boron and carbon atoms competing with hydrogen, for site occupation or for its effectiveness at grain boundaries, have the effect of suppressing the action of hydrogen. However, boron and carbon had no effect on the hydrogen embrittlement of Co$_3$Ti [24,42]. It was also shown that the additions of Fe and Al effectively reduced the hydrogen embrittlement in Co$_3$Ti alloys [14] although its alloying effect has not been clarified yet.

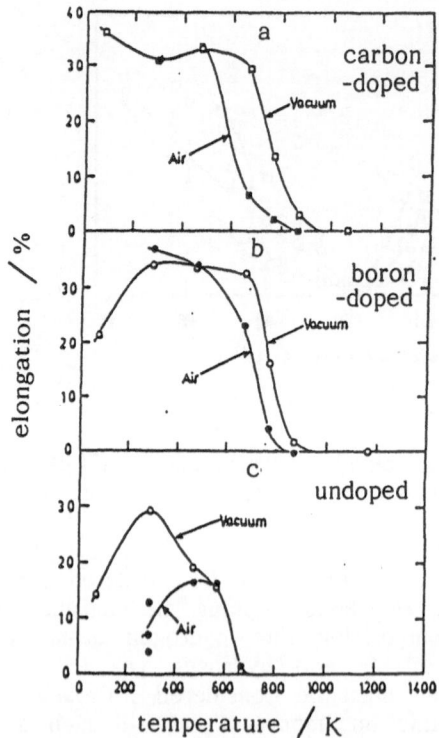

Figure 9. Environmental effect on elongation for undoped, boron-doped and carbon-doped $Ni_3(Si,Ti)$ alloys [30,41].

5. THE MECHANISM CONTROLLING GRAIN BOUNDARY FRACTURE

For understanding the alloying effect on the grain boundary brittleness in $L1_2$ ordered A_3B intermetallics, basically two micro mechanisms have been demonstrated. One is a mechanism based on the grain boundary cohesive strength [16,46,47]. This idea is that the grain boundary fracture directly depends on the bond strength at grain boundary and associated plastic deformation. The other is a mechanism based on the enhancement of plastic flow due to the disordering of the grain boundary region [48]. This idea is that the disordering at grain boundary provides a larger number of slip systems [49] and easier traveling of grain boundary dislocations injected from the grain interior [48], resulting in the accommodation of the stress concentrations near grain boundary. When this argument is confined to the doping effect and alloy stoichiometric effect on the intergranular fracture of Ni_3Al alloys, early studies for chemistry (by scanning transmission electron microscopy and atom probe) [50,51], crystal structure (by lattice image) [52] and the measurement of k_y value in Hall-Petch relation [53] revealed the reduction in order state and enrichment of nickel at grain boundaries in boron-doped Ni-excess Ni_3Al alloys. However, recent studies using the similar technique of Auger [34,35] and high resolution transmission electron microscopy [54] at grain boundaries in boron-doped Ni-excess Ni$_3$Al alloys indicate that ordered structure remained up to the region close to a grain boundary plane and chemical composition of the constituent atoms did not vary so remarkably across a grain boundary plane. Also, the latest measurement of k_y value in Hall-Petch relation in Ni$_3$Al with and without boron did not exhibited a meaningful difference [55]. In addition, the latest studies [56,57], in which the grain boundaries in oriented bicrystals [56] and well-characterized grain boundaries in polycrystals [57] of Ni$_3$Al with and without boron were carefully investigated by high resolution TEM technique, indicate that ordered structure remained up to the region close to a grain boundary plane. Although further studies are required, the latest results thus reveal that an idea based on the cohesive strength is more likely to apply to the results mentioned in the foregoing section and in the following section. Also, if the effect of hydrogen is considered, it is strongly suggested that the cohesive model is very likely as the mechanism controlling the grain boundary fracture. Hydrogen enriched at grain boundaries does not modify the state of ordering but definitely affects the grain boundary fracture through reducing the cohesive strength.

Estimations for the grain boundary cohesive strength of L1$_2$ ordered intermetallics have been recently attempted based on theoretical calculation [58-60]. Cluster calculation of density of states was done for crystal in a bulk and polyhedron in a grain boundary in a number of L1$_2$ ordered intermetallics [58]. Based on this calculation, the relative s-orbital electronegativity of the constituent atoms was predicted to provide a reliable indication of the grain boundary cohesion, being consistent with an experimental data mentioned in the foregoing section.

6. ATOMISTIC COMPOSITION, ELECTRONIC NATURE AND COHESIVE STRENGTH AT GRAIN BOUNDARIES

A number of important feature for the grain boundary of L1$_2$ ordered intermetallics can be argued in the purely geometrical frame, with help of knowledge of the structure and chemistry [46]. The crystal structure at $\Sigma=$ 13/[100] tilt grain boundary in the strongly ordered Ni$_3$(Al,Ti) alloy observed by high resolution transmission electron microscopy indicated that appreciable (translational and local) atomic relaxations can not been found [61]. Also, atomistic calculation using many-body pair potentials on a grain boundary of ordered alloy consisting of high ordering energy drew little atomistic relaxation, resulting in somehow a cavity-like structure [62]. Thus, the bonds at grain boundary in the strongly ordered A$_3$B L1$_2$ alloy appear to be characterized by the 'directionality' and the 'heteropolarity' of electronic charge distribution [46]. As for the 'directionality' [63-65], the A-B bonds (i.e. covalent bonds) prefer to be a 90 degree angle, for which the non-central contribution to the energy of formation is at a minimum. It seems unlikely that the angle between the A-B bonds from a given B atom across a grain boundary will always be close to the desired 90 degree value. This situation is shown in Fig. 10 where the atomic configuration and the bond nature in a Σ=5/[100] CSL boundary is geometrically constructed. Thus, the A-B covalent bonds perpendicular to a grain boundary plane which are supposed to sustain the grain boundary cohesion should be limited, resulting in a lower grain boundary cohesion. Next, from the point of view of the 'heteropolarity' [66-69], the B atom withdraws electron density from the A-A (metal-metal) bonds and forms ionic A(metal)-B bonds. It is then demonstrated that the reduction of charge density in the A-A bonds results in the reduction of the grain boundary cohesive strength. Thus, the 'heteropolarity' of electron charge density due to the A-B bonds at the boundary plane introduces the environment somewhat like a penny-like cavity. Here, the valency and the electronegativity difference criterion can generally be justified by measuring a transfer of electrons between constituent atoms. When the valency difference between A and B atoms is larger (or component atom B is more electronegative with respective to another component atom A), B atoms have a greater tendency to pull electron charge out of the A-A bonds.

O o A atom   ● B atom

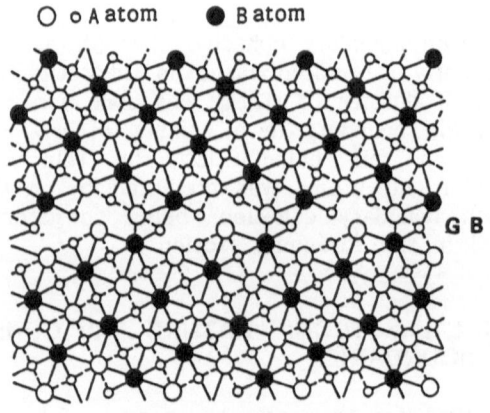

Figure 10. Schematic representation of the bond pairs at a grain boundary of the $L1_2$ ordered A_3B alloy.

The effect of alloying including hydrogen on the grain boundary strength (and resultant fracture) of $L1_2$ ordered intermetallics can be interpreted by the modification of the electronic structure and atomic bonds through compositional modification at the grain boundary [46,70,71]. Based on the geometrical analysis, e.g. as drawn in Fig. 10, the alloying effects of (i) constituent atoms, (ii) ternary additions and (iii) off-stoichiometric atoms on the electronic nature at the grain boundary can be easily understood considering the site occupation of these atoms and their electronic nature. For item (i), in A_3B alloys consisting of VIII group element and b-sub group element the B atom strongly withdraws electron density from the A-A (metal-metal) bonds and then creates the reduction of charge density in the A-A bonds, resulting in reduction of the grain boundary cohesive strength.   Whereas, in alloys consisting both transition elements, the drawing of charge from A-A bonds is less and does not substantially weaken them, resulting in high cohesive strength.   For item (ii), when X in Ni_3X (e.g. X = Al and Si) is replaced with transition element close to VIII group element, more homogeneous electronic charge distribution is created through a grain boundary plane, resulting in enhancement of the cohesive strength.   Whereas, when X in Ni_3X is replaced with b-sub group element or transition element far from VIII group element, more heterogeneous electronic charge distribution is introduced through a grain boundary plane, resulting in reduction of the cohesive strength.   For item (iii), in the A(metal)-rich A_3B alloys, the A(metal)-A(metal) bonds newly introduced by excess A atoms create more homogeneous electronic charge distribution through a grain boundary plane, resulting in enhancement of the grain boundary cohesive strength.   Calculation using pair potential [59] and the tight-binding electronic theory of s, p and d-basis orbitals [60] supports this idea; it was calculated that the grain boundary cohesive strength of Ni-rich Ni_3Al alloys was higher than that of stoichiometric Ni_3Al alloys. Whereas, in the B(b-sub group element)-rich A_3B alloys, the A-B bonds newly introduced by excess B toms create more heterogeneous electronical charge distribution through a grain boundary plane, resulting in reduction of the grain boundary cohesive strength.

The effect of degree of ordering can be interpreted from the point of view of atomistic accommodation at the grain boundary.   When the degree of the ordering is low, extensive atomic relaxations leading to much more homogeneous crystal structure and also electronical structure is expected, resulting in the high cohesive strength, and vice versa.

The effect of boron (or carbon) dopings can be considered in a way that these atoms acted as electronic 'donors' in the region where electronic charge distribution is deficient.   It has been suggested in a grain boundary of pure nickel that boron acts as electronic donors and thereby strengthens the atomic bond at a grain boundary [67,68].   Also, the beneficial effect of boron on the grain boundary cohesive strength of Ni_3Al alloys was indicated by first-principle cluster calculation [72], cluster calculation of density of state [58], tight-binding electron theory [60] and embedded atom calculation [59].

For the hydrogen embrittlement, it is demonstrated that the materials are embrittled by the atomistic and dynamic mechanism by which the intergranular cohesive strength [39,45] and the associated plastic flow around a micro crack [45] were affected.   In this mechanism, hydrogen accumulates at the high stress region near the micro crack tip and thereby lowers the cohesive strength of the lattice at which the bond rupture occurs easier [73-76].   Thus, the effect of hydrogen dopings can be considered in a way that these atoms acted as electronic 'acceptors' in the grain boundary region, in contrast to boron (or carbon) dopings.   It is suggested that hydrogen absorbs much electrons from component atoms and thereby introduces deficient regions of electrons between specific bonds. The details of intergranular embrittlement due to hydrogen can be understood in terms of the electronic concept which has been argued by Eberhart et al. [77].

406

7. SUMMARY

This article, involving a number of studies for the alloying effects on intergranular fracture of $L1_2$ ordered intermetallics, demonstrated that their ductility strongly depended upon atomistic composition and associated crystal and electronical structures at grain boundaries. Dopings (or penetration) of interstitial atoms were extremely effective to enhance or to reduce the grain boundary cohesive strength of these intermetallics. On the other hand, constituent atoms on anti-structure site and on disordered sites, and additive atoms on their preferential sites were, in moderate levels of addition, shown to influence the grain boundary strength and fracture behavior. The understanding of the alloying effect on the grain boundary behavior led to method of alloy design by which the grain boundary cohesion can be manipulated [46,70,71]. One typical example developed by this principle is Ni_3Si alloys; by the addition of Ti elements, control of alloy stoichiometry, doping of boron or carbon and addition of a quaternary elements, extremely tough materials with better corrosion and oxidation resistance were obtained, as shown in Fig. 11 [13].

Most $L1_2$ ordered intermetallics can accept the modification by alloying such as off-stoichiometry, pseudo-binary, addition of ternary (or quaternary) elements, doping and so on. Therefore, alloying technique would be a principle method to control the properties of the ordered intermetallics and to develop useful structural and chemical materials.

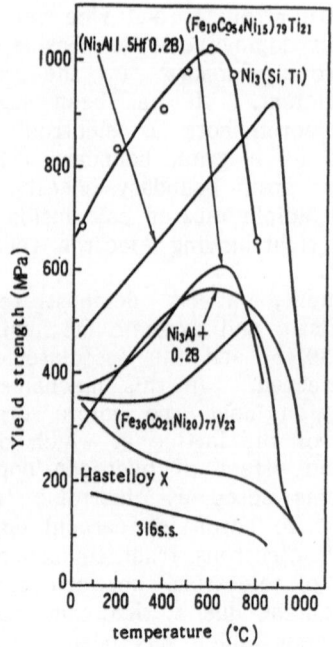

Figure 11. Yield strength as a function of temperature for some intermetallics with ductility, compared with a few conventional alloys [13].

REFERENCES

1. Takasugi, T. and Izumi, O. (1985) 'Electronic and structural studies of grain boundary strength and fracture in $L1_2$ ordered alloys - I. On binary A_3B alloys', Acta Metall. **33**, 1247-1258.
2. Taub, A. I., Briant, C. L., Huang, S. C., Chang K. M. and Jackson, M. R. (1986) 'Ductility in boron-doped, nickel-base $L1_2$ alloys processed by rapid solidification', Scripta Metall. **20**, 129-134.
3. Taub, A. I. and Briant, C. L. (1987) 'Grain boundary chemistry and ductility in Ni-base $L1_2$ intermetallic compounds', MRS Symp. Proc. Publication **81**, 343-353.
4. Takasugi, T. and Izumi, O. (1987) 'Plastic flow of Co_3Ti single crystals', Acta Metall. **35**, 2015-2026.
5. Takasugi, T. and Izumi, O. (1985) 'High temperature strength and ductility of polycrystalline Co_3Ti compound', Acta Metall. **33**, 39-48.
6. Taub, A. I. and Briant, C. L. (1987) 'Composition dependence of ductility in boron-doped, nickel-base $L1_2$ alloys', Acta Metall. **35**, 1597-1603.
7. Briant, C. L. and Taub, A. I. (1989) 'Fracture modes in $L1_2$ compounds', MRS Symp. Proc. Publication **133**, 281-286.
8. Takasugi, T., Izumi, O. and Masahashi, N. (1985) 'Electronic and structural studies of grain boundary strength and fracture in $L1_2$ ordered alloys - II. On the effect of third elements in Ni_3Al alloy', Acta Metall. **33**, 1259-1269.
9. Horton, J. A., Liu, C. T. and Santella, M. L. (1987) 'Microstructure and mechanical properties of Ni_3Al alloyed with iron additions', Metall. Trans. A **18A**, 1265-1277.
10. Oliver, W. C. and White, C. L. (1987) 'The segregation of boron and its effect on the fracture of an Ni_3Si based alloy', MRS Symp. Proc. Publication **81**, 241-246.
11. Takasugi, T., Nagashima, M. and Izumi, O. (1990) 'Strengthening and ductilization of Ni_3Si by the addition of Ti elements', Acta Metall. Mater. **38**, 747-755.
12. Takasugi, T., Izumi, O. and Yoshida, M. (1991) 'Mechanical properties of recrystallized $L1_2$-type $Ni_3(Si,Ti)$ intermetallics', J. Materi. Science **24**, 1173-1178.
13. Takasugi, T. and Yoshida, M. (1991) 'Mechanical properties of the $Ni_3(Si,Ti)$ polycrystals alloyed with various substitutional additions', J. Mater. Science, in press.
14. Liu, Y., Takasugi, T., Izumi, O. and Suenaga, H. (1989) 'Mechanical properties of Co_3Ti polycrystals alloyed with various additions', J. Materi. Science **24**, 4458-4466.
15. Takasugi, T., Masahashi, N. and Izumi, O. (1987) 'Electronic and structural studies of grain boundary strength and fracture in $L1_2$ ordered alloys - III. The effect of stoichiometry', Acta Metall. **35**, 381-391.
16. Liu, C. T., White, C. L. and Horton, J. A. (1985) 'Effect of boron on grain-boundaries in Ni_3Al', Acta Metall. **33**, 213-229.
17. Chiba, A, Hanada, S. and Watanabe, S. (1990) 'Correlation between ductility and ordering energy of Ni_3Al', Mater. Trans. JIM **31**, 824-827.
18. Takasugi, T., Eguchi, T., Yoshida, M. and Izumi, O. (1989) 'Mechanical

408

properties and ordering process in Ni_3Fe polycrystals', J. Japan Inst. Metals **53**, 42-49.

19. Inoue, A., Tomioka, H. and Masumoto, T. (1983) 'Microstructure and mechanical properties of rapidly quenched Ll_2 alloys in Ni-Al-X systems', Metall. Trans. A **14A**, 1367-1377.

20. Carro, G., Bertero, G. A., Witting, J. E. and Flanagan, W. F. F. (1989) 'The effect of anti-phase domain size on the ductility of a rapidly solidified Ni_3Al-Cr alloy', MRS Symp. Proc. Publication **133**, 535-541.

21. Cahn, R. W., Siemers, P. A., Geiger, J. E. and Bardhan, P. (1987) 'The order-disorder transformation in Ni_3Al and Ni_3Fe alloys - I. Determination of the transition temperatures and their relation to ductility', Acta Metall. **35**, 2737-2751.

22. Cahn, R. W., Siemers, P. A. and Hall, E. L. (1987) 'The order-disorder transformation in Ni_3Al and Ni_3Fe alloys - II. Phase transformations and microstructures', Acta Metall. **35**, 2753-2767.

23. Aoki, K. and Izumi, O. (1979) 'Improvement in room temperature ductility of the Ll_2 type intermetallic compound Ni_3Al by boron addition', J. Japan Inst. Met. **43**, 1190-1193.

24. Takasugi, T., Takazawa, M. and Izumi, O. (1990) 'Mechanical properties of Co_3Ti containing boron, carbon and beryllium', J. Mater. Science **25**, 4231-4238.

25. Huang, S. C., Briant, C. L., Chang, K. M., Taub, A. I. and Hall, E. L. (1986) 'Carbon effects in rapidly solidified Ni_3Al', J. Mater. Res. **1**, 60-67.

26. Brenner, S. S. and Ming-Jian, H. (1990) 'Grain boundary segregation of carbon and boron in Ni_3Al+B/C', Scripta Metall. **24**, 667-670.

27. Masahashi, N., Takasugi, T. and Izumi, O. (1988) 'Mechanical properties of Ni_3Al containing C, B and Be', Acta Metall. **36**, 1823-1836.

28. Oliver, W. C. (1989) 'The development of alloys based on Ni_3Si', MRS Symp. Proc. Publication **133**, 397-402.

29. Takasugi, T. and Izumi, O. (1991) 'Design of Ll_2-type intermetallic alloys', to be published in MRS Symp. Proc. Publication.

30. Takasugi, T. and Yoshida, M. (1991) 'Mechanical properties of the $Ni_3(Si,Ti)$ alloys doped with carbon and beryllium', to be published in J. Materi. Science.

31. Masahashi, N., Takasugi, T. and Izumi, O. (1988) 'Atomistic defect structures of Ni_3Al containing B, C and Be', Acta Metall. **36**, 1815-1822.

32. Takasugi, T., Takazawa, M. and Izumi, O. (1990) 'Atomistic defect structures of Co_3Ti containing boron, carbon and beryllium', J. Materi. Science **25**, 4226-4230.

33. Takasugi, T., Masahashi, N. and Izumi, O. (1986) 'Improved ductility and strength of Ni_3Al compound by beryllium addition', Scripta Metall. **20**, 1317-1321.

34. George, E. P., Liu, C. T. and Padgett, R. A. (1989) 'Comparison of grain boundary compositions in B-doped and B-free Ni_3Al', Scripta Metall. **23**, 979-982.

35. Horton, J. A. and Liu, C. T. (1990) 'Does a grain boundary phase exist in Ni-24% Al? Summary of recent experimental work', Scripta Metall. Mater. **24**, 1251-1256.

36. Kuruvilla, A. K., Ashok, S. and Stoloff, N. S. (1982) 'Long range order and hydrogen embrittlement', Proceedings of the Third International Congress on Hydrogen and Material, Pergamon Press, Oxford, p. 629-633.

37. Camus, G. M., Stoloff, N. S. and Dequette, D. J. (1989) 'The effect of order on hydrogen embrittlement of Ni_3Fe', Acta Metall. **37**, 1497-1501.

38. Kuruvilla, A. K. and Stoloff, N. S. (1985) 'Hydrogen embrittlement of Ni_3Al+B', Scripta Metall. **19**, 83-87.

39. Takasugi, T. and Izumi, O. (1986) 'Factors affecting intergranular hydrogen embrittlement of Co_3Ti', Acta Metall. **37**, 607-618.

40. Masahashi, N., Takasugi, T. and Izumi, O. (1988) 'Hydrogen embrittlement of pseudobinary $L1_2$-type $Ni_3(Ai_{0.4}Mn_{0.6})$ intermetallic compound', Metall. Trans. A **19A**, 353-357.

41. Takasugi, T., Suenaga, H. and Izumi, O. (1991) 'Environmental effect on mechanical properties of recrystallized $L1_2$-type $Ni_3(Si,Ti)$ intermetallics', J. Materi. Science **24**, 1179-1186.

42. Takasugi, T., Takazawa, M. and Izumi, O. (1990) 'Environmental effect on the mechanical properties of Co_3Ti containing boron, carbon and beryllium', J. Mater. Science **25**, 4239-4246.

43. Nishimura, C. and Liu, C. T. (1991) 'Environmental embrittlement in $L1_2$-ordered $(Fe,Co)_3V$', Scripta Metall. **25**, 791-794.

44. Kimura, A., Izumi, H., Igarashi, Y., Misawa, T. and Takasugi, T. (1991) 'Mechanism of hydrogen induced intergranular cracking in Co_3Ti' Sixth JIM international symposium on intermetallic compounds - structure and mechanical properties -, to be published in JIM.

45. Liu, Y., Takasugi, T., Izumi, O. and Yamada, T. (1989) 'The influence of hydrogen on deformation and fracture processes in Co_3Ti polycrystals and single crystals', Acta Metall. **37**, 507-517.

46. Izumi, O. and Takasugi, T. (1987) 'Deformability improvements of $L1_2$-type intermetallic compounds', MRS Symp. Proc. Publication **81**, 173-182.

47. Chen, S. P., Voter, A. F., Albers, R. C., Boring, A. M. and Hay, P. J. (1989) 'Theoretical studies of grain boundaries in Ni_3Al with boron or sulfur', Scripta Metall. **23**, 217-222.

48. Schulson, E. M., Weihs, T. P., Baker, I., Frost H. J. and Horton, J. A. (1986) 'Grain boundary accommodation of slip in Ni_3Al containing boron', Acta Metall. **34**, 1395-1986.

49. King, A. H. and Yoo, M. H. (1987) 'Dislocation reactions at grain boundaries in $L1_2$ ordered alloys', MRS Symp. Proc. Publication **81**, 99-104.

50. Baker, I, Schulson, E. M. and Michael, J. R. (1988) 'The effect of boron on the chemistry of grain boundaries in stoichiometric Ni_3Al', Philos. Mag. **B57**, 379-385.

51. Sieloff, D. N., Brenner, S. S. and Ming-Jian, H. (1989) 'The microchemistry of grain boundaries in Ni_3Al', MRS Symp. Proc. Publication **133**, 155-160.

52. Mackenzie, R. A. D. and Sass, S. L. (1988) 'Direct observation of the compositional disordering of Ni_3Al in the vicinity of grain boundaries using high resolution electron microscopy technique', Scripta Metall. **22**, 1807-1812.

53. Weihs, T. P., Zinovev, V., Viens, C. V. and Shulson, E. M. (1987) 'The

strength, hardness and ductility of Ni_3Al with and without boron', Acta Metall. **35**, 1109-1118.

54. Mills, M. J. (1989) 'Determination of compositional ordering at grain boundaries in boron-doped Ni_3Al', Scripta Metall. **23**, 2061-2067.

55. Chiba, A., Hanada, S. and Watanabe, S. (1991) 'Influence of B addition to Pd doped Ni_3Al on the Hall-Petch slope', Scripta Metall. Marter. **25**, 1053-1057.

56. Mills, M. J. and Goods, S. H., (1991) 'The structure and properties of grain boundaries in bicrystals of Ni_3Al', High temperature ordered intermetallic alloys IV, to be published in MRS Symp. Proc. Publication.

57. Kung, H., Rasmussen, D. R. and Sass, S. L. (1991), 'The structure and chemistry of grain boundaries in Ni_3Al', High temperature ordered intermetallic alloys IV, to be published in MRS Symp. Proc. Publication.

58. Eberhart, M. E. and Vvedenski, D. D. (1987) 'Localized grain-boundary electronic states and intergranular fracture', Phys. Rev. Letters **58**, 61-64.

59. Chen, S. P. Voter, A. F. and Srolovitz, D. J. (1986) 'Computer simulation of grain boundaries in Ni_3Al: the effect of grain boundary composition', Scripta Metall. **20**, 1389-1394.

60. Masuda-Jindo, K. (1988) 'Electronic theory for grain boundary segregation and embrittlement of intermetallic compound Ni_3Al', Journal de Physique, Colloque **C5**, 557-562.

61. Sasaki, G., Shindo, D., Hiraga, K., Hirabayashi, M. and Takasugi, T. (1990) 'High resolution electron microscopy of tilt boundary in $Ni_3(Al_{0.6}Ti_{0.4})$ bicrystal', Acta Metall. Mater. **38**, 1417-1421.

62. Ackland, G. J. and Vitek, V. (1989) 'Effect of ordering energy on grain boundary structure in Ll_2 alloys', MRS Symp. Proc. Publication **133**, 105-111.

63. Gelatt Jr. C. D., Williams, A. R. and Moruzzi, V. L. (1983) 'Theory of bonding of transition metals to nontransition metals', Phys. Rev. B **27**, 2005-2013.

64. Miedema, A. R. (1976) 'On the heat of formation of solid alloys. II', J. Less Common Metals **46**, 67-83.

65. Shao, J. and Machlin, E. S. (1983) 'A model analysis of the elastic constants of metals and alloy phases', J. Phys. Chem. Solids **44**, 289-300.

66. Losh, W. (1979) 'A new model of grain boundary failure in temper embrittled steel', Acta Metall. **27**, 1885-1892.

67. Briant, C. L. and Messmer, R. P. (1980) 'Electronic effects of sulphur in nickel A model for grain boundary embrittlement, Phil. Mag. **B42**, 569-576.

68. Briant, C. L. and Messmer, R. P. (1980) 'An electronic model for the effect of alloying elements on the phosphorous induced grain boundary embrittlement of steel', Acta Metall. **30**, 1811-1818.

69. Ishida, Y. et al. (1984) 'Atomistic studies of grain boundary segregation in Fe-P and Fe-B alloys - I. Atomistic structure and stress distribution, II. Electronic structure and intergranular embrittlement, III. Vibrational states of atoms at the grain boundaries', Acta Metall. **32**, 1-11, 13-20, 21-27.

70. Izumi, O. and Takasugi, T. (1988) 'Mechanisms of ductility improvement in Ll_2 compound', J. Mat. Res. **3**, 426-440.

71. Takasugi, T. and Izumi, O. (1988) 'Grain boundary structure and controlling embrittlement in ordered alloys', Materials Forum **12**, 8-25.
72. Painter, G. S. and Averill, F. W. (1987) 'Effects of segregation on grain-boundary cohesion: a density-functional cluster model of boron and sulfur in nickel', Phys. Rev. Lett. **58**, 234-237.
73. Steigerwald, E. A., Schaller, F. W. and Troiano, A. R. (1960) 'The role of stress in hydrogen induced delayed failure', Trans. Met. Soc. AIME **218**, 832-841.
74. Oriani, R. A. and Josephic, P. H. (1974) 'Equilibrium aspects of hydrogen-induced cracking of steels', Acta Metall. **22**, 1065-1074.
75. McMahon Jr., C. J. and Vitek, V. (1979) 'The effects of segregated impurities on intergranular fracture energy', Acta Metall. **27**, 507-513.
76. Beachem, C. D. (1972) 'A new model for hydrogen-assisted cracking (hydrogen "embrittlement")', Met. Trans. **3**, 437-451.
77. Eberhart, M. E., Latanision, R. M. and Johnson, K. H. (1985) 'The chemistry of fracture: a basis for analysis', Acta Metall. **33**, 1769-1783.

FRACTURE AND DUCTILIZATION OF γ-TITANIUM ALUMINIDES

P.A. BEAVEN, F. APPEL, B. DOGAN and R. WAGNER
Institute for Materials Research
GKSS-Research Centre
D-2054 Geesthacht
Germany

ABSTRACT. Essential features of the microstructures which can be produced via phase transformations in two-phase alloys based on γ-TiAl and α_2-Ti$_3$Al are described. Effects of microstructure on the deformation and fracture behaviour are discussed with a view to finding possible pathways towards improved ductility and toughness in such alloys.

1. Introduction

Popular usage of the term "ductilization" is generally associated with improvement in the ambient temperature plastic deformation behaviour of "brittle" monolithic intermetallic phases brought about by a variety of measures either singly or in combination, viz.

(i) compositional control (e.g. of stoichiometry)
(ii) macroalloying (e.g. leading to changes in crystal structure)
(iii) microalloying (e.g. avoidance of grain boundary embrittlement)
(iv) microstructural control via processing methods such as powder metallurgy techniques or thermomechanical treatments (e.g. leading to grain refinement).

The aim of these measures is to ensure that an adequate number of slip systems is not only available, but can also be activated, to allow homogeneous polycrystalline deformation (von Mises), and that processes e.g. planar slip, grain boundary segregation) which lead to premature fracture are avoided. A detailed understanding of the factors controlling crystal structure, dislocation dynamics, and crack initiation and growth is thus necessary in order to elucidate the measures required to improve ductility. An important element in this alloy design philosophy is the use of macroalloying to attain crystal structures with cubic symmetry, e.g. L1$_2$ although this alone does not necessarily guarantee successful "ductilization".

The Titanium Aluminides Ti$_3$Al(DO$_{19}$) and TiAl(L1$_0$) are representative of a class of non-cubic intermetallic phases where according to current knowledge no opportunities exist for producing cubic (e.g. L1$_2$) structures via macroalloying, and where compositional variations (e.g. Ti;Al ratio, alloying element additions)

C. T. Liu et al. (eds.), Ordered Intermetallics – Physical Metallurgy and Mechanical Behaviour, 413–432.

within the ranges of single phase stability (Ti$_3$Al: 22-34 at.% Al, TiAl: 52-60 at.% Al) although affecting dislocation mechanisms (2-7), do not appear to alleviate ambient temperature brittle behaviour. However, the upsurge in interest in these intermetallic phases in recent years has led to the result that via suitable combinations of alloy composition and processing, two-phase alloys based on γ-TiAl and α$_2$-Ti$_3$Al have been produced which show ambient temperature tensile fracture strains up to ~3% [8]. Alloys of current interest are based on compositions close to Ti-48 at.% Al with alloying additions of 1-3 at.% Mn, V, Cr, Nb or combinations of these elements. Such alloys undergo a complex series of phase transformations during solidification and cooling. They can subsequently be thermomechanically treated to give a wide range of microstructures. It is the purpose of this contribution to discuss the influence of microstructure on the deformation and fracture behaviour of such two-phase γ + α$_2$ alloys based on Ti-48 at.% Al, and to discuss the relative importance of the terms "ductilization" and "toughening" as applied to these alloys. Before progressing to these topics a brief summary of phase transformations and microstructural development is given.

2. Phase Transformations and Microstructure

Microstructural development during the solidification, thermomechanical processing and heat treatment of Ti-48Al alloys is best considered by referring to the schematic binary Ti-Al phase diagram shown in Fig. 1 [1]. Under equilibrium conditions the alloy C$_0$ exists as disordered α (hcp) at temperature T$_1$ and enters the α + γ field on cooling at ~1350 °C, and subsequently the α$_2$ + γ phase field below ~1125 °C. It is well established that the transformation from α to a γ + α$_2$ mixture in such an alloy gives rise to a structure of alternating lamellae of the γ and α$_2$ phases (Fig. 2a) with a specific orientation relationship [1,9]. This transformation can be regarded as involving a change in stacking sequence (hcp→fcc) accompanied by an ordering reaction to give the f.c.t. γ-TiAl phase. At T<T$_e$ the formation of γ lamellae takes place from the ordered α$_2$ matrix phase. Over a wide range of cooling rates this transformation occurs at temperatures where little long-range diffusion is needed since the high temperature α phase and the γ phase (T$_3$ - T$_4$) have compositions ~C$_0$. Only at slow cooling rates such that the γ phase is formed at temperatures >T$_3$ can it contain >50 at.% Al.

In reality the alloy with composition C$_0$ does not exist as 100% α at temperature T$_1$, since the peritectic solidification leads to Al enrichment of the melt over a wide range of solidification rates, thus giving rise to interdendritic segregates of Al-rich γ (see Fig. 2b). Typical as-cast microstructures thus consist of relatively equiaxed regions containing the lamellar γ + α$_2$ microstructure (with composition ~47 at.%Al), surrounded by γ phase regions (with composition ~51 at.% Al) at the grain boundaries and triple points. Heat treatment in the single phase α field (at T$_1$) serves to eliminate the γ phase segregates. On subsequent cooling the microstructure which develops is 100% lamellar consisting of ≳90% γ and ≲10% α$_2$ (e.g. Fig. 3).

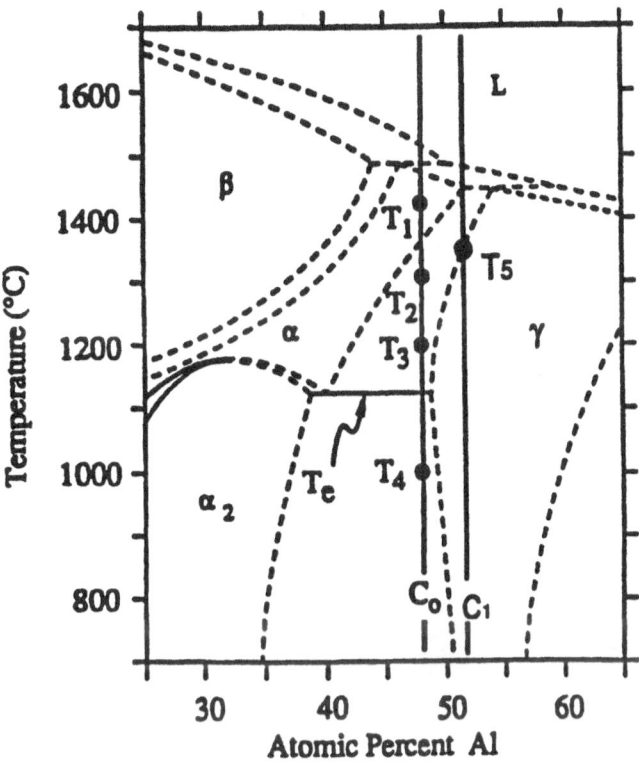

Fig. 1: Schematic binary Ti-Al phase diagram after Kim [1]

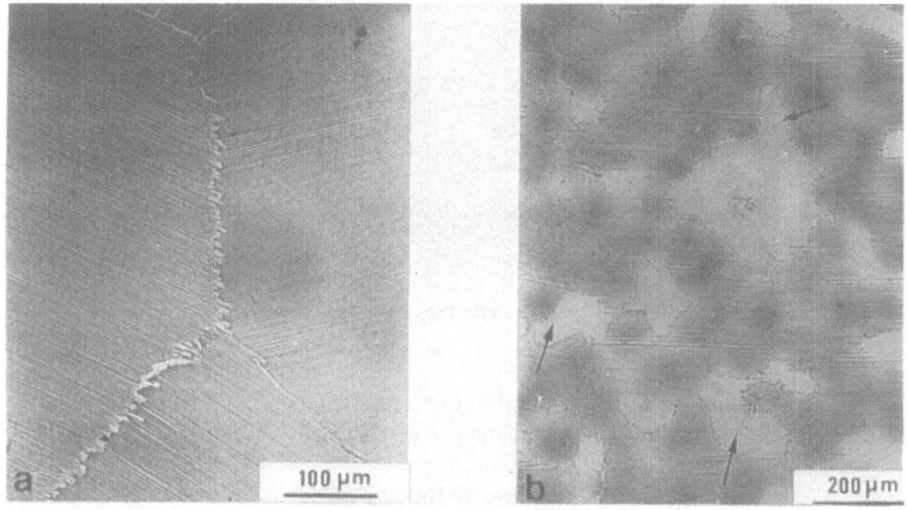

Fig. 2: Optical micrographs of Ti-48Al (a) after homogenization at 1400 °C (b) in the as-cast and HIP'ed (1220 °C/4h) condition; arrows indicate Al-rich γ.

416

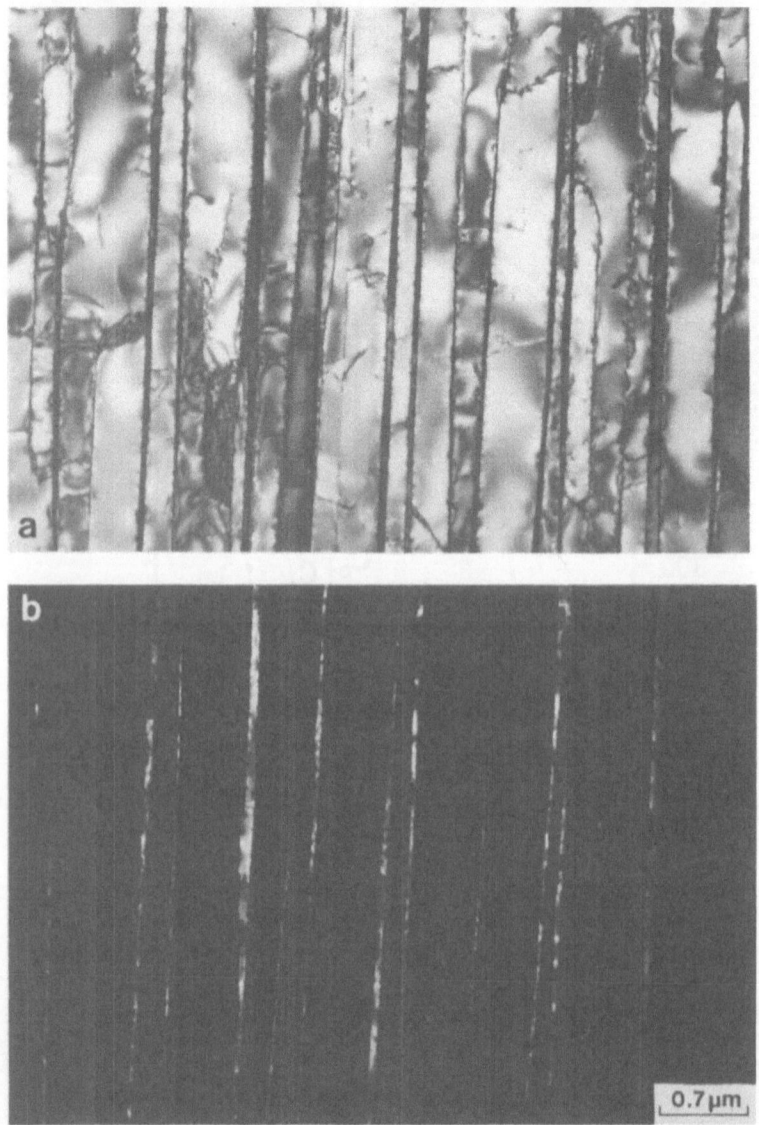

Fig. 3: TEM micrographs of the lamellar ($\alpha_2 + \gamma$) region in a Ti-48 at.% Al alloy.
(a) BF; (b) DF (α_2).

Electron diffraction studies [9] have shown that the γ phase in such lamellar structures adopts an orientation relationship of the form

$$\{111\}\gamma \mid\mid (0001)\alpha_2; <110>\gamma \mid\mid <1120>\alpha_2$$

with respect to the parent α (α_2) phase. Since the <110] and <101] directions in the tetragonal L1$_0$ structure are not crystallographically equivalent, a total of six different variants of the γ phase may form.

This means that in addition to the γ/α_2 interfaces produced during the formation of γ, a variety of γ/γ interfaces is formed, the nature of which depends on the spatial distribution of the six possible variants. Analysis of microdiffraction patterns taken from individual γ lamellae reveals that they consist of three "domains" with mutually perpendicular c-axes (perpendicular transformation twins) such that individual γ lamellae have a "macro-symmetry" equivalent to that of the L1$_2$ structure (Fig. 4a-d). The boundaries between 'c'-domains within the individual γ lamellae are characterized by dislocation networks (Fig. 5a) which accommodate the misfit which results from the tetragonality of the γ phase.

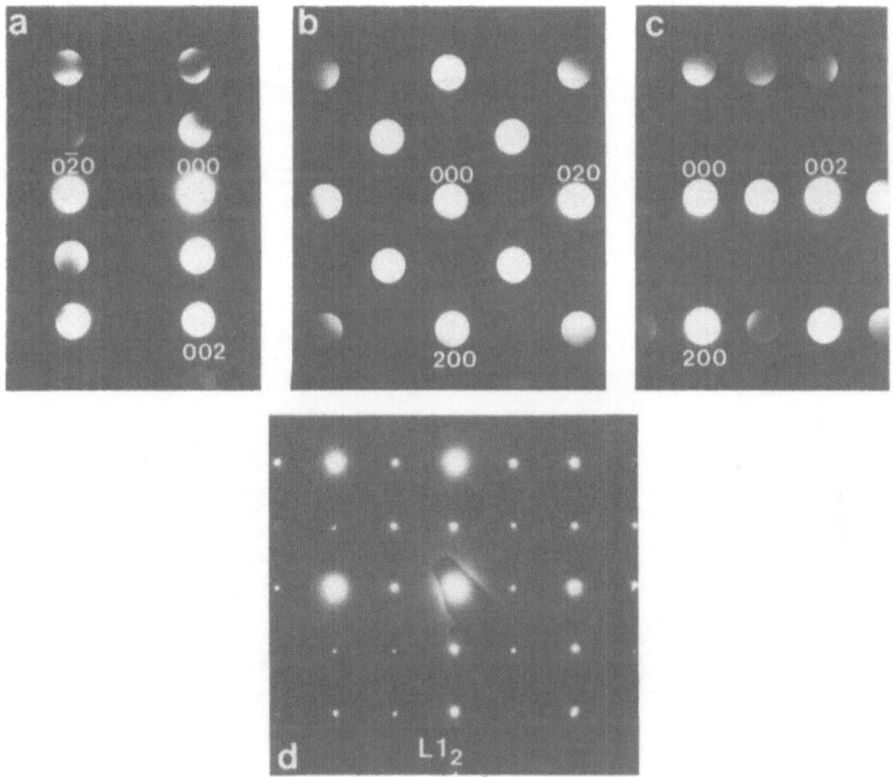

Fig. 4   (a-c) Convergent beam diffraction patterns from individual 'c' domains; (d) composite diffraction pattern showing overall symmetry of the L1$_2$ structure.

418

Since the orientation relationship between γ and α₂ allows a further three "domains" with mutually perpendicular c-axes to be formed, the {111} interfaces between adjacent γ lamellae are thus determined by the particular "pairs" of domains which meet at the boundaries.

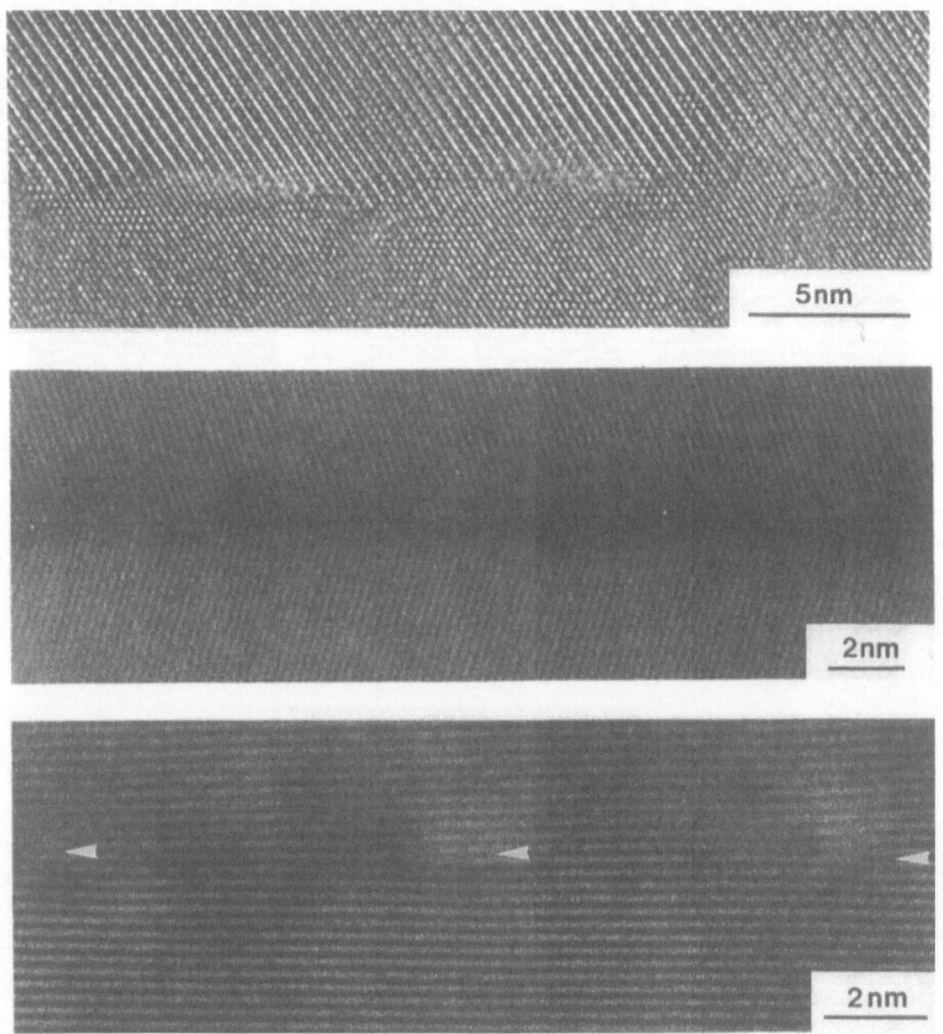

Fig 5.:   Lattice fringe images of frequently observed γ/γ interface types in the α₂ + γ lamellar structure.

(a) domain boundary with misfit dislocations
(b) coherent twin boundary
(c) twin boundary with misfit dislocations

Pairs of domains with c-axes at 70.5° lead to twin relationships of the "order" or "true" twinning type [10] and interfaces which are coherent (Fig. 5b), whereas "pairs" of domains with c-axes at 48.2° lead to "pseudo" twins [10], which as a result of the tetragonality of the γ phase (c/a ~1.02), have interfaces which contain networks of misfit dislocations (Fig. 5c). These γ/γ interfaces together with the α_2/γ interfaces, which consist of smooth facets parallel to $(0001)\alpha_2$ and $(111)\gamma$ separated by "steps" which correspond to the partial dislocations responsible for the $\alpha\rightarrow\gamma$ transformation, are the most frequently observed interface types in the lamellar microstructure [11].

Since the γ component of the lamellar aggregate has a domain structure consisting of twin-related lamellae each having the overall symmetry of the $L1_2$ structure, the diffraction pattern shown in Fig.4 d thus consists of six overlapping patterns from the γ variants together with a pattern due to the α_2 phase. Since c/a $\neq$ 1, the ensuing misfits between adjacent γ crystals lead to small misorientations so that the lamellar aggregate may not be described as being completely "coherent". Effects of these crystallographic characteristics on the deformation mechanisms will be considered in a later section.

The wide variety of microstructures which can be developed via thermomechanical treatments performed on these lamellar microstructures cannot be described in detail here [examples can be found in Refs. 8,12]. Consideration of the phase diagram (Fig. 1) indicates that heat treatments at temperatures T_3-T_2 lead to microstructures consisting of different proportions of α + γ. On subsequent cooling the α phase transforms to lamellar aggregates of α_2 + γ, which are surrounded by equiaxed γ grains. Such microstructures are termed fine-grained "duplex", a typical example of which is shown in Fig. 6.

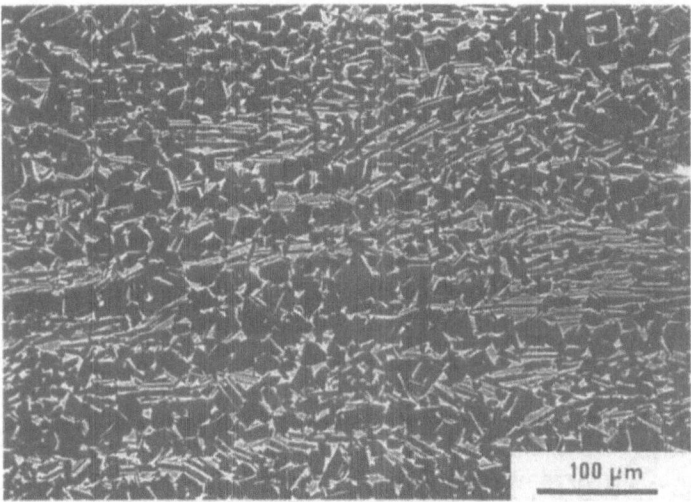

Fig. 6:   SEM micrograph of a Ti-48Al-2Cr alloy after forging at 1025 °C and heat treating at 1300 °C for 4 hours showing "duplex" microstructure.

3. Fracture Behaviour

A Ti-48.5 at.% Al-1 at.% Mn alloy, prepared in a laboratory argon arc furnace, was selected for studies of the deformation and fracture behaviour of lamellar α_2 + γ microstructures. The as-cast microstructure consisted of relatively equiaxed (100 - 300 μm diameter) grains containing the lamellar structure, with γ phase "films" at the grain boundaries. As-cast material was heat treated either at 1000 °C (to relieve residual stresses and ensure that the $\alpha(\alpha_2) \rightarrow \gamma$ transformation was completed) or at 1350 °C (to remove the γ phase "films" and produce 100 % lamellar structure during furnace cooling). Table I summarizes the tensile data and fracture modes for as-cast and heat treated specimens tested at room temperature and at 900 °C.

TABLE I

Tensile Data for Ti-48.5 Al-1 Mn

Heat treatment	Test temp. (°C)	σ_y (MPa)	σ_F (MPa)	ε_F (%)	Fracture mode
As-Cast	R.T.	-	306	0.2	Brittle
As-Cast	900-Argon	214	192	11.3	Brittle
1000 °C/24 hr	R.T.	-	353	0.2	Brittle
1000 °C/24 hr	900-Argon	233	233	8.9	Brittle
1350 °C/2 hr	R.T.	-	208	-	Cleavage
1350 °C/2hr	900-Argon	194	217	3.3	Cleavage

At room temperature the tensile fracture strain of all specimens is low ($\varepsilon_F \leq$ 0.2 %); at 900 °C as-cast material showed a fracture strain of > 11 %. Heat treatment at 1350 °C leads to a reduction in tensile fracture strain at both room temperature and 900 °C, which is associated with a change in observed fracture mode from brittle to cleavage.

Examination of the side surfaces of specimens containing grain boundary γ tested at room temperature showed extensive evidence for preferential microcracking in these γ regions together with some evidence for microcracking within the lamellar structure (Fig. 7a) [13]. Microcracking and brittle fracture facets in grain boundary γ are also apparent on the fracture surfaces (Fig. 7b). The fracture mode at ambient temperature is transgranular brittle showing cleavage facets at both grain and lamellar boundaries. Specimens heat treated at 1350 °C which contained essentially no grain boundary γ failed at room temperature by intragranular cleavage and showed zero ductility.

Polarized light microscopy of the side surfaces of specimens containing grain boundary γ tested at 900 °C showed evidence of plastic deformation via deformation twinning together with extensive 'bending' of the lamellae (Fig. 7c).

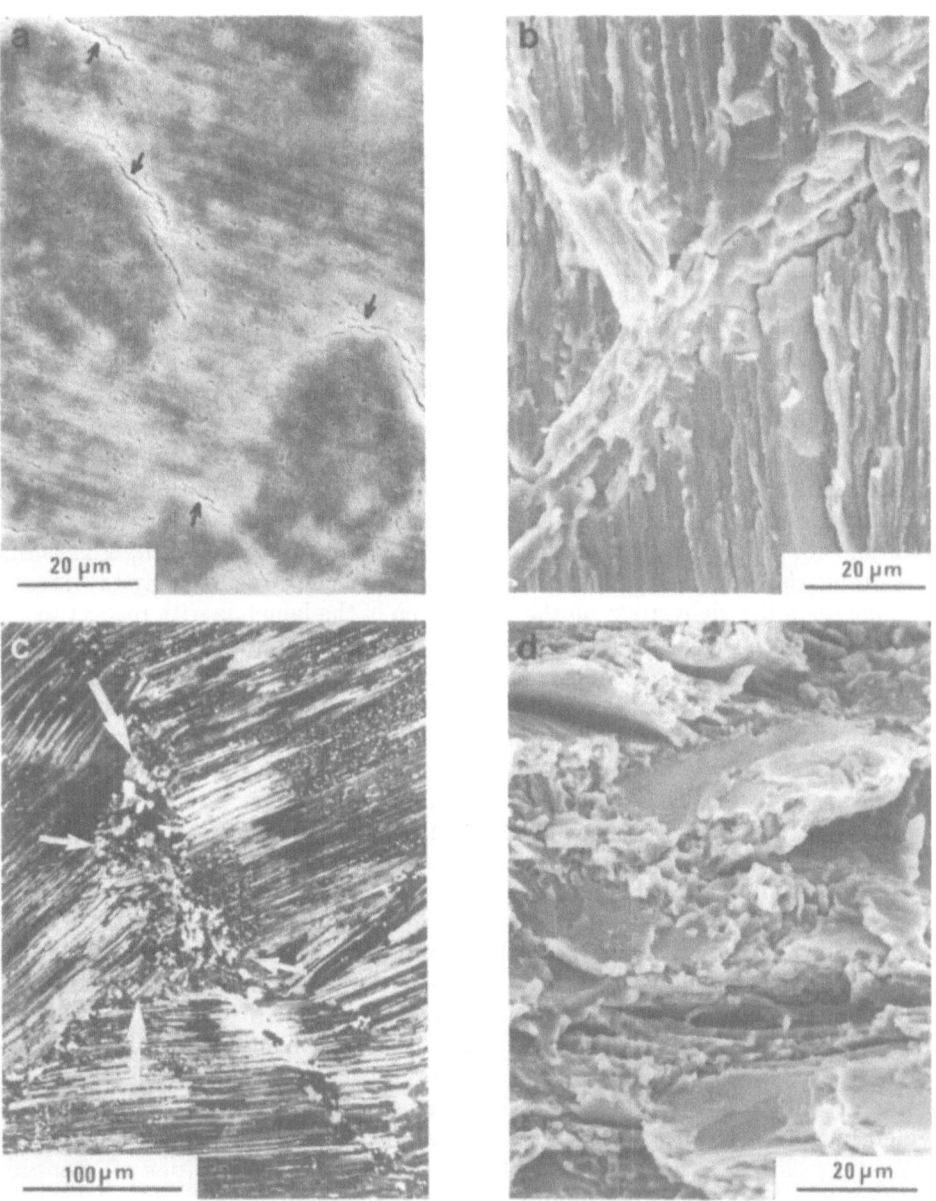

Fig. 7: a) Fracture sites on the side surface of as-cast material tested at room temperature. b) Fracture surface of material heat treated for 24 hrs at 1000 °C, tested at room temperature. c) Microstructure of material tested at 900 °C showing deformation twinning and dynamic recrystallisation of grain boundary γ (at triple points). d) Fracture surface of as-cast material tested at 900 °C.

Although uniform macroscopic tensile elongations of up to about 11 % were measured, the deformed microstructure was inhomogeneous, showing evidence for severe local deformation. Fine-grained γ regions along the grain boundaries and at triple points (Fig. 7c) indicate that dynamic recrystallisation has taken place giving rise to extremely fine (< 5 μm) grains within the grain boundary γ phase. SEM examination of the fracture surfaces showed roughness, which could have been mistakenly interpreted as evidence for a "ductile" fracture mode, but which clearly demonstrates a brittle intergranular fracture path around the fine, dynamically recrystallized γ grains (Fig. 7d). Specimens heat treated at 1350 °C which contained essentially no grain boundary γ failed at 900 °C by cleavage with tensile fracture strains of ~3%.

These results indicate that relatively coarse-grained, polycrystalline material consisting of 100% lamellar structure fails by cleavage at both ambient temperature and 900 °C with zero and ~3% fracture strain respectively. The presence of grain boundary γ thus appears to be essential for measurable (at RT) and appreciable (at high temperatures) tensile strains to be achieved. This suggests that incompatibility effects at grain boundaries resulting from stress concentrations are critical. These could arise as a result of the von Mises' criterion not being fulfilled, or, more likely, as recent work on single crystals containing the lamellar structure would suggest [14], result from the strong dependence of the yield stress on the orientation of the lamellae with respect to the deformation axis; this gives rise to 'soft' and 'hard' deformation modes if slip occurs parallel to the lamellae boundaries or if slip has to cross them respectively. Stress concentrations at grain boundaries which thus develop rapidly are accommodated by the grain boundary γ phase at room temperature via microcracking, and at elevated temperatures via plastic deformation, dynamic recrystallisation, and eventually intergranular fracture.

Although the variation of the yield stress and fracture strain with test temperature (Fig.8) would suggest a classical ductile-brittle transition behaviour (with DBTT > 700 °C) and thus differences in fracture mode above and below the transition, the metallographic and fractographic observations indicate that for the lamellar constituent cleavage is the dominant fracture mode even at 900 °C. Crack initiation appears to be associated with grain boundary sites and crack growth then occurs preferentially along suitably oriented lamellar boundaries (Fig. 9). Since the grain size is relatively coarse this leads to fracture surfaces dominated by intragranular cleavage. Observations of secondary cracking events, however, show that crack propagation "across" the lamellar boundaries is evidently more difficult; crack arrestation and deflection (Fig. 9) thus cause more tortuous crack paths. Such processes lead to the common observation of "stepped" cleavage facets on the intragranular fracture surfaces.

These features of the fracture process are generally observed for the lamellar constituent for all test temperatures, and also appear in the specimens containing the grain boundary γ phase, although in this latter case the fracture process becomes more intergranular in character. Again the fracture mode both

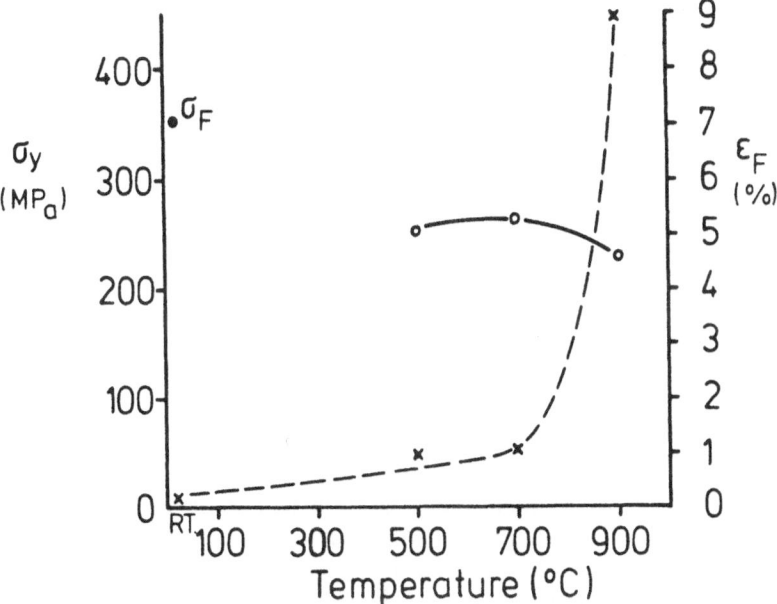

Fig. 8: Variation of yield strength and fracture strain with test temperature.for
Ti-48.5Mn-1Mn heat treated at 1000 °C/24 hr.

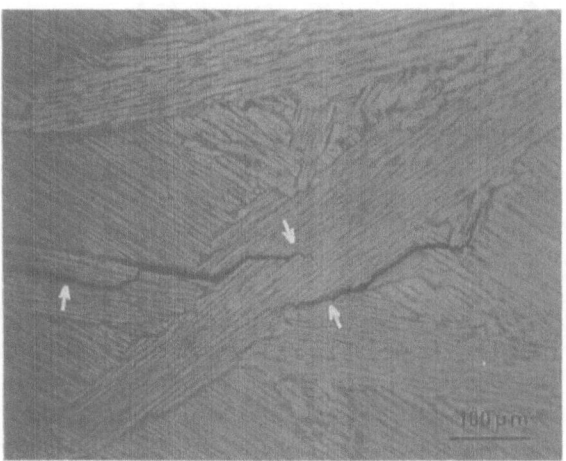

Fig. 9: Fracture paths typically following grain and lamellar boundaries.
Material heat treated for 2 hrs at 1350 °C, tested at 900 °C.

below and above the "apparent" DBTT is brittle. At room temperature this is related to the microcracking and cleavage facets observed in the grain boundary γ phase. At 900 °C, the extensive localized plastic deformation in the grain boundary γ phase, witnessed by the presence of fine dynamically recrystallized grains, leads initially to an apparent 'ductilization' by suppression of grain boundary initiated fracture, but eventually to embrittlement. Although dynamical recovery effects within the lamellar grains may help to reduce the continuing build-up of stress concentrations at these high temperatures (900 °C), it appears that the continuing grain refinement (d < 5μm) of the original grain boundary γ regions leads to a Hall-Petch-type hardening, and under the locally imposed stresses and strain rates the γ grains finally undergo intergranular fracture, once $\sigma_{frac.} < \sigma_{yield}$.

4. Deformation Behaviour

In order to investigate the plastic deformation mechanisms operative below and above the DBTT and to study the role of the lamellar interfaces in more detail, TEM studies were initiated. Constraints arising from the test specimen geometry and the attainment of optimum imaging orientations in the TEM mean that so far our observations have been confined to "hard" deformation modes [14] in which the lamellar boundaries (γ/α$_2$, γ/γ) act as barriers to the propagation of slip or twinning. In all specimens examined evidence was found for plastic deformation occurring via the activitation of <110] (111) slip and 'order' twinning of the type <112] (111) involving the propagation of 1/6 < 112] partial dislocations. Specimens heat treated (at 1350 °C) to remove the grain boundary γ phase which showed nearly zero fracture strain at room temperature even showed evidence for microyielding via fine microtwins (Fig. 10a) which appeared to "cross" the lamellar boundaries. However, these events were extremely localized and sporadic in all samples tested at 20 °C with fracture strains ~0.2%. Due to the early onset of grain-boundary initiated cleavage fracture as discussed above, these events do not significantly contribute to the global fracture strain. Although at this stage it is not possible to quantify the individual contributions of microtwinning, slip and microcracking to the measured global ambient temperature fracture strain, it is probable that the intragranular damage (microcracking) seen in Fig. 7a contributes significantly to the measured total fracture strain.

TEM examination of specimens tested at 900 °C revealed evidence for extensive deformation via twinning (Fig. 10b). The sources of these deformation twins (and of dislocations of the a/2 < 110] type which were also observed) appeared to be associated with the misfit dislocation networks existing at the γ/γ pseudo-twin boundaries described in section 2. Within individual γ lamellae a number of examples of transmission of twinning across boundaries between 'c' domains could be found (e.g. at A in Fig. 10b). The observation that twinning occurs on different systems in the various γ lamellae (see Fig. 10b) is a consequence of the fact that only order twinning is occuring, and that the order twinning system chosen must lead to "extension" rather than "contraction" under the prevailing tensile testing conditions [14]. Further examples of transmission of twinning across γ/γ lamellar boundaries and via deformation twin intersections are shown in Fig. 11. All such examples involved a process akin to cross-slip ("cross-twinning") and the formation of highly strained regions which gave rise to strong strain field contrast. Further propagation of twinning across the lamellar boundaries was generally halted at the γ/α₂ interfaces (Fig. 11).

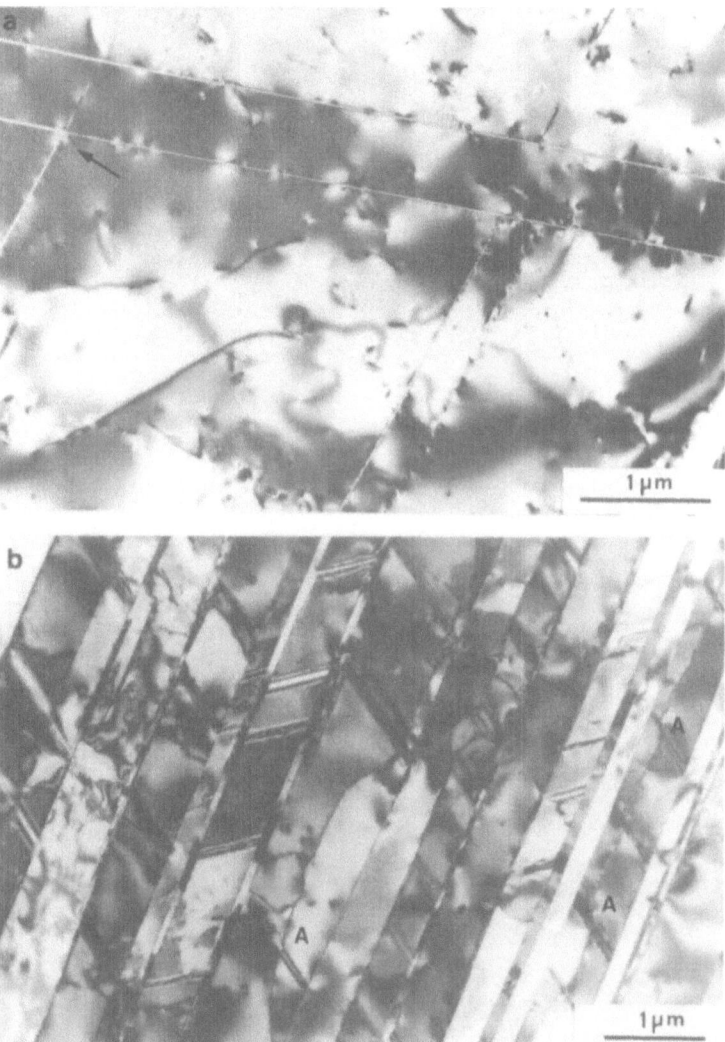

Fig. 10 (a,b) TEM micrographs of the deformation microstructures in Ti-48.5Al-1Mn specimens tensile tested (a) at room temperature, and (b) at 900 °C.

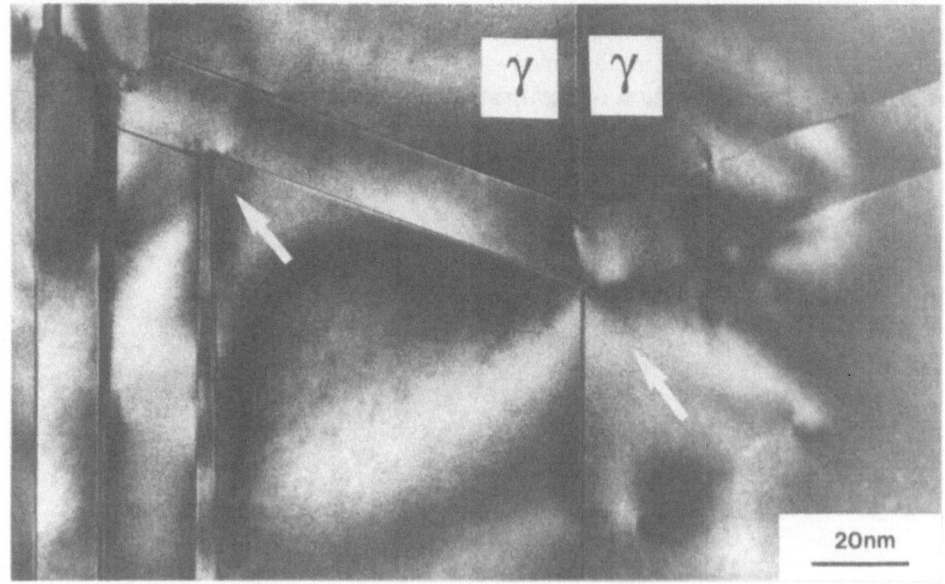

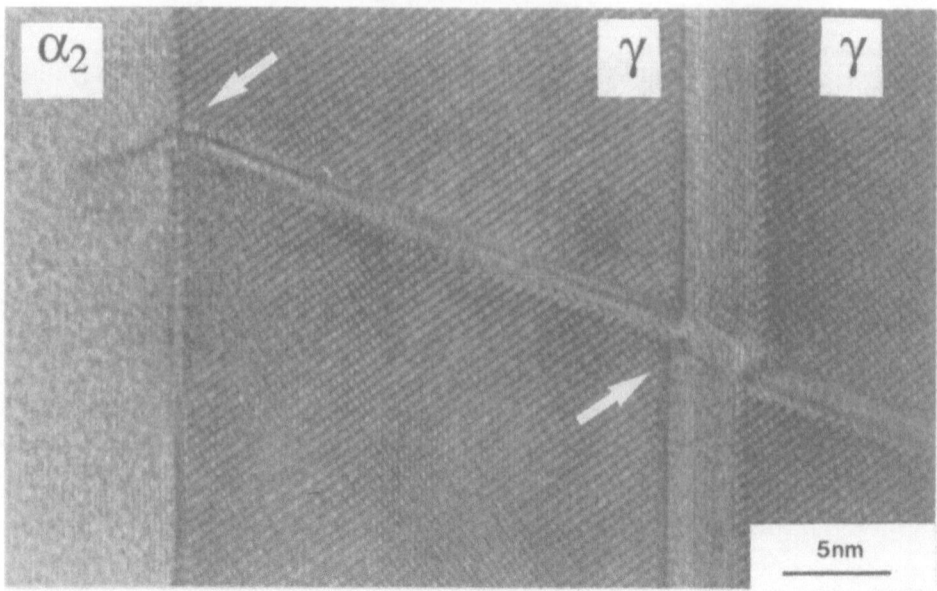

Fig. 11: TEM micrographs showing examples of deformation twin intersections and interactions of deformation twinning with γ/γ and γ/α_2 interfaces.

Although further analysis of these twin-twin interactions is still in progress, we surmise at this stage that it is these processes which generate local stress concentrations which are critical to the plastic deformation behaviour of the γ phase. Relaxation of these stress concentrations at room temperature could involve the nucleation of microcracks; at high temperatures these regions could serve as nucleation sites for dynamic recrystallization (as has been observed under conditions of high temperature creep [15]). Whereas the lamellar structure with its high density of interfaces serves to disperse these events such that they are not so critical to global plasticity, we anticipate that such "cross-twinning" in the single phase γ regions at grain boundaries may be the precursor to microcracking and brittle fracture at room temperature, and to the dynamic recrystallization observed at 900 °C. As a result of the relatively large grain sizes here, no such grain boundary γ regions have yet been subjected to TEM examination.

5. Towards Ductility and Toughness

The observations described thus far for the polycrystalline material, taken together with the results of Inui et al. [16] on single crystals with exclusively lamellar structure, indicate that some degree of plasticity (up to 12.5% tensile fracture strain in suitably oriented single crystals [16]) is possible at ambient temperatures, although the deformation behaviour is extremely anisotropic. Possible measures to improve the properties of the polycrystalline material must therefore be aimed at reducing and accommodating stress concentrations in grain boundary regions. Such measures could include grain refinement, variations in the volume fraction of grain boundary γ, and the use of preferred orientations. In the following sections we describe experiments performed to test these parameters.

5.1 'TEXTURE' EFFECTS

Metallographic characterization of castings of a Ti-48 at.% Al-2 at.% Cr alloy revealed that in addition to the features described in section 2, the lamellae show a high degree of preferred orientation (casting texture) as a result of the radial dendrite growth of the α phase during solidification. It was therefore possible to prepare specimens for tensile testing in which the lamellae interfaces were oriented either parallel, at ~45°, or at 90° to the tensile axis. The tensile data for tests performed at room temperature are shown in Fig. 12 for the orientations I to III.

Although a fuller account of these experiments is to be presented elsewhere [17] a number of preliminary conclusions may be drawn here. The orientation dependence of the strength and fracture strain found in these tensile tests on polycrystalline "textured" specimens appears to be dominated by the fracture behaviour rather than by the plastic deformation behaviour.

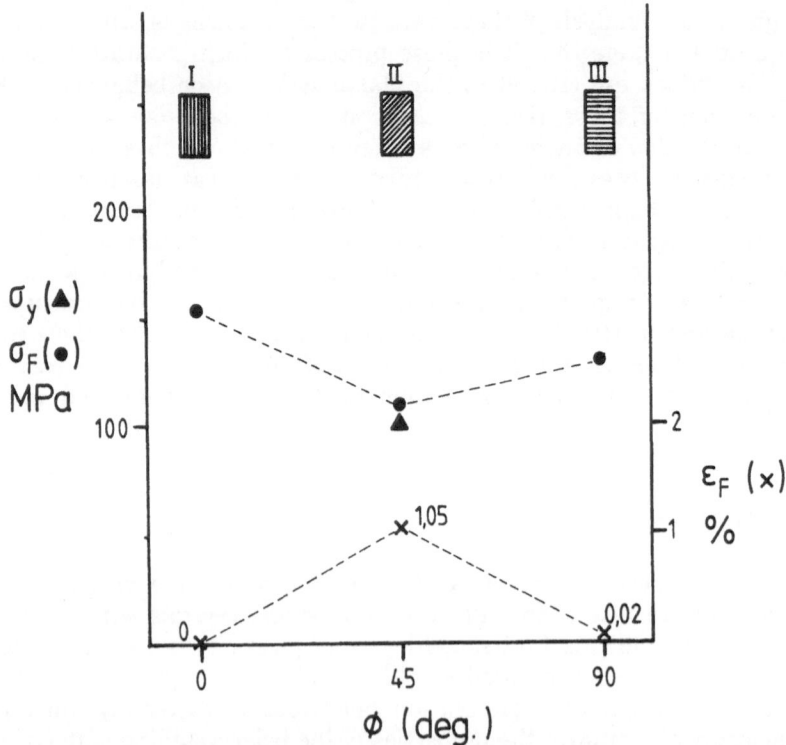

Fig. 12: Tensile data for Ti-48Al - 2Cr tested at room temperature as a function of the angle φ between the lamellar boundaries and the tensile axis.

The orientation II in which the lamellar interfaces are inclined (~45°) to the tensile axis is the only case where a yield stress and fracture strain could be measured. In this orientation, plastic deformation can occur parallel to the lamellar interfaces at low stresses, a result which is consistent with the single crystal studies of Inui et al. [16], and room temperature fracture strains > 1% are reached before failure is initiated in the grain boundary γ phase.

For the orientations I and III ambient temperature yield stresses and fracture strains could not be measured. The fracture stresses determined are significantly lower than the yield stresses found by Fujiwara et al.[14] in compression tests on single crystals with these orientations, in which the lamellar boundaries act as barriers to the propagation of slip events. It is thus evident that in the present tensile tests on polycrystalline "textured" material, failure occurs at stresses lower than those necessary for macroscopic yielding of the lamellar α_2 + γ grains. In the case of orientation I there is a high probability that specimens contain dendrite boundaries which lie at 90° to the tensile axis so that the fracture stress measured pertains to the γ phase grains which are located in these regions (Fig 13). Similarly for orientation III it is thought that the fracture stress measured is associated with the γ phase grains, although the fact that the lamellar boundaries are subjected directly to tensile stresses which cause delamination may also be an important factor. This latter point could be checked using tensile tests on single crystals.

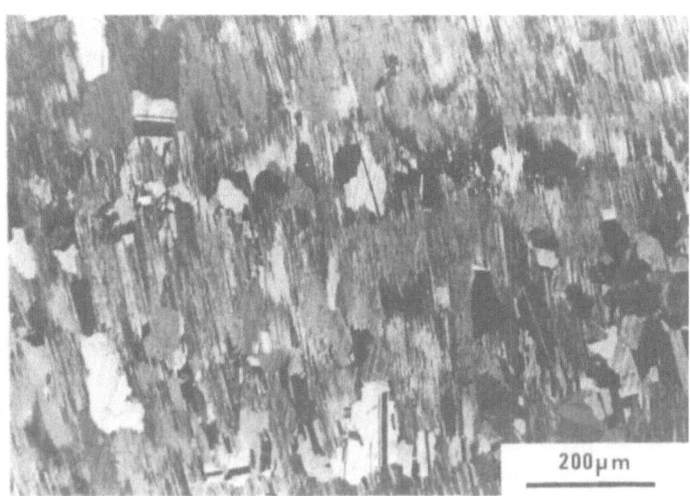

Fig. 13:   Optical micrograph of Ti-48Al-2Cr (cast + HIP) showing the preferred orientation of the lamellar structure and γ grains at dendrite boundaries.

With regard to the use of preferred orientations to improve the ductility of polycrystalline material we conclude at this stage that in coarse-grained materials this can only be achieved if the lamellar grains are so oriented that their yield stress is lower than the fracture stress of the grain boundary γ phase; this is associated with unacceptably low strength levels.

5.2 'MICROSTRUCTURAL REFINEMENT' EFFECTS

To what extent microstructural refinement measures can lead to ductilization at higher strength levels can be demonstrated with the aid of preliminary measurements on investment castings of the Ti-48 at.% Al-2 at.% Cr alloy referred to in the previous section. Such castings show finer microstructures (Fig. 14) as a result of the higher solidification rates; the microstructure which develops during HIP'ing is characterized by a relatively large (~40%) volume fraction of γ phase grains surrounding the "textured" lamellar grains (volume fraction ~60%). Tensile tests [18] at room temperature on specimens with orientation I (lamellar interfaces parallel to the tensile axis) give yield stresses ~380 MPa and tensile fracture strains of ~1.3%. At temperatures above the DBTT, e.g. at 900 °C, yield stresses ~300 MPa and tensile fracture strains >40% could be measured.

Fig. 14:   Optical micrograph showing the microstructure of Ti-48Al-2Cr (investment cast + HIP).

The question as to whether these improvements in ductility are correlated with the increase in volume fraction of γ grains per se, or are associated with the reductions in grain size which ensue when the γ phase consumes the lamellar grains during HIP'ing, cannot be answered at this stage. However, it should be noted that such "mixed" microstructures consisting of fine-grained γ interspersed with lamellar regions are those which can be developed after forging and subsequent heat treatment. Room temperature fracture strains in the range 2-4% have been reported by various workers [8,19] for alloys heat treated after forging such that the proportions of γ grains and α grains (which subsequently transform to the lamellar structure on cooling) are approximately equal, and where grain sizes are of the order of 20-40 μm. It therefore appears that combinations of microstructural refinement and variations in the proportions and distributions of the γ grains and the lamellar component of the microstructure are indeed the measures necessary to achieve a limited "ductilization" in these near γ alloys.

6. Ductilization vs. Toughening

It is instructive to consider whether the route to ductilization of a "brittle" lamellar $\alpha_2 + \gamma$ polycrystalline material brought about by a combination of grain refinement together with the inclusion of a large fraction of single phase γ - a phase known for its brittle behaviour at room temperature - is a result which is deformation or fracture controlled. The ambient temperature (true) fracture toughness of the coarse-grained predominantly lamellar microstructure with <1% tensile ductility investigated in this work is higher (K_{IC} ~10.3 MPa $\sqrt{m}$) than that of the finer-grained investment casting with tensile ductility >1% for which K_{IC} ~7.7 MPa$\sqrt{m}$. This inverse trend between ductility and fracture toughness has also been reported by other workers for forged and heat treated alloys [20]. Fine-grained (20-40 μm) duplex microstructures with tensile ductilities of up to 4% have fracture toughnesses ~12.9 MPa$\sqrt{m}$, whereas "apparent" K_{IC} values >20 MPa$\sqrt{m}$ have been reported [20,21] for coarse-grained lamellar microstructures. Some scatter in the values arises from effects related to the orientation of the crack relative to the orientation of the lamellar structure [21] indicating the role of the lamellar interfaces in producing an anisotropic, composite-like toughening behaviour (see Fig. 9). It is thus tempting to speculate that it is the effect of microstructure on the fracture mechanisms which ultimately determines the tensile ductility, with the lamellar microstructure contributing "toughness" rather than the γ phase contributing "ductility". The strong dependence of strength, ductility, and toughness on the properties and distribution of the lamellar and γ components of the microstructure together with grain size and texture effects, suggests that the primary role of ternary and quaternary (e.g. Mn,Cr) alloying elements is constitutional in nature i.e. causing changes in the locations of the high temperature phase boundaries, thus leading to variations in microstructure. More systematic studies of the effects of well-defined microstructures on the mechanical properties are required in order to establish optimum combinations of both room temperature (strength, ductility, toughness) and high temperature (creep resistance) properties.

Acknowledgements

Thanks are due to J. Rogalla, Th. Pfullmann for helpful discussions, and to U. Lorenz for technical assistance. The support of this work by the Deutsche Forschungsgemeinschaft (G.W. Leibniz-Programm) is also gratefully acknowledged.

References

1. Kim, Y.-W. (1989), "Intermetallic alloys based on Gamma Titanium Aluminide", Journal of Metals 41,24.
2. Greenberg B.A., Aisimov V.I., Gornostirev Yu.N., and Taluts G.G., (1988), "Possible factors affecting the brittleness of the intermetallic compound TiAl. II. Peierls manyvalley relief." Scripta Metall., 22, 859.
3. Court S.A., Vasudevan V.K., and Fraser H.L. (1990) "Deformation mechanisms in the intermetallic compound TiAl", Phil Mag. A61, 141.
4. Hug. C., Loiseau A., and Lasalmonie A., (1986), "A new type of 1/2 <112> slip superdislocation in ordered TiAl strained at room temperature and dissociation of <101> and 1/2 <112> superdislocations." Phil.- Mag. A54, 47.
5. Court S.A., Löfvander J.P.A., Loretto M.H., and Fraser H.L. (1990), "The influence of temperature and alloying additions on the mechanisms of plastic deformation of Ti$_3$Al", Phil. Mag., A61, 109.

6. Vasudevan V.K. Court S.A., Kurath P., and Fraser H.L. (1989), "Effect of purity on the deformation mechanisms in the intermetallic compund TiAl", Scripta Metall., 23, 907.

7. Huang S.-C. and Hall E.L., (1991), "Plastic deformation and fracture of binary TiAl-base alloys". Metall. Trans., 22A, 427.

8. Kim Y.-W., (1991),"Recent advances in gamma Titanium Aluminide alloys", Proc. MRS Symposium "High Temperature Ordered Intermetallic Alloys IV", Vol. 213, 777.

9. Blackburn M.J. (1970), "Some Aspects of Phase Transformations in Titanium Alloys", in "The Science, Technology and Application of Titanium", ed. Jaffee R.I. and Promisel N.E. (Pergamon, Oxford), 633.

10. Christian J.W. and Laughlan D.E. (1988), "The Deformation Twinning of Superlattice Structures Derived from Disordered b.c.c. or f.c.c. Solid Solutions", Acta. Metall. 36, 1617.

11. Appel F., Beaven P.A., and Wagner R., to be published.

12. Beaven P.A., Rogalla J., Pfullmann Th., and Wagner R. (1991), "Phase transformations and microstructural development in TiAl-based alloys", Proc. MRS Symposium "High Temperature Ordered Intermetallic Alloys IV", Vol. 213, 151.

13. Dogan B., Wagner R., and Beaven P.A., (1991), "Fracture behaviour of a Ti-48.5Al-1Mn alloy", Scripta Metall. et Mater., 25, 773.

14. Fujiwara, T., Nakamura, A., Hosomi, M., Nishitani S.R., Shirai Y. and Yamaguchi M. (1990), "Deformation of polysynthetically twinned crystals of TiAl with a nearly stoichiometric composition", Phil. Mag. A. 61, 591.

15. Kinder, J. and Arzt, E. (1991)"Kreuzzwillingsbildung und dynamische Rekristallisation in legierten Titanaluminiden", Hauptversammlung Deutsche Gesell. Materialkde., Graz, to be published.

16. Inui H., Nakamura A. and Yamaguchi M., (1991) "Deformation of Polysynthetically Twinned Crystals of TiAl in Tension and Compression at Room Temperature", Proc. MRS Sympos. High Temperature Ordered Intermetallic Alloys IV, Vol. 213, 569.

17. Dogan B., Wagner. R., Beaven P.A., to be published.

18. Smarsly W., (1991) BMFT-Projekt 03M3030, to be published.

19. Koeppe C., Seeger J., Bartels A., Mecking H., (1991) "Duktilitätssteigerung von intermetallischem Ti48Al2Cr durch thermomechanische Behandlung", Hauptversammlung Deutsche Gesell. Materialkde. Graz, to be published.

20. Chan K.S. and Kim Y.-W., (1991), "Fracture processes in a two-phase gamma Titanium Aluminide alloy", in "Microstructure/Property relationships in Titanium Aluminides and Alloys", ed. Kim A.-W. and Boyer R.R., (Warrendale, PA: TMS), in the press.

21. Reuss S. and Vehoff H., (1990), "Temperature dependence of the fracture toughness of single phase and two phase intermetallics", Scripta Metall. et Mater., 24, 1021.

SOME ASPECTS OF DIFFUSION IN INTERMETALLIC COMPOUNDS

H.BAKKER, D.M.R.LO CASCIO and L.M.DI
Natuurkundig Laboratorium, University of Amsterdam
Valckenierstr. 65, NL-1018 XE Amsterdam
The Netherlands

ABSTRACT. Theoretical aspects of diffusion in intermetallic compounds are discussed. Order-disorder compounds as well as compounds highly ordered at any temperature are considered. Theory is compared to the experiment.

1. Introduction

Diffusion in intermetallic compounds is far from being understood. The number of investigations is scarce compared to diffusion studies in, for example pure metals. It is even very probable that diffusion mechanisms differ from intermetallic to intermetallic and depend strongly on the crystallographic structure, the type and degree of disorder, the ability and the way of vacancy formation etc. So far there are strong indications for a vacancy mediated diffusion mechanism and this will be the starting point of the present paper. A further starting point will be that atomic disorder plays an important role in diffusion. Diffusion between sublattices in intermetallic compounds disorders the material by itself. There are mechanisms imaginable by which this disordering is minimized. This will be the subject of an other paper in these proceedings. We will consider here intermetallic compounds, where there exists at least minimal disorder. In fact we will consider diffusion in two different types of compounds: order-disorder intermetallics such as CuZn and highly ordered compounds such as the A15 compound V_3Ga. It is not the aim of the present paper to review the whole field as was done previously (Bakker (1984)), but to stipulate some essential features. The paper is mainly based on own work.

2. Expression for the diffusion coefficient

For diffusion over longer distances the velocity of the atomic migration is characterized by the diffusion coefficient D. This quantity can be measured, for example by use of radioactive tracers. For the physical interpretation of such

C. T. Liu et al. (eds.), Ordered Intermetallics – Physical Metallurgy and Mechanical Behaviour, 433–448.

measurements knowledge of the diffusion coefficient in terms of the relevant physical parameters is essential. The concept presented here is from Kikuchi and Sato (1969). They argue that the system as a whole must be considered in thermodynamic equilibrium and that in this way a disordering jump at one place of the system is compensated by a reverse jump at another place. Let us take as an example diffusion in the B2 structure. This structure is given in figure 1. It consists of two interpenetrating simple cubic sublattices, which we will call the α and ß sublattice.

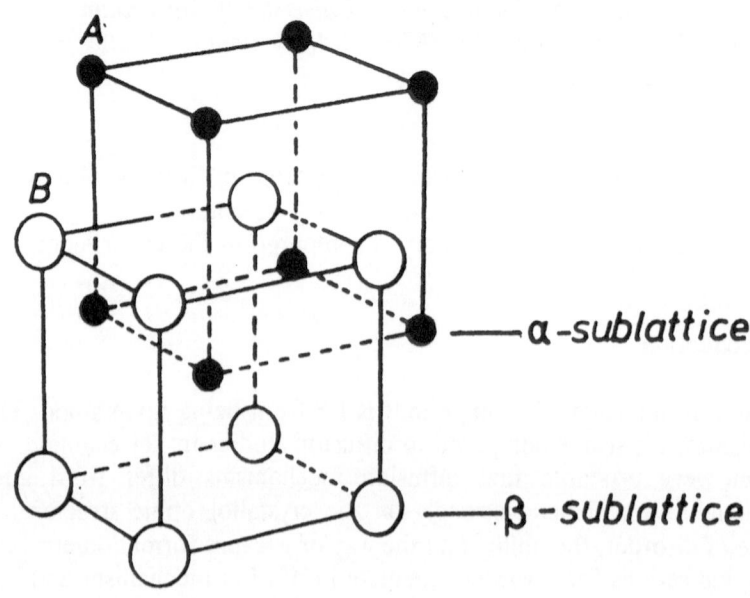

Figure 1. The B2 (or CsCl) structure

The stoichiometric formula of the compound is AB. If there is long-range order, A atoms mainly occupy the α sublattice and B atoms the ß sublattice. We assume atomic jumping to nearest-neighbour positions, i.e. jumping between both sublattices. Moreover we assume a vacancy mechanism of diffusion, i.e. the atoms jump into unoccupied lattice sites. The starting equation for the derivation of an expression for the diffusion coefficient of for example A atoms is

$$D_A = <R_A.R_A>/6t \tag{1}$$

where $<R_A.R_A>$ is the mean squared displacement of an A atom during a time t. The displacement vector R_A can be written in terms of elementary displacement vectors r_{Ai} (i = 1,2,3...) of an atom from lattice position to lattice position:

$$R_A = r_{A1} + r_{A2} + r_{A3} + ... \tag{2}$$

Substitution into the expression for D leads to

$$D_A = f_A . n_A . r^2 / 6t \tag{3}$$

where f_A is called the correlation factor and contains the double products of the elementary jump vectors. The length of the jump vector is r and n_A is the number of jumps during a period t. Assuming a vacancy mechanism for diffusion and thermodynamic equilibrium, the following expression can be derived for the diffusion coefficient of component A (Bakker and Westerveld (1988))

$$D_A = f_A . (c_A^{\alpha} / c_A) . p_{AV}^{\alpha\beta} . w_A^{\alpha} . a^2 \tag{4}$$

where c_A is the fraction of A atoms in the compound, c_A^{α} is the fraction of sites on the α sublattice occupied by A atoms, $p_{AV}^{\alpha\beta}$ the probability of finding a vacancy on the ß sublatice next to an A atom on the α sublattice, w_A^{α} the vacancy-atom exchange rate for an A atom on the α sublattice and where a is the lattice parameter. All quantities in the right-hand side of this equation are strongly dependent on the degree of order and deviation from stoichiometry of the system. Expressions for the diffusion coefficient in a number of more complex cubic structures are given in Bakker and Westerveld (1988).
An alternative form is given by Kikuchi and Sato (1969) and Bakker (1979)

$$D_A = f_A . p_{AV} . w_A . a^2 \tag{5}$$

where p_{AV} is the average probability of finding a vacancy adjacent to an A atom:

$$p_{AV} = (c_A^{\alpha} . p_{AV}^{\alpha\beta} + c_A^{\beta} . p_{AV}^{\beta\alpha}) / 2c_A \tag{6}$$

and where w_A is the harmonic mean of the exchange rates of an A atom on both sublattices

$$1/w_A = (1/w_A^{\alpha} + 1/w_A^{\beta}) / 2 \tag{7}$$

It is interesting to note that eq.(5) is formally the same as the expression for the diffusion coefficient for diffusion in pure metals. This similarity is obtained by the appropriate definition of average quantities. If p_{AV} and w_A can be written as exponentials as in the case of diffusion in pure metals and if f_A is not too much temperature dependent in the temperature range under consideration, D_A takes the form

$$D_A = D_{A0} \exp(-Q_A / kT) \tag{8}$$

This is the well-known Arrhenius equation. Q_A is the activation energy for diffusion. Eq.(8) is found to be valid in most cases (Bakker (1990))

The next task is now to evaluate the various quantities in the diffusion coefficient eq.(5).

3. Diffusion in an order/disorder intermetallic.

A popular model to describe order/disorder phenomena in alloys is the Ising model. We take as a example an alloy AB with the B2 structure. Three energy interaction parameters are defined: v_{AA}, the interaction energy between two neighbouring A atoms, v_{BB} between two B atoms and v_{AB} between an A and a B atom. In order to find the degree of order as a function of temperature the Ising model can be handled in various approximations from the simple mean-field or Bragg-Williams approximation to sophisticated approximations such as the cluster variation method. An 'exact' solution is obtained by computer simulation. The degree of long-range order turns out to vary with the parameter v/kT, where $v = v_{AB} - (v_{AA} + v_{BB})/2$, T is absolute temperature and k is Boltzmann's constant. Below the critical temperature there is long-range order, which is lost above this temperature. The phase transition is second order.

Kikuchi and Sato used the Ising model in their path-probability method in order to calculate the quantities occurring in the diffusion coefficient following eq.(5) (see also Sato and Kikuchi (1983)) and computer simulations were performed among others by Bakker et al.(1976). Kikuchi and Sato used the pair approximation of the Ising model and considered the diffusion coefficient per unit vacancy fraction

$$D_A^* = D_A / c_V = f_A.p_{VA}.w_A.a^2/c_A \qquad (9)$$

Averaging was performed over all possible atomic configurations. By computer simulation (Bakker et al.(1976)) it was demonstrated that by the expressions given by Kikuchi and Sato correct values were obtained for the probability p_{VA} of finding an A atom next to a vacancy and for the exchange frequency w_A. Figure 2 gives the so-called vacancy availibilty factor for the B component $V_B = p_{VB}/c_B$ and figure 3 the reduced B atom-vacancy exchange rate $W_B = w_B.exp(7v_{AB}/kT)$ as a function of the reciprocal of the reduced temperature T_r, where $T_r = T/T'$ with T' the order-disorder temperature. The solid lines are the Kikuchi and Sato values, the points are results of computer simulations (Bakker et al. (1976)). There is close agreement between both results and so these figures demonstrate the success of the Kikuchi and Sato approach for the vacancy availibilty and the atom-vacancy exchange frequency as a function of temperature within their model .

A further quantity that needs attention is the correlation factor f. It is simple to demonstrate that this factor is equal to unity, when the diffusing object performs a random walk over a lattice. However, for an atom diffusing via a vacancy mechanism subsequent jumps into the same vacancy are correlated in direction. Let us consider

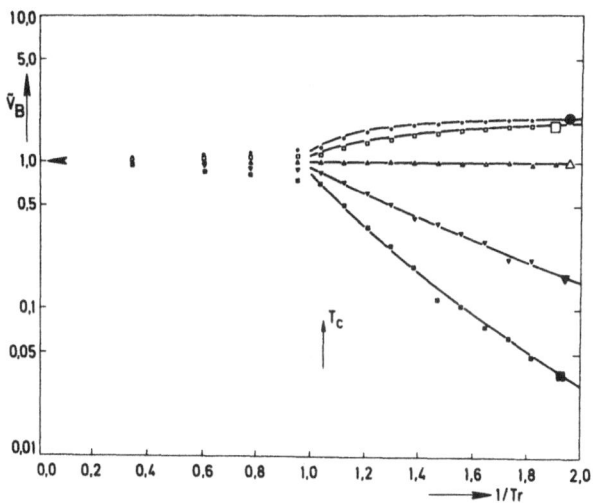

Figure 2. The vacancy availibility factor versus reduced reciprocal temperature for the component B in an alloy with the B2 structure. Solid lines: the Kikuchi and Sato (1969) results; The points: computer simulation results (Bakker et al. (1976)). Curves are given for various values of $U = (v_{BB} - v_{AA})/2v$. From top to bottom: $U = -1.2; -0.6; 0.0; 0.6; 1.2$. The order-disorder temperature is indicated.

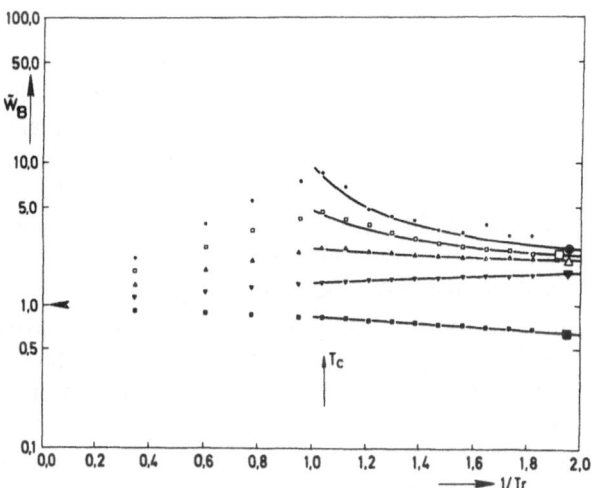

Figure 3. A plot of the reduced B atom-vacancy exchange rate similar to figure 2.

438

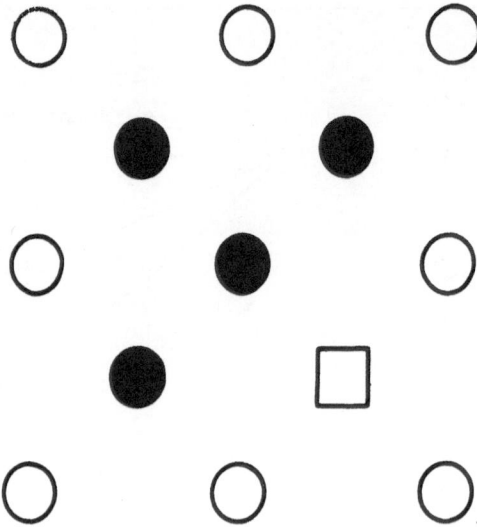

Figure 4. A schematic of an ordered structure, in which the vacancy (square) has just exchanged with a B atom (black circle). The A atoms are drawn as white circles.

a vacancy that has just exchanged position with a vacancy (see figure 4). Then it is clear that the most probable next jump of the atom will be a jump back into the same vacancy, simply because this position is vacant. Then two atomic jumps have taken place, but these jumps were just forwards and backwards and no real migration results. This makes the jumping of atoms via a vacancy mechanism less effective than in a random walk. This effect is incorporated in the correlation factor f. A somewhat different way of expressing the effect of correlation is such: The jumping of the atom under consideration is less effective then one would expect on the basis of the atom-vacancy exchange frequency w. In fact this exchange frequency is reduced to a value f.w, which can be called the effective jump frequency. A calculation of the correlation factor is complicated. Not only may the atom jump back directly after an atom-vacancy exchange, but also may the vacancy dissociate from the atom and return to the atom after some walk through the lattice. In particular in ordered structures, such as intermetallic compounds the effect of correlation between successive jumps is substantial. A jump from the 'right' to the 'wrong' sublattice violates the long-range order and the probability of reversing such a jump into the same vacancy is high. Figure 5, giving results of a computer simulation (Bakker et al.(1976)), makes this clear. At low temperatures (highly ordered state) the correlation factor is low and reduces the diffusion coefficient with an order of magnitude as compared to a random walk.

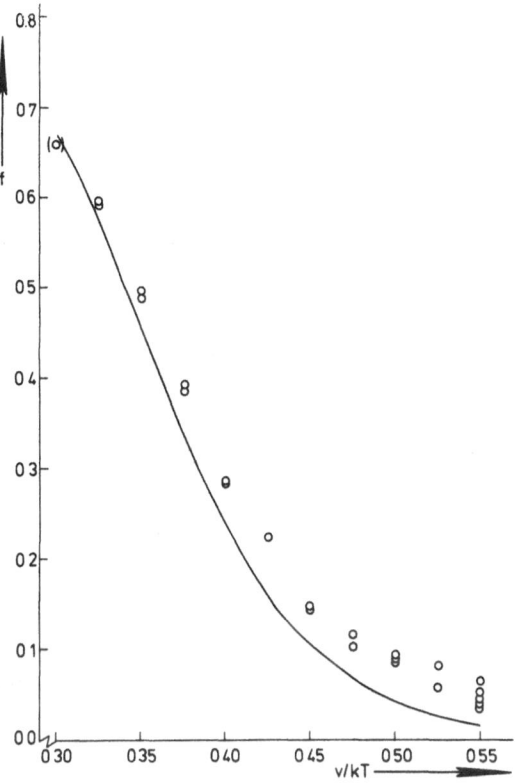

Figure 5. The correlation f for diffusion in an AB alloy with the B2 structure as a function of v/kT. Here $v_{AA} = v_{BB}$ The points are computer simulation results by Bakker et al.(1976). The solid line is from analytical calculations (Bakker (1979)).

A weak point of the theory presented in this section is that the model is far oversimplified. A description in terms of central forces between atoms is rather naive. Moreover, even if such a description is accepted, it is difficult to assign numerical values to the interaction energies. Therefore no more than a qualitative agreement with a real experiment, if any, can be hoped for. Figure 6 gives a classical experiment of a measurement of the tracer diffusion coefficient in the order-disorder compound CuZn (Kuper et al.(1956)). At the order-disorder temperature there is a change of slope in the ln D versus 1/T curve. Such a change of slope is also obtained in a computer simulation study (Bakker et al. 1976, see figure 7). Therefore it turns out that the above model describes the main features of a real measurement rather well. Bakker (1984) speculates that there could be even a quantitative agreement between isotope effect measurements in CuZn and the results of the above model.

440

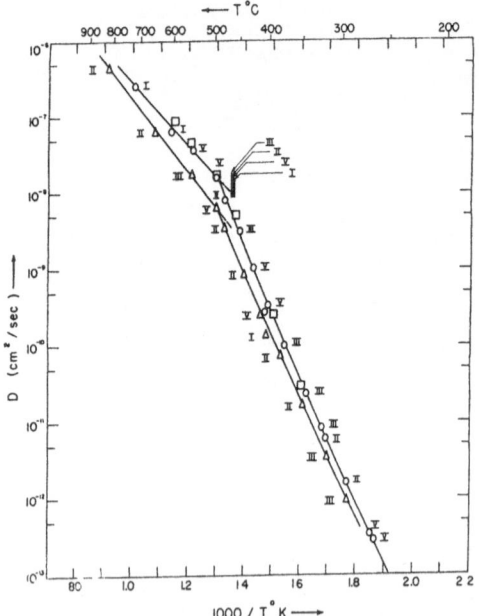

Figure 6. *Arrhenius plots of the diffusion coefficients of Cu (triangles), Zn (circles) and Sb (squares) in β-CuZn.*
(I) 45.65 at % Zn; (II) 47.15at % Zn; (III) 48 at % Zn. (Kuper et al. (1956).

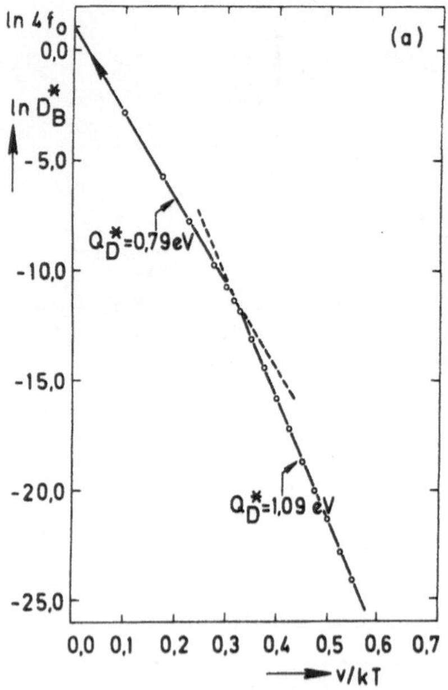

Figure 7. *Arrhenius plot of the diffusion coefficient per unit vacancy concentration in the ordered and disordered structure by computer simulation (Bakker et al. (1976))*

4. Highly ordered intermetallic compounds

In the previous section an order-disorder compound was discussed. In that case there is atomic disorder above the critical temperature. From this fact it is clear that the free-energy increase for accomodating an atom on the 'wrong' sublattice, even in the ordered state, can not be very high. Therefore it is quite logical to assume a diffusion mechanism, where atoms jump between both sublattices, i.e. to nearest-neighbour positions. A quite different situation occurs for example in an ionic compound such as NaCl. The coulombic energy prohibits here the accomadation of atoms on the 'wrong' sublattice and atomic migration takes place over the own sublattice. In completely ordered intermetallics the situation could be similar. Therefore, for ascertaining the diffusion mechanism it is important to know if there is any atomic disorder at higher temperatures. If so, then necessarily migration between both sublattices is possible and plays a role in the diffusion process. To get insight in this problem, we studied a number of intermetallic compounds with the A15 structure. This structure is presented by figure 8. It is the structure in which for example V_3Ga crystallizes. A closer inspection of this structure reveals that the vanadium sublattice

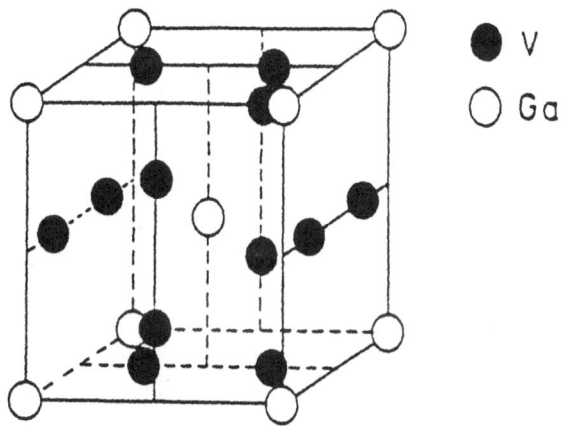

Figure 8. The A15 structure

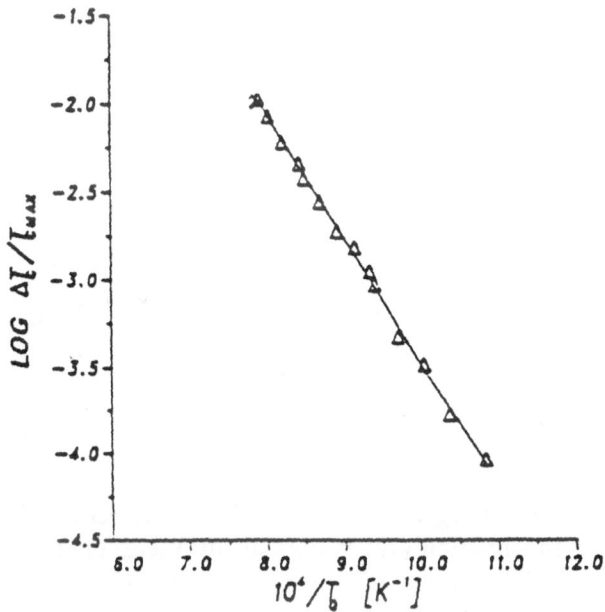

Figure 9. The logarithm of the degradation of the superconducting transition temperature versus the reciprocal quenching temperature for V₃Ga

is build from long linear chains of atoms in three perpendicular directions. Special attention has been given to these chains in relation to the special superconducting properties of a number of compounds with this structure (Labbé and Friedel (1966)). The integrity of these chains is important and disordering would lead to a degradation of the superconducting properties (Labbé and Van Reuth (1970), Junod (1978), Fähnle (1982)). Van Winkel et al. (1984b) quenched V_3Ga from high temperatures and subsequently measured the superconducting transition temperature. This temperature degraded as a result of the quenching. This indicates the existence of at least some atomic disorder at high temperature, which is frozen-in by quenching. Westerveld and Bakker (1986) and Westerveld et al. (1987) made this argument quantitative in the following way. Let us call the two sublattices again the α and β sublattice and the stoichiometric compound A_3B and let us write the rate equation for the disordering reaction (in self-explanatory notation)

$$A^\alpha + B^\beta = A^\beta + B^\alpha$$

as

$$\ln (c_A^\beta . c_B^\alpha / c_A^\alpha . c_B^\beta) = \Delta S/k - \Delta H/k.T_q \tag{10}$$

where ΔS and ΔH are the entropy and enthalpy changes for the disordering reaction, respectively. T_q is the quenching temperature. Since c_A^α and c_B^β are close to unity for the almost completely ordered compound and since c_A^β and c_B^α are proportional, we obtain

$$\ln c_B^\alpha = \text{constant} - \tfrac{1}{2}\Delta H/k.T_q \tag{11}$$

Assuming that for a small degree of disorder the degradation of the superconducting transition temperature ΔT_c is proportional to the fraction of 'wrong' atoms c_B^α, we obtain from eq.(11)

$$\ln \Delta T_c/T_{c\,max} = \ln A - \tfrac{1}{2}\Delta H/k.T_q \tag{12}$$

where A is a constant and $T_{c\,max}$ corresponds to T_c for perfect order.
Figure 9 shows that this relation (12) is obeyed for V_3Ga, quenched from various temperatures. A similar behaviour was found for the 'A15-like' compound $Ca_3Rh_4Sn_{13}$ (Westerveld et al. 1989) and for this compound by X-ray diffraction direct evidence was obtained for disordering in the A15 frame of the structure. Similar results for V_3Ga are given by Flükiger et al. (1976). From the slope of the curve in figure 9 the activation energy for the reaction turns out to be equal to 1.23 eV per pair of wrong atoms. Recent work on Nb_3Au yields 1.72 eV for this compound (Lo Cascio and Bakker 1991). These values scale with the melting temperatures of both compounds.

The above demonstrates that disorder occurs, also in such highly ordered compounds. Therefore it is evident that in A15 compounds the diffusion takes place

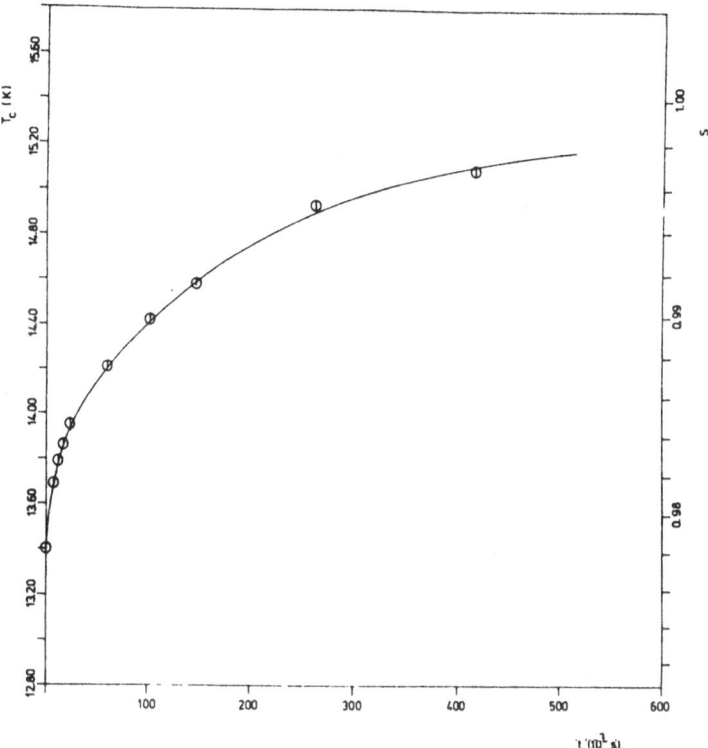

Figure 10. The superconducting transition temperature as a function of annealing time at 580 °C after a quench from 975 °C.

also between the sublattices. We assume atomic migration to occur in the linear chains, between the chains and between both sublattices. The expression for the diffusion coefficient in the A15 structure then becomes (Bakker and Westerveld (1988))

$$D_A = (f_A/16).(c_A^{\alpha}/c_A).(5.p_{AV}^{\alpha\beta}.w_A^{\beta} + p_{AV}^{\alpha\alpha}.w_A^{\alpha\alpha} + 6.p_{AV}^{\alpha\alpha'}.w_A^{\alpha\alpha'}).a^2 \tag{13}$$

where we denoted jumps from a linear chain to a perpendicular linear chain as $\alpha\alpha'$ jumps. A similar expression holds for D_B. Measurements of the tracer diffusion coefficient of V^{48} were performed by Van Winkel et al. (1984a). These measurements reveal an Arrhenius behaviour in the temperature range 1298 K to 1449 K. The activation energy is 4.29 eV and the D_0 value is 1.52 m^2/sec. (Van Winkel et al. (1984a)).

Apart from 'direct' measurements of diffusion by means of radioactive tracers, information about diffusion phenomena can also be obtained in an indirect way, namely by following the recovery of an appropriate physical parameter as a function of time in a system that is brought out of equilibrium. As the physical parameter we used the superconducting transition temperature. First a powdered V$_3$Ga sample is quenched from high temperature, whereby a certain degree of disorder is frozen in. Then the sample is brought to an intermediate temperature, where the atoms are mobile, annealed for some period at this temperature and after that T_c is measured. Since the state of order corresponding to this intermediate temperature is higher than that corresponding to the quenching temperature, the ordering and so the T_c recovers. Such an isothermal measurement of T_c as a function of annaeling time is presented in figure 10.

For the interpretation of such a measurement we first have to realize, that apart from disorder also an excess of vacancies is quenched in. The excess of vacancies will decrease exponentially to zero by annihilation at sinks such as surfaces, dislocations and grain boundaries:

$$\Delta c_V(t) = \Delta c_V(0) \exp(-t/\tau_V) \tag{14}$$

where $\Delta c_V(0)$ is the initial excess of vacancies, t the annealing time and τ_V the relaxation time for the excess of vacancies. Also in first approximation, the excess of anti-site defects $\Delta c_B{}^\alpha$ will decrease exponentially if we assume this decrease to be of first order following

$$d\Delta c_B{}^\alpha/dt = \Delta c_B{}^\alpha/\tau_a(t) \tag{15}$$

$\tau_a(t)$ is the characteristic relaxation time of the excess of anti-site atoms. It is a function of time, because it is proportional to the reciprocal vacancy content. The following relation holds:

$$\tau_a(\infty)/\tau_a(t) = c_V(t)/c_V(\infty) \tag{16}$$

After some algebra and assuming as before that ΔT_c is proportional to the fraction of 'wrong' atoms $c_B{}^\alpha$, we obtain the solution of eq.(15) as

$$\ln(\Delta T_c/\Delta T_{cmax}) = -t/\tau_a + \{\Delta c_V(0)/c_V(\infty)\}.(\tau_V/\tau_a).\{\exp(-t/\tau_V) - 1\} \tag{17}$$

This equation was derived by Westerveld et al. (1989) and is identical with the equation given without derivation by Dew-Hughes (1980).

This type of measurements were performed on V_3Ga in the temperature range from 560 °C to 635 °C after quenching from about 975 °C and analyzed following eq.(17). In figure 11 an Arrhenius plot of $1/\tau_a$ is drawn (Van Winkel and Bakker 1985). The activation energy of the reordering process turns out to be equal to 2.22 eV. The reordering process is obviously determined by the diffusion of the slower component. Since V diffuses with a higher activation energy as measured by tracer diffusion, the slower component must be gallium. Figure 12 presents an Arrhenius plot of the diffusion coefficient for vanadium, measured by tracers and of the diffusion coefficient of the slower component, calculated from τ data. In the whole temperature range the diffusion coefficient for gallium, associated with the slower component is

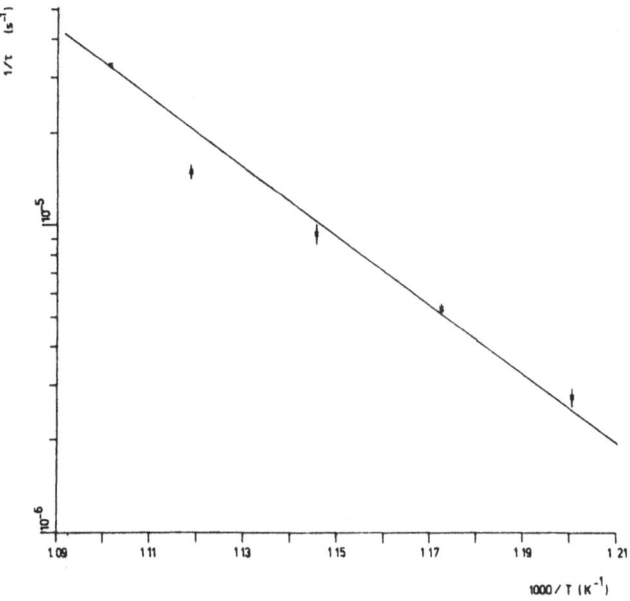

Figure 11. Arrhenius plot of 1/τ.

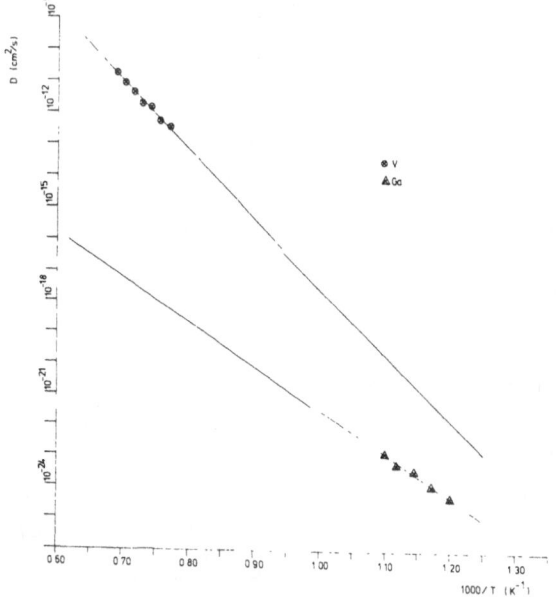

Figure 12. Arrhenius plot of the V tracer diffusion coefficient (circles) and the gallium diffusion coefficient (triangles) calculated from the relaxation time.

lower than the extrapolated values for vanadium diffusion. Similar τ data were obtained for ternary compounds by Westerveld et al. (1989).

At present there is considerable interest in the preparation of non-equilibrium structures by mechanical alloying and mechanical grinding. In this way amorphous alloys, nanocrystalline alloys and non-equilibrium crystalline structures can be synthesized. At the end of this section we will give an example of the use of known diffusion data to understand more about the stability of these structures. Di and Bakker (1991) found recently that mechanical grinding of the A15 compound V_3Ga results in a phase transition from the A15 phase to the A2 high-temperature phase. The latter phase is a disordered solid solution of gallium in b.c.c. vanadium. The material was heated in a differential scanning calorimeter (DSC) at different heating rates and a so-called Kissinger (1957) plot was made. Two peaks were found in DSC and figure 13 presents the Kissinger plots of both. From the slope of these lines activation energies can be derived. The first process reveals the same activation energy as was given above for reordering of V_3Ga. Therefore it is possible to associate this process with reordering and so with diffusion of gallium atoms. This is an example how useful diffusion data may be in the interpretation of what one observes in certain experiments.

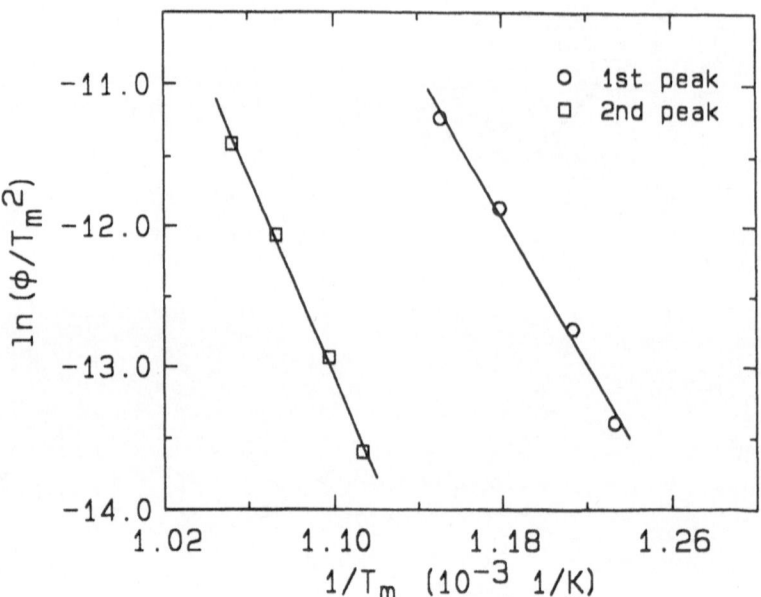

Figure 13. Kissinger (1957) plots corresponding to the two peaks in a DSC scan for ball-milled V_3Ga

5. Conclusions

So far it turns out that in intermetallic compounds diffusion occurs by means of vacancies as in pure metals. The results do not contradict atomic migration between sublattices. That such jumps occur is evident from the disorder at higher temperatures even in highly ordered compounds. The diffusion coefficient as measured, for example by radioactive tracers exhibits an Arrhenius behaviour in the majority of cases. For order-disorder systems a change of slope of the Arrhenius plot is observed close to the order-disorder transition. This behaviour can be reproduced theoretically by computer simulations starting from the Ising model. Information about diffusion can also be obtained by bringing the system out of thermodynamic equilibrium and then following the recovery of an appropriate physical parameter as a function of time at temperatures, where the atoms are mobile.

6. References

Bakker, H. (1979) Phil.Mag.A 40, 525

Bakker, H. (1984) Diffusion in Crystalline Solids, Murch and Nowick eds. (Acad.Press Orlando), p189

Bakker, H. (1990) Landolt-Börnstein New Series Vol.26, Diffusion in Metals and Alloys, H.Mehrer ed. (Springer-Verlag Berlin) p213

Bakker, H., Stolwijk, N.A., Van der Meij, L. and Zuurendonk, T.J. (1976) Nucl.Metall. 20, 96

Bakker, H. and Westerveld, J.P.A. (1988) Phys.Stat.Sol.(b) 145, 409

Dew-Hughes, J. (1980) Journ.Phys.Chem.Solids 41, 851

Di, L.M. and Bakker, H. (1991) Journ.Phys.Condens.Matter 3, 3427

Fähnle, M. (1982) Journ. Low Temp. Phys. 46, 3

Flükiger, R., Staudenmann, J-L and Fischer, P. (1976) Journ. Less Comm.Metals 50, 253

Junod, A. (1978) Journ.Phys.F 8, 1891

Kikuchi, R. and Sato, H. (1969) Journ.Chem.Phys. 51, 161

Kissinger, H.E. (1957) AnalChem. 29, 1702

Kuper, A.B., Lazarus, D. Manning, J.R. and Tomizuka, C.T. (1956) Phys.Rev. 104, 1536

Labbé, J. and Friedel, J. (1966), Journ. de Phys. (Paris) 27, 153

Labbé, J. and Van Reuth (1970) J.Phys.Rev.Lett. 24, 1232

Lo Cascio, D.M.R. and Bakker, H. (1991), accepted by Journ.Phys.Condens.Matter

Sato, H. and Kikuchi, R. (1983) Phys.Rev.B 28, 648

Van Winkel, A., Lemmens, M.P.H. and Bakker, H. (1984a) Journ.Less Common Metals, 99, 257

Van Winkel, A., Weeber, A.W. and Bakker, H. (1984b) Journ.Phys. F 14, 2631

Van Winkel, A and Bakker, H. (1985) Journ.Phys. F 15, 1565

Westerveld, J.P.A. and Bakker, H. (1986) Phil.Mag 54, L15

Westerveld, J.P.A., Lo Cascio, D.M.R. and Bakker, H. (1987) Journ.Phys F 17, 1963
Westerveld, J.P.A., Lo Cascio, D.M.R., Bakker, H., Loopstra B.O. and Goubitz, K. (1989) Journ.Phys. Condens.Matter 1, 5689

DIFFUSION MECHANISMS IN THE B2 TYPE INTERMETALLIC COMPOUNDS

M. KOIWA
Department of Metal Science and Technology,
Kyoto University,
Sakyo-ku, Kyoto 606
Japan

ABSTRACT. The diffusion coefficients of A and B atoms in the B2 type AB alloy are derived for the case that highly correlated vacancy jump cycle (six-jump vacancy cycle) is operative. The effective jump frequency to complete the cycle is expressed in terms of frequencies for individual vacancy jumps; the calculation is made by applying the concept of the mean first passage time known in the theory of stochastic processes. On the basis of the analytical expression for the diffusion coefficients, the ratio of the diffusivities of the two constituent atoms and the isotope effect can be discussed quantitatively.

The correlation effect in the six-jump cycle can be regarded to consist of two effects: a microscopic and a macroscopic effects. The former refers to the correlation between individual jumps involved in the cycle, for which the correlation factor can be neither defined nor calculated.

1. Introduction

Intermetallic compounds or ordered alloys have recently attracted much attention as practical materials. Although much effort is devoted to understand mechanical properties, investigations on diffusion in these materials are scarce. Knowledge of their diffusion behaviour is of interest for the production of these materials or for their use in practical applications.

Apart from its technical importance, diffusion is also interesting from a fundamental point of view. In pure metals or conventional alloys self diffusion occurs by random motion of vacancies. In ordered alloys or compounds, however, random vacancy motion is not possible as it would disrupt the equilibrium ordered arrangement of atoms on lattice sites. Experimental and theoretical investigations on the subject were reviewed by Bakker [1] in 1984, and by Wever, Hünecke and Frohberg [2] in 1989.

Huntington [3] first suggested the possibility of a six-jump vacancy cycle which allows diffusion to take place exclusively by means of nearest neighbour vacancy jumps. The analytical expression for the diffusion coefficient in terms of individual frequencies of vacancy jumps has been derived recently for the six-jump cycle mechanism in a two-dimensional ordered lattice [4], and in a three-dimensional (B2) ordered lattice [5,6].

C. T. Liu et al. (eds.), Ordered Intermetallics – Physical Metallurgy and Mechanical Behaviour, 449–464.
© 1992 Kluwer Academic Publishers.

450

In the present paper, we shall first review briefly experimental observations of the diffusion behaviour in intermetallic compounds, and describe the derivation of the diffusion coefficient for the B2 case.

2. Review of Experimental Observations

Table 1 lists alloys or compounds for which the results of tracer diffusion experiments were reported. Salient features for the alloys of the B2 structure, for which the diffusion behaviour has most extensively been investigated, are summarized as follows:

(1) For alloys with the order–disorder transformation, *e.g.* CuZn, the Arrhenius plot of the diffusion coefficient exhibits a bend near the transformation temperature. The activation energy for diffusion is larger for the ordered- than the disordered state.

(2) For compounds with the stoichiometric composition, the diffusion coefficients of constituent elements, A and B, are of similar magnitudes, as shown in Fig. 1; the ratio, D_A / D_B is not much different from unity.

(3) With increasing the deviation from the stoichiometric composition, the diffusion coefficient increases, *e.g.* AgMg.

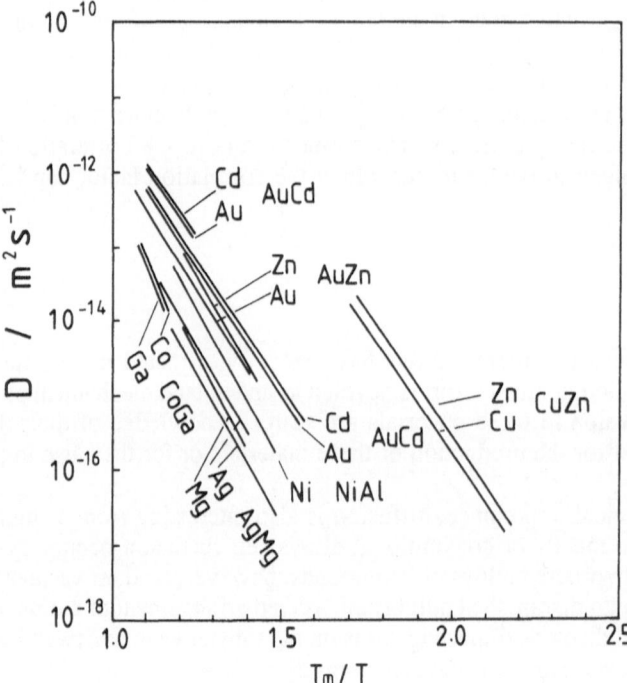

Figure 1. Diffusivities in the B2 type intermetallic compounds as a function of reciprocal temperature. The temperature is normalized to the melting temperatures of the respective compounds.

TABLE 1. Intermetallic compounds for which diffusion measurements have been reported. The underlines indicate that measurements were made for those elements.

Structure type	Intermetallic compounds			
B2	CuZn[1]	AuCd[2,3]	AuZn[4,5,6]	NiGa[7]
	CoGa[8,9]	AgMg[10,11]	NiAl[12,13]	FeCo[14]
	PdIn[15]			
L1$_2$	Ni$_3$Al[16,17,18]	Co$_3$Ti[19]	Pt$_3$Mn[20]	
D0$_3$	Ni$_3$Sb[21]	Cu$_3$Sb[22]	Cu$_3$Sn[23,24]	
A15	V$_3$Ga[25]	Cr$_3$Si[26]		

1. Kuper, A. B. *et al.* (1956) Phys. Rev. **104**, 1536.
2. Huntington, H. B. *et al.* (1961) Acta Met. **9**, 749.
3. Gupta, D. *et al.* (1967) Phys. Rev. **153**, 867.
4. Gupta, D. and Lieberman, D. S. (1971) Phys. Rev. B **4**, 1070.
5. Jeffery, R. N. and Gupta, D. (1972) Phys. Rev. B **6**, 4432.
6. Hilgedieck, R. and Herzig, C (1983) Z. Metallk. **74**, 38.
7. Donaldson, A. T. and Rawlings, R. D. (1976) Acta Met. **24**, 285.
8. Bose, A. *et al.* (1979) Phys. Stat. Sol. (a) **52**, 509.
9. Stolwijk, N. A. *et al.* (1980) Phil.Mag. A **42**, 783.
10. Domian, H. A. and Aaronson, H. I. (1964) Trans. AIME **230**, 44; (1965) Diffusion in Body-Centered Cubic Metals, ASM, 209.
11. Hagel, W. C. and Westbrook, J. H. (1961) Trans. AIME **221**, 951.
12. Hancock, G. F. and McDonell, B. R. (1971) Phys. Stat. Sol. (a) **4**,143.
13. Berkowitz, A. K. *et al.* (1954) Phys. Rev. **95**, 1185.
14. Fishman, S. G. *et al.* (1970) Phys. Rev. B **2**, 1451.
15. Hahn, H. *et al.* (1983) Phys. Stat. Sol. (a) **79**, 559.
16. Hancock, G. F. (1971) Phys. Stat. Sol. (a) **7**, 535.
17. Bronfin, M. B. *et al.* (1975) Fiz. metal. metalloved. **40**, 363.
18. Hoshino, K. *et al.* (1988) Acta Met. **36**, 1271.
19. Nakajima, H. *et al.* (1988) Scripta Met. **22**, 507.
20. Ansel, D. *et al.* (1979) J. Less-Common Met. **65**, 1.
21. Heumann, Th. and Stüer, H. (1965) Phys. Stat Sol. (a) **15**, 95.
22. Heumann, Th. *et al.* (1970) Z. Naturforsch. **25A**, 1883.
23. Prinz, N. and Wever, H. (1980) Phys. Stat. Sol. (a) **61**, 505.
24. Arita, M. *et al.* (1991) Mat. Trans. JIM. **32**, 32.
25. van Winkel, A. *et al.* (1984) J. Less-Common Met. **99**, 257.
26. Jurisch, M. and Bergner, D. (1983) DIMETA-82, 465.

These features indicate that the diffusion process in intermetallic compounds is strongly affected by the ordered arrangement of constituent atoms.

As is evident from Table 1, the diffusion studies have been most extensively made for the B2 compounds. For compounds with the other structures, only the diffusion coefficient of one of the two constituent elements has been reported except for Cu_3Sn; this compound exists only with compositions widely different from the stoichiometry. Needless to say, an accumulation of reliable experimental data is desired before one starts a detailed theoretical analysis. In this paper, we will discuss the mechanism of diffusion in the B2 compounds.

3. Diffusion via Six-Jump Vacancy Cycle

Huntington [3] first suggested a six-jump vacancy cycle as a possible mechanism of diffusion in ordered alloys. Elcock and McCombie [7] and Elcock [8] examined the details of such a cycle for the ordered cubic alloys. Several other authors have considered in detail the case for the B2 and $L1_2$ alloys [9,10]. However, the attempt of deriving the expression of the diffusion coefficient in terms of individual frequencies of vacancy jumps has not been made until recently.

In the case of pure metals or dilute alloys, the diffusion coefficient via the vacancy mechanism is written as

$$D = \alpha f C_v \nu a^2, \tag{1}$$

where α is a numerical factor, f the correlation factor, C_v the concentration of vacancies, ν the jump frequency of a vacancy, and a the lattice constant. In deriving the corresponding expression for intermetallic compounds, the two main tasks are:

(1) to define and calculate an average jump frequency, ν, and
(2) to calculate the correlation factor, f.

The procedures are briefly described below for the B2 case; a detailed account has been given in the previous paper [5].

3.1. Six-jump Vacancy Cycle

The six-jump vacancy cycle in the B2 lattice has been discussed by Wynblatt [9]. Consider the case in which there is a vacancy in the α sublattice in an otherwise perfect B2 crystal as depicted in Fig. 2. The cycle in Fig. 2(a) causes a net transport of the vacancy across a [110] face-diagonal and will be referred to as a 110-type cycle, whereas the cycles in Figs. 2 (b) and (c) cause a net transport of the vacancy across a [100] cube edge and will be referred to as bent 100- and straight 100-type cycles, respectively, depending on whether all six jumps are confined to a single plane or not. The configuration will be specified by the position of the vacancy; the state before the ith vacancy jump is referred to as C_i (configuration i). The numbers of nearest-neighbour wrong bonds A-A and B-B are as listed in Table 2. Figure 3 gives a schematic representation of the energy changes taking place during vacancy motion by the six-jump cycle. The individual jump frequencies of the vacancy are denoted as ν_{i+} (i=1-6) and ν_{i-} (i=1-5), as indicated in the figure. Note that the arrival of the vacancy at C_7 completes the cycle; ν_{6-} does not have to be considered. As seen in the figure, the curve is symmetric with respect to configuration C_4. The following relations hold for the frequencies:

$$\nu_{1-} = \nu_{6+}, \quad \nu_{2-} = \nu_{5+}, \quad \nu_{3-} = \nu_{4+}, \quad \nu_{4-} = \nu_{3+}, \quad \nu_{5-} = \nu_{2+}. \tag{2}$$

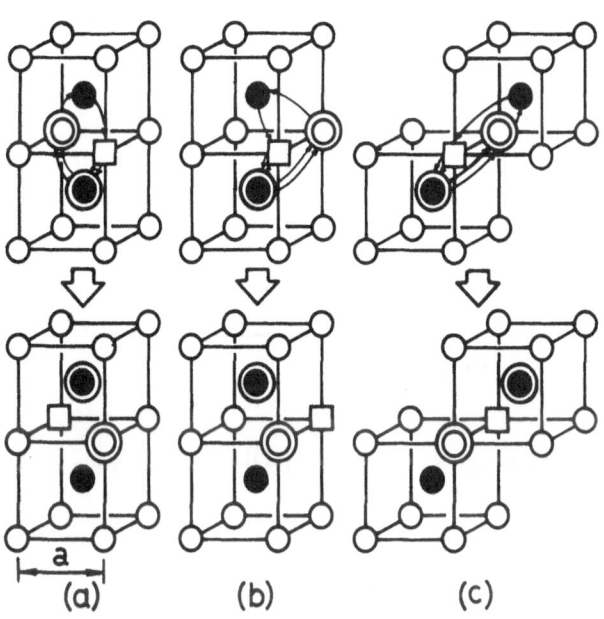

(a)   (b)   (c)

Figure 2. Six-jump vacancy cycle in the B2 lattice. The arrows indicate the path of the vacancy (the lower figures show the atom arrangement after the complete cycle): (a) 110-jump cycle; (b) bent 100-jump cycle; (c) straight 100-jump cycle.

TABLE 2. The number of wrong bonds in each configuration.

Jump cycle	Wrong bond	Number of wrong bonds						
		1	2	3	4	5	6	7
110	B-B	0	7	7	12	7	7	0
	A-A	0	0	6	6	6	0	0
Bent 100	B-B	0	7	7	12	7	7	0
	A-A	0	0	6	6	6	0	0
Straight 100	B-B	0	7	7	12	7	7	0
	A-A	0	0	6	6	6	0	0

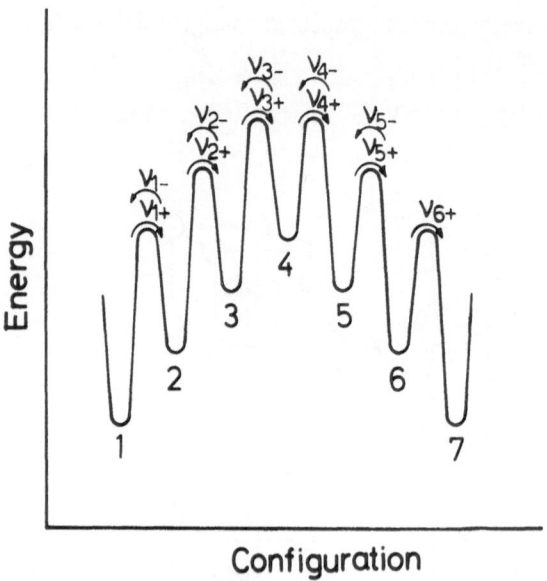

Figure 3.   Schematic diagram of energy changes during the six-jump vacancy cycle.

3.2. Effective Jump Frequency

In order to obtain the expression for the diffusion coefficient in terms of the individual vacancy jump frequencies defined above, we calculate the effective jump frequency for the six-jump cycle.

By denoting the probability of occupation of C_i as P_i, we can write rate equations for P_i. For $i=1$, for example

$$\dot{P}_1 = -8v_{1+}P_1 + 8v_{1-}P_2. \tag{3}$$

By taking into account the number m_i of equivalent configurations, it is more convenient to use new variables $x_i = m_i P_i$; x_i is the probability that any one of the ith configurations is realized. Since $m_1 = 1$ and $m_2 = 8$, the rate equation for x_1 is

$$\dot{x}_1 = -8v_{1+}x_1 + v_{1-}x_2. \tag{4}$$

The simultaneous differential equations for $x_1, x_2, \ldots$ can be written in a matrix form:

$$dX/dt = -KX, \tag{5}$$

where X is column vector with elements $x_1, x_2, \ldots$, and K is a jump frequency matrix.

The mean time $\bar{t}$ for completion of a multibarrier process or the six-jump vacancy cycle as depicted in Fig. 2 can be calculated by applying the concept of the mean first passage time in the theory of stochastic processes [11,12].

$$\bar{t} = \sum_{i=1} \langle i \,|K^{-1}|\, j \rangle , \tag{6}$$

where $<i|K^{-1}|j>$ is the element in the ith row and the j th column of the inverted matrix K^{-1} ; j indicates the initial site of the particle. The effective frequency $\bar{v}$ is given by $\bar{v}=1/\bar{t}$.

In the actual calculation there are some complications to be considered carefully; connectivity or relation between 110, straight 100 and bent 100 cycles. As described in the previous paper [5], it turned out that the number of configurations to be distinguished is ten; the column vector has components $x_1, x_2,..., x_{10}$. The jump frequency matrix K is therefore 10×10 in size.

The relative frequencies of 110 and 100 jumps can also be expressed in terms of the elements of the inverted matrix K^{-1} as

$$P_{110} = v_{6+} \langle 6|K^{-1}|1 \rangle, \tag{7}$$

$$P_{100} = v_{6+} \langle 10|K^{-1}|1 \rangle. \tag{8}$$

P_{100} is a function of individual jump frequencies. The ranges of values permitted for the two quantities are

$$0.25 < P_{110} < 0.5 , \tag{9}$$

$$0.5 < P_{100} < 0.75 . \tag{10}$$

The effective jump frequencies v_1 for 100 and v_2 for 110 jumps are given by

$$v_1 = P_{100} / \bar{t}, \quad v_2 = P_{110} / \bar{t}. \tag{11}$$

The explicit form of $\bar{t}$ in terms of $v_{i\pm}$ is given in the Appendix of Ref. 6.

3.3. Diffusion Coefficients by Means of an α Vacancy Jump Cycle

In the above, we have derived the effective frequencies v_1 and v_2 for an α vacancy. Given v_1 and v_2, the diffusion coefficient of an A tracer atom via α vacancies can be written in a straightforward manner; the diffusion is equivalent to that via a conventional vacancy mechanism in a simple cubic lattice:

$$D_A = \frac{1}{6} f_A C_{V\alpha} a^2 (v_1 + 2v_2) , \tag{12}$$

where f_A is the correlation factor, $C_{V\alpha}$ is the concentration of α vacancies and a is the lattice constant. Note that, in this paper, vacancies are referred to according to their position in their initial configuration C_1. Although an α vacancy passes through β sites during the six-jump cycle, it is called an α vacancy.

The expression for the diffusion coefficient D_B of a tracer B atom via α vacancies can be derived by considering the movement of two B atoms involved in the six-jump cycle; the final result is

$$D_B = \frac{1}{6}f_B C_{V\alpha} a^2 \left(\frac{8}{3}v_1 + 2v_2\right), \tag{13}$$

where f_B is the correlation factor.

The above expressions may be alternatively written in terms of $\bar{v}$ and P_{100}, instead of v_1 and v_2, as

$$D_A = \frac{1}{3}f_A C_{V\alpha} a^2 \bar{v} \left(2 - P_{100}\right), \tag{14}$$

$$D_B = \frac{1}{3}f_B C_{V\alpha} a^2 \bar{v} \left(1 + \frac{1}{3}P_{100}\right). \tag{15}$$

3.4. Correlation Factor

The correlation factor f_A is simply the correlation factor for the diffusion via the vacancy mechanism in the simple cubic lattice, in which the vacancy is allowed to make nearest-neighbour jumps (v_1) and next-nearest-neighbour jumps (v_2). The correlation factor can be calculated analytically, as described in the previous paper [5].

On the other hand, the calculation of the correlation factor f_B is somewhat complicated, because the problem is that of the correlation of B atom movement induced by the atom movement in the counterpart sublattice. The method of calculation has also been described in the previous paper. The values of the correlation factors f_A and f_B for various values of P_{100} are given in Table 3 and in Fig. 4.

TABLE 3.   Correlation factors f_A and f_B as functions of the probability P_{100} of the 100 jump cycle.

P_{100}	f_A	f_B
0.00	0.78145	0.85958
0.10	0.81819	0.85292
0.20	0.83760	0.84584
0.30	0.84717	0.83837
0.40	0.84851	0.83015
0.50	0.84193	0.82226
0.60	0.83701	0.81361
0.70	0.80276	0.80453
0.80	0.76753	0.79501
0.90	0.71887	0.78504
1.00	0.65311	0.77459

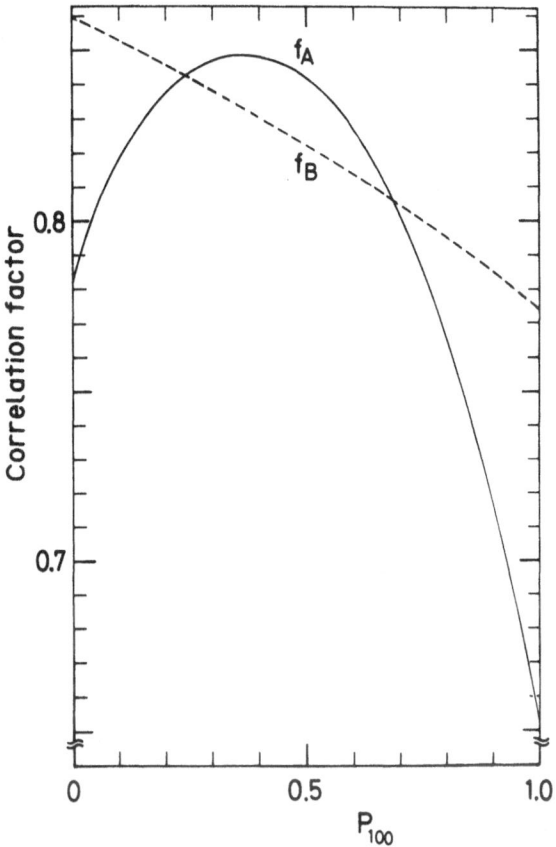

Figure 4. Correlation factor as a function of the probability of the 100 jump cycle.

3.5 Final Expression for Diffusion Coefficients

We have derived the diffusion coefficients of A and B tracer atoms for the case when vacancies exist only in the α sublattice (eqs. (14) and (15)). When vacancies exist in both the α and the β sublattices, the diffusion coefficients D_A and D_B can be written as

$$\left. \begin{aligned} D_A &= \tfrac{1}{6} f_A^\alpha C_{V\alpha} \overline{v}^\alpha a^2 (2 - P^\alpha_{100}) + \tfrac{1}{3} f_A^\beta C_{V\beta} \overline{v}^\beta a^2 \left(1 + \tfrac{1}{3} P^\beta_{100}\right), \\ D_B &= \tfrac{1}{3} f_B^\alpha C_{V\alpha} \overline{v}^\alpha a^2 \left(1 + \tfrac{1}{3} P^\alpha_{100}\right) + \tfrac{1}{6} f_B^\beta C_{V\beta} \overline{v}^\beta a^2 (2 - P^\beta_{100}), \end{aligned} \right\} \tag{16}$$

where α and β indicate that the relevant quantity is related to vacancy belonging to the respective sublattices. For example, the correlation factor f^α_A is that of an A atom with the six-jump cycle of α vacancies.

458

The ratio of the two diffusion coefficients, D_A / D_B, in the binary compounds has attracted much attention in previous theoretical and experimental investigations. For the B2 type compounds, the following relation has been accepted (see, *e.g.*[1]).

$$0.5 < D_A / D_B < 2 . \tag{17}$$

This relation can be understood by noting that, in the six vacancy jump cycle, an A (B) atom exchanges the site with a vacancy while two B (A) atoms exchange the sites each other.

For a rigorous treatment, the correlation effect should be taken into account. By using the expressions (16), the new inequality is [5]:

$$0.4917 < D_A / D_B < 2.034 . \tag{18}$$

3.6. Isotope Effect

The diffusion coefficient is known to depend on the isotopic mass, m, of the diffuser. Quantitatively, the isotope effect, E, is defined by

$$E = \frac{\partial \ln D}{\partial \ln (m^{-1/2})} . \tag{19}$$

Schoen [13] first suggested that isotope effect measurements should be useful in deciding among possible mechanisms in particular cases. For some particular cases, in fact, the isotope effect E can be expressed as

$$E = f \Delta K , \tag{20}$$

where f is the correlation factor and ΔK is the so-called kinetic factor. On the basis of this relation, it is considered that the isotope effect measurement is one of the methods for estimating the magnitude of the correlation factor f. Thus, some researchers have measured the isotope effect of diffusion in intermetallic compounds, and attempted to identify the operative mechanism from the magnitude of E ; ΔK is bounded by zero and unity, and is usually believed to be very close to unity.

Note that, however, the relation (20) is valid only when the following conditions are satisfied:
(1) The jump frequency of the relevant atom, *i. e.* the diffuser, must be unique, and the diffusion coefficient can be written in the form of eq. (1).
(2) The correlation factor is given in the form:

$$f = \frac{u}{u+v} , \tag{21}$$

where v is the jump frequency of the diffuser and u is independent of v.

Evidently, the diffusion via six-jump vacancy cycle in intermetallic compounds satisfies neither of the above two conditions; one cannot use eq. (20) for the interpretation of the isotope effect, E.

The analysis of the isotope effect should be made on the basis of eq. (16), as reported in the separate paper [6]. Salient points are summarized below.

(1) The isotope effect consists of the three contributions: from the effective jump frequency, the correlation factor and the relative probability of 100 jumps. The predominant contribution comes from the first, the effective jump frequency $\bar{v}$.

(2) The isotope effect is smaller than that for self-diffusion in pure metals, provided that the mobilities of A and B atoms are similar. This trend is intuitively understood as a result of inevitable involvement of jumps of atoms other than the tracer atom in the six-jump cycle; two B atoms are involved in addition to the A tracer for α vacancy cycle. The effect of the mass difference of the tracer is diluted by the participation of the other atoms into the unit diffusion process.

(3) The isotope effect can be larger ($E / \Delta K \rightarrow 1$) or smaller ($E / \Delta K \rightarrow 0$), if the mobilities of two atoms are widely different.

3.7. Other Vacancy Jump Cycles

The six-jump vacancy cycle is the process with the formation of the least number of wrong bonds at each step. If one looses the condition so as to allow the formation of some additional wrong bonds, the number of possible vacancy paths evidently increases. Figure 5 illustrates several of these possible paths, and Figure 6 shows the number of wrong bonds vs. jump steps of vacancy. The contribution of such "higher order" paths to the diffusion may not be very large, provided that the wrong bond energy is not too small.

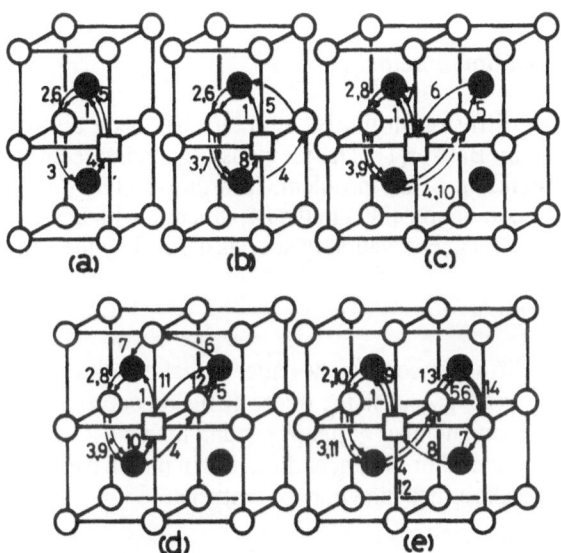

Figure 5. Multi-jump vacancy cycle in the B2 lattice. (a) 6- , (b) 8- , (c) 10- , (d) 12- , and (e) 14-jump cycles.

460

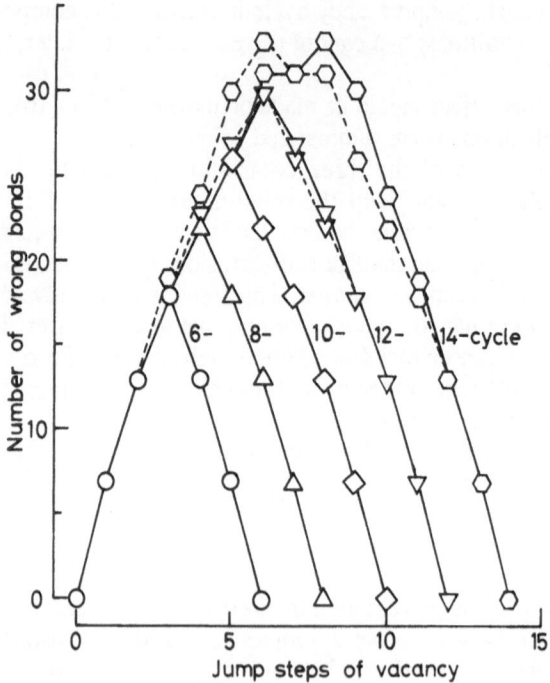

Figure 6. The number of wrong bonds formed in the multi jump vacancy cycles shown in Fig. 5.

4. Remarks on Correlation Factor

The correlation effect for diffusion in the B2 structure has been discussed by several investigators, as reviewed by Bakker[1]. The previous arguments seem to involve some confusions or misunderstanding of the problem. In order to clarify the situation, it is worthwhile to compare the correlation effect in the B2 structure with that in pure metals.

According to Manning [14], the correlation factor f is defined by the equation

$$D_{\text{actual}} = f D_{\text{random}} .$$ (22)

Here D_{actual} is the tracer diffusion coefficient under actual conditions, where the atom follows a correlated walk, and D_{random} is the tracer diffusion coefficient that one would obtain if the atom made the same number of jumps per unit time but successive atom jumps were independent of one another. For the tracer diffusion in pure b.c.c. or f.c.c. metals via the vacancy mechanism, the above equation may be written as

$$D_{\text{actual}} = \frac{1}{6} f\, C_V w d^2 ,$$ (23)

where C_V is the vacancy concentration, w is the jump frequency of the tracer and d is the jump distance.

For diffusion in pure metals, the tracer jump frequency is a unique quantity, and one can rightly write the diffusion coefficient in the above form. In contranst, for diffusion in the B2 structure via the six-jump cycle mechanism, the tracer jumps are specified by several jump frequencies, for example $v_{2\pm}$ and $v_{5\pm}$ for an A tracer in the α vacancy cycle, let alone jump frequencies of B atoms involved in the cycle. It is no longer possible to write the diffusion coefficient in the form of eq. (23) with a single jump frequency. The best that one can do is to define an average frequency $\bar{v}$ for the six-jump cycle and write the diffusion coefficient as

$$D = \frac{1}{6}f_\alpha C_{V\alpha}\bar{v}d^2 , \qquad (24)$$

which is exactly the procedure adopted in the present calculation; the actual form is somewhat more complicated because of the two possible jumps, 100 and 110 (cf. eq. (14)). The correlation factor f_α here has a clear physical meaning; it is the correlation factor for diffusion in a simple cubic lattice via the vacancy mechanism.

It seems appropriate to divide the correlation effect in the B2 structure into two: a macroscopic correlation effect and a microscopic correlation effect. The macroscopic correlation effect is simply that one mentioned above; the six-jump cycle is regarded as a unit process for a long-range diffusion. The microscopic correlation effect is that associated with individual atom (or vacancy) jumps within a six-jump cycle.

The correlation effect which Fishman, Gupta and Lieberman [15] and Arnhold [16] tried to discuss seems to correspond to the latter, the microscopic correlation effect. Fishman et al. presumed intuitively that the isotope effect parameter will be of the order of $f^6 \Delta K$, where f is the correlation factor for the six separate elementary jumps. They assumed that f is of the same magnitude as the correlation factor f_0 for self-diffusion in b.c.c. metals: 0.727... They considered that the correlation factor for the six-jump cycle mechanism is equal to $(0.727)^6 = 0.148$. On the other hand, Arnhold [16] derived a value of 0.83, on the basis of some argument which is difficult to understand. As criticized by Bakker [1], there is no justification for the treatments by Fishman et al. and by Arnhold.

Nonetheless the microscopic correlation effect does exist and is worthy of careful examination. In the six-jump process as depicted in Fig. 3, the vacancy that has jumped from C_1 to C_2 has a higher probability of jumping backwards to C_1 rather than of jumping forwards to C_3; the jumps are highly correlated. For this process, however, it is not possible to conceive a reference case of a hypothetical 'ideal' random walk; the process is inherently highly correlated in its own structure. For such a case, the definition or the calculation of the correlation factor by some formula such as eq. (22) is not possible. All that one can do is to calculate the average time or frequency $\bar{v}$ to complete a full six-jump cycle. The expression for $\bar{v}$ derived in the way described in this paper fully takes into account the microscopic correlation effect.

It is added here that the diffusion kinetics in the B2-type lattice have been discussed by Kikuchi and Sato [17,18] and by Bakker [19]. The former adopted the path probability method, and the latter extended Manning's theory for random alloys. In these treatments the diffusion coefficient is expressed in terms of various average quantities. Although detailed discussions have been made on the correlation factors in their papers, it is difficult to compare them with the present analysis. As argued in the previous paper, the definition of the correlation factor cannot be unique but is necessarily model dependent for diffusion processes involving several jump frequencies. The only significant quantity is the diffusion coefficient as a whole. The correlation factor, average frequency and so on cannot be discussed in isolation.

5. Discussion

If monovacancies are only the defects existing in an otherwise perfectly ordered B2 alloy, the six-jump cycle process is considered to be the most probable path of a vacancy; other types of vacancy movements would bring about configurations of much higher energy.

With the decrease in ordering energy or the wrong bond energy, the atom movement becomes more random creating a larger number of wrong bonds; atoms would move in a variety of ways other than the six-jump vacancy cycle. The diffusion in such cases may be more properly treated on the basis of different models.

With the increase in the ordering energy, the antisite occupation would be more difficult even temporarily. The site exchange of atoms in such alloys would take place by next-nearest neighbour jumps, instead of nearest neighbour jumps of vacancies; the diffusion occurs more or less independently in the respective sublattices of the two constituent atoms.

In this respect, it is interesting to examine the diffusion behaviour of different alloys in comparison with the ordering energies (eV unit) as listed below [20].

CuZn	AgMg	AuCd	AuZn	CoGa	NiGa	PdIn	NiAl
0.12	0.19	0.2	0.27	0.38~0.43	0.45~0.49	0.64	0.69~0.74

If the six-jump cycle is the dominant operative mechanism, the ratio of the two diffusion coefficients, D_A / D_B, must fall in a certain range given by the inequality (18). For alloys with smaller ordering energies: CuZn, AgMg, AuCd, AuZn, the ratios fall in the range (see Fig. 1).Therefore, the six-jump cycle can be the operative mechanism in these alloys; note that, however, the inequality is not the sufficient condition.

The diffusion in the PdIn with a high ordering energy of 0.64 eV has been studied by Hahn, Frohberg and Wever [21]. The inequality is satisfied only for the stoichimetric compound, $Pd_{50}In_{50}$. For other compositions of excess Pd and excess In, the Pd diffusivity is larger by two times or more than the In diffusivity; the Arrhenius plots for the Pd diffusion are straight, whereas the plots for the In diffusion are curved. The authors suggested that the diffusion of Pd is described by next nearest neighbour jumps of Pd into single Pd vacancies, and that the diffusion of In occurs by both the nearest neighbour- and the next nearest neighbour jumps.

For the NiAl with the largest ordering energy among the compounds listed above, only the Ni diffusivity is available, for lack of a suitable Al-radioisotope. Lutze-Birk and Jacobi [22] measured the diffusivity of In, which is known to substitute for Al. For the stoichiometric compound, the ratio D_{Ni} / D_{In} is 6.0; the authours concluded that the next nearest neighbour jumps contribute to the diffusion to some extent.

The diffusion in the CoGa with an intermediate ordering energy has been studied by Stolwijk, van Gend and Bakker [23]. The ratio, D_{Co} / D_{Ga}, for the stoichiometric composition is 6, and is outside the range indicated by (18) also for the other compositions. They proposed a triple defects mechanism in which a divacancy and an antisite atom form an elementary unit for diffusion; with this mechanism, the limits of the range for D_A/D_B are 13.3 and 1/13.3 [24].

The efficiency of the six-jump cycle has been examined by computer simulation [16], as quoted by Bakker [1]. For the stoichiometric compound with the ordering energy 0.05eV and the degree of long-range order S= 0.96, about 45% of the mean squared displacement was due to six-jump cycle with the rest of the migration taking place via normal single vacancy transport. The efficiency of the six-jump cycle mechanism decreases with increasing the deviation from the stoichiometry.

Thus, it is evident that the six-jump cycle process is not the only mechanism operative in the B2 compounds. Further experimental and theoretical efforts are required for a better understanding of the diffusion process in the intermetallic compounds.

References

[1] Bakker, H. (1984) 'Tracer diffusion in concentrated alloys', in G. E. Murch and A. S. Nowick (eds.), Diffusion in Crystalline Solids, Academic Press, 189-256.

[2] Wever, H., Hünecke, J. and Frohberg, G. (1989) 'Diffusion in Intermetallischen Phasen ', Zeitschrift für Metallkunde 80, 389-397.

[3] Huntington, H. B. (1958), private communication to Slifkin, see [7].

[4] Arita, M., Koiwa, M. and Ishioka, S. (1988) 'Diffusion mechanism in ordered alloys —A detailed analysis of six-jump vacancy cycle in a two dimensional ordered lattice—', Trans. JIM. 29, 439-447.

[5] Arita, M., Koiwa, M. and Ishioka, S. (1989) 'Diffusion mechanisms in ordered alloys —A detailed analysis of six-jump vacancy cycle in the B2 type lattice', Acta Metallurgica 37, 1363-1374.

[6] Arita, M., Koiwa, M. and Ishioka, S. (1989) 'Diffusion kinetics in the B2-type lattice: isotope effect and correlation effect ', Philosophical Magazine A 60, 563-580.

[7] Elcock, E. W. and McCombie, C. W. (1958) 'Vacancy diffusion in binary ordered alloys ', Physical Review 109, 605-606.

[8] Elcock, E. W. (1959) 'Vacancy diffusion in ordered alloys', Proc. Physical Society 73, 250-264.

[9] Wynblatt, P. (1967) 'Diffusion mechanisms in ordered body-centered cubic alloys', Acta Metallurgica 15, 1453-1460.

[10] Hancock, G.F. (1971) 'Diffusion of nickel in alloys based on the intermetallic compound Ni_3Al (γ)', Physica Status Solidi (a) 7, 535-540.

[11] Koiwa, M. and MacEwen, S. R. (1972) 'The effect of trapping on the annihilation of a diffusing particle', Philosophical Magazine 26, 173-192.

[12] Koiwa, M. (1974) 'Trapping effect in diffusion of interstitial impurity atoms in b.c.c. lattices', Acta Metallurgica 22, 1259-1268.

[13] Schoen, A. H. (1958) 'Correlation and the isotope effect for diffusion in crystalline solids', Physical Review Letters 1, 138-140.

[14] Manning, J. R. (1968) Diffusion Kinetics for Atoms in Crystals, D. Van Nostrand Princeton, New Jersey.

464

[15] Fishman, S. G., Gupta, D. and Lieberman, D. S. (1970) 'Diffusivity and isotope-effect measurements in equiatomic Fe-Co', Physical Review B **2**, 1451-1460.

[16] Arnhold, V. (1981) Thesis, Westfälische Wilhelms-Universität, Münster.

[17] Kikuchi, R. and Sato, H. (1969) 'Substitutional diffusion in an ordered system', J. Chemical Physics **51**, 161-181.

[18] Kikuchi, R. and Sato, H. (1972) 'Diffusion coefficient in an ordered binary alloy', J. Chemical Physics **51**, 4962-4979.

[19] Bakker, H. (1979) 'On diffusion kinetics in ordered binary alloys', Philosophical Magazine **40**, 525-540.

[20] Neumann, J. P., Chang, Y. A. and Ipser, H. (1976) 'On the relationship between the enthalpy of formation and the disorder parameter of intermetallic phases with the B2 structure', Scripta Metallurgica **10**, 917-922.

[21] Hahn, H. Frohberg, G. and Wever, H. (1983) 'Self-diffusion in the intermetallic B2 electron compound PdIn', Physica Status Solidi (a) **79**, 559-565.

[22] Lutze-Birk, A. and Jacobi, H. (1975) 'Diffusion of ^{114m}In in NiAl', Scripta Metallurgica **9**, 761-765.

[23] Stolwijk, N. A., van Gend, M. and Bakker, H. (1980) 'Self-diffusion in the intermetallic compound CoGa', Philosophical Magazine A **42**, 783-808.

[24] Bakker, H., Stolwijk, N. A. and Hoetjes-Eijkel, M. A. (1981) 'Diffusion kinetics and isotope effects for atomic migration via divacancies and triple defects in the CsCl (B2) structure', Philosophical Magazine A **43**, 251-264.

INTERDIFFUSION IN MULTICOMPONENT SYSTEMS

M. A. DAYANANDA
School of Materials Engineering
Purdue University
W. Lafayette, IN 47907, U.S.A.

ABSTRACT. The phenomenological basis for the description of interdiffusion in multicomponent systems and the experimental approaches employed for the determination of interdiffusion fluxes, interdiffusion coefficients and zero-flux planes are briefly reviewed. Interdiffusion coefficients determined for β (bcc) Fe-Ni-Al alloys at $1000°C$ are presented and the development of zero-flux planes is illustrated with selected Fe-Ni-Al diffusion couples. A new analysis involving the concepts of average, effective interdiffusion coefficients and penetration depths for the individual components in a diffusion couple has been developed. Diffusion paths for a few interdiffusion experiments carried out in the Ti-Al-Nb system at $1100°C$ are also presented. A Ti-Al-Nb diffusion couple that developed B2 phase as a diffusion layer is analyzed for the determination of effective interdiffusion coefficients for the components over selected concentration ranges; these data are discussed in the light of ordering in the B2 phase.

1. Introduction

In the light of the current technological needs for new material systems characterized by high stability and performance in aerospace applications, it has become essential for the materials scientists and engineers to develop a broader understanding of the phenomenon of diffusion in multicomponent systems. Multicomponent diffusion is encountered in the development and use of high temperature alloys, intermetallics, coatings and composites and plays a major role in processes ranging from heat treatments, diffusion bonding, cladding to surface modification.

The phenomenological description of diffusion in multicomponent systems has been based on the Onsager's formalism [1-3] of Fick's law and the major contributions made over the years in this area has been reviewed by Kirkaldy and Young [4]. Systematic isothermal diffusion studies with solid-solid and vapor-solid diffusion couples have been carried out in several multicomponent alloy systems and a compilation of the appreciable data on interdiffusion and intrinsic diffusion coefficients reported in literature for ternary alloys is now available [5]. In addition, a direct method for the determination of interdiffusion fluxes from the concentration profiles of the individual components without the need or use of the interdiffusion coefficients has been developed for the analysis of isothermal diffusion couples [6-8]. Such direct determination of interdiffusion fluxes of the components at any section in the diffusion zone has also helped identify the development of zero-flux planes [ZFP] [6-9].

C. T. Liu et al. (eds.), Ordered Intermetallics – Physical Metallurgy and Mechanical Behaviour, 465–484.
© 1992 *Kluwer Academic Publishers.*

466

At the ZFP the interdiffusion flux of a component goes to zero and undergoes a change in its direction from one side of the plane to the other. The ZFP phenomenon has been identified in several studies of multicomponent diffusion. [9-15].

The main objective of this paper is to review briefly the approaches employed in the determination of interdiffusion coefficients and interdiffusion fluxes and in the identification of ZFPs. Ternary interdiffusion coefficients for β (bcc) Fe-Ni-Al alloys as well as the development of ZFPs are discussed with selected multicomponent diffusion couples in the Fe-Ni-Al system. Interdiffusion studies which were carried out in the Ti-Al-Nb system at $1100°C$ with multiphase couples assembled with commercial alloys (super α_2 and γ) and Nb are presented along with experimental diffusion paths. An analysis based on the concepts of average, effective interdiffusion coefficients and penetration depths for the components is developed and applied to determine effective interdiffusion coefficients in the B2 diffusion layer formed in a Ti-Al-Nb couple.

2. On Interdiffusion Fluxes and Interdiffusion Coefficients

The extended form of Fick's law proposed by Onsager [1-3] for the description of diffusion in an n-component system relates the interdiffusion flux $\tilde{J}_i$ of component i linearly to (n-1) independent concentration gradients, $\partial C_j/\partial x$ and is expressed by:

$$\tilde{J}_i = -\sum_{j=1}^{n-1} \tilde{D}_{ij}^n \frac{\partial C_j}{\partial x} \qquad (i=1,2,...,n) \tag{1}$$

In Eq. (1) $(n-1)^2$ interdiffusion coefficients, $\tilde{D}_{ij}^n$, are defined as functions of composition and the superscript n in $\tilde{D}_{ij}^n$ corresponds to the component taken as the dependent concentration variable. $\tilde{D}_{ii}^n$ correspond to the main coefficients, while $\tilde{D}_{ij}^n$ (i≠j) refer to the cross coefficients. For convenience of analysis, $\tilde{J}_i$ in Eq. (1) is referred to a laboratory-fixed frame and the molar density ρ is assumed invariant with composition. Other possible frames of reference include the volume-fixed frame, the mass-fixed frame and the solvent-fixed frame. For a system of constant density, the interdiffusion fluxes on the laboratory-frame are related by:

$$\sum_{i=1}^{n} \tilde{J}_i = 0 \tag{2}$$

while the concentration gradients satisfy the relation:

$$\sum_{j=1}^{n} \partial C_j/\partial x = 0 \tag{3}$$

2.1. DETERMINATION OF $\tilde{J}_i$ WITHOUT THE NEED FOR $\tilde{D}_{ij}^n$

The concentration profiles developed in solid-solid and vapor-solid diffusion couples assembled

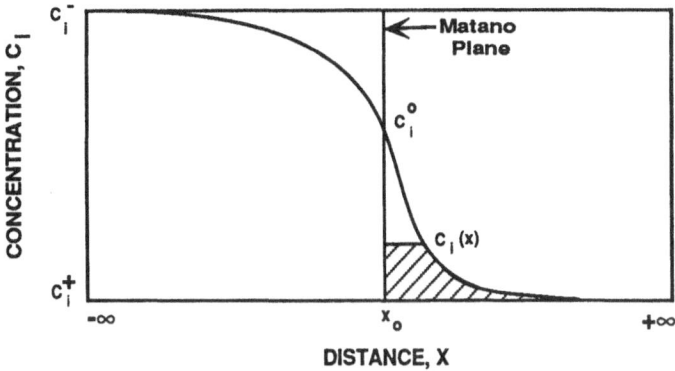

Figure 1. A schematic concentration profile of component i for a solid-solid diffusion couple; the hatched area corresponds to $\int_{C_i^+}^{C_i(x)} (x-x_o)\, dC_i$.

with multicomponent alloys and annealed isothermally can be analyzed directly for the determination of the profiles of interdiffusion fluxes of the various components in the diffusion zone. A schematic concentration profile of a component i developed in an isothermal diffusion couple assembled with disks of two terminal alloys of concentrations C_i^- and C_i^+ is shown in Fig. 1. The initial plane of contact between the alloy disks refers to the Matano plane, x_0. The concentration C_i is a function of the Boltzmann parameter λ expressed by:

$$\lambda = \frac{(x-x_o)}{\sqrt{t}} \tag{4}$$

and a concentration level identified at a given value C_i propagates at a velocity $v(C_i)$ given by [6-8]:

$$v(C_i) = \left. \frac{\partial x}{\partial t} \right)_{C_i}$$

$$= \tfrac{1}{2}\, \frac{\lambda}{\sqrt{t}}$$

$$= \frac{(x-x_o)}{2t} \tag{5}$$

On the basis of the continuity equation,

$$\left[\frac{\partial C_i}{\partial t}\right]_x = -\left[\frac{\partial \tilde{J}_i}{\partial t}\right]_t \qquad (i=1,2,...,n) \tag{6}$$

$$-\left[\frac{\partial x}{\partial t}\right]_{C_i}\left[\frac{\partial C_i}{\partial x}\right]_t = -\left[\frac{\partial \tilde{J}_i}{\partial x}\right]_t$$

$$v(C_i)\left[\frac{\partial C_i}{\partial x}\right]_t = \left[\frac{\partial \tilde{J}_i}{\partial x}\right]_t \tag{7}$$

On substituting Eq. (5) for $v(C_i)$ in Eq. (7) and integrating between C_i^+ at $+\infty$ and C_i at a section x, one gets

$$\tilde{J}_i(x) = \frac{1}{2t} \int_{C_i^+}^{C_i(x)} (x-x_o)\, dC_i \qquad (i=1,2,...,n) \tag{8}$$

since $\tilde{J}_i(\pm\infty) = 0$. The integral $\displaystyle\int_{C_i^+}^{C_i(x)} (x-x_o)dC_i$ corresponds to the hatched area in Fig. 1.

Similarly, by integrating Eq. (7) between C_i^- and $C_i(x)$, one can show:

$$\tilde{J}_i(x) = \frac{1}{2t} \int_{C_i^-}^{C_i(x)} (x-x_o)\, dC_i \qquad (i=1,2,...,n) \tag{9}$$

Eqs. (8) and (9) have been derived without utilizing the Fick's law definition of $\tilde{J}_i$ given by Eq. (1) and by-pass the need for the $(n-1)^2$ interdiffusion coefficients as functions of composition for the determination of interdiffusion fluxes.

2.2. DETERMINATION OF INTERDIFFUSION COEFFICIENTS

On equating Eqs. (1) and (8) for $\tilde{J}_i$, one gets

$$\int_{C_i^+}^{C_i(x)} (x-x_o)\, dC_i = -2t\left[\sum_{j=1}^{n-1} \tilde{D}_{ij}^n \frac{\partial C_j}{\partial x}\right]_{C_i(x)} \tag{10}$$

The use of Eq. (10) for the experimental determination of $\tilde{D}_{ij}^n$ requires setting up $(n-1)$ independent diffusion couples designed to have a common composition develop in their diffusion zones. Such a requirement becomes very difficult to realize experimentally for systems with more than 3 components. For ternary systems, however, a pair of solid-solid couples, A/B

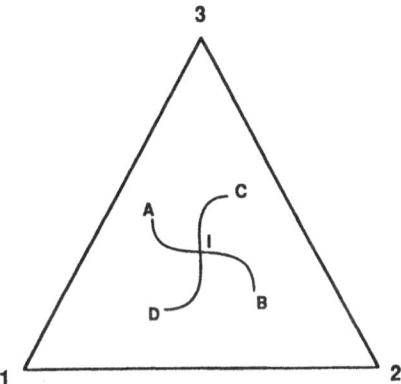

Figure 2. Schematic diffusion paths for a pair of solid-solid diffusion couples intersecting at the common composition I where the four ternary interdiffusion coefficients, $\tilde{D}_{11}^3$, $\tilde{D}_{12}^3$, $\tilde{D}_{21}^3$ and $\tilde{D}_{22}^3$, can be determined.

and C/D, having their diffusion paths intersect at a common composition point I, as shown in Fig. 2, is needed. For n = 3 Eq. (10) becomes:

$$\int_{C_i^+}^{C_i(x)} (x-x_o)\,dC_i = -2t \left[\tilde{D}_{i1}^3 \frac{\partial C_1}{\partial x} + \tilde{D}_{i2}^3 \frac{\partial C_2}{\partial x} \right] \qquad (i=1,2) \qquad (11)$$

From the concentration profiles of the couples A/B and C/D, 4 independent relations on the basis of Eq. (11) are set up for the common composition I and the four ternary interdiffusion coefficients, $\tilde{D}_{11}^3$, $\tilde{D}_{12}^3$, $\tilde{D}_{21}^3$ and $\tilde{D}_{22}^3$ can be determined at that composition. It is apparent that several diffusion couples with as many intersections as possible among their diffusion paths are needed for the determination of concentration-dependent $\tilde{D}_{ij}^3$ over a range of compositions.

A partial determination of $\tilde{D}_{ij}^3$ in ternary systems can be made from a single couple developing relative maxima and minima in the concentration profile of a component. Extrema in the concentrations of a component (say 1) can develop in a diffusion couple assembled with terminal alloys characterized by similar concentrations for component 1. At the concentration extrema, $\partial C_1/\partial x = 0$ and Eq. (11) becomes:

$$\int_{C_i^+}^{C_i} (x-x_o)\,dC_i = -2t \left[\tilde{D}_{i2}^3 \frac{\partial C_2}{\partial x} \right]_{C_i} \qquad (i=1,2) \qquad (12)$$

Eq. (12) allows the calculation of $\tilde{D}_{12}^3$ and $\tilde{D}_{22}^3$ from a single couple with a maximum or a minimum in the concentration profile of component 1. Similarly, $\tilde{D}_{11}^3$ and $\tilde{D}_{21}^3$ can be determined from a couple that develops extrema in the concentration profile of component 2.

Values of ratios of the cross to the main interdiffusion coefficients can be determined at zero-flux planes within the diffusion zone of a ternary diffusion couple. At a ZFP for component i the interdiffusion flux $\tilde{J}_i$ goes to zero and exhibits a change of direction from one side of the plane to the other. At such a plane, Eqs. (8) and (9) yield,

$$\int_{C_i^+ \text{ or } C_i^-}^{C_i(\text{ZFP})} (x-x_o)\, dC_i = 0 \tag{13}$$

From Eq. (11) it follows:

$$\left. \frac{\partial C_1}{\partial C_2} \right]_{\text{ZFP for i}} = - \frac{\tilde{D}_{i2}^3}{\tilde{D}_{i1}^3} \tag{14}$$

It is apparent from Eq. (14) that the ratio of the cross to the main interdiffusion coefficient can be estimated directly from the slope of the diffusion path at a ZFP composition. Such ratios of the interdiffusion coefficients determined at ZFPs have been reported for selected ternary systems [5].

2.3. LOCATION OF ZERO-FLUX PLANES

A ZFP for a component developed in an isothermal diffusion couple can be located or identified directly from the concentration profile of the component. In Fig. 3 is presented a schematic concentration profile with a relative maximum on one side of the Matano plane; $\tilde{J}_i$ profile calculated from the concentration profile from Eq. (8) is also shown in the figure. $\tilde{J}_i$ goes to zero at a section identified as ZFP and has opposite signs on the two sides of the ZFP.

For $\tilde{J}_i$ to go to zero, Eqs. (8) and (9) require:

$$\int_{C_i^+}^{C_i(\text{ZFP})} x\, dC_i = \text{area D} - \text{area C} = 0 \tag{15}$$

and

$$\int_{C_i^-}^{C_i(\text{ZFP})} x\, dC_i = \text{area A} - \text{area B} = 0 \tag{16}$$

Hence, the ZFP is located at the section determined by the equality in the areas A and B as well as by the equality in the areas C and D. This requirement of equality in areas may and may not be observed for a profile that exhibits a relative maximum or a relative minimum in concentration. It is important to note that the presence of an extremum in the concentration profile indicates the development of a ZFP, only if the areal balances identified by Eqs. (15) and (16) are also satisfied. From the viewpoint of mass balance for component i, the region of loss

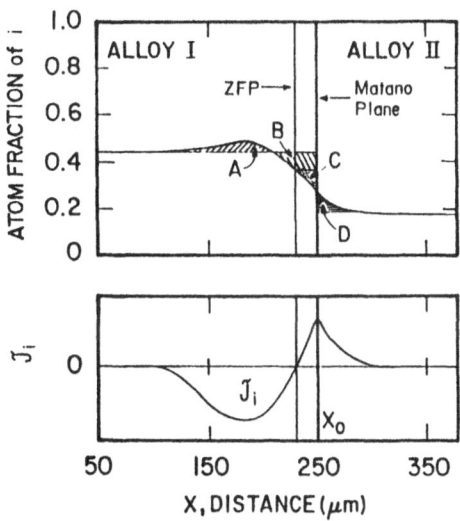

Figure 3. A schematic concentration profile of a component showing a relative maximum on one side of the Matano plane; a zero-flux plane (ZFP) for the component is identified by the requirement: area A = area B and area C = area D.

corresponding to the area (B+C) is balanced by two areas of gain, A and D, one on either side of the Matano plane.

3. Analysis for effective interdiffusion coefficients and penetration depths

A new approach involving the concepts of effective interdiffusion coefficients and penetration depths for individual components in a multicomponent diffusion couple is developed in this section. This analysis will be particularly useful for evaluating an average interdiffusion coefficient for any component over a range of concentrations in the diffusion zone. An integration of the interdiffusion flux $\tilde{J}_i$ over the diffusion zone from one end of the diffusion zone ($+\infty$) to the Matano plane x_0 yields:

$$\int_{+\infty}^{x_0} \tilde{J}_i dx = \int_{+\infty}^{x_0} \tilde{J}_i d(x-x_0)$$

$$= \tilde{J}_i \ (x-x_0)\Big|_{+\infty}^{x_0} - \int_{\tilde{J}_i(+\infty)}^{\tilde{J}_i(x_0)} (x-x_0) \ d\tilde{J}_i \tag{17}$$

Since $d\tilde{J}_i = \dfrac{1}{2t}(x-x_o)\,dC_i$ from Eq. (7) and $\tilde{J}_i(+\infty) = 0$, Eq. (17) becomes:

$$\int_{+\infty}^{x_o} \tilde{J}_i dx = -\frac{1}{2t}\int_{C_i^+}^{C_i^o} (x-x_o)^2 \, dC_i \tag{18}$$

where C_i^o corresponds to the concentration at x_o.

$\tilde{J}_i$ in Eq. (1) can be alternatively expressed by:

$$\tilde{J}_i = -\tilde{D}_i^{\,\text{eff}}\frac{\partial C_i}{\partial x} \qquad (i=1,2,...n) \tag{19}$$

where

$$\tilde{D}_i^{\,\text{eff}} = \tilde{D}_{ii}^{\,n} + \sum_j \frac{\tilde{D}_{ij}^{\,n}\,\partial C_j/\partial x}{\partial C_i/\partial x} \qquad (j\neq i) \tag{20}$$

The second term in Eq. (20) includes the contribution of the cross effects among the diffusing species. Substituting Eq. (19) in Eq. (18), one gets:

$$\int_{C_i^+}^{C_i^o} \tilde{D}_i^{\,\text{eff}}\, dC_i = \frac{1}{2t}\int_{C_i^+}^{C_i^o} (x-x_o)^2 \, dC_i$$

or

$$\tilde{D}_{i,R} = \frac{1}{2t}\frac{\displaystyle\int_{C_i^+}^{C_i^o} (x-x_o)^2 \, dC_i}{\left[C_i^o - C_i^+\right]} \tag{21}$$

where $\tilde{D}_{i,R}$ is the average effective interdiffusion coefficient for component i over the concentration range C_i^+ to C_i^o on the righthand side of the Matano plane. The integral in Eq. (21) can be graphically determined from the cross-hatched area shown in Fig. 4.

On expressing

$$\int_{C_i^+}^{C_i^o} (x-x_o)^2 \, dC_i = \overline{(x-x_o)_{i,R}^2}\left[C_i^o - C_i^+\right] \tag{22}$$

where $\overline{(x-x_o)_{i,R}^2}$ is the mean squared distance from the Matano plane over the concentration range C_i^+ to C_i^o, Eq. (21) yields:

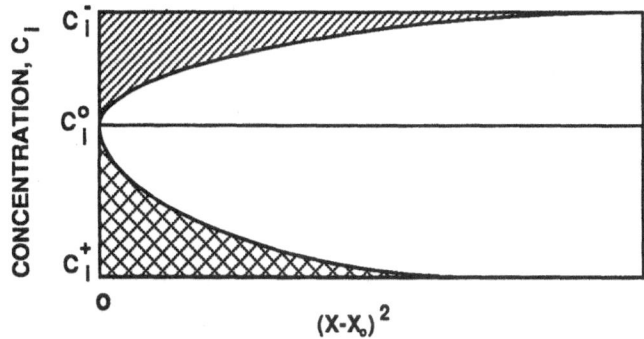

Figure 4. A schematic plot of C_i vs $(x-x_o)^2$ for component i; the integrals, $\int_{C_i^+}^{C_i^o} (x-x_o)^2 \, dC_i$

and $\int_{C_i^o}^{C_i^-} (x-x_o)^2 \, dC_i$ in Eqs. (21) and (24), are determined from the cross-hatched and hatched areas, respectively.

$$\tilde{D}_{i,R} = \frac{\overline{(x-x_o)_{i,R}^2}}{2t} \tag{23}$$

If a similar analysis is applied to the concentration range $\left[C_i^- - C_i^o\right]$ on the left-hand side of the Matano plane, one gets

$$\tilde{D}_{i,L} = \frac{1}{2t} \frac{\int_{C_i^o}^{C_i^-} (x-x_o)^2 \, dC_i}{\left[C_i^- - C_i^o\right]} = \frac{\overline{(x-x_o)_{i,L}^2}}{2t} \tag{24}$$

Effective penetration depths of component i on the two sides of the Matano plane can be determined on the basis of Eqs. (23) and 24) by:

$$x_{i,R} = \sqrt{\overline{(x-x_o)_{i,R}^2}}$$

$$= \sqrt{2\tilde{D}_{i,R}t} \tag{25}$$

and

$$x_{i,L} = \sqrt{\overline{(x-x_o)_{i,L}^2}}$$

$$= \sqrt{2\tilde{D}_{i,L}t} \qquad (26)$$

Similarly, an average effective interdiffusion coefficient for component i over a diffusion layer from x_1 to x_2 can be determined from the expression:

$$\tilde{D}_{i,\,phase} = \frac{\tilde{J}_i(x_1)(x_1-x_o) - \tilde{J}_i(x_2)(x_2-x_o)}{\left[C_i(x_2) - C_i(x_1)\right]} + \frac{\displaystyle\int_{C_i(x_1)}^{C_i(x_2)} (x-x_o)^2 \, dC_i}{2t\left[C_i(x_2) - C_i(x_1)\right]} \qquad (27)$$

This analysis will be applied to the determination of effective interdiffusion coefficients on the two sides of the Matano plane for a diffusion layer developed in a Ti-Al-Nb diffusion couple in section 4.2.

4. Interdiffusion Studies in Selected Multicomponent Systems with Ordered Phases

4.1. Fe-Ni-Al SYSTEM

4.1.1. *Ternary Interdiffusion Coefficients.* Moyer and Dayananda [16] investigated interdiffusion with solid-solid diffusion couples assembled with β (bcc) Fe-Ni-Al alloys at 1004°C and determined ternary interdiffusion coefficients at compositions corresponding to the intersections of diffusion paths and at maxima and minima in concentration profiles. Similar studies were also made at 1000°C by Cheng and Dayananda [17,18] with multiphase couples assembled with β (bcc) and γ(fcc) Fe-Ni-Al alloys at 1000°C. The diffusion paths for the various Fe-Ni-Al diffusion couples from these two investigations are presented in the Figures 5 and 6.

The calculated ternary interdiffusion coefficients, $\tilde{D}_{AlAl}^{Fe}$, $\tilde{D}_{AlNi}^{Fe}$, $\tilde{D}_{NiAl}^{Fe}$, $\tilde{D}_{NiNi}^{Fe}$, for β Fe-Ni-Al alloys are presented in Figures 7-9 as functions of Fe at selected concentration levels of Ni or Al. It is apparent that all the coefficients vary over 1-2 orders of magnitude with composition. From Figures 7 and 8 the variations of the main coefficients with composition for β alloys with less than 5 at.pct Fe can be expressed by:

$$\log_{10} \tilde{D}_{AlAl}^{Fe} = -2.8 \, N_{Ni} - 10.32 \, N_{Al} - 5.08 \qquad (28)$$

and

$$\log_{10} \tilde{D}_{NiNi}^{Fe} = -1.25 \, N_{Ni} - 10.45 \, N_{Al} - 5.95 \qquad (29)$$

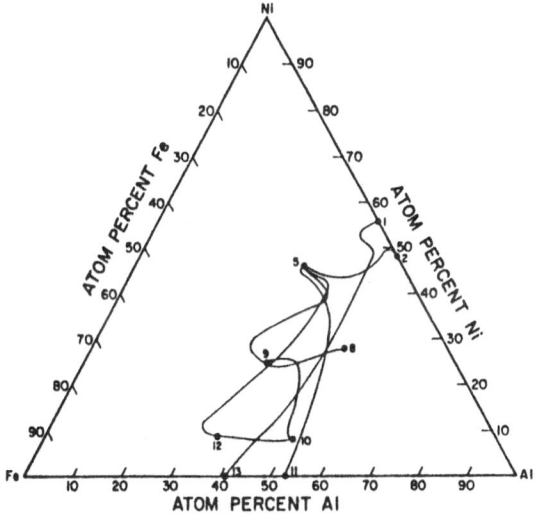

Figure 5. Diffusion paths for β(bcc) Fe-Ni-Al diffusion couples annealed at 1004°C [16].

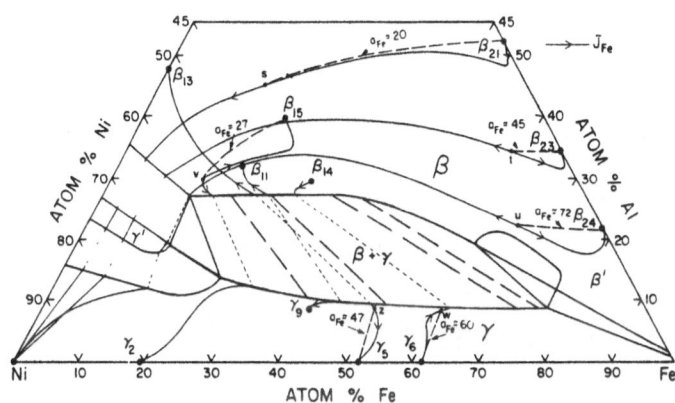

Figure 6. Diffusion paths for multiphase couples assembled with β(bcc) and γ(fcc) Fe-Ni-Al alloys at 1000°C; the composition points, s, t, u, v, w, z correspond to ZFPs developed for Fe. Tie-lines in the two-phase region are indicated by thick and dashed lines, as also a few isoactivity lines passing through terminal alloys [9,17,18].

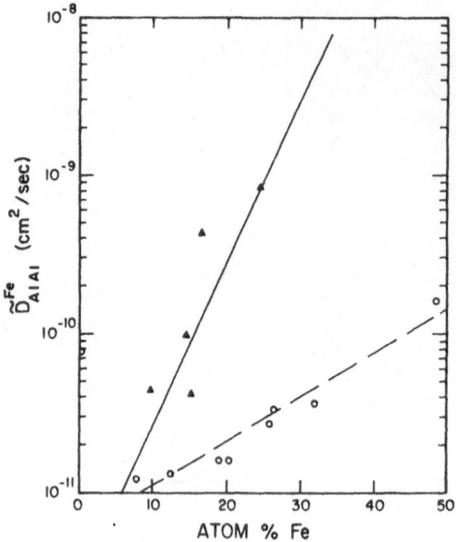

Figure 7. $\tilde{D}_{AlAl}^{Fe}$ as a function of Fe concentration; Δ For alloys with 50 ± 3 at.% Ni [16]; o For alloys with 45 ± 3 at.% Al [18].

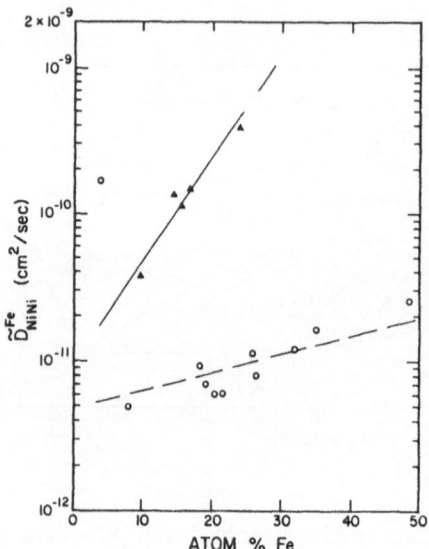

Figure 8. $\tilde{D}_{NiNi}^{Fe}$ as a function of Fe concentration; Δ For alloys with 50 ± 3 at.% Ni; o For alloys with 45 ± 3 at.% Al.

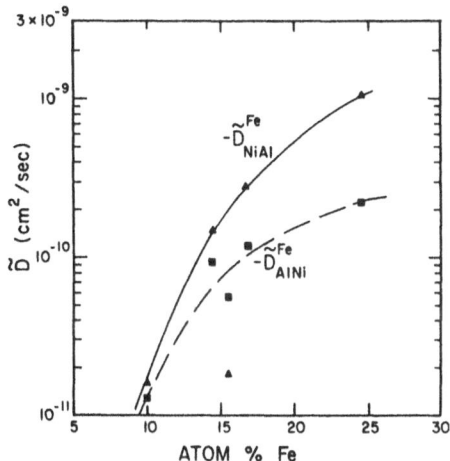

Figure 9. $\tilde{D}_{NiAl}^{Fe}$ and $\tilde{D}_{AlNi}^{Fe}$ as functions of Fe concentration for Fe-Ni-Al alloys with 50 ± 3 at.% Ni at 1000°C.

The cross coefficients $\tilde{D}_{AlNi}^{Fe}$ and $\tilde{D}_{NiAl}^{Fe}$ as indicated in Fig. 9 are mostly negative for alloys at a concentration level of 50 at.pct. Ni, but can become positive at low Fe concentrations and Ni concentrations, respectively [18].

The β (bcc) Fe-Ni-Al alloys are related to both FeAl and NiAl alloys which belong to the CsCl structure. The structure of the (Fe, Ni)Al corresponds to Al atoms at cube corners (sublattice 1) with Fe and Ni atoms at cube centers (sublattice 2). For alloys with less than 50 at. pct. Al, Fe and Ni behave differently with regard to occupying the vacant Al sites; these sites are occupied preferentially by Fe rather than Ni [19]. This is consistent with the degrees of disorder of 4×10^{-4} and 1.25×10^{-2} calculated for NiAl and FeAl, respectively, by Steiner and Komarek [20]. Since NiAl is more strongly ordered than FeAl, Fe is more likely to occupy Al sites than Ni. Hence, the vacancy concentration on sublattice 2 can be expected to increase with increasing Fe concentration at a given Al concentration. This increases the probability of aluminum-vacancy exchange and this expectation is consistent with observed increase of D_{AlAl}^{Fe} with increase in Fe. The increase in $\tilde{D}_{NiNi}$ with decrease in Ni concentration at a constant Al concentration indicates that the increase in vacancy concentration on cube centers favorably influences the jump of Ni atoms from cube centers to cube corners and back to cube centers.

4.1.2. *ZFPs and Flux Reversals for Fe.* The concentration profiles as well as flux profiles for a couple designated by 5/12 are presented in Fig. 10. The flux profiles which were calculated from the concentration profiles on the basis of Eq. 8 clearly show the development of two zero-flux planes for iron, one on either side of the Matano plane. Also, the Al concentration profile shows up-hill diffusion of Al towards the Ni-rich side; this is consistent with the large negative values of $\tilde{D}_{AlNi}^{Fe}$ comparable in magnitude with the values of $\tilde{D}_{AlAl}^{Fe}$.

478

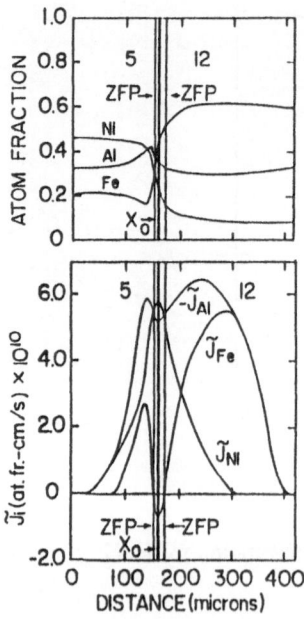

Figure 10. Profiles of concentrations and calculated interdiffusion fluxes for a Fe-Ni-Al couple designated by 5/12 annealed at 1004°C for 96 hr; the couple develops two ZFPs for Fe [9,16].

The concentration profiles and flux profiles for a multiphase couple designated by γ_9/β_{11} are presented in Fig. 11. This couple exhibits a discontinuous flux reversal for iron at the γ/β interface. Such a flux reversal is recognized by the fact that $\int_{C_{Fe}(+\infty)}^{C_{Fe}^{\gamma}\big]_{x_I}} (x-x_0)\, dC_{Fe}$ is negative and $\int_{C_{Fe}(+\infty)}^{C_{Fe}^{\beta}\big]_{x_I}} (x-x_0)\, dC_{Fe}$ is positive, where $C_{Fe}^{\gamma}\big]_{x_I}$ and $C_{Fe}^{\beta}\big]_{x_I}$ refer to Fe concentrations of the γ and β phases, respectively, at the γ/β interface. Fe interdiffuses in both directions away from the interface, as the interface moves towards γ.

ZFPs for Fe were observed for several γ/β two-phase couples at compositions identified by the points s through z on their diffusion paths in Fig. 6. Isoactivity lines passing through the terminal alloys intersect the diffusion paths at compositions very close to those of ZFP. The identification of ZFP compositions being very close to the intersection points of diffusion paths with isoactivity lines drawn through the terminal compositions of the couples on a ternary isotherm has been reported for several systems [7,9-11].

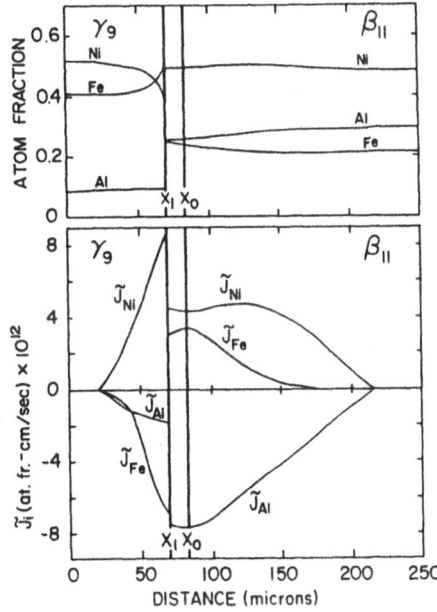

Figure 11. Profiles of concentrations and calculated interdiffusion fluxes for a two-phase Fe-Ni-Al couple annealed at 1000°C for 48 h; Fe exhibits a flux reversal at the interface x_I [9,18].

4.2. Ti-Al-Nb System

Titanium aluminides with additions of Nb, V and Mo are currently of great interest in the development of high-performance structural materials in aerospace applications. Diffusion studies in the Ti-Al-Nb based alloys are limited [21] and no data on interdiffusion coefficients are available. Interdiffusion studies have now been carried out [22] with solid-solid diffusion couples assembled with two commercial alloys, a TiAl(γ) base alloy (Ti - 48Al - 2Nb) and a Ti$_3$Al(α_2) base alloy (Ti-25Al-10Nb-3 V-1Mo), and pure Nb. The γ and α_2 alloys are designated A and B, respectively, and their compositions are given in at. pct. Two sandwich couples, A/Nb/A and B/Nb/B prepared by diffusion bonding at 1000°C and 1000 psi pressure for 2 hrs were supplied by Rockwell International Corporation. The as-received couples were examined by optical microscopy and SEM. A third couple, A/B, was also assembled from disks of alloys A and B held together in a Kovar jig. All couples were sealed in evacuated quartz capsules and annealed at 1100°C for various periods of time and quenched in ice-water. The couples were then sectioned, polished and analyzed for concentration profiles with a scanning electron microscope equipped with energy dispersive X-ray analysis capability.

On the basis of the experimental concentration profiles, diffusion paths for the couples were drawn on a Ti-Al-(Nb+V+Mo) isotherm with Nb, V and Mo grouped together as one concentration variable and are presented in Fig. 12. The various boundaries for the individual

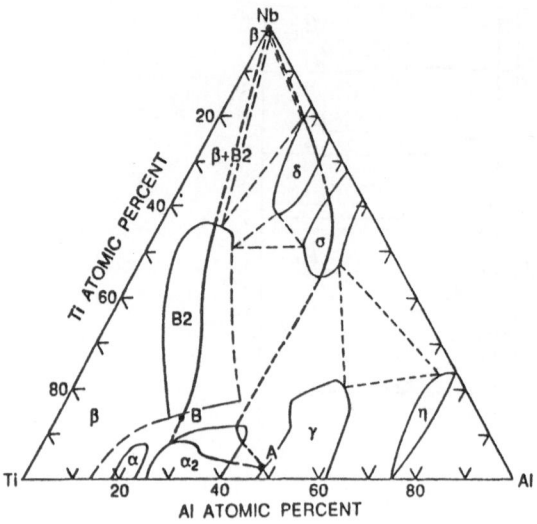

Figure 12. Experimental diffusion paths at 1100°C for diffusion couples assembled with Ti-Al-(Nb + V + Mo) alloys, A and B, and pure Nb; Nb, V and Mo are grouped together as one concentration variable; phase boundaries on the isotherm are approximate and schematic.

single phase regions indicated on the isotherm are schematic but are based on the concentration profiles for the couples and on the Ti-Al, Ti-Nb and Nb-Al binary phase diagrams. Regions of possible three phase equilibria among the various phases are also schematically shown.

The diffusion structure developed for the couple B/Nb/B annealed at 1100°C for 5 days is presented in Fig. 13. A single phase layer (B2) grew at the expense of both Nb and B alloy. The concentration profiles and the calculated profiles of interdiffusion fluxes for the couple are presented in Fig. 14. x_o and x_m correspond, respectively, to the Matano plane and the location of a marker plane after diffusion. B2 phase was identified as an ordered β (bcc) phase on the basis of the superlattice spots observed in selected area diffraction patterns [22]. A B2 (Ti_2NbAl) phase is believed to form during quenching [23].

The B2 phase in the present study is considered to be stable at the annealing temperature of 1100°C, as it forms as a diffusion layer separated by a planar interface on the Nb(β) side of the couple. A development of a planar interface between the β and B2 phases implies a local equilibrium between them and corresponds to a diffusion path crossing of the (B2+β) two-phase field along a tie-line as indicated on the isotherm presented in Fig. 12.

From the concentration profiles for the components presented in Fig. 14, average effective interdiffusion coefficients, $\tilde{D}_{i,L}$ and $\tilde{D}_{i,R}$ (i = Ti, Al or Nb), were determined for the B2 phase on the basis of Eqs. (21) and (24) over the concentration ranges on the two sides of the Matano plane. An average effective interdiffusion coefficient $\tilde{D}_{i,B2}$ over the entire concentration range of the B2 diffusion layer was also calculated from Eq. (27). These data on the effective

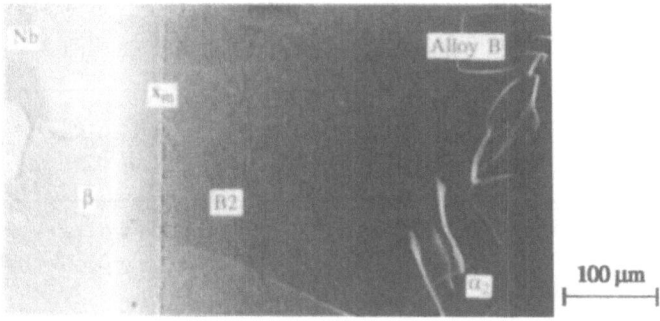

Figure 13. Diffusion structure developed for couple, B/Nb/B, annealed at 1100°C for 5 days.

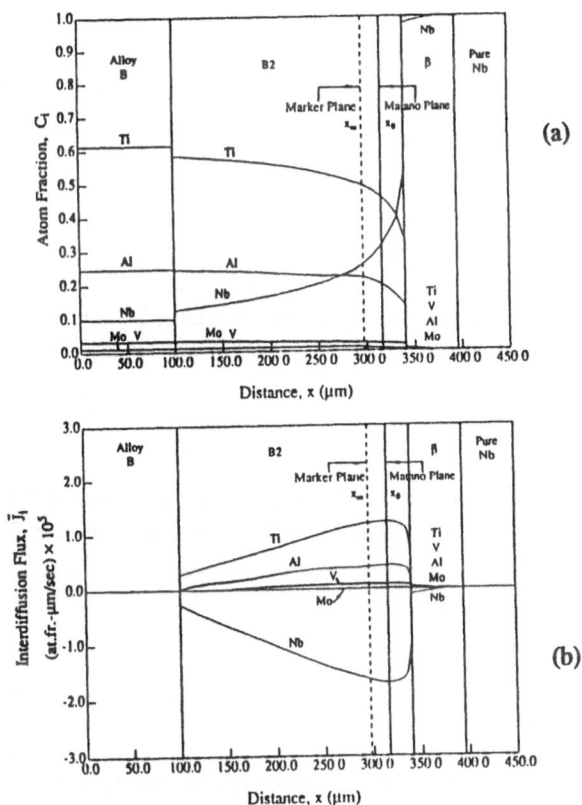

Figure 14. (a) Concentration profiles and (b) calculated interdiffusion fluxes for the B/Nb couple annealed at 1100°C for 5 days.

Table 1. Average effective interdiffusion coefficients and Penetration distances for the components in the B2 phase developed in the couple B/Nb at 1100°C.

i	$\tilde{D}_{i,L}(cm^2/s)$ (10^{-10})	$x_{i,L}$ (μm)	$\tilde{D}_{i,R}(cm^2/s)$ (10^{-12})	$x_{i,R}$ (μm)	$\tilde{D}_{i,B2}(cm^2/s)$ (10^{-10})	$x_{i,B2}(\mu m)$
Ti	1.29	105.7	5.78	22.3	1.77	123.7
Al	1.13	98.6	5.52	21.8	1.49	113.6
Nb	1.17	100.5	5.90	22.6	1.62	118.3

interdiffusion coefficients along with the effective penetration depths are reported in Table 1. It is apparent that all the interdiffusion coefficients vary more than order of magnitude over the composition range of the B2 phase and increase with increase in Ti concentration. However, the compositions that support faster diffusion for all the components cover most of the diffusion layer, as the B2 phase grows faster towards the B alloy side than towards the Nb side. Although the interdiffusion behavior of the Ti, Al and Nb are comparable, Ti interdiffuses the fastest over the B2 layer.

The B2 phase consists of two sublattices, one corresponding to the cube corners (sublattice 1) and the other (sublattice 2) formed by the cube centers. Ti atoms are considered to occupy the sublattice 1, while Nb and Al occupy the sublattice 2. For nonstoichiometric compositions, antisite defects can occur with excess Ti on the sublattice 2; Nb and Al antisite defects can also appear on sublattice 1 with the majority of the Nb and Al atoms staying at the cube centers [24]. This picture appears consistent with the development of a large B2 diffusion layer exhibiting appreciable composition variations for Ti and Nb, as observed for the B/Nb/B couple in this study.

5. Acknowledgement

Sincere thanks are due to Dr. C. Gandhi, formerly at the Science Center, Rockwell International Corporation, Thousand Oaks, California, for supplying the Ti-Al-Nb alloys employed in this study.

6. References

1. Onsager, L. (1931) 'Reciprocal relations in irreversible processes, I', Physical Review, 37, 405-426.

2. Onsager, L. (1931) 'Reciprocal relations in irreversible processes, II', Physical Review, 38, 2265-2279.

3. Onsager, L. (1945) 'Theories and problems of liquid diffusion', Annals, New York Academy of Sciences, 46, 241-265.

4. Kirkaldy, J. S. and Young, D. J. (1987) Diffusion in the Condensed State, The Institute of Metals, London.

5. Dayananda, M. A. (1990) 'Diffusion in ternary alloys', in H. Mehrer (ed.)., Diffusion in Solid Metals and Alloys, Springer-Verlag, Berlin, pp. 372-435.

6. Dayananda, M. A. (1985) 'Zero-flux planes, flux reversals and diffusion paths in ternary and quaternary diffusion', in M. A. Dayananda and G. E. Murch (eds.), Diffusion in Solids: Recent Developments, The Metallurgical Society of AIME, Warrendale, PA, pp. 195-230,.

7. Dayananda, M. A. and Kim, C. W. (1979) 'Zero-flux planes and flux reversals in Cu-Ni-Zn diffusion couples', Metall. Trans. A, 10A, 1333-1339.

8. Dayananda, M. A. (1983) 'An analysis of concentration profiles for fluxes, diffusion depths and zero-flux planes in multicomponent diffusion', Metall. Trans. A, 14A, 1851-1858.

9. Kim, C. W. and Dayananda, M. A. (1983) 'Identification of zero-flux planes and flux reversals in several studies of ternary diffusion', Metall. Trans. A, 14A, 857-864.

10. Kim, C. W. and Dayananda, M. A. (1984) 'Zero-Flux planes and flux reversals in the Cu-Ni-Zn system at 775°C', Metall. Trans. A, 15A, 649-659.

11. Duh, J. G. and Dayananda, M. A. (1985) 'Interdiffusion in Fe-Ni-Cr alloys at 1100°C', Diffusion and Defect Data, 39, 1-49.

12. Kansky, K. E. and Dayananda, M. A. (1985) 'Quaternary diffusion in the Cu-Ni-Zn-Mn system at 775°C', Metall. Trans. A, 16A, 1123-1132.

13. Heaney, III, J. A. and Dayananda, M. A. (1986) 'Interdiffusion in the Ni-Cr-Co-Mo systems at 1300°C', Metall. Trans. A., 17A, 983-990.

14. Nesbitt, J. A. and Heckel, R. W. (1987) 'Interdiffusion in Ni-rich, Ni-Cr-Al alloys at 1100° and 1200°C: Part I, Diffusion paths and microstructures', Metall. Trans. A, 18A, 2061-2073.

15. Dayananda, M. A. (1989) 'Multicomponent diffusion studies in selected high-temperature alloy systems', Mater. Science and Engineering, A121, 351-359.

16. Moyer, T. D. and Dayananda, M. A. (1976) 'Diffusion in β_2 Fe-Ni-Al alloys', Metall. Trans. A, 7A, 1035-1040.

17. Cheng, G. H. and Dayananda, M. A. (1979) 'Multiphase diffusion in Fe-Ni-Al system at 1000°C: I. Diffusion structures and diffusion paths', Metall. Trans. A, 10A, 1407-1414.

18. Cheng, G. H. and Dayananda, M. A. (1979) 'Multiphase diffusion in Fe-Ni-Al system at 1000°C: II. Interdiffusion coefficients for β and γ alloys', Metall. Trans. A, 10A, 1415-1419.

19. Bradley, A. J. and Taylor, A. (1938) 'X-ray study of the Fe-Ni-Al ternary equilibrium diagram', Proc. Royal Soc., Series A, 166, 353-375.

484

20. Steiner, A. and Komarek, K. L. (1964) 'Thermodynamic activities of solid nickel-aluminum alloys', Trans. Metall. Soc. of AIME, 230, 786-790.

21. Perepezco, J. H., Chang, Y. A., Seitzman, L. E., Lin, J. C., Bonda, N. R., Jewett, T. J. and Mishurda, J. C. (1989) 'High temperature phase stability in the Ti-Al-Nb system', in High Temperature Aluminides and Intermetallics, Proc. Symp. Indianapolis, in Press.

22. Ma, Z (1990) 'Multiphase diffusion in Ti-Al-Nb ternary system', M.S. thesis, School of Materials Engineering, Purdue University, W. Lafayette, IN.

23. Strychor, R., Williams, J. C. and Soffa, W. A. (1988) 'Phase transformations and modulated microstructures in Ti-Al-Nb alloys', Metall. Trans. A, 19A, 225-234.

24. Banerjee, D., Nandy, T. K. and Gogia, A. K. (1987) 'Site occupation in the ordered β phase of ternary Ti-Al-Nb alloys', Scripta. Met., 21, 597-600.

DIFFUSION IN EXOTIC INTERMETALLICS

G. FROHBERG, H. WEVER
Institut Metallforschung/TU Berlin
Hardenbergstr.36
D-1000 Berlin 12

ABSTRACT. The experimental data for A-diffusion (A=Ni) and for B-diffusion (B=Sb, Sn, In) in A/B-alloys with the $B8_2$-structure have been compared with a number of possible diffusion mechanisms. The diffusion behaviour is well understood in terms of a main diffusion mechanism, using the double tetrahedral interstices in the case of A-diffusion, and a pure vacancy mechanism in the B-sublattice for B-diffusion. Minority mechanisms explain the temperature variation of the $D_\perp/D_\parallel$-ratio.

1. Introduction

Noncubic intermetallic phases offer some interesting aspects with respect to the diffusion mechanism. Since the diffusion is then anisotropic as a rule, the ratio R of the diffusion coefficients for the direction perpendicular to the c-axis relative to that one parallel to the c-axis is a valuable number to check models for the diffusion mechanism in such phases. The hexagonal B8-phases in the A/B-alloy systems Ni/Sb, Ni/Sn, Ni/In offer this opportunity. In these phases the B-atoms form a compressed hcp-lattice with a c/a-ratio of about 1.25 in which the octahedral sites between the B-atoms are occupied by the A-atoms (Ni), see fig.1a. The A-atoms thus form chains parallel to the c-axis. In such a lattice there is an equal number of A- and B-atoms which is true for the alloy NiSb (NiAs- or $B8_1$-structure). But within this lattice there are still tetrahedral interstices between the B-atoms as shown in fig.1b. Because of the low c/a-ratio two adjacent interstices form one Double Tetrahedral Interstice (DTI). The size of the DTI's allows them to be easily filled with the A-component to 50%. Then there are two energetically different sites for the Ni-atoms: the chain sites (Ni_I) with the lower energy E_1 and the DTI-sites (Ni_{II}) with a slightly higher energy E_2. With this occupation of these sites a $B8_2$-

C. T. Liu et al. (eds.), Ordered Intermetallics – Physical Metallurgy and Mechanical Behaviour, 485–496.
© 1992 *Kluwer Academic Publishers.*

structure is formed and this applies to the alloy Ni_3Sn_2 with a range of homogeneity between 59% and 63% Ni (45 to 70% DTI's filled). With the occupation of every second DTI and the small lattice deformation around the Ni_I-atom on that site the neighbouring unoccupied DTI's get smaller and hence a further occupation of these sites by A-atoms (Ni_{II}) is possible but now again at a slightly higher energy E_3. If all DTI's were filled with Ni-atoms a composition A_2B would result. This applies to the compund Ni_2In with a homogeneity range from 62% to 67% Ni (65 to 75% DTI's filled). It is interesting to note that 75% seems to be the highest degree of filling which is already at 63.4% Ni. For higher concentrations of Ni (up to 67%), Ni-atoms begin to substitute the B-atoms [1],[2]. This results from a combination of density and lattice parameter measurements (Simmons-Baluffi-method).

By the same method also the thermal disorder has been determined. While in all other materials normally thermal vacancies are produced by the creation of new cells of the structure, in the $B8_2$-alloys thermal vacancies are created by the transfer of Ni-atoms from Ni_I-sites to unoccupied DTI's [3]. Thus the free volume in the crystal does not change with temperature, DTI's are just converted into thermal vacancies in the Ni-chains (maximum 2%). Because of problems with the accuracy of such measurements the activation enthalpy of this conversion could not yet be determined. But in the filled-up compounds of Ni_2In the activation energy was found to be 0.85eV and the activation entropy 3.6k [3]. Thermal vacancies in the B-sublattice are produced in the same way: it is only necessary to let a B-atom jump into one of the normally adjacent DTI's. B-antistructure atoms are produced by a jump of a B-atom into a just created thermal vacancy in the Ni_I-chains while A-antistructure atoms are produced by filling one of the just created B-vacancies with an adjacent a-atom.

The crystal data for the three compounds are [7]:
a=0.398nm, c=0.516nm, c/a=1.30 (Ni/Sb)
a=0.412nm, c=0.520nm, c/a=1.26 (Ni/Sn)
a=0.419nm, c=0.516nm, c/a=1.23 (Ni/In).

2. Diffusion Experiments

Up to now there are only three investigations on diffusion in $B8_2$-compounds. P.Schmidt measured the diffusion of [63]Ni in polycrystalline material of Ni_3Sn_2 by the method of residual activity [4]. In the temperature range from 1173K to 1375K the activation enthalpy Q was found to be 2.96eV and the preexponential factor $D_0=678cm^2/sec$. Hähnel et al. have investigated the diffusion of [63]Ni and [124]Sb in single crystals of $Ni_{53}Sb_{47}$ in a- and c-direction [5]. The diffusion coefficient was also measured in polycrystalline $Ni_{50}Sn_{50}$ and $Ni_{53}Sn_{47}$. The results for the single crystals of $Ni_{53}Sb_{47}$ are shown in table 1. H.Schmidt measured the diffusion of ^{63}Ni in $Ni_{61}Sn_{39}$, $Ni_{62}Sn_{38}$ and $Ni_{64}In_{36}$ as well as of ^{113}Sn in $Ni_{61}Sn_{39}$. The results are shown in table 2 and 3. In all cases the activation enthalpies differ only little for the directions parallel and perpendicular to the c-axis, with other words: the ratio $R=D\_/D$ does not vary much with temperature -for A- and B-component as well. The temperature dependence of the ratio R is shown in fig.2 for $Ni_{62}Sn_{38}$ as an example (all others are very similar). This is a strong indication that for each component there is a common main diffusion mechanism acting in both directions. The small variation of R with temperature indicates that there might be a minority mechanism involved.

3. Theoretical Considerations and Discussion

Generally the diffusion coefficient D for a tracer in one direction is given by

$$(1) \qquad D = \frac{f}{2} l^2 w \quad , \quad w = 1/\tau \quad ,$$

where f is the correlation factor, l the component of the jump vector in the considered direction, w the effective jump frequency and τ the total time between two such jumps.

From the structure of the material, the size of the atoms, the distance of neighbouring atoms and the energy considerations given above the probability of possible atom jumps can be estimated. The (covalent) atom diameters are $D_A=0.230nm(Ni)$, $D_B=0.280nm(Sb)$, $=0.282nm(Sn)$, $=0.288nm(In)$. Hence the size d_c of a DTI in c-direction is

$$(2) \qquad d_c = c - D_B$$

i.e. $d_c=0.236nm(Ni/Sb)$, $=0.238nm(Ni/Sn)$, $=0.228nm(Ni/In)$. The size in a-direction is

$$(3) \qquad d_a = 2\ a/\sqrt{3} - D_B$$

i.e. $d_a=0.180nm(Ni/Sb)$, $=0.194nm(Ni/Sn)$, $=0.196nm(Ni/In)$. Thus Ni-atoms fit well in c-direction of the DTI's but tend to widen the lattice in a-direction.

3.1.SELF-DIFFUSION OF Ni

For the diffusion of the A-component (Ni) we propose a mixture of three mechanisms D, C, A.

3.1.1. *The D-Mechanism*

Let us consider a vacancy in the Ni_I-sublattice. A vacancy strictly moving along a chain does not contribute to the diffusion of Ni-tracers because the correlation factor is zero. Hence we must allow the vacancy to move aside from the chain which often will include reconversion of a Ni_I-vacancy into a DTI. Hence let us consider a tracer atom A^* in the chain (site 1 in fig.3). The probability is V_2 that there is an unoccupied DTI on a neighbouring site (site 2 in fig.3). V_2 is the molefraction of vacant DTI's in the DTI-sublattice. Then the tracer will jump into the DTI (site 2) with a jump-rate w_{12}. The just created vacancy in the chain at site 1 will normally disappear via site 6 or 3, 7. Then normally the tracer must wait at site 2 until in one of the 3 neighbouring Ni-chains (e.g. along sites 4, 5 or 6, 1, 3, 7) another (thermal) vacancy in the Ni_I-sublattice will appear to enable a return of the tracer into its sublattice with jump frequency w_{12}. Since the energy E_2 is somewhat larger than E_1 we conclude that

(4) $\quad w_{12} < w_{11} < w_{21}$ and $\quad w_{12}\,w_{21} = w_{11}{}^2$

Hence the molefraction V_1 of thermal vacancies in the chains is not as low as in element crystals (up to 2%). The just described "D - m e c h a n i s m" (vacancy mechanism via DTI's) is supposed to be the m a i n mechanism for Ni-diffusion. The mechanism contributes to diffusion parallel to the c-axis ($D_{\parallel}^{(D)}$) as well as to diffusion perpendicular to it ($D_{\perp}^{(D)}$). Contribution $\parallel$c is e.g. by jumps 1-2-3 or 1-2-5 (3 cases) and to $\perp$c by jumps 1-2-4 (2 cases). There is a total of 6 cases in each of the 3 diffusion planes and so the average square of the total jump length is

(5) $\quad l_{\parallel}{}^2 = (\,0 + 3\;c^2/4\,)/6 = c^2/8$

(6) $\quad l_{\perp}{}^2 = ((0+2\;3a^2/4)2/3 + (0+1\;3a^2/4)4/3) = a^2/3$

Thus for the effective jump frequency of the D-mechanism we get

(7) $\quad w_D = V_1\,V_2\,w_{12}\,w_{21}\,/\,(\,w_{12} + w_{21}\,)$

and for the diffusion coefficients

(8) $\quad D_{\perp}^{(D)} = f_D\,l_{\perp}^2\,w_D\,/\,2\;,\;\;D_{\parallel}^{(D)} = f_D\,l_{\parallel}^2\,w_D\,/\,2$

Hence for the ratio R of the diffusion coefficients we get

(9) $\quad R_D = D_{\perp}/D_{\parallel} = (l_{\perp}/l_{\parallel})^2 = 8/3\;(a/c)^2$

if the diffusion occurs by the D-mechanism only.

3.1.2. *The C-Mechanism*

The C-mechanism also starts with a tracer jump 1-2 as in the D-mechanism (fig.2). But then there is also the possibility for the tracer to change places with the same just created chain vacancy (now still at site 1) again. The largest probability is for a jump back (2-1) because of (4). In such a case there is no contribution to diffusion because the first jump is cancelled. But there is also a certain possibility p_A that the vacancy meanwhile moves to site 3 and then the tracer may jump from site 2 to site 3 (compare discussion under 3.1.2). In this case there is only a contribution parallel to the c-axis with

$$(10) \quad l_{||}^2 = c^2/4 \quad , \quad w_C = p_A V_2 w_{12} w_{21} / (w_{12} w_{21})$$

$$(11) \quad D_{||}^{(C)} = 1/8 f_C c^2 w_C \quad , \quad D_{\perp}^{(C)} = 0 \quad , \quad R_C = 0$$

Thus in the case of a combination of D- and C-mechanisms we get for the total diffusion coefficients

$$(12) \quad D_{||} = D_{||}^{(D)} + D_{||}^{(C)} \quad , \quad D_{\perp} = D_{\perp}^{(D)}$$

$$(13) \quad R = D_{\perp}/D_{||} = D_{\perp}^{(D)} / (D_{||}^{(D)} + D_{||}^{(C)}) < 8/3 \ (a/c)^2$$

3.1.3. *Discussion of A-Diffusion*

Comparing the different temperature dependences of $D_{||}^{(D)}$ and $D_{||}^{(C)}$ we note that w_D according to (7) contains an additional factor V_1 while w_C contains p_A i.e. w_{11} (diffusion along the chains). Because of (4) the activation energy in V_1 is thought to be small and even smaller than the activation energy of w_{11}. Hence, as usual, we expect the minority mechanism to have a higher activation energy, i.e. it will come more into play at higher temperatures. Indeed the measured values for $R=D_{\perp}/D_{||}$ are lower than the theoretical value of the pure D-mechanism and the departure is larger at higher temperatures as is expected from (13), see fig.4, where the theoretical R-values are given at for the pure main mechanisms (100% at low temperatures) and the corresponding parts of the minority mechanisms at higher temperatures.

The relative good agreement between the experimental data and the theoretical predictions of a combination of the D-mechanism as the main mechanism and the C-mechanism as the minority mechanism for all the three investigated alloys seems to indicate that this interpretation of the Ni-diffusion process is very probable.

A recent investigation of Vogel et al.[6] on diffusion in the B8$_2$-compound Ni/Sb by means of quasielestic neutron scattering with one of our single crystals confirms our conclusion for the D-mechanism as the majority mechanism. Moreover, the authors observe that about 30% of the jumps are double-jumps, i.e. during the interaction with the neutron the Ni-tracer jumps not only from one octahedral site to the DTI but also back to a tetrahedral site, which confirms our C-mechanism and is in agreement with our theoretical interpretations for the C-mechanism.

Inspite of this we still like to discuss another mechanism.

3.1.4. *The A-Mechanism*

Since there is a small window between adjacent DTI-sites - as can be seen from a model of spheres of the B8$_2$-structure - it might be supposed that Ni-atoms may diffuse directly (with a frequency w_{AA}) to neighbouring DTI-sites. This would mean a direct diffusion within the DTI-sublattice which is equivalent to the B-sublattice. Hence the squares of the diffusion lengths are

(14) $l_{||}^2 = c^2/4$, $l_{\perp}^2 = 1/3 \ (0 + 2 \ a^2/4) = a^2/6$

and so we get

(15) $D_{||}^{(A)} = 1/8 \ f_A \ c^2 \ A_2 \ V_2 \ w_{22}$, $D_{\perp}^{(A)} = 1/8 \ f_A \ a^2 \ A_2 \ V_2 \ w_{22}$

f_A is the correlation factor, A_2 the molefraction of Ni_{II}-atoms and w_{22} the jump frequency of Ni in the Ni_{II}-sublattice. Thus for a pure A-mechanism

(16) $R_A = (l_{\perp}/l_{||})^2 = 2/3 \ (a/c)^2$

The migration energy of this process is expected to be high and hence the contribution to diffusion should be very small. So we think that we can neglect the contribution of the A-mechanism except for the diffusion in Ni/In where the R-value of the pure A-mechanism is larger than the experimental values. Hence the A-mechanism cannot be excluded in this case as a minority mechanism but it then contributes with a higher R-value (similar to that one of the D-mechanism) than the C-mechanism. Hence there might be a combination of D-, C- and A-mechanisms working.

If there is a significant amount of Ni-antistructure atoms as in the Ni/In-alloys it is also possible that such Ni-atoms migrate within the B-sublattice via a vacancy mechanism. The diffusion geometry then is the same as for the vacancy mechanism for B-diffusion (V-mechanism) and the same as the A-mechanism. Hence (14) and (16) are again valid and so these two mechanisms cannot be separated since they have the same ratio R which should be thought to comprise both mechanisms, taking into account, that the diffusion of vacancies in the B-sublattice is even higher activated than the direct A-mechanism in the Ni_{II}-sublattice and hence we may neglect this contribution as well.

3.2.THE DIFFUSION OF Sb, Sn, In.

The diffusion of the B-component is also thought to be a mixture of three mechanisms V, R and I. As in the case of the diffusion of Ni all proposed mechanisms use the DTI's as a reservoir to create vacancies in another sublattice.

3.2.1.*The V-Mechanism*

This is a normal vacancy mechanism within the B-sublattice and is thought to be the m a i n B-mechanism. The B-vacancies (molefraction V_B) are created when a B-atom jumps from a B-site into a neighbouring DTI (e.g.1'-2', fig.3) where it rests as a B-antistructure atom. For the motion of the vacancy through the B-sublattice we get

$$(17) \quad l_{||}^2 = c^2/4 \ , \ l_\perp^2 = 1/3 \ (0 + 2 \ a^2/4) = a^2/6$$

$$(18) \quad D_{||}^{(V)} = 1/8 \ f_V \ c^2 \ V_B \ w_B \ , \ D_\perp^{(V)} = 1/8 \ f_V \ a^2 \ V_B \ w_B$$

where w_B is the vacancy jump frequency and f_V the correlation factor for tracer diffusion by a vacancy in the B-sublattice. Hence, if only the main mechanism (V-mechanism) is contributing

$$(19) \quad R_V = D_\perp^{(V)}/D_{||}^{(V)} = 2/3 \ (a/c)^2$$

3.2.2.*The R4-Mechanism*

Similar as in the case of the Ni-diffusion there is a certain chance with probability p_B that the just created B-antistructure atom on site 2' in the V-mechanism will exchange again with the just created B-vacancy, starting from site 1'(fig.3). Again, the jump back (2'-1') does not contribute to diffusion. But when the vacancy has moved to e.g. site 3' then a jump 2'-3' with frequency w_{21}' would contribute to diffusion. Then effectively the B-atoms from sites 1', 3' would have changed places in a type of a 3-step vacancy-ring-mechanism

(R3-mechanism) within the B-sublattice. Hence the jump components are the same as given by (17) and so

(20) $\quad D_{||}^{(R)} = 1/8 \; f_R \; c^2 \; p_B \; w_{BB} \; , \; D_{\perp}^{(R)} = 1/12 \; f_R \; a^2 \; p_B \; w_{BB}$

where f_B is the correlation factor and

(21) $\quad w_{BB} = 2 \; w_{12}' \; w_{21}' \; / \; (\; w_{12}' + w_{21}' \;)$

The ratio R is the same as given by (19) and cannot be separated from the V-mechanism. Hence we regard this as a part of the main mechanism.

But there is also the possibility that after the jumps 1'-2', 3'-1' the vancy moves to 4' (tracer: 4'-3'), giving tracer B at 2' the chance to jump to 4'. Then in this "4-step vacancy-ring-mechanism" (R4) effectively 3 atoms have changed sites : 1'-4', 3'-1', 4'-3'. Since a B-atom has 5 neighbouring DTI's, 2 in c-direction and 3 in a-direction, and if we assume the jump frequencies to be equal, then

(22) $\quad l_{||}^2 = 2/5 \; (c^2 + 2 \; c^2/4) + 3/5 \; (0 + 2 \; c^2/4) = 9/10 \; c^2$

$\quad\quad\; l_{\perp}^2 = 2/10 \; (0 + 2 \; a^2/3) + 3/10 \; (a^2 + 2 \; a^2/3) = 19/30 \; a^2$

and thus

(23) $\quad R_4 = 19/27 \; (a/c)^2$

for the R4-mechanism. The values are very close to those of the V-mechanism, but this is not represented by the experimental values (compare fig.4). From this fact it must be concluded, that the jumps $||$ and $\perp$ to the c-direction are not at all equal and that $w_{||} << w_{\perp}$. Then

(24) $\quad l_{||}^2 = c^2/2 \quad , \quad l_{\perp}^2 = 5/6 \; a^2$

and hence

(25) $\quad R_4 = D_{\perp}/ \; D_{||} = 5/3 \; (a/c)^2$

for the R4-mechanism.

3.2.3. *The B-Mechanism*

There is also the possibility that the B-antistructure atoms move directly between DTI's. Since the structure of the DTI-sublattice is the same as for B, the l-values are given again by (17) and hence

(26) $\quad D_{||}^{(B)} = 1/8 \; c^2 \; B_2 \; w_B' \quad , \quad D_{\perp}^{(B)} = 1/12 \; a^2 \; B_2 \; w_B'$

where w_B' is the jump frequency of the B-atom and B_2 is the molefraction of antistructure B-atoms. Again $R_B = D_\perp/D$ is given by (19). Since B_2 is expected to be very small and the migration energy large, the contribution of (25) should be negligible or at least very small.

3.2.4. *The I-Mechanism*

In the case of the V-mechanism the B-antistructure atom will wait on it's site at 2' in the Ni_{II}-lattice until any vacancy will appear in its neighbourhood in the B-sublattice in order to return into it's sublattice (fig.3). The contribution to diffusion by this return is small since it is proportional to V_B^2, but will give rise to a minority mechanism, the I-mechanism (interstice mechanism), which uses the DTI-(interstice-) position in a similar way as the D-mechanism for A-tracers. Then if we assume w_{12}' equal for jumps $\parallel$ and $\perp$ to the c-axis (similar for w_{21}'):

(27) $l_\parallel^2 = 3/25 \; 2 \; c^2/4 + 2/25 \; 7 \; c^2/4 = 1/5 \; c^2$

$l_\perp^2 = 2/25 \; (2 \; a^2/12 + a^2/3) + 2/25 \; 3a^2/4 + 1/25 \; 2 \; 3a^2/4$

$= 4/25 \; a^2$

and hence

(28) $D_\parallel^{(I)} = 1/10 \; f_I \; c^2 \; V_2 \; V_B \; w_I, \quad D_\perp^{(I)} = 2/25 \; f_I \; a^2 \; V_2 \; V_B \; w_I$

where f_I is the correlation factor and $w_I = w_{BB}$ is given by (21). Thus for a pure I-mechanism we would get

(29) $R_I = D_\perp/D_\parallel = 4/5 \; (a/c)^2$

which gives slightly larger values than (19) for V- and R4-mechanism.

3.2.5. *Discussion of B-Diffusion*

According to the discussions given above the contributions of the B- and I-mechanisms seem to be small and thus may be neglected in a first approximation. Hence the experimental results are interpreted in terms of the V-mechanism as the main mechanism and the R4-mechanism as the leading minority mechanism. The experimental data and the theoretical predictions for the main mechanism are given in fig.4. For higher temperatures the parts of the minority mechanism R4 are also given at the curves. But these percentages are not as low as for the diffusion of Ni and in the case of Ni/Sb

494

the part of the R4-mechanism even exceeds 50% (65%), i.e. the "minority" mechanism is getting the main mechanism. So the situation for the diffusion of the B-tracer seems to be not so simple as for the Ni-diffusion and the question of a participation of the B- and I-mechanism cannot really be answered. There might be such contributions.

4. Summary

The experimental data for A-diffusion (A=Ni) and for B-diffusion (B=Sb, Sn, In) in A/B-phases with the $B8_2$-structure have been compared with a number of possible diffusion mechanisms. As a result of this comparison and the discussions of the theory we conclude that

- The experimental data for Ni-diffusion as indicated by the parallel Arrhenius-lines for diffusion $\parallel$ and $\perp$ to the c-axis may be well interpreted by the D-mechanism as the main mechanism (>90%) and the C-mechanism as a minority mechanism with contributions up to 10%. In the case of Ni/In the main mechanism may include contributions from the A-mechanism.

- The experimental data for B-diffusion (B=Sb, Sn, In) as also indicated by parallel Arrhenius-lines for the diffusion $\parallel$ and $\perp$ to the c-axis may be interpreted for low temperatures by the V-mechanism as the main mechanism and the R4-mechanism as the minority mechanism but for higher temperatures the "main" mechanism is changing to a minority mechanism which only contributes with 35% (1500K) while the "minority mechanism contributes with 65%. In this case contributions of other mechanisms (B-, I-mechanism) may as well exist.

5. References

1. Laves F., Wallbaum J.(1941), Z.angew.Mineral.9,p.17

2. Brand P.(1967),Z.anorg.allg.Chemie 353,p.280

3. Hünecke J.,Frohberg G.,Wever H.(1989),Defect and
   Diffusion Forum 66-69,p.483

4. Schmidt P.(1965),Dissertation FB17,TU Berlin

5. Hähnel R.,Miekeley W.,Wever H.(1986),
   phys.stat.sol.(a)97,p.181

6. Randl O.G.(1990),Magister Thesis,Univ.Wien

7. Schmidt H.(1990) "Selbstdiffusion in den Phasen Ni_3Sn_2,
   Ni_2In",Dissertation FB17,TU Berlin

6.Tables

Table 1: Experimental diffusion results for $Ni_{53}Sb_{47}$.
$D = D_o \exp(- Q / k T)$

tracer:	Ni-63	Sb-124
$Do_\perp$ (cm^2/s)	0.095	3.4
$Q_\perp$ (eV)	1.69	2.91
$D_{o\parallel}$ (cm^2/s)	0.069	5.6
$Q_\parallel$ (eV)	1.71	2.94

Table 2:Experimental Ni-diffusion results for Ni_3Sn_2, Ni_2In.

alloy:	$Ni_{64}In_{36}$	$Ni_{62}Sn_{38}$	$Ni_{61}Sn_{39}$
$Do_\perp$ (cm^2/s)	63.2	0.61	0.51
$Q_\perp$ (eV)	2.31	2.16	2.14
$D_{o\parallel}$ (cm^2/s)	108	0.70	0.95
$Q_\parallel$ (eV)	2.34	2.21	2.24

Table 3:Experimental Sn-diffusion results for Ni_3Sn_2

tracer:	Sn-113
$Do_\perp$ (cm^2/s)	9.9
$Q_\perp$ (eV)	3.07
$D_{o\parallel}$ (cm^2/s)	4.55
$Q_\parallel$ (eV)	2.93

7.Figures

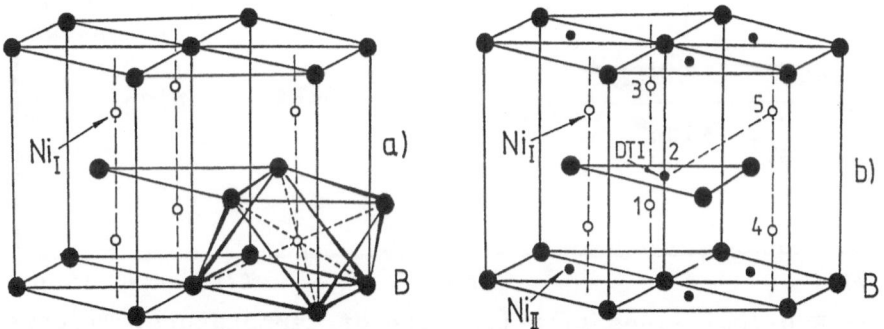

Fig.1: Octahedral (a) and tetrahedral (b) interstices of the $B8_2$-structure.

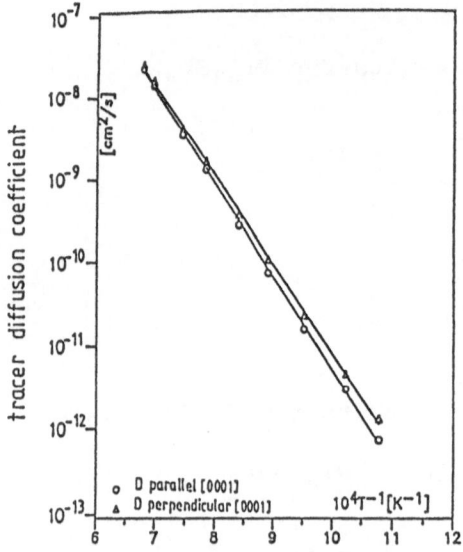

Fig.2:Arrhenius plot ‖ and ⊥
to c-axis:63Ni in Ni62Sn38.

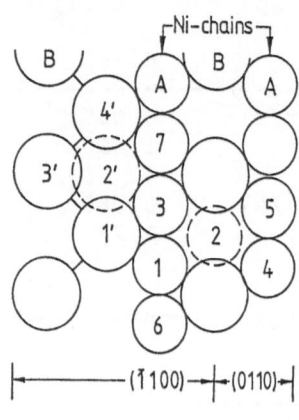

Fig.3: Planes (0110),(-1100):
diffusion paths in B8$_2$-phase.

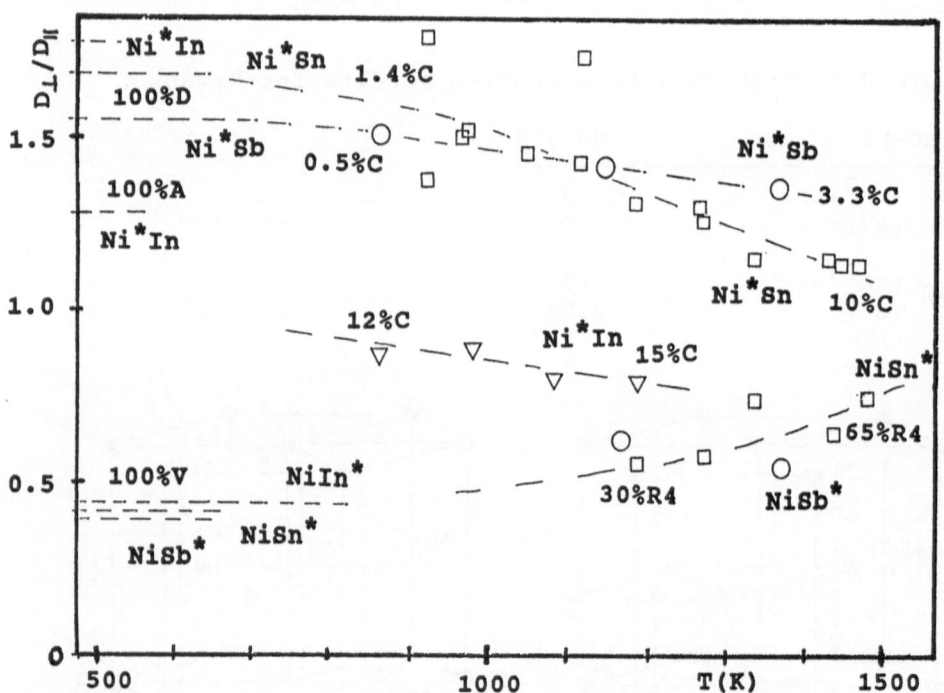

Fig.4: R=D$_\perp$/D$_\parallel$-values as a function of temperature. Tracers:
Ni* in Ni/Sb (o), in Ni/Sn (□), in Ni/In (▽);
Sb* in Ni/Sb (o); Sn* in Ni/Sn (□).

USE OF ATOMISTIC TECHNIQUES TO STUDY DIFFUSION IN INTERMETALLICS

G. VOGL, CH. KARNER, O. RANDL, B. SEPIOL*
and D. TUPPINGER
Institut für Festkörperphysik der Universität Wien
Strudlhofgasse 4
A-1090 Wien
Austria

ABSTRACT. Quasielastic neutron scattering (QNS) and quasielastic Mössbauer spectroscopy (QMS) permit to deduce the jump vector of diffusing atoms. This is possible by comparing the angular dependence of quasielastic line broadening with model predictions. Results are reported for Ni diffusion in NiSb (B8) and Ni_3Sb (DO_3) and for Fe diffusion in FeAl (B2). For NiSb we conclude that the Ni atoms jump alternately between regular and interstitial sites. Ni_3Sb and FeAl contain high concentrations of vacancies; conclusions on the possibilities for jumps via the vacancies are drawn.

1. Introduction

All methods studying diffusion by measurements of penetration profiles are *macroscopic* in the sense that they determine the net result of a great number of jumps. Their result is a proportionality constant, i.e. the diffusivity D. To conclude on the elementary diffusion jump is only possible if flanking measurements are performed (e.g. isotope effect, pressure dependence) and further assumptions, e.g. on the behaviour of a jumping atom at the saddle point between two consecutive positions are made. A *microscopic* way to determine the *diffusion mechanism* on an atomistic scale is therefore highly desirable.

For intermetallic phases various reviews and original papers report on the models invented for the jump mechanism (e.g. Bakker 1987, Bakker and Westerveld 1988, Bakker, this Workshop, Arita et al. 1989, Koiwa, this Workshop). The crucial point is that in most cases of simple intermetallic structures (e.g. B2) the nearest neighbour sites are occupied by atoms of the other species. Therefore it is not immediately evident why rather high diffusivities appear, even though the degree of disorder remains low. In many cases the explanation is a surprisingly high vacancy concentration, in others – as we shall see – we conclude that interstitial sites are involved.

497

C. T. Liu et al. (eds.), Ordered Intermetallics – Physical Metallurgy and Mechanical Behaviour, 497–509.
© 1992 Kluwer Academic Publishers.

2. Methods

We report about investigations of diffusion by help of radiation emitted, absorbed or scattered from/at the nucleus of a jumping atom. These methods are quasielastic neutron scattering (QNS) and quasielastic Mössbauer spectroscopy (QMS). The energy width Γ of the emitted (absorbed, scattered) radiation is a direct measure of the atom's jump frequency: an increase in the frequency of the diffusion jumps leads to an *increase in energy width* of the radiation interacting with a nucleus of a moving atom. This is an effect based on Heisenberg's uncertainty principle: the energy uncertainty is proportional to the reciprocal of the time uncertainty, i.e. the residence time τ at *one* position

$$\Gamma \propto 2\hbar/\tau. \tag{1}$$

The positions scanned by an atom during its interaction with the radiation enter into the *angular dependence* of the energy width if orientated single crystals are used as specimens,

$$\Gamma(\mathbf{Q}) \propto [1 - (1/N) \Sigma_n \exp(i\,\mathbf{Q}\,\mathbf{R}_n)] \tag{2}$$

with $\mathbf{R}_n$ the N jump vectors between these positions and $\mathbf{Q}$ the momentum transfer to the scattered neutron or the momentum of the emitted (or absorbed) gamma quantum. The word "quasielastic" signalizes that the energy change of the radiation is very small, the scattering (emission, absorption) is very close to elastic. The energy width Γ of the radiation, i.e. the broadening of its linewidth caused by diffusion, can be measured by high resolution neutron or gamma spectroscopy, the latter commonly known as Mössbauer spectroscopy.

What is interesting in the present context are not the physical details of the diffusional quasielastic line broadening. They have been discussed e.g. by Petry and Vogl (1987) and by Vogl et al. (1989) on the basis of the theory by Singwi and Sjölander (1960), and a review on the applicability of Mössbauer spectroscopy for diffusion has recently been given by Vogl (1990). We shall rather list the abilities and limitations of the atomistic methods compared with diffusivity studies by the measurement of penetration profiles e.g. of radioactive tracers.

2.1. PARTICULAR ABILITIES OF ATOMISTIC TECHNIQUES

2.1.1. From both QNS and QMS atomistic details of diffusion, namely the *jump frequency and the jump vector*, can be deduced by comparing the measured angular dependence of the diffusional line broadening with predictions from models (equations (1) and (2)). For neutron scattering the scattering vector $\mathbf{Q}$ can also be varied and thus additional information be derived from the $\mathbf{Q}$ dependence of the line broadening.

2.1.2. The atomistic techniques permit *non-destructive* investigation of diffusion and are performed in *thermal equilibrium.*

2.2.1. *Limitation of accessible range of diffusivity.* Present day techniques set limitations on the diffusivity range accessible to QNS measurements: the still leading technique, i.e. neutron backscattering spectroscopy at a cold neutron beam, needs diffusivities of at least a few 10^{-13} m²/s in order to cause significant line broadening, i.e. for systems with "normal" diffusivities QNS measurements of the jump vector must be performed within a narrow range below the melting point. It has to be added that such investigations can presently be done only in two or three places in the world.

With present day Mössbauer spectroscopy on ^{57}Fe, diffusivities between 10^{-14} m²/s and 10^{-10} m²/s can be studied. Slower diffusivity leads to line broadening below experimental resolution, larger diffusivity (appearing for fast diffusion) to line broadening as large as to be difficult to scan with present day standard equipment, though there is no principal limit.

2.2.2. *Limitation to elements.* Again this is principally mainly a question of measuring techniques, but for experiments as complicated from inherent reasons as studying diffusion at high temperatures in the intermetallics (preparation of single crystals, phase stability etc.) it is advisable not to accumulate further problems.

For QNS the criterion for choosing an isotope is the incoherent scattering cross section. Among the more popular metallic elements the best ones are, in range of decreasing scattering cross section, Co, Ni, V, Cs, Ti, Cr, W, Cu with Cu already a factor of 10 worse than Ni, which means that measurements on Cu must last 10 times longer, i.e. several days, which is presently impossible at the very few nuclear reactors with their heavy overload of experiments.

The first criterion for the choice of an isotope for Mössbauer spectroscopy (nuclear resonance absorption) is a high resonance effect and this is best realized in the case of the "milk cow" of Mössbauer spectroscopy ^{57}Fe. Investigations of diffusion with other relatively easy candidates as ^{119}Sn have not been performed until now, and with more delicate isotopes as the principally most attractive ^{181}Ta with its much higher energy resolution will only be performed if the interest in the field should increase considerably.

3. Study of the Diffusion Jump in three Ordered Structures

As Bakker (1987) has put it nearly 5 years ago in his review paper for the Berlin conference on defects in metals: For understanding material properties of intermetallics, the knowledge on an atomistic scale of the state of a sample is essential, but disorder, defects and defect mobility in intermetallics form still a nearly unexplored field. One of the reasons is that already the problem of preparing good samples may be enormous, another difficulty is the interpretation of experimental results in terms of atomistic quantities. So far Bakker.

We claim that through a completely different experimental access the problem of interpreting experimental data in terms of defect mobility is reduced, namely by measuring directly the atomic jump vector. The other problem, however, i.e. preparing good samples, is even more serious in our experiments, since for receiving full information on the atomic jump vector we need single crystal specimens which a priori are not available for many interesting systems.

We, therefore, have entered the field with studying a system where the single crystals were already available. This was NiSb with the "exotic" NiAs structure (B8 phase), the single crystals being provided by the Wever-Frohberg group (compare the paper by Frohberg and Wever at this Workshop). The study was performed with quasielastic neutron scattering. After having become familiar with the problems of studying intermetallics with QNS we dared to attack the high-temperature phase of Ni_3Sb (DO_3-structure) where we had to grow the crystals by ourselves directly at the neutron beam. Simultaneously a system with B2-structure, namely FeAl, was studied with Mössbauer spectroscopy.

We shall start with the terminated study of the B8 phase NiSb.

3.1. ATOMIC JUMP OF Ni ATOMS IN $Ni_{53}Sb_{47}$ (B8 STRUCTURE)

This system is ideal for investigating the Ni diffusion jumps by quasielastic neutron scattering (QNS) by the following reasons:
- Tracer studies have been performed by the Berlin group (Hähnel et al. 1986).
- Ni atoms diffuse sufficiently fast in order to produce a reasonable quasielastic line broadening, whereas the diffusivity of Sb atoms is 2 to 3 orders of magnitude smaller.
- Ni has one of the highest incoherent scattering cross sections of all elements, thus spectra can be taken in reasonable time.
- The incoherent scattering cross section of Sb is extremely low, thus Sb is practically invisible for QNS and there is no problem with deconvoluting the signals from the two partners. (Of course, therefore, only Ni diffusion jumps can be studied.)

Fig.1 shows the arrangement of Ni and Sb atoms in the B8 (NiAs) structure which represents principally a hexagonal close-packed lattice (compressed along the c-axis) of Sb atoms (●) with the Ni atoms on its octahedral interstitial sites (O). The surplus Ni atoms above equiatomic concentration occupy double tetrahedral interstices (●), and even below that concentration there is some occupancy of these sites (Leubolt et al. 1986). At least three different types of Ni diffusion jumps are immediately conceivable:
(a) between the regular Ni sites via vacancies in the Ni sublattice,
(b) between the double tetrahedral interstitial sites,
(c) alternately between regular and interstitial sites (Frohberg model, see Hähnel et al. 1986).
One appealing jump variant, however, appears to be excluded from the tracer results, namely Ni jumps alternating between the two sublattices, since Sb diffusivity is so much smaller than Ni diffusivity and it is hardly conceivable how Ni atoms could use Sb vacancies a factor of 500 times more often than the Sb atoms themselves.

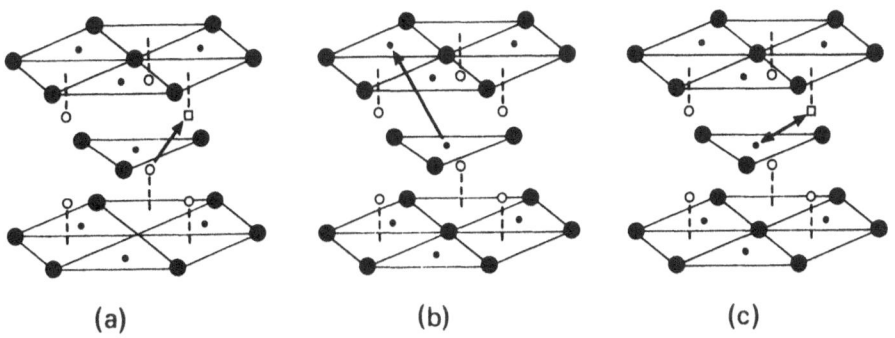

(a)   (b)   (c)

Figure 1. Exemplary jump vectors of Ni atoms in NiSb for three different jump models. In model (a) the Ni atoms jump between their regular sites, i.e. the octahedral interstices in the hcp Sb sublattice. The DTI model (b) assumes jumps between the double tetrahedral interstices (DTI) that are occupied at alloy concentrations above (and, to a certain extent, even below) stoichiometry. According to the Frohberg model (c) the Ni atoms jump alternately between regular and DTI sites.

The principle of a QNS measurement has been described e.g. by Vogl et al. (1989) for self-diffusion in ß–titanium. Fig.2 from Vogl et al. 1991 compares measured QNS data from $Ni_{53}Sb_{47}$ (left column) with theoretical predictions for some of the above-mentioned models and various orientations of the NiSb single crystal relative to the neutron beam. The middle column shows the line widths calculated with equation (2) for various variants of jumps exclusively between regular sites. The inappropriateness of the models is clear, most evident for the orientation indicated (-30,90,90): the models expect a minimum with no line broadening for scattering wave numbers Q=1.8 $Å^{-1}$, whereas the experimental data only show a small relative minimum there. The third column gives model predictions for jumps between interstitial sites (model b) and alternately between regular and interstitial sites (model c). Whereas the clear maxima predicted by the interstitial model donot find their correspondence in the experimental data, the deviation from the model c (Frohberg model) is not too striking but nevertheless cannot be denied.

502

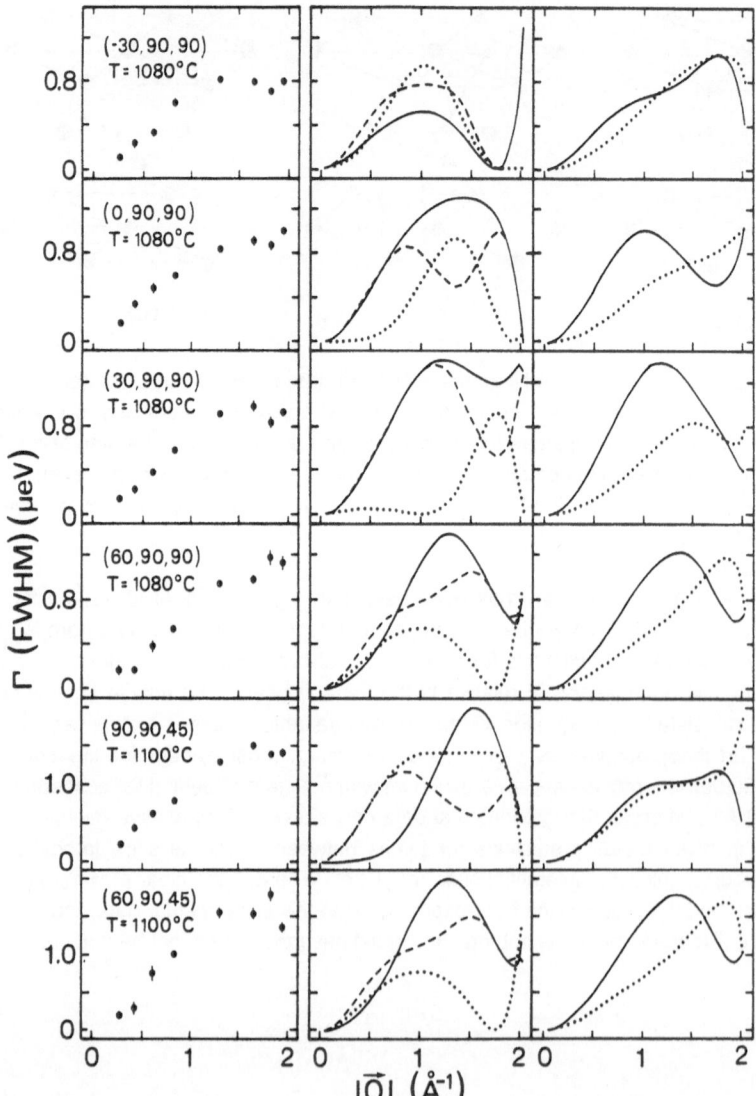

Figure 2. Ni$_{53}$Sb$_{47}$: Comparison of experimental QNS diffusional line broadening Γ for a set of Euler angles (left column) and calculations for various models (middle and right column) for the same Euler angles.

Left column: Results of measurements with the crystal orientations defined by Euler angles (Φ, Θ, Ψ). Middle column: Model calculations of line broadening for jumps between the regular ("octahedral") sites (solid line: jumps parallel to the c–axis, dashed line: diagonal jumps, dotted line: jumps in the plane normal to the c–axis). Right column: model calculations for the DTI model (solid line) and the Frohberg model (dotted line). Note that the experimental values are given in μeV, whereas the theoretical values are normalized to one and the same jump frequency for all spectra. That frequency was chosen so that an easy comparison with the experimental values is guaranteed.

The way out, and obviously perfect fits (fig.3) were possible by remembering that for diffusion in non-ordered alloys second and further jumps during the "encounter" with one vacancy have to be considered. In our case of model c (fig.1), too, further jumps appear possible if the vacancy left back by the jumping Ni atom changes its place by just one NN jump and offers itself now again to the Ni atom which accepts that new chance. The double jump eventually can be described as a jump between two regular sites. The optimum fit indicates that in 30% of all cases Ni atoms continue jumping in this way immediately after the first jump.

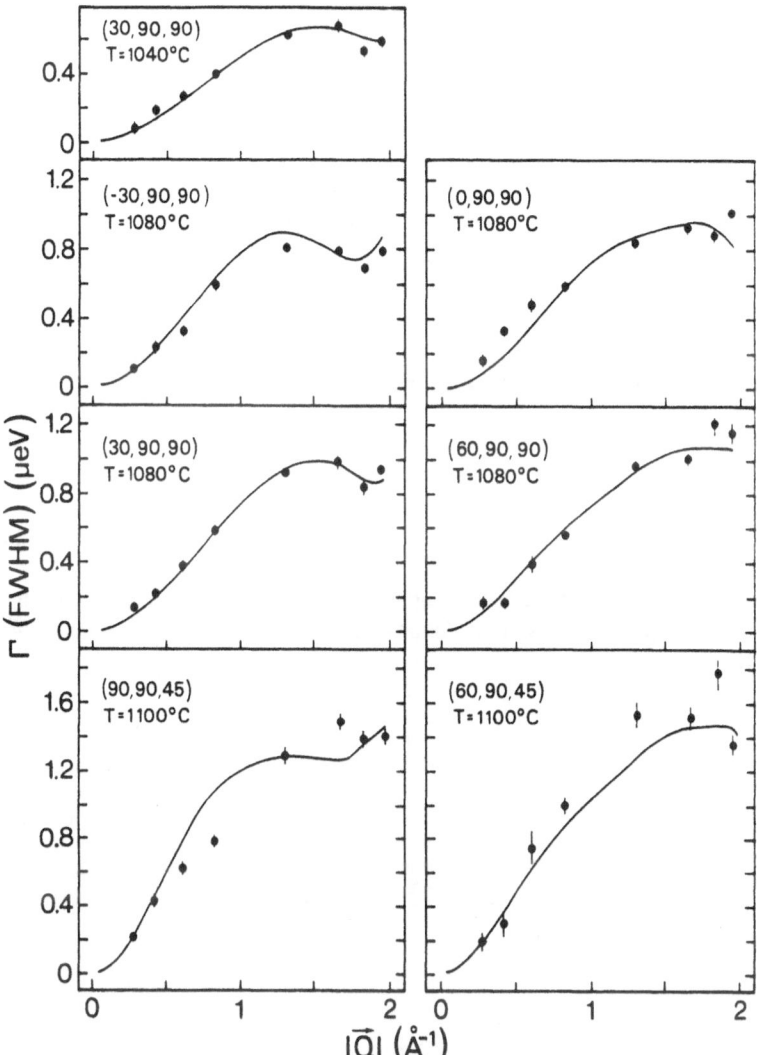

Figure 3. Ni$_{53}$Sb$_{47}$: Results of the optimum fits with our model described above (Frohberg model + 30 % double jumps into vacancy).

3.2. DIFFUSION JUMP OF Ni ATOMS IN Ni$_3$Sb (DO$_3$ STRUCTURE)

A very recent application of QNS to the atomistic study of diffusivity and an obvious extension of the work on NiSb is the investigation of Ni diffusion in the DO$_3$ phase Ni$_3$Sb (fig.4).

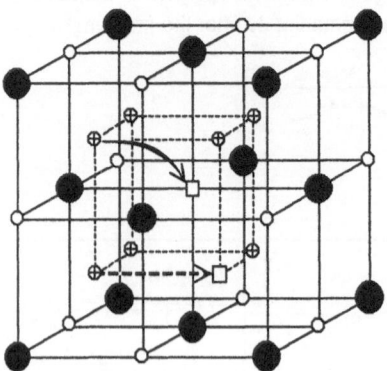

Figure 4. DO$_3$ structure of Ni$_3$Sb. The full arrow indicates jumps between Ni1 and Ni2 sites, the broken arrow between Ni2 and Ni2 sites. O Ni1, ⊕ Ni2, ● Sb, ☐ vacancy.

Tracer studies several years ago by Heumann and Stüer (1966) had proved that Ni diffuses very fast (D of the order of 10^{-10} m²/s at 900°C). For a QNS study the same advantages as for NiSb prevail, but sample preparation is a problem since single crystals of the high-temperature phase have to be prepared, the low-temperature phase having a complicated non-cubic structure.

We solved the problem by growing and orienting the crystals of Ni$_{71}$Sb$_{29}$ (not exactly Ni$_3$Sb) in-situ at the neutron spectrometers in a furnace specially constructed for such problems (Flottmann et al. 1987).

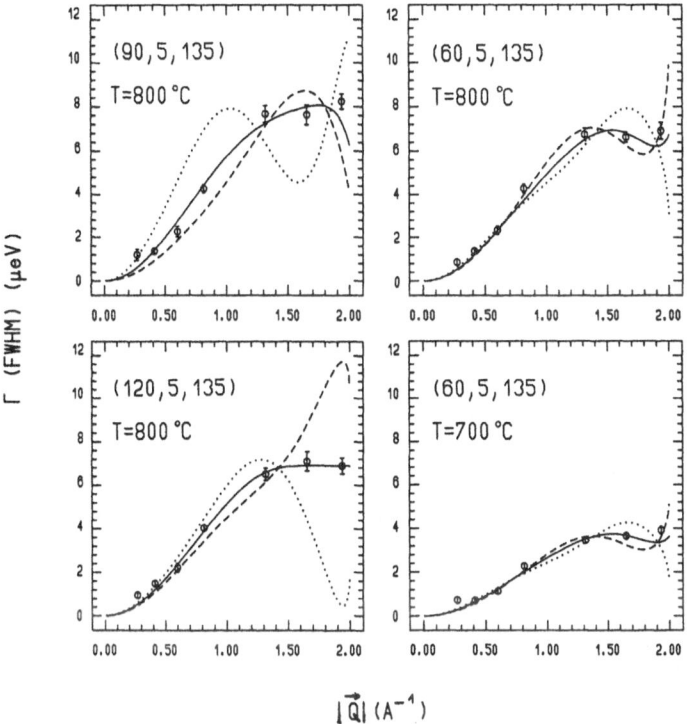

Figure 5. Ni$_{71}$Sb$_{29}$: QNS diffusional line broadening Γ at two different temperatures for a set of Euler angles (Φ, Θ, Ψ). Comparison with calculations for various models. Broken line: jumps between Ni1 and Ni2 sites, dotted line: jumps between Ni2 and Ni2 sites. Full line: best fitting combination of Ni1↔Ni2 and Ni2↔Ni2 jumps.

Figure 5 shows the results of preliminary fits of a first measuring series. The representation is the same as chosen for figures 2 and 3. Comparison with two possible models for diffusional jumps of the Ni atoms (fig. 4). It is evident that the (shorter) Ni1↔Ni2 jumps alone cannot explain the experimental results, but that an appropriate combination with Ni2↔Ni2 jumps yields a satisfactory fit. The conclusion is: Ni atoms jump via vacancies with nearly equal mobility on Ni1 and Ni2 sites.

3.3. DIFFUSION JUMPS OF Fe ATOMS IN $Fe_{50}Al_{50}$ (B2 STRUCTURE)

This system is ideally suited for a Mössbauer study of Fe diffusion if single crystals can be grown which – as expected – proved to be a considerable problem.

From measurements of lattice parameter, length and density by Ho and Dodd (1978) and Riviere (1977) it is known that FeAl can contain large numbers of vacancies, namely several percent. The best studied B2 face is CoGa which according to van Ommen et al. (1981) also contains several percent of vacancies. For CoGa Stolwijk et al. (1980) proposed that vacancies equally distributed on Co and Ga nearest neighbour (NN) sites may transform in the course of the diffusion jump to a configuration of two Co vacancies on next nearest neighbour (NNN) sites and an antisite Co atom as NN to both vacancies, called "triple-defect" (fig.6a). The authors assume that in the further course of the elementary diffusion event the triple defect retransforms to a Co and a Ga vacancy, but the Co vacancy has eventually exchanged its position with a Co atom originally one lattice parameter apart, i.e. it has jumped along an edge of the cubic unit cell.

If we admit (Bakker, private discussion) that in FeAl one member of the divacancy in the course of the jump can be separated from the other vacancy further than just a NNN distance, or that a second vacancy may arrive at a NN site of an Fe atom which has jumped into a single vacancy, then alternatively to jumping along the cube edge, the Fe atom may jump via the face diagonal of the cubic cell (fig.6b) (even jumps via the body diagonal appear possible, though with lower statistical probability).

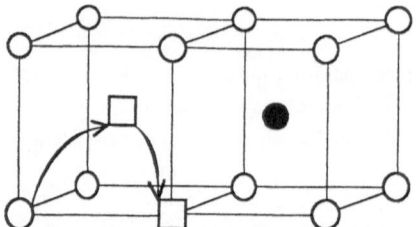 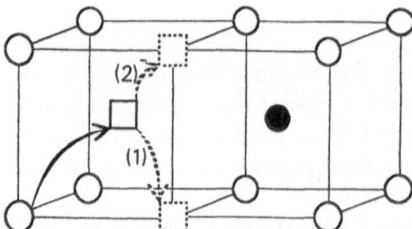

Figure 6. B2 structure. Model a (left): Triple defect jump along edge of cubic cell according to Stolwijk et al. (1980). Model b (right): jumps via cube edge (1) or via face diagonal (2) of cubic cell. O Fe atom, ● Al atom, ☐ vacancy.

We have studied Fe diffusion in $Fe_{50}Al_{50}$ by Mössbauer spectroscopy of ^{57}Fe. A single crystal of FeAl was cut so that the surface was close to a (111) plane. The directions <110> and <111> were in the measuring plane, i.e. the plane containing the direction of the gamma rays. The sample was the Mössbauer absorber, i.e. the furnace containing the sample was midway between Mössbauer gamma source and detector. The set-up allowed to scan essentially all directions between <110> and <111>. We report here preliminary results of first experiments with a single crystal of FeAl. As stressed in the Introduction what we can deduce from our experiments is, at least until now, not the details of an encounter with a defect acting as means of diffusion, but rather the vector between the position of the diffusing atom *before arrival* and *after departure* of the defect. What we find is shown in fig.7, together with simulated curves corresponding to certain jump vectors.

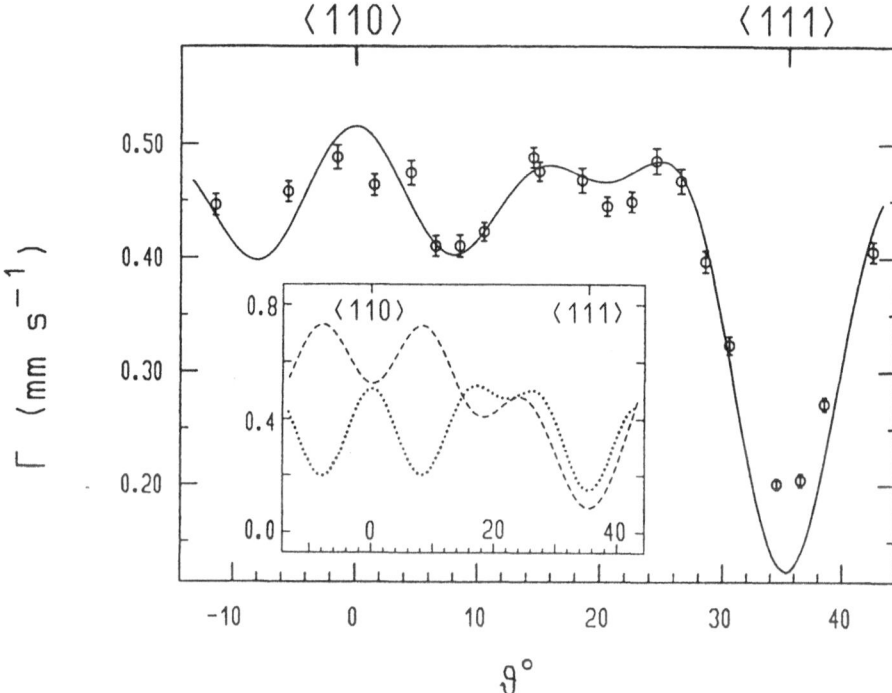

Figure 7. $Fe_{50}Al_{50}$: Diffusional broadening Γ of Mössbauer line as a function of observation direction at $T = 1080°C$. ϑ denotes the angle between the <110> crystal axis and the observation direction in the measuring plane. Full line: computer simulation for 40% NNN jumps along the cube edge (model 1) and 60% jumps across the face diagonal (model 2) calculated according to equation (2). Insert: Calculated angular dependences of line width according to jumps into NNN vacancy along cube edge (broken line) and via the cubic face diagonal (dotted line).

The angular dependence of the diffusional line broadening shows a dramatic minimum close to the direction <111>. This is exactly what is expected for all jumps in a cubic structure and thus represents only a check on the correct orientation of the sample. More important for deciding between different jump models are the smaller minima near the angles $\vartheta = 10°$ and $\vartheta = 20°$. The broken line in the insert of figure 7 shows the angular dependence of the linewidth as expected for Fe jumps along cube edges to NNN sites (model a). The dotted line is for jumps across the face diagonal to third nearest neighbour sites (model b). For $\vartheta = 10°$ the predictions of the two models are completely opposite ones: a maximum linewidth for the jumps along the cube edge

508

and a minimum for jumps across the face diagonal. It is evident that model b yields the much better description of the measurement, though the experimental minimum is softer than predicted by the model. Best fit is therefore received with 40% admixture of model-a jumps along the cube edge, the majority (60%) jumping across the face diagonal. Such a combination also gives a satisfactory fit to the shallower minimum at $\vartheta = 20°$.

4. Conclusions

Atomistic techniques like quasielastic neutron scattering (QNS) or quasielastic Mössbauer spectroscopy (QMS) can measure the energy broadening of radiation scattered, emitted or absorbed at nuclei of diffusing atoms. Different diffusion models predict different anisotropies of the broadening, thus a comparison of experimental data and theory permits to induce directly on the elementary diffusion jump. Of course, uniqueness is not guaranteed and sometimes it will only be possible to exclude models, but not to conclude with certainty on the correct model. It must be stressed that, at least up to now, the net jump vector during an "encounter" with a lattice defect is determined, but not the "internal details". Thus, e.g. for FeAl, we can say that between arrival and departure of a vacancy type defect Fe has been displaced along a cube edge or via the cubic face diagonal, but we cannot decide which jumps have occured during the encounter.

For the B8, the DO_3 and the B2 structures systems suitable for QNS or QMS, respectively, have been found and the atomic diffusion jumps have been induced. That has been possible due to sufficiently high diffusivities and isotopes appropriate for QNS (Ni) or QMS (Fe).

Acknowledgement

We thank H. Wever for numerous valuable discussions and him and J. Hünecke for providing the $Ni_{53}Sb_{47}$ samples. We further acknowlegde discussions with P. Terzieff and D. Schicketanz on the NiSb system, M. Mantler and A. Heiming for help with the determination of the Ni_3Sb structure, W. Petry, A. Heiming and J. Trampenau for their cooperation in the QNS measurements, E. Seidl for support in producing FeAl single crystals and P. Willbacher for technical assistance.

This work was financially supported by the Austrian Fonds zur Förderung der Wissenschaftlichen Forschung (project P7426).

References

* On leave from Institute of Nuclear Physics, Cracow.

Arita, M., Koiwa, M., and Ishioka, S. (1989) "Diffusion mechanisms in ordered alloys – a detailed analysis of six-jump vacancy cycle in the B2 type lattice", Acta metall. 37, 1363–1374.

Bakker, H. (1987) "Diffusion ordering and vacancies in intermetallic compounds", Materials Science Forum 15–18, 1155–1182.

Bakker, H. and Westerveld, J. P. A. (1988) "Expression for the diffusion coefficient in intermetallic compounds with cubic structures", phys. stat. sol. (b), 145, 409–417.

Flottmann, T., Petry, W., Serve, R., and Vogl, G. (1987) "A combined furnace for crystal growth and neutron scattering", Nucl. Instr. and Meth. in Phys. Res. A 260, 165–170.

Hähnel, R., Miekeley, W., and Wever, H. (1986) "Diffusion studies on the B8 phase of the Ni/Sb system", phys. stat. sol. (a) 97, 181–190.

Heumann, T. and Stüer, H. (1966) "Bestimmung der Diffusionskoeffizienten des Nickels und der Leerstellenkonzentration in der ß-Phase des Systems Nickel-Antimon", phys. stat. sol. 15, 95–105.

Ho, K. and Dodd, R. A. (1978) "Point defects in FeAl", Scripta Metall, 12, 1055–1058.

Leubolt, R., Ipser, H., Terzieff, P., and Komarek, K. L. (1986) "Nonstoichiometry in B8-Type NiSb", Z. anorg. allg. Chem. 533, 205–214.

Ommen, A. H. Van, Waegemaekers, A., Molemann, A. C., Schlatter, H., and Bakker, H. (1981) "Vacancies in CoGa", Acta Metall. 29, 123–133.

Petry, W. and Vogl, G. (1987) "Potential and limits of nuclear methods in diffusion studies", Materials Science Forum 15–18, 323–348.

Riviere, J. P. (1977) "Structural defects in ß phase Fe-Al", Mat. Res. Bull. 12, 995–1000.

Singwi, K. G. and Sjölander, A. (1960) "Resonance absorption of nuclear gamma rays and the dynamics of atomic motions", Phys. Rev. 120, 1093–1102.

Stolwijk, N. A., Van Gend, M., and Bakker, H. (1980) "Self-diffusion in the intermetallic compound CoGa", Philos. Mag. A42, 783–808.

Vogl, G. (1990) "Diffusion studies", Hyperfine Interactions 53, 197–212.

Vogl, G., Petry, W., Flottmann, Th., and Heiming, A. (1989) "Direct determination of the self-diffusion mechanism in bcc ß-titanium", Phys. Rev. B 39, 5025–5034.

Vogl, G., Randl, O., Petry, W., and Hünecke, J. (1991) "Direct determination of the Ni diffusion mechanism in the intermetallics NiSb", submitted to J. Physics: Condensed Matter.

KINETICS OF ORDERING AND DISORDERING OF ALLOYS

R.W. CAHN
Department of Materials Science & Metallurgy
Cambridge University
Pembroke Street
Cambridge CB2 3QZ
England

ABSTRACT. In the first part, the ordering kinetics of reversibly ordered phases such as Ni_4Mo and Ni_3Fe are examined; these can be initially disordered either by quenching from above T_c (which $< T_m$ for a 'reversibly' ordered alloy) or by cold-working the ordered alloy. The relationship of ordering kinetics to diffusivities is outlined. The interrelation of ordering kinetics and the kinetics of recovery and recrystallization of reversibly ordered alloys is briefly discussed. — In the second part, the reordering of Ni_3Al, an alloy of the 'permanently' ordered variety, is analysed. This alloy is ordered, in equilibrium, up to the melting temperature; it can be substantially disordered either by severe cold work or by ultrarapid quenching. Some recent work on disordering of Ni_3Al by ball-milling is described; the process is associated with unexpected changes in lattice parameter. The effect of annealing and gradually reordering such a mechanically disordered Ni_3Al upon its hardness is briefly treated.

1. Introduction

This concise overview is primarily concerned with changes in the degree of long-range order (LRO) in stoichiometric alloy phases, either with reordering after LRO has first been eliminated by appropriate heat-treatment or plastic deformation, or with the stages of that elimination of LRO as an interesting process in its own right. There is a range of measurement techniques that can be used to follow the change of S, the long-range order (or Bragg) parameter: x-ray diffraction line intensities, lattice parameter, dilatometry, electrical resistivity, differential scanning calorimetry, magnetic properties. Several of these will be exemplified here.

Attempts have been made to correlate the ordering kinetics with the known diffusivity values in various alloy phases and also with the ordering energies of the systems concerned. For 'reversibly' ordered alloys (those that, on heating, disorder before they melt) there is evidence that as the holding temperature is changed, the isothermal ordering rate reaches a maximum some way below the critical temperature at which LRO disappears, resulting in a characteristic 'nose' on the time-temperature-ordering transformation diagram.

Undoubtedly the most interesting aspect of recent researches is the discovery that 'permanently' ordered phases (those that melt before they can disorder) can be entirely disordered by intense plastic deformation (or irradiation) or even, if

511

C. T. Liu et al. (eds.), Ordered Intermetallics – Physical Metallurgy and Mechanical Behaviour, 511–524.

512

special techniques are used, by effectively ultrarapid quenching, and it is then possible to study how the physical and mechanical properties of such artificially disordered alloys change as order is progressively reintroduced. This now offers the possibility of a better understanding of the properties of strongly ordered alloys, which are also those that in equilibrium remain ordered up to the melting temperature. It is probable that much more will be heard of such studies in the near future.

2. Ordering Kinetics of Reversibly Ordered Alloy Phases

2.1 THERMALLY DISORDERED PHASES

The available experimental information on isothermal ordering kinetics mainly of $L1_2$ type alloy phases has been reviewed by Cahn [1]. Many of these studies have been of a special kind in which an alloy is equilibrated at one temperature and then the temperature is abruptly changed to another, both below T_c. These are easier to interpret theoretically, which accounts for the preference of many physicists for this kind of experiment. True ordering kinetics, starting from the initially disordered (or short-range ordered) condition, have also been performed, notably with Cu_3Au and Ni_3Fe (both $L1_2$), also with the tetragonal $D1_2$ type alloy Ni_4Mo. Characteristic data are shown in Figs. 1 and 2. Fig. 1 is based upon direct diffractometric measurements of S vs time for Ni_3Fe [2], while Fig. 2 is derived from resistometric measurements on Ni_4Mo [3]. On each set of curves

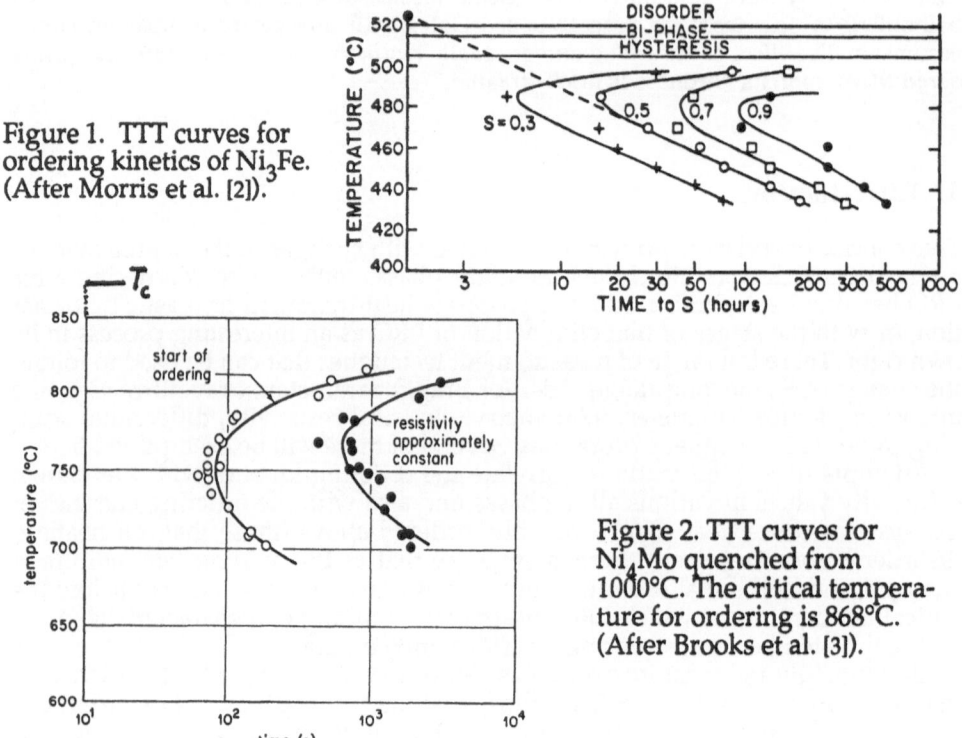

Figure 1. TTT curves for ordering kinetics of Ni_3Fe. (After Morris et al. [2]).

Figure 2. TTT curves for Ni_4Mo quenched from 1000°C. The critical temperature for ordering is 868°C. (After Brooks et al. [3]).

it can be seen that ordering is fastest at a temperature some way below T_c. This is interpreted by noting that as the temperature is lowered beyond T_c, the driving force for ordering increases but the diffusivities in the alloy decrease, so that two opposing factors are in play. This maximum ordering rate, or minimum relaxation time, below T_c is superficially similar to but has a physical origin distinct from the 'critical slowing down' phenomenon close to T_c, as shown in Fig. 3 which refers to Ni₃Mn [4]. Measurements were made by neutron diffraction. Here the transition is from a partial state of LRO to a more complete state. The matter is further discussed in Cahn's review [1].

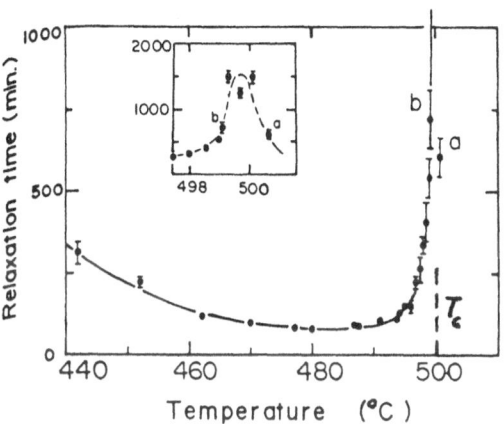

Figure 3. Relaxation time of the LRO parameter of Ni₃Mn, held isothermally at various temperatures. (After Collins and Teh [4]).

Cahn [1] also sought to correlate measured self-diffusivities in several L1₂ type alloys with the corresponding ordering kinetics. Since in a binary alloy there are two distinct self-diffusivities, it is difficult to know which one to use for the correlation and the best procedure was thought to be to choose the slower of the two diffusivities, D_s. It was concluded that the value of D_s corresponding to a given ordering rate is the smaller, the higher the critical ordering temperature T_c. In other words, diffusion can be regarded as more '"efficient" in bringing about LRO if the driving force, resulting from a higher ordering energy (associated with a higher value of T_c), is raised. To reach this conclusion, ordering kinetics and diffusivities for Ni₃Al were included; this phase has a much higher ordering energy than the other (reversible) phases used in Cahn's analysis and orders correspondingly faster. The reordering of Ni₃Al is discussed in the next Section. Ni₃Al orders at much lower reduced temperatures than, e.g., Ni₃Mn or Ni₃Fe.

Ordering kinetics have recently been examined in an alloy of composition $(Co_{78}Fe_{22})_3V$, usually code-named LRO-1 following the early studies of this, and related, alloys by Liu [5]. The disordered alloy is f.c.c. and the superlattice is L1₂; special interest attaches to this phase because the ordering temperature (910°C) is high for a reversibly ordered alloy. Precision resistivity measurements were made during isothermal anneals [6] and attempts were made to fit equations to the isotherms. An example is shown in Fig. 4, which refers to a temperature of 770°C (this is close to the 'nose' temperature for this alloy). There is an initial, rapid reduction of resistivity followed by a much slower process; we believe the rapid process to be linked with the elimination of excess point defects. It is interesting to compare this figure with Fig. 2, above. The 'nose temperature' in Fig. 2 is identical with the temperature corresponding to Fig. 4, but the (Co,Fe,V) alloy orders much more slowly than does Ni_4Mo. The two alloys have very similar T_c values, but the superlattices are quite different. This indicates that kinetic omparisons between different ordered crystallographies are not likely to be fruitful. — Measurements of resistivity were also made during continuous heating and matched against corresponding DSC runs, which allowed the release of enthalpy to be related to resistivity changes. There was again in these runs an indication of an early stage corresponding to the removal of excess point defects. — Comparisons between specimens disordered by quenching, as for Fig. 4, and others disordered by cold-rolling lent support to the point defect hypothesis.

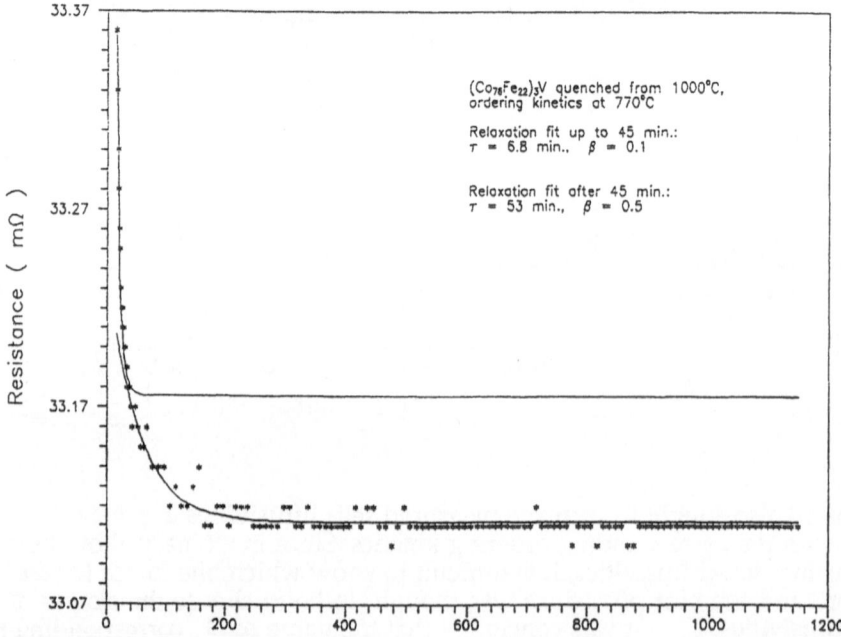

Figure 4. Resistivity isotherm at 770°C for $(Co_{78}Fe_{22})_3V$, quenched from 1000°C. Two distinct functional relationships have been fitted to the early and late stages of ordering, with relaxation times of 6.8 and 53 min, respectively. (After Soltys et al. [6]).

A similar separation of ordering into separate stages during continuous heating was found for an alloy of composition Cu_4Pt; the first stage was found to be more pronounced if the alloy had been disordered by cold-rolling, which was consistent with its being due to the removal of excess point defects [7,8].

2.2 BEHAVIOUR OF COLD-WORKED ALLOYS

The comparison of ordering kinetics in an alloy disordered by quenching from above T_c, and the same alloy partially disordered by rolling the initially fully ordered material, has already been alluded to in the immediately preceding remarks. The ordering kinetics in the second case is intimately linked with the recovery and recrystallization of the deformed alloy. This aspect has been studied by Cahn et al. [9] and by Gialanella and Cahn [10], using the $(Co_{78}Fe_{22})_3V$ alloy, alias LRO-1. Comparisons were made in these two studies between the alloy rolled in the originally ordered state and alternatively in the initially disordered state, and the behaviour was radically different in the two cases. In the first case, the alloy softened rapidly, well before recrystallization had got under way; in the second case (Fig. 5) the alloy hardened initially, reaching a maximum well before ordering was complete, and then softened substantially. Even at 850°C recrystallization took ≈150 minutes to go to completion, and here again the changes in properties are due to a form of 'recovery', not to recrystallization. Fig. 6 shows how the order parameter changes with time during isothermal annealing, for comparison with Fig. 5; it can incidentally be seen that the ordering rate is virtually identical at 770°C and 850°C, suggesting that the TTT nose temperature is around 800°C, ≈ 100 deg C below T_c.

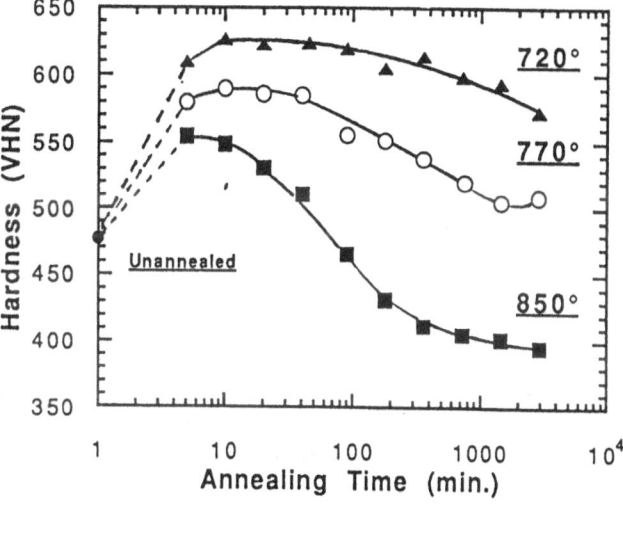

Figure 5. Hardness of LRO-1 alloy, initially disordered, rolled to 50% reduction and isothermally annealed at three temperatures. (After Gialanella and Cahn [10]).

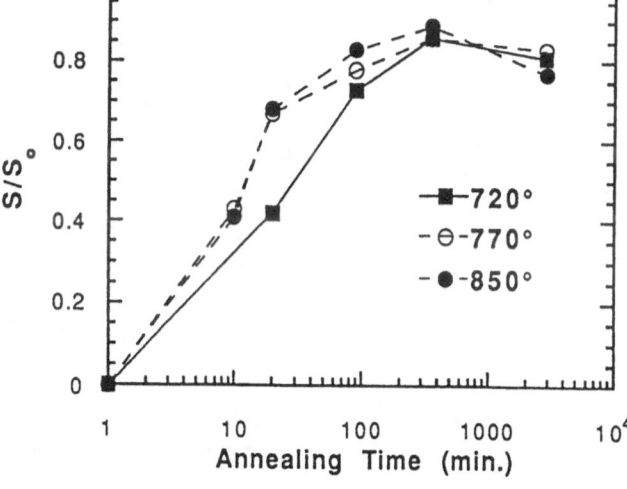

Figure 6. As Fig. 5, showing increase of order parameter S with annealing time. (After Gialanella and Cahn [10]).

We will return to the reason for the radically different changes in mechanical properties during annealing according to the initial state of order of the alloy, after we have discussed the behaviour of Ni_3Al. In this review, nothing further will be said specifically about *recrystallization* in relation to order, which is a large subject in its own right. The reader wishing to pursue this topic is referred to references [9], [10] and [11].

3. Disordering and Reordering of Ni_3Al

3.1 INTRODUCTION

The phase Ni_3Al is of such fundamental importance to modern metallurgy that any new findings concerning its behaviour are bound to be of interest to the intermetallics community, even if they refer to metastable states which would not last for a second at the kind of temperature at which alloys incorporating this phase are actually used. Accordingly, we discuss here recent researches (and a few not so recent) which bear upon ways of disordering Ni_3Al and the properties of wholly or partly disordered Ni_3Al.

3.2 WAYS OF DISORDERING A 'PERMANENTLY' ORDERED PHASE, Ni_3Al

3.2.1 *Mechanical Disordering*. Many years ago, Corey and Potter [12], followed by Clark and Mohanty [13], showed that Ni_3Al can be substantially disordered by intense deformation; filing is particularly effective. Clark and Mohanty examined the gradual return of order when such a batch of filings was isothermally annealed at various temperatures (Fig. 7). The first stages of order return quickly but order does not return all the way to $S = 1$ even after long periods at 280°C. Whether or not recrystallization (i.e., removal of dislocations) is needed to achieve this was not examined.

Figure 7. Reordering of mechanically disordered Ni_3Al filings at 220-280°C. (After Clark and Mohanty [13]).

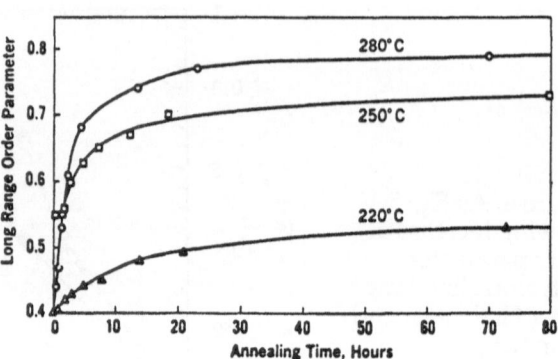

Recently, Jang and Koch in a fascinating series of experiments [14] showed that Ni_3Al can be efficiently disordered by ball-milling in a vibrating ball-mill, using a tool-steel vial and hardened stainless steel balls. Continued milling eventually turned the powder into a partially amorphous form, but this part of their investigation does not concern us here. They were able to make microhardness measurements on polished sections of the powder at various stages of milling, with the striking results shown in Fig. 8. There was a very pronounced hardness peak when the order parameter had reached 0.5. This kind of peak had been observed years ago for other, reversibly ordering, alloys which had been brought to various levels of partial order by purely thermal treatment, but not nearly as high and sharp.

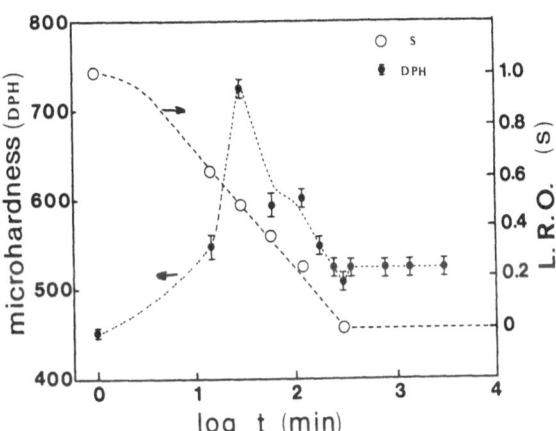

Figure 8. Microhardness of Ni_3Al powder as a function of ball-milling time at room temperature. At the hardness peak, $S \approx 0.5$. (After Jang and Koch [14]).

Recent experiments in the author's laboratory which have taken this study of disorder induced by milling a stage further will be discussed in the next Subsection.

The characteristics shown in Fig. 8 serve to interpret the results shown in Fig. 5 taken in conjunction with Fig. 6, if it is assumed that what applies to Ni_3Al applies to the alloy LRO-1 likewise. Initially disordered, rolled alloy on annealing will progressively become ordered and, presumably, climb over a steep hardness peak before softening again, as seen in Fig. 5. In that figure, the peak hardness approximately corresponds to $S = 0.5$. A discussed more fully in ref. 9, at the degree of order corresponding to peak hardness, the motion of single dislocations gives way to the motion of paired (super-)dislocations.

Ni_3Al can also be disordered by irradiation and alternatively by Al ion implantation into Ni films, but that falls outside the scope of this survey.

3.3.2 *Recent Experiments on Mechanical Disordering of Ni_3Al*. The behaviour of powders of stoichiometric and slightly off-stoichiometric Ni_3Al during prolonged ball-milling in planetary mill with balls and vials of different materials has been systematically examined by Gialanella et al. [15], working in Cambridge and Barcelona; order and lattice parameters were determined at various stages of the process. Fig. 9 shows the most surprising result of this study. The lattice parameter increases steadily as the order parameter diminishes to zero, but then continues to increase to almost double the value it had when S had dropped to

518

zero. There undoubtedly is some contamination of the powder by the material from which the grinding balls and vials are made, but experiments with different grinding media and with reannealing after grinding demonstrated that most of the unexpected supplementary rise of the lattice parameter was genuine and not to be attributed to contamination.

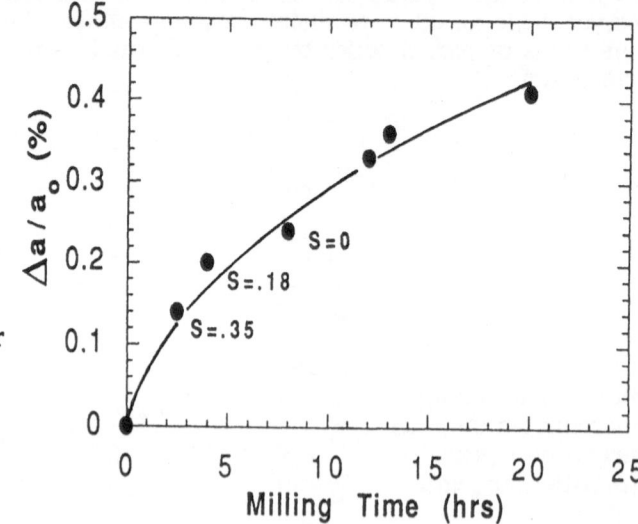

Figure 9. Atomized Ni_3Al+B powder, ball-milled in steel vial with hardened steel balls. Lattice parameter and order parameter, S, are shown. (After Gialanella et al. [15]).

The x-ray analysis was complemented by extensive differential scanning calorimetry (DSC) which showed that the stored energy due to ball-milling continued to increase as milling time increased (as demonstrated by measurements of energy released on heating in the DSC after various milling times, Fig. 10, [15] [25]). There is a small low-energy heat release at $\approx$ 400K (125°C), which does *not* correspond to reordering but may be linked with the annealing out of point defects. At present we think that the continued expansion of the lattice on milling beyond $S = 0$ is due to the progressive destruction of short-range order.

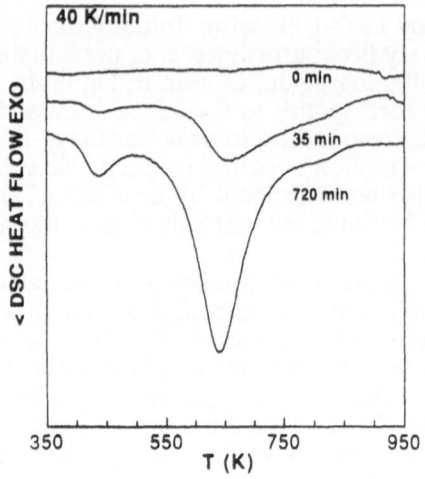

Figure 10. DSC curves for Ni_3Al+B powder milled for 3 different durations. DSC heating rate = 40K/min. (Courtesy of M.D. Baró; see also ref. 15).

3.2.3 Thermal Disordering of Ni$_3$Al.

Repeated attempts have been made over the years to disorder Ni$_3$Al by rapidly quenching it from the melt. Until recently, these were substantially unsuccessful, in spite of the fact that the (virtual) critical ordering temperature for stoichiometric Ni$_3$Al is only slightly above the melting temperature, and for off-stoichiometric (Ni$_{78}$Al$_{22}$), T_c is in fact just below T_m [16, 17]; from this one might deduce that the ordering energy is not so high as to exclude totally the possibility of disordering by quenching. However, a typical melt-quenching experiment using arc-hammer quenching (an effective technique) yielded a fully ordered alloy [18].

An even faster melt-quenching technique based on the use of a spark-erosion machine [19] did generate partially disordered powder when a liquid argon dielectric fluid was used, but with more normal dielectric fluids, the alloy's composition was changed, which spoiled the experiment.

Very recently, however, three groups were independently successful in disordering Ni$_3$Al completely, either by even more drastic quenching methods or else by reducing the ordering energy through ternary alloying. Yavari and Bochu [20] followed the second strategy: by alloying Ni$_3$Al with 13at.% Fe they reduced the ordering energy sufficiently [16] to produce entirely disordered f.c.c. alloy after melt-spinning to make ribbons only 10 µm thick; melt-spinning pure Ni$_3$Al in the same way created partial disorder only. —The other approach was followed by West et al. [21] and by Harris et al. [22]. West et al. transiently melted thin films of Ni$_3$Al, made by evaporation from separate sources of the two metals, by means of a pulse from a XeCl excimer laser with a pulse duration of $\approx$ 30 ns; resolidification was estimated to take place at a rate of $\approx$ 4 m/s, which sufficed to trap the resolidified film in the fully disordered state. Further x-ray studies established that not only the laser-treated films but the original evaporated films themselves (with $\approx$ 15 nm grain size) were disordered. (Fecht et al. [23] produced a disordered film of Ni$_3$Al by co-evaporation from Ni and Al sources, in 1989).

Harris et al. also evaporated Ni$_3$Al, this time from a prealloyed ingot, on to substrates at either ambient or liquid nitrogen temperatures; they concentrated on the evaporated films themselves. With either substrate temperature they found the films (average grain size $\approx$ 5 nm) to be entirely devoid of atomic (i.e., chemical) order. It may well be that the phenomenon is necessarily linked to the formation of nanometre-sized grains.

In the light of these recent findings, the way is now open for physical and even (by use of a nanoindenter) mechanical properties of disordered or partly ordered Ni$_3$Al to be examined, and that opportunity surely will lead to extensive research. It would be extremely interesting to discover whether the characteristics seen in Fig. 8 (and particularly the sharp peak at $S = 0.5$), which pertain to heavily deformed Ni$_3$Al, will also be found for the undeformed, partially ordered alloy. Even the elastic properties of disordered Ni$_3$Al would be interesting to know because the ratio μ/K has a bearing on the ductility of an alloy and so one would like to know the value of this ratio for disordered as well as ordered Ni$_3$Al [24].

3.3 REORDERING OF THERMALLY OR MECHANICALLY DISORDERED Ni$_3$Al

In Fig. 7, the reordering, at several temperatures, of mechanically disordered Ni$_3$Al filings has already been displayed. The Cambridge and Barcelona groups, working together, have examined the reordering of Ni$_3$Al powders ball-milled for different times, both during continuous heating in the DSC and by isothermal anneals. The full results and discussion will be found in a joint paper by the two groups [25]. Here (Fig. 11) we show only the results of isothermal anneals of powder ball-milled for three different durations: 2.5 and 4 h, which achieved only partial disordering, and 12 h, which disordered the powder completely.

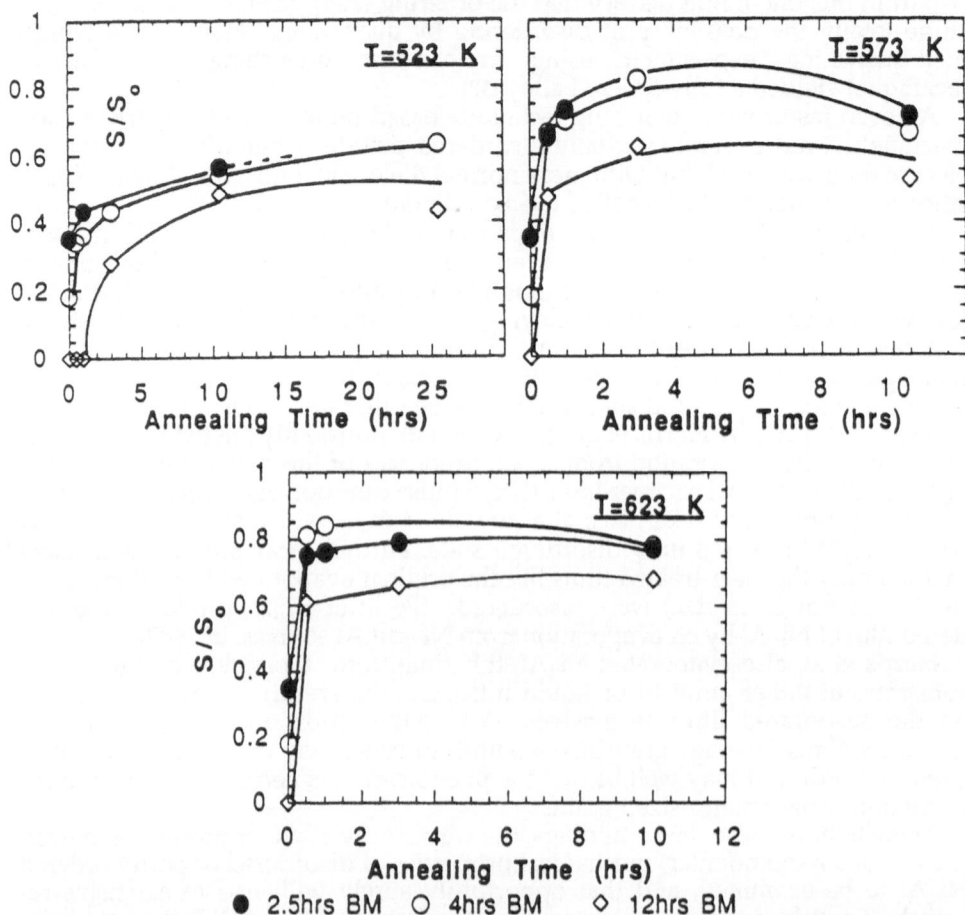

Figure 11. Variation of the relative LRO parameter (normalised to the value of the fully ordered, unmilled powder) for three different premilling times, at three different annealing temperatures, 523, 573 and 623K. (After Malagelada et al. [25])

Attention is specially drawn to the curve for fully disordered powder (ball-milled for 12 h) annealed at 523K (250°C). This material does not begin ordering at once, but requires to be annealed for more than an hour for the process to start. Alloys premilled for shorter times, which are not yet wholly disordered, reorder sluggishly but the process begins at once. At present we do not know the reason for this curious behaviour. — Comparison with Fig. 7, above, which refers to the annealing of filings and includes a 250°C isotherm, is appropriate by taking the curves in Fig. 11 top left, for the partially disordered powders 2.5 or 4 h milling). These specimens reorder at about the same rate as the filings at the same temperature, though the latter reach $S = 0.7$ while our specimens only reached $S \approx 0.6$; the difference is not significant. (Filings made from the fully ordered alloy will certainly not be wholly disordered, in spite of the fact that the isotherms in Fig. 7 are drawn as starting from $S = 0$). — It is striking that in both Fig. 7 and Fig. 11, even at 623K (350°C) full order is not achieved. This may well be because this temperature is not high enough to recrystallize the alloy, intense though the deformation has been; this is clearly because of the inhibition against recrystallization known to be conferred by LRO [9] [11]. It is to be presumed that complete reordering is only possible when the temperature is high enough to recrystallize the milled alloy. Malagelada et al. [25] found that annealing for 2 h at 773K (525°C) reordered the alloy completely. This is certainly high enough to recrystallize such an intensely deformed alloy. The linkage between complete reordering and recrystallization remains to be rigorously examined. — We do not believe that the *apparent* fall of S with prolonged annealing at 573K is significant; it is probably due to a change of extinction coefficient with progressive ordering.

Comparison of the reordering isotherms just presented with the behaviour of thermally disordered material is difficult because none of those who have made thermally disordered material have undertaken proper isothermal reordering measurements as yet. Quenched and partially ordered Ni_3Al, annealed for 2 h at 550°C, examined by Yavari and Bochu [20], proved to have $S = 0.9$, slightly higher than the approximately corresponding value in the bottom righthand graph of Fig. 11. Here, the difference between thermally and mechanically disordered material does not seem to be great. In the study by West et al. [21], reordering was found to have reached a steady state (but S was not measured) on the basis of resistivity measurements, after 15 min at 300°C plus 15 min at 350°C. This again is consistent with the bottom righthand graph of Fig. 11. Harris et al. [22] simulated heating in a DSC with intermediate measurements of LRO. After heating steps of a few minutes in successively higher temperature ranges, no reordering was found until 300°C had been reached, whereupon it became very rapid. A few (≈ 5) minutes at 350°C produced $S = 0.8$, once again approximately consistent with Fig. 11.

Insofar as valid comparisons can be made, reordering kinetics seem to match reasonably well as between mechanically and thermally disordered Ni_3Al; the uncertainty concerning the role of recrystallization in enabling complete reordering remains to be resolved,but it is interesting to note that Harris et al. found that reordering was always accompanied by growth of the nanocrystalline grains.

One other study remains to be cited. Because it has proved possible to disorder a Ni_3Al+Fe ternary alloy completely by melt-spinning [20], Yavari et al. [26] compared the reordering during steady heating (no isothermal runs were made) of thermally and mechanically disordered alloy samples of identical compostion. Clear evidence of two-stage reordering, as first observed in Cu_4Pt by Mitsui et al. [7], was obtained from DSC records, and the degree of reordering could also be estimated from magnetization maeasurements, since the ferromagnetic properties are intimately affected by order in this alloy. No evidence of different reordering behaviour was reported as between the two types of specimen, but it appeared that mechanical disordering gave a more homogeneous disordered state than did quenching. One curious observation [20] has yet to be interpreted; the meltspun, disordered Ni_3Al+Fe alloy reordered slightly over a period of some months at room temperature, which is very surprising.

4. Conclusions

Ordering characteristics of reversibly ordered alloys like those discussed in the earlier part of this paper are interesting in their own right and a number of issues remain to be resolved, but their main current interest is as background for the behaviour of mechanically and thermally disordered forms of phases of current practical interest. Hitherto, the only such phase examined has been Ni_3Al, and there is now much to do in examining in detail the properties of disordered Ni_3Al. Since the difference in properties between ordered and disordered forms of 'permanently' ordered phases should be the greater, the higher the ordering energy, it will be interesting to examine whether other such phases, with higher ordering energy than Ni_3Al, can be effectively disordered by mechanical means.

5. Acknowledgments

I am deeply indebted to my collaborators in a Cambridge/Barcelona/Grenoble research project, notably Dr. S. Gialanella, Prof. M.D. Baró and Dr. A.R. Yavari, for their readiness to provide material for this overview. We are also indebted to the Commission of the European Communities for financial support for this research. I am obliged to Dr. J.A. West and Prof. B. Fultz for providing information about their researches on thermal disordering of Ni_3Al prior to publication, and to Prof. C.J. Humphreys for making the facilities of the Materials Science Department in Cambridge available to me. Finally, I wish to express my appreciation to NATO and its co-sponsors for making possible the Workshop for which this overview was prepared.

6. References

[1]   Cahn, R.W. (1987) 'Ordering kinetics and diffusion in some $L1_2$ alloys', in: Cargill III, G.S., Spaepen, F. and Tu, K.-N. (eds.), Phase Transitions in Condensed Systems — Experiment and Theory, (Mat. Res. Soc. Symp. Proc. Vol. 57), Materials Research Society, Pittsburgh, pp. 385-404.

[2]   Morris, D.G., Brown, G.T., Piller, R.C. and Smallman, R.E. (1976),'Ordering and Domain Growth in Ni_3Fe', Acta Metall. 23 , 21-28.

[3]   Brooks, C.R., Spruiell, J.E. and Stansbury, E.E. (1984), 'Physical metallurgy of nickel-molybdenum alloys', International Metals Reviews 29, 210-248.

[4]   Collins, M.R. and Teh, H.C. (1973), 'Neutron scattering observations of critical slowing down of an Ising system', Phys. Rev. Lett. 30 , 781-784.

[5]   Liu, C.T. (1984), 'Physical metallurgy and mechanical properties of ductile ordered alloys $(Fe,Co,Ni)_3V$', International Metallurgical Reviews 29, 168-194.

[6]   Soltys, J., Abdank-Kozubski, R., Gialanella, S., Malageda, J., Suriñach, S., Baró, M.D. and Cahn, R.W. (1991), 'Ordering kinetics of a $(Co,Fe)_3V$ alloy', to be submitted to J. Mater. Sci.

[7]   Mitsui, K., Mishima, Y. and Suzuki, T. (1986),'Composition dependence of ordering kinetics in Cu_4Pt', Phil. Mag. A 54, 501; (1989), 'The role of excess vacancies in two-stage ordering in ordered alloys', ibid 59, 123.

[8] Yavari, A.R., Baró, M.D. and Suriñach, S. (1991), 'Ordering of L1$_2$, Cu$_4$Pt disordered by cold-rolling and by melt-spinning, Anales de Física (Spain), in press.

[9] Cahn, R.W., Takeyama, M., Horton, J.A. and Liu, C.T. (1991), 'Recovery and recrystallization of the deformed, orderable alloy (Co$_{78}$Fe$_{22}$)$_3$V', J. Mater. Res. 6, 57-70.

[10] Gialanella, S. and Cahn, R.W. (1991), 'Recovery and recrystallization of rolled (Co,Fe)$_3$V as a function of the initial state of the alloy', in: Proc. Mat. Res. Soc. Symposium on High-Temperature Ordered Intermetallic Alloys IV, MRS Symp. Proc. Vol. 213, in press.

[11] R.W. Cahn, 'Recovery, strain-age-hardening and recrystallization in deformed intermetallics', in: Whang, S.H., Liu, C.T., Pope, D.P. and Stiegler, J.O. (eds.) High Temperature Aluminides and Intermetallics, TMS, Warrendale, PA, pp. 245-270.

[12] Corey, C.L. and Potter, D.I. (1967), 'Recovery processes and ordering in Ni$_3$Al', J. Appl. Phys. 38, 3894-3900.

[13] Clark, J.P. and Mohanty, G.P. (1974), 'Anomalous recovery phenomena during annealing of cold-worked Ni$_3$Al', Scripta Metall. 8, 959-964.

[14] J.S.C. Jang and C.C. Koch, 'Amorphization and disordering of the Ni$_3$Al ordered intermetallic by mechanical milling', J. Mater. Res. 5, 498-510.

[15] Gialanella, S., Cahn, R.W., Malagelada, J., Suriñach, S. and Baró, M.D. (1991), 'Disordering of Ni$_3$Al by ball-milling', in: Yavari, A.R. (ed.), Proc. European Workshop on Ordering and Disordering, Grenoble, July 1991, to be publishedby Elsevier.

[16] Cahn, R.W., Siemers, P.A., Geiger, J.E. and Bardhan, P. (1987), 'The order-disorder transformation in Ni$_3$Al and Ni$_3$Al-Fe alloys', Acta Metall. 35, 2737-2751.

[17] Bremer, F.J., Beyss, M. and Wenzl, H. (1988), 'The order-disorder transition of the intermetallic phase Ni$_3$Al', phys. stat. sol. (a) 110, 77-82.

[18] Koch, C.C., Horton, J.A., Liu, C.T., Gavin, O.B. and Scarborough, J.O. (1983), in: Mehrabian, R. (ed.), Proc. 3rd Conf. on Rapid Solidification Processing, Gaithersburg, MD. National Bureau of Standards, Washington DC, p. 264.

[19] Cahn, R.W., Walter, J.L. and Marsh, D.W. (1988),'Characteristics of Ni$_3$Al+Fe powders produced by spark erosion quenching', Mater. Sci. Eng. 98, 33-37.

[20] Yavari, A.R. and Bochu, B. (1989), 'L1$_2$ ordering in Ni$_3$Al-Fe disordered by rapid quenching', Phil. Mag. A 59, 697-705.

[21] West, J.A., Manos, J.T. and Aziz, M.J. (1991), 'Formation of metastable disordered Ni$_3$Al by pulsed laser-induced rapid solidification', presented at MRS Symposium on High Temperature Intermetallic Alloys IV, Boston, November 1990; to be published in Mat. Res. Soc. Symp. Proc. Vol. 213.

[22] Harris, S.R., Pearson, D.H., Garland, C.M. and Fultz, B. (1991), 'Chemically disordered Ni$_3$Al sysnthesized by high vacuum evaporation', J. Mater. Res. 6, in press.

[23] Fecht, H.J., Hellstern, E., Fu, Z. and Johnson, W.L. (1989), Adv. Powder Metallurgy 1-3, 111.

524

[24] Cahn, R.W. (1990), 'The nature of grain boundaries in ordered alloys', in: Charles, J.A., and Smith, G.C. (eds.), Advances in Physical Metallurgy, The Institute of Metals, London, pp. 67-76.

[25] Malagelada, J., Suriñach, S., Baró, M.D., Gialanella, S. and Cahn, R.W. (1991), 'International Conference on Mechanical Alloying, Kyoto, Japan, May 1991. Proceedings to be published in Materials Science Forum, Aedermannsdorf, Switzerland: Trans Tech Publications.

[26] Yavari, A.R., Baró, M.D., Fillion, G., Suriñach, S., Gialanella, S., Clavaguera-Mora, M.T., Desré, P. and Cahn, R.W. (1991), 'L1$_2$ ordering in disordered Ni-Al-Fe alloys', presented at the MRS Symposium on High Temperature Ordered Intermetallic Alloys IV, Boston, November 1990. To be published in Mat. Res. Soc. Symp. Proc. Vol. 213.

Creep Behaviour and Creep Mechanisms in Ordered Intermetallics

G. SAUTHOFF
Max-Planck-Institut für Eisenforschung GmbH.
Postfach 140 444
D-4000 Düsseldorf
Germany

ABSTRACT. In the first section of this overview the creep behaviour of single-phase intermetallic alloys is discussed with respect to stress dependence, temperature dependence and effects of composition. The second section refers to two-phase intermetallic alloys, and both particulate and non-particulate alloys are regarded. Data are presented for single-phase and two-phase NiAl-base alloys, and the prospects for materials developments for application temperatures above those of the superalloys are briefly discussed.

1. Introduction

Intermetallics - i.e. intermetallic ordered alloys, intermetallic phases or compounds (IMCs) - are regarded as most promising for developing new structural materials for applications at high temperatures and various materials developments are under way (e.g. (Sauthoff, 1986; Sauthoff, 1989; Sauthoff, 1990a; Engell et al.1991)). The aim is on the one hand to develop new lightweight materials and for this the titanium aluminides are in the centre of interest. On the other hand, the gap between metallic materials and ceramics is to be closed, i.e. materials are to be developed with application temperatures above those of the superalloys and with toughness values above those of the ceramics. In view of the latter aim the nickel aluminides and less common phases with sufficiently high melting temperatures are studied, and indeed various intemetallic phases and alloys show sufficiently high strengths at high temperatures. This is exemplified by Fig. 1 which shows flow stress data for various intermetallic phases and alloys in comparison to the advanced superalloy MA 6000 and the engineering ceramic SiN.

However, the situation is less favourable with respect to long-term behaviour, i.e. with respect to creep strength as is illustrated by Fig. 2. The creep resistance which is the stress for producing a secondary creep rate of 10^{-7} s^{-1} is lower than that of the superalloy MA 6000 or only reaches it for the phases and alloys in Fig. 1 and similar phases and alloys. In particular the advanced Ni$_3$Al is even less creep resistant than the lightweight phase TiAl. Only the hard and brittle C14 Laves phase NbNiAl is slightly more creep restistant than MA 6000. It is noted that the curve for advanced Ni$_3$Al is continued by that of the perovskite-type Fe$_3$AlC with L'1$_2$ structure which can be regarded as a complex carbide or an L1$_2$ Fe$_3$Al phase that is stabilized by dissolved C.

C. T. Liu et al. (eds.), Ordered Intermetallics – Physical Metallurgy and Mechanical Behaviour, 525–539.
© 1992 *Kluwer Academic Publishers.*

526

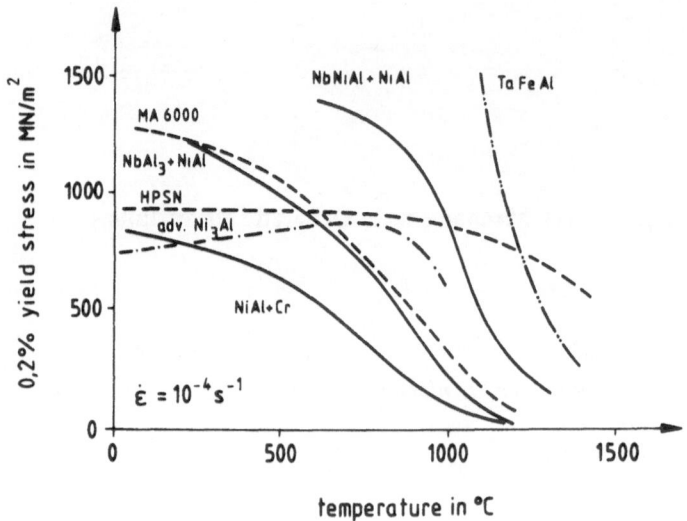

Figure 1. 0.2 % proof stress (strain rate 10^{-4} s^{-1}) as a function of temperature for the C14 Laves phase TaFeAl, the two-phase intermetallic alloys NbNiAl-NiAl and Al₃Nb-NiAl, the eutectic alloy NiAl-Cr (all in compression, see (Sauthoff, 1990a)) and the Ni₃Al-base Advanced Aluminides (Liu, 1988) in comparison to the advanced ODS superalloy MA 6000 (in tension) (Inco, 1982) and the hot-pressed SiN-base ceramic HPSN (in bending) (Porz and Grathwohl, 1984).

The reason for the comparatively low creep resistance of the intermetallics is not a particular weakness of the intermetallics at high temperatures, but the exceptionally high strength of the superalloys which results from the specific interaction of the dislocations with the Ni₃Al precipitate interfaces. The strength at high temperatures, i.e. above 1/3 of the melting temperature, is controlled by creep processes as was discussed in (Sauthoff, 1990b). In the following the characteristics of creep are overviewed and the problems in understanding are noted. The presented exemplifying data refer to B2 phases and related intermetallic alloys.

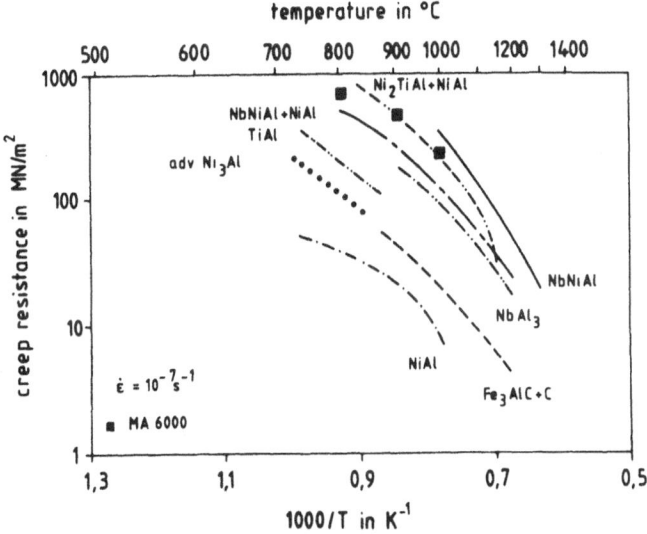

Figure 2. Creep resistance (10^{-7} s^{-1} secondary creep rate) as a function of temperature for the C14 Laves phase NbNiAl (Sauthoff, 1991a), the $D0_{22}$ phase Al_3Nb (Reip, 1991; Sauthoff, 1991a), the $L1_0$ phase TiAl (Martin et al.1983), the B2 phase NiAl (Rudy, 1986; Rudy and Sauthoff, 1985; Jung et al.1987), advanced Ni_3Al (Schneibel et al.1986; Nicholls and Rawlings, 1977) and the two-phase intermetallic alloys Ni2TiAl-NiAl (Polvani et al.1976), NbNiAl-NiAl (Sauthoff, 1991a) and Fe_3AlC-C with strengthening graphite precipitates (Jung and Sauthoff, 1989a) (all in compression with the exception of (Schneibel et al.1986) which is in tension) in comparison to MA 6000 (in tension) (Inco, 1982).

2. Single-Phase B2 Alloys

2.1. STRESS-STRAIN RATE DEPENDENCE OF (NI,FE)AL

The stoichiometric B2 phase NiAl contains 50 at.% Al and is completely ordered up to the melting point (see (Sauthoff, 1986) for references). In this phase Al can be substituted partially by Ni, i.e., excess Ni atoms occupy Al sites, whereas Ni can be substituted completely by Fe and Co without affecting the B2 order, but not by Al, i.e., with excess Al atoms Ni vacancies are created. It is noted that the crystal structure of the resulting ternary phases is designated as $L2_1$, too.

Figure 3 shows representative creep data for the binaries FeAl and NiAl and for two ternaries which result from alloying the binaries with each other. The data follow straight lines in the double-logarithmic plot - at least above $10^{-8}\,s^{-1}$ strain rate - which indicates the familiar power-law behaviour. This stress-strain rate dependence is well understood for conventional disordered alloys and is described by the Dorn equation for dislocation creep:

$$\dot{\varepsilon} = A\,(DGb/kT)\,(\sigma/G)^n \qquad\qquad\qquad\qquad\qquad \text{Eq.} \quad 1$$

where $\dot{\varepsilon}$ = secondary strain rate, A = dimensionless factor which considers the microstructural effects, dislocation reactions etc., D = effective diffusion coefficient, G = shear modulus, b = Burgers vector, k = Boltzmann's constant, T = temperature, σ = applied stress, and the exponent n is between 3 and 5 depending on the rate controlling creep mechanism (for references see e.g. (Rudy and Sauthoff, 1985)).

Dislocation creep of conventional disordered alloys is produced by gliding and climbing dislocations. If climb is the slower step as in pure metals, the creep rate is controlled by dislocation climb which gives rise to a well-defined subgrain structure, and the stress exponent is 4 or 5. This is characteristic for the so-called class-II alloys (see (Sauthoff, 1991b)). Otherwise viscous dislocation glide is rate controlling which leads to dislocation tangles without subgrain formation and to a stress exponent 3 (class-I alloys). The data in Fig. 3 indicate n = 3 for FeAl and the Fe-rich ternary and indeed only dislocation tangles were observed in these alloys (Rudy, 1986; Rudy and Sauthoff, 1986). For NiAl and the Ni-rich ternary the stress exponent was found between 4 and 5 and indeed subgrains were formed during creep. The different behaviour of the Fe-rich and Ni-rich phases is not caused by different dislocation types since in both cases only <100> dislocations were observed (Rudy, 1986; Rudy and Sauthoff, 1986). Obviously the driving force and the atom mobility which is necessary for subgrain formation is sufficient only in the Ni-rich phases (Jung et al.1987).For a quantitative discussion of these effects further studies are necessary.

At strain rates below $10^{-8}\,s^{-1}$ the data for the Ni-rich ternary which showed the highest creep resistance clearly deviate from a straight line in Fig. 3 which indicates a contribution of diffusion creep to the observed total creep rate. The constitutive equation for diffusion creep is

$$\dot{\varepsilon}_{diff} = A_{diff}(\Omega D/kTd^2)\sigma \qquad\qquad\qquad\qquad\qquad \text{Eq.} \quad 2$$

where A_{diff} is a dimensionless factor which considers the diffusion geometry (usually A_{diff} = 14), Ω is the atomic volume, D is the effective diffusion coefficient which considers both the diffusion through the grain (Nabarro-Herring creep) and along the grain boundaries (Coble creep), and d is the effective diffusion length which is usually approximated by the grain size (see (Jung et al.1987)). Dislocation creep and diffusion creep are independent creep processes which act in parallel in the grains, and the total creep rate is given by the sum of the partial rates. Because of the stronger stress dependence of dislocation creep (Eq.1) the contribution of diffusion creep becomes more prominent with decreasing stress.

In view of these and other examples (Schneibel et al.1986; Kampe et al.1991; Hayes, 1991; Hayashi et al.1991; Miura et al.1991; Rowe et al.1991; Takahashi and Oikawa, 1991; Albert and Thompson, 1991; Argon et al.1991; Polvani et al.1976; Nicholls and Rawlings, 1977) it is concluded that the creep of intermetallic phases is controlled by the same creep mechanisms and is described by the same constitutive equations as the creep of the familiar disordered alloys though

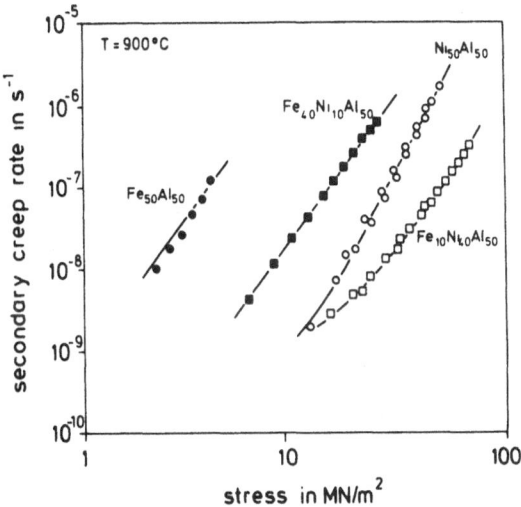

Figure 3. Secondary creep rate (in compression with stepwise loading) as a function of stress for some binary and ternary B2 phases at 900 °C (Rudy, 1986; Rudy and Sauthoff, 1986).

the crystal structures of the intermetallics are much less symmetric than the simple structures of the disordered alloys. The lower crystal symmetry leads to particular dislocation types and higher dislocation energies, including special superlattice dislocations and to a reduced number of slip systems. However, dislocation creep is controlled by non-conservative dislocation motions, i.e., dislocation motions are coupled with local diffusion fluxes, and the latter are rate controlling. Then the particular dislocation mechanisms are not of primary importance for the total macroscopic creep rate which explains the similar behaviour of metallic and intermetallic alloys

2.2. EFFECT OF DIFFUSION COEFFICIENT

Figure 4 shows a marked dependence of the creep resistance of the ternary B2 phases (M',M'')Al with M' and M'' = Fe, Co, Ni and stoichiometric Al content. In the case of (Ni,Fe)Al it could be shown that this composition dependence results from the composition dependence of the diffusion coefficient which could be estimated by using data in (Cheng and Dayananda, 1979; Moyer and Dayananda, 1976) (see (Rudy, 1986; Rudy and Sauthoff, 1986)). The maximum in the creep resistance is directly related to the minimum in the diffusion coefficient. It has to be noted that the analysis of creep data with respect to diffusion coefficients poses problems because - apart from the scarcity of data - the diffusion coefficients in Eqs. 1 and 2 are effective ones depending on the particular creep process which determines the coupling of partial diffusion fluxes, i.e., the needed effective diffusion coefficients are not those which are measured in diffusion experiments. For binary solid solutions expressions for the effective diffusion coefficients are available for

530

various creep mechanisms (Fuentes-Samaniego and Nix, 1981; Chin et al.1977), whereas for multinary phases an expression for the effective diffusion coefficient is known only for the case of diffusion creep (Herring, 1950). Here more theoretical and experimental work is necessary.

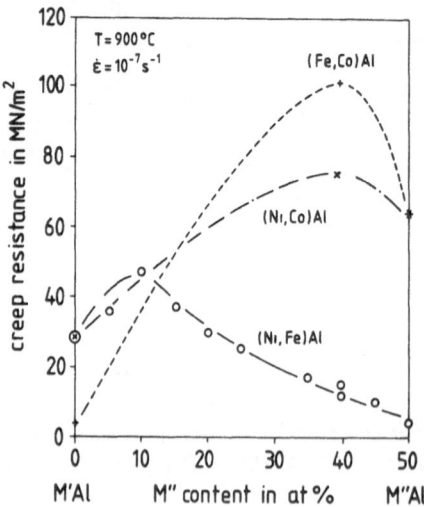

Figure 4. Composition dependence of the creep resistance of the ternary stoichiometric B2 phases (M',M'')Al with M', M'' = Fe, Co, Ni at 900 °C (in compression with 10^{-7} s^{-1} strain rate) (Rudy, 1986; Jung et al.1987; Sauthoff, 1989).

Figure 5 shows the correlation between the creep resistance of various B2 phases and the respective diffusion coefficient as found in the literature or estimated on the basis of literature data (Rudy, 1986; Rudy and Sauthoff, 1986; Sauthoff, 1991b). It can be seen that the correlation is surprisingly good and corresponds to Eq. 1 in spite of the problems with the appropriate determination of the effective diffusion coefficients. The question now is why the diffusion coefficient depends in the shown way on the composition of the B2 phases. As other properties the diffusion coefficient depends on the crystal properties, i.e. on the character and strength of atomic bonding and indeed the activation energy of diffusion increases with increasing total heat of phase formation as is illustrated by Fig. 6. However, for understanding the effects of composition variations on the respective diffusion coefficients the correlation with the character and strength of bonding must be studied in more detail. Here again much more theoretical and experimental work is necessary.

Figure 7 shows the effect of variations of Al content, i.e. deviations from stoichiometry on the creep resistance of such B2 aluminides. As discussed in (Sauthoff, 1990b), deviations from stoichiometry introduce constitutional disorder, i.e. point defects which increase the diffusion coefficient and thereby the creep rate. This is well understood in the case of the binary NiAl (Hancock and McDonnell, 1971; Vandervoort et al.1966). However, it has to be noted that at lower temperatures these point defects become immobile and act as obstacles to dislocation movements, i.e., they increase the low-temperature strength. At intermediate temperatures the two effects balance each other and deviations from stoichiometry do not affect the creep resistance as was found for NiAl (Vandervoort et al.1966). In Fig. 7 the Ni-rich phases show a strong stoichiometry effect whereas the effect is significantly smaller for the Fe-rich phases, i.e., the test temperature of 900 °C leads to high-temperature behaviour of the Ni-rich phases and to intermediate-temperature behaviour of the Fe-rich phases in spite of their lower melting temperatures. The reasons for this behaviour are still to be clarified.

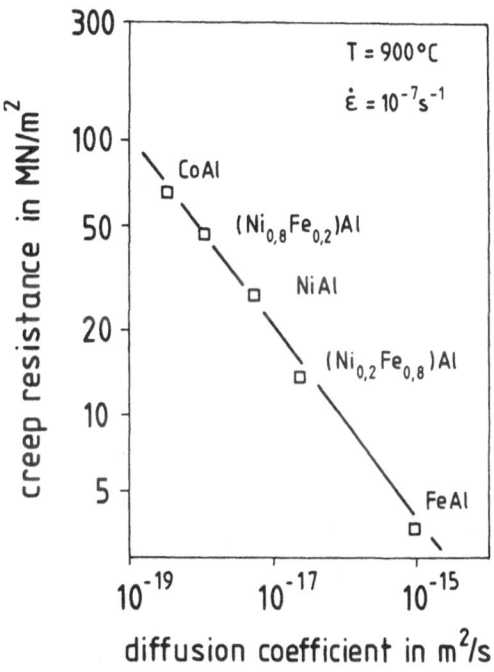

Figure 5. Creep resistance of binary and ternary stoichiometric B2 aluminides at 900 °C (in compression with 10^{-7} s^{-1} strain rate) as a function of diffusion coefficient as found in the literature or estimated from available data (Rudy, 1986; Jung et al.1987; Sauthoff, 1991b; Shankar and Seigle, 1978; Akuezue and Whittle, 1983; Cheng and Dayananda, 1979; Moyer and Dayananda, 1976).

In view of the discussed composition dependence of the creep resistance it is concluded that the effective diffusion coefficient is of primary importance for controlling the creep resistance. This of course does not mean that the other parameters in Eq. 1 can be neglected. This is demonstrated by the temperature dependence of creep of B2 (Ni,Fe)Al which has been discussed in (Sauthoff, 1991b). In view of Eqs. 1 and 2 the apparent activation energy of creep is expected to correspond to that of diffusion since the other parameters depend less sensibly on temperature and indeed this has been confirmed repeatedly in the case of conventional disordered alloys. However, in the case of B2 (Ni,Fe)Al the apparent activation energy of creep corresponds to that of diffusion only at not too high temperatures up to 900 °C whereas at higher temperatures the apparent activation energy of creep is much higher. This means that here the microstructure-dependent parameter A in Eq. 1 shows a strong temperature dependence, too, since the shear modulus G still depends only weakly on temperature (Harmouche and Wolfenden, 1991).

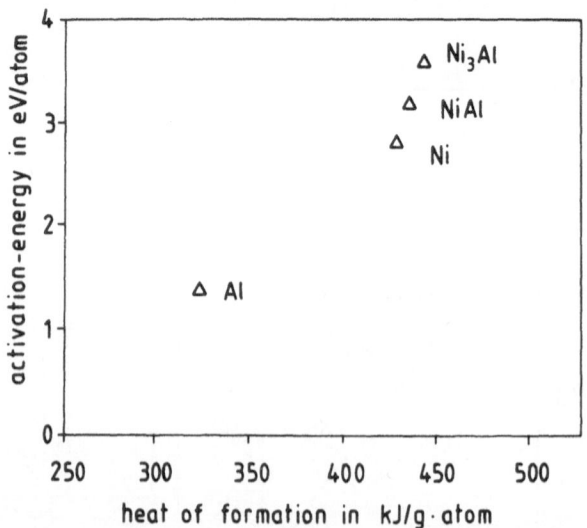

Figure 6. Activation energy of diffusion (Schneibel et al.1986; Frost and Ashby, 1982) as a function of total heat of phase formation for Ni, Al, NiAl and Ni₃Al (Engell et al.1991; Von Keitz and Sauthoff, 1991).

In summary it is concluded that the creep behaviour of intermetallic phases can be described as that of the conventional disordered alloys by the familiar phenomenologic constitutive equations, and indeed deformation maps have been calculated as a function of available data (Jung et al.1987). However, the physical mechanisms which control the creep behaviour are only partially understood and much more work is necessary.

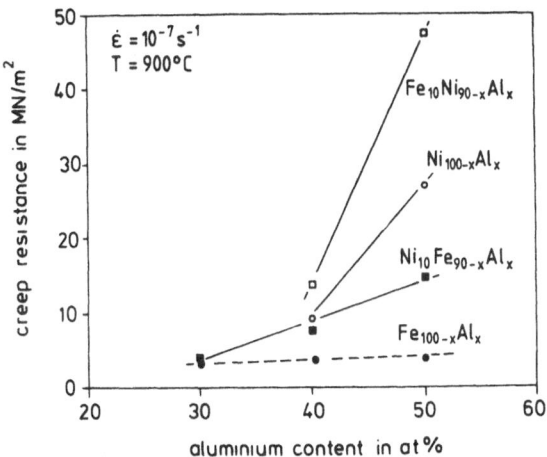

Figure 7. Creep resistance of binary and ternary B2 aluminides at 900 °C (in compression with 10^{-7} s^{-1} strain rate) as a function of Al content (Rudy, 1986; Rudy and Sauthoff, 1986).

3. Multiphase Alloys

3.1. EFFECT OF PARTICLES

Precipitated particles are usually used in conventional alloys for increasing the creep resistance. The strengthening effect of precipitated particles was studied in NiAl-Fe alloys with B2 NiAl matrix and α-Fe particles (Jung and Sauthoff, 1989b). It was found that the effect of the α-Fe particles was quite analogous to that of B2 NiAl particles in Fe-NiAl alloys with α-Fe matrix which were studied in (Jung and Sauthoff, 1987; Jung, 1986). In both cases the particles act as dislocation obstacles because of the dislocation-particle interaction, and indeed adhering dislocations were observed at particles before detachment in the NiAl-Fe case. Such obstacles are surmounted by climb which gives rise to a threshold stress σ_{th} and increases the creep resistance:

$$\dot{\varepsilon} = A\,(DGb/kT)\,[(\sigma - \sigma_{th})/G]^n \qquad\qquad \text{Eq. 3}$$

This threshold stress is proportional to the Orowan stress as was shown theoretically for various climb processes (Arzt and Rösler, 1988) and thus is proportional to the reciprocal particle distance in agreement with the experiments (Jung and Sauthoff, 1987; Jung and Sauthoff, 1989b). The observed secondary dislocation creep can be described by Eq. 3 and deformation maps can be calculated on the basis of the experimental data (Jung and Sauthoff, 1989b).

It is noted that a double-logarithmic plot of stress-strain rate data which follow Eq. 3 produces curves which indicate apparent stress exponents above 5 according to Eq. 1. Indeed such high stress exponents have been reported for alloys with strengthening precipitates or dispersoid, and thus a stress exponent of 6 or higher may be taken for an indication of the presence of a second strengthening phase.

3.2. NON-PARTICULATE ALLOYS

In particulate alloys one phase - the matrix - is distributed continuously whereas the second phase is distributed discontinuously. In non-particulate alloys all phases are distributed continuously as is the case in lamellar or fibrous composites. The effect of phase distribution on the creep behaviour was studied in detail in (Klöwer, 1989; Klöwer and Sauthoff, 1991). For this a Ni-40at.%Fe-18at.%Al alloy was chosen which was produced by directional solidification to produce a lamellar microstructure with the phases NiAl (B2) and γ-Fe-Ni (fcc) with equal volume fractions. It was found that the creep resistance of such a lamellar alloy is related to the creep resistances of the constituent phases according to a rule of mixtures as long as the lamellae spacing is larger than a critical spacing which is of the order of the free dislocation path. If the lamellae spacings are smaller than this critical value then the lamellae interfaces give rise to an additional strengthening effect. This strengthening effect can be described as in Eq. 3 by a threshold stress which again is proportional to the reciprocal lamellae spacing.

Similar effects have been observed in other intermetallic NiAl-base alloys with less regular distributions of hexagonal C14 Laves phases and have been discussed in (Sauthoff, 1991c). In those alloys with coarse phase distributions the observed secondary creep rates follow a rule of mixtures in first approximation and additional strengthening effects are only observed for alloys with fine phase distributions. From this it is concluded that particulate and non-particulate intermetallic alloys creep in similar ways and can be described by the same constitutive eqations as conventional multiphase alloys.

However, the above discussion has referred to secondary creep only. The situation is less clear for the transient primary creep stage which precedes secondary creep. As discussed in (Sauthoff, 1991c) the primary creep strain is reduced significantly by the presence of second phases and it decreases with increasing stress and with decreasing interface spacing at least in some NiAl-base alloys (Reip, 1991; Klöwer, 1989). Such a behaviour is known for conventional alloys, too, but there are other alloys which show an opposite behaviour, i.e. an increasing primary strain with increasing stress (see (Sauthoff, 1991c)). Such effects are not yet understood even for disordered alloys. It is further noted that not only the normal primary creep with decelerating creep rate is observed, but also inverse primary creep with accelerating creep rate e.g. in the case of the ternary Laves phases (Machon and Sauthoff, 1991) as has already been reported for Ni$_3$Al (Hazzledine and Schneibel, 1989). Inverse creep results from an insufficient number of mobile dislocations, and a classic example for such a deformation behaviour is silicon where the deformation behaviour has been analysed in detail (Alexander and Haasen, 1968; Alexander, 1986).

4. Conclusions and Prospects

The creep behaviour of intermetallic phases and alloys is quite similar to that of conventional disordered alloys and can be described phenomenologically by the known constitutive equations. The particularities of the various intermetallic phases find expression in the respective materials parameters which are the effective diffusion coefficient D, the shear modulus G and the microstructure parameter A in the case of dislocation creep. The physical understanding of the rate controlling processes is, however, far from complete and much more experimental and theoretical work is necessary. In particular the fundamental data with respect to diffusion and elasticity are lacking in most cases.

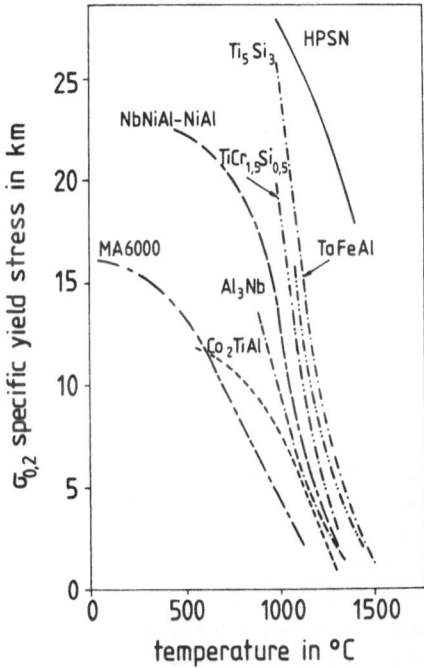

Figure 8. Specific yield strength (0.2 % proof stress per unit weight density) at $10^{-4} s^{-1}$ strain rate in compression) as a function of temperature for the $D0_{22}$ phase Al_3Nb (Reip, 1991; Sauthoff, 1990b), the Heusler-type phase Co_2TiAl (Machon and Sauthoff, 1991; Sauthoff, 1990b), the Laves phases $TiCr_{1.5}Si_{0.5}$ and $TaFeAl$ (Machon and Sauthoff, 1991; Sauthoff, 1990b), the two-phase alloy NbNiAl-NiAl with 15 vol.% NiAl in the Laves phase NbNiAl (Machon and Sauthoff, 1991; Sauthoff, 1990b), and the hexagonal $D8_8$ phase Ti_5Si_3 (Frommeyer et al.1990) in comparison to the superalloy MA 6000 (in tension) (Inco, 1982) and the hot-pressed silicon nitride HPSN (upper limit of flexural strength) (Porz and Grathwohl, 1984).

536

As to materials developments which aim at temperatures above the service temperatures of superalloys, there are phases which are promising for these high temperatures and there are possibilities for increasing the creep resistance significantly as has been discussed here and in (Sauthoff, 1991b). Figure 8 shows specific strength data as a function of temperature for various phases in comparison to the superalloy MA 6000 and the ceramic SiN. The specific strength which is the density-compensated strength (strength divided by weight density) and which is an important design parameter for rotating components is favourable for the aluminides and silicides because of the respective low densities. It can be seen in Fig. 8 that there are various phases with cubic and hexagonal crystal structures in the stress-temperature range between the superalloy and the ceramic. The best possibility for improving both the creep strength and the low-temperature fracture toughness is to combine different phases to form multiphase alloys by proper alloying since interfaces can increase creep strength and hinder growing microcracks (see (Sauthoff, 1991b)).

5. Acknowledgements

The financial support by Deutsche Forschungsgemeinschaft und Bundesminister für Forschung und Technolgy is gratefully acknowledged.

6. References

Akuezue, H.C. and Whittle, D.P. (1983) "Interdiffusion in Fe-Al System: Aluminizing", Met. Science 17, 27-33.

Albert, D. and Thompson, A. (1991) "The Effect of Microstructure on the Creep Properties of Ti-24Al-11Nb", in L. Johnson, D.P. Pope and J.O. Stiegler (eds.), High Temperature Ordered Intermetallic Alloys IV, MRS, Pittsburgh, in print.

Alexander, H. (1986) "Dislocations in Covalent Crystals", in F.R.N. Nabarro (ed.), Dislocations in Solids, North Holland, Amsterdam, pp.113-234.

Alexander, H. and Haasen, P. (1968) "Dislocations and Plastic Flow in the Diamond Structure", in F. Seitz, D. Turnbull and H. Ehrenreich (eds.), Solid State Physics, Academic Press, New York, pp.27 ff.

Argon, A.S., Haubensak, F.G. and Pollock, T.M. (1991) "Sources of Creep Resistance in CMSX-3 Single Crystals", in L. Johnson, D.P. Pope and J.O. Stiegler (eds.), High Temperature Ordered Intermetallic Alloys IV, MRS, Pittsburgh, in print.

Arzt, E. and Rösler, J. (1988) "The Kinetics of Dislocation Climb over Hard Particles - II. Effects of an Attractive Particle-Dislocation Interaction", Acta Metall. 36, 1053-1060.

Cheng, G.H. and Dayananda, M.A. (1979) "Multiphase Diffusion in Fe-Ni-Al System at 1000 °C: II. Interdiffusion Coefficients for β and γ Alloys", Metall. Trans. 10A, 1415-1419.

Chin, B.A., Pound, G.M. and Nix, W.D. (1977) "The Role of Diffusion in Determining the Controlling Creep Mechanisms in Al-Zn Solid-Solutions: Part I", Metall. Trans. 8A, 1517-1522.

Engell, H.-J., Von Keitz, A. and Sauthoff, G. (1991) "Intermetallics - Fundamentals and Prospects", in W. Bunk (ed.), Advanced Structural and Functional Materials, Springer Verlag, Berlin, pp.91-132.

Frommeyer, G., Rosenkranz, R. and Lüdecke, C. (1990) "Microstructure and Properties of the Refractory Intermetallic Ti5Si3 Compound and the Unidirectionally Solidified Eutectic Ti-Ti$_5$Si$_3$ Alloy", Z. Metallkde. 81, 307-313.

Frost, H.J. and Ashby, M.F. (1982) "Deformation Mechanism Maps", Pergamon Press, Oxford.

Fuentes-Samaniego, R. and Nix, W.D. (1981) "Appropriate Diffusion Coefficients for Describing Creep Processes in Solid Solution Alloys", Scr. Metall. 15, 15-20.

Hancock, G.F. and McDonnell, B.R. (1971) "Diffusion in the Intermetallic Compound NiAl", phys. stat. sol. a 4, 143-150.

Harmouche, M.R. and Wolfenden, A. (1991) "Temperature and Composition Dependence of Young's Modulus in Polycrystalline B2 NiAl", Journal of Testing and Evaluation 15, 101-104.

Hayashi, T., Shinoda, T., Mishima, Y. and Suzuki, T. (1991) "Effect of Offstoichiometry on the Creep Behavior of Binary and Ternary Ni$_3$Al", in L. Johnson, D.P. Pope and J.O. Stiegler (eds.), High Temperature Ordered Intermetallic Alloys IV, MRS, Pittsburgh, in print.

Hayes, R.W. (1991) "On the Creep Behavior of the Ti$_3$Al Titanium Aluminide Ti-25Al- 10Nb-3V-1Mo", Acta Metall. Mater. 39, 569-578.

Hazzledine, P.M. and Schneibel, J.H. (1989) "Inverse Creep in Ni$_3$Al", Scr. Metall. 23, 1887-1892.

Herring, C. (1950) "Diffusional Viscosity of a Polycrystalline Solid", J. Appl. Phys. 21, 437-443.

Inco (1982) "Inconel Alloy MA 6000", Inco Pamphlet.

Jung, I. (1986) "Untersuchung des Verformungsverhaltens ferritischer zweiphasiger Fe-Ni-Al-Legierungen mit großen Anteilen der intermetallischen (Fe,Ni)Al-Phase bei hohen Temperaturen", Dr.rer.nat. thesis, RWTH Aachen, pp. 1-120.

Jung, I., Rudy, M. and Sauthoff, G. (1987) "Creep in Ternary B2 Aluminides and Other Intermetallic Phases", in N.S. Stoloff, C.C. Koch, C.T. Liu and O. Izumi (eds.), High-Temperature Ordered Intermetallic Alloys II, Materials Research Society, Pittsburgh, pp.263-274.

Jung, I. and Sauthoff, G. (1987) "The Effect of Ordered Precipitates on Creep in Ferritic Fe-Ni-Al Alloys", in B. Wilshire and R.W. Evans (eds.), Creep and Fracture of Engineering Materials and Structures, The Institute of Metals, London, pp.257-270.

Jung, I. and Sauthoff, G. (1989a) "High-Temperature Deformation Behaviour of the Perovskite-type Phases Fe$_3$AlC and Ni$_3$AlC", Z. Metallkde. 80, 490-496.

Jung, I. and Sauthoff, G. (1989b) "Creep Behaviour of the Intermetallic B2 Phase (Ni,Fe)Al with Strengthening Soft Precipitates", Z. Metallkde. 80, 484-489.

Kampe, S.L., Bryant, J.D. and Christodoulou, L. (1991) "Creep Deformation of TiB$_2$-Reinforced Near-γ Titanium Aluminides", Metall. Trans. 22A, 447.

Klöwer, J. (1989) "Untersuchung des Kriechverhaltens gerichtet erstarrter, lamellarer Eisen-Nickel-Aluminium-Legierungen", Dr.Ing. thesis, RWTH Achen, pp. 1-112.

Klöwer, J. and Sauthoff, G. (1991) "Creep Behaviour of Directionally Solidified Lamellar Nickel-Iron-Aluminium Alloys - Part I: Effect of Lamellae Spacing and Orientation", Z. Metallkde. 82, July issue, in print.

Liu, C.T. (1988) unpubl.

Machon, L. and Sauthoff, G. (1991) unpubl.

538

Martin, P.L., Mendiratta, M.G. and Lipsitt, H.A. (1983) "Creep Deformation of TiAl and TiAl+W Alloys", Metall. Trans. 14A, 2171-2174.

Miura, S., Mishima, Y., Hayashi, T. and Suzuki, T. (1991) "The Compression Creep Behavior of Ni_3Al-X Single Crystals", in L. Johnson, D.P. Pope and J.O. Stiegler (eds.), High Temperature Ordered Intermetallic Alloys IV, MRS, Pittsburgh, in print.

Moyer, T.D. and Dayananda, M.A. (1976) "Diffusion in β_2 Fe-Ni-Al Alloys", Metall. Trans. 7A, 1035.

Nicholls, J.R. and Rawlings, R.D. (1977) "Steady-State Creep of an Alloy Based on the Intermetallic Compound Ni3Al (γ')", J. Mater. Sci. 12, 2456-2464.

Polvani, R.S., Tzeng, W.-S. and Strutt, P.R. (1976) "High Temperature Creep in a Semi-Coherent $NiAl$-Ni_2AlTi Alloy", Metall. Trans. 7A, 33-40.

Porz, F. and Grathwohl, G. (1984) "Nichtoxidische keramische Werkstoffe für hohe Beanspruchungen", KfK-Nachr. 16, 94-108.

Reip, C.-P. (1991) "Untersuchung des Verformungsverhaltens der DO_{22}-geordneten intermetallischen Phase Al_3Nb", Dr.rer.nat thesis, RWTH Aachen, pp. 1-118.

Rowe, R.G., Konitzer, D.G., Woodfield, A.P. and Chesnutt, J.A. (1991) "Tensile and Creep Behavior of Ordered Orthorhombic Ti_2AlNb-based Alloys", in L. Johnson, D.P. Pope and J.O. Stiegler (eds.), High Temperature Ordered Intermetallic Alloys IV, MRS, Pittsburgh, in print.

Rudy, M. (1986) "Untersuchung des Verformungsverhaltens der intermetallischen Phase (Fe,Ni)Al bei hohen Temperaturen", Dr.Ing. thesis, RWTH Aachen, pp. 1-122.

Rudy, M. and Sauthoff, G. (1985) "Creep Behaviour of the Ordered Intermetallic (Fe,Ni)Al Phase", in C.C. Koch, C.T. Liu and N.S. Stoloff (eds.), High-Temperature Ordered Intermetallic Alloys, MRS, Pittsburgh, pp.327-333.

Rudy, M. and Sauthoff, G. (1986) "Dislocation Creep in the Ordered Intermetallic (Fe,Ni)Al Phase", Mater. Sci. Engg. 81, 525-530.

Sauthoff, G. (1986) "Intermetallic Phases as High-Temperature Material", Z. Metallkde. 77, 654-666.

Sauthoff, G. (1989) "Intermetallic Phases - Materials Developments and Prospects", Z. Metallkde. 80, 337-344.

Sauthoff, G. (1990a) "Intermetallic Alloys - Overview on New Materials Developments for Structural Applications in West Germany", Z. Metallkde. 81, 855-861.

Sauthoff, G. (1990b) "Mechanical Properties of Intermetallics at High Temperatures", in S.H. Whang, C.T. Liu, D.P. Pope and J.D. Stiegler (eds.), High-Temperature Aluminides and Intermetallics, TMS, Warrendale, pp.329-352.

Sauthoff, G. (1991a) "Hochtemperatureigenschaften von NiAl-Basislegierungen", in F.J. Bremer (ed.), Intermetallische Phasen als Strukturwerkstoffe für hohe Temperaturen, Forschungszentrum Jülich GmbH, Jülich, pp.53-63.

Sauthoff, G. (1991b) "High Temperature Deformation and Creep Behaviour of BCC Based Intermetallics", in O. Izumi (ed.), Proceedings of the International Symposium on Intermetallic Compounds - Structure and Mechanical Properties - (JIMIS-6), The Japan Institute of Metals, Sendai, pp.371-378.

Sauthoff, G. (1991c) "Creep Behaviour of Intermetallics", in Microstructure and Mechanical Properties of Materials, Deutsche Gesellschaft für Materialkunde, Oberursel, in print.

Schneibel, J.H., Petersen, G.F. and Liu, C.T. (1986) "Creep Behaviour of a Polycrystalline Nickel Aluminide: Ni-23.5at.%Al-0.5at.%Hf-0.2at.%B", J. Mater. Res. 1, 68-72.

Shankar, S. and Seigle, L.L. (1978) "Interdiffusion and Intrinsic Diffusion in the NiAl (δ) Phase in the Al-Ni System", Metall. Trans. 9A, 1467-1476.

Takahashi, T. and Oikawa, H. (1991) "Structure Dependence of Tensile Creep Behavior in Thermomechanically-Treated Ti-50mol%Al Intermetallics", in L. Johnson, D.P. Pope and J.O. Stiegler (eds.), High Temperature Ordered Intermetallic Alloys IV, MRS, Pittsburgh, in print.

Vandervoort, R.R., Mukherjee, A.K. and Dorn, J.E. (1966) "Elevated-Temperature Deformation Mechanisms in β' NiAl", Trans. ASM 59, 930-944.

Von Keitz, A. and Sauthoff, G. (1991) unpubl.

CREEP DEFORMATION OF B2 ALUMINIDES

M. V. Nathal
NASA Lewis Research Center
Cleveland, OH USA 44135

ABSTRACT. The creep resistance and elevated temperature deformation mechanisms in CoAl, FeAl, and NiAl are reviewed. The stress and temperature dependencies of the steady state creep rate, the primary creep behavior, the dislocation substructure, and the response during transient tests are used as the main indicators of the deformation processes. In single phase intermetallics, the influence of grain size, stoichiometry, and solid solution hardening have been examined. In addition, the effect of adding dispersoids, precipitates, and other types of reinforcements to improve creep strength are compared.

1. INTRODUCTION

The B2 structure aluminides CoAl, FeAl, and NiAl have several attributes which provide a basis for interest as high temperature structural materials. They have low densities, relatively high melting points, and the potential for excellent oxidation resistance. In addition, their simple cubic crystal structures and large solubility ranges allow alloy design flexibility for improved properties over those of the binary compounds. Creep resistance is one of the important properties required for extended use at high temperatures. Of course, in most conditions a balance of high temperature creep strength, low temperature fracture toughness, and environmental resistance is required. This paper will not address the difficult challenge of achieving the required balance in properties; however, it should be kept in mind that in many cases improvements in one area result in a degradation in other properties.

The review is divided into two main parts. First, the creep behavior of the binary, single phase compounds is presented, and both diffusional and dislocation creep are discussed. The second part will describe the attempts at improving the creep strength of these intermetallics. The vast majority of these strategies involves adding a reinforcing second phase. It will become readily apparent that the amount of work performed on NiAl far exceeds that of both CoAl and FeAl combined. CoAl appears to be hampered by its low toughness values coupled with no advantages over NiAl in either density or melting point. FeAl, despite its higher levels of ductility, is limited to lower temperatures by its strength, melting point, and oxidation resistance.

2.0 CREEP OF BINARY COMPOUNDS

Creep deformation of the B2 aluminides follows that for metals and alloys and can be divided into primary, secondary and tertiary stages. The shape of the primary creep curve provides an important clue for determining the deformation mechanisms, although the secondary stage is usually of more interest because it tends to comprise the majority of

C. T. Liu et al. (eds.), Ordered Intermetallics – Physical Metallurgy and Mechanical Behaviour, 541–563.
© 1992 Kluwer Academic Publishers.

the creep life. The tertiary stage has rarely been examined in these
materials, primarily because most testing to date has been in
compression, where tertiary creep is suppressed or eliminated.

The second stage or steady-state creep rate $\dot{\epsilon}$ is usually expressed
as a form of the Dorn equation(1):

$$\dot{\epsilon} = A\sigma^n \exp(-Q/RT) \qquad [1]$$

Here σ is the applied stress, n is the stress exponent, Q is the
activation energy for creep, R is the gas constant, T is the absolute
temperature, and A is a constant which includes microstructural
variables such as grain size and stacking fault or anti-phase boundary
energy. The values for the various parameters in Eqn. 1 are dependent
on the operative deformation mechanisms within a given temperature and
stress regime. Although the existence of a true steady state is
difficult to substantiate, it is usually a satisfactory approximation;
as an alternative, a minimum creep rate can usually be substituted with
consistent results. Another important point implied by Eqn. 1 is that
the steady-state creep rate is independent of test mode. Although creep
tests have traditionally been performed under constant load or constant
stress conditions, with the steady state strain rate measured as the
dependent variable, it is equally valid to impose a constant strain rate
and measure the steady state flow stress as the dependent variable.
This has been experimentally verified for a number of systems(2,3) with
known exceptions that can be traced to microstructural instabilities(4).

2.1 Dislocation Creep

Dislocation creep mechanisms are well described by the above semi-
empirical equation, although the details of the mechanisms themselves
are being continuously refined(5). For single phase metals and alloys,
dislocation creep can be classified as either of two main types, known
as Class M, or pure metal type, and Class A, or alloy type(5-7). Class
M creep is characterized by glide being much faster than climb, and thus
creep becomes controlled by the rate of climb past substructural
obstacles. Class A creep is often called viscous glide controlled
creep, since the glide of dislocations is restricted by solute atoms or
perhaps by a high lattice friction stress due to long range order. This
reduced glide mobility is the limiting creep process, while climb can
occur readily. Although these categories represent limiting conditions,
and many materials exhibit an intermediate behavior comprised of a
mixture of the defining traits, these two types of behavior can be
distinguished by several criteria(1,5-7), as listed in Table I. The
most commonly used indicator for determining the deformation behavior is
the stress exponent, which takes on values close to 5 for Class M and 3
for Class A. The activation energy for creep is equal to that for
lattice diffusion regardless of the mechanism, although corrections for
the temperature dependence of the elastic modulus are frequently
necessary(1,8). In terms of dislocation structure, Class M materials
tend to form subgrains during creep, while in Class A materials, a
uniform distribution of dislocations is developed. The two types also
have different primary creep behavior. For Class M materials in a
constant load (or constant stress) test, a normal primary creep region
is exhibited, in which the creep rate is initially high and decreases to
the steady state value as the subgrains form. This is in contrast to an

inverted primary region seen in Class A materials, where the creep rate
is initially low but increases to the steady value. If the test is
performed under constant crosshead speed (or strain rate) conditions
instead of constant load, the analogous behaviors of work hardening or
yield points are observed(2). Finally, there are tests which are
important for discriminating between mechanisms, whereby sudden changes
in applied stress or strain rate are made and the responses are unique
to the material and the type of test(5,9,10). For example, the
instantaneous strain increment after a small stress increase would be
primarily elastic for Class A materials but would have a large plastic
component in Class M behavior.

TABLE I
Criteria For Classifying Dislocation Creep Behavior

	Class M	Class A
Controlling Mechanism	Climb	Viscous Glide
n	5	3
Q	Diffusion	Diffusion
Dislocation Structure	Subgrains	Homogeneous
Primary Creep const.σ const.$\dot{\epsilon}$	Normal Work Hardening	Inverted Yield Point
Response To Transients	Class M	Class A

A summary of the stress exponents and activation energies
determined by various authors(11-23) is presented in Table II. For NiAl
and CoAl, the best choice for activation energy appears to be close to
310 kJ/mol, which is reasonably close to the activation energies
determined in diffusion experiments(24-28). For FeAl, however, an
activation energy for creep of ~450 kJ/mol is much higher than that for
diffusion, ~300 kJ/mol(28-30). Correction for the temperature
dependence of the elastic modulus, using dynamic moduli(31,32) results
in reductions in Q ranging from 15 Kj/mol in CoAl to 40 kJ/mol in FeAl,
which is not sufficient to explain the discrepancies observed for FeAl.
Certainly, the larger number of both diffusion and creep studies which
have been conducted on NiAl allows for more confident analyses compared
to either FeAl and CoAl.

Fig. 1 presents a summary of the creep data for NiAl at 1175 K,
a temperature chosen because it required the least amount of
interpolation. Most of the data from the various studies cluster with
reasonable agreement, within about a factor of 5 in creep rate at a
given stress level, and the stress exponents are also similar, between
about 4.5 to 6. The early data by Vandervoort et al.(16) appears to be
abnormally weak for no known reason. Fig. 2 shows the stress exponent
as a function of temperature and compiles data generated from many
studies on NiAl. Included in this plot are data from materials having a
wide variety of grain sizes, including single crystals. Below about

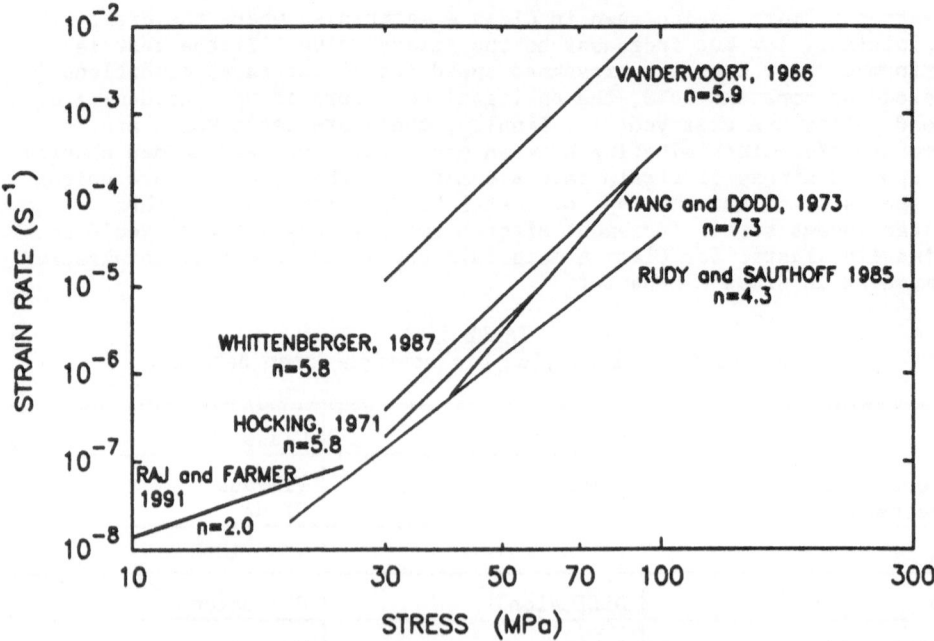

Figure 1.   Compressive creep data for stoichiometric NiAl at 1175 K. Data are taken from refs. 11,14-16,18,33.

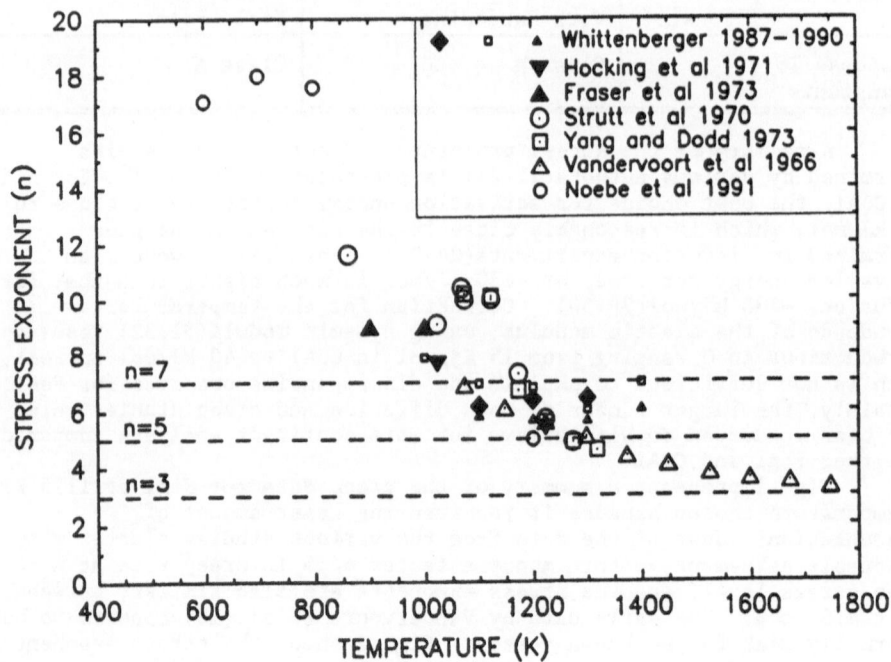

Figure 2.   Temperature dependence of the stress exponent for stoichiometric NiAl.   Data are taken from refs. 11-19,34.

TABLE II
Summary of Creep Parameters for B2 Aluminides

At% Al	Grain Size (μm)	T (K)	n	Q (kJ/mol)	Reference
NiAl					
48.25	5-9	1000-1400	6-7.5	313	11
49.2	15-20	1100-1400	5.75	314	12
50	12	1200-1300	6	350	13
50	450	1073-1318	10.2-4.6	283	14
50	500	1173	4.7	--	15
50.4	1000	1075-1750	7-3.3	230-290	16
50	SX[123]	1023-1223	7.7-5.4	--	17
50	SX	1023-1328	4-4.5	293	18
50	SX[001]	1000-1300	6	440	19
CoAl					
44	8	1100-1400	2.7	200	20
49.9	10	1100-1400	4.5	345	20
50	20	1200-1300	3	384	21
50	SX[123]	1323	2.6	--	17
FeAl					
27	70	810-1010	6	370	22
45.7	40	1300-1400	3.5	410	23
48.7	40	1100-1200	6.5	470	23
50	500	1173	4.0	--	15

1000-1100 K, the stress exponent starts to rise drastically, indicating a transition between high temperature creep and lower temperature behavior which can perhaps be called power-law breakdown. Discounting the abnormally weak material of Vandervoort et al., this figure reveals that between ~1100-1400 K, the values for n cluster between about 5 and 7. Although a stress exponent of 7 is higher than typical, such high values have been observed in several Class M materials(35,36). Combined with the stress exponent data, the other criteria listed in Table I have also been examined. Observations of subgrain formation after high temperature deformation have been made by numerous workers(11-15,37-40), and normal primary creep behavior, under both constant load and constant crosshead speed conditions, has also been observed(11-13,39,41). Additionally, the strain rate transient tests performed by Yaney and Nix(41) have also been consistent with Class M behavior. So in summary, the vast majority of results from a wide variety of sources indicate that high temperature creep in NiAl is climb controlled.

The 1300 K creep data for CoAl is summarized in Fig. 3. For this compound, the n values range between 2-5, which indicates a possible tendency towards Class A behavior. In fact, most of the data appear to show a transition from Class A to M between 1200 and 1400 K. At 1200 K, inverted primary creep(20,21,41), homogeneous dislocation structures(17, 41), and Class A response to strain rate changes(41) have been observed. In contrast, normal primary creep(20,41) and Class M response during strain rate changes(41) are observed at 1400 K. Dislocation structures during creep have not been reported, although subgrains have been observed after extrusion at ~1500 K(42). The fact that the stress exponent has not shown a corresponding increase at higher temperatures is probably the result of additional contributions to deformation by diffusional creep mechanisms.

FeAl also appears to creep by at least two different mechanisms, as shown in Fig. 4 and Table II. At high stresses and at temperatures below 1200-1300 K, the powder metallurgy materials studied by Whittenberger exhibit stress exponents which cluster near 6(23,44). In addition, normal primary creep is reported(23,44), but subgrains have not been observed(15,44). More conclusive determination of the mechanism(s) in this regime requires further experiments. At low stress levels at 1200 and 1300 K, and at higher temperatures, deformation appears to follow a stress exponent of 3-4(23). In this regime, normal primary creep was observed(23,43), but a lack of subgrains was again reported(43). The evidence of temperature dependent n values(23), a failure to reach the typical steady-state deformation conditions(23), observations of grain growth and dynamic recrystallization during deformation(23,45), and improved creep strength with coarser grain sizes(23) all lead to the conclusion that a superposition of both dislocation and grain boundary mechanisms are contributing to the creep response in this regime. Sauthoff and co-workers(15,43,46) have also observed such a transition in n values, although at a lower temperature, ~1025 K. Thus, the data from their studies in Fig. 4 show n ~4 at the same temperatures where the data of Whittenberger tend to display n ~6. Because their material was made by casting, grain boundary mobility is probably much higher than in the powder metallurgy material, which has oxide particles that restrict grain growth.

The next topic regarding creep of these compounds is the effect of stoichiometry. Fig. 5 is a plot of steady state creep rate at a given temperature and stress, as a function of Al content for all three B2 compounds. The testing temperatures indicated in the figure are approximately 70% of the absolute melting point, T_m. For NiAl, there is a broad range in composition ranging between about 45 and 52% Al, where the creep rate is roughly independent of composition. The largest difference in creep rates within this range of compositions is only a factor of 5. Only at very low Al contents is NiAl noticeably weaker(47, 48). This is most easily explained by the lower melting points of these compositions, which in turn implies a higher diffusivity, although the diffusivity data(24) would imply a more significant effect. These trends as a function of stoichiometry are reversed from those observed at lower temperatures(49), where defect hardening predominates over the effects of diffusion. As seen in Fig. 5, a much stronger dependence on stoichiometry is observed in CoAl, with the best creep strength close to the equiatomic composition. The creep behavior of FeAl follows that of the melting point, where a slight decrease in creep rate is observed as

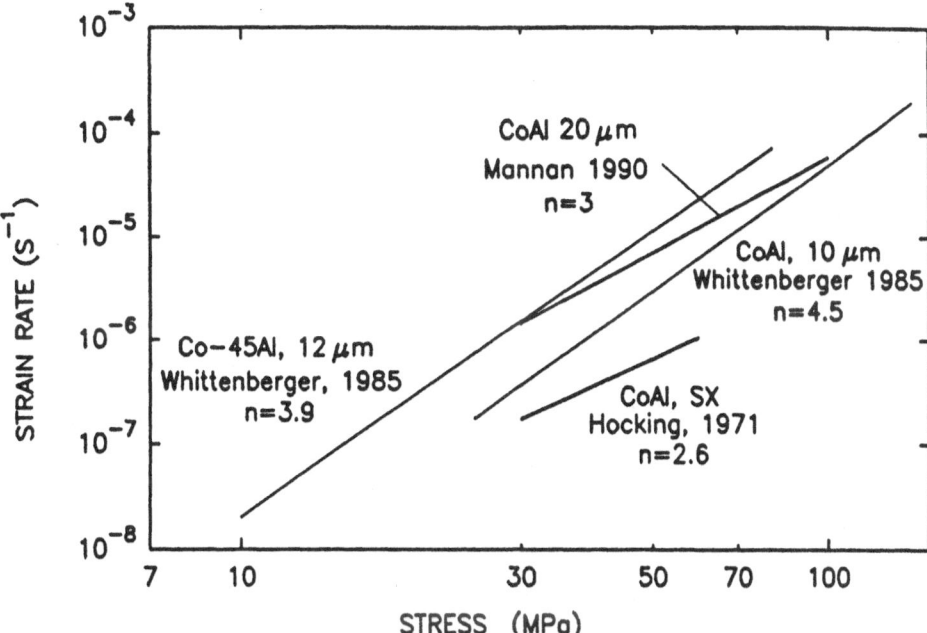

Figure 3.   Compressive creep data for CoAl at 1300 K.   Data are taken from refs. 17,20,21.

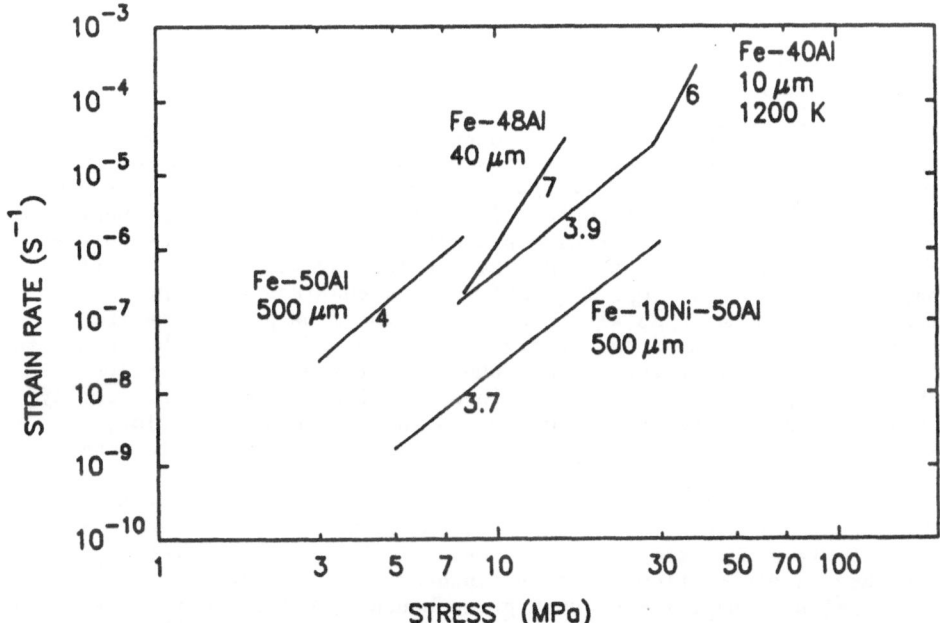

Figure 4.   Compressive creep data for FeAl materials at 1175 K and Fe-40Al at 1200 K.   Data are taken from refs. 15,23,43.

Al level is decreased. Diffusion data for FeAl also follow a weak dependence on Al content(28-30).

Although dislocation creep mechanisms are generally considered to be independent of grain size, studies of NiAl and FeAl have demonstrated that elevated temperature creep strength can sometimes be improved by decreasing the grain size. For example, Fig. 4 shows that the strength of the finer grained binary FeAl alloys is superior to the coarser grained forms, and such behavior was explained as a Hall-Petch effect (23). In the case of NiAl, the data in Fig. 6 illustrate that material with a grain size below ~10μm is capable of improved creep resistance. Since NiAl is a Class M material which exhibits subgrain boundaries that act as obstacles for dislocations, this behavior can be expected when the grain size is finer than the equilibrium subgrain size(11). Unfortunately, the effectiveness of fine grain size is restricted to lower temperatures and/or higher stresses, where diffusional creep mechanisms have less of an influence.

2.2 Diffusional Creep

Time dependent deformation can occur by stress assisted vacancy flow at stresses which are too low for dislocation processes to be significant. Creep by these diffusional mechanisms such as Nabarro-Herring or Coble creep is attributed solely to movement of vacancies from sources to sinks, which are usually grain boundaries of different orientations with respect to the applied stress. These mechanisms are characterized by stress exponents n - 1 and very strong dependencies on grain size, with large grained materials being more creep resistant. Rudy and Sauthoff(46) provided the most convincing evidence for diffusional mechanisms in a Ni-20Fe-50Al alloy, namely a stress exponent of 1. Additionally, there is some evidence in binary NiAl at temperatures above 1300 K and at low strain rates, where grain growth during the creep test resulted in coarser grained material having higher strengths(12). Fig. 1 also displays some new data generated by Raj and Farmer(33) showing a low stress exponent that may indicate some grain boundary assisted mechanism operating at low stresses. Despite these examples, the large body of literature regarding creep of NiAl provides little direct experimental support for creep by diffusional mechanisms over large regimes of temperature and stress. In fact, the calculated creep rates by this type of mechanism are much greater than the actual experimental data(50), implying that these mechanisms are suppressed compared to disordered metals. In Fig. 6, the fine grained material begins to lose its advantage at stresses below ~30 MPa, as the change in slope indicates. In general, the tendencies for diffusional creep mechanisms begin to become noticeable around 1300 K and are not really very strong until ~1400 K. As discussed in the previous section, evidence for diffusional mechanisms in FeAl and CoAl appear to commence at approximately $0.7T_m$ also.

Some qualifications should be mentioned with regard to the influence of grain size. The trends seen above are restricted to single phase materials that are relatively weak at these temperatures. If the resistance to dislocation creep is improved, for example by precipitation strengthening, then grain boundary effects may become more influential and coarse grained materials may be more desirable. Furthermore, since the majority of the creep data has been obtained by compression testing, grain boundary cavitation is rarely observed. Thus it is possible that different trends with grain size would be seen in

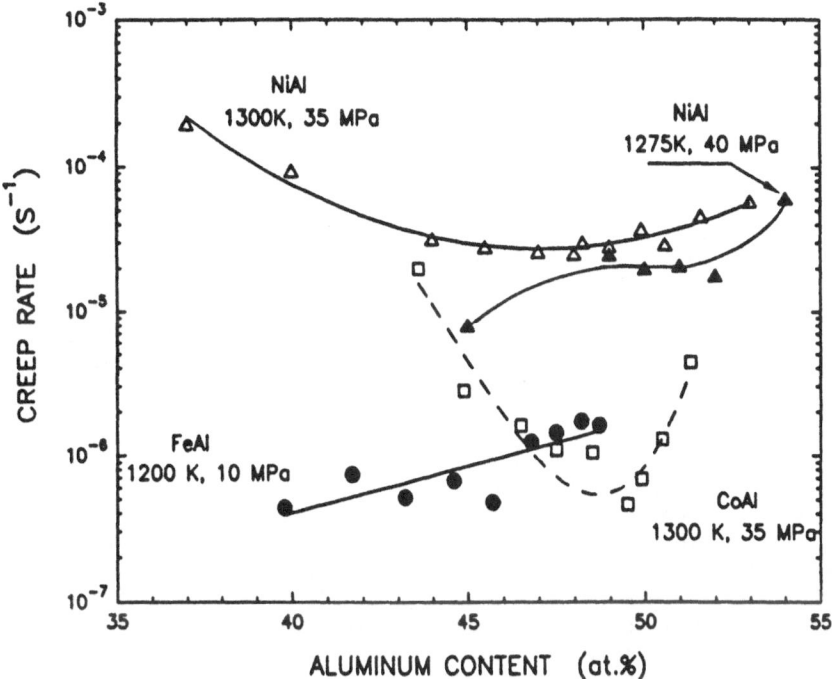

Figure 5. Creep rate as a function of stoichiometry for NiAl(12,14,47, 48), FeAl(23), and CoAl(20).

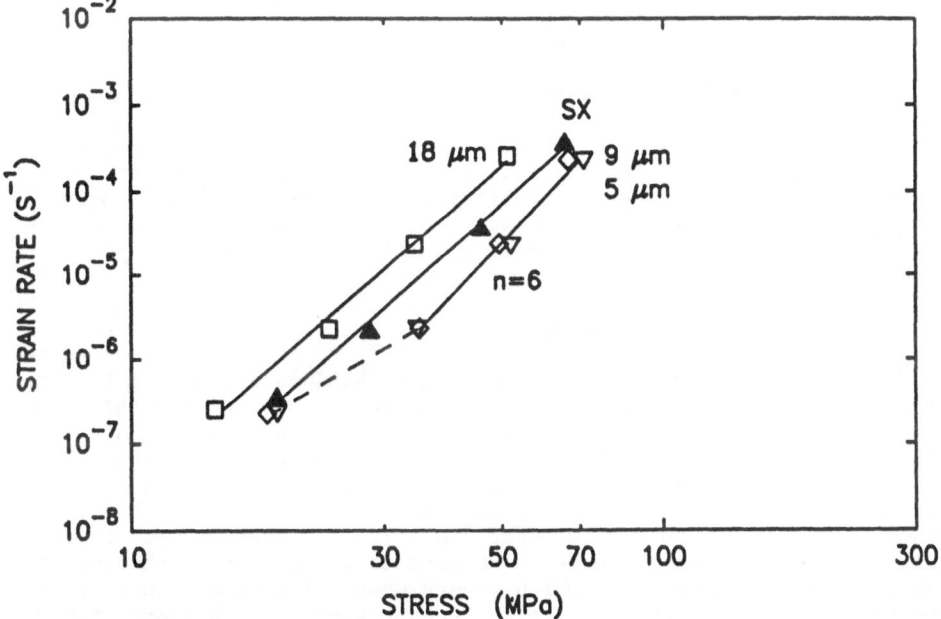

Figure 6. Compressive creep resistance as a function of grain size for NiAl at 1300K. Data are taken from refs. 11,19.

tension tests. One example of this is illustrated in Fig. 7 for FeAl.
At 1100 K, the creep rates measured in tension and compression were in
good agreement for binary Fe-40Al, but not for the alloy which was
precipitate strengthened by Zr and B additions. This difference was
traced to the onset of cavitation after a few percent strain in tension
(51). Also note that the precipitate strengthened alloy had a 2-3 order
of magnitude improvement in creep rate over the binary alloy in
compression, but only a factor of 15 improvement in tension tests.

3. STRATEGIES FOR INCREASING CREEP RESISTANCE

This section considers various concepts for improving creep
strength, primarily for NiAl. As an introduction, Fig. 8 is an example
of how creep strength is built in a single crystal Ni-base superalloy
(2). Starting with pure Ni, one can see that solid solution hardening
with a heavy element such as W provides about 2 orders of magnitude
improvement in creep rate. Creep data for a solid solution alloy that
has a combination of Cr, Co, and W in a composition similar to that of
the γ phase of the superalloy, follows the creep response of the Ni-2W
binary. In other words, the effects of combining several solid solution
elements are not additive. The presence of a long range ordered
structure in Ni_3Al provides another 2 orders of magnitude improvement in
creep rate, and solid solution hardening of Ni_3Al to form the
composition of the γ' phase in the superalloy gives an additional factor
of ten. However, the largest improvement in creep resistance, about a
factor of 1000, is obtained by the superalloy with a two phase mixture
of about 50% γ and γ'. So in total, there is a decrease of about 8
orders of magnitude in creep rate as progressive changes are made from
pure Ni to the Ni-base superalloy, and the addition of a second phase is
the major reason for this improvement.

Similar improvements are necessary for the aluminides to compete
with current materials such as superalloys and Ni_3Al base alloys. Plots
similar to Fig. 8 will be used for most of the comparisons of the
various strengthening concepts which are discussed in the following
sections. A rough criteria of success is obtained by judging how close
the creep strength is to the superalloy, without compensating for the
lower density of NiAl. Creep properties of the intermetallics, which
were generated primarily in compression, will be compared to the tensile
creep response of NASAIR 100, a first generation single crystal
superalloy(52).

3.1 Solid Solution Strengthening

The role of solid solution hardening in NiAl is summarized in Fig.
9 at 1200 K, where the bulk of the data has been generated and the least
amount of extrapolation was needed. Two data sets for binary NiAl are
shown which cover the range in creep strength seen in the numerous
studies, with both having a high stress exponent, n ~5. All of the
solid solution alloys show some improvements in strength, but they also
exhibit a change in n to a value near 3-4. Thus it appears that these
solute additions change the creep mechanism to viscous drag behavior, in
a manner very similar to that which occurs when alloying elements are
added to pure metals. In fact, in one recent study(55), a transition
from Class M to Class A behavior as a function of applied stress was
observed, and the transition appeared to be well described by current
theories developed for disordered solid solutions(6,7). However,

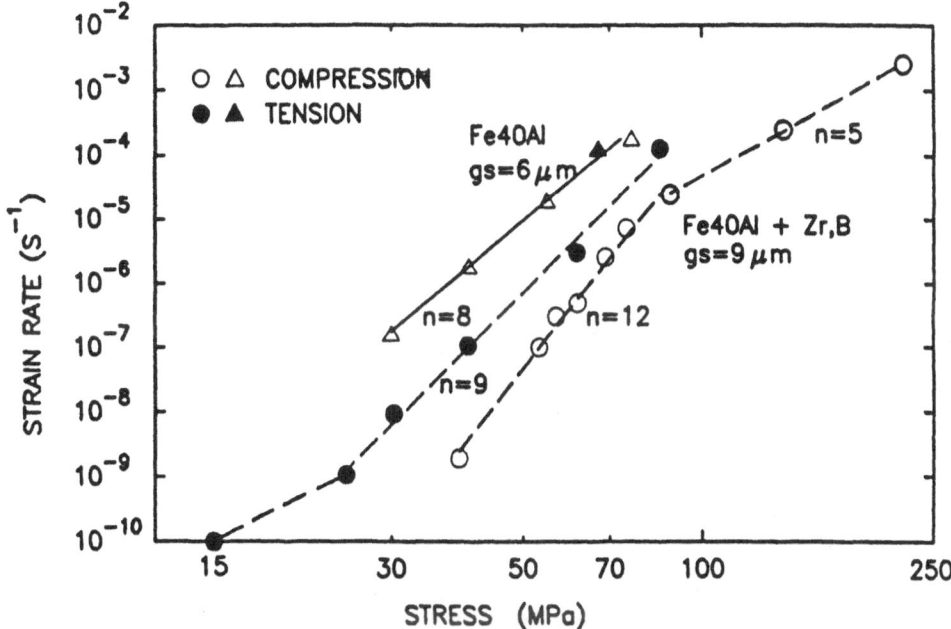

Figure 7. Comparison of creep response under tension and compression for Fe-40Al and Fe-40Al-0.1Zr-0.5B at 1100 K (51).

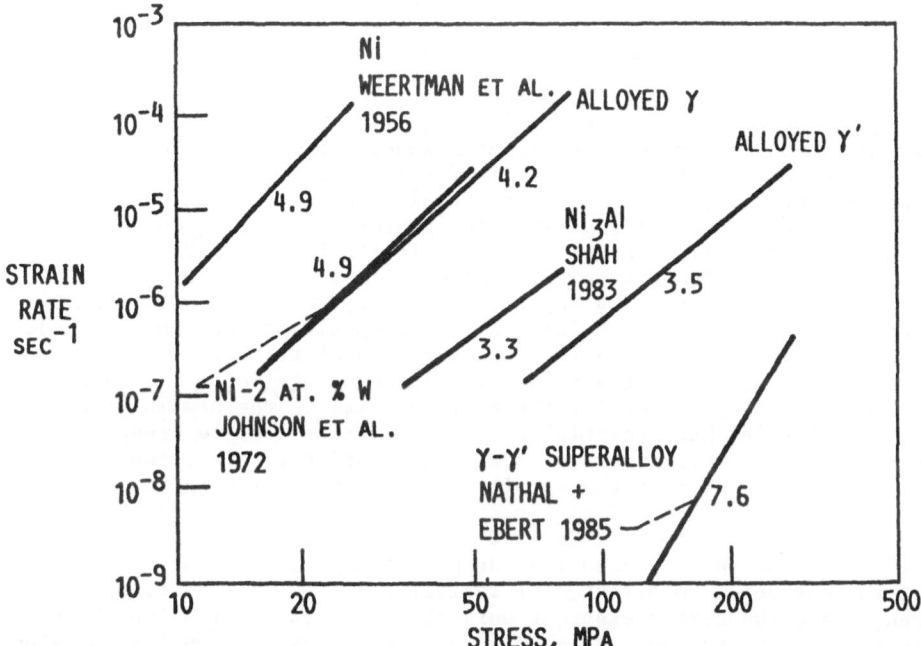

Figure 8. Building creep strength at 1275 K in steps from pure Ni to a Ni-base superalloy(2).

because of this new stress exponent, the strength improvements over the binary are only large at high stresses, and in the lower stress regime the advantage is reduced or eliminated. Finally, it is of interest to note in Fig. 9 that one of the largest strengthening effects was produced by an addition of only 0.05 at.% Zr. This sensitivity to small differences in composition might be the main reason for discrepancies in mechanical properties among various published results. Fig. 4 shows data for FeAl that indicate significant improvements in creep resistance with solid solution additions of Ni. Solid solution hardening of CoAl by Fe and Ni has also been reported(56). Since both of these binary compounds have tendencies towards Class A behavior, it would not be surprising that the ternary alloys would also show similar trends. In summary, it appears that solid solution hardening does provide some creep strength improvements over the binary compounds, but this concept is inadequate by itself and must eventually be used in combination with other strengthening mechanisms.

3.2 Precipitation Strengthening

Significant improvements in creep strength of NiAl by precipitation hardening were first demonstrated by Polvani et al.(57) by adding Ti to form a two phase mixture of NiAl and Heusler phase Ni_2AlTi. Similar Heusler phases and other phases such as Laves (eg., NiAlTa) can be formed with many ternary additions such as Nb, Ta, Hf and V. The creep properties of some Ti and Ta containing alloys are presented in Fig. 10, where it is evident that these materials are reasonably strong, but again, extrapolation to low stresses shows the advantages diminishing. The reasons for the low stress exponent in these alloys is not entirely clear, since most creep resistant, two-phase alloys exhibit significantly higher stress exponents than the matrix phase. In most cases, the microstructures of the ternary alloys were probably not optimized: for example if the second phase is not fine enough, effective strengthening would not be expected. Equally valid explanations may be that the observed n values represent a superposition of several deformation mechanisms, including diffusional creep, and/or that coarsening of the precipitate phase results in less strengthening in the low stress/long life regime. TEM micrographs of the Ti-rich Heusler phase containing alloys after creep testing(53,57) exhibit semi-coherent β-β' interfaces which are reminiscent of microstructures in γ-γ' superalloys(52). Again, by analogy to the superalloys, optimizing the creep strength requires a balance of the compositions of the two phases, the precipitate volume fraction, the size and distribution of the precipitates. The use of single crystal technology would also seem necessary in order to realize the full benefits of the precipitates. In fact, Darolia(58) has recently reported promising tensile creep-rupture properties for single crystals containing Heusler precipitates.

Another example of the sensitivity of creep strength to microstructure in these alloys is shown in Fig. 11. The alloys with Nb-rich Laves phase behave similarly to the other alloys in Fig. 10, but it is also evident that by changing the processing of the same alloy from cast and extrusion to directional solidification, both the creep strength and the stress exponent were changed. It is also important to note that the directionally solidified material was not optimized, since the second phase was only partially aligned(60). Other options involving precipitation hardening also exist, such as the use of different phases which may have better strengthening or microstructural

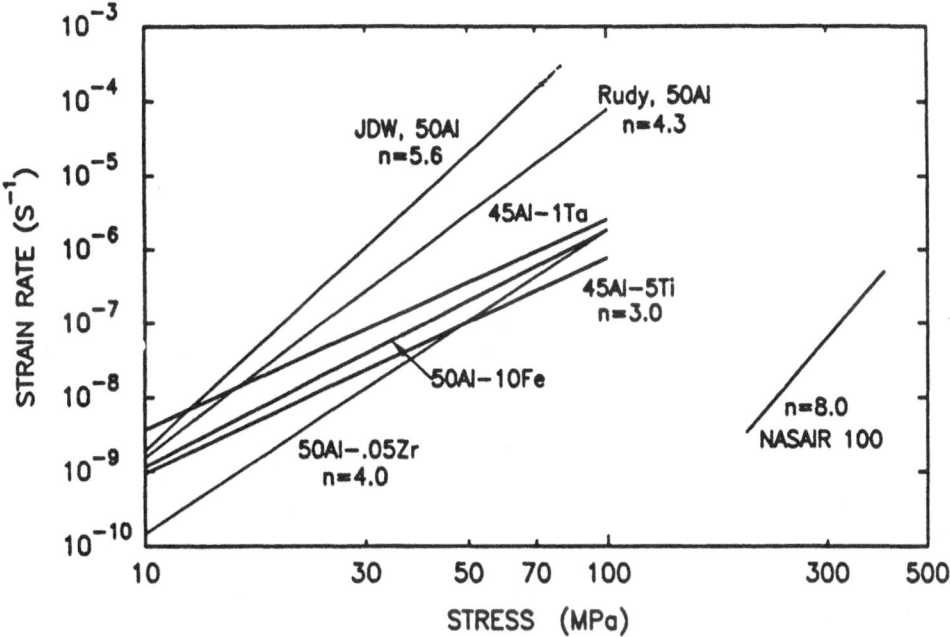

Figure 9. Solid solution hardening of polycrystalline NiAl in compressive creep at 1200 K. Data are taken from refs. 12,15,19,46,52-54.

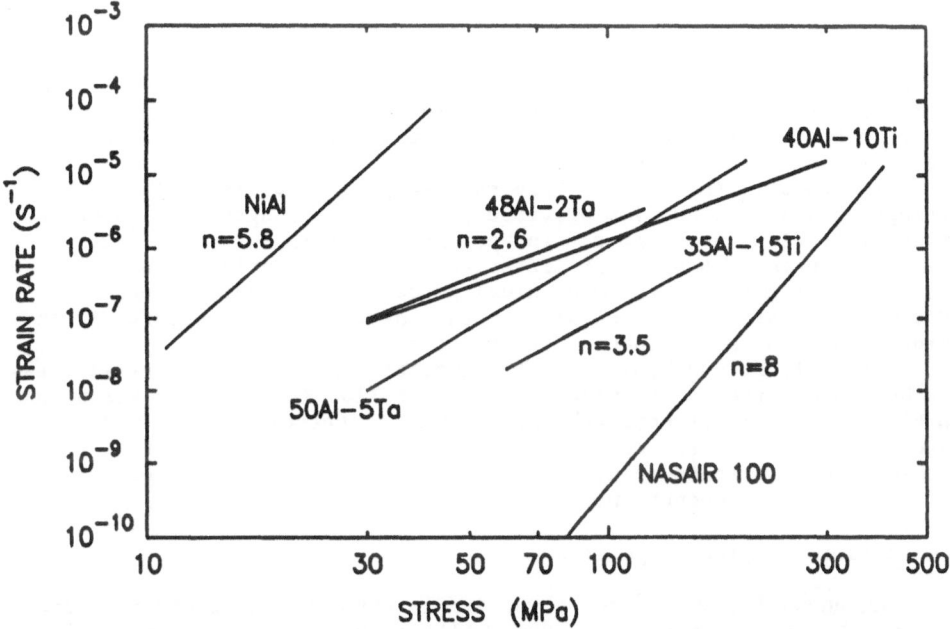

Figure 10. Precipitation hardening of NiAl by Heusler and Laves phases in compressive creep at 1300 K. Data are taken from refs. 12,52-54,57.

stability characteristics(61).

3.3 Dispersion Strengthening

This section addresses the use of rapid solidification to add fine dispersoids. This concept takes advantage of the fact that many elements or compounds are soluble in liquid NiAl but insoluble in the solid. By rapid solidification, very fine dispersions of second phases, with diameters on the order of 20-50 nm, can form that are resistant to coarsening due to the very low solubility. For NiAl, both pure elements such as W and Mo and various carbides or borides are candidates for this type of strategy. These dispersoids also tend to pin grain boundaries and result in significant grain refinement(62-64). Fig. 12 reveals that the additions of pure W(64) and TiC(62) had very little strengthening effect, and TiB_2(62) showed about an order of magnitude improvement over binary NiAl. However, this degree of strengthening can be achieved simply due to grain refinement similar to that shown in Fig. 6, so dispersoid/dislocation interactions are not expected for these alloys.

More interesting are the improvements achieved with HfB_2(65) and HfC(62,63), which are both considerably stronger than binary NiAl. In the case of HfC there was some indication of a threshold stress at ~50 MPa below which no creep occurs(63). However, more recent work has shown that this apparent threshold stress is the result of dynamic grain growth which is a function of testing conditions(4). These new data indicate that for HfC strengthened NiAl, coarse grained material is more creep resistant than the finer grained product, which is the opposite of the trends observed in the binary. Thus the grain interiors have been strengthened sufficiently such that diffusional creep mechanisms are occurring at similar rates. The mechanism for the improved strength are related primarily to the interaction of the dispersoids with mobile dislocations(62,63) and subgrain boundaries(65). However, it is probably not a coincidence that the two most effective dispersoids contained Hf, and it is quite possible that small amounts of Hf in solution could strengthen NiAl in a manner similar to that observed in Fig. 9 for small Zr additions.

In summary, the use of rapid solidification to dispersion strengthen NiAl has not yet provided sufficient strengthening to compete effectively with superalloys. The potential for improvement exists, primarily in the areas of optimizing the dispersoid volume fractions, and in devising the thermo-mechanical processing schedules needed to produce the desired grain structures that have proven successful in the oxide dispersion strengthened Ni-base alloys(66).

In FeAl, strengthening by additions of Fe_6Al_6Zr and ZrB_2 particles has been demonstrated at temperatures up to 1100 K(51,67), as was seen in Fig 7. Additional strengthening by Y_2O_3 dispersoids has also been observed(68). However, the relative contributions to creep strength from elements in solid solution, precipitates, dispersoids, and grain refinement remain uncertain.

3.4 Reaction Milled Composites

An unusual example of a composite is the AlN dispersoid-reinforced NiAl which was produced by reaction milling in liquid nitrogen(69,70). This process produces very fine dispersoids, on the order of 20-50 nm, at relatively high volume fractions of ~10%. These particles are not uniformly distributed, but are clustered along prior particle

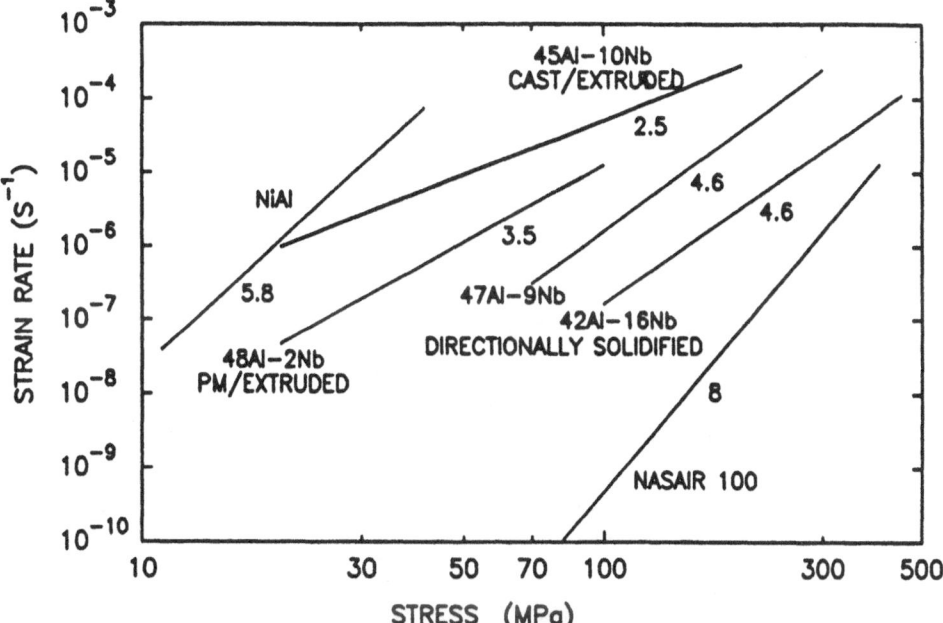

Figure 11. Strengthening NiAl by NiAlNb Laves phase in compressive creep at 1300 K, showing the effect of variations in microstructure. Data are taken from refs. 12,52,54,59,60.

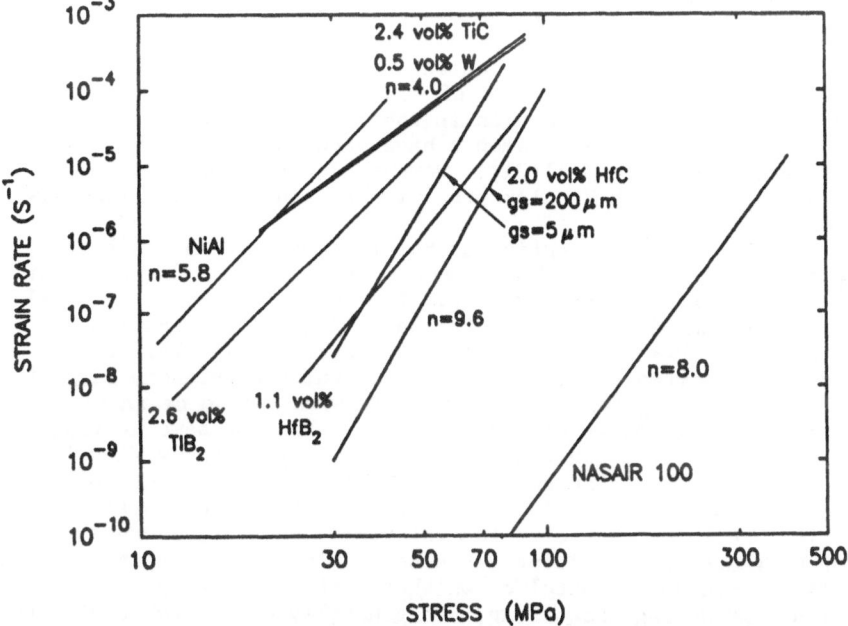

Figure 12. Effect of dispersoids added by rapid solidification techniques on compressive creep of NiAl at 1300 K. Data are taken from refs. 4,12,52,62,63,65.

boundaries. The properties of this material are very promising, since as seen in Fig. 13, creep strengths approaching that of NASAIR 100 were obtained. After correcting for density, the deformation resistance of the superalloy and AlN/NiAl are actually equivalent. Also of interest is that the properties of extruded material, where the particle-rich regions are strung out along the extrusion direction, were roughly equivalent to HIP-consolidated material, where the particles are not aligned but still segregated. The reasons for the exceptional properties of this type of second phase reinforcement are not currently understood, although the promising results in compression certainly warrant more extensive testing in tension.

3.5 Discontinuous Reinforced Composites

Another approach to strengthening is through the use of composites with discontinuous reinforcements. These reinforcements are typically larger in size and present in higher concentrations than that found in dispersion strengthened materials. One example is TiB_2 particulate reinforced NiAl, where composites with $1\mu m$ diameter particles were produced by an exothermic reaction process(13). Fig. 14 reveals that composites made in this manner do show improvements in strength that scale with the amount of reinforcement. It is important to note that the stress exponents are all high, which leads to a creep strength advantage that is maintained or improved at lower stresses. When the stress exponent is increased, it usually indicates that the dislocation substructure is refined and stabilized by the second phase when compared to the same matrix without the reinforcement. Evidence for this has been provided by transmission electron microscopy(13), where the creep deformation structure was characterized by subgrain boundaries which are usually pinned by the particles, in combination with a much higher dislocation content within the subgrains. It is clear that the TiB_2 particles are effective in stabilizing a higher dislocation density which results in the observed strengthening.

Similar strength improvements have been observed when TiB_2 particles were added to CoAl(21). However, an attempt at combined strengthening from both Heusler precipitates and TiB_2 particles was unsuccessful(53).

Another type of discontinuous reinforcement which has been produced in NiAl is Al_2O_3 whiskers(71). The whiskers, which had an average aspect ratio of approximately 7.5, were added by mechanical blending at volume fractions ranging up to 25%. The mechanical properties from those composites are presented in Fig. 15. Some improvements in creep resistance are seen, but the whiskers are not as effective as the TiB_2 particles described above. Also of interest is that the stress exponents of the composites were the same as that of the matrix, which indicates that deformation is controlled by flow in the matrix, as predicted by several models of composite strengthening(72). These models would predict further improvements in creep strength by increasing the aspect ratio of the whiskers. However, because some whisker breakage after testing was observed, higher strength whiskers will also be needed. Control of whisker distribution and alignment, and whisker damage during processing, are generally major problems in this type of composite. Finally, hybrid composites containing both TiB_2 particulates and Al_2O_3 whiskers have been made and have shown that these strengthening concepts were additive.

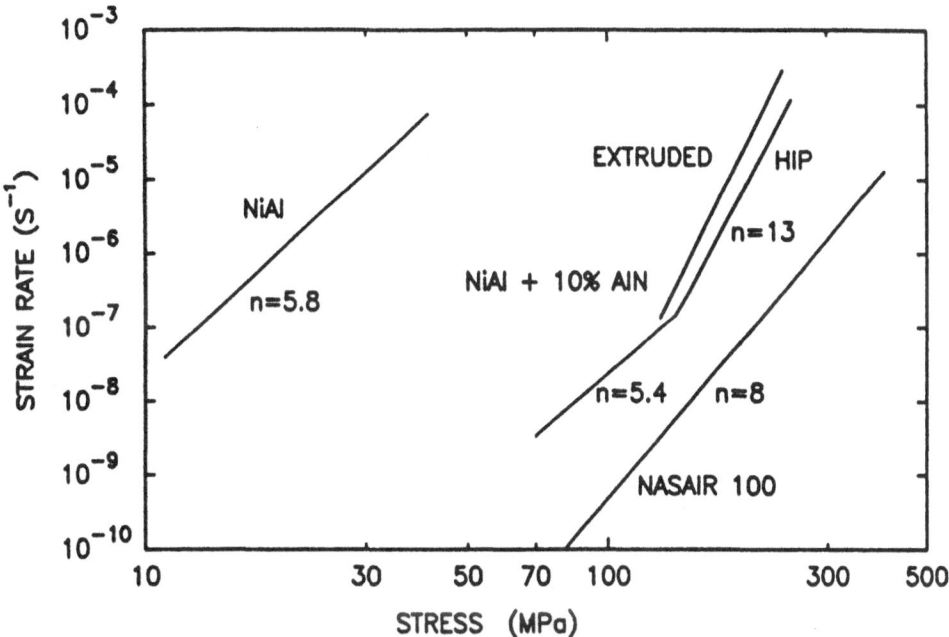

Figure 13. Effect of AlN dispersoids added by a reaction milling process on compressive creep of NiAl at 1300 K. Data are taken from refs. 12,52,69,70.

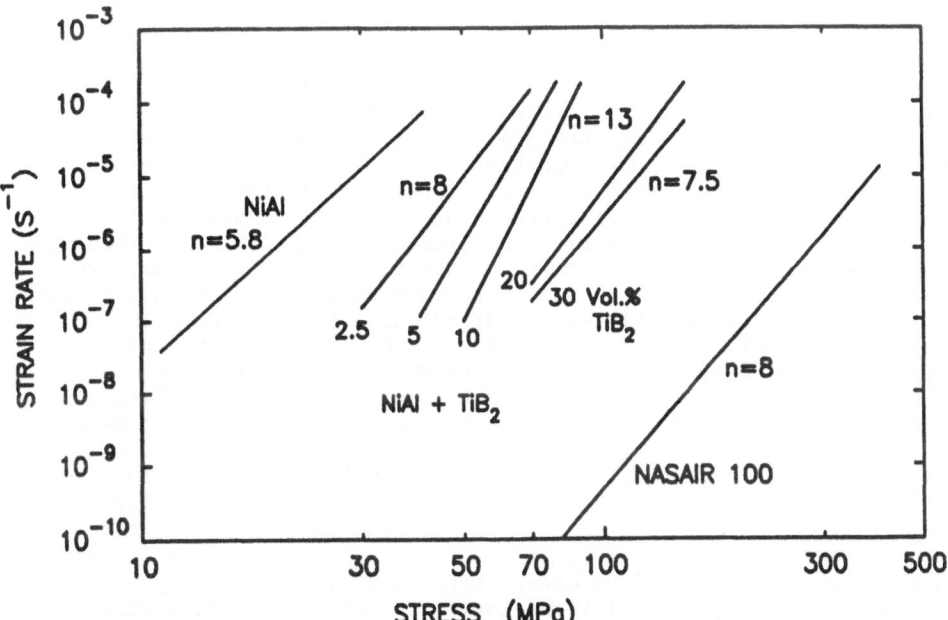

Figure 14. Effect of TiB$_2$ particles added by exothermic reaction synthesis on compressive creep of NiAl at 1300 K. Data are taken from refs. 12,13,52.

3.6 Continuous Reinforcement Composites

The final strategy to discuss is the reinforcement of NiAl with continuous fibers. Such composites can either be natural, such as directionally solidified eutectics, or artificially fabricated, in order to use fiber-matrix combinations not achievable through eutectic solidification. The early work by Walter and Cline(73) has shown that a eutectic consisting of α-Cr rods in a NiAl matrix possessed some very promising creep properties, as indicated in Fig. 16. Similar eutectic microstructures can be produced with Mo(74), W(64), Re(75), and NiAlNb (60) phases, and these may also prove to be advantageous.

Some artificial composites have been made by a powder metallurgy approach of hot pressing matrix powders around either W or Al_2O_3 fibers and have been tested in bending(76). There were substantial strength improvements in the W/NiAl system over the stand-alone matrix, whereas the Al_2O_3/NiAl composites did not show any strengthening. This was traceable to different degrees of bonding, where load could be transferred to the strongly bonded W fibers but not to the weakly bonded Al_2O_3. However, the composite with weakly bonded fibers did show evidence of toughening, and thus a hybrid concept of using two types of reinforcement is one way to achieve a balance of properties. Further testing of these composites is required to ascertain whether the high strengths will be maintained in creep tests, and whether they can survive in an environment involving thermal cycling.

4.0 SUMMARY

High temperature creep deformation in NiAl appears to be satisfactorily described by pure metal type, or dislocation climb controlled creep, as all of the major defining characteristics of this class of creep have been observed. Stoichiometry variations appear to be relatively unimportant, especially between 45 and 52 % Al, a fact which is surprising based on the relatively large effect of stoichiometry on diffusion characteristics. Diffusional creep mechanisms appear to become important at low stresses and above about $0.7T_m$, although they appear to be more prominent in ternary alloys and in materials which have been strengthened against dislocation creep. CoAl has exhibited mixtures of both climb and glide controlled behavior, and may in fact undergo a transition from Class A to M as temperature increases. Creep in CoAl is much more dependent on stoichiometry, with a maximum in creep strength seen near 50% Al. FeAl has also shown intermediate behavior with a mixture of both Class A and M traits, and exhibits the opposite effect of being more creep resistant at substoichiometric Al levels. Both CoAl and FeAl have shown larger contributions from diffusional creep mechanisms that appear to be acting concurrently with dislocation creep. Of the three aluminides, only FeAl has exhibited an activation energy significantly higher than that for diffusion. An additional note of importance is the heavy reliance on compression testing that exists in the literature. Compression tests are very valuable in isolating deformation mechanisms and for providing an indication of the maximum creep strength achievable in a given material. However, several technologically important topics such as grain boundary cavitation, necking, and tertiary creep can best be examined in tension creep experiments.

The various strategies to improve the creep resistance of NiAl are summarized in Table III, along with suggested areas for future work

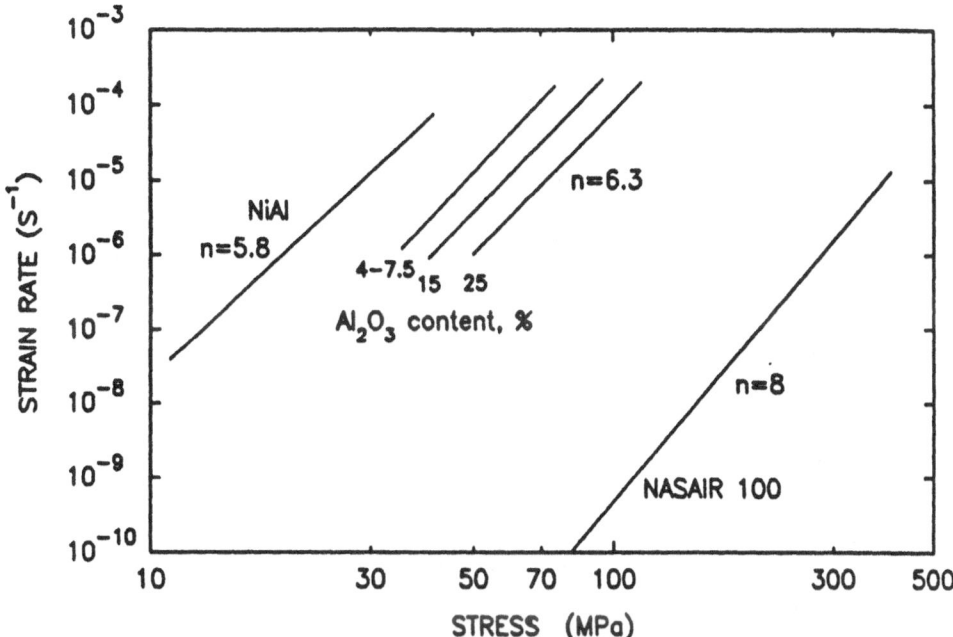

Figure 15.   Effect of Al_2O_3 whiskers added by a mechanical blending on compressive creep of NiAl at 1300 K. Data are taken from refs. 12,52,71

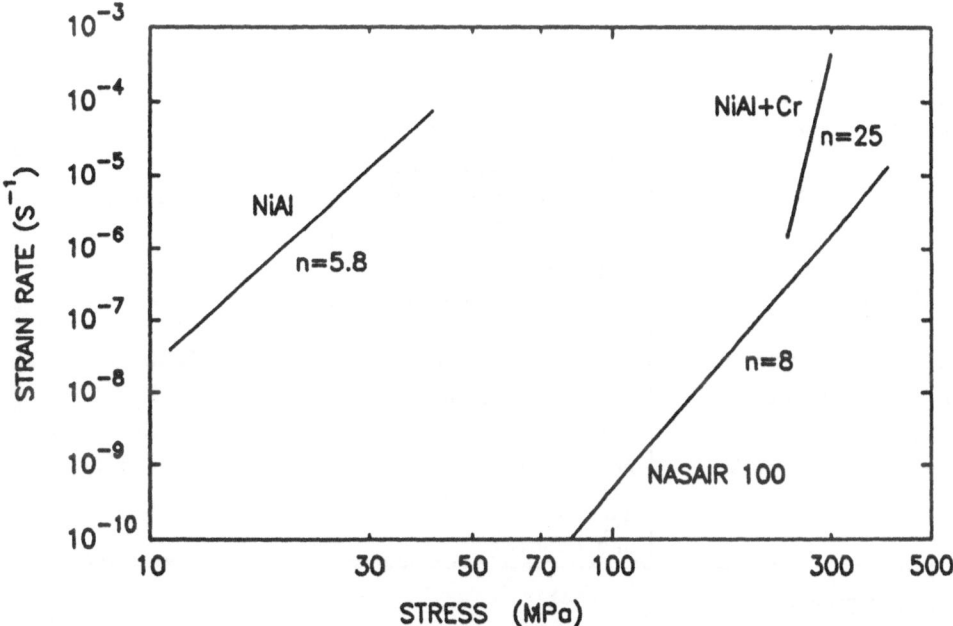

Figure 16.   Tensile creep of directionally solidified NiAl-34Cr eutectic at 1300 K.   Data are taken from refs. 12,52,73.

aimed at further improvements. Of these attempts, both solid solution and precipitation hardening have shown progress, but the low stress exponents of these materials results in less attractive properties at lower stresses and creep rates. However, recent advances using precipitate strengthening in single crystals have shown some very good promise. Rapid solidification has shown only small improvements in strength, whereas TiB_2 particulates and Al_2O_3 whiskers showed larger but still insufficient advances. The AlN/NiAl composite has some of the best creep properties to date, comparable to the Cr containing directionally solidified eutectic. Finally, the strengthening which can be achieved with continuous fibers is dependent on the choice of the fiber and the bonding with the matrix.

In most of the above strategies, options exist for exploring further improvements in creep strength. New types of precipitates or other reinforcements in combination with microstructural control may produce more significant results. Furthermore it is important to acknowledge that a balance of properties is required for most applications, and low temperature toughness, density, environmental resistance, fabricability and cost must eventually be considered.

TABLE III
Strategies For Improving Creep Strength of NiAl

STRATEGY	CURRENT STATUS	SUGGESTED MEANS OF IMPROVEMENT
Solid Solution + Precipitates	Limited benefits in low ϵ regime	Single crystal + possible new phases
Dispersoids by RST	Only small improvements in strength	New vol. % + mechanical working
AlN Dispersoids	Largest strengthening to date	Processing
Particulates	TiB_2 significant benefit but not enough	Hybrid composites
Whiskers	Al_2O_3 small benefit	Whisker availability + alignment, processing
D.S. Eutectics	Large strengthening with Cr	
Continuous Fibers	Potential strengthening but dependent on fiber and bonding	Fiber availability

5.0 REFERENCES

1. W.D. Nix and B. Ilschner, in 5th Int. Conf. Strength of Metals and Alloys, ed. by P. Hassen et al., Pergamon Press, 3, p. 1503 (1980).

2. M.V. Nathal, J.O. Díaz, and R.V. Miner, in High Temperature Ordered Intermetallic Alloys III, Proc. MRS, ed. by C.T. Liu et al., 133, p. 269 (1989).

3. W.C. Oliver and W.D. Nix, Acta Metall., 30, p. 1335 (1982).

4. J.D. Whittenberger, R. Ray, and S.C. Jha, in High Temperature Ordered Intermetallic Alloys IV, Proc. MRS, ed. by L.A. Johnson et al., 213, p. 581 (1991).

5. M.J. Mills, J.C. Gibeling and W.D. Nix, Acta Metall., 34, p. 915, (1986).

6. F.A. Mohamed and T.G. Langdon, Acta Metall., 22, p. 779 (1974).

7. M.S. Soliman and F.A. Mohamed, Mat. Sci. Eng., 55, p. 111 (1982).

8. A.M. Mukherjee, J.E. Bird, and J.E. Dorn, Trans. ASM, 62, p. 155 (1969).

9. D.L. Yaney, J.C. Gibeling and W.D. Nix, Acta Metall., 35, p. 1391 (1987).

10. T.G. Langdon and P. Yavari, in Creep and Fracture of Engineering Materials and Structures, ed. by B. Wilshire and D.R.J. Owen, Pineridge Press, Swansea, UK, p. 71 (1981).

11. J.D. Whittenberger, J. Mat. Sci., 23, p. 235 (1988).

12. J.D. Whittenberger, J. Mat. Sci., 22, p. 394 (1987).

13. J.D. Whittenberger, R.K. Viswanadham, S.K. Mannan, and B. Sprissler, J. Mat. Sci., 25, p. 35 (1990).

14. W.J. Yang and R.A. Dodd, Met. Sci. J., 7, p. 41 (1973).

15. M. Rudy and G. Sauthoff, in High-Temperature Ordered Intermetallic Alloys, Proc. MRS, ed. by C.C. Koch et al., 39, p. 327 (1985).

16. R.R. Vandervoort, A.K. Murkerjee, and J.E. Dorn, Trans. ASM, 59, p. 930 (1966).

17. L.A. Hocking, P.R. Strutt, and R.A. Dodd, J. Inst. Metals, 99, p. 98 (1971).

18. J. Bevk, R.A. Dodd, and P.R. Strutt, Met. Trans., 4, p. 159 (1973).

19. R.D. Noebe and J.D. Whittenberger, unpublished research, NASA-Lewis Research Center, Cleveland, OH (1991).

20. J.D. Whittenberger, Mat. Sci. Eng., 73, p. 87 (1985).

21. S.K. Mannan, K.S. Kumar, and J.D. Whittenberger, Metall. Trans., 21A, p. 2179 (1990).

22. A. Lawley, J.A. Coll, and R.W. Cahn, Trans. ASM, 218, p. 166 (1960).

23. J.D. Whittenberger, Mat. Sci. Eng., 77, p. 103 (1986).

24. G.F. Hancock and B.R. McDonnell, Phys. Stat. Sol., 4(a), p. 143 (1971).

25. A. Lutze-Birk and H. Jacobi, Scripta Metall., 9, p. 761 (1975).

26. S. Shankar and L.L. Seige, Metall. Trans, 9A, p. 1467 (1978).

27. A.E. Berkowitz, F.E. Jaumot, and F.C. Nix, Phys. Rev., 95, p. 1185 (1954).

28. H.C. Hagel, in Intermetallic Compounds, ed. by J.H. Westbrook, p. 377 (1977).

29. K. Hirano and A. Hishinuma, Nippon Kinzoku Gakkaishi, 32, p. 516 (1968); Diffus. Data, 3, p. 270 (1969).

30. K. Nishida, T. Yamamoto, and T. Nagata, Nippon Kinzoku Gakkaishi, 34, p. 591 (1970); Diffus. Data, 5, p. 26 (1971).

31. M.R. Harmouche and A. Wolfenden, Mat. Sci. Eng., 84, p. 35 (1986).

32. M.R. Harmouche and A. Wolfenden, J. Testing and Evaluation, 15, p. 101 (1987).

33. S.V. Raj and S. Farmer, unpublished research, NASA-Lewis Research Center, Cleveland, OH (1991).

34. H.L. Fraser, R.E. Smallman, and M.H. Loretto, Phil. Mag., 28, p. 651 (1973).

35. M.F. Ashby, Acta Metall., 20, p. 887 (1971).
36. J.E. Bird, A.K. Mukherjee, and J.E. Dorn, in Int. Conf. Quantitative Relation Between Properties and Microstructure, ed. by D.G. Brandon and A. Rosen, Israel Univ. Press, p. 255 (1969).
37. A. Ball and R.E. Smallman, Acta Metall., 14, p. 1517 (1966).
38. E.P. Lautenschlager, T.C. Tisone, and J.O. Brittain, Phys. Stat. Sol., 20, p. 443 (1967).
39. P.R. Strutt, R.A. Dodd, and G.M. Rowe, in 2nd Int'l. Conf. Metals and Alloys, ASM, Metals Park, OH, III, p. 1057 (1971).
40. W.R. Kanne Jr., R.R. Strutt, and R.A. Dodd, Trans. AIME, 245, p. 1259 (1969).
41. D.L. Yaney and W.D. Nix, J. Mat. Sci., 23, p. 3088 (1988).
42. D.L. Yaney, A.R. Pelton, and W.D. Nix, J. Mat. Sci., 21, p. 2083 (1986).
43. M. Rudy and G. Sauthoff, Mat. Sci. Eng., 81, p. 525 (1986).
44. J.D. Whittenberger, Mat. Sci. Eng., 57, p. 77 (1983).
45. I. Baker and D.J. Gaydosh, Metallography, 20, p. 347 (1987).
46. I. Jung, M. Rudy, and G. Sauthoff, in High-Temperature Ordered Intermetallic Alloys II, Proc. MRS, ed. by N.S. Stoloff et al., 81, p. 263 (1987).
47. J.D. Whittenberger, R.D. Noebe, C.L. Cullers, K.S. Kumar, and S.K. Mannan, Metall. Trans., in press (1991).
48. J.D. Whittenberger, K.S. Kumar, and S. K. Mannan, J. Mat. Sci., 26, p. 2015 (1991).
49. R.T. Pascoe and C.W.A. Newey, Met. Sci. J., 2, p. 138 (1968).
50. J.D. Whittenberger, NASA-TM 101382 (1987).
51. J.D. Whittenberger, D.J. Gaydosh, and M.V. Nathal, unpublished research, NASA-Lewis Research Center, Cleveland, OH (1991).
52. M.V. Nathal and L.J. Ebert, Metall. Trans., 16A, p. 1863 (1985).
53. J.D. Whittenberger, R.K. Viswanadham, S.K. Mannan, and S.K. Kumar, J. Mat. Res., 4, p. 1164 (1989).
54. V.M. Pathare, PhD Thesis, Case Western Reserve University, Cleveland, OH (1987); NASA CR-182113 (1988).
55. S.V. Raj, I.E. Locci, and R.D. Noebe, unpublished research, NASA-Lewis Research Center, Cleveland, OH (1991).
56. J.D. Whittenberger, Mat. Sci. Eng., 85, p. 91 (1987).
57. R.S. Polvani, W.S. Tzeng, and P.R. Strutt, Metall. Trans., 7A, p. 33 (1976).
58. R. Darolia, J. Metals, 43, no. 3, p. 44 (1991).
59. J.D. Whittenberger, L.J. Westfall, and M.V. Nathal, Scripta Metall., 23, p. 2127 (1989).
60. B. Oliver, R.D. Noebe, and J.D. Whittenberger, unpublished research, Univ. of Tennessee, Knoxville, TN (1991).
61. I.E. Locci, R.D. Noebe, R.R. Bowman, R.V. Miner, M.V. Nathal, and R. Darolia, in High Temperature Ordered Intermetallic Alloys IV, Proc. MRS, ed. by L.A. Johnson et al., 213, p. 1013 (1991).
62. J.D. Whittenberger, D.J. Gaydosh, and K.S. Kumar, J. Mat. Sci., 25, p. 2771 (1990).
63. S.C. Jha, R. Ray, and J.D. Whittenberger, Mat. Sci. Eng., A119, p. 103 (1989).
64. I.E. Locci, R.D. Noebe, J.A. Moser, D.S. Lee, and M.V. Nathal, in High Temperature Ordered Intermetallic Alloys III, Proc. MRS, ed. by C.T. Liu et al., 133, p. 639 (1989).
65. J.D. Whittenberger, R. Ray, S. Jha, and S.L. Draper, Met. Sci. Eng., in press (1991).
66. R.C. Benn, in MiCon 86: Optimization of Processing, Properties, and

Service Performance through Microstructural Control, ed. by B.L. Bramfitt et al., ASTM STP 979, p. 238 (1988).
67. M.A. Morris and D.G. Morris, Acta Metall. Mater., 38, p. 551 (1990).
68. S. Struthers, PhD Thesis, Case Western Reserve University, Cleveland, OH (1991).
69. J.D. Whittenberger, E. Arzt, and M.J. Luton, J. Mater. Res., 5, p. 271 (1990).
70. J.D. Whittenberger, E. Arzt, and M.J. Luton, J. Mater. Res., 5, p. 2819 (1990).
71. J.D. Whittenberger, K.S. Kumar, and S.K. Mannan, Materials at High Temperature, 9, p. 3 (1991).
72. A. Kelly and K.N. Street, Proc. Roy. Soc. London, A 328, p. 283 (1972).
73. J.L. Walter and H.E. Cline, Metall. Trans., 1, p. 1221 (1970).
74. J.L. Walter and H.E. Cline, Metall. Trans., 4, p. 33 (1973).
75. D.P. Mason, D.C. Van Aken, R.D. Noebe, I.E. Locci, and K.L. King, in High Temperature Ordered Intermetallic Alloys IV, Proc. MRS, ed. by L.A. Johnson et al., 213, p. 1033 (1991).
76. R.D. Noebe, R.R. Bowman, and J.I. Eldridge, in Intermetallic Matrix Composites, Proc. MRS, ed. by D.L. Anton et al., 194, p. 323 (1990).

CREEP IN L1$_2$-INTERMETALLICS

J. H. SCHNEIBEL
Oak Ridge National Laboratory
P.O. Box 2008
Oak Ridge, TN 37831-6116
U.S.A.

P. M. HAZZLEDINE
University of Oxford
Department of Materials
Parks Road
Oxford, OX1 3PH
United Kingdom

ABSTRACT. Several intermetallic compounds with the L1$_2$-structure exhibit a yield strength anomaly (YSA), i.e., the yield stress increases as the temperature increases. As shown for Ni$_3$Al, no corresponding "creep strength anomaly" exists. The creep strength decreases in a normal manner with increasing temperature. Its orientation dependence is different from that observed in the YSA regime. The transient creep of Ni$_3$Al is composed of a normal ("work hardening") and an inverse ("work softening") part. The reasons for such transients are discussed. Specific models for the dependence of strain rate vs. strain during inverse creep as well as for the orientation dependence of the creep strength are reviewed and compared with experiments. Experiments and models indicate that the strain rate $\dot{\varepsilon}$ depends on the strain ε approximately as ε or $\varepsilon^{2/3}$. The orientation dependence of the creep strength is particularly pronounced at low temperatures and changes at high temperatures and low stresses. At low creep stresses, Ni$_3$Al is not as strong as high-strength superalloys, due to its low stress exponent. However, further optimization may be possible. As to the creep of other L1$_2$-compounds, very little information is available and no distinct evidence for inverse creep has been found. This may be due to the lower yield stresses and less pronounced YSA in these compounds.

1. Introduction

Many intermetallics with the L1$_2$-structure show an anomalous dependence of yield stress vs. temperature. The yield stress anomaly (YSA) of the technologically important Ni$_3$Al system has attracted the most attention. The orientation dependence and tension-compression asymmetry of the yield stress of Ni$_3$Al have been studied very carefully both experimentally and theoretically [1,2]. While the YSA is a very important feature of Ni$_3$Al, the strain rates at which it is typically measured are much

C. T. Liu et al. (eds.), Ordered Intermetallics – Physical Metallurgy and Mechanical Behaviour, 565–581.

higher than those likely to be encountered in practical applications, where high-temperature creep, and not the yield stress, are life-time controlling. In high temperature creep there is no creep strength anomaly which corresponds to the YSA - as the temperature is raised, the creep strength decreases. Nevertheless it will be seen that Ni_3Al exhibits a much higher creep strength than would be expected from the solution strengthening of aluminium in nickel. It is in this sense that the creep strength of single-phase Ni_3Al may be called anomalous.

We will start out with a general discussion of creep deformation, focussing on creep transients. Then the most important observations concerning slip planes and dislocations in Ni_3Al, the most intensively investigated $L1_2$-compound, will be briefly summarized. Mechanisms responsible for the inverse creep behavior of Ni_3Al will be outlined and compared to available creep data. Predicted orientation dependences of the creep strength will be compared to measured ones. Finally, some results on other $L1_2$-intermetallics will be discussed.

2. Phenomenology of Normal and Inverse Creep

Creep refers to time-dependent deformation at a constant applied stress. The mechanisms which govern creep deformation are usually studied under conditions of constant stress or constant load, although they may be important in other testing conditions as well. In particular in the case of an externally applied constant extension or strain rate, creep mechanisms may be stress-controlling for low strain rates and high temperatures. Unless otherwise stated we assume in the following creep mechanisms which are based on dislocations, i.e., we are concerned either with single crystals, or with polycrystals with sufficiently large grain sizes such that contributions from Nabarro-Herring and Coble creep can be ignored.

Upon applying a constant stress, creep deformation usually begins with a transient, i.e., the strain rate changes with strain and time. Assuming that the dislocation density at the start of a creep test is small, we can distinguish between two different creep regimes. One is usually associated with pure metals (M), whereas the other one is observed in alloys (A). The two regimes show different transient behavior during the initial stages of creep (see Fig. 1 and Table I). If the applied stress σ_a is higher than the yield stress σ_y, then the material will deform significantly on initial loading. Due to work hardening (dislocation-dislocation interaction) the initially high strain rate will usually decrease monotonically with time and strain. Such a "normal" transient is often observed in pure metals. If the applied stress is smaller than the yield stress, the strain upon initial loading will be very small. The available dislocations will move slowly by thermally activated processes. Correspondingly dislocation nucleation will be a slow process. With increasing strain and time more dislocations are generated and the strain rate increases. As long as the dislocation density ρ is well below $(\sigma/Gb)^2$, interactions between adjacent dislocations can be ignored and the creep rate increases according to $\dot{\varepsilon} = \rho\, b\, v$. Such "inverse" creep transients, which are also called work-softening transients [3] are observed in some solid solution alloys such as Al-Mg [4]. For more detailed discussions of creep behavior in pure metals and alloys we refer to Poirier [5] and Blum et al. [3]. It should be noted that we deviate from conventional practice in so far as we define regimes M and A in terms of their creep transients, and not in terms of the stress exponent under steady-state creep conditions.

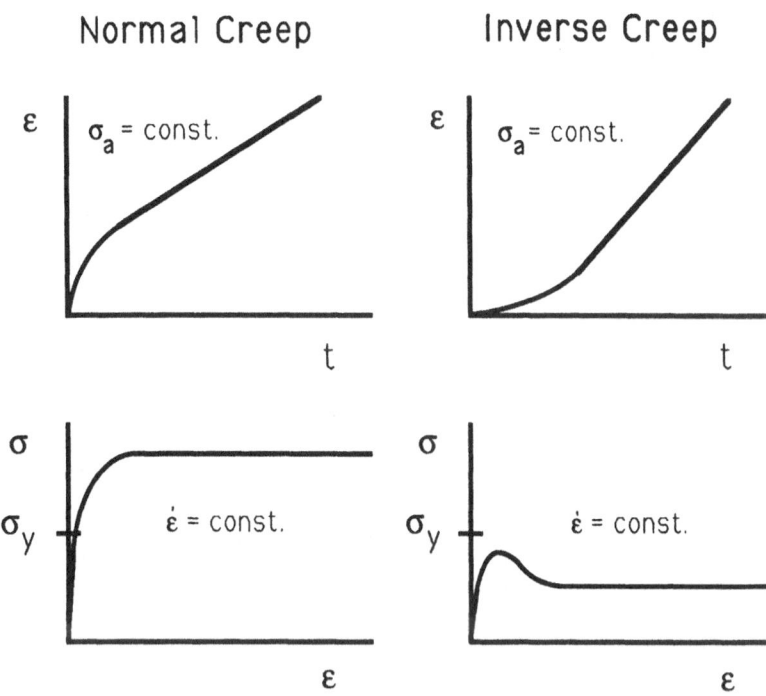

Fig. 1. Schematic comparison of normal and inverse transient creep for constant stress and constant strain rate conditions.

TABLE 1
Characteristics of the two creep regimes M and A

Creep Regime	Condition	Type of Transient	Dislocation Interaction
M (Class II)	$\sigma_a > \sigma_y$	Normal (Work Hardening)	Yes
A (Class I)	$\sigma_a < \sigma_y$	Inverse (Work Softening)	No

Our classification of creep behavior is only as precise as our definition of the yield stress σ_y. If the yield stress is a rate-independent stress required for propagation of individual dislocations, then the above classification is rigorous. If the yield stress is not well defined, for example if it consists of a micro- and a macro-yield, the situation becomes more complicated. In this case the distinction between regimes M and A in terms of applied and yield stresses requires some care.

If no macroscopic instabilites or failure processes occur, primary creep is often followed by a secondary, steady-state creep regime in which the strain rate remains constant, for a given applied stress and temperature. In true steady-state dislocation creep a dynamic dislocation configuration is formed which is independent of strain or time. Secondary creep is followed by tertiary creep, which involves internal damage processes like void formation as well as external instabilites such as necking.

Inverse creep curves may look very similar to tertiary ones. In contrast to tertiary creep, inverse creep may be followed by a steady-state (secondary) regime. Inverse creep is therefore a primary transient. In practice inverse and tertiary creep may occur at the same time and their experimental separation then requires microstructural observations or, alternatively, testing techniques designed to eliminate tertiary processes.

3. Slip Systems and Dislocations in Ni$_3$Al

It is well established that at strain rates around 10^{-4} to 10^{-3} s^{-1} and at temperatures below that of the peak yield stress, dislocations glide on the primary octahedral slip system. At higher temperatures glide occurs usually on the primary cube plane. Due to the rate sensitivity of the yield stress on cube planes the temperature of the peak yield stress decreases with decreasing strain rate. For sufficiently low strain rates, cube glide may therefore occur at relatively low temperatures. Hemker and Nix [6, 7, 8] showed that at relatively low temperatures and high stresses (903 K, 745 MPa) slip in Hf-alloyed Ni$_3$Al occurs on the secondary, instead of the primary, cube cross slip plane. In this case dislocations generated on the primary octahedral slip system aid slip on the secondary cube slip system.

For orientations very close to the [001] orientation, for which the resolved shear stress on cube planes is close to 0, slip on {111} planes may occur at temperatures above the peak yield stress [9]. Under creep conditions, glide on [011] planes has also been observed [10].

Under creep conditions, screw dislocations on {111} planes are fully cross-slipped with a typically 5 nm wide antiphase boundary (APB) ribbon on {001}. Some degree of mobility on {111} planes is achieved by the lateral propagation of superkinks [11]. Dislocations gliding on {100}, on the other hand, lie mostly in this plane, with the exception of narrow (0.5 nm) complex stacking fault ribbons on {111}. When the temperature is high enough to reduce the high Peierls stress on {100} these dislocations can glide relatively easily on {111} as demonstrated by yield stress measurements in this regime. Dislocation gliding on {110} have been observed by Caron et al. [10] in the creep of Ni$_3$Al. These dislocations are dissociated perpendicular to their glide plane. Caron et al. suggested that the movement of such dislocations involves diffusive APB dragging. A similar process might also occur for screw dislocations on {111}.

When cube slip occurs, many dislocation reactions resulting in many types of locks may occur. For more information we refer to work by Hazzledine et al. [12] as well as Sun et al. [13]. The situtation is likely to become even more complicated if slip on {011} occurs.

4. Models of Inverse Creep in Ni_3Al

4.1. INVERSE CREEP OF SUPERALLOYS

Although superalloys do not consist of single-phase Ni_3Al, their creep behavior shows certain similarities with that of Ni_3Al. In particular, the creep rate of superalloys often increases with strain and time, i.e., inverse creep is observed. Ghosh et al. [14] assume therefore that the rate of dislocation generation in superalloys is proportional to the strain rate. If interactions between gliding dislocations can be ignored, then it follows from $\dot{\varepsilon} = \rho$ bv that the strain rate increase during creep is proportional to the strain. Ghosh et al. refer to this process as tertiary creep. Since the model of Ghosh et al. does not involve fracture or damage processes and since steady-state creep, if it occurs, will only be found in later stages of creep, this model refers, strictly speaking, to primary creep.

4.2. THE MODEL OF HEMKER

Hemker found [6,7,8] that the small amount of octahedral slip generated in the first stages of creep is quickly exhausted due to locking of the screw dislocations by cube cross-slip. In his model for inverse creep [8] he assumes that the octahedral slip creates a constant density N_s of dislocation sources for slip on the cube cross slip planes. A new dislocation is nucleated whenever that previously generated has reached a distance λ from the source, i.e., a loop diameter of 2λ. Since the dislocations are assumed to expand with a constant velocity v, their nucleation rate is constant. The strain rate calculated from this model is [8] :

$$\dot{\varepsilon} = \pi\, N_s\, b\, v^2\, [t + (v/\lambda)\, t^2] \tag{1}$$

Assuming the initial strain to be 0, the strain as a function of time is given by:

$$\varepsilon = \pi\, N_s\, b\, v^2\, [(1/2)\, t^2 + (1/3)\, (v/\lambda)\, t^3] \tag{2}$$

The relationship between strain rate and strain is characterized by two limiting cases:

$$t\, v \ll \lambda: \qquad\qquad \dot{\varepsilon} = (2\, \pi\, N_s\, b)^{1/2}\, v\, \varepsilon^{1/2} \tag{3}$$

and

$$t\, v \gg \lambda: \qquad\qquad \dot{\varepsilon} = 3^{2/3}\, (\pi\, N_s\, b/\lambda)^{1/3}\, v\, \varepsilon^{2/3}\,. \tag{4}$$

The first case corresponds to the expansion of a single dislocation loop without nucleation of new dislocations, whereas the second case involves continuous nucleation at time intervals $\Delta t = \lambda/v$.

The density N_s of the sources is assumed to remain constant during deformation. As dislocation density builds up during deformation, the source density might increase, and this could result in an exponent higher than 2/3 in the equation relating $\dot{\varepsilon}$ and ε. A close to linear relationship similar to that observed in superalloys might then result.

570

For t v >> λ we obtain for the time t to obtain a certain strain ε:

$$t = (3 \varepsilon /\pi b)^{1/3} (\lambda/N_s)^{1/3}(1/v) .$$

If λ and N_s are assumed to be stress-independent and if dislocation glide is a viscous process obeying v ∝ σ then the time required to reach a given strain is proportional to 1/σ. In order to obtain a more realistic, i.e. stronger, dependence of t on the applied stress the term λ/N_s must in practice depend significantly on the stress.

Inverse creep in Hemker's model takes place on cube planes. Correspondingly, the creep strength decreases as the resolved shear stress on the cube plane increases, i.e., as the sample orientation moves away from [001]. This trend is schematically shown in the stereographic triangle in Fig. 2(a).

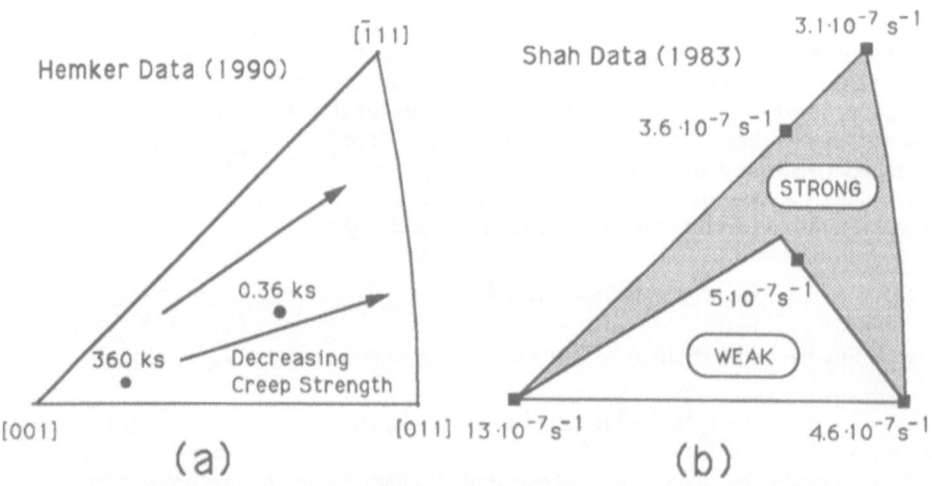

Fig. 2. (a) Standard stereographic triangle indicating the orientation dependence of the creep strength if cube slip is the important factor. Times-to-fracture measured by Hemker [8] for Al-22Al-1Hf-0.2B (at.%) crept at 916 K and 745 MPa are indicated. (b) Standard stereographic triangle with the orientation dependence of creep in Ni₃Al predicted by Hazzledine and Schneibel [15]. Creep rates determined by Shah [23] for Ni-23.5 at. % Al at 1255 K and 69 MPa are indicated.

4.3 THE MODEL OF HAZZLEDINE AND SCHNEIBEL

Based on Hemker's experimental results, Hazzledine and Schneibel [15] proposed an alternative model to explain inverse creep in Ni₃Al which also makes predictions about the orientation dependence of the creep strength. In this model it is assumed that dislocations are generated on two intersecting octahedral slip systems which interact on a common cube cross slip plane according to

$$[\bar{1}01] + [101] = [001] + [001] .$$

The assumption made here is that the [001] dislocations are more glissile on the cube plane than the <101> dislocations. Hazzledine et al.´s [12] work suggests that this may be the case. As viscous glide continues on the cube cross slip plane, more dislocations are injected from the octahedral slip systems. The dislocation density on the cube plane increases and inverse creep ensues. This mechanism applies to those orientations for which (a) the cube cross slip plane is supplied with dislocations from two intersecting octahedral planes and (b) for which the Schmid factor of the cube plane is larger than 0. For all other orientations the above dislocation reaction cannot occur and the creep rate will be lower (provided that a high Schmid factor for the active slip plane does not increase the creep rate excessively). That region where the creep rate is likely to be high and which should be avoided is indicated in the standard stereographic triangle in Fig. 2(b). For [001] orientations the model would predict, like Hemker´s model, a negligable creep rate. In practice, however, {111} and {011} slip may then result in a finite creep rate.

5. Creep Experiments With Ni₃Al and Their Interpretation

5.1. SOLID SOLUTION STRENGTHENING

We will show first that Ni_3Al has a much higher creep strength than would be expected from an extrapolation of solid solution strengthened nickel-aluminium. In Fig. 3 we show the creep strengths of nickel [16], Ni-10 at.% Al [17] and Ni-24 at.% Al [18] at 1033 K. Solid solution strengthening due to a solute concentration c is approximately proportional to $c^{1/2}$. Extrapolation of the creep strengths of nickel and of the solid-solution alloy Ni-10 at.% Al with an upper bound indicated by the broken line shows clearly that Ni_3Al is much stronger in creep than expected from solid solution strengthening alone. One reason for this is the expected reduction in the diffusivities due to ordering. However, this effect is generally small in $L1_2$-intermetallics [19]. We expect therefore that, in a manner similar to the YSA, dislocation dissociations and transformations to be the cause for the high creep strength. In this sense the creep strength of Ni_3Al may be called anomalous. As the aluminum concentration is increased such that the compound Ni_3Al forms, the creep strength increases discontinuously to very high values

5.2. INVERSE CREEP IN Ni₃Al

In Table II the results of creep experiments for a number of alloys based on Ni_3Al are compiled. Quite clearly, inverse creep is very common in this system. It occurs in single crystals as well as in polycrystals, in tension as well as in compression, and over a wide range of temperatures (916 - 1273 K). In a number of cases a small normal transient is observed upon application of the load which is then followed by an inverse transient. The small normal transient was shown by Hemker and Nix [6,7,8] to be due to the exhaustion of octahedral slip (the creep stress was only 75% of the yield stress). Subsequent inverse creep in these experiments was controlled by glide on the cube cross slip system and not by glide on the primary cube slip system, i.e., that with the highest Schmid factor. TEM observations suggest that the inverse transient occurs **prior** to the formation of a stationary creep structure [6,7,8,21].

TABLE 2
Compilation of pertinent creep data for Ni₃Al - alloys

Authors	Composition (at. %)	Orientation	Temp. (K)	Stress(MPa)	Primary I (normal)	Primary II (inverse)	Min. Creep Rate (1/s)	Steady State Rate (1/s)
Davies [20]	Ni-14.5Al-10.5Ti	Polycrystal	1198	205	N	Y	$1.5 \cdot 10^{-7}$	-
Schneibel [21]	Ni-23.5Al-0.5Hf-0.2B	Polycrystal	1033	250	0.0-0.3%	Y	$2 \cdot 10^{-8}$	-
Hemker [8]	N-22.2Al-1.0Hf-0.24B	[-0.05,0.20]	916	745	0-0.25%	Y	$2.6 \cdot 10^{-8}$	-
Hemker [8]	a/a	a/a	1273	312	-	0-1%	$9.1 \cdot 10^{-5}$	$3.5 \cdot 10^{-4}$
Hemker [8]	a/a	[-0.65,1.75,3]	1275	131	-	-	$3.3 \cdot 10^{-6}$	$3.3 \cdot 10^{-6}$
Khan [22]	Ni-Co-Cr-Mo-W-Al-Ti-Ta	within 5° of [001]	1033	650	0-4%	?	$3 \cdot 10^{-7}$	-
Khan [22]	Ni-22Al-2Hf	[001]	1033	280	-	Y	$2.1 \cdot 10^{-8}$	-
Shah [23]	Ni-23.5Al	[-123]	1255	69	-	-	$5 \cdot 10^{-7}$	$5 \cdot 10^{-7}$
Anton [24]	Ni-17Al-6Ta	within 5° of [001]	1033	552	0-0.3%	Y	0	-
Nathal [25]	Ni-Al-Cr-Ti-Ta-W	within 2° of [001]	1273	160	N	0.5%	10^{-6}	$3 \cdot 10^{-6}$

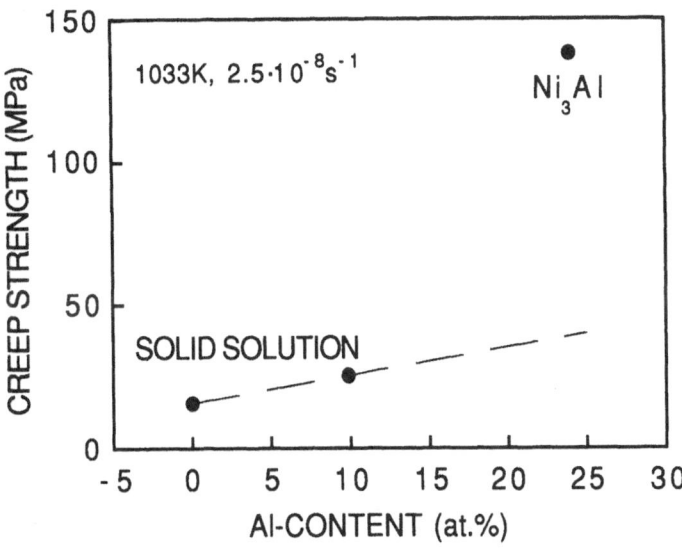

Fig. 3. Creep strengths in the Ni-Al system as a function of Al-concentration. The creep strengths for Ni and a Ni-10 at.% Al solid solution were extrapolated from references [16] and [17]. The creep strength for Ni₃Al is that given by Liu [18].

This transient is therefore a primary transient, just like the initial normal transient. We call these two transients Primary I and Primary II. As described by Hazzledine and Schneibel [15], two transients occur since two different yield stresses exists. The lower one corresponds to the movement of mobile edge dislocations on octahedral planes whereas the higher one corresponds to the more difficult propagation of cross-slipped screw dislocations. The creep stress lies between these two yield stresses. Upon applying the creep stress we obtain first a normal (work hardening) transient and then an inverse (work softening) transient (see also Fig. 1 for σ_a = const.). An example for these two transients is given in Fig. 4. The minimum creep rate at the transition between the two transients does not correspond to steady-state creep since the dislocation structure continues to evolve during inverse creep [21]. The measured dislocation densities suggest that dislocation interaction during inverse creep is not a major factor. After a strain of 0.9% the dislocation density is approximately 10^{12} m^{-2} (Fig. 5 in Ref. [21]). The dislocation density calculated from the relation $\rho = (\sigma/Gb)^2$, on the other hand, has a much higher value of 10^{14} m^{-2}. The requirements for inverse creep outlined in section 2 are therefore fullfilled.

At low temperatures (e.g. 1000 K) there is usually no evidence for steady-state creep. Tertiary creep sets in before steady-state conditions are reached. At high temperatures (e.g. 1270 K) steady-state creep is readily found [25], but then there is no evidence for Primary I, and Primary II is not very pronounced. Since the stresses applied at high temperatures are significantly lower than those at low temperatures, the small strains associated with Primary I are more difficult to verify, and Primary I may nevertheless exist.

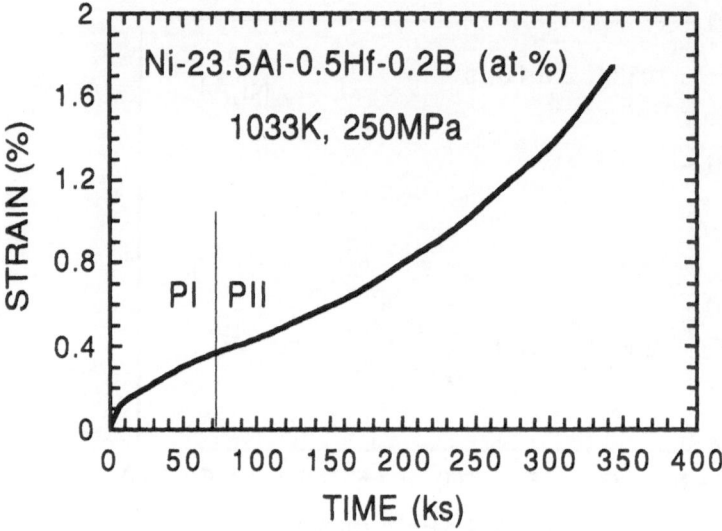

Fig. 4. The primary creep in Ni-23.5Al-0.5Hf-0.2B (at. %) [21] consists of a
normal transient (Primary I) and an inverse transient (Primary II).

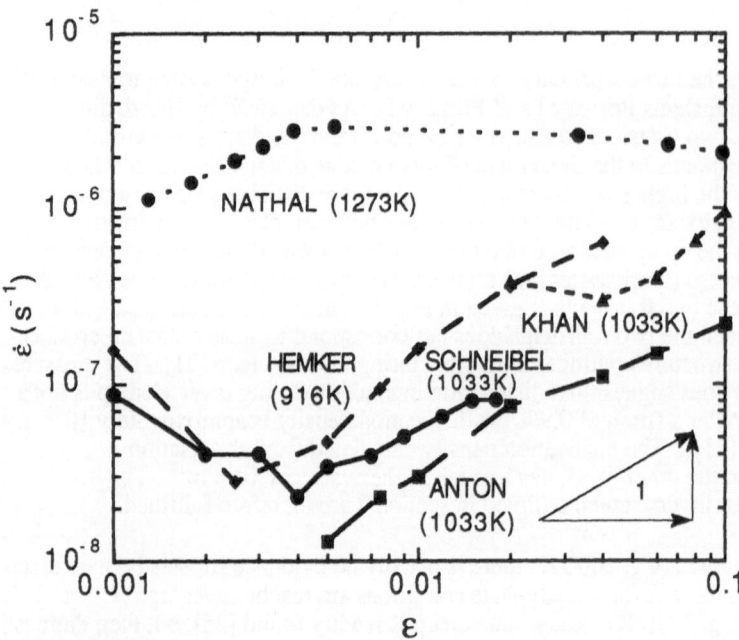

Fig. 5. Strain rate as a function of strain for several Ni$_3$Al alloys at various
temperatures (for details see Table II).

In Fig. 5 the inverse creep behavior of some of the Ni$_3$Al alloys listed in Table II is visualized in a plot of log $\dot{\varepsilon}$ vs. log ε. In the inverse creep regime the data is described fairly well by ($\dot{\varepsilon} \propto \varepsilon$ or $\dot{\varepsilon} \propto \varepsilon^{2/3}$. These dependences agree both with the Ghosh et al [14] and Hemker [8] models. The scatter in the data does not however permit a discrimation between the two models.

Figure 6 shows compression experiments performed with constant extension rates by Miura et al [9]. The orientation was near [100] and a range of temperatures and strain rates was examined. For relatively low temperatures and high strain rates, i.e., in the region of the YSA, work hardening is observed. For high temperatures and low strain rates, on the other hand, work softening occurs. As pointed out by Nathal et al. [25] and shown schematically in Fig. 1, the stress peaks in such experiments correspond to the inverse creep behavior found under constant stress conditions.

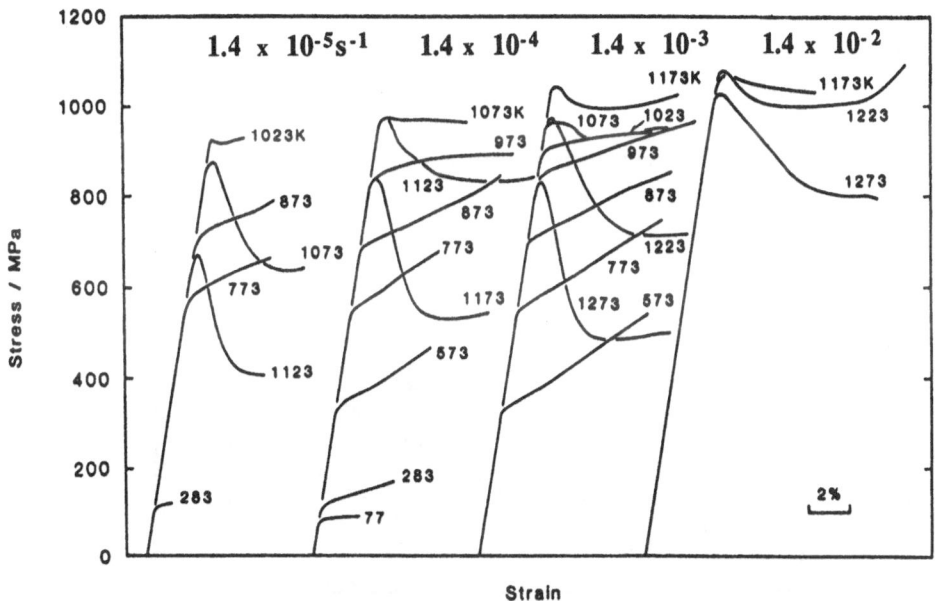

Fig. 6. Stress vs. strain for Ni$_3$(Al,Ti) deformed with different rates in compression [9]. Reproduction by permission of the Deutsche Gesellschaft für Materialkunde.

At high temperatures (typ. 1250 K) the dislocations are homogeneously distributed during creep [8,25]. At lower temperatures and higher stresses this is not the case. Generally, there are distinct inhomogeneities in the distribution of the dislocation [8,21]. This is plausible since deformation, once started in a particular location, tends to concentrate there due to strain softening. The dislocation arrangement in polycrystalline Ni$_3$Al after 1 - 2% inverse creep at 1033 K can be described by dislocation-free and dislocation-containing regions [21]. The volume fraction of dislocation-containing regions increases with increasing strain. This suggests an increase in the average dislocation density as required for inverse creep. In a control experiment dislocations were introduced by room-temperature deformation prior to creep-testing. The initial creep rate increased and inverse creep disappeared. Similarly, if creep is not controlled by dislocation movement but rather by diffusional creep, the inverse creep behavior is not observed [21]. All these findings are qualitatively consistent with the various models of inverse creep. Unfortunately, neither the models nor the experiments are presently precise enough to give any of the models a clear preference.

5.3. STEADY-STATE CREEP IN Ni3Al

The experiments by Miura et al [9] in Fig. 6 show that for sufficiently high temperatures and low strain rates the flow stress of Ni₃(Al,Ti) becomes approximately constant after a strain of about 5%. This indicates steady-state creep. The appropriate strain rates are plotted in Fig. 7 as a function of the flow stress. The stress exponents reach values as low as 3. Similar results were obtained by Hsu et al. [26] for creep of single crystal Ni₃Al containing chromium. They determined stress exponents varying between 2.9 and 3.2. The fact that steady-state may be achieved in Ni₃Al is another indicator that the inverse creep observed prior to steady-state is a primary transient.

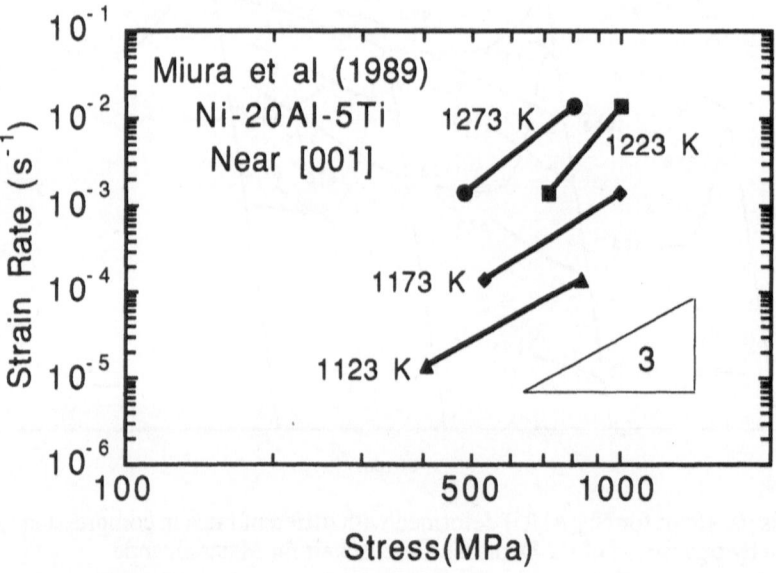

Fig. 7. Dependence of strain rate on stress for steady-state creep in Ni₃(Al,Ti), evaluated from data by Miura [9] in Fig. 6.

5.4 ORIENTATION DEPENDENCE OF CREEP IN Ni3Al

The orientation dependence of the yield strength of Ni₃Al changes as one increases the temperature while keeping the strain rate constant [27]. In compression experiments in the YSA regime, orientations close to [001] are weaker than those close to [111], since the tendency for cross slip is smaller. At temperatures above the YSA, however, the [001] orientation is strongest, since in this regime cube slip becomes important and since orientations close to [001] have a small Schmid factor for cube slip [27]. Since creep is associated with cube slip, one therefore expects that the orientation dependence of the creep strength is different from that in the YSA-regime.

In his examination of the orientation dependence of the creep strength, Hemker creep-tested specimens with orientations close to [001] and [$\bar{1}$23] at 916 K and 745 MPa. The Schmid factors for the cube planes of the [$\bar{1}$23] orientation were 0.437 and 0.372, respectively. The time to failure for the near [001] orientation with cube plane Schmid factors of 0.17 and 0.152 was 1000 times longer [see Fig. 2(a)]. This large effect is difficult to rationalize on the basis of Schmid factors alone since a very high dependence of the strain rate on the stress would then be required. An additional factor indicated in the work of Bellows and Tien [28] probably becomes important. These authors noted in creep-fatigue experiments with Ni-11.7Al-0.51B-3.5Hf (wt.%) at 1033 K that specimens more than 7° off [001] showed erratic behavior. This irreproducibility creates some uncertainty as to the exact magnitude of the orientation effect in creep fatigue. It is likely that similar irreproducibilities occur in creep experiments. A possible explanation for the erratic behavior is the inverse creep behavior. Once deformation sets in it may quickly become localized, with a concomitant increase in the strain rate and eventual failure. In spite of these complications Bellows and Tien´s experiments are generally consistent with Hemker´s work.

At temperatures around 1300 K the orientation dependence of the creep strength is much less pronounced than at lower temperatures. A rationalization in terms of cube plane Schmid factors is in this case unsuccessful. This is readily seen from Fig. 2(b) which contains creep rates measured by Shah [23] for different orientations. It is however seen that those orientations which straddle the "weak" regime in the Hazzledine and Schneibel model are those with the highest creep rates. The orientations in the "strong" regime, on the other hand, exhibit the lowest creep rates. The particularly high creep rate for the [001] orientation cannot be rationalized in terms of cube slip, since the corresponding Schmid factor is 0. Presumably {111} and/or {011} slip become active for this orientation. Although Shah´s results do not prove conclusively the Hazzledine and Schneibel model they are nevertheless consistent with it.

6. Creep in Other Ll$_2$ Intermetallics

Considering the fact that there has been relatively little interest in the creep behavior of a much investigated material like Ni$_3$Al it is not surprising that there has not been a lot of activity regarding the creep behavior of other Ll$_2$-intermetallics. One notable exception is the work of Liu on (Fe, Co, Ni)$_3$V alloys [29]. This work shows very clearly the increase in creep strength on ordering. On ordering the creep rate decreases by over two orders of magnitude. Inverse creep transients have however not been reported. This is not surprising for two reasons. First, the YSA in Fe-38Ni-25V-0.5Ti (at.%) is much less pronounced than that in Ni$_3$Al. The peak yield stress is significantly lower than in Ni$_3$Al. Second, the creep stresses employed in these creep experiments were close to the yield stress, whereas they should be significantly below the yield stress for inverse creep to be observed. Provided that suitable parameters are chosen, inverse creep may very well be observable in FeNiVTi.

Trialuminides based on Al_3Ti have recently received some attention as high-temperature, low-density structural materials. These systems show generally only a small YSA. Two compression creep tests performed by us with Al-8Cr-25Ti [30] have not shown conclusive evidence for inverse creep. Although inverse creep may exist in this system, it will be harder to identify than in Ni_3Al. The reason for this are the relatively low yield stresses which correspond to lower creep stresses. For low creep stresses small strains are required to produce the relatively low dislocation densities corresponding to steady-state creep. Therefore inverse creep is likely to be less pronounced.

7. Discussion

The inverse creep behavior in Ni_3Al is now well established. The main reason for the distinctiveness of inverse creep in this material is the high critical resolved shear stress required for extensive dislocation motion. Typical creep stresses are substantially below these levels and therefore creep rates are controlled by some form of viscous dislocation glide. The slow glide motion corresponds to a gradual build-up of the low initial dislocation density. To a first approximation the increase in dislocation density is proportional to the creep strain. Therefore, as long as the interaction between individual dislocations can be ignored the creep rate increases steadily with strain and time. High creep strengths are obtained only for low dislocation densities.

On a more microscopic basis the work of Hemker and Hemker and Nix [6,7,8] has significantly increased our understanding of the creep behavior of Ni_3Al. These authors showed that at relatively low temperatures and high stresses, cube slip does not occur on the cube slip system with the highest Schmid factor but rather on the cube cross slip system for the primary octahedral system. Dislocation generation is therefore linked to activity on the primary octahedral system. It appears that dislocation generation on the cube system without aid from octahedral slip is a much more difficult process for these particular experimental conditions. The situation is different at higher temperatures (e.g. 1250 K) at which slip occurs also on the primary cube plane. Similarly, the orientation dependence of the creep strength changes at higher temperatures, suggesting different creep mechanisms. It should however be realized that the applied creep stresses are usually quite different at different temperatures. It is conceivable that the creep mechanisms are independent of temperature as long as they compared at the same stress.

The explanations for inverse creep in Ni_3Al proposed to date are based on cube slip. Inverse creep is however more generic in nature - in particular it occurs in solid solution alloys. Miura et al [9] found inverse creep (strictly speaking strain softening) for [001] orientations exhibiting octahedral slip. Cube slip is therefore not a necessary condition for inverse creep to occur in Ni_3Al. The main requirement is the viscous motion of dislocations which may occur not only on {001}, but also on {111} and {011} slip planes.

Inverse creep as it is observed in Ni$_3$Al is, in principle, well suited to study the viscous glide behavior of dislocations since interaction between individual dislocations is not a major factor and since most dislocations can be assumed to be mobile. If dislocation densities are determined, the average velocity of individual dislocations can be assessed. This velocity will depend on such processes as superkink motion [11] or APB dragging [10]. Ultimately it may be possible to calculate, from information on the glide plane, dislocation structure, dislocation density and self-diffusivity the rate with which individual dislocations propagate during high temperature deformation and compare it to experimental values.

The special creep behavior of Ni$_3$Al has implications for future uses. As shown by Khan [22], single-phase Ni$_3$Al alloys can have high temperature yield strengths similar to those of high-strength superalloys. However, due to its low stress exponent, the single-phase Ni$_3$Al is significantly weaker under creep conditions. If creep strength is the only criterion, Ni$_3$Al can therefore not compete with advanced superalloys exhibiting significantly higher stress exponents. However, there are several points which require more investigation and which may be of importance in future uses of Ni$_3$Al and possibly other L1$_2$-intermetallics. For example, it is not clear how the inverse creep behavior of Ni$_3$Al influences its creep crack growth behavior, as compared to superalloys. On the one hand, inverse creep is a destabilizing process, since local deformation increases the local creep rate and remains therefore localized. Creep deformation at a crack tip may thus weaken the material at the tip of the crack. The stress exponent for steady-state deformation, on the other hand, has a relatively low value. This translates into a high resistance towards crack growth and neck formation. Finite element calculations to assess the net crack growth behavior of Ni$_3$Al would be desirable.

In the past, several comparisons have been made between the creep strength of superalloys and the corresponding γ and γ'-phases [22,25]. In all cases the single-phase γ' is distinctly weaker in creep than the superalloy. It should be pointed out, however, that these comparisons between γ and γ' creep behavior are not as clear as they seem at first sight. In superalloys creep deformation at low stresses takes place primarily in the γ-phase, i.e., γ' does not deform significantly [31]. Also, the optimization of the creep strength of superalloys involves a combination of γ and γ'. Since the γ' in the superalloy is not directly involved in the creep deformation there is no reason to assume that the optimum composition of γ' in a superalloy will be the optimum composition of single-phase γ'. Similarly, the orientation dependence of the creep strength of γ' has not been examined in sufficient detail. It is clear by now that that orientation which maximizes the yield strength in the regime of the YSA is not that with the highest creep strength. The optimum orientation for maximum creep strength in Ni$_3$Al has not yet been found. This is not an easy task since the orientation dependence is stress and/or temperature dependent.

The creep behavior of L1$_2$-intermetallics other than Ni$_3$Al has not been studied in great detail. If the mechanism of the YSA is similar to that in Ni$_3$Al and if cube glide occurs at high temperatures or low strain rates, respectively, then inverse creep behavior qualitatively similar to that of Ni$_3$Al is expected. Ni$_3$Ga falls into this category. The situation is different if no or only a weak YSA exists. The creep stresses employed are than relatively low and only small strains are required to build up a sufficiently high dislocation density for dislocation interactions and steady-state creep to occur. This means that inverse creep will be less easily observed.

580

Acknowledgments

This research was sponsored by the Division of Materials Science, U.S. Department of Energy under contract DE-AC05-84OR21400 with Martin Marietta Energy Systems, Inc. J. H. Schneibel acknowledges stimulating discussions with Dr. J. Rösler. The manuscript was prepared by G. Sims.

References

1. Pope, D. P. and Ezz, S. S. (1984), 'Mechanical properties of Ni_3Al and nickel-base alloys with high volume fraction of γ''', Int. Met. Rev. **29**, 136-167.
2. Paidar, V., Pope, D. P. and Vitek, V. (1984), 'A theory of the anomalous yield behavior in $L1_2$ ordered alloys', Acta Metall. **32**, 435-448.
3. Blum, W., Straub, S. and Vogler, S. (1991), 'Creep of pure materials and alloys', to be published in ICSMA-9, Haifa, Israel.
4. Horiuchi, R. and Otsuka, M. (1972), 'Mechanism of high-temperature creep of aluminum-magnesium solid-solution alloys', Trans. JIM **13**, 284.
5. Poirier, J.-P. (1985) Creep of Crystals, Cambridge University Press, Cambridge.
6. Hemker, K. J. and Nix, W. D. (1989), 'An investigation of the creep of $Ni_3Al(B,Hf)$ single crystals at intermediate temperatures', in Mat. Res. Soc. Symp. Proc. Vol. 133 on 'High-temperature ordered intermetallic alloys III,' C. T. Liu, A. I. Taub, N. S. Stoloff and C. C. Koch (eds.), Materials Research Society, Pittsburgh, PA, pp. 480-486.
7. Hemker, K. J. and Nix, W. D. (1990), 'An investigation of creep in $Ni_3Al(B,Hf)$', in B. Wilshire and R. W. Evans (eds.), Creep and fracture of engineering materials and structures, The Institute of Metals, London, pp. 51-63.
8. Hemker, K. J. (1990), 'A study of the high temperature deformation of the intermetallic alloy Ni_3Al', Doctoral Dissertation, Stanford.
9. Miura, S. , Mishima, Y. and Suzuki, T. (1989), 'The CRSS for octahedral slip in $Ni_3(Al,X)$ single crystals at elevated temperatures', Z. Metallkde. **80**, 164-169.
10. Caron, P., Khan, T. and Veyssière, P. (1989), '{110} slip in Ni_3Al crept at 760°C along [001]', Philos. Mag. A **60**, 267-281.
11. Sun, Y. Q. and Hazzledine, P. M. (1988), 'A TEM weak-beam study of dislocations in γ' in a deformed Ni-based superalloy', Philos. Mag. A **58**, 603-618.
12. Hazzledine, P. M., Yoo, M. H. and Sun, Y. Q. (1989), 'The geometry of glide in Ni_3Al at temperatures above the flow stress peak', Acta Metall. **37**, 3235-3244.
13. Sun, Y. Q., Hazzledine, P. M. and Crimp, M. A. (1991), 'Dislocations with non-planar locked structures in the $L1_2$ ordered alloys', in Mat. Res. Soc. Symp. Proc. Vol. 213 on 'High-temperature ordered intermetallic alloys IV', L. A. Johnson, D. P. Pope and J. O. Stiegler (eds.), Materials Research Society, Pittsburgh, PA, pp. 311-316.
14. Ghosh, R. N., Curtis, R. V. and McLean, M. (1990), 'Creep deformation of single crystal superalloys - modelling the crystallographic anisotropy', Acta Metall. **38**, 1977-1992.
15. Hazzledine, P. M. and Schneibel, J. H. (1989), 'Inverse creep in Ni_3Al', Scr. Metall. **23**, 1887-1892.
16. Weertman, J. and Shahinian, P. (1956), 'Creep of polycrystalline nickel,' Trans. AIME **206**, 1223-1226.

17. Dobes, F. and Cadek, J. (1977), 'Steady state creep in two Ni-Al alloys,' Metall. Trans. **8A**, 1809-1816.

18. Liu, C. T. (1988), 'Development of nickel and nickel-iron aluminides for elevated-temperature structural use,' in ASTM STP 979, B. L. Bramfitt, R. C. Benn, C. R. Brinkman and G. F. Vander Voort (eds.), American Society for Testing and Materials, Philadelphia, PA, pp. 222-237.

19. Cahn, R. W. (1991), MRS Bulletin XVI, No. 5, 18-23.

20. Davies, R. G. and Johnston, T. T. (1979), in 'Ordered Alloys,' B.H. Kear (eds.), Claitor's Publishing Division, Baton Rouge, LA, pp. 447-474.

21. Schneibel, J. H. and Horton, J. A. (1988), 'Evolution of dislocation structure during inverse creep of a nickel aluminide: Ni-23.5Al-0.5Hf-0.2B (at.%), J. Mat. Res. **3**, 651-655.

22. Khan, T., Caron, P. and Naka, S. (1990), 'Mechanical Behaviour of Ni_3Al-based intermetallics and the need for designing multiphase alloys', in High Temperature Aluminides and Intermetallics, S. H. Whang, C. T. Liu, D. P. Pope and J. O. Stiegler (eds.), The Minerals, Metals & Materials Soc., Warrendale, PA, pp. 219-241.

23. Shah, D. M. (1983), 'Orientation dependence of creep behavior of single crystal γ' (Ni_3Al)', Scr. Metall. **17**, 997-1002.

24. Anton, D. L. Pearson, D. D. And Snow, D. B. (1987), 'Creep deformation of Ta modified gamma prime single crystals', in Mater. Res. Soc. Symp. Proc. Vol. 81 on 'High-Temperature Ordered Intermetallic Alloys II,' N. S. Stoloff, C. C. Koch, C. T. Liu and O. Izumi (eds.), Materials Research Society, Pittsburg, PA, pp. 287-295.

25. Nathal, M. V., Diaz, J. O. and Miner, R. V. (1989), 'High temperature creep behavior of single crystal gamma prime and gamma alloys,' in Mat. Res. Soc. Symp. Proc. Vol. 133 on 'High-Temperature Ordered Intermetallic Alloys III,' C. T. Liu, A. I. Taub, N. S. Stoloff and C. C. Koch (eds.), Materials Research Society, Pittsburg, PA, pp. 269-274.

26. Hsu, S. E., Lee, T. S., Yang, C. C., Wang, C. Y., Tong, C. H. and Wu, S. K., 'Intermetallics research and development in Taiwan, R. O. C.,' This Symposium.

27. Ezz, S. S., Pope, D. P. and Paidar, V. (1982), 'The tension/compression flow stress asymmetry in Ni_3(Al,Nb) single crystals', Acta Metall. **30**, 921-926.

28. Bellows, R. S. and Tien, J. K. (1987), 'Orientation effects on the creep-fatigue behavior of single crystal Ni_3Al(B,Hf),' Scr. Metall. **21**, 653-656.

29. Liu, C. T. (1984), 'Physical metallurgy and mechanical properties of ductile ordered alloys (Fe,Co,Ni)$_3$V,' Int. Metals Rev. **29**, 168-194.

30. Schneibel, J. H., Horton, J. A., and Porter, W. D., 'Bend ductility and physical properties of extruded chromium-modified Al_3Ti,' to be published in Proc. of 'International Conference on High-temperature aluminides and intermetallics,' San Diego, 16-19 Sept. 1991.

31. Mukherji, D., Jiao, F., Chen, W. and Wahi, R. P., 'Stacking fault formation in γ' phase during monotonic deformation of IN738LC at elevated temperatures,' accepted for publication in Acta Metall.

Basic Research on Intermetallic Compounds
Supported by the U.S. Department of Energy

Joseph B. Darby, Jr.
Division of Materials Sciences
Office of Basic Energy Sciences
U.S. Department of Energy
Washington, DC   20585
U.S.A.

Abstract

This contribution is a review of the programs at the National
Laboratories and various U.S. Universities that are engaged in
research on intermetallic compounds and alloys that are either
ductile or ductility can be induced by small additions of alloying
elements. The coverage ranges from the growth of single crystals
to the investigation of grain boundary effects on the atomic level.

Introduction

The expanding interest in intermetallic compounds is driven by the
quest to achieve higher operating temperatures and to reduce the
weight of systems that will yield higher energy efficiency. Many
intermetallic compounds have high melting temperatures that reflect
their stability and some of the compounds, particularly aluminides,
have relatively low densities. Further, small additions of boron
have been found to improve room temperature ductility in certain
compounds and others have some ductility that potentially could be
improved further. The research on intermetallic alloys funded by
the Division of Materials Sciences (DMS), U.S. Department of Energy
(USDOE) has the goal of providing the knowledge base on which
technology can build. Hence, the DMS program[1] covers a broad
range, as indicated below, from theoretical studies to experimental
research.

C. T. Liu et al. (eds.), Ordered Intermetallics – Physical Metallurgy and Mechanical Behaviour, 583–589.
© 1992 Kluwer Academic Publishers.

584

Discussion

A need for single crystal materials was recognized early in the program and an experimental effort was initiated at the Materials Preparation Center, Ames Laboratory, to produce single crystals of Ni_3Al and other feasible intermetallic compounds upon request. Several experiments employing a Bridgeman technique to produce single-crystal Ni_3Al indicated that the binary Ni-Al phase diagram was incorrect. It appeared that the temperature-composition coordinates of the peritectic and eutectic points were not correct. A new phase diagram for the Ni-Al binary system has been published[2] and improved the successful growth of single-crystal Ni_3Al by the Bridgeman technique.

The results of the directional solidification study and phase diagram determination conducted during FY 1989 and FY 1990 has led to the growth of large crystals of Ni_3Al. With the correct form of the phase diagram identified, a small range of composition from 75.5 a/o to 76 a/o Ni were identified over which the Ni_3Al phase precipitates directly from the liquid, a necessary requirement for the growth of single crystals. Crystals of Ni_3Al within this composition range (with and without boron additions) have been successfully and routinely prepared by the Bridgman technique with sizes ranging up to 15 mm diameter and 75 mm long. Prior to this study, maximum sizes commonly prepared were limited to approximately 6 mm diameter and 25 mm long. Furthermore, additions of boron and other ternary elements enhance single crystal preparation. Efforts at Ames Laboratory for the preparation of intermetallic single crystals, including NiAl and $MoSi_2$, are continuing.

The center piece of the DMS research program on intermetallic compounds is carried out in the Metals and Ceramics Division of the Oak Ridge National Laboratory in the High-Temperature Alloy Design Group, under the leadership of C. T. Liu. The activity includes both theoretical (M. H. Yoo) and experimental activities (J. A. Horton, E. P. George, W. C. Oliver and J. H. Schneibel). Also, theoretical support from C. L. Fu and G. M. Stocks of the Theoretical Studies of Metals and Alloys Group is a strong component of the intermetallic compound research. The research activities include a range of aluminides and this proceedings has a thorough coverage of this work such that it will not be repeated here.

Alloy theory on intermetallic phase diagrams is underway at the Lawrence Berkeley Laboratory under the leadership of D. DeFontaine. The cluster variation method (CVM) is employed to calculate phase equilibria without empirical parameters for intermediate phases. The current status of this work is a part of this proceedings.

The Mechanical Properties of Intermetallic Compounds are under investigation by C. Loxton, I. M. Robertson and H. K. Birnbaum and they are investigating solute effects on mechanical properties of grain boundaries in the Materials Research Laboratory that is supported by the USDOE. This activity [3-5] is primarily concerned with dislocation-grain boundary (DGB) interactions in Ni_3Al including the effects of structure and chemistry. This research group studied the response of grain boundaries in $Ni_{76}Al_{24}$ with and without boron, using the in-situ TEM deformation technique,[5] and concluded the following:

a. the incorporation of lattice dislocations into grain boundaries was the same in boron-doped and boron-free material and dislocations remained at the point at which they entered the grain boundary, yielding a point of stress concentration.

b. the matrix dislocations did not dissociate into grain boundary dislocations. Hence, lack of mobility of grain boundary dislocations.

c. The slip system which created the residual grain boundary dislocation with the smallest Burgers vector was preferred.

d. In boron-face specimens, the stress was relieved by the nucleation and propagation of a crack along the grain boundary. In the presence of boron, the stress was relieved by slip in the adjoining grain.

The authors concluded, based on the above observations, that the enhancement of ductility was caused by boron increasing the boundary cohesive energy.

All of the above programs are underway at the major laboratory facilities support by the USDOE. A sizeable effort on intermetallic compounds is maintained through the grant program to universities. Also, two of the university grant programs are exploring non-aluminide compounds. The Hall-Patch Relationship and Mechanisms of Fracture in B2 Compounds is the focus of an investigation by Ian Baker at Dartmouth College [6]. TEM in-situ straining studies have been performed on NiAl and FeAl and both materials show intense activity ahead of transgranular cracks in their foils. Studies of the fracture behavior as a function of stoichiometry have been completed for both compounds. The addition of V and Cr to the two compounds have been studied to determine the site occupancy of the ternary addition. Other binary and ternary B2 compounds are under investigation. Baker's research on Ll_2 compounds are part of a contribution to this proceedings.

The potential for Laves Phase Compounds are being investigated [8,9] in a program entitled "Slip, Twinning and Transformations in Laves Phases," by S. M. Allen and J. D. Livingston at Massachusetts Institute of Technology. The compounds of interest include HfV_2 based systems and $TiCr_2$, $Ti-TiCr_2$ systems, and $(Zr, Ti) Fe_2$ near the C15-C14 transitions.

The program "Intergranular Fracture and The Accommodation of Slip at Grain Boundaries," is lead by E. M. Schulson at Dartmouth College. In-situ TEM deformation studies are underway on Ni-rich, stoichiometric and Ni-lean Ni_3Al both with and without boron. Grain boundary sliding is being investigated in Ni_3Al by systematic experiments on the effects of grain size on high-temperature deformation with and without boron. A similar investigation is in progress on Ni_3Si, Ni_3Ge and Ni_3Ga. Progress on this program is reported in this Proceedings and will not be repeated here.

Experimental and theoretical studies of the structure of grain boundaries at Cornell University with S. L. Sass as the principal investigator, includes studies on the structure and chemistry of grain boundaries in Ni_3Al. The current progress on the Ni_3Al studies are reported within this proceedings.

A theoretical effort on atomistic studies of grain boundaries in alloys and compounds is performed at the University of Pennsylvania by V. Vitek and computer simulation studies are carried out on a variety of materials including $NiAl$, Ni_3Al Cu_3Au, etc. The current progress on this program is a part of this proceedings.

A micromechanical analysis of intermetallic alloys using the field-ion microscope atom probe was an activity by S. S. Brenner and associates at the University of Pittsburgh. The thrust of the effort was to determine the microchemistry in grain boundaries of intermetallics [10]. They noted that B does segregate more strongly to grain boundaries than to external surfaces and B segregation to the grain boundaries increased the cohesive energy of the grain boundaries to the extent that stress concentrations at boundaries are relieved by initiation of slip rather than fracture. Another finding was that the average grain boundary concentration of boron is independent of the stoichiometry of the Ni_3Al alloy. Preliminary results show that Co_3Ti and $Ni_3(Si+Ti)$ respond to boron doping in a similar fashion as Ni_3Al.

A program entitled "Irradiation Induced Disordering, Ductility and Phase Transformations in Ordered Intermetallic Compounds," is underway at the University of California Los Angeles with A. J. Ardell as the principal investigator[11] of the yield strengths of Ni_3Al of various stoichiometries using the miniaturized disk-bond test technique, are in excellent agreement with those obtained from uniaxial tensile tests on the exact same material. It was found that the addition of boron greatly enhances the ductility of the 24% Al alloy, moderately enhances that of the 25% Al alloy, but has no influence on the 26% Al alloy. Also the fracture loads of all three alloys are increased by the addition of boron. It is of interest to note that while the addition of boron to the 26% Al alloy did not enhance the ductility, the observed increase in fracture load is due to the increased cohesive strength of grain boundaries.

Irradiation studies in this program of Zr_3Al by 3.8 MeV Zr^{3+} ions at 250°C are in progress and will emphasize the correlation of fracture observations, long-range order measurements and defect analysis with the mechanical behavior.

A research program at the University of Connecticut, A Coherent Model of Martensitic Nucleation and Growth, P. C. Clapp, principal investigator, has been investigating martensitic transformations in NiAl (B2 type). The molecular dynamic studies of 10,000 atom arrays of stoichiometric NiAl continuing a crack under external stress in Model I loading has indicated that a martensitic transformation generally occurs, starting in the vicinity of the crack tip, prior to the generation of dislocations and/or the propagation of the crack. The thermally induced cubic to tetragonal martensitic transformation in NiAl occurs by the Bagers-Burgers double shear mechanism.

Another University of Connecticut grant program lead by J. I. Budnick employees synchrotron radiation to study a variety of materials including transition metal aluminides. The goal of the effort on aluminides is to explore local structural properties and determine their relevance to the mechanical properties of binary transition metal aluminides and the role of small ternary additions that appear to enhance ductility. The studies use temperature dependent EXAFS to obtain mean squared relative displacement of atomic sites as a function of temperature for NiAl with substitution of Fe, Ga, etc. on Al sites.

Mechanisms of high temperature rupture under multiaxial stress is a grant program at the University of California at Irvine under the leadership of J. C. Earthman and F. A. Mohamed. Investigation is underway of the mechanisms of high temperature rupture and damage under different multiaxial stress rates in Ni_3Al and other elements and alloys. This is a relatively new program and the experimental portion of the effort is underway.

588

Summary

A broad based program on intermediate phases is supported at
national laboratories and through the grant program at several
universities. The range of activities extend from phase equilibria
and materials preparation to a variety of property measurements
with emphasis on grain boundary effects. The focus of the overall
program is to understand the mechanism by which intermetallics can
be made to have some ductility and the deformation mechanisms that
operate under such conditions.

Acknowledgements

The author acknowledges the support of the U.S. Department of
Energy and partial support by the NATO Advanced Research Workshop.
Special appreciation to the Organizers of the Workshop for an
excellent meeting is acknowledged.

References

(1) Materials Sciences Programs - Fiscal Year 1990; January 1991,
     U.S. Department of Energy, Report DOE/ER-0483P.

(2) Verhoeven, J. D., Lee, J. H., Laabs, F. C., and Jones, L. L.,
     "The Phase Equilibria of Ni_3Al Evaluated by Directional
     Solidification and Diffusion Couple Experiments," Journal of
     Phase Equilibria, Vol. 12, #1, 1991.

(3) Lee, T. C., Subramanian, R., Robertson, I. M. and Birnbaum, H.
     K. "Dislocation Grain Boundary Interactions in Ni_3Al; Effects
     of Structure and Chemistry," Scripta Met. Vol. 25, p. 1265,
     1991.

(4) Subramanian, R., Robertson, I. M. and Birnbaum, H. K.,
     "Nickel Segregation at Characterized Grain Boundaries in Ni_3Al
     Alloys," Scripta Met. to be published.

(5) Lee, T. C., Robertson, I. M. and Birnbaum, H. K.,
     "Interactions of Dislocations with Grain Boundaries in Ni_3Al,"
     Acta Metall. accepted for publication.

(6) Baker, I. and Munroe, P. R. "Properties of B2 Ordered Alloys,"
     Proc. Symp. High Temperature Aluminides and Intermetallics,"
     (Ed. S. H. Whang, C. T. Liu, D. P. Pope and J. O. Stiegler),
     TMS, 1990, pages 425-452.

(7) Munroe, P. R. and Baker, I., "An ALCHEMI Investigation of
     Ternary Site Occupancy in NiAl-based Alloys," , Proc. XIIth
     International Congress for Electron Microscopy, (1990), pages
     448-449.

(8) Hall, E. L. and Livingston, J. D., "Deformation Modes in Laves
     Phase Intermetallics," Proc. 47th Ann. Meeting, Electron
     Microscopy Society of America, G. W. Bailey, ed., San
     Francisco Press (1989), p. 318.

(9) Livingston, J. D. and Hall, E. L., "Room Temperature
     Deformation in a Laves Phase," Journal of Materials Research,
     Vol. 5, p. 5, 1990.

(10) Sieloff, D. N., Brenner, S. S. and Ming-Jian, H.,
     "Microchemistry of Grain Boundaries in Ni$_3$Al," Matls. Res.
     Soc., Symp., p. 133, p. 155 (1989).

(11) Li, H. and Ardell, A. J. "Miniaturized Disk-Bend Testing of
     Ni$_3$Al: Effect of Stoichiometry and Boron Content," High
     Temperature Ordered Intermetallic Alloys IV, accepted for
     publication.

UTILIZATION OF COMPUTATIONAL MATERIALS DESIGN TO IMPROVE HIGH TEMPERATURE INTERMETALLICS

ALAN H. ROSENSTEIN
Air Force Office of Scientific Research
Directorate of Electronic and Material Sciences
Bolling Air Force Base - Building 410
Washington DC 20332-6448
USA

ABSTRACT.

The quest for advanced materials has renewed interest in documenting relationships between atomistic considerations and macroscopic properties. This effort in computational materials design spans the spectrum from angstroms to millimeters, from physics to materials science to mechanics, from first-principles to design criteria. Establishment of relevant relationships will often run into a gap of existing knowledge at some point of the spectrum. The pull across this gap must often be supplied by requirements from the more macroscopic community. One class of materials that has received much attention through this approach is the high temperature intermetallics. Ductility issues have been explored and some progress has been made. The gap of knowledge appears to be between the atomic scale properties that may be modeled and microstructural and micromechanical mechanisms that are influenced. Defining the gap helps to identify and prioritize key areas of research. The use of computational materials design for structural materials research and development is in its infancy. It is hoped that this discussion will encourage additional systems and problems to be identified and attacked by this approach.

1. Introduction

There is currently a considerable amount of interest in utilizing fundamental, first-principles, computational approaches in the microstructural design and development of advanced materials. This concept is not new, but has been given impetus recently by several factors: better theoretical understanding of fundamental phenomena; better experimental means to observe nano/microstructures and material processes, and to validate prediction of such processes; and better and faster computational tools such as the supercomputer. This interest in what is sometimes called "Computational Materials Design" has been further nurtured by the need to understand and solve some very difficult materials science issues if better materials are to be obtained.

One example of an advanced material system that lends itself to this approach is the intermetallic materials. The basis for the viability of computational materials science is that specific research problems can be identified that relate to specific fundamental processes. In turn, these processes can be linked to macroscopic properties. In the case of the intermetallics, the specific research problems isolated

C. T. Liu et al. (eds.), Ordered Intermetallics – Physical Metallurgy and Mechanical Behaviour, 591–595.
© 1992 *Kluwer Academic Publishers.*

include deformation processes and interface phenomena issues. These relate to fundamental processes that can be modeled on the atomic scale such as bonding, cohesion, lattice symmetry and phase stability. It is hoped to be able to link these fundamental processes to measurements of macroscopic properties such as ductility, toughness and strength.

It must be recognized that in proceeding from atomistics to macroscale some linkages will be difficult, and some may not currently be able to be made. A "gap" of knowledge or understanding may be found. Gaps should be expected and serve to identify fruitful areas of research.

The concept of linkages from the nanoscale to the macroscale is shown schematically in Figure 1. This is basically a trip from the angstrom world of the physics community, through the micron world of the materials science/metallurgy community, and into the millimeter and continuum world of the mechanics community - eventually leading into ties with structural design considerations. It is a long journey that will require time, and those awaiting advanced structural materials for applications must be patient. The potential, if success is achieved, is high and the payoffs great.

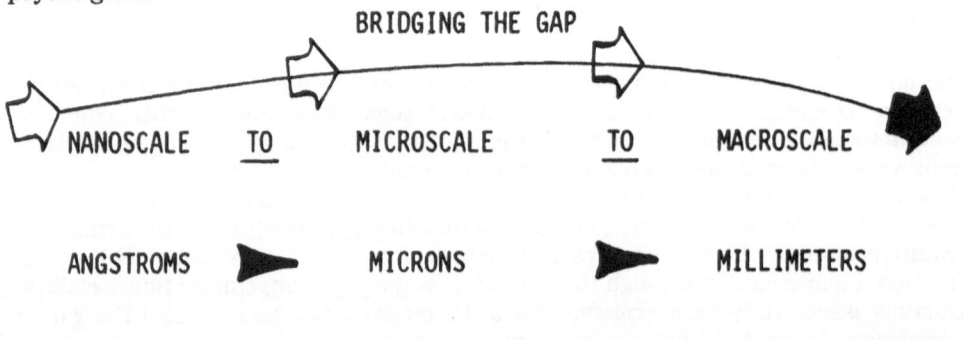

Figure 1 - BRIDGING THE GAP

2. Discussion

The representation of computational materials design by a schematic diagram having some generality, yet sufficient detail to suggest validity, is a tricky task. Such a representation is merely a flow chart or roadmap and will certainly change with time. The contents of such a diagram are a function of the perspective of the individual drafting it, and do not provide mutually exclusive or unique solutions. Nor does it provide the only connections that can be made across the spectrum of activities.

With a certain amount of trepidation, such a representation is shown in Figure 2 for high temperature intermetallic materials. The relationships shown are based on the premise that macroscopic properties are related to atomic scale properties. The research vehicles available include fundamental physics approaches (such as being pursued by A.J. Freeman and coworkers at Northwestern University) coupled with extensive use of supercomputers to perform an otherwise formidable amount of calculations. These techniques deal nicely with the periodic lattice. Attempts are being made to extend the methods to handle larger clusters of atoms to deal with problems of broken symmetry caused by impurities, defects, microalloying, stacking faults, antiphase boundaries and twins. Current methods can handle 20-70

COMPUTATIONAL MATERIALS DESIGN: HIGH TEMPERATURE INTERMETALLICS

PREMISE: MACROSCOPIC PROPERTIES ARE RELATED TO ATOMIC-SCALE PROPERTIES.

ATOMIC SCALE PROPERTIES MODELED	"BRIDGING THE GAP"	MECHANISMS INFLUENCED: (MICROMECHANICAL/ MICROSTRUCTURAL)	EXPERIMENTAL VERIFICATION OF MECHANICAL PROPERTIES
Bonding	THE PULL ACROSS	Bonding Directionality, Strength, Type	Deformation Processes
Cohesion	THE "GAP" MUST	Interfacial Phenomena Grain Boundary Sliding Void Formation	Fracture, Creep
Phase Stability Cubic Structure (L1$_2$)	BE SUPPLIED BY REQUIREMENTS FROM THE MORE	Slip Systems Available [111] Planes Promote Easier Slip/Deformation	Ductility, Toughness
Alloying Microalloying	MACROSCOPIC	Solid Solution Strengthening Solute/Dislocation Interactions	Strength, Ductility Fatigue
	COMMUNITY.	Control of Microstructure Role of Ternary Additions Transformation Toughening	All Properties
Phase Transformations		Dislocation Mobility Interfacial Energies Activation of Specific Slip Systems	Deformation, Fracture
Fault Energies (APB, Twins, Stacking, etc.)			

RESEARCH VEHICLE

FUNDAMENTAL PHYSICS
All-electron total energy quantum mechanical approach
• Periodic Lattice
• Large Clusters
  Broken Symmetry

COMPUTATIONAL TOOL
Use supercomputers to perform otherwise formidable amount of calculations

Electronic Materials   Experience

Materials Science .... Metallurgy
Microstructure/Property Relationships
Lots of Experience!!

Figure 2 - COMPUTATIONAL MATERIALS DESIGN: HIGH TEMPERATURE INTERMETALLICS atom clusters, but clearly larger cluster methods are required for treating local atomic conditions that affect macroscopic properties.

Some of the atomic scale properties that may be modeled are shown in the second column of Figure 2. Much success has been obtained in relationships between fundamental physics and atomic scale properties by the electronic materials community, and this experience must be built upon by the structural materials community.

The gap of knowledge in the high temperature intermetallics area appears to be between the atomic scale properties and mechanisms influencing microstructural and micromechanical phenomena. Some of these mechanisms are shown in the fourth column of Figure 2. Certainly, dislocation theory has helped to bridge this gap for mechanical properties, but much additional effort is required. Current research is actively pursuing relationships between crystal structure, broken symmetry and deformation processes that control ductility and toughness. For high temperature intermetallics, this is a critical issue.

On the atomistic side of the gap, much detailed information may be generated, a great deal of which is limited to idealized conditions. Adjustments for non-perfect lattices, temperature dependence and non-equilibrium conditions may introduce severe restraints on interpretation. However, as one moves to more macroscopic considerations, at each step one must select that information that appears to be controlling at the next step. This leaves most information behind. In may cases, particularly where relationships are not well documented, all information possible must be generated in order to determine what is important and controlling.

The mechanistic side of the gap must help in the selection and focusing of information. Mechanisms must be defined based on microstructural and experimental observations. The setting of priorities must be done with an eye to applications. Key and limiting problem areas must be suggested as well as appropriate material systems to be explored. The pull across the gap must often be supplied by requirements from the more macroscopic community.

The connections between microstructural and micromechanical mechanisms and macroscopic mechanical properties are more easily made based on extensive metallurgical experience. Although all information required is not currently available, directions for research are usually easily discernable.

A key concept as one moves through this spectrum of activities is to do research at the appropriate scale to understand and control phenomena of interest.

3. Closure

It is hoped that this discussion will encourage additional material systems and research problems to be identified and attacked by a "computational materials design" approach. It is reasonable to anticipate some research successes in areas such as: assessment of the efficacy of a new material system - interrogation of its feasibility; evaluation of a material's chemical balance and the role of specific elements in a more quantitative and scientific manner than currently available; separation of the role of chemistry as it affects atomic level phenomena from the role of microstructure; and, uncovering of some of the more fundamental aspects of cleavage fracture which may depend heavily on atomic level phenomena associated with bonding characteristics and surface energies. These successes will take time. Meaningful models must be developed that will require intensive research efforts. One might expect atomic level approaches to produce advances for some intermetallic systems over the next five years, however, major impact on structural material systems in general is probably at least ten years away.

4. Acknowledgements

I would like to thank Art Freeman for his insights and encouragement while preparing this paper. I would also like to acknowledge fruitful discussions with Dennis Dimiduk and Dave Srolovitz. In addition, I would like to recognize the foresight of Harry Lipsitt who in the early 1980's encouraged me to become involved in computational materials science. This area of research has been fruitful due to the efforts of a great many researchers, and to them the materials science community is grateful.

INTERMETALLICS RESEARCH AND DEVELOPMENT IN TAIWAN, R.O.C.

S.E. Hsu, T.S. Lee, C.C. Yang, C.Y. Wang and C.H. Tong
Materials R&D Center, CSIST
P.O. Box 90008-8, Lung-Tan, Taiwan, Taiwan
S.K. Wu
National Taiwan University, Taipei, Taiwan, R.O.C.

ABSTRACT. In this article, research and development on intermetallics in Taiwan, R.O.C. is reviewed. Starting from 1984, R&D on intermetallics has been aiming at nickel aluminide, titanium aluminide and titanium nickel shape memory alloys. For Ni_3Al intermetallics, a series of effort has been focused on ductilization and strengthening by alloying and processing. Single crystal study and creep resistance for Ni_3Al intermetallics and its alloy modifications were emphasized in this topic. A mechanism of strengthening for long-range ordered $L1_2$ matrix by a dispersoid of disordered $Ni_3(AlCr)$ is proposed. For the titanium aluminide systems, microstructure of $\alpha_2+\gamma$ two phase TiAl and texture-strengthening of Ti_3Al-Nb has been investigated. For titanium nickel SMA intermetallics, recent research interests on TiNi intermetallics are centered on the premartensitic transformation, transformation sequences, high temperature SMA and hydrogenation of NiTi alloys. Lastly, an innovative process for preparing any kind of intermetallics was developed and will be described briefly in this article.

1. Introduction

In late 1970's, some prominent results showed that ductility of ordered intermetallic Ni_3Al could be effectively improved by alloying with micro-amount of boron [1]. This result demonstrated the feasibility of achieving high tensile ductility in other intermetallic alloys [2]. These promising achievements have stimulated world-wide interest including Taiwan in studying on ordered intermetallics. Since thin, here in Taiwan, initiated from MRDC/CSIST, a few projects have been set up to focus on understanding the brittle fracture and ductilization in ordered intermetallics. The research and development work has been concentrated in MRDC/CSIST (Materials R&D Center/ Chung Shan Institute of Science and Technology) and NTU.(National Taiwan University). The topics have been centered primarily on aluminides of nickel and titanium because these intermetallics possess a number of attributes that make them attraction for aero-space applications.

This article will report some of the results of the development work in Taiwan, especially in nickel-aluminides, titanium aluminides and titanium nickel alloys. Alloying and processing have been emphasized to control all the parameters such as microstructure, grain boundary, and compositions in order to improve both the ductility and strength at elevated temperatures.

Single crystals of Ni_3Al and multi-element modified Ni_3Al have been

C. T. Liu et al. (eds.), Ordered Intermetallics – Physical Metallurgy and Mechanical Behaviour, 597–616.
© 1992 *Kluwer Academic Publishers.*

successfully grown in MRDC/CSIST. A parallel effort, two innovative processes to sythesize intermetallics, has been contributed to fabricated various alloy systems. Some of these results will be presented in this article.

2. Nickel Aluminide Intermetallics

2.1. DUCTILIZATION AND STRENGTHENING OF POLY–CRYSTALLINE NI₃AL INTERMETALLICS

There has been tremendous interest in Ni_3Al development in recent years. The anomalous temperature dependence of the yield strength, the inherent good fatigue strength, the oxidation resistance, and a relative low density in comparing with nickel base superalloys, make Ni_3Al a potential alloy for high temperature applications among various promising properties which are important to engineering materials, the high temperature creep and ductility are major concerns in current development. The following articles are to summarize our results.

2.1.1. *Recrystallization and Grain Growth Kinetics* [3]. Addition of boron has been found capable of increasing low temperature ductility of Ni_3Al [1]. But the mechanism is operative only in off–stoichiometric, containing less than 25 at% aluminum [4,5], in which nickel rich $\gamma-\gamma'$ eutectic structure is very stable and renders homogenization processing very time consuming. In our study, recrystallization and grain growth behavior of Ni_3Al were studied attempting to speed up the homogenization process so as to improve creep resistance. A repeatedly thermal mechanical treatment (TMT) process followed by a high temperature annealing were employed. It was found that the homogenization time for eliminating γ–phase from $\gamma-\gamma'$ eutectic structure of off–stoichiometric Ni_3Al (23 at% Al) can be reduced from 75 hrs to 3 hrs at homogenization temperature of 1100°C. From kinetic study of recrystallization and grain growth, it is concluded that the apparent activation energy for grain growth is close to 220 KJ/mole for Ni_3Al single phase, no matter the extent of departure from stoichiometric composition. Comparing with the activation energy for Ni_3Al and inter–diffusion in Ni_3Al ordered structure, it is judged that the grain boundary structure of the single phase Ni_3Al is partially ordered. However, off stoichiometric aluminides which contains 15 vol% coarsen eutectic structure with activation energy around 280 KJ/mole indicating that both recrystallization and grain growth were delayed by the eutectic structure. It is also concluded that TMT process is not only the effective procedure to speed up homogenization for the off–stoichiometric structure but also a workable process to improve the ductility of near stoichiometric Ni_3Al even contain without boron.

2.1.2. *Effects of Zr Addition on the Mechanical Behavior of Ni_3Al* [6]. In 1980's, there were many studies on macro–alloying with third element X addition for Ni–Al–X system [7].
Fig. 1 shows ternary phase diagram of Ni–Al–X at 1000°C [7]. As shown in Fig. 1, Zr is in the region of Al–substitution. Experiments indicated that Zr–addition within the solubility of 1.5 at% could substantially reduce the activation energy for (100) cross–slip and possess the potential for solution strengthening at elevated temperatures. Yield strength and elongation as a function of atomic percent of Zr–content are plotted in Fig. 2 [8]. Yield strengths of $Ni_{77}Al_{23-x}Zr_x$ vs. testing temperature with x=0, 1.0, 1.5 at% are plotted in Fig. 3.

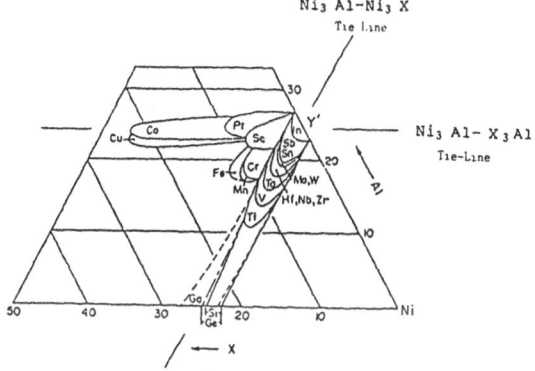

Fig. 1. Solubility lobe for Ni–Al–X phase diagram at 1000°C [7]

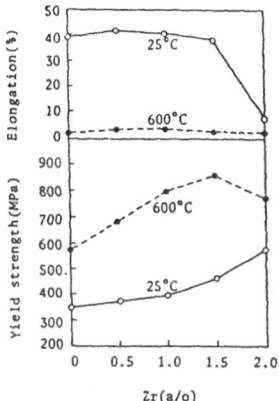

Fig. 2. Yield strength and elongation as a function of Zr content

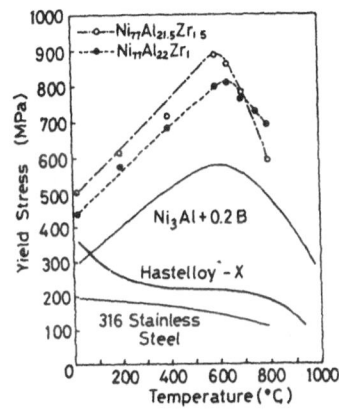

Fig. 3. Y.S. vs. T curve for $Ni_{77}Al_{23-x}Zr_x$, x=1, 1.5

As illustrate in Figs. 2 and 3, the effect of Zr addition on mechanical behavior of Ni_3Al intermetallics can be summarized as follows:

(a) The yield strength increases as a function of Zr–addition up to the solubility of Zr, 1.5%

(b) Peak strength in Y.S. vs. T curves for $Ni_{77}Al_{23-x}Zr_x$ alloy increases as x increased.

(c) Stress rupture tests demonstrated that the rupture strength at elevated temperature increased as Zr–content increased.

(d) Effect of Zr–addition on ductility of Ni_3Al is negligible even SEM fractography shows ductile fracture.

From the above encouraging results, most of alloy design for Ni_3Al intermetallics in this laboratory, either polycrystal or single crystals will contain Zirconium as a part of aluminum equivalent in the following work.

2.1.3. *High Temperature Mechanical Behaviors of Zr/Cr Modified Ni_3Al*

Intermetallics [8]. Following the research on toughening mechanical behavior, especially the rupture strength of the Zr/Cr modified Ni$_3$Al appreciably on the phase presented in the alloy, only those alloys which have a structure very close to the ordered L1$_2$ phase have been examined. Zr–content was limited to its solubility of 1.5 at% and 500 ppm of boron was added to the alloy in most cases, while Cr–addition was rather macroscopically changed from 2 at% to 8 at%.

It is observed that additions of Zr and Cr notably refine the as–cast dendritic structure. As mentioned in §2.1.1, the TMT process has effectively broken up the dendrites which would otherwise take very long time to homogenize.

The tensile properties of Zr/Cr modified Ni$_3$Al are shown in Fig. 4. It is observed that Zr/Cr–addition is capable of increasing both the high temperature strength and ductility as well.

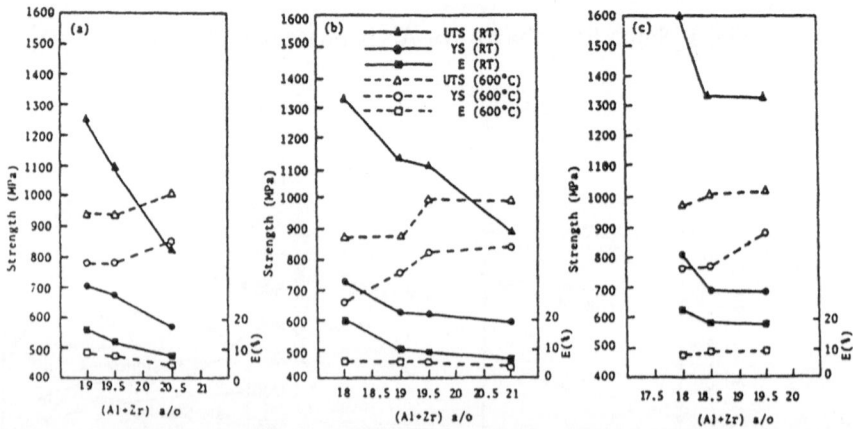

Fig. 4. Tensile properties of Ni–Al–Zr–Cr quarternary alloys as function of (Al+Zr) content, (a) 4 at% Cr, (b) 6 at% Cr, (c) 8 at% Cr.

When the yield strengths of alloys under investigation are plotted against the test temperatures, as already shown in Fig. 3, the anomolous behavior of yield strength increases with temperature is again preserved. Moreover, a combination of Zr and Cr addition has the effects of improving not only the yield strength, but also the Tp, temperature of peak strength, from a yield strength of about 550 Mpa at 600°C for Ni$_{77}$Al$_{23}$ shifting to a yield strength of 900 Mpa at 700°C for Ni$_{73.5}$Al$_{17}$Cr$_8$Zr$_{1.5}$ alloy.

The rupture strength at 760°C of some Zr/Cr modified Ni$_3$Al have been plotted in Fig. 5. The rupture strength increases with Cr–addition as measured at the same Al+Zr content. It is to be emphasized that the Al+Zr content is an important factor in rupture strength at off–stoichiometric composition, for instance, Ni$_{73.5}$Al$_{19}$Cr$_6$Zr$_{1.5}$ alloy has a 760°C, 400 Mpa rupture life of over 600 hours.

2.1.4. *High Temperature Creep Behavior of Cr–Modified Ni$_3$Al* [9]. Effects of Cr–addition on high temperature creep for Ni$_3$Al intermetallics have been studied in MRDC/CSIST. Two sets of specimens, Ni$_{77.97}$Al$_{20.16}$Zr$_{1.05}$B$_{0.38}$, designated Cr–0 and Ni$_{72.64}$Al$_{18.2}$Zr$_{1.04}$Cr$_{7.72}$B$_{0.4}$, designated as Cr–8 were creeped at temperatures of 760°C, 790°C, 810°C and 830°C, respectively with strain rate of 10^{-8}sec^{-1}. Creep data were fitted into a power law of creep equation of the following form.

$$\dot{\epsilon}_s = \frac{C}{kT} \sigma^n \exp^{-Q/kT}$$

where C is constant; k is Bolzmann constant; σ is stress in psi; Q is activation energy for creep; n is stress exponent.

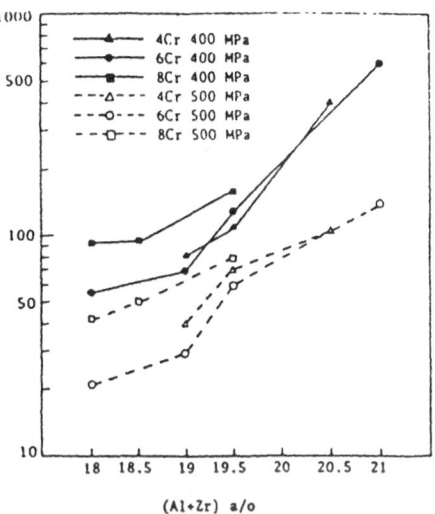

(Al+Zr) a/o

Fig. 5. Rupture life vs. temperature plot for various Zr/Cr strengthened Ni₃Al

The values of n and steady state activation energy for creep are listed in Table I.

Table I. Creep data of Cr-Modified Ni₃Al

Material	Stress Exponent		Activation Energy for Creep	
	n	Δn	Q(KJ/mode)	ΔQ(KJ/mole)
Cr-0 (a)	2.9	0.1	339	4
Cr-0 (b)	2.9	0.1	346	16
Cr-8 (a)	3.3	0.2	391	17
Cr-8 (b)	3.3	0.1	400	31

(a) σ = 300 Mpa   (b) σ = 400 Mpa

It is noted that the stress exponent of Cr-0 and Cr-8 are close to each other and to those of the other Ni₃Al intermetallics [10] under the similar test conditions. However, the activation energy for creep of Cr-8, (400 KJ/mole), is much higher than those reported in other experiments, 300 KJ/mole for Ni diffusion in Ni₃Al [11]. Apparently, Cr-addition might change the structure of the anti-phase boundary on the glide-plane, an additional energy required for the mobile superpartial dislocation to pull-away from the distorted anti-phase boundaries.

2.1.5. *Effects of Ta–addition on Mechanical Behavior of the Cr/Zr Modified Ni_3Al Intermetallics* [12]. Effects of Ta on Mechanical behavior of superalloy and intermetallics have been extensively studied [13]. In this studied, various amount of Ta, 0.7 at%, 1.3 at% and 2.0 at% were added into $Ni_{73}Al_{18}Cr_8$ intermetallics which has been discussed in §2.1.3. Following a series processes of TMT treatment, the four sets of specimens were mechanically tested at two temperatures, 20°C and 600°C respectively. Table II listed the results:

Table II. Mechanical Behavior of Ta-Modified Ni₃Al

Test Temperature	Ta-0 *		Ta-0.7		Ta-1.3		Ta-2.0	
	Y.S(Mpa)	E%	Y.S(Mpa)	E%	Y.S(Mpa)	E%	Y.S(Mpa)	E%
25°C	620	18	951	6	820	14	1077	11
600°C	879	13	961	6	1010	7	910	2

*Base alloy : $Ni_{73}Al_{18}Zr_1Cr_8$

Obviously, Ta–1.3 alloy exhibits optimum properties for toughness. Therefore, Ta–1.3 alloy was selected as the material for alloy design. Fig. 6 shows yield strength and elongation plots as a function of test temperature for Ni₃Al intermetallics which has been strengthened by Zr, Zr+Cr and Zr+Cr+Ta separately. These curves reveal that a tremendous increase in yield strength by Zr–addition. However, Ta–addition improve little on ductility for $Ni_{73}Al_{18}ZrCr_8$ at temperature below 600°C. Combining the effects of Zr, Cr and Ta–additions, both yield strength and ductility for the Ni₃Al can be improved. Hopefully, Ni–Al–Zr–Cr–Ta, the penta–element intermetallics will become a candidate of wrought alloys for high temperature applications. In the following article, single crystal study will be emphasized mainly based on the penta–element modification.

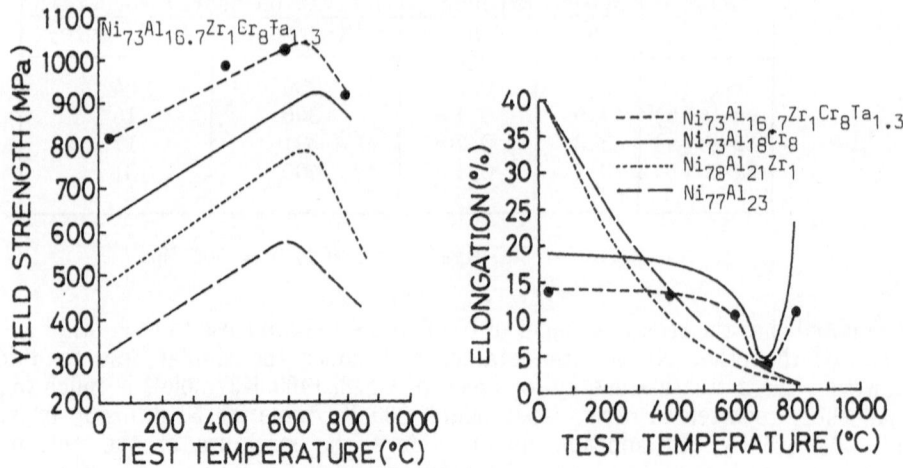

Fig. 6. The yield strength and elongation vs. temperature curves for Zr/Cr/Ta modified Ni₃Al intermetallics

2.2. SINGLE CRYSTAL GROWTH AND MECHANICAL BEHAVIOR OF NI₃AL INTERMETALLICS

2.2.1 *Single Crystal Growth.* It is realized that single crystals of Ni₃Al are ductile at ambient temperature, but polycrystalline Ni₃Al fails by brittle grain boundary fracture with little ductility [14]. Startling discoveries found [1] that Ni₃Al could be converted to ductile at room temperature by addition of boron. Apparently, grain boundaries play fatal important role for controlling the ductility of poly–crystalline Ni₃Al. Study of ductilization led to a conceptual attempt to investigate single crystals that could directly simplify the complexity of problems concerning grain boundaries, such as grain boundary structure, grain boundary ordering, grain boundary segregation as well as grain boundary energies.

Cylindrical and plate–type single crystal of Ni₃Al with various addition elements were successfully grown by directional solidification technique with withdrawal speed of 10 cm/hr. Three groups of Ni₃Al single crystals were grown within past two years. Chemical compositions are listed in the following Table III.

Table III. Compositions of Ni₃Al Single Crystals

Group	Alloy I.D.	Compositions (at%)						
		Ni	Al	Cr	Ta	Zr	B	Al(Equiv)
1	SC–530	76	24					24
	SC–531	76	24				500ppm	24
	SC–532	76	23			1	500ppm	24
2	SC–515	75	21	4				23
	SC–516	75	23		2			25
	SC–517	77	19	4				21
3	SC–510	71	17	6	5	1		26
	SC–550	71.5	18	4.6	5.5	0.4		26.2
	SC–560	74	17	4.6	4	0.4		23.7

The first group is for verification of inherent ductility with and without addition of boron. Second group is for comparison of creep resistance between single and polycrystals when Ni₃Al is modified with Cr or Ta. The third group is to investigate the mechanical behavior and microstructure of Zr/Cr/Ta modified penta–element Ni₃Al single crystal.

2.2.2 *Ductility of Ni₃Al With and Without Boron.* The off–stoichiometric–Ni₃Al single crystals with boron (SC–#531) and without boron (SC–#530) as listed in table III have been tested at room temperature and various elevated temperatures. The results of mechanical behavior are listed in Table IV and plotted in Fig. 7. The data for polycrystal Ni₇₆Al₂₄ is also shown for comparison. It is interesting to note that elongation of Ni₇₆Al₂₄ single crystal is as high as 96% for tensile test at room temperature reveals its inherent ductility and independence of boron addition. Effect of boron addition, on the other hand, improves yield strength at elevated temperatures. However, it is hardly to prove, in the time being, whether the strengthening is caused by solution alloying or orientation difference due to the shortage of specimen with the identical orientation. In comparing with the data of Poly–$Ni_{76}Al_{24}$–$B_{0.2}$ which were tested at temperature of 600°C and 760°C, the yield strengths are close to that of single crystal. However, the ductility is incomparable. It is further proved that grain boundaries are the main source of brittleness for the Ni_3Al intermetallics.

Table IV. Mechanical Behaviors of Single Crystal $Ni_{76}Al_{24}$

Alloy#/Cond.	T.Temp.	U.T.S.(Mpa)	Y.S(Mpa)	Elong.(%)	Oriention
S.C. #530	R.T.	541	235	96	
(Ni-24Al,	400°C	780	324	63	111
Aleq = 24	600°C	679	439	53	with ±10
1050°C/72hr Homog.)	800°C	682	604	23	off
S.C. #531	R.T.	552	193	59	
(Ni-24Al-500ppmB	400°C	657	360	68	100
Aleq = 24	600°C	655	592	27	within ±10
1050°C/72hr Homog.)	800°C	718	689	32	off
	1000°C	362	211	121	
Poly. Ni-24Al-0.2B	600°C	549	531	0.6	
(Equiaxed,G.S.=90 m)	760°C	590	567	0.7	

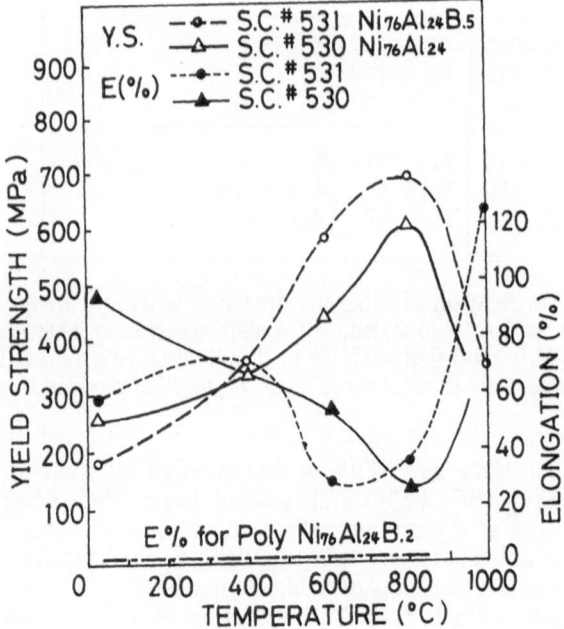

Fig. 7 Yield strength and elongation as a function of testing temperature for single crystal Ni_3Al with and without boron

2.2.3. *The Creep Behavior of Cr–Modified Ni₃Al Single Crystals* [9]. The steady state as well as transient creep tests for Cr–modified single crystal, as listed in Table III, group 2, were conducted at temperatures range from 760°C to 860°C. The steady state creep rate of these two specimens, S.C. #517 and S.C. #515, are plotted as a function of T^{-1} by using Arrhenius–type equation as shown in Figs. 8 and 9.

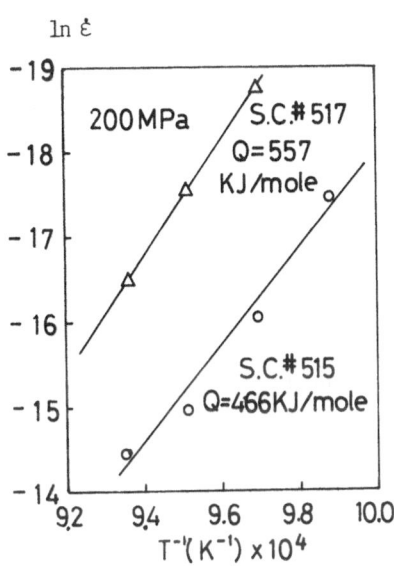

Fig. 8. A plot of $\ln \dot{\varepsilon}$ vs. T^{-1} for SC#517 and SC#515 under 200Mpa

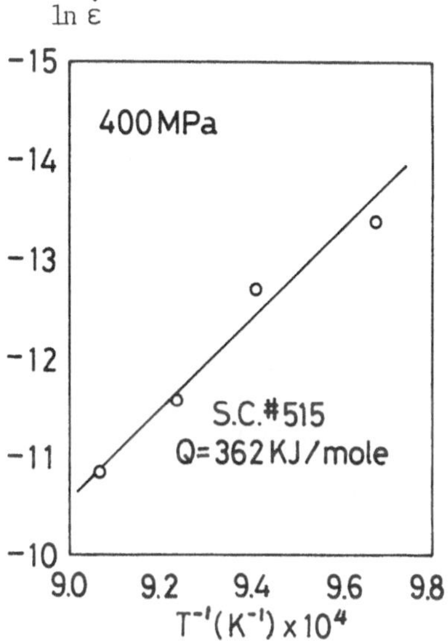

Fig. 9. A plot of $\ln \dot{\varepsilon}$ vs. T^{-1} for SC#515 under 400Mpa

The activation energy for creep were obtained by least mean square fitting the data with straight lines. Note that the activation energy, Qc, are much higher than that reported for Ni–atoms diffusion in Ni₃Al and even more higher than that for poly–crystal Cr–8 as previously discussed in §2.1.4. (listed in Table I).

In Cr–modified single crystals, the addition of Cr coincides with the positive deviation in the activation energy for creep. Judging from the bilateral nature of Cr on substituting for Ni and Al, the anisotropy of planar faults may be changed due to Cr–addition, subsequently the dissociation of Shockley partials may also be affected.

2.2.4. *Development of Ni₃Al Single Crystal with Zr/Cr/Ta Modification* [15]. The mechanical behavior and microstructure analysis of Ni₃Al–single crystal with Zr/Cr/Ta modification, as listed in Table III, group 3, will be discussed in this paragraph. The ideal stoichiometric composition for long–range ordered Ni₃Al–intermetallics is 25 at% Al and 75 at% Ni. It exists stably in non–stoichiometric composition range of about 77.5 at% to 77 at% nickel. The penta–element alloy could be regarded as a composite material composed of Ni₃Al (γ') and nickel base superalloy ($\gamma+\gamma'$). The interdendritic structure ($\gamma+\gamma'$) will present in the as–cast single crystal, as shown in Fig. 10.

Fig. 10. The as—cast Zr/Cr/Ta dendritic structure for
modified Ni₃Al single crystal

The dendritic structure was very stable and could not be removed even be
homogenized at very high temperature for many hours. All the specimens, SC#510,
#550 and #560, listed in Table III were homogenized at 1200°C for 72 hours and
tested at five different temperatures. The mechanical data is listed in Table V.

Table V. Mechanical Properties of Ni₃Al Single Crystals

Alloy#/Cond.	T. Temp.	UTS(Mpa)	YS(Mpa)	Elong.(%)	R.A.(%)
#510 (Ni-17Al-6Cr-5Ta-1Zr Aleq = 26 1200°C/72hr Homog.)	R.T. 400°C 600°C 780°C 860°C	859 1176 1152 1204 856	841 1176 1113 1106 841	1.1 Nil Nil Nil Nil	-- -- -- -- --
	*Fracture time = 6.5 hours at 980°C and 245Mpa				
#550 (Ni-18Al-4.6Cr-5.5Ta- 0.4Zr Aleq = 26.2 1200°C/72hr Homog.)	R.T. 400°C 600°C 800°C	1010 1260 1380 1275	970 1260 1380 1275	2.6 Nil Nil Nil	-- -- -- --
#560 (NI-17Al-4.6Cr-4Ta- 0.4Zr Aleq = 23.7 1200°C/72hr Homog.)	R.T. 400°C 600°C 800°C 1000°C	669 772 1015 1105 505	568 659 930 1057 505	3.0 8.7 2.8 3.1 22	5.8 17 12 11 26
	*Fracture time = 29.1 hours at 980°C and 245Mpa				

The Y.S vs. temperature and elongation vs. temperature curves for these crystals
are illustrated in Fig. 11.

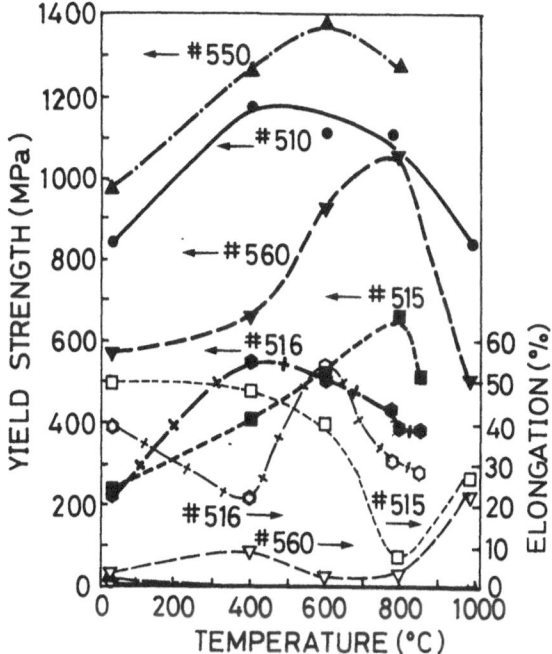

Fig. 11. The yield strength and elongation vs. temperature
curves for Zr/Cr/Ta modified Ni₃Al single crystals

Few conclusions can be drawn after investigating the mechanical data.
(a) Both yield strength and ultimate tensile strength of the Ni–Al–Zr intermetallics
increase with temperature and reach a maximum at peak temperature (Tp) which
tends to shift to high temperature side as the Ni₃Al alloy single crystal was modified
with Cr/Ta.
(b) Either chromium or tantalum in effective strengthener for Ni₃Al intermetallics
no matter the Ni₃Al alloy is single or poly crystal. However, the combined
strengthening effect makes Ni₃Al single reaching an optimum yield strength, as high
as 1380 Mpa at Tp 600ºC which has never seen in literatures.
(c) Deficit of Al or Al–equivalent from stoichiometric composition 25 at%, is
probably a criterion of ductilization for Ni₃Al intermetallics. From the statistical
elongation data as listed in Table IV and V, all the ductile specimens exhibit
off–stoichiometric composition with deficit Al or Al–equivalent which is based on
Ni–Al–X phase diagram [7]. Al–atom can be substituted for alloying elements of Ta
and Zr. While Cr substitute both Al and Ni sites, we count it half for
Al–equivalent.
    In order to investigate the intensive strengthening mechanism by addition of
Cr/Ta in Ni₃Al single crystal, micro–structure analysis was conducted by using
JEM–4000 FX STEM. Single crystal specimen, SC–#550, was cut along growing
director which was identified by Laue diffraction to be <001>.
    Figs. 12 and 13 illustrate a cluster and an isolated tiny particle (<0.1μ) which
were found in both as–cast and 1200ºC/72hr homogenized SC–#550 single crystal
specimens.

608

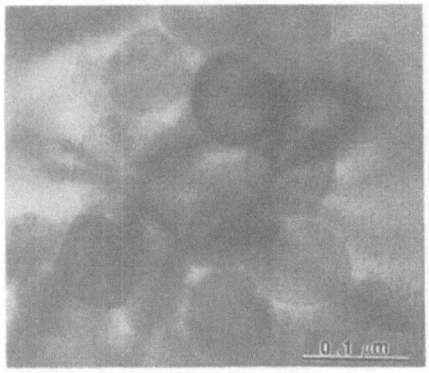 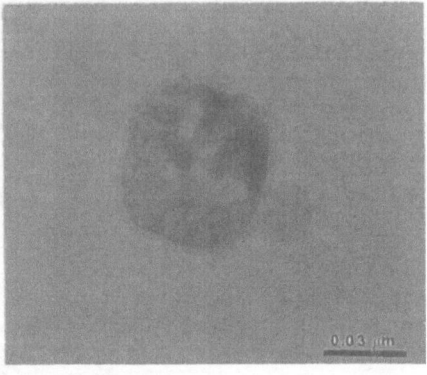

Fig. 12. A cluster of tiny particles from the as–cast and homogenized Ni–Al–Zr–Cr–Ta single crystals

Fig. 13. An isolated tiny particle from Fig. 12

Figs. 14 and 15 illustrated the SADP diffraction patterns which were taken from the matrix and the isolated tiny–particle in the SC–#550 crystal.

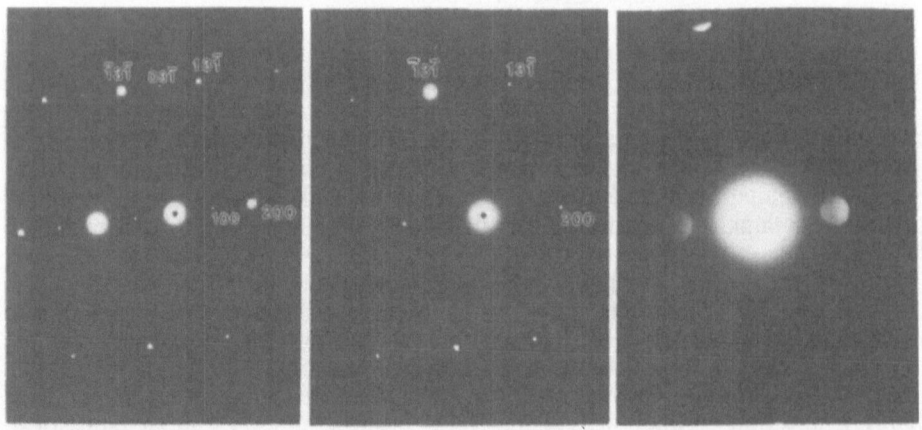

Fig. 14. SADP taken from the matrix of Ni–Al–Zr–Cr–Ta single crystal

Fig. 15. SADP taken from the tiny particle

Fig. 16. Micro–Micro diffraction pattern taken from the tiny particle

From Fig. 14, SADP reveals that the matrix is indeed a single crystal with an ideal $L1_2$ ordered structure including a superlattice point of (100). However, the diffraction orientation is [013] instead of [001] which was primarily identified by Laue diffraction. It is interesting to note that the SADP taken from the tiny particle, Fig. 15, exhibits exactly the same orientation and pattern with the matrix but without the super–lattice point. The extinction of (100) lattice point can be further verified by the micro–micro diffraction by focusing on the tiny particle, which has been shown in Fig. 16. Principally, the superlattice point (100) should be simultaneously presented with the matrix due to the fact that the minimum aperture for SADP is 10μ in diameter which is much larger than the micro–structure of the tiny particle ($<0.1\mu$). The only explanation is that the microstructure of the tiny particle is a disordered structure which is embedded in the ordered matrix due to the non–uniformly substitution of Cr for Al sites. Fig. 17(a) and (b) show the EDS spectrum for the matrix and the tiny particle. Apparently, the Cr–rich tiny particles exist as–cast with disordered structure which

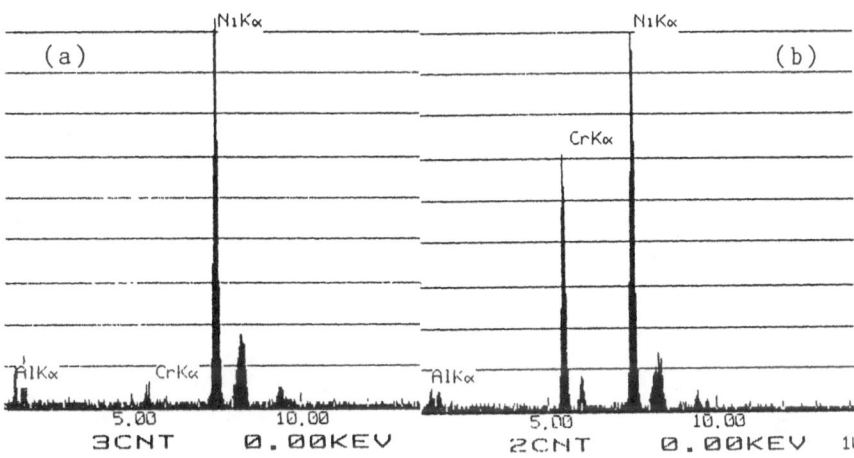

Fig. 17. EDS of Ni–Al–Zr–Cr–Ta single crystal
(a) matrix,   (b) tiny particle

dispersed in the $L1_2$ matrix of ordered Ni_3Al–crystal co–existing with a coherent epitaxial orientation. This prediction can be verified from the photo image as shown in Fig. 18 which was taken from the coherent interface between the tiny particle and the matrix. Therefore, we propose that a disordered Cr–rich phase $Ni_3(CrAl)$

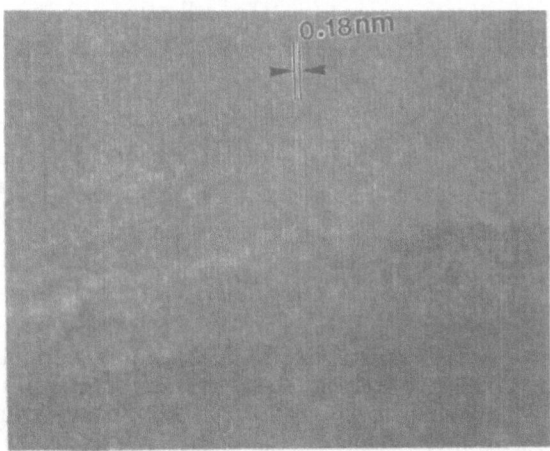

Fig. 18. Photo image taken from the coherent interface between the tiny particle and the matrix

dispersed in an ordered L1$_2$ matrix which disturb the homogeneity in the region of anti–phase boundaries to block the motion of superpartial dislocations, resulting an effective mechanism to strengthen the Ni$_3$Al single crystals.

3. Recent Development of TiAl and Ti$_3$Al Alloys

3.1. ORIENTATION FAULTS OF γ LAMELLAE IN A TI–40 AT% AL ALLOY [16]

A two phase region of $\alpha_2 + \gamma$ has been reported to exist with aluminum contents between 34 and 50 atomic% at temperatures below 1250°C. Microstructures of the two–phase titanium aluminides have been investigated extensively because their mechanical properties are superior to the single α_2 or γ. Experimental results show that the adjacent two γ plates, as shown in Fig. 19, are found to have four different orientation relationships.

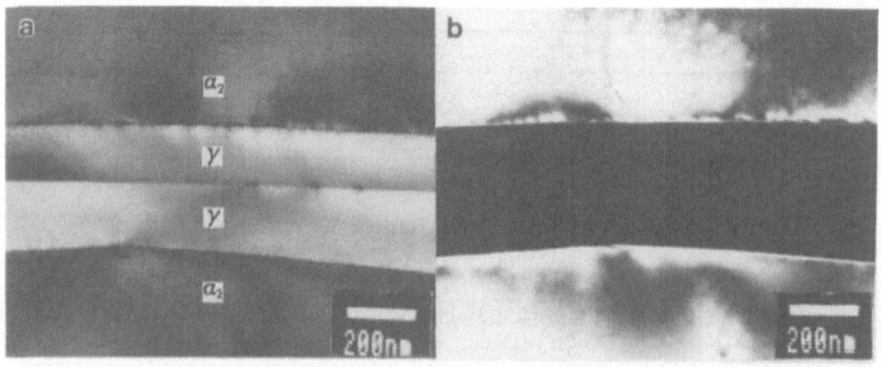

Fig. 19. The $\alpha_2/\gamma/\gamma/\alpha_2$ microstructure. (a) bright field image, (b) dark field image using (2021)α_2 reflection revealing that there was no α_2 layer existing between the adjacent γ plates.

Each orientation relationship is composed of two different variants. The reason for this characteristic is suggested to be that different variants can accommodate the strain induced by the γ plates when they precipitate from the α_2 matrix. Stacking faults in α_2 with opposite sign of $a/3$ [10$\bar{1}$0] Shockley partial dislocations are proposed as the nucleation sites of the different γ variants.

At the same time, differently orientated γ regions are frequently observed to coexist within the same γ lamella. The atomic arrangements of these regions are suggested to be similar to the α_2 matrix. Figs. 20 and 21 are two examples. In Fig. 20, regions 1 and 3 are one variant, and region 2 is another variant. These two variants are found to have the same stacking sequence with the APB–induced fault existing between them. This is due to the fact the stacking sequence in the same γ lamella is created from the same Shockley partial and therefore should be identical. In Fig. 21, regions 1 and 2 in the same γ lamella are found to have the same characteristic, as shown in Fig. 20. However, the boundary between region 1 and 2 is the (011)γ/(101)γ twin planes, instead of the APB–induced fault plane in Fig. 20.

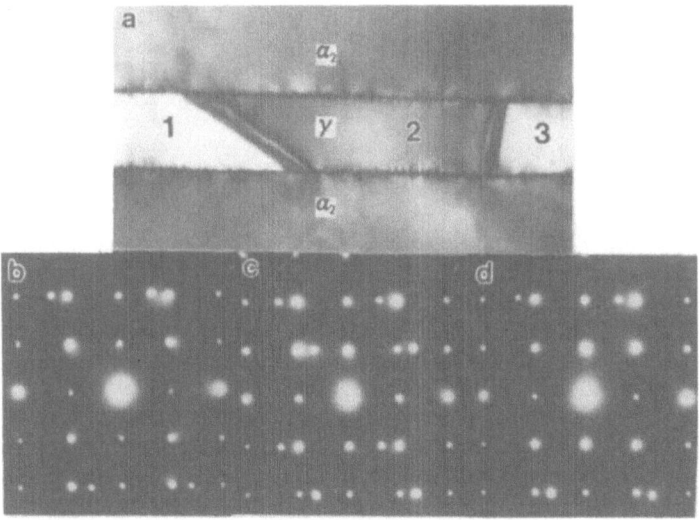

Fig. 20. TEM micrograph shows (a) differently orientated regions 1, 2 and 3 coexisting within the same γ lamella. SADPs of regions 1, 2 and 3 are shown in (b), (c) and (d), respectively.

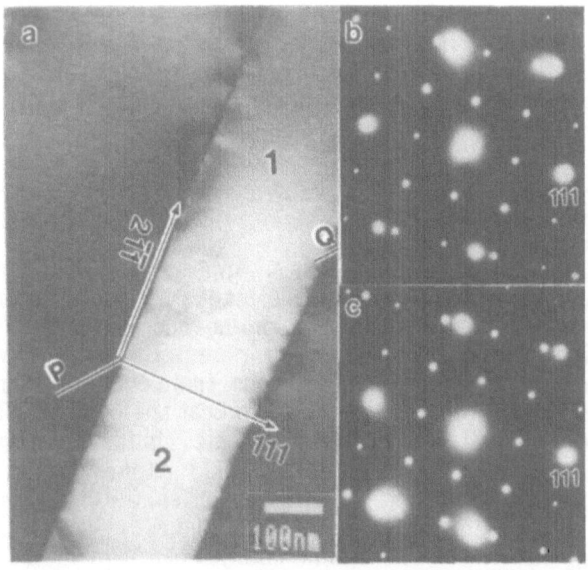

Fig. 21. TEM micrographs show (a) region 1 and 2 in the same γ lamella with $(011)\gamma/(101)\gamma$ twin plane PQ (b) and (c) SADPs of region 1 and 2

The mechanism for the formation of $(011)\gamma/(101)\gamma$ twin planes is suggested to grow from the "APB–induced fault" formed in the γ lamella during the precipitation transformation of $\alpha_2 \longrightarrow \alpha_2+\gamma$. Total strain induced by the structure change (hcp $\longrightarrow$ fcc) and by the tetragonality of γ phase during the precipitation transformation can be minimized by a composite of $\{111\}\gamma$ and $\{011\}\gamma$ twins coexisting in γ 0.5cm.

3.2. THE ROLLING TEXTURE AND IT'S RELATED MECHANICAL PROPERTIES OF THE TI₃AL–NB ALLOY [17]

The Ti₃Al ordered alloy with α_2 phase was found brittle in the temperature range from room temperature to about 600ºC. A significant ductility improvement is found by Nb–modifying Ti₃Al alloy, especially the alloy with the composition of Ti₆₅Al₂₅Nb₁₀. A rolling textured with α_2 basal type can be formed by hot rolling process in a Ti₆₅Al₂₅Nb₁₀ alloy. Experimental results show that the elongation in the rolling direction of the rolling–textured Ti₆₅Al₂₅Nb₁₀ alloy can be improved significantly (12.7% elongation at room temperature) compared with the Ti₃Al alloys. It is suggested that the improvement in elongation in this textured alloy results from the lower average orientation factor and from the smaller grain boundary angles present in the textured alloy. Both factors contribute a lower blocking energy of the grain boundaries for dislocation slipping.

4. Titanium Nickel Intermetallics

The most famous titanium–nickel intermetallics is the equiatomic titanium–nickel alloy, or TiNi alloy, due to its shape memory effect (SME) and pseudoelasticity (PE) properties. The characteristics of SME and PE are associated with the thermoelastic martensitic transformation of TiNi alloys. Recent research interests on TiNi alloys in Taiwan are focused on the premartensitic transformation and its transformation sequence, the high–temperature shape memory alloys, and the hydrogen in TiNi alloys.

4.1. THE PREMARTENSITIC TRANSFORMATION AND ITS TRANSFORMATION SEQUENCE [18]

It is well known that the premartensite R–phase appears if TiNi alloys are cold worked and annealed, or thermal cycled, or thermal aged (for Ni$\geq$50.4 at%), or ternary element being added. The R–phase is a rhombohedral structure which can be formed by elongating <111> directions in the parent B2 phase. Owing to B2 phase having four independent <111> direction, thus, there exist four variants of R phase. In 1989, Wu et. al. [19] constructed the reciprocal lattice of the R phase by using electron diffraction and proposed the atomic shuffles associated with premartensitic R–phase transformation as follows. (1) Atoms on the elongated <111>$_{B2}$ direction are slightly elongated. (2)Other atoms shuffle slightly inward the same distance along <112>$_{B2}$ direction on $\{111\}_{B2}$ planes which are the planes normal to the <111>$_{B2}$ directions. The atomic shuffles can explain the facts that the observed 1/3–reflections appear in the electron diffraction pattern, the rhombohedral angle α is slightly less than 90 degrees, and α can vary with temperature.

As mentioned above, the SME and PE are associated with the martensitic transformation of TiNi alloys (the martensite is a B19' structure). It is also confirmed that the premartensite R–phase has the SME and PE properties. This means that the premartensitic transformation is also a thermoelastic transformation. We are interested to know that, when the premartensite appears, what is the transformation sequence between martensite and premartensite transformations. One research result is shown in Fig. 22 which indicates that, by internal friction test, the $Ti_{49}Ni_{51}$ alloy aged at 400oC conducts the transformation sequence of B_2 <—> R–phase <—> B19'. The well separated premartensitic and martensitic transformations indicate one can utilize either one of these two transformations in the engineering applications.

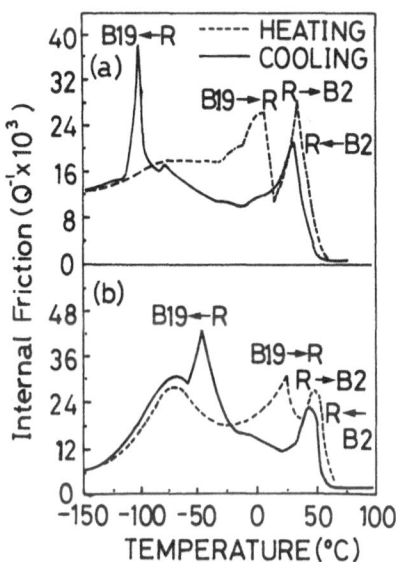

Fig. 22. Internal friction of $Ti_{49}Ni_{51}$ as a function of transformation sequence at 400oC

4.2. THE HIGH–TEMPERATURE SHAPE MEMORY ALLOYS [20]

The TiNi binary alloy shows martensitic transformation around room temperature, and therefore, can not be used as high–temperature shape memory alloy (SMA). An intensive study has been conducted to look for the high–temperature SMA. The $Ti_{50}Ni_{50-x}Pd_x$ and $Ti_{50}Ni_{50-x}Au_x$ alloys are found to be the potential ones. Fig. 23 shows the start temperature of the martensitic transformation, Ms, vs. the Pd composition in $Ti_{50}Ni_{50-x}Pd_x$ alloys by electrical resistivity test. By using DSC measurement, a transformation heat ΔH associated with martensitic transformation in $Ti_{50}Ni_{50-x}Pd_x$ alloys, with x : 20 − 50 at%, has been proved to possess the characteristics of thermoelastic martensitic transformation. The similar situation also occur in $Ti_{50}Ni_{50-x}Au_x$ alloys. Both $Ti_{50}Ni_{50-x}Pd_x$ and $Ti_{50}Ni_{50-x}Au_x$ alloys have the one–way SME and rather good shape memory recovery.

4.3. HYDROGEN IN TINI ALLOYS [21]

TiNi alloys can absorb hydrogen and become TiNi–H hydrides. Wu et. al. [21] observed the 1/2 − or − 1/4 − extra electron reflection spots between B2 fundamental spots, and proposed that TiNi–H hydride is a Ti_8Ni_8H composition with interstitial hydrogen in the octerhedral sites. Recent research indicates that hydrogen permeability in TiNi alloys is faster than that in steels. TiNi–H with unidentified composition can easily form on the surface if the hydrogen concentration charged into TiNi alloys is over 2000 ppm. This surface hydride serves as the diffusion barrier for hydrogen and expands in volume during its formation. A preliminary results show that there is a hydride which has hydrogen concentration higher than Ti_8Ni_8H. The research on characteristics of TiNi–H hydrides is still undertaken.

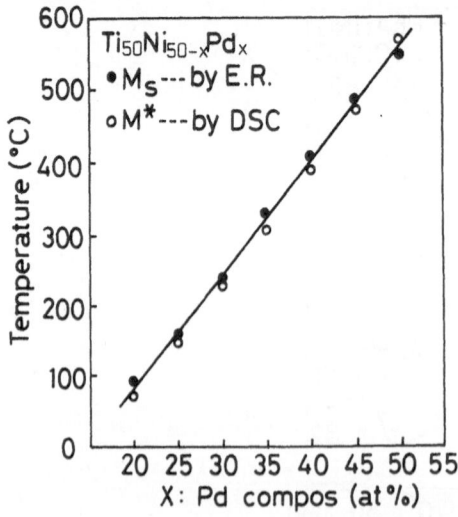

Fig. 23. Ms, vs. Pd at% in $Ti_{50}Ni_{50-x}Pd_x$ SMA alloy

5. An Innovative Process to Prepare Intermetallics [22]

An innovative spontaneous reaction synthesis (SRS) process was developed in MRDC to prepare many kinds of intermetallics including nickel aluminide, titanium aluminide and nickel titanide. SRS process is carried out in a controlled environment in which the premixed and consolidated powders with any designed stoichiometric compositions were firstly ignited by heating element and then exthermally reacted spontaneously. The whole process to fabricate a TiAl, for instance, intermetallics powder with 500 gr in weight can be completed within a few minutes.

Utilizing this process, TiAl, Ti_3Al, NiAl, Ni_3Al and NiTi intermetallics have been successfully synthesized. The synthesized powders were attrited down to $10\mu m$ and cold compacted with pressure of 137 Mpa in a die. Then the specimen can either be sintered by a conventional P/M process or remelted by a vacuum arc furnace. Fig. 24 shows the EDS spectrum of SRS–synthesized TiAl powder.

It is found that the SRS process exhibits many unique advantages:
(a) Extremely pure ordered phase can be obtained.
(b) Any desired stoichiometric composition can be synthesized.
(c) To smelt H.T. intermetallics without furnace.

(d) Minimum cost.
(e) The quickest process to fabricate intermetallics.

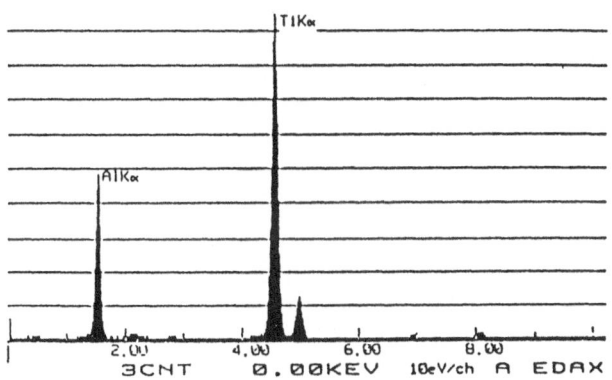

Fig. 24. EDAX spectrum for SRS synthesized TiAl intermetallic powder

6. Summary

Ordered intermetallics based on aluminide and titanide constitute a unique family of metallic materials that have promising mechanical behaviors for temperature applications.

For the past few years, Taiwan has devoted a substantial effort to the research and development on intermetallic compounds and appreciable progress has been made in alloy design and in improving their high temperature strength as well as ductility at room and elevated temperatures.

The alloy design work has been centered on the multi–element modified Ni_3Al both for poly and single crystals. A strengthening mechanism was proposed that an ordered $L1_2$ matrix could be strengthened by a disordered Cr–rich $Ni_3(AlCr)$ dispersoids. The development work has resulted in substantial improvement in the mechanical properties and ductility which have been considered for future elevated temperature applications.

An innovative SRS process has been developed. Based on this process, many desired stoichiometric intermetallic compounds can be fabricated in a quickest way.

Acknowledgement

The authors wish to thank Mr. J.Y. Ma, S.Y. Lee, T.S. Liu, S.C. Yang and Dr. B.C. Fu for their contribution on the experimental work. The authors are grateful to Miss Y.C. Su and Mr. C.R. Lee for the manuscript preparation.

Reference:

1. Aoki, A. and Izumi, O. (1979) Nippon Kinzoku Gokkaishi, Vol. 39, p.1190.
2. Koch, C.C., Liu, C.T., and Stoloff, N.S. (1985) 'High–Temperature Ordered Intermetallic Alloys' MRS Symp. Proc. Vol.39.

616

3. Hsu, S.E., Hsu, N.N., Tong, C.H. and Ma, J.Y. (1986) Chinese Journal of Materials Science, Vol.18A, No.1, pp.40–53.
4. Liu, C.T., White, C.L. and Horton, J.A. (1985) Acta Metall., Vol.33, pp.213–219
5. Taub, A.I., Huang, S.C. and Chang, K.M. (1985) 'Stoichiometry Effects on the Strengthening and Ductilization of Ni_3Al by Boron Modification and Rapid Solidification' in Early, J.G., Shives, T.R. and Smith, J.H. "Failure Mechanisms in High Temp. Materials", Cambridge Univ. Press. pp.57–65
6. Hsu, S.E., Hsu, N.N., Tong, C.H., Fu, P.C. and Ma, J.Y. (1987) Proc. Chinese Soc. of Mat. Sci.
7. Ohiai, S., Oya, Y. and Suzuki, T. (1984) Acta Metall., vol.32, 2, p.289.
8. Hsu, S.E., Hsu, N.N., Tong, C.H., Ma, C.Y. and Lee, S.Y. (1987) Mat. Res. Soc. Proc., 81, pp.506–512.
9. Hsu, S.E., Tong, C.H., Lee, T.S. and Liu, T.S. (1988) Mat. Res. Soc. Proc.
10. Flinn, P.A. (1960) AIME Trans., 218, 145.
11. Hancock, G.F. (1971) Phys. Stat. Sol. A7, 535.
12. Ma, C.Y., Tong, C.H., Fu, B.C. and Hsu, S.E. (1989) Proc. of the 1989 Annual Conf. of the Chinese Soc. for Mat. Sci., pp.145–149.
13. Mishima, Y., Ochiai, S., Yadogawa, M. and Suzuki, T. (1984) Trans. of the Japan Inst. of Metals, 27, 1, p.41.
14. Liu, C.T. and Koch, C.C. (1983) 'Trend in Critical Materials Requirement for Steel of the Future' NBSIR–83–2679–2 National Bureau of Standards.
15. Hsu, S.E., Lee, T.S., Yang, C.C. and Tong, C.H. (1991) Morris Fine Symp. Edited by Liaw, P.K., Weertman, J.R., Marcus, H.I. and Santner, J.S., TMS published. p.101.
16. Yang, Y.S. and Wu, S.K. (1991) Scripta Metall, et Materialia, Vol.25, p.255.
17. Hou, W.P., Wu, S.K. and Koo, H.K. (1990) MRS Symp.
18. Wu, S.K. (1990) Invited Paper in Proc. of 1990 Annual Conf. of Chi. Soc. for Mat. Sci., pp.86–95.
19. Wu, S.K. and Wayman, C.M. (1987) Acta Metall. England, Vol.37, pp.83–88.
20. Wu, S.K. and Wayman, C.M. (1987) Scripta Metall., Vol.21, pp.75–77 and pp.83–88.
21. Wu, S.K. and Wayman, C.M. (1988) MRS Int'l Meeting , Tokyo.
22. Hsu, S.E., Wu, H.D., Li, C.M., Chou, H.Y. and Wang, K.L. (1991) Presented in JIMIS–6, Tohoku, Japan.

INTERMETALLIC RESEARCH IN THE UK

M H LORETTO,
IRC in Materials for High Performance Applications
The University of Birmingham
Edgbaston
Birmingham B15 2TT, UK

ABSTRACT. The brief for this paper is to outline the intermetallics work in the UK, which is not covered in the presentations from Cambridge, Oxford and Imperial College at this meeting. The total research effort in the UK into intermetallics is not large and in this brief review an attempt is made to summarise other University-based work on defects in intermetallics, before describing some of the more industrially-based work. The industrially-supported work is aimed at producing alloys based on intermetallics and is centred on the new plasma melting facility, processing and testing facilities at the IRC. Finally the direction of future UK research into alloys based on intermetallics is briefly discussed.

1. Introduction

Papers at this meeting will cover much of the intermetallics research effort in Oxford, Cambridge and Imperial College and this brief review will cover work elsewhere in the UK. There is in fact very little other work going on in this area, with the main effort being in the School of Metallurgy and Materials at Birmingham and at the Interdisciplinary Research Centre (IRC) in Materials for High Performance Applications in Birmingham and Swansea, together with some additional work at Imperial College not presented at this meeting. Much of the work at the IRC is supported by industry through IMI and Rolls Royce. Other work in the area of Ti aluminides is underway at RAE Farnborough and at Harwell. This paper will be concerned therefore with the effort outside Oxford, Cambridge and will be divided into two areas; firstly the work which is essentially physical metallurgy (which is the main interest at this meeting) and the second area, which will include work involving processing. The physical metallurgy work on intermetallics is being done at Imperial College (in the Department of Metallurgy and Materials Science) at the IRC and in the School of Metallurgy and Materials at Birmingham and covers work in L12, L10 and DO19 intermetallics. Some examples of recent work in these areas will be presented briefly before looking at the broader scene.

C. T. Liu et al. (eds.), Ordered Intermetallics – Physical Metallurgy and Mechanical Behaviour, 617–622.
© 1992 *Kluwer Academic Publishers.*

2. Defect Studies in Intermetallics

2.1. L12

A detailed study which is underway of the factors that control the plastic deformation of Fe_3Ge, (which shows no flow stress anomaly with increase of temperature) has shown [1] that all deformation appears to take place via <110>{001} slip. This is despite the fact that the annealed structure consists almost exclusively of faulted defects on {111}. Consideration of the various factors which could lead to the preference of this slip system over the <110>{111} suggests that the Peierls stress on {111} is too high to allow dislocation mobility at the stresses that allow the {001} systems to operate.

2.2. L10

The influence of alloying additions and of temperature on the slip systems in TiAl has been studied in a recent research programme [2]. In agreement with other work it has been found that the addition of Mn increases the room temperature plasticity both by increasing the amount of twinning and by increasing the contribution of 1/2<110> dislocations. The Mn appears to substitute mainly for the Al and in so doing it has been suggested that it reduces the effective Peierls stress for certain segments of line direction of the dislocations. The limited plasticity found in two phase alloys with compositions near to 48 at%Al has been shown to be associated with deformation in the TiAl phase rather than in the Ti_3Al. The evidence, both from this study and from other work, is that several factors may be important in explaining this somewhat surprising observation. Two of these suggestions seem to be accepted; the higher solubility for oxygen of Ti_3Al may result in the effective scavenging of oxygen from the TiAl phase with a consequent increase in the mobility of dislocations [3]; the interface between the TiAl and the Ti_3Al appears to act as an effective dislocation source [2].

Interestingly it has been found that even when samples are deformed at liquid nitrogen temperature, there are some grains which contain a high density of faulted dislocation dipoles of the sort analysed earlier by Hug et al [4,5]. The density can be so high that it is difficult to believe that there were a sufficient number of forest dislocations to give rise to the number of intersection events, required for the mechanism discussed by Hug et al, to be solely responsible. On that basis it has been suggested [6] that there may be other mechanisms which give rise to dipoles and to other extended defects observed in intermetallics and this is discussed below.

2.3. DO19

In conjunction with colleagues in the USA the influence of alloying additions and of deformation temperature on slip systems has also been studied in Ti_3Al. It has been found [7,8] that the main slip systems which operate during compressive strain at room temperature are $<1\bar{2}10>\{1\bar{1}00\}$, and in grains where the resolved shear stress on these systems is small the $<11\bar{2}6>\{11\bar{2}1\}$ systems are found to operate. The same systems operate at temperatures up to 600°C, where some evidence for climb has also been noted. The limited ductility found at room temperature has been attributed to the difficulty of operating the $<11\bar{2}6>\{11\bar{2}1\}$ systems and the increase in ductility at 600°C has been associated with climb, as well as with an increase in the mobility of the relatively sessile $<11\bar{2}6>$ dislocations.

It has been found that the addition of either about 4at%Nb in solid solution, or the removal of oxygen (and carbon) by gettering with Er, results in the <11$\bar{2}$0> dislocations gliding virtually exclusively on (0001) rather than on {1$\bar{1}$00} and also in a very significant increase in the number of <11$\bar{2}$6> dislocations. These observations have been interpreted in terms of the directionality and relative strengths of the interatomic bonds, but at this stage there is no conclusive evidence for this suggestion although it is undoubtedly important. Evidence for the importance of near neighbour interactions is clear from the fact that displacement fringes are commonly observed when dislocations of <11$\bar{2}$6> glide on {11$\bar{2}$1}; these fringes are visible when using fundamental diffracting vectors contained in rows which do not contain superlattice vectors so that the contrast is not APB contrast, but is a rigid body displacement associated with wrong bonds across the {11$\bar{2}$1} planes [7]. The fact that two identical 1/6<11$\bar{2}$6> partials which make up the 1/3<11$\bar{2}$6> dislocations, can be pulled apart effectively to infinity, suggests that there is a strong pinning mechanism which allows the stress to separate these identical partials. This possibility is discussed below.

2.4. MOBILITY OF DISLOCATIONS IN INTERMETALLICS

It has been suggested [6], on the basis of the above work, that in addition to the importance of bonding effects, the influence of dislocation dissociations, is likely to be a significant factor at all temperatures. This is relevant, whatever the ordering energies on the different glide planes, since many of the possible dislocation reactions do not produce either stacking faults or anti phase boundaries. These dissociations are possible because many of the dislocations in intermetallics are made up of vectors which are simply the sum of two perfect lattice translations and although the dissociations do not lead to energy reductions the influence of the stresses acting on gliding dislocations will commonly be in the sense to cause the dissociation.

If dislocations in TiAl are considered the following dissociations are possible:

$$1/2<112> = 1/2<110> + <001>.$$
$$<101> = 1/2<112> + 1/2<1\bar{1}0>$$
$$<101> = <100> + <001>$$

To a first approximation there is no reduction in energy associated with any of these reactions but it is clear that under appropriate stress conditions the dislocations produced by all three of these reactions could glide apart. A full anisotropic calculation could be carried out, but because it is clear that the applied stress could drive the component dislocations apart even if the energy balance is (slightly) unfavourable, the calculation has not been done.

If a 1/2[112] dislocation is in edge orientation on ($\bar{1}\bar{1}$1) it will lie along [$\bar{1}$10] and the applied stress could act on the [001] component of the dislocation and extend the core on (110). Similarly if the dislocation were in screw orientation and lay along [112] the [001] segment could again extend on (1$\bar{1}$0). Thus both edge and screw dislocations of b = 1/2<112> would be pinned and the dislocation would be difficult to move. Similarly the glide of a screw dislocation of b = [101] on (11$\bar{1}$) will be impeded if the core dissociates into the two dislocations [100] and [001] if these can extend on (010). These very simple considerations thus suggest that because the dislocations can dissociate in this manner there may be consequences on the mobility of various segments. There are several observations of the defect structure in deformed TiAl which are consistent with the dissociations discussed here; firstly the relative mobility of the various segments of the superlattice dislocations and secondly the formation of the very high density of extrinsically faulted dipoles generated by the movement of <101> and 1/2<112> dislocations.

The formation of extended dipoles in TiAl was noted initially by Lipsitt et al [9] who pointed out that dislocations of b = <101> were pinned by unknown pinning centres which gave rise to the formation of long dipoles. In a series of detailed papers these dipoles have been shown to be extended extrinsic dipoles (eg[4], [5]) and are an extremely common feature of the deformation structure in TiAl. The basic idea behind the mechanism put forward by Hug et al is that a gliding dislocation is jogged by a forest dislocation and that the glide of this jogged dislocation produces a dipole which transforms to a faulted dipole on the active glide plane. Typically these dipoles are about 40nm wide and it is a tacit assumption that they are single atom spacing jogs.

A dislocation of b= 1/2<112> may be pinned if it dissociates to produce a 1/2<110> and <001> dislocation as discussed earlier. Subsequent glide of the unpinned portion of the dislocation of b= 1/2<112> will drag out a dipole on the active slip plane. The width of this dipole on {111}, the glide plane, would then be determined by the length along which the 1/2<112> dislocation dissociated to form the 1/2<110> and <001> dislocation. The direction of the extended dipole would generally be determined by the direction in which the dislocation continued to glide and in principle there is no constraint on this direction other than that it lie in the appropriate {111}. As noted by Hug et al if cross slip is involved then the dislocation must locally align along a <101> direction in the active glide plane. If this idea is taken up and the part of the 1/2<112> dislocation forming the dipole is assumed to dissociate according to the following reaction

$$1/2<112> = 1/2<101> + 1/2<011> + APB$$

it is then possible for these segments to cross slip as suggested by Hug et al [4,5] who describe the subsequent dissociations in detail.

A <101> dislocation can dissociate according to the following type of reaction

$$<101> = 1/2<112> + 1/2<110>$$

and the reaction giving rise to the faulted dipole involves the dissociation of the 1/2<112> component. Hence whether or not the dipole originates from <101> or from 1/2<112< dislocations the cross slip event is associated with the 1/2<101> partials that make up the 1/2<112> dislocation.

In the case of Ti_3Al it has been pointed out above that the $1/6<11\bar{2}6>$ partials which make up the superlattice dislocation are pulled apart during deformation. It is apparent that without some locking mechanism the two identical $1/6<11\bar{2}6>$ partials would always respond to the applied stress identically and thus be impossible to separate beyond the separation defined by the APB energy on $\{11\bar{2}1\}$. It is suggested then that the following reactions occur under the influence of stress:

$$1/3[11\bar{2}6] = 1/6[11\bar{2}6] + 1/6[11\bar{2}6]$$

$$1/6[11\bar{2}6] = 1/6[11\bar{2}0] + [0001]$$

If a segment of the trailing $1/6[11\bar{2}6]$ dislocation were in screw orientation then under appropriate stress conditions the $1/6[11\bar{2}0]$ partial could be extended onto $(1\bar{1}00)$ (where the fault energy is low) whereas if the partial were in edge orientation the [0001] could be extended onto $(11\bar{2}0)$ - this would lead to the generation of no fault. In either case the $1/6<11\bar{2}6>$ partial would be locked and the leading partial could continue to glide on $\{11\bar{2}1\}$. The force needed to pull two $1/6<11\bar{2}6>$ dislocations apart under these conditions would have to be equal to the APB energy and calculations show that a stress of about 0.3% of the shear modulus would be needed [6].

The type of dissociation which has been discussed could be very common in intermetallics but in most cases the conditions for the formation of such obvious debris will not be fulfilled. Thus the evidence that this type of core dissociation occurs along the length of the gliding superdislocations may not be as apparent in other systems (eg if the fault energy is high) although any influence on the critical resolved shear stress will be felt. Interestingly it has recently been reported [10] that dislocation loops of b = <100> are left behind dislocations of b = <101> in NiAl.

Since the possibility of dissociation will be influenced by the direction of the local stress on the dislocation it is unlikely that there will be any method of limiting this type of dislocation locking whereas alloying can be used to influence the extent of covalency effects. This type of dislocation locking will occur under conditions which drive the component dislocations apart and since the product dislocations concerned in these reactions all have Burgers vectors at right angles to each other the condition that one of the resultant dislocations feels a larger force acting on it than does the other will be fulfilled for virtually all directions of the applied stress.

2.5. TRANSFORMATION AND CONSTITUTIONAL STUDIES

The work in the Department of Metallurgy and Materials Science at Imperial College is concentrated in the areas of phase transformations and constitutional studies of the titanium aluminides[11]. The main technique being used is surface alloying using laser melting and powder injection. This approach is yielding both equilibrium data and data on the kinetics of the complex transformations in these systems. The technique can be applied to other more exotic intermetallic systems and this work will be continued in such areas.

3. Other Aspects of Intermetallic Research in the UK

The main effort, other than that described elsewhere at this meeting, is concerned with the production of alloys based on intermetallics. Not surprisingly it is in this area that UK industry is more closely involved - and of course details of all industrial work is not available. IMI are carrying out melting work on TiAl and Ti_3Al-based alloys in conjunction with Rolls Royce, with the Royal Aircraft Establishment at Farnborough and with the IRC at Birmingham.

The work at the IRC makes use of a new twin torch plasma melting facility funded by SERC which allows the production of ingots up to 15cm in diameter and 150cm long. A range of ingots of Ti-based intermetallic alloys has already been produced. These alloys are being thermo-mechanically processed in the IRC. In addition to the actual processing both the plasma melting and the subsequent forging are being modelled within the IRC. Ingot-route materials are currently being tested over a range of conditions using the extensive mechanical testing facilities available in the IRC and this programme is part of a major effort aimed at relating processing route to properties and to microstructure. Interestingly it appears that the TiAl-based alloys do not require HIPping before forging. The longer term aim is to produce some demonstrator components using the machining and forging facilities in the IRC.

The plasma melter is currently being modified to allow the production of both spray-formed materials and powders of a range of intermetallics by introducing a bottom pouring cold wall induction furnace into the melter - this facility enables a controlled stream of 25Kg of intermetallic to be processed. A smaller single torch plasma melter at Imperial College is also being used to produce powder of Ti-based intermetallics.

622

The focus of the work in the IRC is to produce alloys of defined composition from elemental feedstock, so that expensive master alloys do not limit the range of alloys which can be produced and work carried out in conjunction with IMI has shown that the ingots have good homogeneity even after a single melting operation and the loss of volatile elements is very small. This represents an obvious advantage of plasma melting over vacuum-based routes, but of course there are other areas where vacuum technology, although being more expensive is to be preferred. An important part of the programme in the IRC is to determine which are the intermetallic-based alloys where plasma rather than vacuum-based melting is to be preferred.

The extent of coring is extremely small in the alloys assessed so far - presumably because the ingots are only 15cm in diameter and are water cooled and stirred during solidification. The pick-up of interstitial elements is negligible and it is clear that this facility should allow production of a range of intermetallic alloys of controlled and reproducible composition and controlled microstructure. Attempts are underway to integrate the research programme in the UK so that with the limited support available the facilities are used as efficiently as possible.

In addition to the programme outlined above there are small efforts aimed at assessing oxidation resistance of Ti-based and other intermetallics and some small programmes investigating some of the more exotic intermetallics alloys. Work in these areas is going on at Oxford, Cambridge, Birmingham and at Imperial College.

The other papers at this conference have made it clear that the extent of funding in the USA and in Japan is far greater than in the UK. On that basis it is apparent that there is the need both to increase the support in the UK and to integrate the effort more efficiently. Actions are being taken along both of these lines and it is hoped that by the time the next intermetallics workshop is held (which will surely include processing) the level of UK activity will have increased and the integration improved.

References

[1] Ngan A, Smallman R E and Jones I P. Submitted for publication.

[2] Wardle S and Jones I P (Submitted for publication)

[3] Vasudevan V K, Court S A, J Kurath P and Fraser H L. 1989 Scripta Met 23, 907

[4] Hug G, Loiseau A and Lasalmonie A. 1986 Phil Mag A54 47.

[5] Hug G, Loiseau A and Veysiére P. Revue Appl Phys 23, 673.

[6] Loretto M H (Submitted for publication).

[7] Court S A, Löfvander J P A, Loretto M H and Fraser H L. Phil Mag 1990, 61, 109.

[8] Court S A Loretto M H and Fraser H L Phil Mag. 1989, 59, 379

[9] Lipsitt H A, Schetman D and Schafrik R E 1975 Metall Trans A 6 1991.

[10] Field R D, Lahrman D F and Dariola R. Submitted for publication.

[11] Flower H M, Private Communication.

THE ORDERED INTERMETALLIC RESEARCH IN P.R.CHINA

Dongliang Lin (T.L.Lin)
Institute of Materials Science and Engineering
Shanghai Jiao Tong University
Shanghai 200030, P.R.China

ABSTRACT. This paper will provide an overview of the current status of the research and development of ordered alloys for high temperature structural applications, which were conducted in P.R.China. The physical and mechanical properties of ordered alloy will be reviewed with enphasis on aluminides. Current research programs in P.R.China will be outlined, and the expections of the future will be discussed.

Introduction

The interest in intermetallic research and development was rewakened in P.R.China from 1985 after the first Materials Reseach Society Symposium on High-Temperature Ordered Intermetallic Alloy, held in Boston, November 26-28, 1984. The potential uses of intermetallic alloys for structural applications attracted the researchers in the universities and the research institutes as well as the people in industry. A five-year (1986-1990) project on high temperature ordered intermetallic sponsored by National Advanced Materials Committee of China (NAMCC) was promoted as a part of the long term project for advanced materials and technology. This project, together with the financial support from National Nature Science Foundation of China (NNSFC), have in the past five years stimulated much work on intermetallics, largely in the area of improving low temperature ductility and increasing high temperature strength for aluminides in P.R.China, see Table I .

Table I Compilation of Research Efforts on Ordered Intermetallic Alloys in P.R.China

Laboratory	Source of Founding	System Studied	Principal Investigators
AMRI	NAMCC	Ti₃Al	Chunxiao Cao
CISRI	NAMCC	Ni₃Al,Ti₃Al	Zhenyong Zhong
CSUT	NAMCC	TiAl	Baiyun Huang
IMR	NAMCC, NNSFC	Ni₃Al,Ti₃Al,Fe-Al	Jianting Guo
NUT	NNSFC	Ti-Al	Shiming Hao
SJTU	NAMCC, NNSFC	Ni₃Al,TiAl₃,Fe₃Al,FeAl	Dongliang Lin(T.L.Lin)
UST	NAMCC	Nb-Ti-Al	Guoliang Chen

AMRI – Aeronautic Materials Research Institute, Beijing
CISRI – Central Iron and Steel Research Institute, Beijing
CSUT – Central South University of Technology, Changsha
IMR – Institute of Metal Research, Shenyan

C. T. Liu et al. (eds.), Ordered Intermetallics – Physical Metallurgy and Mechanical Behaviour, 623–643.
© 1992 Kluwer Academic Publishers.

624

NUT – Northeast University of Technology, Shenyang
SJTU – Shanghai Jiao Tong University, Shanghai
UST – The University of Science and Technology, Beijing

1. Nickel Aluminide Ni₃Al

Following the discovery that the pure, polycrystalline Ni₃Al can be ductiled at room temperature by doping with boron [1], a flood of research began, much of it either carried out or stimulated by Liu's team at Oak Ridge, U.S.A.[2].

The ductility and fabricability of Ni₃Al containing 24 at%Al were dramatically improved by adding a few hundred parts per million of boron, which tends to segregate strongly to grain boundaries and suppresses the brittle intergranular fracture in Ni₃Al [3]. The solid-solution hardening of Ni₃Al depends on the substitutional behavior of alloying elements, atomic size misfit and the non-stoichiometry of the alloy. Hafnium additions are very effective in improving high-temperature properties of ternary Ni₃Al (Al + Hf = 24at%) doped with boron.

The Alloy Development Team at Oak Ridge has succeeded in establishing a series of 5 Ni₃Al derivatives which are finding rapidly increasing terrestrial applications. For stuctural applications, mechanical and metallurgical properties of Ni₃Al alloys can be further improved by alloy design. Among them are high-temperature ductility and fabricability, creep resistance and low-temperature strength. For this purpose a great effort has been made in the laboratories of Institute of Materials Science and Engineering, Shanghai Jiao Tong University (SJTU) and Institute of Metals Research, Shenyan (IMR) to systematically investigate the effects of boron content and alloy additions of Mg, Ca, Si, transition metal elements and rare-earth elements on the mechanical properties of directionally solidified (DS), mono- and poly-crstalline Ni₃Al. Some results are summerized as follows:

1.1. The Effect of Boron Content

The microstructres and the mechanical properties at room- and high-temperature of mono-, poly- and micro-crstalline Ni₃Al containing 0, 0.52, 1.37 and 2.22 at%B [Table II] have been studied [4].

Table II   Chemical Composition of Ni₃Al + B alloys, at%

Alloy	Ni	Al	B
2	77.11	22.89	0.0
3	75.65	23.83	0.52
4	76.30	22.33	1.37
6	79.44	23.84	2.22

Microcrystalline Ni₃Al ribbons were obtained by melt spinning method with the cooling rate of 5×10^6 K/s. Mono-crystalline specimens of 16 mm diameter and 75 mm length were produced by high rate solidification (HRS) method with the temperature gradient of about 50K/cm and the solidification rate of about 1 mm/min.

Both conventional casting rods of polycrystalline and monocrystalline Ni₃Al were annealed at 1473K for 4h, furnace cooled to 1373K for 1h, and then air cooled.

The results have shown that the ultimate and yield strength at 293K and 1073K of poly-, mono- and micro-crystalline Ni₃Al as well as the stress rupture life of polycrystalline Ni₃Al alloys at 1123K and 100MPa steadily increase with the increase of B content up to 1.37at%B. With the increase of B content beyond this value they sharply drop for polycrystalline Ni₃Al, slowly drop for monocrystalline Ni₃Al and continuously increase for microcrystalline Ni₃Al. The stress rupture life of monocrystalline Ni₃Al continuously decreases with the increase of B content. Conclusion has been obtained that optimum B content might be 0.52-1.37 at% in Ni₃Al for compromising mechanical properties both at room and high temperature, instead of 0.1-0.5 at% for optimum room temperature ductility.

1.2. Effects of Alloy Addition on the mechanical properties of DS Ni₃Al

From previous work [3,5-8], it is known that the mechanical properties of Ni₃Al are strongly affected by the tertiary elemental additions. To make best use of Ni₃Al alloys as a potential engineering materials, the proper choice of elemental additions is anticipated, and the mechanical behavior of Ni₃Al alloy with each tertiary alloy element should be systematically investigated at first.

A project sponsored by the NAMCC has been conducted in our laboratory since 1989, aimed toward the development of Ni₃Al-based alloys for practical uses. In present study, the mechanical properties of DS Ni₃Al of ⟨001⟩ orientation with additions of tertiary microalloy elements such as B, C, Mg, Ca and rare-earth elements, and tertiary macroalloy elements such as Si and transition metal elements were systematically investigated by tensile test in a temperature range from 123K to 1223K [9-12].

1.2.1. Effects of Alloy Additions on the Temperature Dependence of Yield Stress [9]

(1) Temperature Dependence of Yield Stress

The temperature dependence of yield stress in B-doped DS Ni₃Al alloys with additions of Si and tranisition metal elements such as Ti, V, Mo, W, Nb, Ta, Hf and Zr has been determined and yield stress v.s. temperature curves for the alloys with additions of Ti are shown in Fig.1 as an example. Results show that the yield stress of DS Ni₃Al alloys increases progressively with temperature to a peak value at an elevated temperature, and beyond this peak the yield stress decreases with temperature. It is also found that the yield stress level of B-doped Ni₃Al at test temperatures below the peak increases with increasing concentration of each alloy element. It is noted that the stress level drastically increases especially with addition of such transition metal elements as Zr, Hf, Ta, Nb and W, and

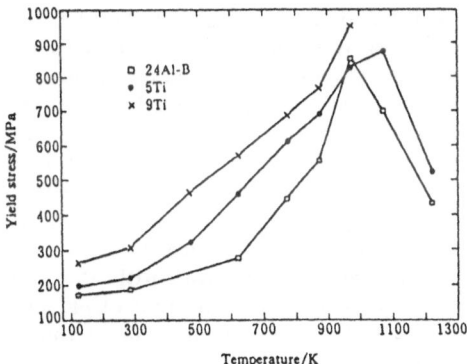

Fig.1 Temperature dependence of the yield stress in B doped DS Ni₃Al with addition of Ti

that the addition of certain amount of Mo and V reduces the yield stress at higher temperatures. On the other hand, the strenghening effect of the additive elements at sufficiently high temperatures above the peak weakens or disappears. The peak temperature for every alloy is found to be higher than or identical to that of B-doped Ni₃Al, i.e., 973K.

(2) Effect of Alloy Additions on the Positive Temperature Dependence of the Yield Stress

It has been proposed [13] that the observed yield stress, σ_y, can be expressed as

$$\sigma_y = \sigma_0(1 - BT) + A\exp(-U/RT)$$

where σ_0 is yield stress at 0 K, U is the activation energy for a thermally activated process, R is gas constant, T is temperature, and A and B are constants. According to the equation, Arrhenius type curves for each alloy were plotted for $\ln[\sigma_y - \sigma_0(1\text{-}BT)]$ with reciprocal test temperature, 1/T, in which a well-defined linear relation is obtained. Furthermore, the effect of each alloy element on the activation energy for the thermally activated process to cause the positive temperature dependence of the yield stress for Ni₃Al is shown in Fig.2, as a function of concentration of alloy elements. It is demonstrated that every alloy element in the B-doped Ni₃Al significantly reduces the value of the activation energy and the slope of U vs. alloy element content becomes steeper as a sequence of Ti < V < Mo < W < Nb < Ta < Hf < Zr for the addition of transition metal elements, while the slope for the Si is close to that for Mo.

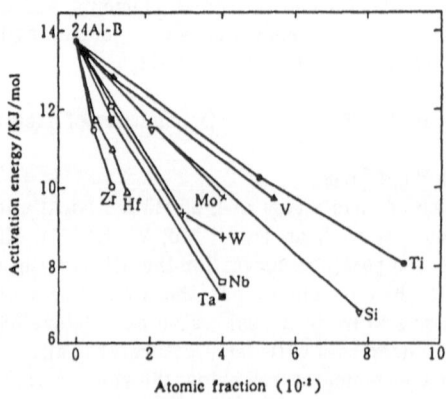

Fig.2 Effect of alloy elements in the activation energy of B-doped DS Ni₃Al

As for the mechanism of the anomalous mechanical behavior in Ni₃Al affected by the elemental additions, previous work [7,8] shows that adding such atoms into L1₂-type A₃B ordered phase as to increase its electron concentration e/a and/or atomic radius ratio R_B/R_A would enhance the anomalous yield stress behavior for Ni₃Al. In present work, it is found that among the alloy elements of 5th or 6th period there exists a similar variation of rate of activation energy change (dU/dC) with the difference of Pauling electronegativity between Ni and the alloy elements (δX_{AB}^P). Namely, with increasing δX_{AB}^P in a sequence of Mo < Nb < Zr of 5th period, and W < Ta < Hf of 6th period, the dU/dC decreases in its negative value. In the case of the addition of the elements of 3rd or 4th period, if the rate of activation energy change (dU/dC) is correlated with the valence difference (δVE) between the alloy element and Al, it can be found that dU/dC decreases in its negative value with increasing δVE for Ni₃Al with the addition of Al and Si of 3rd period having δVE equal to 0 and 1, and Ti and V of 4th period having δVE equal to 1 and 2, respectively. Therefore, present results imply that in the Ni₃Al alloy an increase in the positive temperature dependence of yield stress with the addition of Si and those transition metal elements is obtained mainly through increasing e/a for the alloy.

(3) Effect of Alloy Additions on the Solid Solution Strengthening

The effect of each alloy element on the solid solution strengthening is evaluated with the yield stress at 123K, where the contribution of Aexp($-$U/RT) is negligible. The yield stress obtained at 123K as a function of concentration of the alloy elements in the B-doped DS Ni₃Al is shown in Fig.3. Almost linear correlation is observed for each alloy element in the Ni₃Al alloys. It is evident in this figure that the strengthening capacity in the B-doped DS Ni₃Al is rather strong for the addition of Zr, Hf, Ta, Nb and W, while weak for addition of Ti and V. Again, it is found that as medium strengtheners in Ni₃Al, Si and Mo are close to each other in their strengthening capacity.

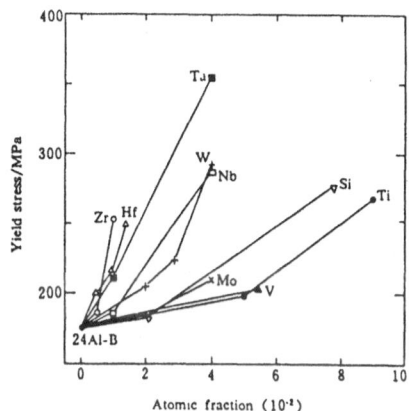

Fig.3 Variation of the yield stress at 123K with the additive element content in B-doped DS Ni₃Al

As for the capacity of solid solution strengthening with the addition of substitutional atoms in the Ni₃Al, the atomic size effect in the elastic interaction would be a main factor to be accounted [5,6,14]. If the yield stress increment per atomic fraction of alloy element ($\delta\sigma_y/\delta C$) is correlated with the lattice strain increment per atomic fraction of alloy element ($\delta\varepsilon/\delta C$), it is found that the observed solid solution strengthening induced by the transition metal elements, like Hf, Ta and Nb, can not be interpreted only by the size effect in the elastic interaction between the solute atom and the dislocation. By applying the concept of the electronegativity of alloy element to this phenomenon, it is found that the larger the difference of δX_{AB}^p is, the stronger the solid solution strengthening would be for the additions of the transition metal elements, Hf, Ta and Nb, in the B-doped DS Ni₃Al. Such tendency suggests that the electrical effect would be another main factor to respond to the solid solution strenghening in Ni₃Al alloy with the additions of transition metal elements.

1.2.2 Effects of Alloy Additions on the Ductilities of DS Ni₃Al [11]

(1) Effect of alloy additions on the ductility of DS Ni₃Al at ambient temperature

In present work, the magnitude of the ductility of Ni₃Al alloy is measured with the tensile elongation percentage at fracture. Some test results at ambient temperature are listed in Table III. It is shown that at ambient temperature the ductility of DS Ni₃Al intermetallics significantly increases with B additions, and with increasing B content (300-1000 ppm) the ductililty increases progressively (see alloy No.1-4 in Table Ⅲ). For example, by adding 1000ppm B its tensile elongation reaches 31% at ambient temperature. When 9 at%Fe or Mn is added, the ductility of DS Ni₃Al with or without B sharply increases, reaching about 40% (see alloy No.5-7 in Table Ⅲ). This result demonstrates that Fe and Mn are the most effective ductilisers, and the effects of alloying with both B and Fe on the ductility of Ni₃Al would belong to a competing but a combining mechanism. In the case of alloying with Cr, Hf, Nb, Mg and Y, Table Ⅲ shows that the ductility of B-doped DS Ni₃Al can be further improved with additions of certain amount of them at ambient temperature (see Alloy No.8-12 in Table Ⅲ).

Table Ⅲ Effects of alloy additions in the ductility of DS Ni₃Al at ambient temperature

№	1	2	3	4	5	6
Composition	24Al	24Al 300ppm B	24Al 600ppm B	24Al 1000ppm B	19.5Al 9Fe	19.5Al 9Fe 300ppm B
Enlongation(%)	9.0	19.5	22.0	31.0	39.5	39.0

№	7	8	9	10	11	12
Composition	16Al 9Mn 300ppm B	24Al 300ppm B 100ppm Y	23.5Al 0.5Hf 300ppm B	23Al 1Nb 300ppm B	19Al 5Ti 300ppm B	20.5Al 7Cr 300ppm B
Elongation(%)	42.0	26.0	26.0	28.0	30.0	28.5

The ductilizing effects of these solutes in Ni₃Al at ambient temperature could be mainly attributed to two factors, i.e., the electrical effect of the substitutional elements such as Fe, Mn and Cr on grain boundaries and the effect of grain boundary segregation of other alloy elements. If the two effects would homogenize the chemical bonding environment on grain boundaries, this will result in an increase in cohesion and deformability of grain boundaries, thereby enhancing the ductility of Ni₃Al intermetallics.

(2) Effect of alloy additions on the intermediate temperature brittleness of DS Ni₃Al alloys

The Ni₃Al intermetallics is brittle, accompanied by intergranular fracture in the intermediate temperature range. The ductility values of DS Ni₃Al with or without alloy additions are shown in Table Ⅳ in the intermediate temperature range from 773K to 1073K. In the intermediate temperature range the ductility of Ni₃Al intermetallic is lower. Especially at 973K and 1073K, its ductility becomes zero (see Alloy No.1 in Table Ⅳ). With addition of small amount of B (300 ppm) the ductility of Ni₃Al appreciably increases and reaches 9.0% and 6.0% at 973K and 1073K, respectively. From the table it can be seen that adding certain amount of Cr (4-9 at%) can further alleviate the brittleness in B-doped DS Ni₃Al especially at lower temperatures of 773K and 873K (see Alloy No.3-5 in Table Ⅳ). At higher temperatures of 973K and 1073K, it is found that the ductility of B-doped DS Ni₃Al can further increase through adding 9 at% Fe together with increasing Al content (see Alloy No.6 in Table Ⅳ).

Table IV Effects of alloy additions on the brittleness of DS Ni$_3$Al in the intermediate temperature range from 773K to 1073K

No.	Alloy Composition	Elongation (%)			
		773K	873K	973K	1073K
1	Ni$_{76}$Al$_{24}$	11.5	13.0	0	0
2	(Ni$_{76}$Al$_{24}$)$_{99.97}$B$_{0.03}$	10.0	8.0	9.0	6.0
3	(Ni$_{74}$Al$_{22}$Cr$_4$)$_{99.97}$B$_{0.03}$	24.0	10.0	6.5	9.0
4	(Ni$_{72.5}$Al$_{20.5}$Cr$_7$)$_{99.97}$B$_{0.03}$	29.0	6.5	9.5	8.5
5	(Ni$_{71.5}$Al$_{19.5}$Cr$_9$)$_{99.97}$B$_{0.03}$	20.5	28.0	12.5	–
6	(Ni$_{69}$Al$_{22}$Fe$_9$)$_{99.97}$B$_{0.03}$	12.0	7.5	14.0	13.0

Compared with that at lower temperatures, the cohesion and deformability of grain boundary of Ni$_3$Al is lower in the intermediate temperature range. Through segregation of B and substitution of Cr or Fe for the constituents of Ni$_3$Al on the grain boundary, the cohesion and deformability may increase. The beneficial effects of Cr may also respond to its resistance to the penetration of O along the grain boundaries; besides the effects of adding both Fe and Al may also respond to its resistance to the grain boundary cracking by precipitation of second phase in Ni$_3$Al alloy, and this may be an effective way to alleviate the brittle intergranular fracture in Ni$_3$Al at higher temperatures.

(3) Effect of alloy additions on the ductilities of DS Ni$_3$Al at a high temperature of 1223K

Table V Effect of alloy additions on the ductility of DS Ni$_3$Al at a high temperature of 1223K

No	1	2	3	4	5	
Composition	24Al	24Al 300ppm B	24Al 600ppm B	24Al 1000ppm B	24Al 9Fe 300ppm B	
Elongation(%)	8.5	69.5	53.8	65.2	84.4	
No	6	7	8	9	10	11
Composition	22Al 2.1Si 300ppm B	23.5Al 0.5Hf 300ppm B	23Al 1Hf 300ppm B	22.5Al 1.4Hf 300ppm B	16Al 9Mn 300ppm B	24Al 300pp B 200ppm Mg
Elongation(%)	98.4	51.5	58.6	43.0	42.5	38.0

Table V shows the ductility of DS Ni$_3$Al affected by alloy additions at 1223K. It is noted that the ductility of Ni$_3$Al intermetallics is low, only about 8.5% at 1223K (see Alloy No.1 in Table V). The addition of various levels of B (300-1000 ppm) sharply enhances the ductility of DS Ni$_3$Al, and the maxium value is about 70% (see Alloy No.2-4 in Table V). Besides, the ductility of the B-doped DS Ni$_3$Al can be further improved with addition of certain amount of Ca (200 ppm) and Si (2.1 at%), and the ductilities reach 84% and 98%, respectively (see Alloy No.5-6 in Table V). From the table it is also indicated that the B-doped DS Ni₃Al alloys retain higher ductilities at 1223K when certain amount of Hf, Mn, or Mg is added (see Alloy No.7-11 in Table V). Additions of other elements significantly reduce the ductility of B-doped Ni₃Al alloy at 1223K.

SEM fractographic analyses show that for Ni₃Al alloys exhibiting higher ductilities at 1223K, the fracture is transcrystalline in nature and the dynamic recrystallizing process occurs in the tensile specimen, while for those exhibiting lower ductilities at 1223K, the above features are not observed. This phenomenon predicts that an increase in grain boundary cohesion and mobility of dislocations arising from such elements as B, Ca, Si, and Mg would be a main factor to respond to the high ductility behavior in Ni₃Al alloy.

It should be noted here that with Mg, Y, or Ce additions the high temperature ductilities of polycrystalline Ni₃Al alloy can be improved, which is reported by other two studies [15,16]. In the work, a number of Ni₃Al-Cr-Zr-B ordered alloys (IC218) containing various levels of Mg, Y, or Ce were prepared by arc-melting, or induction melting and conventional casting. Compression test results in a temperature range from 1273K to 1423K indicate that adding moderate levels of Mg, Y, or Ce significantly increases the high temperature ductilities of Ni₃Al-Cr-Zr-B ordered alloy (IC218). In the work, it is suggested that the ductility improvement of the Ni₃Al alloy is attributed to an increase in grain boundary cohesion resulting from moderate Mg, Y, or Ce additions, which is in agreement with the predictions of our work mentioned above.

1.3. The Investigation of Grain Boundary Structure, Segregation and Deformation Characteristics of the Ni₃Al Alloys

A great effort has been devoted to elucidating mechanisms responsible for boron-enhenced ductility. One suggestion is that boron increases the cohesive strength of grain boundaries [19], which is supported by the intergranular segregation of boron. An alternative suggestion [20] is that boron appears to increase the mobility of grain-boundary dislocation and, thereby, to increase the grain-boundary slip accommodation, which makes propagation of slip across grain boundary easier. Besides, there exists the interpretation [21] that boron lowers the stress for dislocation generation from sources along the boundary.

In order to understand the intrinsic characteristics of brittleness and mechanism of boron induced ductility in Ni₃Al alloys, systematic work has been done in our laboratory, which includes three related parts of research work, that is, the grain-boundary (GB) structure, the GB segregation and deformation behavior of Ni₃Al alloys.

1.3.1. Atomistic Simulations of Grain Boundaries in Ni₃Al

To systematically study such problems as GB atomic structures, energies, and other related problems, as well as their relationships with the GB composition, has the significant meaning for understanding the intrinsic characters of GBs in Ni₃Al. Because it is not easy to combine the chemical composition with the structural observation on the atomic scale, and also difficult to obtain the relationships between the various structures and their properties, so computer atomic simulation as an 'experimental technique' becomes an effective method to study the GB atomic structures and their properties.

By now the publications dealing with the simulation of Ni₃Al GB structures are mostly given by Chen, et al. [22-24]. They have performed a series of simulations on symmetric tilt [100] GB structures and energies. They have also studied the effect of boron, sulfur and nickel segregation on the structure and strength of GBs, and computed the GB (and bulk) cohesive properties by the frozen and slow straining methods. Vitek et al. [25] have also calculated and compared the GBs in pure f.c.c. metals and in L1₂ alloys and analyzed their similarities and differences in the view of ordering tendency and formation of atomic size cavities in the boundaries, and discussed their relations with the intergranular fracture.

On the basis of above previous studies, the GBs in Ni₃Al alloys have been more systematically studied with the consideration of the problems from some other aspects. The interaction between the dislocation and GB has also been investigated in our laboratory, about which no publication has been seen by now as we know. The details of computational procedure were described elsewhere [26,27] and only main results or conclusions will be presented here.

The contents of computer simulations include a series of symmetric tilt [100], [110] and [111] GB structures, each of them was selected with several GB geometrical indexes (the reciprocal density of coincidence site Σ and misorientation angle θ^0) with the different GB composition (Al-rich, Ni-rich and stoichiometry). The relationships between the GB structure, GB composition, GB energy, GB electron density distribution, GB stress field and the formation energy of vacancy in the GB region have been studied. Besides, due to the reason that the core structure of the dislocation close to the GB and the interaction between the GB and the dislocation, e.g., dislocation absorption and emission at GB, will affect the capacity of deformation accommodation of GBs and resistances against the intergranular fracture, so the core structures and its transformations of the dislocation close to the GB under the effect of applied shear stress have also been investigated. Furthermore, in all above studies, the influence of boron addition has been inspected, which is most important for understanding the mechanism resposible for boron enhanced ductility.

From all above simulated results, the following conclusions can be inferred:

(1) There is a strong dependence of energies on the local GB composition. Ni-rich boundaries generally exhibit lower GB energies and higher cohesive energies than those of stoichiometric or Al-rich boundaries for all three [100], [110], and [111] symmetric tilt axes boundary system. Moreover, $\Sigma 3$ GB has much lower GB energy but higher GB cohesive energy and fracture work than those of other GBs, which indicates that it has very strong resistance against intergranular fracture, and this phenomenon has also been confirmed by Lin, H. et al. [28].

(2) The boron addition in Ni₃Al greatly increases GB cohesive energies and fracture work, thus the intergranular cracking tendency will be decreased. Moreover, the most striking feature is that such effect in Ni-rich GBs is higher than that in other GBs.

(3) The segregation of boron makes the distribution of electron density more homogeneous at the GB region. The more homogeneous the distribution is, the more stronger the cohesion between atoms is, and furthermore the increment of homogeneous level due to boron addition is higher in Ni-rich GBs than those in Al-rich GBs, which will make the GBs stronger in former case.

(4) The vacancy concentrations are higher in Al-rich alloys (bulk) than those in Ni-rich alloys (bulk), and the hydrostatic stresses are more tensile in Ni-rich GBs than those in Al-rich GBs. These two factors all make boron atoms have higher tendency for segregation in Ni-rich GBs than that in other GBs.

(5) In Ni₃Al alloys, the superdislocation dissociates on the (010) plane into the 1/2[$\bar{1}$01] superpartials separated by an APB, and on the (1$\bar{1}$1) plane the superdislocation dissociates into two 1/2[$\bar{1}$01] dislocation bounding an APB, or into a 1/3[$\bar{2}\bar{1}$1] and a 1/3[$\bar{1}$12] superpartial separated by a superlattice intrinsic stacking fault (SISF). The dislocation

close to GBs leads to the distortion of GB structure. The addition of boron decreases all the corresponding distortion difference between neighboring structural units for three types of the interaction, which implies that the intergranular boron segregation eases the spreading of the dislocation core and may reduce the stress concentration at the GB.

(6) As results of interaction with the GB, all the three types of screw dislocation cores are non-planar in Ni_3Al with and without boron. Although, the segregation of boron can not change the spatial extension nature of the dislocation cores which is responsible for their sessile character, it decreases the extension width of the corresponding cores, thus the non-planar cores are easy to change into planar configuration and then to transform under applied stress. As the stress is applied on the dsilocation, the core splits along the GB plane gradually in boron-doped Ni_3Al, while the core in boron-free Ni_3Al does not.

(7) The addition of boron eases the core spreading along the GB plane, i.e., reduces the motion resistance of dislocation near GBs, and also eases the dislocation absorption by GBs and makes the GBs an effective dislocation sinks, therefore, improves the capacity of deformation accommodation of GBs and reduces the intergranular fracture in Ni_3Al.

1.3.2. The Segregation Behavior of Boron in Ni_3Al

Because of the fact that the segregation of boron plays a decisive role on the ductility of Ni_3Al, so to systematically study the effects of various factors on the segregation state has a siginificant meaning for understanding the mechanical behaviors of Ni_3Al and the mechanism of boron induced ductility.

Choudhury et al. [29] have studied the effect of thermal history on intergranular boron segregation with Auger Electron Spectroscopy (AES) and conclusion was made that the effect of thermal history is entirely reversible and appears to be consistent with equilibrium segregation. However, AES does not actually look at random boundaries but those can, by hydrogen charging, be induced to intergranular failure. The particle-tracking autoradiograph (PTA) technique does look at all boundaries and it can be used for an application involving grain boundary [30].

In our laboratory, the PTA technique has been used to study the effect of such factors as aluminium content (24-26 at% Al), the level of added boron (700-1200 ppm) and different heat treatments (including quenching from different temperatures, quenching plus step annealing, and step annealing plus quenching, etc.) on the GB segregation behaviors of boron in Ni_3Al alloys. Besides, the effect of heat treatment, and hence the amount of segregated boron on the ductility has also been inspected by the tensile testing and corresponding fractographs observed in SEM. The details of experimental procedure is described in [31].

From the PTA results, the following conclusions can be obtained:

(1) The amount of segregated boron is affected distinctly by the aluminium content of matrix, while the aluminium content increases from 24 to 26 at%, the amount of boron at the GB in latter alloy is only about one third of that in former alloy.

(2) The amount of segregated boron increases with the increase of bulk boron amount in the alloy, but these two increments are out of proportion, and the former is less than the latter.

(3) The thickness of segregated region is about several atomic layers, furthermore, inhomogeneity and site selection of segregation are existed.

(4) The higher of the temperature, the less of the segregated boron, and the effect of heat history on the segregation is reversible, therefore, the segregation of boron in Ni_3Al is equilibrium in character.

(5) The ductility decreases with the increasing of quenching temperature, which is consistent with the variation of segregated boron as the quenching temperature changing, and failure mode changes from transgranular to mixed and then to completely intergranular as the quenching temperature from room temperature to 1423K.

1.3.3. The Deformation Behavior in Ni₃Al Alloys

As mentioned previously, there are several explanations to interpret the mechanism of boron induced ductility, in order to understand this problem more deeply in the aspect of GB behavior in Ni₃Al alloys during deformation, the present study was therefore undertaken to obtain direct observation of the slip process, microcrack propagation and dislocation/grain-boundary interactions, and the results obtained are compared for boron-doped and undoped Ni₃Al alloys with different aluminium contents (24-26 at%Al), the amount of added boron (500-700 ppm) and third alloy addition (1 at%Hf and 9 at%Mn). The influence of heat treatment, and hence the amount of segregated boron is also considered. The behavior of deformation slip bands, microcrack propagation, and their interactions with GBs, were in-situ observed in SEM. Furthermore, dislocation arrangements near the GB were also observed in TEM. The details of experimental procedure is presented in [32].

From the SEM and TEM observations, the following conclusions can be deduced:
(1) With the boron addition, there exists a thin intermediate transition region close to the GB, within which slip is reoriented or other slip systems operated. In this case a local strain accommodation at and close to the GB could lead to reduce the stress concentration at the GB and therefore decrease the tendency for intergranular cracking.
(2) During plastic deformation, dislocations are trapped by GBs and held immobile. The number trapped corresponds to a plastic strain at GBs is rather less than the macroscopic strain. Boron additions may also serve to increase the number of dislocation sources and lower the stress for dislocation generation from sources close to and along the boundary or to ease cross-slip close to the boundary.
(3) The trapped dislocations in GBs do not subsequently move within the boundary, nor do they reemerge from the boundary.
(4) The stoichiometry, heat treatment and tertiary alloy element additions will strongly affect the boron segregation behavior, thus change the local strain accommodation close to the boundary.

Summarizing the above simulated and experimental results it can be deduced that several mechanisms resposible for boron-enhanced ductility may take their effects at the same time. That is, in one aspect boron enhances the GB cohesive strength, which is in agreement with the suggestion proposed by Liu et al.[19]. Accompany this effect, in the other aspect, boron reduces the motion resistance of dislocation near GBs, which makes the GBs become effective dislocation sinks and also sources, therefore, the capability of deformation accommodation of GBs is improved, thus the decreasing of intergranular fracture tendency can be expected. In our view the latter effect is controlled by the GB structural aspect, which will be the decisive factor for boron-enhanced ductility in Ni₃Al. Although, the boron-enhanced GB cohesive strength will play an role on the decrease of intergranular fracture tendency, at the same time the GB structure has been changed because of the boron segregation, in other words, the effect of boron-enhanced GB cohesive strength is also taken, in more intrinsic point of view, through the structural change of GBs. According to Vitek [25], L1₂ compounds with near stoichiometric composition will have high tendency to preserve the ordering, which leads to formation of atomic size cavities in the boundaries where can be the sites for the nucleation of micro-cracks, and ensures low dislocation mobility in the boundary region, however, ordering tendency and cavities will greatly decrease if the composition sustantially deviates from the stoichiometry. They suggested that boron attracts nickel into GBs, which promots the deviation from stoichiometry in the boundary region, then the cavities are reduced and the dislocation mobility is enhanced. So the problem of whether boron-enhanced ductility in Ni₃Al is caused by the increment of GB cohesive strength or by the more effectiveness for the dislocation absorption to and emission from the GB, is no contrary to each other, rather two sides to one coin, however, at last these two interpretations can be atrributed to the aspect of structural change if considering the problem more intrinsically as we think.

1.4. The Dislocation Structure of Ni₃Al During Deformation

The flow stress of Ni_3Al alloys has an anomalous flow behavior. The flow stress increases with increasing temperature until a maximum value is reached at the peak temperature (for a review see [33]). Slip line analyses show that around the peak temperature a transition in slip systems occurs, i.e., from $<110>\{111\}$ below the peak to $<110>\{001\}$ above the peak. It has been concluded that the anomalous increase in the flow stress below the peak temperature is due to the cross-slip of dislocations from $\{111\}$ planes onto $\{010\}$ planes, to form the sessile Kear-Wilsdorf (KW) configurations [34].

The present form of cross slip pinning (CSP) model was suggested by Takeuchi and Kuramoto (TK) [35] and modified by Paidar, Pope and Vitek (PPV)[36] to include the so-called "core width effect". The CSP model can adquately explain the temperature and orientation dependence of flow strength as well as the tension/compression asymmetry of the yield strength. However, in recent years, a number of studies using the weak beam TEM techniques showed that within $\{111\}$ slip at intermediate temperatures, the KW locks sometimes deviate from exact screw orientations by expansion in the $\{010\}$ plane, which can not be interpreted by the CSP model [37]. Therefore, the CSP model should undergo further development. Because in the [001] orientation samples, which behaves the same flow anomaly as the other oriented samples do, no transformation in slip systems takes place around the peak temperature [33], it could be expected that a change in the manner of the dislocation motion occurs.

The dislocation structure in a directionally solidified (DS) Ni₃Al-based alloy (of nominal composition 8.5Al, 0.8Zr, 7.8Cr, 0.02B and balanced Ni, wt%) deformed along the [001] orientation at temperatures below and above the peak in flow stress has been studied by the weak-beam TEM technique in order to gain some insight into the modification of CSP model and the $<110>\{111\}$ slips in the region above the peak temperature. The results are summized as follows [38,39]:

At room temperature, the screw superdislocations are partly dissociated on $\{010\}$ and transformed into Kear-Wilsdorf configurations. With increasing temperatures, the transformed KW parts increase until the temperature reaches 723K, at which the screw superdislocations are wholly transformed. The segments of mixed superdislocations dissociated on $\{111\}$. At 450°C, screw dislocation segments are connected by mixed dislocation segments which appear to be superkinks. Except the bending dislocation structure in the $\{010\}$ plane the TEM observations clearly indicate that $1/2<110>$ dislocation pairs move according to the CSP model in which the cross lip distance on $\{010\}$ is not only b/2 or b, but mb at 450°C, m will be ~20. The bending of KW configurations in $\{010\}$ planes is the result of interaction between KW segments and superkinks, and is aided by thermal activation. The contribution of the superkink-aided motion to deformation is not important.

At 850°C, most dislocations are screw or near-screw in character and dissociated on the $\{111\}$ slip planes. This reflects that the mixed dislocations move freely on their slip planes.

Therefore, a model for the ⟨110⟩{111} slips above peak temperature for ⟨001⟩ orientated specimens of Ni₃Al was proposed. The 1/2⟨110⟩ dislocation composed of a screw and a mixed segment. The mixed segment dissociates on the (I̅11) slip plane, and the screw segment on the (001) cross-slip plane. The screw segment is less mobile and can only proceed on the (001) plane a very short distance by local stress and by thermal activation. As the mixed segment slip freely on the (I̅11) plane a distance, a new screw segment is created. This screw segment immediately cross slips onto the (001) plane. In order for the mixed segment to slip further under the effect of the applied stress, the deviated screw segment on the (001) plane has to be drawn back towards the exact screw orientation, and finally onto the original (I̅11) slip plane. This drawing back process is thermally activated.

2. Ti-Al System

Adding third element is an effective method to improve the ductility of TiAl-based alloys at room temperature and elevated temperature. As we know that the addition of manganese can reduce the axial ratio (c/a) of TiAl crystal lattice and vanadium is the largest one among β-forming elements in Ti-based alloys. Therefore, manganes and vanadium are largely used as the third element to improve the mechanical properties of TiAl-based alloys [40,41].

2.1. The Microstructures of TiAl Based Alloy

The effects of manganese and vanadium on microstructures have been investigated by Hao [40]. The nominal compositions are: the near-stoichiometric TiAl (Ti-36.6 wt%Al), TiAl + Mn (Ti-34.5 wt%Al-1.5 wt%Mn) and Ti-30-20-V (Ti-30 wt%Al-20 wt%V). No evidence of the peritectic reaction has been found in the microstructure of cast TiAl + Mn alloy. It means that only L→β transformation takes place during solidification for this alloy. The amount of α_2 phase in the microstructure of TiAl + Mn alloy is more than in TiAl alloy. After annealing at 1273K for 7 days the average grain size of the TiAl + Mn alloy is approximately 30μm which is obviously finer than those of TiAl alloys. The twinning structure which existed in casting alloy disappears. Microstructures and electron probe microanalysis for cast and annealed (1273K for 24hrs) Ti-30Al-20V alloy show that the microstructure consists of γ phase (rich in Al and poor in V) and second phase β (rich in V and poor in Al) distributed along grain boundaries.

It has been reported that faulted dipoles observed in TiAl are one of the most important factors, which result in the room-temperature brittleness. Huang and his coworkers [41] studied the nature and the formation mechanism of the faulted dipoles in Ti-50 at% Al alloy. The faulted dipoles observed in Ti-50 at%Al alloy are intrinsic and bounded by a/6[112] type partial dislocations. They formed from the discontinuous dissociation of a/2[112] and [101] types of superdislocations on the same slip plane when a/6 [112] type partial dislocations are pinned by some obstacles.

2.2. The Deformation Behavior of TiAl Based Alloy

Recently, much work has been done on the brittle nature of TiAl compound and the results revealed that alloying vanadium was benificial to the ductility of TiAl. Pu and his coworkers [42] studied the effect of vanadium on the high temperature deformation behavior using double-notch shear test. The nominal composition (wt%) of materials used in their investigation is:

No.	Ti	Al	V
1	53.4	46.6	0.0
2	52.2	46.7	1.1
3	49.4	46.7	3.7
4	45.5	46.9	7.6

It is found that the fracture strain increases with the increasing temperature. This is because the sessile superdislocations were operated by the thermal activation process as the temperature increased gradually. The fact that obvious ductility can be observed for TiAl compound unless the temperature was higher than 773K shows that the element V can improve the ductility and decrease the ductile-brittle transition temperature effectively. The ductility of TiAl compound with addition of V shows the trend of increasing first and decreasing finally with increasing V and the best content is 3.7at.%.

The shear yield strength of TiAl decreases with the increase of temperatuire. However, the strength of TiAl with addition of V increases with increasing temperature up to the peak value of 973K approximately, and then shows a normal temperature dependence at above 973K.

For all specimens, the crack direction is along 45° relative to the load direction. This shows that the brittle crack direction is vertical to the maximum main stress, and exhibits that the controlling factor for brittle facture of TiAl compound is the maximum main stress. The nominal fracture stress τ_f increases with the increase of temperature. The plastic fracture strain shows a minimum value at 773K, this is because the value of $\tau_f - \tau_s$ has a minimum at 773K.

According to the work of Hao [43], the compressive proof stress of the poly-crystalline TiAl (Ti-36.6wt%Al) is found to be positive temperature dependence as same as the single crystal one (Fig.4). The correlation of the flow stress together with strain rate and deformation temperature is in good agreement with the exprssion

$$\dot{\varepsilon} = A(\sigma_p)^n \exp(-Q/RT)$$

where $\dot{\varepsilon}$ is strain rate, Q is deformation activation energy, R is gas constant, σ_p is maximum flow stress, n is stress index, and A is constant.

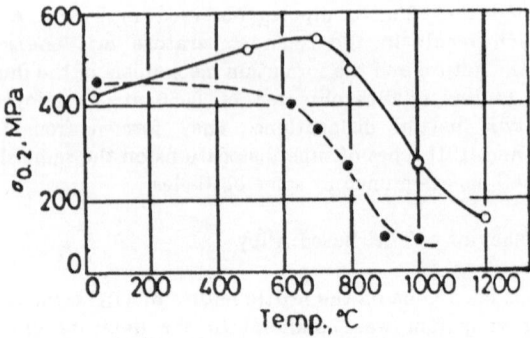

Fig.4 Temperature dependence on proof stress for TiAl compound. ——O—— this study, compressive; ——●—— after Lipssit, tensile

However, Pu and his coworkers [44] show that the stress-temperature dependence of TiAl based alloys depends upon their chemical composition. For TiAl−X (V, Mn, Ga) ternary compound, the stress vs temperature curves can be divided into three regions. In region 1, the strength decreases with increasing temperature below 473K and the deformation process is controlled by P-N mechanism. In region 2, the strength increases with increasing temperature between 200 and 973K, i.e., the positive stress-temperature dependence. In region 3, the strength decreases with increasing temperature above 973K. For TiAl binary compound, the yield strength decreases with increasing temperature between 25 and 1073K, i.e., the normal stress-temperature dependence.

It is well known that a small amount of Mn addition can improve the room temperature ductility of TiAl alloys. Huang and his coworkers [41] investigated the effects of Mn addition on the plastic deformation of TiAl based on the comparison of the dislocation structure of TiAl and TiAl + Mn alloy. The pinning effect of a/6[112] partial dislocation is eliminated by Mn addition, as a result, the movement of a[101] and a/2[112] superdislocations and twinning deformation were activated. In other words, the ductility improvement of TiAl by Mn-addition is associated with enhancement of the movement of a/6[112] partial dislocation and superdislocations as well as the twinning process.

In order to understand the brittle nature and to establish a guideline for advanced alloy design, much work about the dislocation structure and deformation mechanism for TiAl based alloys has been done by Pu et al [45]. The results show that the deformation is mainly contributed by the glide of ⟨011⟩ superdislocation rather than ⟨110⟩ ordinary dislocation in Al-rich $Ti_{46}Al_{54}$ alloy at room temperature and 77K. However, in Ti-rich TiAl based binary alloy and ternary alloy, the deformation is mainly contributed by the glide of ⟨110⟩ ordinary dislocation and 1/6⟨112⟩ twinning.

3. Ti-Al-Nb Ternary Intermetallic Alloy

There has been an interest in Ti-Al-Nb ternary intermetallic alloy for high temperature application. Chen and his cowokers [46] systematically investigated the Nb-Ti-Al ternary system. They took the concept of the phase-constitution map, the density-composition map and the high temperature strength map as the guideline for selecting composition. The phase-constitution map summarizes all the results for phase identification. The bonds shown on the density map the composition of iso-density. For the single phase γ_1 alloy the density varies from 4.1 to 4.7g/cm³ as the composion changes. In the multi-phase regions the density depends on the relative amounts of the intermetallics involed and the density difference between the intermetallics.

The isothermal section of Ti-Al-Nb ternary phase diagram at 1273K has been determined using $(Ti-Nb)/TiAl_3$, $(Ti-Nb)/NbAl_3$, $(Ti-TiAl)/NbTiAl_3$ and $(TiAl-Nb)/Ti$ diffusion couples [47] and a new ternary intermetallic compound (Ti_5Al_3Nb) with large solubility range in Ti-Al-Nb systems has been discovered [48].

4. Al_3Ti-Based $L1_2$ Intermetallic Alloys

Trialuminide alloys based on Al_3Ti are recently of significant interest as a result of their low density, good oxidation resistance, and relatively attractive mechanical properties. Binary Al_3Ti is very brittle at room temperature. The main deformation mode was found to be $(111)\{11\bar{2}\}$ twinning[49]. Efforts have been made to ductilize the Al_3Ti by substituting some amounts of Al with Cu, Ni, Fe, Mn or Cr [50-53] to bring about a change to the $L1_2$ structure which may have a sufficient number of slip systems for homogeneous deformation. Following such an idea, it is expected that the ductility of modified

Al₃Ti-based alloys will be greatly improved. However, these materials having L1₂ structure remain brittle, falling by cleavage[52,54]. Therefore, investigations both in improving the ductility via chemistry modification and processing, and in exploring the deformation mechanism are necessary before the practical engineering applications of these alloys.

A project sponsored by the NAMCC has been conducted by Hu and his coworkers, aimed toward the development of ductile Al₃Ti-based L1₂ intermetallic alloys. The results to date look promising and are summarized as below.

4.1. Alloying Effect on Ductility [55,56]

A number of Al₃Ti alloys containing various substituting elements were prepared by arc-melting, MF induction melting, or melt-spinning. X-ray and electron diffraction analyses verified that Al₃Ti-xM (where $M = Ni$, Fe, or Mn; $x = 8$-10 at%) alloys are of ordered L1₂ structure, in either as-cast or homogenized conditions. Optical and SEM micrographs further show a few second phase particals in irregular shapes, distributed mainly at grain corners. Compression tests indicated that these ternary alloys exhibit appreciable ductility, compared to the binary Al₃Ti. Table VI lists some results of the compressive tests at room temperature.

Table VI Compressive properties of Al₃Ti-based alloys containing various elemental additions at room temperature

Alloy	8Ni	4Ni/4Mn	9Fe	6Fe/3Mn	8Mn
$\sigma_{0.2}$(MPa)	312±12	261±7	272±14	226±8	218±2
ε_p(%)	6.4±0.2	9.7±0.4	11.0±0.4	13.0±0.7	17.0±0.8

Among Ni, Fe and Mn additions, the alloy cotaining Mn shows the best ductility at room temperature, and its plastic strain at fracture, ε_p, reaches 17%. When certain amounts of Ni or Fe are replaced by Mn, the ductility of the quarternary alloys is also moderatedly improved.

The strain distribution in the compressed specimen is inhomogeneous as the observed flow markings in relief on the specimen surfaces. This also indicates that slip propagated from one grain to the next throughout a region extending right across the specimen. Final fracture occurred by a shear-off process along a surface oriented about 45° to the compression axis. SEM fractographic analyses showed that the fracture is transcrytalline in nature and the fracture mode is mainly quasicleavage plus tearing. The quasicleavage regions are associated with dense steps of slip bands and curved subsurfaces. These feature together with the tear ridges observed illustrate that these regions were plastically deformed before fracture. Therefore, the observed compression ductility of the L1₂-structured Al₃Ti alloys is attributed to a sufficient number of operating slip systems during deformation.

4.2. Deformation Substructures [55,57,58]

For the purpose of exploring the deformation mechanisms, some of the specimens were compressed only to a small amount of plastic strain (below 3%) and the dislocation substructures were analyzed by TEM. The present alloys in the homogenized state contain a few straight free dislocations. In contrast, the dislocation density is greatly increased after deformation. The moving dislocations introduced through deformation are locally pinned, as indicated by the bowing-out. They appear to be in pairs, showing that the a⟨101⟩

superdislocations dissociated into two partials. These observations confirmed that dislocation slip must be the main mode of twinning in the DO_{22}-type Al_3Ti.

Contrast analysis of the partial dislocations has been carried out by using different reflections. The results confirmed that the a⟨101⟩ superdislocation in the Al_3Ti-9Fe alloy dissociates into two superpartials of a/3⟨112⟩-type, bounding a superlattice intrinsic stacking fault (SISF), for instance, in the (1$\bar{1}$1) slip plane:

$$[\bar{1}01] \rightarrow 1/3[\bar{2}\bar{1}1] + SISF + 1/3[\bar{1}12]$$

Such an a⟨101⟩ dislocation dissociation and slip mechanism was further verified by the following observations in the deformed specimens: (1) the dipoles of the extended dislocations, (2) the planar arrays of slipping dislocations and (3) the elongated dislocation loops multiplied from subboundaries. It has been found that the temperature dependence of yield stress for the present alloy resembles that of a group of $L1_2$ alloys such as Co_3Ti. It was suggested that the superdislocation with SISF would be responsible for the anomalous increase of yield stress in the low temperature region [59], although there has been no appropriate mechanism proposed to explain such yield behavior.

In the Mn-modified Al_3Ti alloy, the dissociation of a⟨101⟩ superdislocations were more frequently observed than was in the Al_3Ti-Fe. The seperation between two a/3⟨112⟩ partials was found to appear slightly wider. These may also related to the ductility difference among the alloys with various additions.

5. Summary and Remarks

This paper has reviewed the current status of research and development of ordered intermetallics during past five years in P.R.China. Progress has been made in mechanical properties and physical metallurgy of ordered intermetallics with the achievements in the improvements of ductility, fabricability and high temperature strength for aluminides, which encourages the scientists and engineers to pursuit the further research and development of ordered alloys for high temperature structural applications in the future.

The next five-year project sponsored by National Advanced Materials Committee of China will start from 1991, with special enphasis upon aluminides. The alternative intermetallics, e.g., Ti-Al-Nb, Ti_5Si_3-Ti_3Al, etc. also included in the project. The intermetallics matrix composite will also be conducted in the next five years. Material processing and high temperature structral applications will be emphasized in the future. One project for the applications of Ti_3Al-based alloys to aircraft engins parts will be carried on by the areonautic industry.

Acknowledgements   The auther is grateful to his former and present graduate students, Da Chen, Yun Zhang, Mao Wen, Bo Yang and Min Lu for their contributions to our understanding of the behavior of intermetallic compounds, and his colleages Gengxiang Hu and Shipu Chen, from whose work many of the results are taken. The auther acknowledge the financial support of National Advanced Materials Committee of China and National Nature Science Foundation of China.

References

1   Aoki, K. and Izumi, O. (1979) 'Improvement in room temperature ductility of the $L1_2$ type intermetallic compound Ni_3Al by boron addition', J.Japan Inst.Met.43, 1190-1196.

640

2　Liu, C.T. and White, C.L. (1985) 'Design of ductile polycrystalline Ni₃Al alloys', in Koch, C.C., Liu, C.T., and Stoloff, N.S. (eds.), High Temperature Ordered Intermetallic Alloys,MRS Symp.Proc.,Vol.39, Materials Research Society, Pittsburgh, PA, pp.365-380.

3　Liu, C.T., White, C.L., and Horton, J.A. (1985) 'Effect of boron on grain-boundaries in Ni₃Al', Acta Metall. 33, 213-229.

4　Guo, J.J., Zhu, Y.X, Li, H., Sun, C., Wan, S.H., and Hu, Z.Z. (1990) 'The effect of boron content on mechanical properties and microstructure of poly-, mono- and micro-crystalline Ni₃Al', Chin. J. Met. Sci. Technol. 7, 113-121.

5　Guard, R.W. and Westbrook, J.H. (1959) 'Alloying behavior of Ni₃Al (γ' phase) ', Trans. AIME 215, 807-814.

6　Rawlings, R.D. and Staton-Bevan, A.E. (1975) 'The alloying behavior and mechanical properties of polycrystalline Ni₃Al (γ' phase) with ternary additions', J. Mater. Sci.10, 505-514.

7　Ochiai, S., Mishima, Y., Yodogawa, M., and Suzuki, T. (1986) 'Mechanical properties of Ni₃Al with ternary addition of B-subgroup elements', Trans. JIM 27, 32-40.

8　Mishima, Y., Ochiai, S., Yodogawa, M., and Suzuki, T. (1986) 'Mechanical properties of Ni₃Al with ternary addition of transition metal elements', Trans. JIM 27, 41-50.

9　Zhang, Y. and Lin, T.L. (1990) 'The effects of alloy additions of Si and transition metal elements on the mechanical properties of B-doped DS Ni₃Al', to be published in Mater. Res.Soc.Symp.Proc. on High Temperature Ordered Intermetallic Alloys IV, November 26-December 1, 1990, Boston, MA, U.S.A.

10　Lin, T.L. and Zhang, Y. (1991) 'Mechanical properties of B-doped DS Ni₃Al containing C, Mg, Ca and rare-earth elements', to be published in Int.Conf. on High-Temperature Aluminides and Intermetallics, Septemper 16-19, 1991, San Diego, CA, U.S.A.

11　Lin, T.L. and Zhang, Y. (1991) 'The effects of alloy additions on mechanical properties of DS Ni₃Al', ibid.

12　Lin, T.L. and Zhang, Y. (1990) 'Effects of alloying with transition metal elements on the ductility of B-doped DS Ni₃Al', to be published in Symp.Proc. on Intermetallic alloys for high-temperature structural use, December 23-24, 1990, Beijing, P.R.China.

13　Suzuki, T., Oya, Y., and Wee, D.M. (1980) 'Transition from positive to negtive temperature dependence of the strength in Ni₃Ge solid solution', Acta Metall. 28, 301-310.

14　Mishima, Y., Ochiai, S., Hamao, N., Yodogawa, M., and Suzuki, T. (1986) 'Solid solution hardening of Ni₃Al with ternary additions', Trans. JIM 27, 648-655.

15　Ma, P.L., Han, L., Yuan, Y., Lu, Y., and Zhang, Z.Y. (1991) 'Improvement of hot workability of Ni₃Al-Cr-Zr-B ordered alloy', to be published in Int. Conf. on High-Temperature Aluminides and Intermetallics, September 16-19, 1991, San Diego, CA, U.S.A.

16　Li, H., Guo, J.T., Shun, C., Wang, S.H., and Tan, M.H. (1991) 'The effect of yttrium and cerium on the compression properties of Ni₃Al-base alloys', Journal of the Chinese Rare Earth Society (in Chinese) 9, 209-215.

17　Stoloff, N.S. (1984) 'Ordered alloys-physical metallurgy and structural applications', Int. Met. Rev. 29, 123-135.

18　Liu, C.T., Taub, A.I., Stoloff, N.S., Koch, C.C., and Izumi, O. (1985, 1987 and 1989) High-Temperature Ordered Intermetallic Alloys I, II, III (Proc. Mater. Res. Soc. Symp., Vol.39, Vol.81 and Vol.133), Materials Research Society, Pittsburgh, PA, U.S.A.

19　Liu, C.T. and White, C.L. (1985) 'Design of ductile polycrystalline Ni₃Al alloys', in Koch, C.C., Liu, C.T., and Stoloff, N.S. (eds.), Mat. Res. Soc. Symp. Proc. Vol.39, Materials Reacher Society, Pittsburgh, PA, U.S.A., pp.365-380.

20　Schulson, E.M., Weihs, T.P., Baker, I., Frost, H.J., and Horton, J.A. (1986) 'Grain boundary accommodation of slip in Ni₃Al containing boron', Acta Metall. 34, 1395-1399.

21 Bond, G.M., Robertson, I.M., and Birnbaum, H.K. (1987) 'Effect of boron on the mechanism of strain transfer accross grain boundaries in Ni₃Al', J. Mater. Res. 2, 436-440.

22 Chen, S.P., Voter, A.F., and Scrlovitz, D.J. (1986) 'Computer simulation of grain boundaries in Ni₃Al: the effect of grain boundary composition', Scr. Metall. 20, 1389-1394.

23 Chen, S.P., Voter, A.F., and Scrolovitz, D.J. (1987) 'Atomistic simulation of [001] symmetric tilt boundaries in Ni₃Al', in Stoloff, N.S., Koch, C.C., Liu, C.T., and Izumi, O. (eds.), Mat. Res. Soc. Symp. Proc., Vol.81, Materials Research Society, Pittsburgh, PA, pp.45-50.

24 Chen, S.P., Voter, A.F., Albers, R.C., Boring, A.M., and Hay, P.J. (1990) 'Investigation of the effects of boron on Ni₃Al grain boundaries by atomistic simulations', J. Mater. Res. 5, 955-970.

25 Vitek, V., Chen, S.P., Voter, A.F. Kruisman, J.J., and DeHosso, J.Th.M. (1989) 'Grain boundary structure and intergranular fracture in L1₂ ordered alloys', Materials Science Forum 46, 237-252.

26 Lin, T.L. and Chen, D. (1990) 'Atomistic simulations of [100], [110] and [111] symmetric tilt grain boundaries in Ni₃Al', Journal De Physique, Colloque, C-1, 227-232.

27 Lin, T.L. and Yang. B. 'Computer simulation of the interaction between the grain boundary and dislocation in Ni₃Al', to be published.

28 Lin, H. and Pope, D.P. (1990) 'The effect of grain boundary geometry on intergranular fracture in Ni₃Al at room temperature', in Johnson, L., Pope, D.P., and Stiegler, J.O. (eds.), High Temperature Intermetallic Alloy IV, in process, (MRS Symp. Proc.) Materials Research Society, Pittsburgh, PA., U.S.A.

29 Choudhury, A., White, C.L., and Brooks, C.R. (1986) 'The effect of thermal history on intergranular boron segregation and fracture morphology of substoichiometric Ni₃Al', Scr. Metall. 20, 1061-1066.

30 He, X.L. and Chu, Y.Y. (1983) 'The application of the ¹⁰B(n,α)⁷ Li fission reaction to study boron behavior in materials', J. Phys. D: Appl. Phys. 16, 1145-1158.

31 Lin, T.L., Chen, D., and Lin, H. (1990) 'Investigation of the segregation behavior of boron in Ni₃Al alloys by PTA technique', Acta Metall. 39, 523-528.

32 Lin, T.L. and Chen, D. (1989) 'In-situ observation of grain boundary behavior in Ni₃Al alloys during tensile deformation in SEM', in Liu, C.T., Taub, A.I., Stoloff, N.S., and Koch, C.C. (eds.), Mat. Res. Soc. Symp. Proc., Vol.133, Materials Research Society, Pittsburgh, PA, U.S.A., pp.217-223.

33 Pope, D.P. and Ezz S.S. (1984) 'Mechanical Properties of Ni₃Al and nickel-based alloys with high volume fraction of γ′', Int. Metals Rev. 29, 136-167.

34 Kear, B.H. and Wilsdorf, H.G.F. (1962) 'Dislocation configurations in plastically deformed polycrystalline Cu₃Au alloys', Trans. TMS-AIME, 224, 382-386.

35 Takeuchi, S. and Kuramoto, E. (1973) 'Temperature and orientation dependence of the yield stress in Ni₃Ga single crystle', Acta Metall. 21, 415-425.

36 Paider, V., Pope, D.P., and Vitek, V. (1984) 'A theory of the anomalous yield behavior', Acta Metall. 32, 435-447.

37 Veyssiere, P. (1989) 'Transmission electron microscope obveration of dislocation in ordered intermetallic alloys and the flow stress anomaly', in Liu, C.T., Taub, A.I., Stoloff, N.S., and Koch, C.C. (eds.), High Temperature Ordered Intermetallic Alloys III (MRS Symp. Proc.), Materials Research Society, Pittsburgh, PA, U.S.A., pp.175-188.

38 Wen, M. and Lin, T.L. (1991) 'A TEM study of the dislocation structure in a Ni₃Al-based alloy at temperature below the peak in flow stress', in Johnson, L., Pope, D.P., and Stiegler, J.O. (eds.), High Temperature Ordered Intermetallic Alloy IV, in Press, MRS Symp. Proc., Materials Research Society, Pittsburgh, PA, U.S.A.

642

39 Wen, M. and Lin, T.L. (1991) 'Slip behaviors in a directionally solidified Ni$_3$Al at 850°C along [001]', to be published in Scr. Metall. Mater.

40 Hao, S.M. and Zheng, Z.Q. (1990) 'Effects of third element on the microstructure and properties of TiAl-based alloys', in Yan, D.S., Shi, C.X., and Li, H.D. (eds.), C-MRS'90 International Conference Proc., Vol.2, Advanced Structural Materials, Elsevier Science Publishers, Amsterdam, the Netherland, pp.831-834.

41 Huang, B.Y., Qu, X.H., Wen, J.H., and Lei, C.M. (1990), 'Plastic deformation of TiAl intermetallic with Mn addition', to be presented at the Sixth JIM Symp. on Intermetallic Compound Structure and Mechnical Properties, June 17-20, 1991, Sendai, Japan.

42 Pu, Z.Q., Cai, Q.G., and Zhu, J. (1990) 'Deformation behavior of TiAl-V intermetallic compound at elevated temperature', Acta Metall.Sinica (English Edition) 3, 142-144.

43 Hao, S.M. and Zheng, Z.Q. (1990) 'High temperature deformation behavior of TiAl intermetallic compound', Acta Metall. Sinica (English edition) 3, 415-415.

44 Pu, Z.Q., Zhu, D.,Zhu, J., Cai, Q.G., and Zou, D.X. (1990) 'The plastic deformation behavior of TiAl-based alloys', Chinese Science Bulletin, 35, 1397-1400.

45 Pu, Z.J., Zhu, Q., Zou, D.X., and Cai, Q.G. (1990) 'The deformation mechanism and alloying behavior of TiAl intermetallic compound', in Yan, D.S., Shi, C.X., and Li, H.D. (eds), C-MRS' 90 International Conf.Proc., Vol.2, Advanced Structural Materials, Elsevier Science Publishes, Amsterdam, the Netherland, pp.731-736.

46 Chen, G., Sun, Z., Xie, X., Ren, Y., Yao, K., Zhou, X., Sha, L., Fu, L., and Yang, W. (1990) 'Nb-Ti-Al ternary intermetallic alloys-a potential high temperature ordered alloy system', in Yan, D.S., Shi, C.X., and Li, H.D. (eds.), C-MRS'90 International Conf. Proc., Vol.2, Advanced Structural Materials, Elsevier Science Publishers, Amsterdam, the Netherland, pp.803-806.

47 Hao, S.M., and Zhao, Q. (1990) 'Investigation on the 1000°C isothermal section of Ti-Al-Nb ternary phase diagram', Proc. 6th National Symp. on Phase Diagrams, Shenyang, China, pp.142-143.

48 Hao, S.M., and Zhao, Q. (1990) 'A new ternary intermetallic compound in Ti-Al-Nb system', Proc.6th National Symp. on Phase Digrams, Shenyang, China, pp.144-145.

49 Yamaguchi, M., Umakoshi, Y., and Yamane, T. (1987) 'Deformation of the intermetallic compound Al$_3$Ti and some alloys with an Al$_3$Ti base', in Stolloff, N.S., Koch, C.C., Liu, C.T., and Izumi, O. (eds.), High-Temperature Ordered Intermetallic Alloys II, MRS Symp. Proc. Vol.81, Pittsburgh, PA, U.S.A., pp275-285.

50 Kumar, K.S., and Pickens, J.R. (1988) 'Compression behavior of the L1$_2$ intermetallic Al$_{22}$Fe$_{38}$', Scr. Metall. 22, 1015-1018.

51 Tarnacki, J., and Kim, Y.W. (1988) 'A study of rapidly solidified Al$_3$Ti intermetallics with alloying additions', Scr. Metall. 22, 329-334.

52 Turner, C.D., Powers, W.O., and Wert, J.A. (1989) 'Microstructure, deformation and fracture characteristics of an Al$_{67}$Ni$_8$Ti$_{25}$ intermetallic alloy', Acta Metall. 23, 2635-2643.

53 Zhang, S., Nic, J.P., and Mikkola, D.E. (1990) 'New cubic phases formed by alloying Al$_3$ Ti with Mn and Cr', Scr.Metall. 24, 57-62.

54 George, E.P., Porter, W.D., Henson, H.M., and Oliver, W.C. (1989) 'Cleavage fracture in an Al$_3$Ti-based alloy having the L1$_2$ structure', J. Mater. Res. 4, No.1, 78-84.

55 Hu, G.X., Chen, S.P., Wu, X.H., and Chen, X.F. (1991) 'Plastic deformation and fracture behavior of a Fe-modified Al$_3$Ti-base L1$_2$ intermetallic alloy', J. Mater. Res. 6, No.5, 1-7.

56 Wu, X.H., Chen, X.F., Chen, S.P., and Hu, G.X. (1990) 'Al$_3$Ti-based L1$_2$-type intermetallic alloys containing Fe and Mn', to be published in Proc. Symp. on Intermetallic Alloys for High-Temperature Structural Use, November 23-24,1990, Beijing, P.R.China.

57   Chen, S.P., Wu XH., and Hu G.X. (1990) 'TEM study on dislocation structures of Al$_{66}$Fe$_9$Ti$_{24}$ intermetallic alloy deformed at room temperature', in Proc. XIIth Intern. Congr. for Electron Microscopy, August 12-18, 1990, Seattle, WA, U.S.A., Vol.4: Advances in Microscopy in the Material Sciences.

58   Chen, S.P., Rong, Y.H., Wu, X.H., and Hu, G.X. (1990) 'TEM observations of dislocation multiplication mechanism in an Al$_3$Ti-based L1$_2$ alloy', to be published in Proc. Symp. on Intermetallic Alloys for High-Temperature Structural Use, November 23-24, 1990, Beijing, P.R.China.

59   Liu, Y., Takasugi, T., Izumi, O., and Takahashi, T. (1988) 'TEM investigation on dislocation dissociation and planar faults in deformed (Co,Ni)$_3$Ti single crystal', Acta Metall. 36, 2959-2966.

POTENTIAL AND PROSPECTS OF SOME INTERMETALLIC COMPOUNDS FOR STRUCTURAL APPLICATIONS

S. NAKA, M. THOMAS and T. KHAN
Office National d'Etudes et de Recherches Aérospatiales
29 Av. de la division Leclerc
92322 Châtillon CEDEX
France

ABSTRACT.   In the present paper, potential and prospects of some intermetallic compounds for structural applications are discussed by emphasizing two specific points which seem to be of prime importance for the future research and development activities. The first point stresses upon the benefits associated with the development of multi-phase intermetallics and describes a novel approach for creating the two-phase γ-γ' type microstructure in different alloy systems. The second point deals with the problem of stability of the constituent phases in these multi-phase materials. Views expressed in this paper are frequently corroborated by referring to our most recent results obtained in various alloy systems based on Ni_3Al, iron, niobium, Ti_3Al+X (X=Nb, V and Mo) and NiAl.

1. Introduction

The potential of current conventional materials has been exploited to a large extent. For example, both the nickel-base superalloys and the titanium alloys have reached their upper temperature limit of utilization. The aerospace industry is therefore looking forward to the development of "alternative" materials which would be lighter, stronger and having a higher temperature potential, compared to the presently available alloys. In this context, research and development activities are underway on a wide range of materials that may find eventual applications. Among the various potential high temperature materials, intermetallic-based alloys are prime candidates for "alternative" materials.

However, the lack of both ductility and toughness is often the "Achilles' heel" in intermetallics. The successful development of new materials based on intermetallic compounds will therefore strongly depend on the improvement of these properties. Ductilization of intermetallics is now extensively being tried through different approaches. Increase in the number of easy slip systems through alloying with solid solution elements, modification of the crystallographic structure by macroalloying, strengthening of the grain boundaries by microalloying, grain size refinement and single crystal approach are well-known examples. Toughning of intermetallics is also tried, although the success is still limited. Suitable microstructural control by introducing a second phase of lamellar morphology is an example. It is worth noting that toughening observed through martensitic transformations in certain ceramics (yttria stabilized zirconia) is an interesting approach, although such an ap-

C. T. Liu et al. (eds.), Ordered Intermetallics – Physical Metallurgy and Mechanical Behaviour, 645–662.
© 1992 *Kluwer Academic Publishers.*

proach has not yet been applied to intermetallic-based materials.

The strength of the single-phase intermetallic compounds is often mediocre. For example, the stoichiometric or Ni-rich NiAl has a yield strength of 100-200MPa which is very low from the engineering viewpoint, despite the fact that this strength corresponding roughly to $10^{-3}\mu$ (μ being the shear modulus) is quite high in view of the plasticity theory. Hardening of intrinsically strong crystals which have a high lattice friction stress is not very easy. There is, however, ample experimental evidence available today to show that the mechanical strength (especially the creep strength) of multi-phase alloys is higher than that of single-phase intermetallics [1]. In the particular case of NiAl, the introduction of a high volume fraction of the Ni_2AlTi phase was reported to be very successful in strengthening NiAl, in particular at high temperatures [2].

The present paper does not aim at presenting an extensive review on the state of the art of the intermetallic-based materials, which has already been presented elsewhere [1]. Here we will rather focus our attention on some selected cases, often by referring to the most recent results of our activities in this field. An important purpose of this paper is to suggest some useful directions that may be fruitful to follow in the future research and development activities.

2. Ni₃Al vs. γ-γ ' nickel-base superalloys

A considerable amount of research effort has been devoted to the compound Ni_3Al. The main reason is that Ni_3Al (γ ') constitutes the principal strengthening phase in many so-called Ni-base γ-γ ' superalloys. This intermetallic with the cubic ordered $L1_2$ structure is important because it exhibits an increasing flow stress with increasing temperature up to 600-800°C.

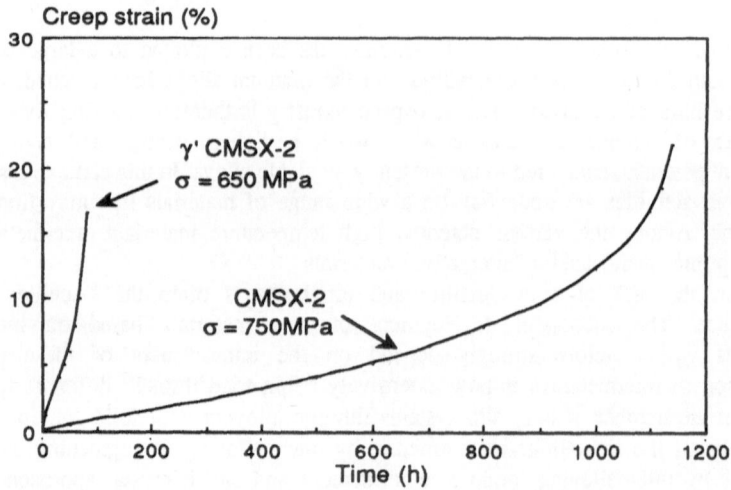

Figure 1. Creep curves at 760°C of [001] CMSX-2 and γ' CMSX-2 single crystals.

Much attention has been payed to this particular behaviour by many investigators specializing in plasticity theory and the recent explanation seems to be very convincing [3]. Although this property is useful as a base for high temperature structural applications, the binary compound in the polycrystalline state exhibits brittle intergranular fracture without any ductility at room temperature. Since the discovery of a very strong ductilizing effect of boron made by Aoki and Izumi [4], extensive investigations undertaken in particular by Liu and co-workers [5,6] led to the development of some semi-industrial polycrystalline γ' alloys. However, the relatively poor mechanical strength of such alloys, especially at high temperatures, seem to seriously restrict their field of application.

It is worth recalling, here, our recent experiments [7] which illustrate the advantages of Ni-base γ-γ' superalloys over single-phase γ' alloys. Fig.1 shows the creep curves of one of the most advanced γ-γ' superalloys, designated "CMSX-2", and of a single-phase γ' alloy whose composition corresponds to that of the constituant γ' phase of this superalloy. Both these materials were directionally solidified as [001] oriented single crystals and then tested in creep at various temperatures. It is clear from this figure that the two-phase γ-γ' superalloy is by far much stronger than the single-phase γ' alloy at 760°C in spite of the fact that the tensile strengths of the two materials measured at the same temperature were almost identical. These results practically rule out the possibility of replacing high strength γ-γ' superalloys by γ' based alloys for high temperature-high stress applications.

3. Advantages of the multi-phase intermetallic-based materials

Advantages of multi-phase alloys over single-phase ones are also evident in many other intermetallic-based materials. In order to illustrate this point, we will be considering the following three examples : Ti_3Al-based alloys, TiAl-based alloys and NiAl-based alloys.

3.1. Ti_3Al-BASED ALLOYS

Due to its strongly brittle nature, the hexagonal DO_{19} Ti_3Al intermetallic compound per se is not considered useful for industrial applications. Since the discovery of a certain ductilizing effect of a BCC phase formed through additions of β stabilizing elements (e.g. Nb, Mo, V), a considerable effort has been devoted to enhance the mechanical properties (ductility and strength) of the two-phase α_2 (Ti_3Al)+β alloys, the most advanced example of which is the so-called Super α_2 (Ti-25Al-10Nb-3V-1Mo) of TIMET. Depending on the thermal and thermomechanical treatments, a large variety of microstructures can be generated in the alloys of this category, as shown in Fig.2. Their mecahnical properties are strongly influenced by the resultant microstructure, in particular by the degree of its refinement, by the amount of β (or β_2) and α_2 phases, and by their distribution and morphology [8]. When the alloys contain a relatively high amount of β stabilizing elements, the ordered β_2 phase, which has the cubic B2 structure, is likely to be stabilized at the expense of the disordered form. However, an important point to be kept in mind is the fact that in these alloys, the α_2 (Ti_3Al) phase is the strengthening phase and their deformability is ensured by either the β phase or the ordered B2 phase of cubic symmetry.

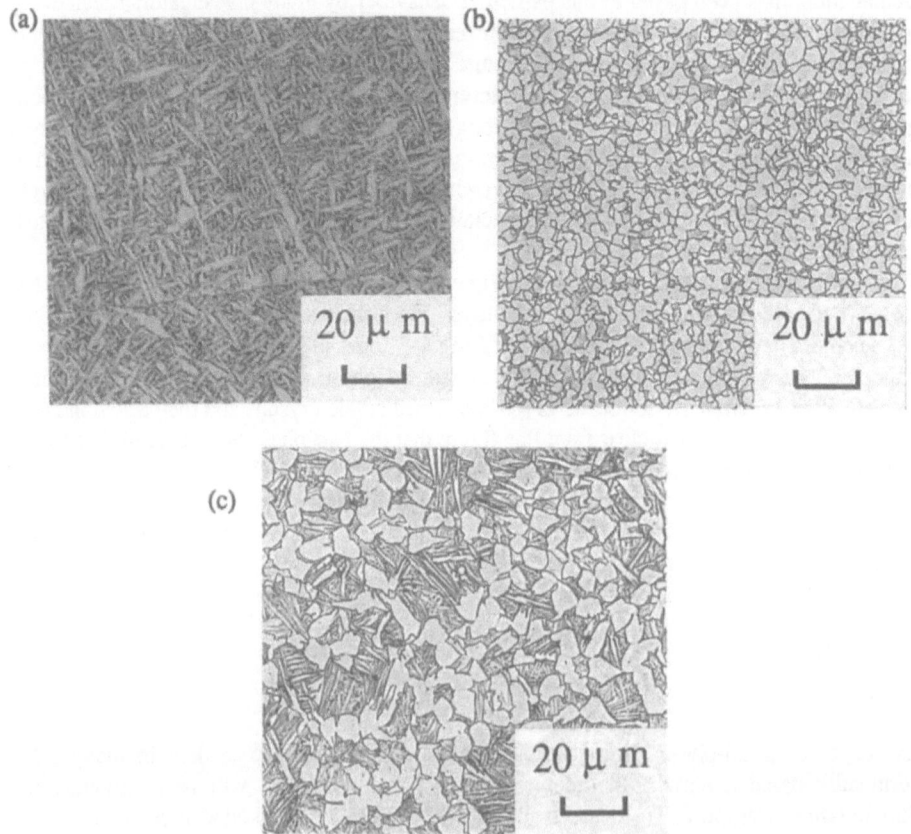

Figure 2. Various microstructures observed in Super α_2.
(a) intermediate Widmanstätten, (b) globular and (c) mixed.

3.2. TiAl-BASED ALLOYS

Recently, an increasing attention is being payed to the tetragonal $L1_0$ TiAl (γ) intermetallic compound, which has a great potential because of its high specific strength, high modulus and oxydation resistance. The alloys based on this compound can be divided into two categories : single-phase γ ($L1_0$) alloys and two-phase γ ($L1_0$)+α_2 (DO_{19}) alloys. The single-phase alloys show a very limited plasticity, even if some dislocations seem to have a certain mobility [9]. The two-phase alloys exhibit on the contrary some macroscopic ductility; for example, about 2% of ductility was reported for the binary Ti-48at.%Al alloy [10]. The origin for this ductility has not been clearly identified, although several explanations were suggested [11,12]. Indeed, contrary to the case of Ti$_3$Al-based alloys where one of the two constituent phases possesses a cubic symmetry and is therefore likely to promote the ductility, none of the two phases (γ and

Figure 3. Lamellar microstructure observed in the two-phase $\gamma+\alpha_2$ alloys.

α_2) in the TiAl-based alloys is cubic. One of the possible explanations is that the presence of a small volume fraction of α_2 which is believed to scavenge interstitial impurities from the γ matrix may enhance the ductility; in this case, the very pure γ phase should be intrinsically ductile. This point has not yet been elucidated. Another major explanation is based on some microstructural refinement which is promoted through the formation of a lamellar structure (Fig.3) typical of the two-phase $\gamma+\alpha_2$ alloys. The origin of formation of this lamellar morphlogy is now attributed to the movement of Shockley partial dislocations which locally modify stacking sequences both in α_2 and γ phases [13,14,15], but neither the mechanism nor the kinetics of the lamella growth is elucidated up to now. Some ternary or quaternary alloys with alloying elements such as V, Mn, Nb and Cr have recently been reported to improve the ductility (e.g. 3% for Ti-48Al-2Cr-2Nb [16]). In all cases, such an improvement seems to be possible only if alloys have a two-phase microstructure.

3.3. NiAl-BASED ALLOYS

The NiAl compound is recognized as an attractive intermetallic which may constitute the basis of an "alternative" material replacing some superalloys. This compound has a cubic B2 structure, melts at 1638°C and has a low density (6g/cm^3). Furthremore, it has an excellent oxydation resistance. As alredy mentioned, its intrinsically low mechanical strength can be remedied by incorporating the second phase Ni$_2$AlTi in the NiAl matrix [2]. Crystallographically, the structure of Ni$_2$AlTi (L2$_1$) is a superstructure of that of NiAl (B2), and in two-phase alloys, these two phases show the cube-cube orientation relationship. The lattice coherency between these two phases is not however very high. The lattice parameter of stoichiometric NiAl (a_{NiAl}) is 2.886×10^{-10}m while that of stoichiometric Ni$_2$AlTi (a_{Ni_2AlTi}) is 5.87×10^{-10}m.

The $L2_1$ structure has a lattice parameter which is twice that of the B2 structure, so the difference between $2a_{NiAl}$ and a_{Ni_2AlTi} is about 1.7%. This relatively high mismatch may explain why the two-phase alloys are much more brittle than the single-phase NiAl compound. However, the creep strength of two-phase NiAl-Ni$_2$AlTi alloys, superior to that of the Mar-M200 Ni-base superalloy currently used for turbine blades opens the way for new developments.

4. Attempts to design multi-phase alloys having a high phase compatibility

As emphasized above, the mechanical performance of multi-phase alloys is much better than that of the monolithic intermetallics and the most prominent example is certainly provided by the so-called Ni-base γ-γ' superalloys. Some of the Ni-base γ-γ' superalloys now being used in modern aero-engines have been pushed to a temperature of about $0.8T_m$ (T_m is the melting point). These alloys contain a very high volume fraction (up to 70%) of γ' phase in the γ-matrix, and the γ' ($L1_2$) phase is crystallographically a super-structure of the γ (A1) phase. These two phases not only have a cube-cube orientation relationship but also a very small lattice misfit; in other words, the compatibility of the two lattices is very high. This compatibility combined with the high volume fraction of γ' phase certainly explains the outstanding performance (creep strength and ductility) of the alloys of this category.

Assuming that a two-phase microstructure with a good lattice compatibility and a high volume fraction of the second phase are the key factors for obtaining useful high-temperature materials, we have recently attempted to create a γ-γ' type microstructure in several alloy systems. The approach and the results of some of these experiments are briefly described hereafter.

4.1. Fe-Ni$_2$AlTi

The first case deals with the pseudo-binary system Fe-Ni$_2$AlTi and its derivatives. In this case, the initial idea was to incorporate the NiAl phase (B2) in a BCC (A2) matrix. Among the ternary systems X-Ni-Al with X=BCC transition metals such as Fe, Cr, V, Mo, W, Nb and Ta, three systems Fe-Ni-Al, Cr-Ni-Al and V-Ni-Al possess a two-phase (A2+B2) field in a certain composition range. It is interesting here to compare the lattice parameter of each of the above metals (Table 1) with that of NiAl (a_{NiAl}=2.886x10^{-10}m). Both Fe and Cr have a parameter

TABLE 1. Lattice parameter of some BCC transition metals.

Element	Ta	Nb	W	Mo	V	Cr	Fe
a (x10^{-10}m)	3.30	3.30	3.16	3.15	3.03	2.89	2.87

close to that of NiAl while V has a slightly larger parameter ((a_V-a_{NiAl})/$a_V \approx 0.04$). All the others (Mo, W, Nb and Ta) have a much larger parameter, especially in the case of both Nb and Ta (e.g. (a_{Nb}-a_{NiAl})/$a_{Nb} \approx 0.13$). It is important to note the difference between the pseudo-binary phase diagramme Fe-NiAl on the one hand and Cr-NiAl and V-NiAl diagrammes on the other hand (Fig.4). For the Fe-NiAl system, two single-phase solid solution fields, either

(a)   (b)

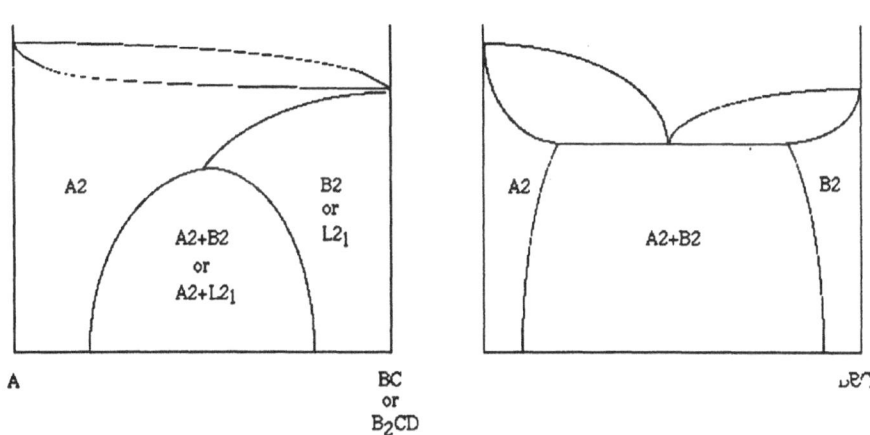

Figure 4. Two types of phase diagramme, schematically depicted.

disordered (A2) or ordered (B2), occupy the whole composition range at high temperatures and these solid solutions decompose into two phases (A2+B2) at low temperatures; a schematic phase diagramme is shown in Fig.4a. The other two systems show a eutectic reaction, like that depicted in Fig.4b.

In general, a system showing the phase diagramme of the Fig.4a type offers a great flexibility for controlling different microstructural parameters (size and distribution of the second phase and its volume fraction) through a suitable choice of composition and heat treatments. Our experiments showed that the microstructure observed in alloys of the Fe-NiAl system after solidification was indeed characterized by a very fine two-phase morphology probably due to a spinodal-like decomposition but that this fine microstructure is very stable during heat treatments. These alloys were therefore always very hard and brittle.

When a part of Al is replaced by Ti (the composition of the alloys studied lies on the Fe-Ni_2AlTi tie-line), the microstructure observed after uni-directional solidification was found to show a morphology very similar to that of nickel-base $\gamma-\gamma'$ alloys and characterized by a very regular distribution of cuboidal particles (size $\approx 0.2\mu m$) of $L2_1$ phase in the BCC matrix with a volume fraction of about 50% (Fig.5). The $L2_1$ phase is a super-structure of the BCC (A2) lattice and these two phases have a cube-cube orientation relationship but their lattice coherency is not very high (mismatch $\approx 1\%$). Interestingly, these alloys show a recrystallized grain morphology with a grain size of about 100μm after uni-directional solidification. The recrystallization is presumably induced by phase decomposition (solid solution $\Rightarrow$ A2+$L2_1$). Although these alloys exhibit a certain macroscopic ductility at room temperature (tensile elongation $\approx$ 1-2%) with a yield stress level of 1000MPa at 25°C and 800MPa at 600°C, they show a sharp

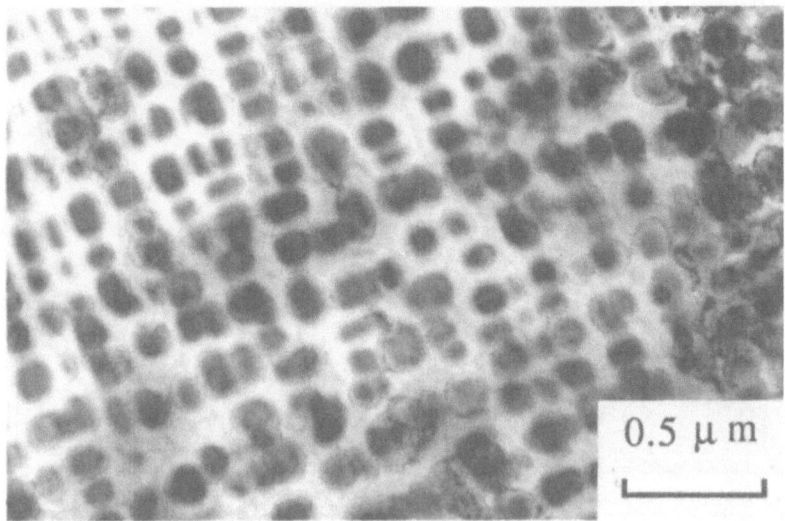

Figure 5. Two-phase microstructure (A2+L2$_1$) obtained in the Fe-Ni$_2$AlTi system.

drop in strength beyond this temperature, partly because of their polycrystalline character. Attempts to suppress the recrystallization by reducing the mismatch between the two phases A2 and L2$_1$ through further alloying have not been successful up to now.

4.2.Nb-BASE ALLOYS

The second example concerns the Nb-base alloys where an attempt is made to create a γ-γ ' like microstructure. Niobium alloys are now considered as possible candidates for high temperature applications [17,18], because of both their refractory character and reasonably low density (T_m=2468°C and d=8.6g/cm^3 for pure Nb). However, they have a very poor oxydation resistance and their strength level is low at low and intermediate temperatures (25-900°C). The purpose of the present investigation was to obtain niobium alloys having a substantial strength level in this temperature range.

Since Nb has a BCC (A2) lattice, a suitable second phase may be of the B2-type. However, among about 300 binary B2 compounds inventoried in the literature [19], no binary B2 compound can be formed with Nb. The experiments were therefore extended both to ternary Nb-XY and to quaternary Nb-X$_2$YZ systems. The choice of B2-type XY or X$_2$YZ compounds were made by taking into account the difference in the lattice parameters between Nb (a_{Nb}=3.30x10^{-10}m) and these B2 compounds. If the difference is too large, the system cannot permit a coexistence of two phases A2 and B2. Furthermore, even if the system accepts such a coexistence, it will tend to take the shape of a phase diagramme with a eutectic reaction (like that schematized in Fig.4b) rather than that with a solid-state decomposition (see Fig.4a). The compatibility of two lattices may also be very low.

4.2.1. *Nb-CoZr system.* When the B2 compound CoZr ($a_{CoZr}=3.197\times10^{-10}$m) was chosen for XY (($a_{Nb}-a_{CoZr})/a_{Nb} \approx 0.03$), the as-cast $Nb_{60}Co_{20}Zr_{20}$ alloy (one of the compositions of the pseudo-binary system Nb-CoZr) were found to be nearly two-phased but showed a very coarse dendritic structure constituted by two phases : Nb (A2) in dendrite cores and CoZr (B2) in interdendritic regions. The microanalysis showed that the Nb phase contains only a small amount of Co and the solubility of Nb in CoZr is low. Subsequent microstructural examinations after heat treatments suggested that this pseudo-binary system has a eutectic reaction like that schematically shown in Fig.4b. Although this alloy shows a ductile behaviour at room temperature, its strength is quite low, probably because of the coarse microstructure. Only a rapid solidification processing technique, for example, by using the electron beam surface melting led to a fine-scale two-phase microstructure.

4.2.2. *Nb-Ti₂AlMo system.* A survey of the literature [19, 20] indicates that there are some ternary B2 compounds of the Ti₂AlX ((X=Mo, Fe, Cr, Nb) type. Their lattice parameters are not well known but seem to be of the order of $3.10\text{-}3.15\times10^{-10}$m. The field of existence of these phases both in concentration and in temperature is also not well-known. Since our preliminary experiments indicated that Ti₂AlMo was the most stable among the above compounds, alloys of the quaternary system Nb-Ti-Al-Mo along the tie-line Nb-Ti₂AlMo (Fig.6) were in-

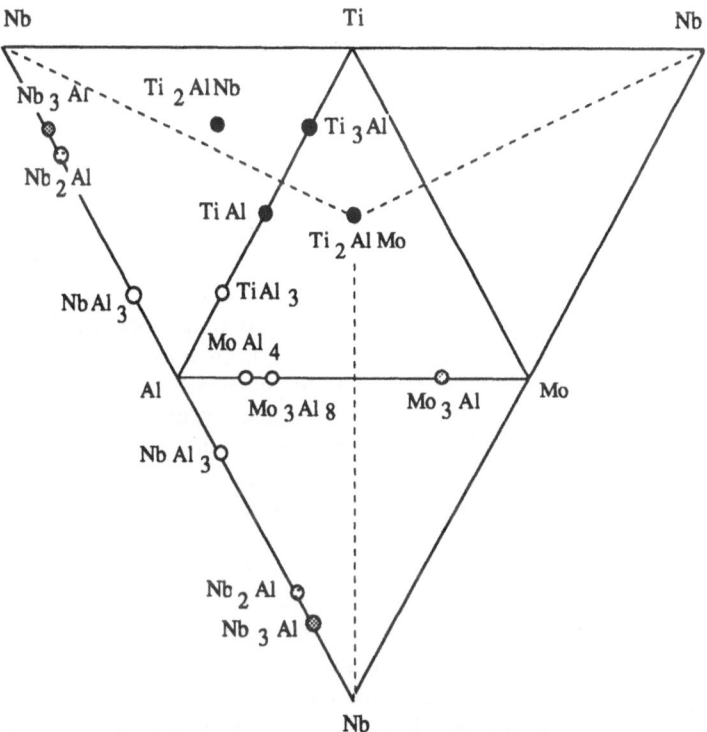

Figure 6. Various phases existing in the quaternary Nb-Ti-Al-Mo system.

654

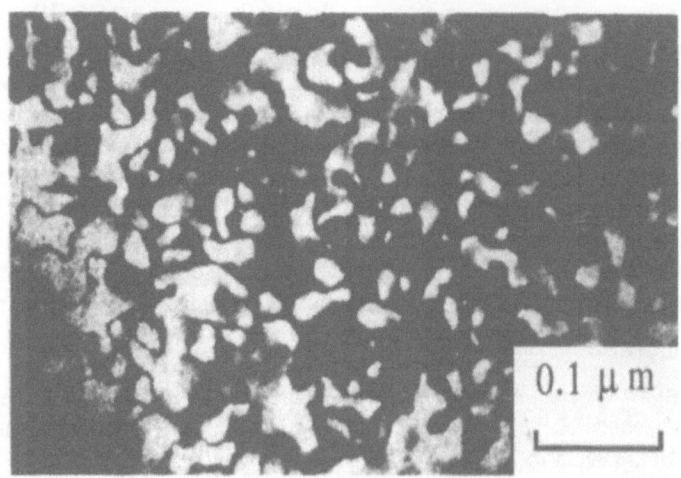

Figure 7. Anti-phase domains observed in ordered alloys of the Nb(+Cr)-Ti$_2$AlMo system.

vestigated. Niobium, β-titanium and molybdenum are totally miscible in binary couples, while the solubility of aluminium in niobium, in titanium and in molybdenum is limited to about 10at.%. In the Nb-Al phase diagramme, there are three intermetallic phases : Nb$_3$Al (cubic A15), Nb$_2$Al (tetragonal D8$_b$) and NbAl$_3$ (tetragonal DO$_{22}$); in Ti-Al, Ti$_3$Al (hexagonal DO$_{19}$), TiAl (tetragonal L1$_0$), TiAl$_3$ (DO$_{22}$) etc.; in Mo-Al, Mo$_3$Al (A15), Mo$_3$Al$_8$ (monoclinic), MoAl$_4$ etc.. These phases are shown in the diagramme comprising four ternary systems : Ti-Mo-Al at the center, and Nb-Mo-Ti, Nb-Al-Mo, Nb-Ti-Al in the corners.

Microstructural examinations of three alloys of the Nb(+Cr)-Ti$_2$AlMo system both in the as-cast state and after heat treatments indicated that they were totally single-phased, but electron diffraction studies showed that, at room temperature, they were either disordered (A2) or ordered (B2) depending on the alloy composition. Small quantities of Cr addition were totally soluble and not supposed to modify the nature of the phases present. The size of the anti-phase domains (Fig.7) in the as-cast ordered alloys also depends on the compositions. The results of these microstructural observations suggest that the pseudo-binary system Nb-Ti$_2$AlMo shows the second order (or continuous) order-disorder transition (Fig.8). In this schematic phase diagramme, both A2 and B2 phases have a wide solid-solution range, but there is no two-phase field.

Fig.9 shows the temperature dependence of yield stress for these alloys (ordered or disordered). For the sake of comparison, this figure comprises also the yield stress vs. temperature curves of some other alloys (F-48 : conventional Nb alloy; IN100 : nickel base superalloy; Super α$_2$: Ti$_3$Al-based alloy). The three alloys selected for this study possess a surprisingly high yield strength in the temperature range 25-800°C (about 1200MPa at 25°C and 650-900MPa at 800°C). The yield strength of these alloys is therefore higher than (or at least comparable to)

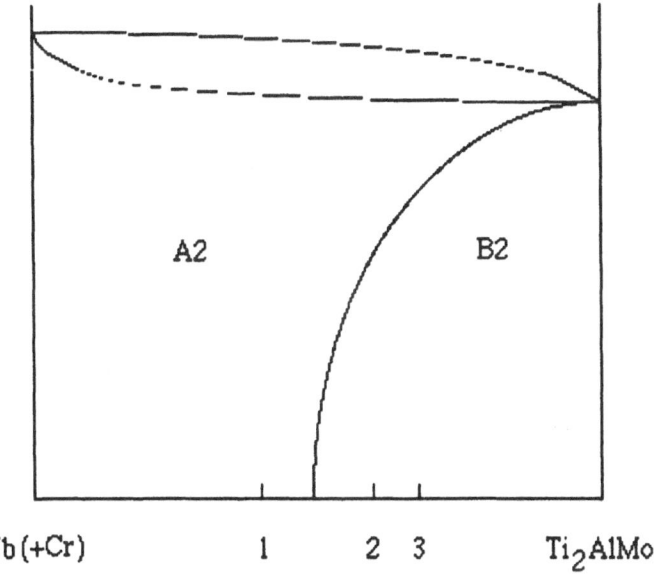

Figure 8. Schematic phase diagramme of the pseudo-binary Nb(+Cr)-Ti$_2$AlMo, suggested from microstructural observations.

that of the currently used IN100 superalloy.

However, based on our previous experience, it is anticipated that a single-phase microstructure which we have been able to generate up to now would not have an adequate creep resistance and/or a balance of other pertinent engineering properties. Creep tests are now in progress to clarify this hypothesis. Concurrently, work is in hand to identify suitable additional alloying elements which would make the Nb(+Cr)-Ti$_2$AlMo alloy system show the first order order-disorder transition, such as the one illustrated in Fig.4a. Such attempts are now underway through both experimental and theoretical approaches. Finally, a theoretical study [21] using the so-called CVM (Cluster Variation Method) is worth mentioning. The results of this study show that both the first order and second order transitions are possible between A2 and B2 phases depending on the nature of interaction between atoms of both the first and the second neighbours.

4.3. DISCUSSION

We have attempted to generate the two-phase γ-γ ' type microstructure in different alloy systems. It is implicit in this approach that the considered alloy must contain a fairly high volume fraction of relatively small (<1μm) and regularly distributed second phase particles which have, crystallographically, a superstructure of the matrix phase and that these two phases possess a high lattice compatibility. The approach adopted in this study is of an empirical "crystallo-chemistry" type. However, such an approach may be justified due to the fact that reliable

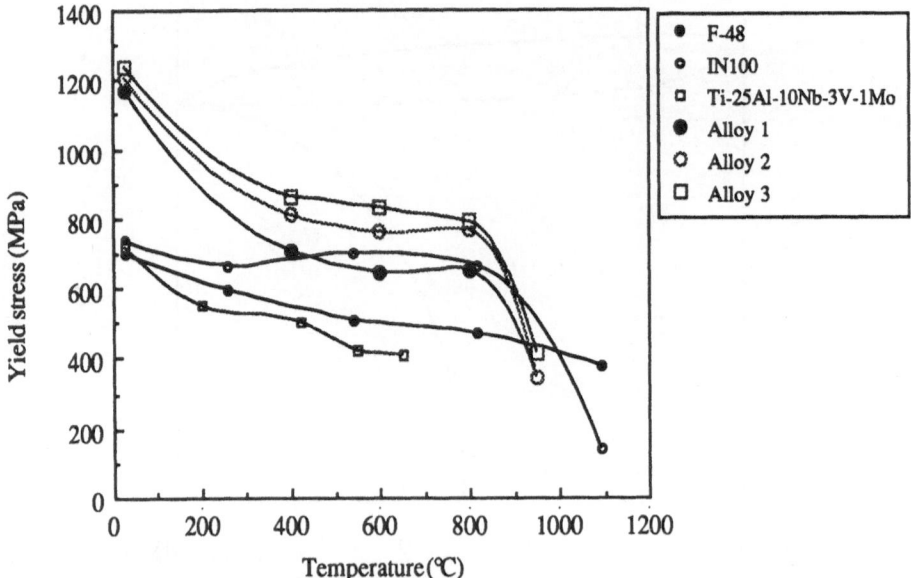

Figure 9. Temperature dependence of yield stress observed in three alloys of the Nb(+Cr)-Ti$_2$AlMo system. Compare these alloys withe three other industrially available alloys.

ternary or more complex phase diagrammes are often unavailable in the literature. In view of our experimental results, a rough screening of promising alloy systems seems to be possible through this approach, in which only crystal structures, size effect (lattice parameter) and mutual miscibility between constituent elements are taken into account.

Among the three schematic phase diagrammes shown in Fig.4(a, b) and Fig.8, only the case of Fig.4a (e.g. Fe-NiAl and Fe-Ni$_2$AlTi) is suitable to achieve the desired γ-γ' like microstructure. If the phase diagramme is of the Fig.4b type (e.g. Nb-CoZr and Cr-NiAl), no solutioning heat treatment is possible for the alloys of the composition located beyond the solubility limit observed at the eutectic temperature. It is therefore practically impossible to control the size and distribution of the second phase.

It is worth emphasizing here the need for a more theoretical approach in order to achieve, by alloying, the decomposition into two phases in alloys based on the Nb(+Cr)-Ti$_2$AlMo system (phase diagramme of the Fig.8 type), because of the promising mechanical strengths of the alloys of this system, that we have shown previously.

According to Sauthoff [22], a two-phase alloy based on a Laves phase NbNiAl containing about 15 vol.% of NiAl has a very high yield strength level (900MPa at 1000°C and 500MPa at 1100°C) with a "reduced brittleness" compared to the monolithic Laves compound. The reduced brittleness is probably promoted by the presence of less brittle NiAl phase. However, these two phases are likely to have practically no compatibility between them. In this regard, a comparative study of the mechanical properties (degree of brittleness and creep strength) between the NbNiAl-NiAl alloy and alloys of the NiAl-Ni$_2$AlTi system which are also brittle would be very interesting. As already mentioned, the phase compatibility is much higher (L2$_1$ is a superstructure of B2) in the latter system.

5. Stability of the constituent phases in multi-phase intermetallic based-alloys

When designing new alloys, especially multi-phase alloys, the stability of the constituent phases is of prime importance not only during thermo-mechanical processing but also during long-term use. This problem is now discussed in the following two cases, by using the results of our most recent investigations.

5.1. B2 PHASE IN TERNARY Ti-Al-X ALLOYS (X=Nb,V,Mo)

As already mentioned, when Ti₃Al-based alloys contain a relatively high amount of β stabilizing elements, the ordered β_2 phase, which has the cubic B2 structure, is stabilized at the expense of the disordered form. It is therefore expected that the alloys which contain a higher amount of refractory β stabilizing elements than in the Super α_2 show an improved high-temperature strength [8]. An increasing amount of data is now available concerning the phase transformations in the Ti-Al-Nb system. In particular, the question of whether the high temperature B2 phase can be preserved or not down to room temperature has been discussed by various investigators [23,24,25]. Much less information is available about the β stabilizing elements other than Nb.

In order to preserve the benefit brought about by additions of β stabilizing elements, it is clear that the possible transformation of the B2 phase must be controlled, and this requires an

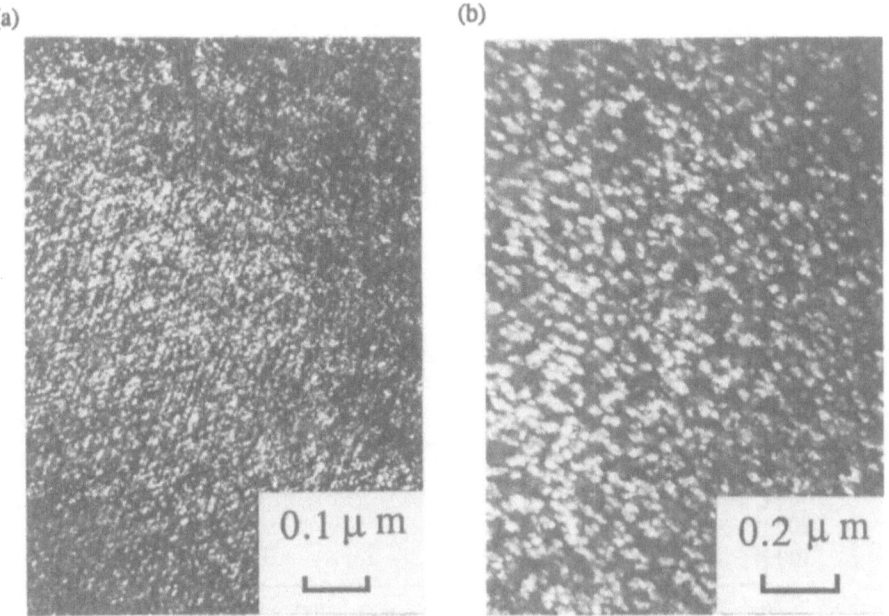

(a) (b)

0.1 μm

0.2 μm

Figure 10. (a) O-type phase observed in Ti₂AlNb,

(b) ω-type phase observed in Ti₆₂Al₂₅V₁₃.

understanding of the prevailing metallurgical factors and the knowledge of the phase transitions. We have recently examined three alloys $Ti_{50}Al_{25}Nb_{25}$ (=Ti_2AlNb), $Ti_{62}Al_{25}V_{13}$ and $Ti_{50}Al_{25}Mo_{25}$ (=Ti_2AlMo). The amount c of β stabilizing elements varies from 13 to 25 at.% in the formula $Ti_{(75-c)}Al_{25}X_c$. In the Super α_2 (Ti-25Al-10Nb-3V-1Mo) of TIMET, c is equal to 14 (=10+3+1). If the value c is 25, the alloy composition corresponds to Ti_2AlX. It is the case for both X=Nb and X=Mo. The composition $Ti_{62}Al_{25}V_{13}$ lies on the tie-line Ti_3Al - Ti_2AlV. While the B2 phase was always quite stable in the Mo containing alloy, it was less stable in the other two alloys (Nb or V containing) and various phase transitions were observed. A particular attention must however be payed to both ω-type and O(orthorhombic)-type phases. These phases appear both on slow cooling from the high temperature field and during subsequent exposure (or aging) in the low temperature region (around 500°C) (Fig.10). The formation of these phases are probably initiated through the so-called "displacive" shearings as well as "replacive" chemical rearrangements. As it is well known, the displacive atom movement is due to the lattice instability of the B2 phase. It is also important to note that these two phases are not formed simultaneously. Our results suggest that each of these decompositions takes place in two separate composition ranges of the ternary systems.

When these three alloys at the single-phase state (B2) were deformed at various temperatures, quite an anomalous temperature dependence of yield stress was observed (Fig.11) in Ti_2AlNb and $Ti_{62}Al_{25}V_{13}$. In these two alloys, a slight decrease of the yield stress from 1100MPa at room temperature to 900MPa at 400°C is followed by a drastic rise up to 1500-1600MPa at 500-550°C. Subsequent drop above the peak temperature is also very spectacular. Ti_2AlNb appears to harden more quickly than $Ti_{62}Al_{25}V_{13}$, and as a consequence the peak stress of the former alloy is shifted slightly toward a lower temperature. The most striking as-

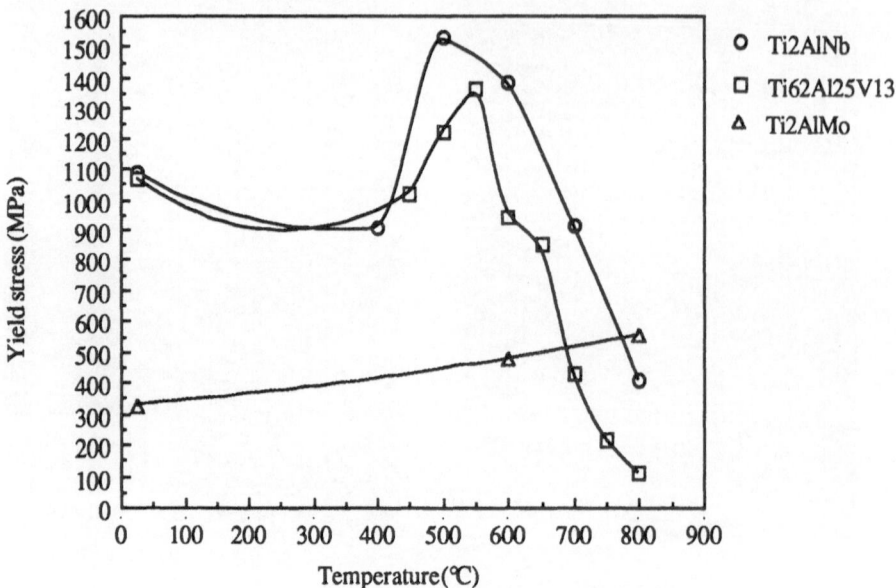

Figure 11. Temperature dependence of yield stress observed
in Ti_2AlNb, $Ti_{62}Al_{25}V_{13}$ and Ti_2AlMo.

pect, however, is that the behaviour is very similar in both these alloys. On the contrary, Ti_2AlMo exhibits a totally different behaviour. The temperature dependence of yield stress is very weak (from 330 MPa at room temperature up to 550 MPa at 800°C). While the yield stress of this alloy at room temperature is unexpectedly lower than that of both the others, Ti_2AlMo becomes the most heat-resistant at 800°C.

The anomalous variation of yield stress with temperature in Ti_2AlNb and $Ti_{62}Al_{25}V_{13}$ is probably related to a strong hardening effect of the precipitation of second-phase particles (O-type in the former case and ω-type in the latter) observed in the specimens deformed at temperatures just around the yield stress peak. It is here worth comparing our results with those of a recent study made by Rowe on the mechanical strength of the Ti_2AlNb alloy after both β (or $β_2$) and $α_2 + β$ (or $β_2$) heat treaments [8]. Rowe's results are very different from our present results. According to him, the yield strength estimated from Vickers Hardness measurements varies normally with temperature (monotonic and slight decrease of yield stress up to 815°C). He also showed the possibility of keeping high strength (600MPa-750MPa) up to temperatures as high as 815°C after the two heat treatments. His samples were argon cooled, which is not fast enough to preserve the high temperature B2 structure during the cooling from the β (or $β_2$) field. The β (or $β_2$) heat treated alloy therefore transforms through a sequence Ti_2AlNb (B2) → Ti_2AlNb (Orthorhombic) and retains a constant composition. In the $α_2 + β$ (or $β_2$) heat treated alloy, it can be supposed that the B2 phase is sufficiently stabilized to be maintained at room temperature. But, in this case, the subsequent decomposition of this stabilized B2 phase during low temperature aging cannot be totally excluded.

An important conclusion can be drawn from these experiments. In order to find out suitable alloy compositions which could correspond to an "improved" version of Ti_3Al-based alloys, it is crucial to verify the stability of the B2 phase over a wide temperature range, because decomposed products such as the ω-type phase are often harmful to the ductility. More generally speaking, only those alloys which will not exhibit instability would be serious candidates for a long-term utilization.

5.2. B2 PHASE IN TERNARY Ni-Al-Ti ALLOYS

More recently, we also investigated some uni-directionally solidified ternary Ni-Al-Ti alloys. The alloys studied were both two-phase (β+γ') and three-phase (β+β'+ γ') in the equilibrium state at 900°C [26]. β corresponds to NiAl, β' to Ni_2AlTi and γ' to Ni_3Al. The purpose of this study was to examine whether or not the incorporation of the γ' phase in the β matrix improves the low temperature ductility.

Depending on the heat treatment conditions, these alloys show a large variety of microstructures, including often metastables phases (martensitic phase and ω-type phase). In one of these alloys ($Ni_{68}Al_{30}Ti_2$), a certain room temperature ductility was observed in the as-solidified state. The corresponding microstructure is dendritic; dendrite cores are formed by the β matrix and coarse γ' precipitates while the interdendritic regions are γ' single-phase. While the γ' precipitates of the dendrite cores can be totally dissolved in the β matrix by homogenizing heat treatment at 1300°C/24h, the interdendritic γ' phase remains unchanged. After this heat treatment followed by air cooling (AC), the β matrix showed a microstructure typical of the martensitic transformation (Fig.12); in this state, the alloy completely lost its ductility.

Assuming that the ductility observed in the as-solidified state can be explained by the defor-

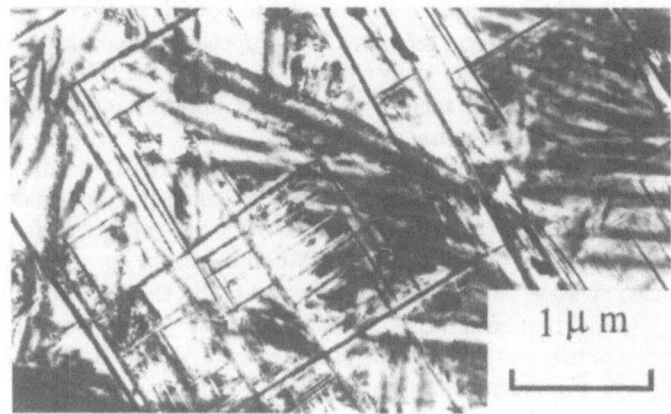

Figure 12. Microstructure typical of the martensitic transformation observed in
Ni$_{68}$Al$_{30}$Ti$_2$ after heat treatment at 1300°C/24h.
followed by air cooling (AC).

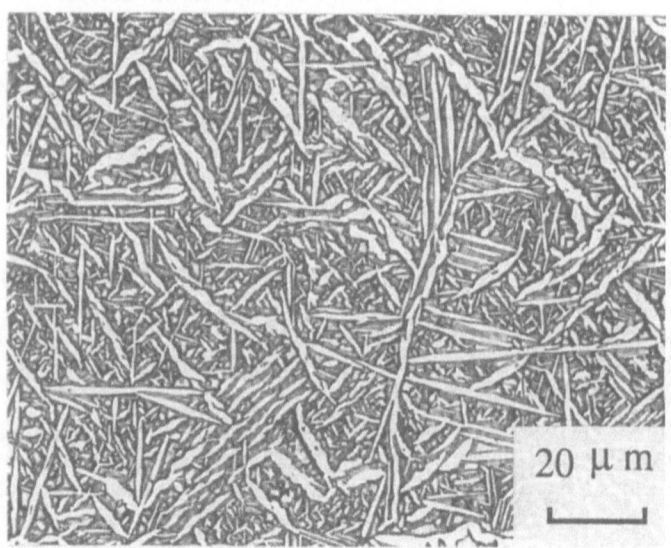

Figure 13. Homogeneous precipitation of γ in the β matrix
after heat treatment at 1000°C/24h.(+AC).

mation induced due to the martensitic transformation of the β matrix under applied stress and that the martensite transformation start temperature (M_S) is, in this case, below room temperature, it is possible to understand the brittleness observed after the homogenizing heat treatment. Indeed, the matrix is enriched in nickel through the dissolution of γ'. This increases the M_S temperature. In such a situation, the martensitic transformation takes place during cooling and the alloy loses the deformability, hence its ductility.

Heat treatment at 1000°C/24h (+AC) following the homogenizing heat treatment resulted in a relatively fine and very homogeneous precipitation of the γ' phase (Fig.13). In this state, a certain ductility could be expected because of the decrease in M_S. However, only a very low plasticity (0.1%) was observed. This may be explained either by an insufficient decrease in M_S ($M_S > 25°C$) or by a hardening due to the γ' precipitation, reducing the ductility.

6. Conclusions

The present paper emphasizes two particular points which, in the authors' opinion, seem to be of prime importance for the future research and development activities on intermetallic-based materials : (1) advantages of the multi-phase intermetallic-based alloys over the monolithic single-phase intermetallics, (2) stability of the constituent phases in multi-phase intermetallic based-alloys.

The superiority of the Ni-base γ-γ' superalloys over the single-phase γ' alloy was demonstrated by comparing their creep behaviours. Advantages of multiphase alloys were also briefly reviewed and discussed in three other alloy categories : Ti_3Al-based, TiAl-based and NiAl-based alloys. Finally, we report the approach and experimental results in an effort to create the two-phase γ-γ' like microstructure in different alloy systems.

Problems related to the stability of the constituent phases in multi-phase intermetallic based-alloys were also discussed in two cases (alloy systems Ti-Al-X with X=Nb, V and Mo and NiAl-based alloys) by using our most recent results.

Acknowledgements

The authors are grateful to Dr. F. Ducastelle for his very useful advice. Thanks are also due to A. Bachelier and J.L. Raffestin for their technical assistance.

References

[1]   T. Khan, S. Naka, P. Veyssière and P. Costa, in "High Temperature Materials for Power Engineering", COST 501 and COST 505 Conf. Proc., Liège Belgium, September (1990).
[2]   R.S. Polvani, W.S. Tzeng and P.R. Strutt, Metall. Trans., 7A (1976) 33.
[3]   C. Bontemps and P. Veyssière, Phil. Mag. Let., 61 (1990) 259.
[4]   K. Aoki and O. Izumi, J. Japan Inst. Metals, 43 (1979) 1190.
[5]   C.T. Liu, C.L. White and J.A. Horton, Acta Metall., 33 (1985) 213.
[6]   C.T. Liu and V.K. Sikka, JOM, 38. N°5 (1986) 19.

662

[7]   T. Khan, P. Caron and S. Naka, Proc. Joint ASM/TMS Symp. "High Temperature
      Aluminides and Intermetallics", ed. SH. Whang, C.T. Liu, D.P. Pope and J.O. Stiegler,
      TMS, Warrendale, PA USA, (1990) p. 219.
[8]   R.G. Rowe, Proc. Joint ASM/TMS Symp. "High Temperature Aluminides and
      Intermetallics", ed. SH. Whang, C.T. Liu, D.P. Pope and J.O. Stiegler, TMS,
      Warrendale, PA USA, (1990) p. 375.
[9]   G. Hug., PhD Thesis, University of Paris-Sud, Orsay France, (1988).
[10] Y. Nishiyama, T. Miyashita, S. Isobe and T. Noda, Proc. Joint ASM/TMS Symp. "High
      Temperature Aluminides and Intermetallics", ed. SH. Whang, C.T. Liu, D.P. Pope and
      J.O. Stiegler, TMS, Warrendale, PA USA, (1990) p. 557.
[11] Y.W. Kim, JOM, 41, N°7 (1989) 24.
[12] K. Hashimoto, H. Doi, K. Kasahara, O. Nakano, T. Tsujimoto and T. Suzuki, J. Japan
      Inst. Metals, 52 (1988) 1159.
[13] C.R. Feng, D.J. Michel and C.R. Crowe, Scripta Metall., 23 (1989) 241.
[14] C. McCullough, J.J. Valencia, C.G. Levi and R. Mehrabian, Acta Metall., 37 (1989)
      1321.
[15] Y.S. Yang and S.K. Wu, Scripta Metall. et Mater., 24 (1990) 1801.
[16] S.C. Huang, U.S. Patent 4 879 092, General Electric Company (1989).
[17] E.A. Loria, JOM, 39, N°7 (1987) 22.
[18] M.R. Jackson, European Patent 4 879 092, General Electric Company (1989).
[19] W.B. Pearson, "A Hand Book of Lattice Spacings and Structures of Metals and Alloys",
      Pergamon Press, Oxford (1967).
[20] "Titanium : Physico-chemical Properties of its compounds and Alloys",
      ed. K.L. Komarek, International Atomic Energy Agency, Vienna Austria, (1983)
[21] H. Ackermann, G. Inden and R. Kikuchi, Acta Metall., 37 (1989) 1.
[22] G. Sauthoff, Z. Metallkde., 80 (1989) 337.
[23] R. Strychor, J.C. Williams and W.A. Soffa, Metall. Trans., 19A (1988) 225.
[24] D. Banerjee, A.K. Gogia, T.K. Nandi and V.A. Joshi, Acta Metall., 36 (1988) 871.
[25] L.A. Bendersky, W.J. Boettinger, B.P. Burton, F.S. Biancaniello and C.B. Shoemaker,
      Acta Metall., 38 (1990) 931.
[26] P. Nash and W.W. Liang, Metall. trans., 16A (1985) 319.

RESEARCH AND DEVELOPMENT OF TITANIUM ALUMINIDES IN GERMANY

A. BARTELS AND H. MECKING
Technical University of Hamburg–Harburg
Eissendorfer Str. 42
W–2100 Hamburg 90
Germany

ABSTRACT. A survey of recent and current research on lightweight intermetallic titanium aluminides in Germany is presented. The main activities are centered around γ–TiAl based alloys and multi–phase alloys based on Ti_3Al and Ti_5Si_3 with Nb additions. The focus is on the interrelation between microstructure and mechanical properties.

1. Introduction

In West Germany a programme of research on intermetallic phases and alloys was begun six years ago. The aim of these activities was to explore the potential of a variety of intermetallic phases as advanced structural materials. The results obtained in this period have been overviewed by Sauthoff [1] and have been partly reported in the proceedings of a national conference [2].

As a result of this fundamental survey, new programs were established, aimed at applications of some selected intermetallic alloys. As promising light-weight alloys the titanium aluminides are being considered for high temperature applications. γ–TiAl ($L1_0$ structure) based alloys with a small amount of α_2– Ti_3Al (DO_{19} structure) and additions of Cr, Mn, Nb und Si are regarded as possible structural materials in turbines and turbo–chargers and also as skin and heat protection shields of hypersonic spacecraft (Sänger project).

Recently a two phase alloy containing the intermetallic phases α_2–Ti_3 (Al, Si) (DO_{19}) and Ti_5 (Si, Al)$_3$ ($D8_8$ structure) was developed with better high temperature properties than the γ–TiAl based alloys. With an alloying addition of Nb the oxidation resistance is substantially improved and applications in turbines up to 1000°C seem to be feasible.

These projects are being carried out in cooperation with industry. As scientific institutes the Institute for Materials Research of the GKSS Research Centre and the Division of Physics and Technology of Materials of the Technical University of Hamburg–Harburg are taking part in this research work. Besides these activities the development of γ–TiAl alloys is being carried out at the Max–Planck–Institut für Eisenforschung in Düsseldorf.

C. T. Liu et al. (eds.), Ordered Intermetallics – Physical Metallurgy and Mechanical Behaviour, 663–678.
© *1992 Kluwer Academic Publishers.*

2. γ–TiAl based alloys

2.1 THE SCREENING PHASE

During the initial screening phase investigations on a series of γ–TiAl based alloys were performed. Initial difficulties in melting these alloys with high purity had to be overcome. For a long time only laboratory scale material (about 100g) was available, too small for serial standard tests. Therefore, material saving methods like bending tests were applied.

The single phase γ–TiAl gave promise for applications at high temperatures ($T_M \geq 1470^oC$). But brittleness and inclination to grain growth and to micro—cracking turned out to be unsurmountable obstacles. A way had to be found to make the γ–phase applicable. Similar to developments in other countries the Al content was reduced. The resulting two phase alloy $\gamma-TiAl/\alpha_2-Ti_3Al$ possesses superior mechanical properties, but reduces the application temperature below the eutectoid temperature of about 1125°C. The best ductility was reported in Ti—48Al [3]. The fracture toughness of these dual phase alloys is rather poor. $K_{IC} = 8$ MPa m$^{1/2}$ at room temperature and an increased value of about 12 MPa m$^{1/2}$ at 600°C were reported [4]. An improved fracture toughness of about 20 MPa m$^{1/2}$ was observed in Ti–enriched alloys with 42–45 at% Al [4]. A lack of data on the high temperature phases in TiAl around stoichiometry initiated work at the GKSS Research Centre to determine this part of the phase diagram [5]. The results are consistent with those of other authors, e.g. Valencia et al [6].

Further development continued on two phase alloys with 46 − 49 at% Al and minor additions of alloying elements such as Cr,Mn,V, Nb, Si and C [3,7 −10]. Most of the alloys were prepared as 80 − 100g ingots in an Argon–arc melting facility and also by powder metallurgy. The microstructure in TiAl (Nb) was studied extensively by optical and electron microscopy [8]. Only small improvements in the mechanical properties in comparison to binary alloys were achieved [7,9]. A study of the fracture behaviour of Ti–48.5Al–1Mn leads to a better understanding of the ductility in a mixed microstructure consisting of lamellar $\gamma-TiAl/\alpha_2-Ti_3Al$ regions with single–phase γ regions at the grain boundaries [10]. This is discussed in detail by Beaven et al. in this workshop [11].

The largest improvement in ductility was reported by Wunderlich et al. [3]. Elongations to failure of 4% at room temperature and an increasing ductility above 100°C are outstanding properties in a Ti–46Al–1Cr–0.2 Si alloy, which were explained by a preferred glide of [121] partial dislocations and subsequent twin formation. Additionally Cr and Si stabilize the Ti$_3$Al phase and increase the strength [3]. New research programs in cooperation with industry have been started with Ti–48Al and Ti–48Al–2Cr. The first results of the investigations are reported in the next chapters.

2.2. UPSCALING OF CAST ALLOYS

First upscaling was performed with the alloys Ti–48Al and Ti–48Al–2Cr by casting ingots of about 18 cm in diameter and a weight of more than 100 kg. The material was HIPed at 1220°C to romove porosity. The microstructure consisted of very coarse grains of lamellar α_2–Ti$_3$Al and γ–TiAl. The lamellae interfaces were strongly aligned perpendicular to the radial dendrite growth directions and therefore parallel to the surface of the cylindrical ingot.

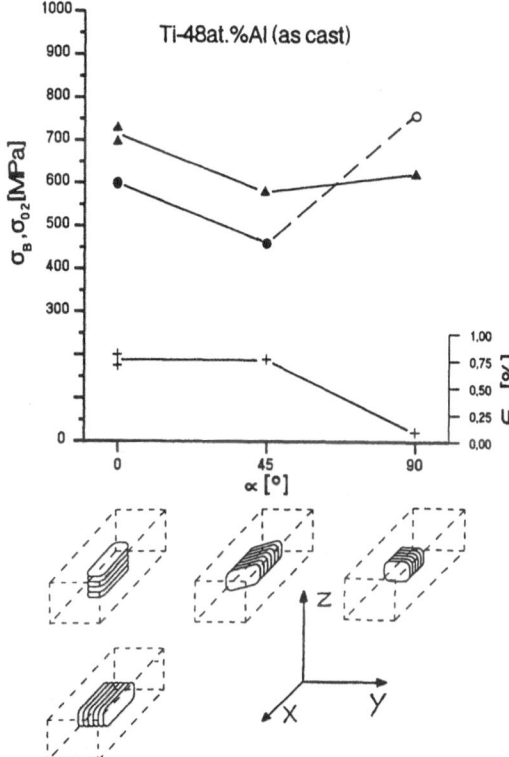

Fig. 1.
Yield stress σ_{02} ($\bullet$), fracture stress σ_B ($\triangle$) and fracture strain ϵ_B (+) as a function of α the orientation angle of γ/α_2–lamellae to the X–axis, i.e. the direction of the maximum tensile stress in the bending tests [12,14].

Specimens for four–point bending tests were cut by spark erosion [12–14]. As outlined in fig. 1, four different orientations were chosen. The bending force acts in the z–direction and x was the direction of maximum tensile stress during bending. In the two cases where the maximum tensile stress acts parallel to the lamellar interfaces ($\alpha = 0°$, see Fig. 1), a significantly higher yield stress and fracture stress were observed, compared to those determined on specimens with $\alpha = 45°$. With $\alpha = 90°$ the tensile stress acts perpendicularly to the lamellae and the bending bars broke at very small elongations before reaching the yield stress [12–14]. The 0° and 90°–orientations acted as hard modes and the 45°–orientation as a soft mode.

A recent paper [15] reports this behaviour in tensile tests of "polysynthetically twinned crystals" of Ti–49.3Al, i.e. a single crystal with the alternating layers of γ–TiAl twins and α_2–Ti$_3$Al laths. Here a substantial decrease of the yield stress was measured in the soft mode combined with a maximum strain to fracture (12.8%). In the case of our bending tests the soft mode was not accompanied with an increase in ductility. In our coarse grained polycrystalline cast material the fracture behaviour was determined mainly by the grain boundaries of the lamellar grains and by the occurrence of some very coarse single phase γ–grains.

2.3. THERMOMECHANICAL TREATMENT

Normally isotropic mechanical properties are required for structural materials. Therefore, a thermo–mechanical treatment is necessary to eliminate the cast microstructure. A first step in this direction was to carry out unidirectional compression tests on stoichiometric TiAl polycrystals with only a small amount of the α_2–Ti$_3$Al phase [16]. The initial material was produced in two different ways: by powder metallurgy (PM) and by arc melting. In both cases grain refinement caused by dynamic recrystallization during compression was observed. The micro- structure consists of equiaxed grains with a homogeneous size distribution de- pending on temperature and strain rate during compression. Some isolated coarse grains of the single phase γ–TiAl or lamellar α_2–Ti$_3$Al/ γ–TiAl regions are found which have not undergone dynamic recrystallization.

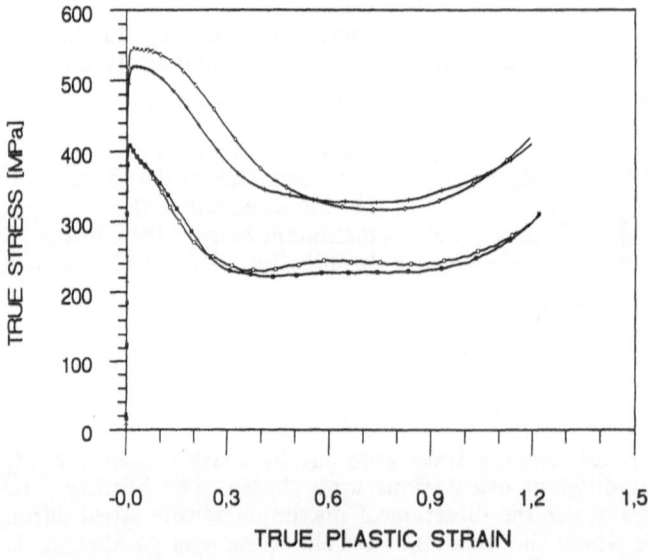

Fig. 2.
Stress–strain–curves of compression tests at 950°C (hard mode) with strain rates of $3 \cdot 10^{-4} s^{-1}$ (lower curves) and $3 \cdot 10^{-3} s^{-1}$ (upper curves).

The new cast material with a lower concentration of Al (Ti–48Al–2Cr) and Cr additions contains a larger volume fraction of α_2–Ti$_3$Al predominantly in the form of thin lamellar laths. Fig. 2 shows the stress vs. strain curves for compression tests performed on a lamellar specimen of "hard" orientation, i.e. the stress acted mainly parallel to the lamellar layers of γ–TiAl/α_2–T$_3$Al. Due to dynamic recrystallization the stress decreases immediately after reaching a maximum stress value at very low plastic strain. A high strain rate sensitivity was observed. The decrease of stress is followed by stationary deformation conditions ending in a final increase of stress possibly due to frictional effects. In the microstructure after a strain of 1.2 in fig. 3, zones affected by dynamic recrystallization and those with deformed and bent lamellae and also coarse single phase γ–grain with deformation twins can be distinguished. Obviously this inhomogeneous behaviour is caused by fluctuations of composition in the cast alloy [11].

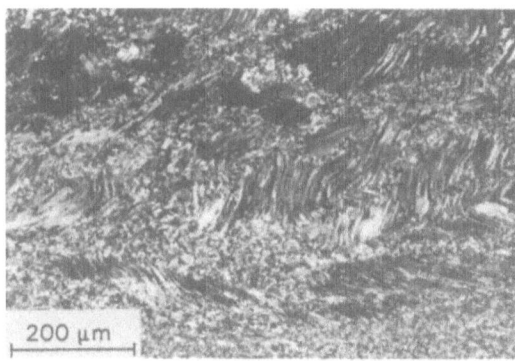

Fig. 3.
Microstructure after compression test as seen in fig. 2 (symbol □).

Fig. 4.
Microstructure of Ti–48Al–2Cr after forging (1050°C, 80%), annealing (2h, 1180°C) and ageing (4h, 1000°C).

Heat treatments in the $(\alpha_2 + \gamma)$–phase region (i.e. 1125°C − 1350°C) resulted in "striped" microstructures containing zones of coarse grained nearly single phase γ–TiAl and more fine grained γ–TiAl stabilized by second phase α_2–Ti$_3$Al at the grain boundaries [12,17]. As an example fig. 4 illustrates the microstructure after compression at 1050°C (80%), annealing at 1180°C (2h) and ageing at 1000°C (4h). In previous work measurements of the mechanical properties with four–point bending tests were performed after forging at 1050°C and systematic variations of the annealing temperature fig. 5. The yield stress drops from a very high value (850 MPa) in the forged alloy to 510 − 630 MPa after heat treatment due to a change in the microstructure as mentioned above [18]. (It should be noted that the bending test tends to overestimate the yield stress and work is in progress to correct these values by FEM calculations.)

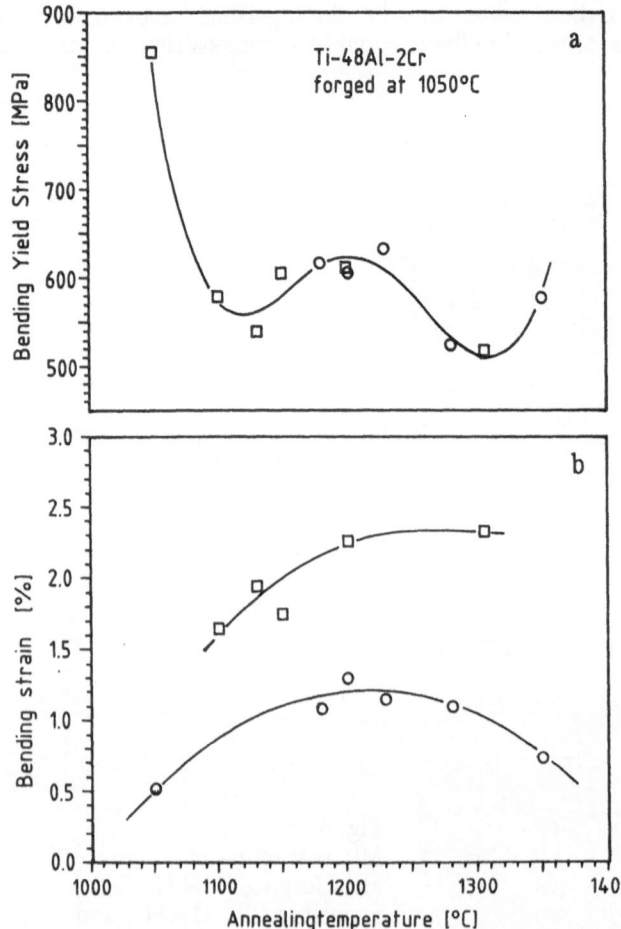

Fig.5.
Yield stress (a) and strain to fracture (b) (four–point bending test of Ti–48Al–2Cr as a function of annealing temperature (2h) after forging at 1050°C (about 80%) [18]. The two curves in (b) correspond to two different parts of the same ingot.

The shape of the yield stress vs. annealing temperature curve can be explained as follows – The first minimum corresponds to behaviour dominated by the zones of large single phase γ–grains. The subsequent increase is due to a reduction of this grain size. A larger amount of α_2–Ti$_3$Al (as predicted by the phase diagram) inhibits grain growth during annealing. Above 1200°C the fine grained parts of the microstructure, stabilized by α_2–phase at the grain bounderies, determines the mechanical behaviour. A slight increase of grain size with temperature causes a decrease in yield stress. The tendency follows a Hall–Petch behaviour, although a quantitative analysis is complicated due to the bimodal character of the microstructure. The increase above 1300°C is due to the appearance of new lamellar grains, i.e. the so–called duplex microstructure. The ductility attained the largest values after annealing in the range 1180°C to 1300°C. The two curves in fig. 5 present results obtained with specimens of the same ingot, but cut out from two different parts. Obviously this entails different flaw contents such as micro–cracks.

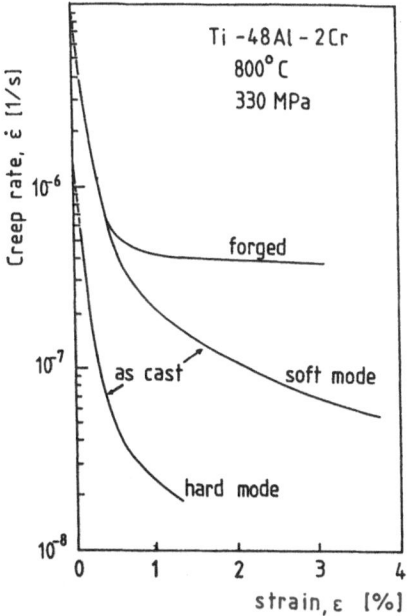

Fig. 6.
Creep behaviour (four–point bending test) of cast and thermomechanically treated Ti–48Al–2Cr (forged at 1050ºC, 80%, annealled 2h at 1200ºC, aged 3h at 1000ºC). Hard mode indicates the γ/α_2–lamellae to be parallel to maximum tensil stress, soft mode under 45º, respectively.

The thermomechanical treatment improves the ductility from 0.8% (as cast) up to about 2.2%. First results of a combination of homogenization in the α–phase region together with an additional forging step suggest a further improvement in the room temperature ductility by avoiding a striped microstructure described above. However, the globular microstructure produced without lamellar grains loses the outstanding creep resistance of a lamellar microstructure. Fig. 6 demonstrates the difference in the creep behaviour of cast and thermomechanically treated materials (four–point bending test). The creep rates in the cast material are significantly lower than those in a globular microstructure.

In summary, it can be stated that the thermomechanically processed material with a globular and homogeneous microstructure, which exhibits isotropic mechanical behaviour, seems to be ideally suited for further shaping operations such as rolling or forging. But for applications as a structural material the microstructure should be transformed into one containing the lamellar structure. This will result in an improved creep resistance, but will be connected with a loss in ductility (see also ref. [11]).

2.4. TEXTURES

One aim of the work at the Technical University of Hamburg–Harburg is the explanation of the plastic behaviour of polycrystals by means of texture measurements and texture simulations. In γ–TiAl these methods have some special features. With the (110)–Bragg reflection (L1$_0$–lattice) it is possible to detect the distribution of the c–axis. Furthermore the weak tetragonality of the L1$_0$ structure in TiAl (c/a = 1.02) causes a separation of (002)– and (200)– or (220)– and (022)–reflexes but the difference in the reflexion angle of about 1º is

670

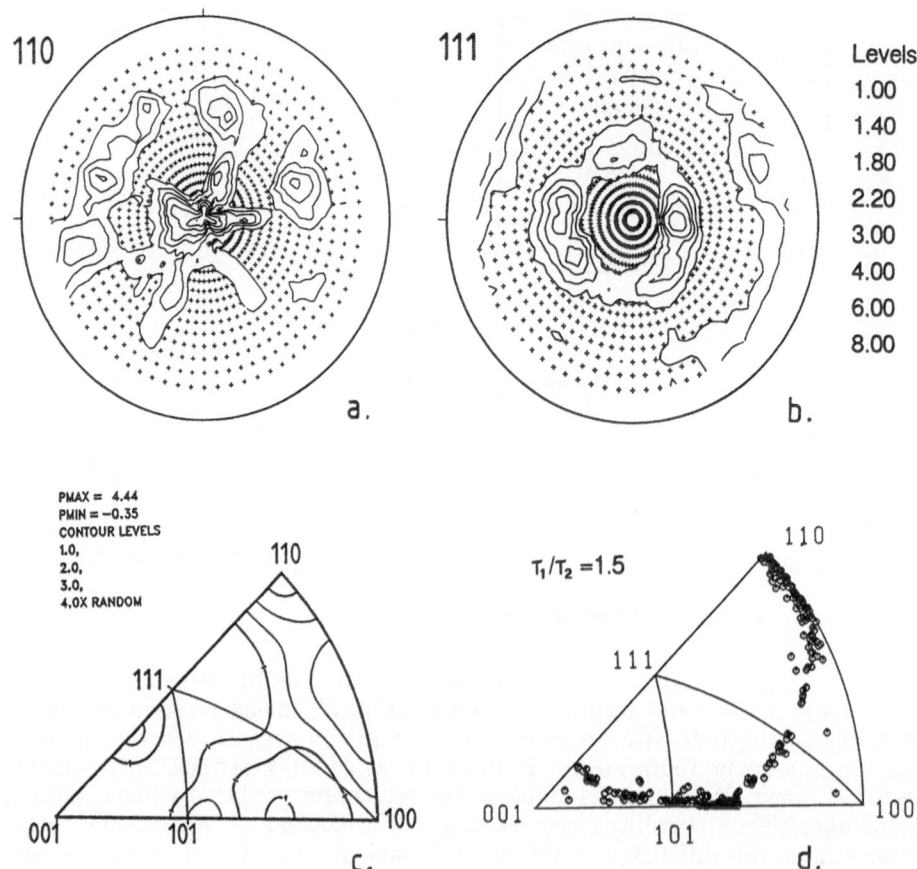

Fig. 7.
Pole figures (110) (a) and (111) (b) in Ti–50Al after compression at 450°C (46%).
The inverse polefigure (c) shows the distribution of the compression axis in view
of the crystal orientations of the grain. Best fit of the simulations (see d) is
achieved with a ratio CRSS({111}<110>)/CRSS({111}<101>)= 1.5 [19].

too small for experimental resolution. Therefore, only superimposed pole figures
can be measured in these cases which, complicates the analysis of the textures.

The possibilities of the texture analysis will be demonstrated by the following example. Polycrystalline Ti–50Al was deformed at 450°C, in compression (ϵ
= 46%). The (110) and (111) pole figures shown in fig. 7a,b (for a complete
texture analysis at least three pole figures are used) exhibit a typical fibre texture
with two components. The intensity at $\alpha = 0°$ in the (110) pole figures indicates
a (110)–component while the intensity at $\alpha = 60°$ is typical for an (011)–component or (101), respectively. Both superimposed in the (111) pole figure at $\alpha =$
35°.

From a complete ODF–analysis the inverse pole figure was calculated
(fig. 7c). It shows the distribution of the compression axis with respect to the
crystallographic directions of the grains. The highest concentration of compression
axis is located around (110). The part near (101) is shifted towards (100). This

result was compared with texture simulations (fig. 7d) based on Taylor's approach to polycrystal deformation. Two different slip modes were taken into account: $\{111\}<110>$ slip due to ordinary dislocations and $\{111\}<101>$ slip by superdislocations. A free parameter in the simulations is the ratio of the critical resolved shear stresses (CRSS) for these two kinds of slip systems. The best agreement with the experimental texture could be achieved by assuming a higher CRSS for the slip mode $\{111\}<110>$ (ordinary dislocation) than for the $\{111\}<101>$ slip mode (superdislocation). This implies preferential motion of superdislocations. (A more detailed discussion is given in ref. [19]).

Further work in this field is in progress, in particular with hot rolled materials.

2.5. POWDER METALLURGY

In developing powder metallurgy techniques for the production of titanium aluminides, two methods were chosen. Rapidly solidified alloyed powders are produced by a facility which has been assembled at the GKSS Research Centre. The intention is to produce γ–TiAl based alloys, as well as Ti–Al–Si alloys which are considered in the next section.

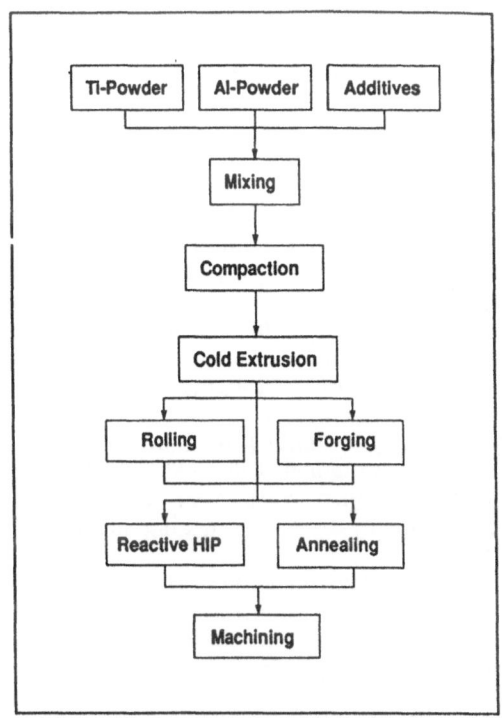

Fig. 8.
Flow diagram of powder
metallurgical procedure [21].

The other method proceeds from elemental powders [20,21]. The schematic course of the treatment is demonstrated in fig. 8. Very pure powders of Ti and Al (d < 120 μm) are mixed and, if required, additives can be introduced easily. The powder mixture is compacted by cold isostatic pressing (CIP). The subsequent cold extrusion with a ratio of at least 9 is important. In this way oxide layers on the particle surfaces are ruptured, metallic contact occurs and the particles are deformed to fibres. The density exceeds up to 99.5% of the theoretical value. The extruded material is still a mixture of elemental grains. To transform the material into intermetallic alloys, a reaction treatment is required, which enables the interdiffusion, preferably under isostatic pressure, to avoid Kirkendall pores (reactive HIP). During interdiffusion several intermediate phases occur like Al_3Ti and Al_2Ti. The final result, however, is a microstructure consisting of γ–TiAl together with islands of α_2–T_3Al aligned in the extrusion direction. No lamellar grains are formed. Therefore, this material can be deformed at lower stresses than the cast material with lamellar microstructure. Also a lower brittle–ductile transition temperature is observed [21]. A further advantage is the possibility of forming (rolling or pressing) the material in the extruded state into a nearly final shape with subsequent reactive HIP.

The material exhibits promising properties but the room temperature ductility achieved is still rather poor. Probably, a reduction of the oxygen content is indispensable to improve the ductility.

3. The two phase alloys based on $Ti_3(Al,Si)$ and $Ti_5(Si,Al)_3$

In the United States, the intermetallic phase α_2–Ti_3Al with a hexagonal DO_{19} structure was one starting point for alloy development. To improve the room temperature ductility, Nb additions were used to stabilize the β–phase which is rather ductile due to its bcc structure. The development culminated in the so–called Super Alpha 2 (Ti–25Al–10Nb–3V–1Mo; in at %) as an α_2–base metal matrix alloy. The Metal Science Group at the Technical Universität Hamburg–Harburg has reported on thermomechanical processing of a commercial Super Alpha 2 alloy with the aim of optimizing the microstructure and mechanical properties for applications up to temperatures of 600°C and higher [22].

In order to extend the development to higher application temperatures the α_2–Ti_3Al phase was combined with Ti_5Si_3 (hexagonal $D8_8$). The outstanding properties of Ti_5Si_3 are a low density combined with very high melting temperature (2130°C). Brittleness and a tendency to microcracking prevent the application as single phase material [23].

Promising properties could be achieved in two phase materials with either Ti [23] or Ti_3Al [24] as the matrix material. In a first step the ternary Ti–Al–Si phase diagram was determined in the eutectic region of the phases $Ti_3(Al,Si)$ and $Ti_5(Si,Al)_3$ [24–26]. The eutectic microstructure consists of fine needles of $Ti_5(Si,Al)_3$ embedded in a matrix of $Ti_3(Al,Si)$. Hypereutectic alloys with a surplus of Si form needle–shaped primary $Ti_5(Si,Al)_3$ containing micro–cracks. These alloys are brittle and not suited as structural materials. Lack of Si in the hypoeutectic alloys results in the formation of additional primary $Ti_3(Al,Si)$ embedded in the eutectic microstructure [24–26].

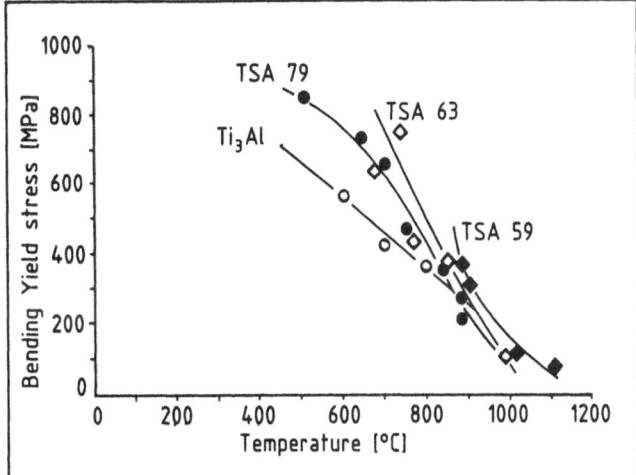

a.

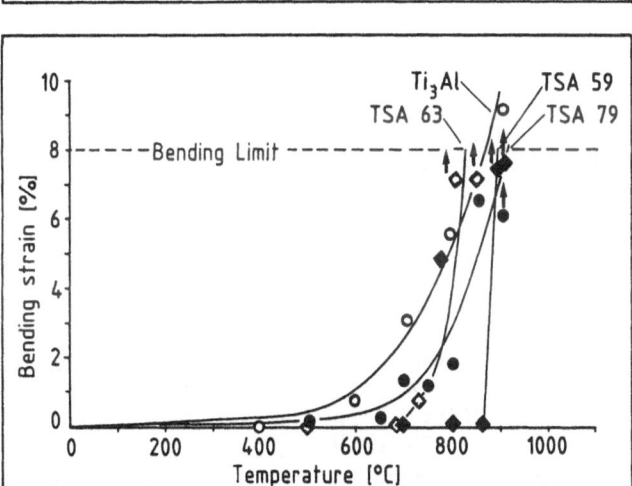

Fig. 9a and b.
Yield stress (a) and
strain to fracture (b)
(four—point bending
test) as function of
temperature for the
$Ti_3(Al,Si)$ —
$Ti_5(Si,Al)_3$ alloys
TSA 59
(Ti—6.5Si—29Al), TSA
63 (Ti—6.5Si—23.5Al)
and TSA 79
(Ti—8.5Si—20.5Al)
[26] in comparison to
those of Ti_3Al [27].

b.

    The true bending strain to fracture and the bending yield stress are shown
in fig. 9a and fig. 9b as a function of the test temperature for the eutectic alloys
TSA 59 and TSA 79 (Ti—6.5Si—29Al and Ti—8.5Si—20.5Al (at %)) and the hypoeutectic alloy TSA 63 with the composition Ti—6.5Si—23.5Al. Values for single
phase Ti_3Al from Ref. [27] are given for comparision. At room temperature no
measurable ductility was found. Above 600oC the alloys show a brittle to ductile
transition. TSA 59 becomes ductile with a sharp transition temperature between
860 and 885oC. In the temperature range 700—900oC, which is of particular
interest for technical applications, TSA 63 and TSA 79 show higher yield stresses
than Ti_3Al, although the composition of the matrix phase in TSA 79 is comparable with that of Ti_3Al. A solution hardening effect stems from Si in solid solution. The fracture toughness of TSA 63 ist found to be between 11.2 and 12.9

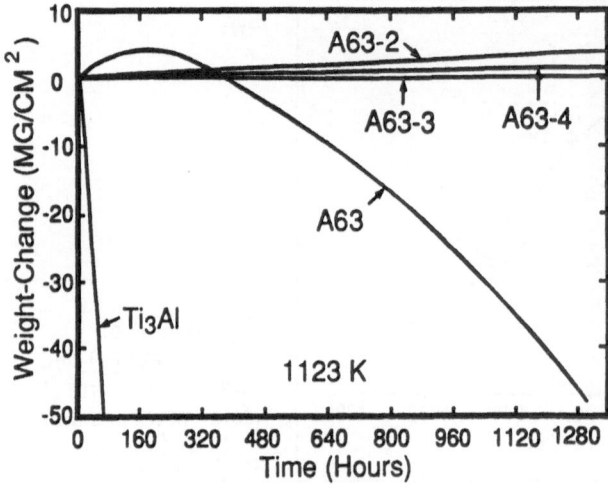

Fig. 10.
Weight change vs. time in cyclic oxidation test for Ti₃Al, alloy 63 (Ti–6.5Si–23.5Al) and Nb–modified alloy 63(63–2 with 5at%Nb, 63–3 with 10at%Nb and 63–4 with 15at%Nb) [28].

MPa m$^{1/2}$. These properties make TSA 63 and, with some restrictions, also TSA 79 promising candidates for structural applications in the temperature range 700–1000°C.

Similar to Ti₃Al based Super Alpha 2, the oxidation resistance of the Ti–Al–Si alloys is poor as compared to Ni–based superalloys. Therefore, a basic requirement is the improvement of the oxidation resistance at elevated temperatures besides that of the room temperature ductility. A drastic reduction of the high temperature oxidation was achieved with Nb additions [26,28,30]. Fig. 10 demonstrates the results of the oxidation tests carried out under cyclic conditions in air (cycling between room temperature and 850°C, 1 hour, more than 1000 cycles). The weight increment is plotted vs. time. The Nb–modified hypoeutectic alloys show a slow growth of a stable oxide layer following classical parabolic oxidation kinetics. The smallest rate constant was achieved in TSA 63–3: a hypoeutectic alloy Ti–6.5Si–23.5Al–10Nb (at %). The rate constant is superior to that of γ–TiAl and of all Nb–modified α_2 – Ti₃Al alloys [28].

The microstructure of the quaternary Ti–Si–Al–Nb alloys is more complicated than that of the ternary alloy. Nb shows an extensive solubility in the β–phase and in Ti₅Si₃. After solidification the hypoeutectic or eutectic character of the microstructure basically remains. The matrix phase is (Ti, Nb)₃ (Al, Si), but now in a fine grained Widmannstätten microstructure. The second phase is (Ti,Nb)₅ (Si,Al)₃ (Nb₅Si₃ is isomorphous with Ti₅Si₃). A further discussion of the complex microstructure is beyond the scope of this paper and is given in detail elsewhere [28–29].

In TSA63, a Nb content of 15 at % increases the temperature of the brittle to ductile transition as well as the yield stress above this temperature [28]. This behaviour might be due to the difference in the microstructure as compared to TSA63. The creep behaviour does not change significantly by Nb additions [28]. All tested alloys show a superior creep resistance as compared to Nb modified α_2–Ti₃Al alloys. The results of steady state creep for the arc melted alloys TSA–63–3 and TSA79–3 (both with 10at% Nb) are shown in Fig. 11 (four–point bending test). The hypoeutectic alloy (TSA 63–3) shows a higher creep resistance than the eutectic alloy (TSA 79–3) [30]. The stress exponent n of the power law,

$$\dot{\epsilon} \propto \sigma^n ,$$

is between 3.44 and 3.88 indicating a dislocation creep mechanism.

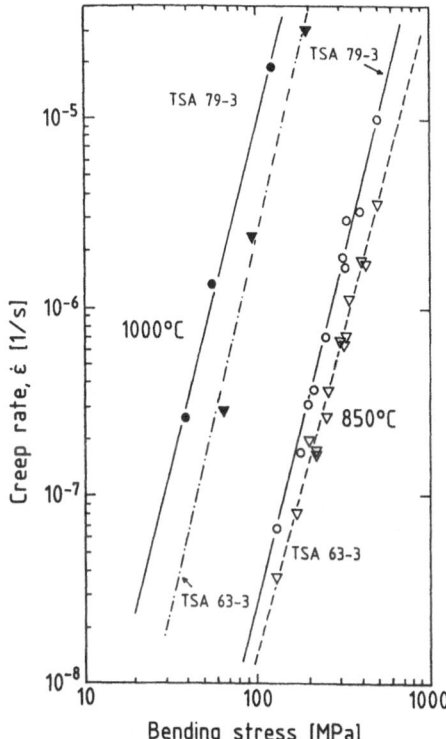

Fig. 11.
Creep rate vs. maximum stress in four–point bending tests at 850°C and 1000°C for the alloys TSA63–3 (Ti–6.5Si–23.5Al–10Nb) and TSA 79–3 (Ti–8.5Si–20.5Al–10Nb).

    Work on microstructure optimization is in progress, with the main goal to improve the mechanical properties, such as ductility and the fracture toughness at lower temperatures. In addition to heat treatments of cast alloys, powder metal-lurgical techniques are applied as well, [31]. Especially with rapidly solidified powders, a finer distribution of the silicide phase can be produced which is ex-pected to improve the fracture toughness.

Acknowledgements
The authors are grateful to M. v. Schwerin and K. Koeppe for providing their unpublished results. Thanks are also due to J. Seeger and Ch. Hartig for valuable discussions. Part of this work was financially supported by the Bundesministerium für Forschung und Technologie (BMFT 03M 3029 F2) and by the GKSS Forschungszentrum through research grants. Close cooperation with the research group of Prof. R. Wagner on many aspects of the project is gratefully acknowledged.

676

References

1. Sauthoff, G. (1990) "Intermetallic Alloys—Overview on New Materials Developments for Structural Applications in West Germany", Z. Metallkde 81. 855–861.
2. "Intermetallische Phasen als Strukturwerkstoffe für höhere Temperaturen",Bremer, F.J., (ed), (1991), Konferenzen des Forschungszentrums Jülich, Band 6, Post box 1913, W–5170 Jülich.
3. Wunderlich, W., Kremser, T. and Frommeyer, G. (1990) "Enhanced Plasticity by Deformation Twinning of Ti–Al–Base Alloys with Cr and Si", Z. Metallkde. 81, 802–808.
4. Reuss, S. and Vehoff, H. (1990) "Temperature dependence of the Fracture Toughness of single Phase and Two Phase Intermetallics", Scripta Metall. Mater. 24, 1021–1026.
5. Pfullmann, Th. (1988) "Konstitution von TiAl–Legierungen im Konzentrationsbereich 30–55 At% Al," Diploma Thesis, University of Hamburg and GKSS Research Centre
6. Valencia, J. J., McCullough, C., Levi, C.G. and Mehrabian , R. (1987) Scripta Metall. 21. 1341.
7. Hartig, Ch., Chen, S., Beaven, P.A. and Fukotomi, H., (1988) "Mechanical Properties and Microstructure of Powder and Ingot Metallurgy Intermetallic TiAl Alloys," Proc. Sixth World Conference on Titanium, France 1021.
8. Chen, Shipu (1988) "Microstructural Analysis of Intermetallic TiAl (Nb)–Compounds Prepared by Arc Melting and by Powder Metallurgy" Thesis Technical University of Hamburg–Harburg and GKSS Research Centre.
9. Mecking, H., Hartig, Ch. and Seeger, J. (1991) "Intrinsic Mechanical Properties and Strengthening Methods in Inorganic Crystalline Materials" J. Physique III, 1, 829–849.
10. Dogan, B., Wagner, R. and Beaven, P.A. (1991) "Fracture Behaviour of a TI–48.5 Al–1Mn Alloy", Scripta Metall. Mater. 25, 773.
11. Beaven, P.A., Appel, F., Dogan, B., and Wagner, R. (1992) "Fracture and Ductilization of Titanium Aluminides", this workshop
12. Seeger, J., Hartig, Ch., Bartels, A. and Mecking, H. (1991) "Microstructure of Titanium–Aluminides After Thermomechanical Treatment" in "High Temperature Ordered Intermetallic Alloys IV, Material Research Society Sympos. proc. Series 213, Boston 1990, 157.
13. Seeger, J., Bartels, A. and Mecking, H. (1991) "Auswirkung der Lamellenorientierung auf die mechanischen Eigenschaften der intermetallischen Basislegierung Ti–48% Al", presented at the DGM 1991 Annual Meeting in Graz, Austria.
14. Mecking, H., Seeger, J., Hartig, Ch. and Bartels, A.,(1991) "Zur thermomechanischen Behandlung von Intermetallischen Phasen, in ref. [2].
15. Inui, H., Nakamura, A. and Yamaguchi, M. (1991) "Deformation of Polysynthetically Twinned (PST) Crystals of TiAl in Tension and Compression at Room Temperature", in "High Temperature Ordered Intermetallic Alloys IV", Materials Research Society Sympos. Proc., Serie 213, Boston 1990, 569.

16. Fukutomi, H., Hartig, Ch. and Mecking, H. (1990) "Change of Microstructure in a TiAl Intermetallic Compound during High Temperature Deformation", Z. Metallkde. 81, 272–277.

17, Beaven, P.A., Pfullmann, Th., Rogalla, J. and Wagner, P. (1991) "Phase Transformation and Microstructural Development in TiAl–Based Alloys in "High Temperature Ordered Intermetallic Alloys IV", Material Research Society Sympos. Proc. Series 213, Boston 1990, 151.

18. Koeppe, C., Seeger, J., Bartels, A. and Mecking, H. (1991) "Duktilitätssteigerung von intermetallischem Ti48Al2Cr durch thermomechanische Behandlung", poster presented at the DGM 1991 Annual Meeting in Granz, Austria.

19. Hartig, Ch., Fang, X.F., Mecking, H. and Dahm, M. (1991) "Textures and Plastic Anisotropy in γ–TiAl", Acta Metall. Mater. (submitted)

20. Dahms, M. (1989) "Formation of titanium aluminides by heat treatment of extruded elemental powders", Mat. Sci. Eng. A 111, L5–L8.

21. Dahms, M., Seeger, J., Smarsly, W. and Wildhagen, B. (1991) "Titanium–Aluminides by Hot Isostatic Pressing of Cold Extruded Titanium–Aluminium Powder Mixtures", ISIJ International, The Iron and Steel Institute of Japan (to be published).

22. Luetjering, G., Proske, G., Albrecht, J., Helm, D. and Daeubler, M. (1991) "Microstructure and Mechanical Properties of Super Alpha 2", Proc. Sixth Jap. Inst. Metals Sympos. (JIMIS–6) on Intermetallic Compounds, 537.

23. Frommeyer, G., Rosenkranz, R. and Lüdecke, Ch. (1990) "Microstructure and Properties of the Refractory Intermetallic Ti_5Si_3 Compound and the Unidirectionally Solidified Eutectic $Ti–Ti_5Si_3$", Z. Metallkde. 81, 307–313.

24. Wu, J. S., Beaven, P.A. and Wagner, R. (1990) "The Ti_3 (Al,Si) + $Ti_5(Si,Al)_3$ Eutectic Reaction in the Ti–Al–Si System". Scripta Metall. Mater. 24, 207–212.

25. Wu, J. S., Beaven, P.A., Wagner, R., Hartig, Ch. and Seeger, J. (1989) "Microstructures and Mechanical Properties of Dual Phase Alloys Consisting of the Intermetallic Phases $Ti_3(Al,Si)$ and $Ti_5(Si,Al)_3$" in Liu, C.T., Taub, A.I., Stoloff, N.S., Koch, C.C. (Eds) "High Temperature Ordered Intermetallic Alloys III." Materials Research Society, Boston, MA, USA, p.761–766.

26. Beaven, P.A., Wu, J.S., Dogan, B. Hartig, Ch., Seeger, J. and Wagner, R. (1989) "Entwicklung von intermetallischen Verbindungen für neue Hochtemperaturwerkstoffe", GKSS Jahresbericht 1989, p. 49–61.

27. Lipsitt, H.A. Shechtman, D. and Schafrik, R.E. (1980) "The deformation and fracture of Ti_3Al at elevated temperatures". Metall. Trans. 11A, 1369–1375.

28. Wagner, R., Es–Souni, M., Chen, D., Dogan, B., Seeger, J. and Beaven, P.A. (1991) "Microstructure and Properties of Eutectic Composites Based on $(Ti,Nb)_3$ (Al,Si) and $(Ti,Nb)_5$ (Si,Al)$_3$". "In High Temperature Ordered Intermetallic AlloysIV", Materials Research Society Sympos. proc. Series 213, Boston 1990, 1007.

29. Chen, D., Es–Souni, M., Beaven, P.A. and Wagner, R. (1991) "Microstructure of Ti_3Al Based Alloys Containing Niobium and Silicon". Scripta Metall. Mater. 25 (6).

30. Es–Souni, M., Chen, D., Dogan, B., Wagner, R., Beaven, P.A. and
     Bartels, A. (1991) "Microstructure and Properties of Dual–Phase Inter-
     metallics, Based on $(Ti,Nb)_3(Al,Si)$ and $(Ti,Nb)_5$–$(Si,Al)_3$. Proc. Sixth
     Jap. Inst. Metals Sympos. (JIMIS–6) on Intermetallic Compounds, Sendai
     1991, 525.
31. Wagner, R., Gerling, R., Schimansky, F.P., Beaven, P.A. and Dahms, M.
     (1991) "Pulvermetallurgie von Titan–Aluminiden und –Siliziden" in ref.
     [2], p 139–154.

OVERVIEW OF NiAl ALLOYS FOR HIGH TEMPERATURE STRUCTURAL APPLICATIONS

R. Darolia, D.F Lahrman, R. D. Field, J. R. Dobbs, K. M. Chang,
E H. Goldman and D.G. Konitzer
GE Aircraft Engines, 1 Neumann Way, Cincinnati, Ohio 45215 USA

ABSTRACT. NiAl alloys offer significant payoffs in gas turbine applications. Excellent progress has been made in understanding their mechanical behavior and improving low temperature ductility and high temperature strength. Significant improvements in mechanical properties have been obtained with microalloying. The next challenge is to develop an alloy which has the required balance of ductility, toughness, strength and other properties such as fatigue and impact resistance. Development of design, processing and test methodology for components made out of low ductility and anisotropic materials will also be required. While significant challenges remain, the prognosis for using NiAl alloys as high temperature structural materials is good.

1. Introduction

The development of low density, high strength intermetallic alloys will enable the design and production of higher performance, lighter (high thrust-to-weight) engines for future military aircraft and supersonic commercial transport such as the National Aerospace Plane (NASP) and the High Speed Civil Transport (HSCT). Nickel aluminide (NiAl) has been the subject of several recent development programs [1-7]. This overview paper will describe the physical and mechanical properties of NiAl, its application potential, processing issues, and the progress made in improving its mechanical properties. Considerable progress has been made in identifying alloying additions to improve the low temperature ductility and the high temperature strength of NiAl, as described in this paper, such that the prospects for using NiAl as a replacement for Ni base superalloys are good.

2. Advantages and Potential Applications of NiAl

NiAl offers six key advantages: 1) density of 5.95 gm/cc which is approximately 2/3 those of state-of-the-art nickel base superalloys, 2) thermal conductivity which is 4 to 8X (dependent on composition and temperature) those of nickel base superalloys, 3) excellent oxidation resistance, 4) simple ordered body centered cubic (CsCl) crystal structure and small slip vectors for potentially easier plastic deformation compared to many other intermetallic compounds, 5) lower ductile to brittle transition temperature

679

C. T. Liu et al. (eds.), Ordered Intermetallics – Physical Metallurgy and Mechanical Behaviour, 679–698.
© 1992 *Kluwer Academic Publishers.*

relative to other intermetallics, 6) high melting temperature (1640°C) which is approximately 300°C higher than the nickel base superalloys. The thermal expansion characteristics are similar to those of the nickel base superalloys.

One of the potential applications of NiAl is as a high pressure turbine blade material. Low density NiAl turbine blades can reduce the turbine rotor weight (blades and disk) by as much as 40% as shown in Figure 1. A major portion of this reduction in weight comes from reduction in the weight of the rotating disk due to reduced blade weight and centrifugal stresses. Besides the advantage of the low density, an equally important payoff comes from the high thermal conductivity of NiAl. The thermal conductivity of NiAl is compared with a single crystal nickel base superalloy in Figure 2. Because of the higher conductivity, the temperature distribution in a turbine blade is much more uniform and the life limiting "hot spot" temperature is reduced by as much as 50°C, as shown in Figure 3. When the density and the thermal conductivity advantages are combined, NiAl alloys with strengths equivalent to older superalloys such as Rene' 80 can compete with the newer single crystal superalloys such as Rene' N5. The oxidation resistance of NiAl is well proven, having served as the major component of the protective 'aluminide' coating on virtually all high pressure turbine blades and vanes of aircraft engines.

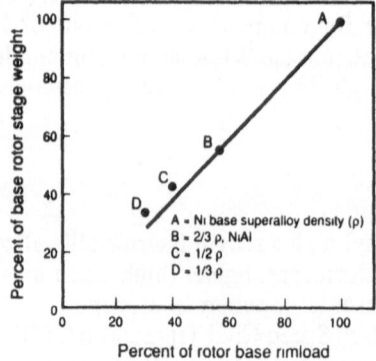

Figure 1 Reduction in rotor weight with reduced density blade materials.

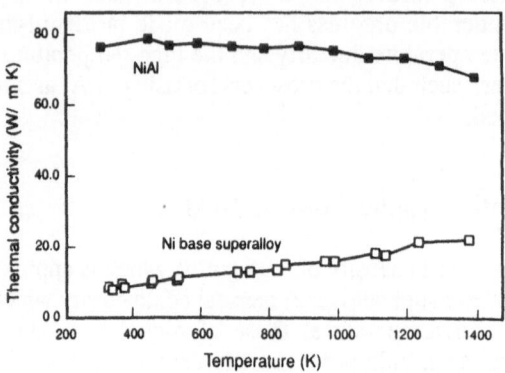

Figure 2 NiAl has a thermal conductivity advantage of 3X to 8X over Ni base superalloys.

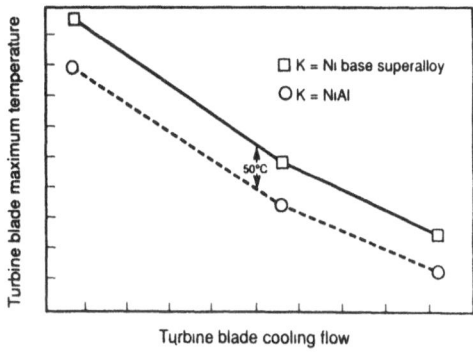

Figure 3 A 50°C reduction in the blade temperature can be achieved with higher conductivity NiAl.

3. General Characteristics

Several features in the binary Ni-Al phase diagram in Figure 4 are worth mentioning. NiAl melts congruently at ≈ 1640°C and has a wide single phase field which extends from 45 to 60 at% nickel. This wide phase field allows deviations from stoichiometry and alloying for improvements of the mechanical properties without entering a two phase field. This feature is different from the majority of other intermetallic compounds which are either line compounds or have a very narrow phase field. The CsCl crystal structure of NiAl is shown in Figure 5, which also shows the slip vectors generally observed in NiAl. The crystal structure of the Heusler, Ni_2AlTi, phase is also shown, which will be discussed when addressing the strengthening approaches for NiAl. NiAl has an ordered body centered cubic crystal structure where Ni can be thought of as occupying the corner sites and Al the body centered site. The ordering energy is believed to be very high which makes dislocation mobility rather difficult. The elastic modulus is highly anisotropic; the modulus values are 94.46, 184.51 and 270.43 GPa for the <100>, <110> and <111> orientations, respectively [8].

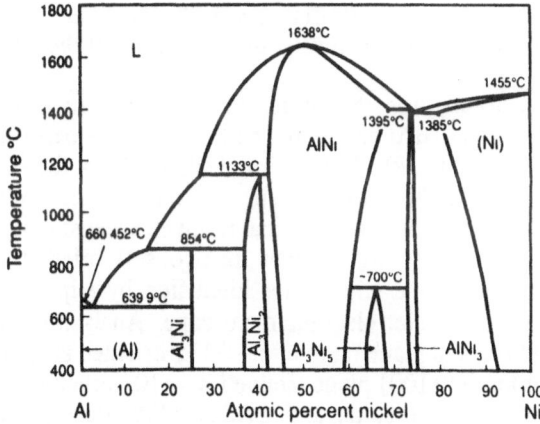

Figure 4 Ni-Al Phase Diagram

682

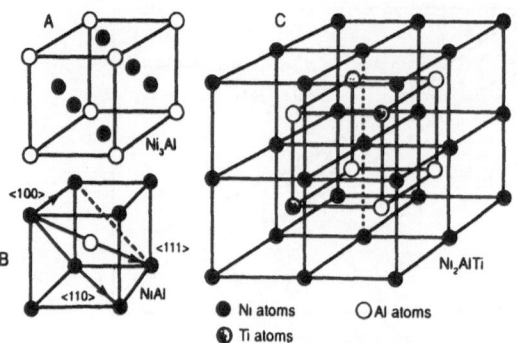

Figure 5 Crystal Structures of a) Ni₃Al, b) NiAl and c) Ni₂AlTi.

4. Slip Mechanisms

Binary NiAl, in both polycrystalline and single crystal forms, is brittle in tension at room temperature in most cases and ductile at high temperatures. It is interesting to note that the ductile-to-brittle-transition temperature (DBTT) is only 350°C to 400°C in <100> oriented single crystals and only around 200°C for the <110> oriented crystals as shown in Figure 6. These temperatures are only ~ 0.35 of the melting point of NiAl. The majority of other intermetallic compounds have higher ductile-to-brittle-transition temperatures. There is an abrupt increase in ductility in NiAl over a small range of temperatures near the DBTT. Large plastic elongations (>20%) are obtained just above this temperature. The reasons for such abrupt change in ductility are not fully understood. Onset of dislocation climb, thermally activated slip, unlocking of dislocations from interstitial impurities and additional slip systems are possible explanations proposed for this behavior. It is also interesting to note that a reduction in the degree of ordering in NiAl has been proposed around this temperature range [10]. Recent results have provided strong evidence that the <110> slip vector becomes active in 'hard' <100> oriented specimens above the DBTT [11 - 13]. These leave behind loops with <100> Burgers vectors, believed to result from dissociation at superjogs[11]. In Figure 7, **b**=<110> dislocations, with trailing **b**=<100> loops, are shown in a <001> specimen tested in tension at 471°C. A further understanding of the DBTT behavior in NiAl is required.

The shortest lattice translation vector in the B2 structure is <100>, and the predominance of <100> slip in NiAl has been well established experimentally (14 - 19) and justified by calculations [20 - 22]. The preferred slip plane for the <100> slip vector is generally considered to be {110}. Most investigators have considered the {100} to be an alternate slip plane with a slightly higher critical resolved shear stress (CRSS) [16-18].

In a recent study [23], yield strength values for two 'soft' orientations, <111> and <110>, were measured and active slip planes identified by slip trace analysis and dislocation studies in the transmission electron microscope. Analysis of the data revealed that the CRSS for the {110} slip plane (active for <111> oriented crystals) was actually slightly higher than that for the {100} plane (active for <110> oriented crystals) at room temperature. The yield stress results, and CRSS values derived from them, are summarized in Table 1. Orientation ranges within a standard triangle in which each of the two slip planes is expected to be active at RT are shown in Figure 8.

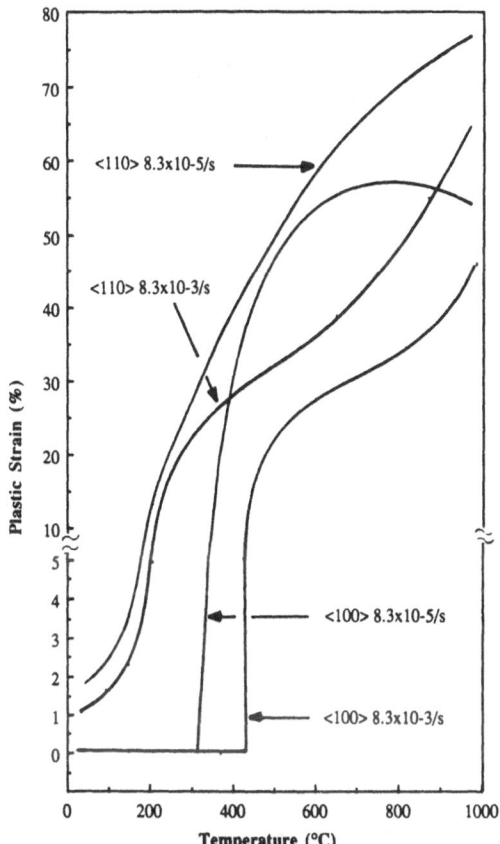

Figure 6   Plastic strain to failure in tension test in <100> and <110> oriented stoichiometric NiAl as a function of temperature and strain rate.

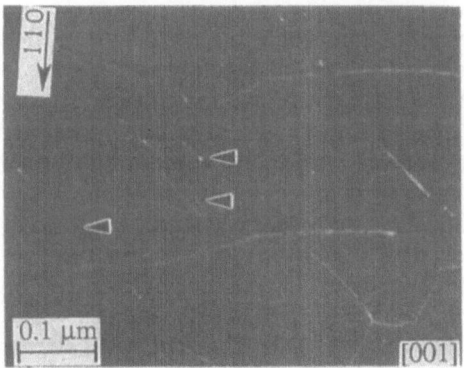

Figure 7   Weak beam TEM micorgraph of an [100] oriented tensile specimen tested at 427°C to 0.5% plastic strain. Loops with **b**=[100], marked with arrows, are seen trailing **b**=[10$\bar{1}$] dislocations which are bowed out on (101) planes.

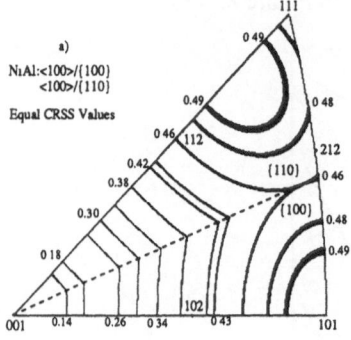
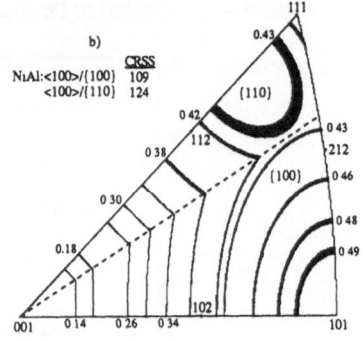

Figure 8 Iso-Schmid factopr plots showing the regimes in which {110} and {100} slip planes are expected to be active for the <100> slip vector in NiAl. In (a) the actual Schmid factors are plotted, while in (b) the factors for {110} have been adjusted to reflect the higher CRSS measured for this plane.

TABLE 1. Resolved Shear Stress Values for {110} and {100} Slip Planes from "Soft" Orientation NiAl Crystals

Orientation	0.2% YS (MPa)	Resolved Shear Stress (MPa) {110}	{100}	Source
111	264±4.6	124±2.2	87±1.5	Current Study
110	217±0.7	76±0.2	109±0.4	Current Study
111	145*	68	48	Wasilewski et al [18]
110	290*	100	145	Wasilewski et al [18]
112	200*	94	66	Wasilewski et al [18]

Data represent average and spread for two tests.
* Estimated from stress-strain curves in Figure 1 of Wasilewski et al [18]

One of the factors believed to limit ductility in NiAl is the availability of only three independent slip systems with the <100> slip direction, and thus the failure to meet Von Mises criterion of a minimum of five systems to accommodate an arbitrary shape change. Although this criterion is generally associated with polycrystalline ductility, plastic anisotropy in NiAl is also a concern in single crystal applications where components are subjected to multiaxial stresses. In addition, multiaxial stresses which develop around crack tips may not be relieved by the three independent slip systems in NiAl. The introduction of dislocations with <111> Burgers vectors (the slip direction for the disordered bcc structure on which the B2 structure of NiAl is based) would relieve this limitation. Superpartials based on the bcc 1/2<111> Burgers vector, separated by an anti-phase boundary (APB), are possible in the B2 structure and do represent the majority of dislocations in some B2 compounds (eg. CuZn [24]). Due to the very high CRSS for <111> versus <100> slip in NiAl, <111> slip has been clearly demonstrated only when compression testing is performed at or below room temperature along a <100> axis, where the Schmid factor for <100> slip is zero [25]. Even then, bending stresses and local inhomogeneities can result in kinking, which has been established as a <100> slip phenomenon [26, 27].

Several alloying research activities have focussed on promoting <111> dislocations to meet Von Mises' criterion of five independent slip systems. For example, in an Air Force Office of Scientific Research sponsored program, first principle modeling and alloying studies were carried with the aim of reducing anti - phase boundary energy to promote <111> dislocations [28, 29]. Calculations demonstrated that the substitution of Cr on Al sites on the APB plane significantly lowers the APB energy. Lowering of the APB energy could result in the formation of <111> superdislocations providing an ample number of slip systems and, provided the Peierls stress was low enough to give sufficient mobility, should promote ductility in NiAl.

Experimentally, the addition of 1 at% Cr, the solubility limit for this element in NiAl, was shown to promote the activation of <111> slip over deformation by kinking during compressive testing of <001> oriented specimens [30]. However, the measured CRSS for <111> slip was found to increase, not decrease as would be expected from a lowering of the APB energy. No evidence of separation of the 1/2<111> superpartials was observed in these specimens, as shown in Figure 9. The promotion of <111> slip was interpreted as a result of differential proportional hardening of the <100> versus <111> slip systems as opposed to actual softening of the <111> slip direction. CRSS data are given in Table 2. Note that the addition of 1%Cr results in a decrease in the ratio of CRSS values for <111>/<100> slip, thus inhibiting kinking, which is a <100> slip phenomenon [26, 27], and promoting <111> slip. The alloy displayed uniform yielding by <111> dislocations in RT compression without kinking. No increase in RT tensile elongation is observed in this alloy. It appears that the activation of the <111> slip direction is not a sufficient criterion for room temperature ductility improvement in NiAl alloys. Even though the addition of <111> slip systems would satisfy the Von Mises' criterion, increased ductility is not observed, since the CRSS for the additional slip system is too high to accommodate yielding prior to fracture. In general, the lack of ductility in most of the intermetallics can be attributed to their fracture stresses being lower than their yield stresses.

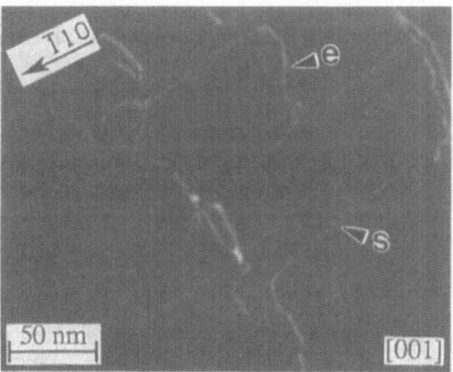

Figure 9 Weak beam TEM micrograph showing **b**=[$\bar{1}$11] dislocations on a (1$\bar{1}$2) slip plane in a [100] NiAl+1%Cr compression specimen tested at room temperature. Screw(s) and edge(e) segments are marked. No dissociation of the dislocations is

observed.

Table 2. CRSS Ratios For NiAl Alloys

	CRSS (MPa) [010]/(001)	CRSS (MPa) [Ī11]/(1Ī2)	Ratio [Ī11]/(1Ī2) to [010]/(001)
Stoichiometric NiAl	114	623	5.5
NiAl+1% Cr	178	721	4.1

4.1. STRAIN RATE EFFECTS

Intermetallic compounds are generally strain rate sensitive. Strain rate dependence in single crystal NiAl was investigated by performing tensile tests as a function of temperature and two strain rates (8.3×10-5/s and 8.3×10-3/s) [9]. Three crystallographic orientations, <100>, <110> and <111> were studied. The strain rate dependence of both yield stress and ductility in this material is surprisingly small for the two strain rates evaluated as shown in Figure 6. For the <110> and <111> orientations, the effect on ductility is minimal below the DBTT. The DBTT for the <110> orientation is strain rate independent and occurs at about 200°C. The yield strengths at the the high strain rate are somewhat higher at low temperatures but this difference diminishes with increasing temperature. For the <100> orientation, there is no ductility below the DBTT. The DBTT occurs at about 350°C for the low strain rate, and is increased to about 425°C for the higher strain rate.

Thus, the differences in properties between two strain rates, which would encompass the typical strain rates encountered in a turbine during normal operation, are relatively small. However, at much higher strain rates, which might be encountered in an impact situation, most intermetallics, including NiAl, can fail in a brittle manner even at temperatures above the DBTT. The sensitivity of intermetallics to impact damage needs to be addressed.

5. The Effect of Microalloying on Room Temperature Ductility

Polycrystalline binary NiAl is reported [31] to have about 0 to 2% tensile ductility at room temperature. The ductility is highly dependent on aluminum content, grain size [32], impurity content and texture. The fracture is generally intergranular. Attempts to strengthen grain boundaries by boron addition have not succeeded in improving ductility [33]. The fracture mode was changed to transgranular fracture with the addition of boron, as has been observed in the case of Ni_3Al. However, boron was found to be a potent solid solution strengthener in NiAl so that the yield strength exceeded the fracture strength.

For single crystal NiAl, the plastic deformation behavior is highly anisotropic. No plastic elongation is seen in the <100> oriented specimens, whereas for <110> and <111> oriented NiAl ('soft' orientations), plastic elongations up to 2% can be obtained [9]. Like most other intermetallics, ductility in compression is much higher than in tension because cracking during compression testing does not lead to immediate failure. Again, the plastic behavior is highly sensitive to composition and interstitial content.

Extensive alloying studies conducted at GE Aircraft Engines have shown beneficial effects of microalloying on the room temperature tensile ductility of single crystal NiAl [34]. In room temperature tensile tests, up to 6% plastic elongation to failure has been obtained in <110> oriented NiAl alloys containing 0.1 and 0.25 at% Fe as shown in Figure 10. Such large increases in tensile ductility by microalloying are exciting and enhance the outlook for ductile intermetallic alloys. Very significant improvements in ductility have also been made by alloying with Ga or Mo. In the case of <110> oriented NiAl alloys, up to 4.5% plastic elongation in room temperature tensile tests can be obtained with a 0.1at% Ga addition. One very important observation made from Figure 10 is that while ductility enhancements are obtained at microalloying levels, the beneficial effects are either reduced or disappear at higher levels. The mechanisms by which these elements enhance room temperature ductility of <110> oriented NiAl are currently being studied. Preliminary investigations have shown that, although the yield stress is somewhat reduced by Fe and Ga additions, the relative values of CRSS for {100} and {110} slip planes, as well as the work hardening rates, remain the same. Also, no decrease in CRSS for the <111> slip vector is observed. Thus, the slip system is unchanged with such microalloying additions. Attempts to significantly increase ductility in the <100> orientation by alloying have, thus far, been unsuccessful.

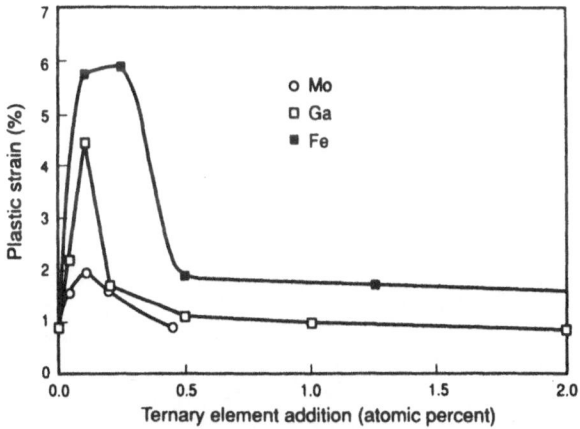

Figure 10 Significant improvement in room temperature tensile ductility of <110> NiAl is achieved by microalloying with Fe, Ga or Mo.

6.0 Fracture Toughness

Like some other intermetallic compounds, the fracture toughness of binary NiAl is low. The reported fracture toughness values are dependent on the specimen configuration, crystallographic direction with respect to loading direction, and the geometry of the notch. Typically a value of 8 MPa√m is obtained [35] from a specimen oriented in the <100> direction in a four point bend test with a simple notch (Figure 11), decreasing to 4 MPa√m in the <110> orientation. In a Chevron notch configuration, a value as high as 20 MPa√m is obtained in a <100> oriented specimen. Reasons for such differences are not clearly understood. Therefore, specimen and notch configuration and the test method should be specified when quoting fracture toughness values. Like ductility, the fracture toughness increases with temperature due to increased plasticity at the tip of the growing crack [2]. Preliminary data obtained at GE Aircraft Engines indicate that the alloying elements which improve ductility also improve fracture toughness. An alternative approach for improving fracture toughness is ductile reinforcements in an NiAl matrix. The fracture toughness has been doubled with NiAl + Cr and NiAl + Mo eutectic microstructures (See below). Unlike other B2 materials such as FeAl, which cleaves on {100} planes, the cleavage plane for NiAl is not well defined. Prior to final cleavage on the {110} plane, which appears to be the cleavage plane in NiAl, cleavage on "transient" fracture planes such as {115} or {117} is seen in NiAl, as shown in Figure 12 [35]. The occurrence of the {115} or the {117} cleavage planes has been confirmed under a variety of test conditions. The cleavage phenomena on such high index planes are not yet understood.

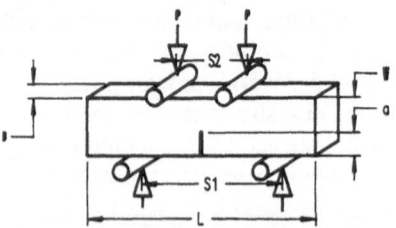

Figure 11   Four point bend test with a simple notch for toughness measurement.

Figure 12   Prior to final cleavage on the {110} plane, cleavage on "transient" fracture planes such as {115} or {117} is seen in NiAl.

6.1 NiAl EUTECTIC ALLOYS FOR IMPROVED FRACTURE TOUGHNESS AND STRENGTH

An alternate approach for improving the toughness of brittle materials, such as intermetallic alloys or ceramics, is to include a ductile phase. Usually inclusion of the ductile phase requires that a built up composite be produced, adding extra processing steps. In the case of NiAl alloys, the inclusion of the ductile phase can be done through directional solidification of eutectic ternary systems. Examples of these systems are NiAl + Cr, NiAl + Mo or NiAl + Re. Directional solidification of the composition 33Ni-33Al-34Cr (atomic %) produces a microstructure which consists of long fibers of the alpha-Cr (with Ni and Al in solution) in the matrix of NiAl (with Cr in solution). This microstructure is shown for both the longitudinal and transverse sections in Figure 13 and is similar to a continuous fiber composite, although it does not require the large number of steps which would normally be necessary to produce a composite. This microstructure can be varied significantly by varying the processing conditions. Figure 14 shows the effect of increasing the growth rate on the size and spacing of the fibers for an NiAl + Mo composite.

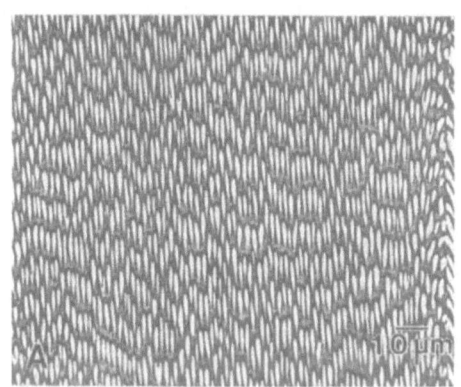
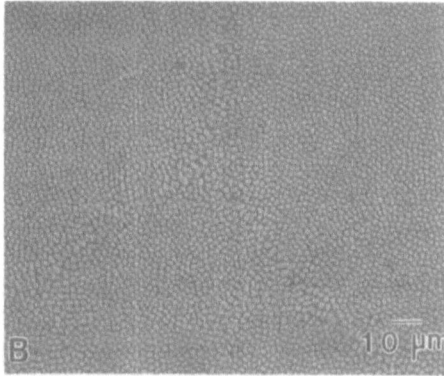

Figure 13 Longitudinal a), and transverse b) section of a NiAl+Cr eutectic alloy which has been directionally solidified.

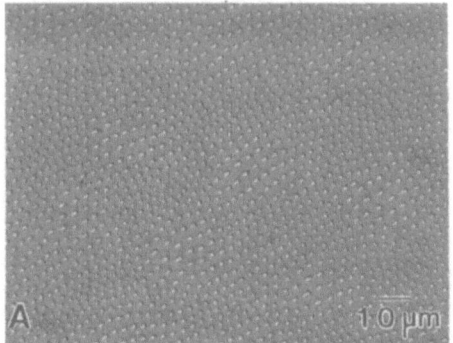

Figure 14 NiAl+Mo Eutectic grown at a) 0.5 in/hr and b) 4.0 in/hr growth rates showing the differences in microstructures resulting from the difference in processing.

The objective of producing the microstructures described above is to improve both the strength and damage tolerance properties of the intermetallic matrix. The effect of the fibers has been shown previously [36] to increase the strength over that of the binary NiAl matrix. The increase in strength was larger for the material which was grown at a faster rate. This faster growth rate corresponds to a finer structure as shown in Figure 14. The fibers also act to increase the toughness of the material as shown in Figure 15, which compares the toughness of stoichiometric NiAl with the eutectic materials. From this figure, it can be seen that the toughness for the pseudo-eutectic is about double over that of binary NiAl.

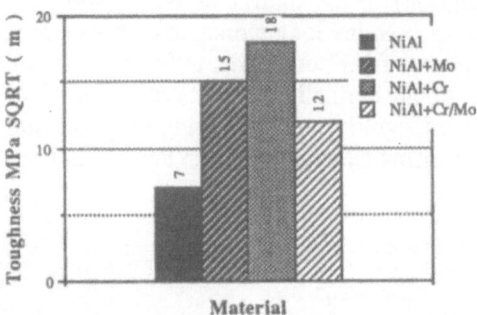

Figure 15 Toughness comparison of NiAl based alloys with NiAl eutectic materials.

In the eutectic material there is an appreciable amount of solubility for NiAl in the Cr and for Cr in the NiAl. This may result in solid solution strengthening effects, or after heat treating in precipitation strengthening effects. An example of a heat treated structure in the NiAl+Cr system is shown in Figure 16, from which it can be seen that there is a significant amount of precipitation in both the fiber and matrix. Therefore, a more accurate comparison to illustrate the composite toughening effect would be between an NiAl alloy saturated with Cr, rather than binary NiAl. Further development of such NiAl composites is being pursued.

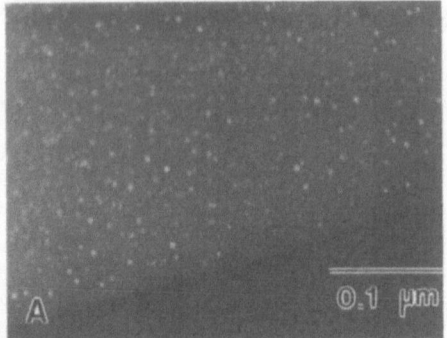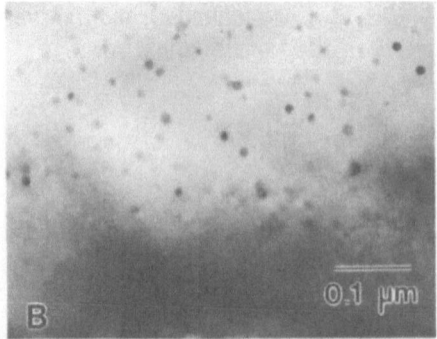

Figure 16 TEM micrographs showing a) precipitation of B2 and the Cr fiber and b) precipitation of Cr in the NiAl matrix following heat treatment at 1400°C/1hr+1200°C/24hrs.

7. High Temperature Strength

The high temperature strength of unalloyed NiAl needs improvement to be competitive with the superalloys[19, 29, 37, 38]. There are five ways by which the high temperature strength of NiAl has been improved: 1) elimination of the grain boundaries, 2) solid solution strengthening, 3) metallic precipitate strengthening, 4) intermetallic precipitate strengthening, and 5) composite strengthening.

Elements such as cobalt, iron and titanium have a high solubility in NiAl and can provide significant solid solution strengthening [39]. It appears that solid solution strengthening can be quite significant even at low levels of the ternary element, especially with addition of group IVB and VB elements such as Ti, Hf and Ta. Elements such as Cr, Mo and Re have limited low temperature solubility (<1%). These elements, when added beyond their solubility limit, precipitate metallic a Cr, - Mo or - Re particles which strengthen NiAl [7]. Microstructure control is achievable with proper solution and aging treatments.

The addition of the group IVB or VB elements beyond their solubility limits to NiAl produces several ternary intermetallic compounds including the Heusler phase (50Ni-25Al-25X) and the Laves phase (33Ni-33Al-33X) as shown on the Ni-Al-Hf ternary diagram in Figure 17. Such phases can contribute significantly to the strengthening of the NiAl alloys [2, 39, 40]. The Heusler phase is of particular interest because it is simply a further ordering of the NiAl B2 phase caused by atoms of the added element substituting for every other Al atom. The BiF$_3$ (L2$_1$) type crystal structure for this is compound shown in Figure 5. The close relationship between the B2 and L2$_1$ crystal structures results in coherent or semi coherent precipitates of Heusler phase in NiAl. The Laves phase is primitive hexagonal, and will therefore not be coherent with the β phase matrix. The Heusler phase, also referred to as the β' phase, typically precipitates coherently in the β phase on cooling, but becomes incoherent as the material is aged.

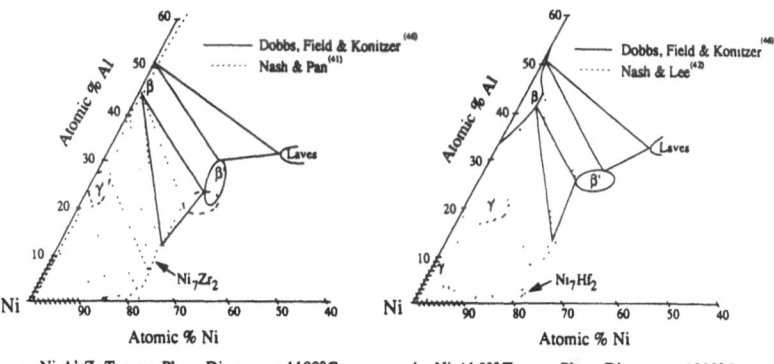

a. Ni-Al-Zr Ternary Phase Diagram at 1100°C   b. Ni-Al-Hf Ternary Phase Diagram at 1200°C

Figure 17 Ternary phase diagrams of the Ni-Al-Zr at 1100°C a) and Ni-Al-Hf at 1200°C.

An understanding of the phase relationship between the β and the surrounding phases is needed to correctly develop alloys in this family. Several existing ternary phase diagrams were used to conduct analysis of the phase relationships around NiAl [41- 45]. The existing diagrams have predominantly focused on Ni rich compounds in support of γ-γ' superalloys. In a more recent study [46], alloys were produced around NiAl to reflect the compounds of interest, and these alloys were subjected to the temperatures listed on the diagrams for up to 100 hours. The specimens were then analyzed using electron microprobe and DTA, and ternary diagrams were constructed around NiAl. Two of these diagrams are shown in Figures 17a and 17b. Figure 17a is an 1100°C section from the Ni-Al-Zr ternary phase diagram proposed in [41] and revised in the more recent study [46]. Notice the shift of the β-β'-Ni$_7$Zr$_2$ three phase field and the existence of the Laves phase field. Figure 17b is a section of the Ni-Al-Hf ternary phase diagram at 1200°C, proposed in [42] and updated in the more recent study. Again the β-β'-Laves three phase field is defined and the β-β' two phase field is verified. Notice from these two diagrams that Zr and Hf behave similarly in the two systems, although Zr has a much lower solubility in NiAl. These diagrams also show that nickel rich (>50 at%) compositions are required to avoid the formation of the NiAlTa phase. This hexagonal NiAlTa Laves phase can also be used to strengthen NiAl [2]. However, NiAl alloys strengthened by the Laves phase have been shown to be more brittle than the alloys strengthened by the β' phase. In general, the strengthened NiAl alloys are less ductile than binary NiAl and have higher brittle to ductile transition temperatures. Strengthening by a different type of intermetallic phase, referred to as a G phase ($Ni_{16} X_6 Si_7$, where X is Zr or Hf) has also been studied [47]. The strengthening potential of this phase is being further explored.

Figure 18 presents a summary of progress made at GE Aircraft Engines in strengthening NiAl alloys. In this figure, tensile rupture stress for single crystal NiAl alloys is plotted against a Larson Miller Parameter which combines both time and temperature on the same axis. The rupture behavior of the conventionally cast nickel base superalloy Rene' 80 is also plotted. It can be seen that the rupture strengths of NiAl alloys have been improved to the Rene' 80 strength levels. Figure 19 shows that the yield strengths of NiAl alloys also have been improved significantly to Rene' 80 strength levels. Stresses in Figures 18 and 19 are not corrected for density. When specific strengths are considered, and taking into account the higher thermal conductivity benefits, these NiAl alloys can be competitive with the current single crystal nickel base superalloys. As was pointed out earlier, the strongest alloys are less ductile than binary NiAl. A proper balance of strength and ductility is currently being pursued.

NiAl composites consisting of an NiAl matrix and dispersoids such as Al_2O_3, TiB_2 and HfC are also being developed by various research groups [48 - 50]. These composites are in early stages of development and do not yet compete with the β' strengthened NiAl, as shown in Figure 18 where the tensile stress rupture properties of TiB_2 dispersed NiAl are compared to β' strengthened alloys.

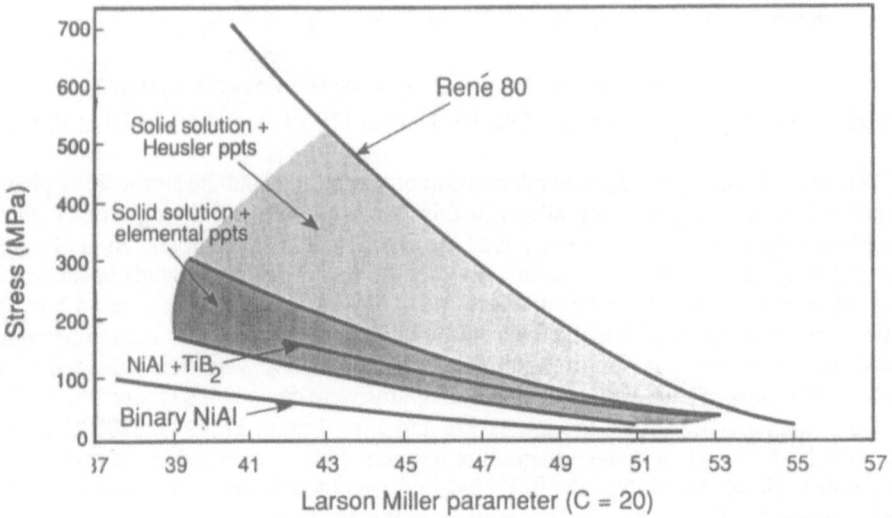

Figure 18   Stress rupture properties of NiAl alloys and composites compared with superalloy Rene'80.

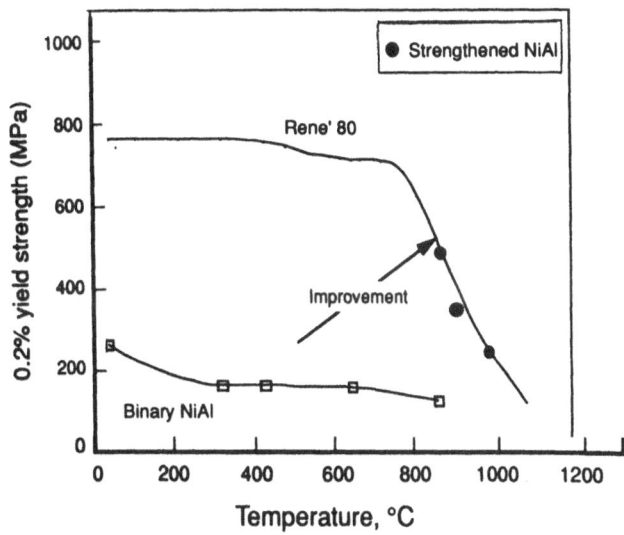

Figure 19 Tensile yield strengths of NiAl alloys in the <110> orientation. Alloying produces yield strengths equivalent to Rene'80. No yielding occurs below 800°C.

8. Processing of NiAl Alloys

Processing of NiAl is relatively easy, in spite of its high melting temperature. Casting, powder processing and extrusion have been successfully used. The preferred processing route for turbine blades and vanes is, however, directional solidification which can be used to produce single crystal structures. The elimination of grain boundaries not only allows a study of the intrinsic properties of NiAl, but is necessary to obtain high temperature creep strengths comparable to current superalloys. Directional solidification is currently used to produce hollow, thin-wall turbine blades and vanes from advanced Ni-base superalloys (Figure 20). Columnar-grain and single crystal structures are routinely produced using this process which is derived from that practiced by Bridgman [51]. The directional solidification process and the benefits of columnar and single crystal structures are well-known; surveys of these technologies can be found in many publications [52, 53].

Building on this existing technology, GE Aircraft Engines has successfully produced large (up to 4 cm square) single crystal bars of NiAl alloys; Figure 21 shows a 12 cm x 2.5 cm x 4 cm typical single crystal. The strength and ductility properties discussed earlier were obtained from these single crystal bars. Solid near net shape blades have also been cast, as shown in Figure 22. The major challenge in producing single crystals of NiAl alloys is the high melting temperature, which is up to 300°C higher than the most advanced superalloys. In addition to making furnace-related issues more complex (temperature capability, furnace durability, temperature measurement and control, etc), the higher processing temperatures present challenges to existing ceramic mold and core materials, which are already approaching their limits for current superalloys in terms of structural capability and reactivity with the molten metal (Figure 23). The very high

694

thermal conductivity of NiAl relative to superalloys also requires attention.

Figure 20 Hollow thin-wall directionally solidified turbine blades.

Figure 21 Typical 12x2.5x4 cm single crystal produced by a Bridgman process.

Figure 22 Solid near net shape single crystal NiAl high pressure turbine blade produced by Bridgman process.

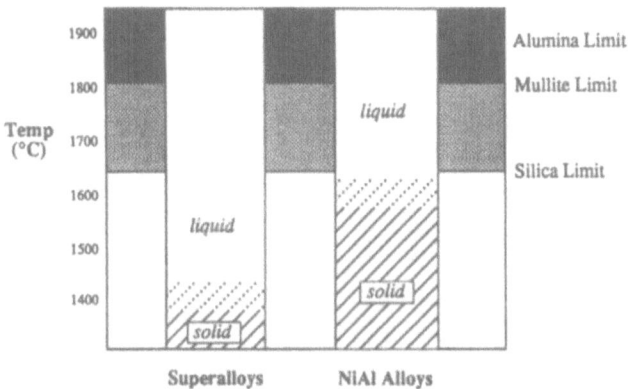

Figrue 23 Temperature limits for investment casting mold and core materials.

Melt growth processes, such as Czochralski and modified Edge-Defined Film-Fed Growth, as well as float zone processes, have also successfully yielded NiAl single crystals, although size limitations have not yet been overcome.

While large single crystal bars and blade shapes can be made using the modified Bridgman process, casting of hollow, thin-wall structures has not been demonstrated. The limited low-temperature plasticity of most NiAl alloys, combined with the strains generated when the metal shrinks around the ceramic core during cooling from the casting temperature, results in severe cracking of the metal. Also, core-related problems such as shift, sag, breakage and reactivity are a major source of low yields in casting *superalloy* turbine blades in advanced designs. This experience, along with the additional challenges associated with the high melting temperatures and low RT ductility of NiAl, suggests that an alternate approach be used to produce complex, thin-wall NiAl alloy turbine airfoils.

Such an approach, shown schematically in Figure 24, involves machining a component from a single crystal preform, and contains the following major elements:
• produce a single crystal preform (block or near net shape) and split by wire-EDM (electrodischarge machining) or produce two or more single crystal "matching halves",
• machine internal cooling passages into each half by electrode EDM or other methods,
• bond the two halves by a process which produces a single crystal across the joint,
• finish machine the airfoil contours, other external blade features, and film-cooling holes by a combination of mechanical and electrochemical processes.

This fabricated blade approach precludes the need for producing thin wall castings and can result in improved dimensional tolerances and improved inspectability, while allowing the freedom to incorporate advanced cooling designs.

Many conventional and non-conventional material removal techniques can be used on NiAl alloys, although some must be modified to accommodate low RT plasticity. Some of the machining processes required for producing a fabricated blade were used to produce the small high pressure turbine blade shown in Figure 25. This blade was machined from a standard 2.5 cm x 4 cm rectangular single crystal bar using a ternary NiAl alloy, and does not contain internal features other than the machined-in radial holes.

Single crystal bond joints, with properties equivalent to the base metal, can be produced in superalloy single crystals using an activated diffusion bonding process [54]. Initial results at GE Aircraft Engines suggest that similar results can be achieved in NiAl.

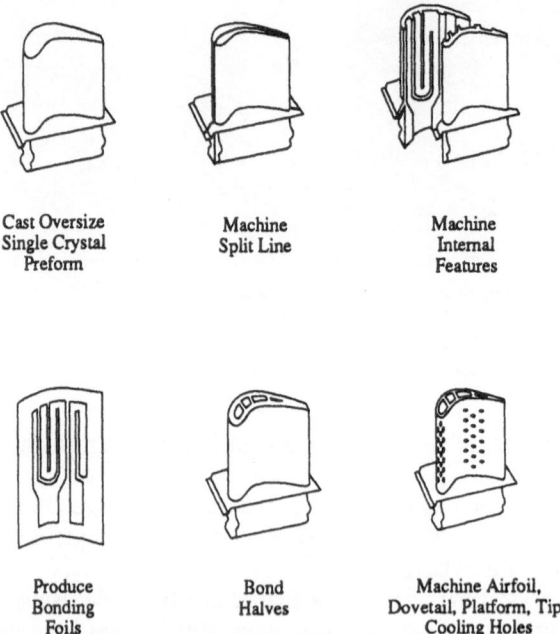

Cast Oversize Single Crystal Preform	Machine Split Line	Machine Internal Features

Produce Bonding Foils	Bond Halves	Machine Airfoil, Dovetail, Platform, Tip, Cooling Holes

Figure 24 Fabricated blade processing sequence.

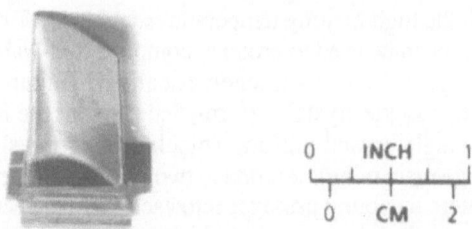

Figure 25 Small NiAl high pressure turbine blade that was machined from a single crystal bar.

ACKNOWLEDGEMENT

We would like to acknowledge the discussions with D. B. Miracle of the Air Force Wright Laboratory, M. V. Nathal of NASA Lewis Research Center and K. Vedula of the Case Western Reserve University. In addition to the financial support under the Independent Research and Development projects at GE Aircraft Engines, the work was supported by the Air Force Office of Scientific Research (A. H. Rosenstein), Air Force Wright Laboratory (W. A. Troha, M. J. Kinsella and R. H. Lilly), and the Naval Air Propulsion Center (A. S. Culbertson).

REFERENCES

1. R. Darolia, *JOM* **43**, 44 (1991).
2. G. Sauthoff, *Z. Metallkde* **80**, 337 (1989).
3. G. Sauthoff, in *High Temperature Aluminides and Intermetallics*, ed. S. H. Whang, et. al. (Warrendale, PA: TMS), 329 (1989).
4. C. C. Law and M. J. Blackburn, "Rapidly Solidified Lightweight Durable Disk Material" (Final Report AFWAL - TR - 87 - 4102, 1987).
5. R. D. Noebe, R. R. Bowman, C. L. Cullers and S. V. Raj, in *HIGHTEMP Review 1990* (NASA CP 10051 20-1), 20-1 (1990).
6. R. D. Noebe, R. R. Bowman, J. T. Kim, M. Larsen and R. Gibala, in *High Temperature Aluminides and Intermetallics*, ed. S. H. Whang, et. al. (Warrendale, PA: TMS), 271 (1989).
7. K. Vedula, V. Pathare, I. Aslandis and H. H. Titran, in *High Temperature Ordered Intermetallics Alloys*, ed. C. C. Koch et al. (MRS Symposium Proceeding **39**), 411 (1985).
8. R. J. Wasilewski, *Metall.Trans* **236**, 455 (1966).
9. D. F. Lahrman, R. D. Field and R. Darolia, in *High Temperature Ordered Intermetallics Alloys IV,* ed. L. A. Johnson, et al. (MRS Symposium Proceeding **213**) 603 (1991).
10. L. A. Kuchererenko and V. A. Troshkina, *IZVEST AKAD NAUK SSSR METAALLY* **1**, 115 (1971).
11. R. D. Field, D. F. Lahrman and R. Darolia, accepted for publication in *Acta Metallurgica*.
12. D. Miracle, "The Deformation of NiAl Bicrystals" (Ph.D. Dissertation, Ohio State University 1990).
13. J. T. Kim, "On the Slip Behavior and Surface Film Effects in B2 Ordered NiAl Single Crystals" (Ph.D. Dissertation, University of Michigan, 1990).
14. A. Ball and R. E. Smallman, *Acta Met.* **14**, 1349 (1966).
15. A. Ball and R. E. Smallman, *Acta Met.* **14**, 1517 (1966).
16. R. T. Pascoe and C. W. A. Newey, *Met. Sci. J.* **2**, 138 (1968).
17. R. T. Pascoe and C. W. A. Newey, *Phys. Stat. Sol.* **29**, 357 (1968).
18. R. J. Wasilewski, S. R. Butler and J. E. Hanlon, *Trans. AIME* **239**, 1357 (1967).
19. R. R Bowman, R. D. Noebe and R. Darolia, in *HIGHTEMP Review 1989* (NASA CP - 10039), 47-1 (1989).
20. M. H. Yoo, T. Takasugi, S. Hanada and O. Izumi, *JIM Mater. Trans.* **31**, 435 (1990).
21. D.I. Potter, *Mater. Sci. Eng.* **5**, 201 (1969).
22 M.G. Mendiratta and C.C. Law, *J. Mat. Sci.* **22**, 607 (1987).
23. R.D. Field, D.F. Lahrman, and R. Darolia, *High Temperature Ordered Intermetallic Alloys IV*, ed. L. A. Johnson et al. (MRS Symposium Proceeding **213**), 255 (1991).
24. H. Saka, M. Kawase, A. Nohara and T. Imura, *Phil Mag.* **50**, 65 (1984).
25. M. H. Loretto and and R. J. Wasilewski, *Phil Mag.* **23**, 1311 (1971).
26. H. L. Fraser, R. E. Smallman and M. H. Loretto, *Phil Mag.* **28**, 651 (1973).
27. H. L. Fraser, R. E. Smallman and M. H. Loretto, *Phil Mag.* **28**, 667 (1973).
28. R. Darolia, D. F. Lahrman, R. D. Field and A. J. Freeman, in *High Temperature Ordered Intermetallics Alloys III*, ed. C. T. Liu et al. (MRS Symposium Proceeding **133**), 113 (1989).

698

29. R. D. Field, R. Darolia, D. F. Lahrman and A. J. Freeman, "Alloy Modeling and Experimental Correlation for Ductility Enhancement in Near Stoichiometric Single Crystal Nickel Aluminide" (AFOSR contract F49620-88-C-0052, Final Report, July 1991).

30. R. D. Field, D. F. Lahrman and R. Darolia,"The Effect of Alloying on Slip Systems in [100] Oriented NiAl Single Crystals," accepted for publication in *Acta Metallurgica*.

31. K. Vedula, K. H. Hahn and B. Boulogne, in *High Temperature Ordered Intermetallics Alloys III*, ed. C. T. Liu et el. (MRS Symposium Proceeding **133**), 299 (1989).

32. E. M. Schulson and D. R. Barker, *Scripta Met.* **17**, 519 (1983).

33. E. P. George and C. T. Liu, *J. Mater. Res.* **5**, 754 (1990).

34. R. Darolia, D. F. Lahrman and R. Field, in preparation for publication in *Scripta Metallurgica*.

35. K.-M. Chang, R. Darolia and H. A. Lipsitt, *High Temperature Ordered Intermetallic Alloys IV*, ed. L. A. Johnson et al. (MRS Symposium Proceeding **213**), 597 (1991).

36. J. L. Walter and H. E. Cline, *Metall. Trans.* **1**, 1221 (1970).

37. S. V. Raj, R. D. Noebe and R. R. Bowman, *Scripta Met.* **23** 2049 (1989).

38. J. D. Whittenberger, *J. Mat. Sci.* **22**, 394 (1987).

39. M. Rudy and G. Sauthoff, *Mat. Sci. Eng.* **81**, 525 (1986).

40. R. S. Polvani, W. S. Tzeng and P. R. Strutt, *Metall. Trans.* **7A**, 33 (1976).

41. P. Nash and Y. Y. Pan, "The Al - Ni - Zr System," To be published in *J. of Alloy Phase Equilibria*.

42. K. J. Lee and P. Nash, "The Al - Ni - Hf System," To be published in *J. of Alloy Phase Equilibria*.

43. K. J. Lee and P. Nash, "The Al - Ni - Nb System," To be published in *J. of Alloy Phase Equilibria*.

44. K. J. Lee and P. Nash, "The Al - Ni - Ti System," To be published in *J. of Alloy Phase Equilibria*.

45. Y. Y. Pan and P. Nash, "The Al - Ni - Ta System," To be published in *J. of Alloy Phase Equilibria*.

46. J. R. Dobbs, R. D. Field and D. G. Konitzer, "Phase Relationships in NiAl + X (X = Zr, Hf, Nb, Ta) Alloys", (presented at the Fall 1990 TMS meeting, Detroit 1990).

47. I. E. Locci, R. D. Noebe, R. R. Bowman, R. V. Minor, M. V. Nathal and R. Darolia, *High Temperature Ordered Intermetallic Alloys IV*, ed. L. A. Johnson et al. (MRS Symposium Proceeding **213**), 1013 (1991).

48. J. D. Whittenberger, R. K. Viswanadham, in *High Temperature Ordered Intermetallics Alloys III*, ed. C. T. Liu et el. (MRS Symposium Proceeding **133**), 621 (1989)

49. S. C. Jha, R. Ray and D. J. Gaydosh, *Scripta Met.* **23**, 805 (1989).

50. J. D. Whittenberger, E. Arzt and M. J. Luton, *J. Mat. Res.* **5**, 271 (1990).

51. P. W. Bridgman, *Proceedings of American Academy of Arts and Sciences* **60** #6, 304 (1925).

52. C. T. Sims, N. S. Stoloff, W. C. Hagel, *Superalloys II*, John Wiley & Sons (1987).

53. M. McLean, *Directionally Solidified Materials for High Temperature Service*, The Metals Society (1983).

54. R. Darolia, W. D. Grossklaus, D. M. Matey, P. W. Stanek, M. Smith, US Patent Pending.

A BRIEF SUMMARY OF THE NATO ADVANCED RESEARCH WORKSHOP ON ORDERED INTERMETALLICS

C. T. Liu
Oak Ridge National Laboratory
P. O. Box 2008
Oak Ridge, TN 37831-6115

G. Sauthoff
Max-Planck-Institute für Eisenforschung GmbH, Germany

R. W. Cahn
University of Cambridge, Great Britain

The Advanced Research Workshop on Ordered Intermetallics—Physical Metallurgy and Mechanical Behavior provided a scientific forum to discuss and assess the research and development of ordered intermetallic alloys. The workshop consisted of four and a half days of feature presentations and discussions on recent advances in these areas. The last half day was devoted to assessment of current intermetallic research and recommendations of critical areas for future studies. The workshop focused on three specific areas: (1) electronic structure and phase stability, (2) deformation and fracture, and (3) high-temperature properties (including creep, diffusion, grain growth, etc.).

Recently, multidisciplinary approaches have been used to study electronic structure and phase stability of intermetallic phases, including first-principles total-energy calculations, atomistic simulations, statistical and classic thermodynamic calculations, phenomenological analyses, and experimental verification. Several attempts have been made to construct phase diagrams using statistical models with interatomic potentials calculated from first-principles quantum-mechanical calculations. In addition to prediction of intermetallic phase stability, the first-principle calculations have advanced to the stage that they are able to calculate atomic bonding, elastic moduli, defect energies [e.g., antiphase boundary (APB) energies] and fracture strengths. A general agreement between theoretical calculations and experimental measurements of these properties has been established in some intermetallic systems including TiAl, Ti_3Al, FeAl, Ni_3Al and NiAl.

Low ductility and brittle fracture at ambient temperatures are the major concern for structural use of ordered intermetallics. Considerable efforts have been devoted to understanding brittle fracture and improving ductility and toughness in aluminides and silicides. Both intrinsic and extrinsic factors are identified as major causes of brittleness. Intrinsic causes include insufficient number of

C. T. Liu et al. (eds.), Ordered Intermetallics – Physical Metallurgy and Mechanical Behaviour, 699–701.
© 1992 *All Rights Reserved.*

deformation modes, high hardness due to difficulty in generation and glide of dislocations, poor cleavage strength or low surface energy, grain-boundary weakness, high strain-rate sensitivity at crack tips, and planar slip and localized deformation. Recent studies have demonstrated that environmental embrittlement - an extrinsic factor - is a major cause of brittle fracture and low ductility in ordered intermetallics containing reactive elements. The embrittlement involves decomposition of moisture in air and generation of atomic hydrogen which embrittles crack tips. The understanding of the brittleness together with alloy design efforts has led to significant improvement in ductility in a number of intermetallic alloys, including Ni_3Al, Ni_3Si, Ti_3Al, TiAl, Co_3V, Ni_3V, Fe_3Al, FeAl and NiAl.

Many ordered intermetallics show unusual mechanical behavior at elevated temperatures. The most striking one is that their yield strength increases rather than decreases with temperature. Several dislocation models based on thermally activated processes have been proposed to explain the positive temperature dependence of the yield strength. The advantages and disadvantages of these models were discussed, and their predictions were compared with experimental observations of dislocation activities using transmission electron microscopy.

Creep properties of aluminides and silicides were studied as a function of test temperature and applied stress. The intermetallics with high crystal symmetries (such as $L1_2$ and B2) showed limited creep resistance while those with complex crystal symmetries (such as $L2_1$) exhibited excellent creep properties. The reason for this difference is not well understood, but it may be related to different diffusion behavior in these materials. Attempts to correlate the creep properties with diffusion data have been made among several $L1_2$ and B2 intermetallics. Atomistic modeling of diffusion in ordered lattices was presented and discussed, and additional work is clearly needed in this area.

The direction for future intermetallics research was discussed extensively in the last session, "Assessment of Current Research and Recommendation for Future Work". A general consensus was that there was a lack of precise information on elastic moduli, shear fault energies, and dislocation mobility in intermetallics. Both experimental data and theoretical calculations are required to address these areas. Atomistic-simulation modeling and, in particular, first-principles quantum-mechanical calculations have demonstrated their ability to help in understanding and solving some real material problems. An integrated approach (including electronic, atomic, microscopic, and macroscopic scales) toward understanding the brittleness and improving the ductility of intermetallics was discussed. Some emphasis was placed on the role of deviations from stoichiometry and on that played by minor alloying additions in relieving extreme brittlenesss. A lack of general understanding in the areas of crack-tip plasticity, fracture toughness, and notch sensitivity was specifically mentioned in the discussion. At present, only limited data on high-temperature creep are available. A need for establishing a correlation between the creep properties and microstructural features and diffusion data was stressed. Several groups emphasized the need of single crystals for fundamental studies. The possibility of setting central facilities for crystal growth at national laboratories was also suggested.

_____

This research was partially sponsored by the Division of Materials Sciences, United States Department of Energy under contract DE-AC05-84OR21400 with Martin Marietta Energy Systems, Inc.